Lecture Notes in Artificial Intelligence 7520

Subseries of Lecture Notes in Computer Science

LNAI Series Editors

Randy Goebel
  University of Alberta, Edmonton, Canada
Yuzuru Tanaka
  Hokkaido University, Sapporo, Japan
Wolfgang Wahlster
  DFKI and Saarland University, Saarbrücken, Germany

LNAI Founding Series Editor

Joerg Siekmann
  DFKI and Saarland University, Saarbrücken, Germany

Eyke Hüllermeier   Sebastian Link
Thomas Fober   Bernhard Seeger (Eds.)

Scalable Uncertainty Management

6th International Conference, SUM 2012
Marburg, Germany, September 17-19, 2012
Proceedings

 Springer

Series Editors

Randy Goebel, University of Alberta, Edmonton, Canada
Jörg Siekmann, University of Saarland, Saarbrücken, Germany
Wolfgang Wahlster, DFKI and University of Saarland, Saarbrücken, Germany

Volume Editors

Eyke Hüllermeier
Thomas Fober
Bernhard Seeger
Marburg University
Department of Mathematics and Computer Science
Hans-Meerwein-Straße 6, 35032 Marburg, Germany
E-mail: {eyke, thomas, seeger}@informatik.uni-marburg.de

Sebastian Link
The University of Auckland
Department of Computer Science
38 Princes St., Auckland 1010, New Zealand
E-mail: s.link@auckland.ac.nz

ISSN 0302-9743   e-ISSN 1611-3349
ISBN 978-3-642-33361-3   e-ISBN 978-3-642-33362-0
DOI 10.1007/978-3-642-33362-0
Springer Heidelberg Dordrecht London New York

Library of Congress Control Number: 2012946351

CR Subject Classification (1998): I.2, H.4, H.3, H.5, C.2, H.2

LNCS Sublibrary: SL 7 – Artificial Intelligence

Typesetting: Camera-ready by author, data conversion by Scientific Publishing Services, Chennai, India

Printed on acid-free paper

Springer is part of Springer Science+Business Media (www.springer.com)

Preface

In many applications nowadays, information systems are becoming increasingly complex, open, and dynamic. They involve massive amounts of data, generally issued from different sources. Moreover, information is often inconsistent, incomplete, heterogeneous, and pervaded with uncertainty. The annual International Conference on Scalable Uncertainty Management (SUM) has grown out of this wide-ranging interest in the management of uncertainty and inconsistency in databases, the Web, the Semantic Web, and artificial intelligence applications.

The SUM conference series aims at bringing together researchers from these areas by highlighting new methods and technologies devoted to the problems raised by the need for a meaningful and computationally tractable management of uncertainty when huge amounts of data have to be processed. The First International Conference on Scalable Uncertainty Management (SUM 2007) was held in Washington DC, USA, in October 2007. Since then, the SUM conferences have taken place successively in Naples (Italy) in 2008, again in Washington DC (USA) in 2009, in Toulouse (France) in 2010, and in Dayton (USA) in 2011.

This volume contains the papers presented at the 6th International Conference on Scalable Uncertainty Management (SUM 2012), which was held in Marburg, Germany, during September 17–19, 2012. This year, SUM received 5 submission. Each paper was reviewed by at least three Program Committee mbers. Based on the review reports and discussion, 41 papers were accepted gular papers, and 13 papers as short papers.

addition, the conference greatly benefited from invited lectures by three eading researchers: Joachim Buhmann (ETH Zürich, Switzerland) on t Sensitive Information: Which Bits Matter in Data?", Minos Garo-Technical University of Crete, Greece) on "HeisenData: Towards Next-on Uncertain Database Systems", and Lawrence Hunter (University of o USA) on "Knowledge-Based Analysis of Genome-Scale Data". More-kinstian Destercke (CNRS, Université de Technologie de Compiègne) was ceivgh to accept our invitation for an introductory talk that was con-infor overview of different approaches to uncertainty modeling in modern systems.

In , we would like to express our gratitude to several people and insti-tutions, all helped to make SUM 2012 a success:

- all the thors of submitted papers, the invited speakers, and all the conference pticipants for fruitful discussions;
- the members of the Program Committee, as well as the additional reviewers, who devoted time to the reviewing process;
- Alfred Hofmann and Springer for providing continuous assistance and ready advice whenever needed;

- the European Society for Fuzzy Logic and Technology (EUSFLAT) and the Marburg Center for Synthetic Microbiology (SYNMIKRO) for sponsoring and financial support;
- the Philipps-Universität Marburg for providing local facilities;
- the creators and maintainers of the conference management system Easy-Chair (http://www.easychair.org).

July 2012

Eyke Hüllermeier
Sebastian Link
Thomas Fober
Bernhard Seeger

Organization

General Chair

Bernhard Seeger   Philipps-Universität Marburg, Germany

Local Chair

Thomas Fober   Philipps-Universität Marburg, Germany

Program Committee Chairs

Eyke Hüllermeier   Philipps-Universität Marburg, Germany
Sebastian Link   The University of Auckland, New Zealand

Topic Chairs

Michael Beer	Uncertainty, Reliability, and Risk in Engineering
Djamal Benslimane	Managing Preferences in Web Services Retrieval
Mohamed Gaber	Massive Data Streams
Allel Hadjali	Managing Preferences in Web Services Retrieval
Daniel Keim	Visual Analytics
Ioannis Kougioumtzoglou	Uncertainty, Reliability, and Risk in Engineering
Anne Laurent	Scalable Data Mining
Edwin Lughofer	Evolving Fuzzy Systems and Modeling
Edoardo Patelli	Uncertainty, Reliability, and Risk in Engineering
Volker Roth	Biological and Medical Data Analysis
Moamar Sayed-Mouchaweh	Evolving Fuzzy Systems and Modeling
Ingo Schmitt	Computational Preference Analysis
Steven Schokaert	Logic Programming
Guillermo Simari	Computational Argumentation
Martin Theobald	Ranking and Uncertain Data Management
Lena Wiese	Uncertainty, Inconsistency, and Incompleteness in Security and Privacy
Jef Wijsen	Consistent Query Answering

International Program Committee

Leila Amgoud
Michael Beer
Nahla Ben Amor
Leopoldo Bertossi
Isabelle Bloch
Abdelhamid Bouchachia
Reynold Cheng
Carlos Chesñevar
Laurence Cholvy
Jan Chomicki
Alfredo Cuzzocrea
Anish Das Sarma
Thierry Denoeux
Jürgen Dix
Zied Elouedi
Avigdor Gal
Lluis Godo
Fernando Gomide
John Grant
Gianluigi Greco
Sven Hartmann
Jon Helton
Anthony Hunter
Gabriele Kern-Isberner
Ioannis Kougioumtzoglou
Vladik Kreinovich
Weiru Liu
Jorge Lobo
Peter Lucas
Thomas Lukasiewicz
Zongmin Ma

Thomas Meyer
Cristian Molinaro
Charles Morisset
Guillermo Navarro-Arribas
Zoran Ognjanovic
Francesco Parisi
Bijan Parsia
Simon Parsons
Gabriella Pasi
Edoardo Patelli
Olivier Pivert
Henri Prade
Andrea Pugliese
Guilin Qi
Matthias Renz
Daniel Sanchez
Kai-Uwe Sattler
Ingo Schmitt
Prakash Shenoy
Guillermo Simari
Umberto Straccia
V.S. Subrahmanian
Shamik Sural
Karim Tabia
Vicenc Torra
Sunil Vadera
Aida Valls
Peter Vojtáš
Jef Wijsen
Ronald Yager
Vladimir Zadorozhny

Additional Reviewers

Silvia Calegari
Chang Liu
Niccolo Meneghetti
Giovanni Ponti
Antonino Rullo

Moamar Sayed-Mouchaweh
Ammar Shaker
Francesca Spezzano
Yu Tang
Wangda Zhang

Sponsors

We wish to express our gratitude to the Sponsors of SUM 2012 for their essential contribution to the conference:

Philipps-Universität Marburg

European Society for Fuzzy Logic and Technology (EUSFLAT)

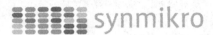

LOEWE-Zentrum für Synthetische Mikrobiologie (SYNMIKRO)

Table of Contents

Regular Papers

Short Papers

Xtream: A System for Continuous Querying over Uncertain Data Streams

Mohammad G. Dezfuli and Mostafa S. Haghjoo

Computer Engineering Department,
Iran University of Science and Technology,
Tehran, Iran
{mghalambor,haghjoom}@iust.ac.ir

Abstract. Data stream and probabilistic data have been recently considered noticeably in isolation. However, there are many applications including sensor data management systems and object monitoring systems which need both issues in tandem. The existence of complex correlations and lineages prevents Probabilistic DBMSs (PDBMSs) from continuously querying temporal positioning and sensed data. Our main contribution is developing a new system to continuously run monitoring queries on probabilistic data streams with a satisfactory fast speed, while being faithful to correlations and uncertainty aspects of data. We designed a new data model for probabilistic data streams. We also presented new query operators to implement threshold SPJ queries with aggregation (SPJA queries). In addition and most importantly, we build a java-based working system, called Xtream, which supports uncertainty from input data streams to final query results. Unlike probabilistic databases, the data-driven design of Xtream makes it possible to continuously query high-volumes of bursty probabilistic data streams. In this paper, after reviewing main characteristics and motivating applications for probabilistic data streams, we present our new data model. Then we focus on algorithms and approximations for basic operators (select, project, join, and aggregate). Finally, we compare our prototype with Orion the only existing probabilistic DBMS that supports continuous distributions. Our experiments demonstrate how Xtream outperforms Orion w.r.t. efficiency metrics such as tuple latency (response time) and throughput as well as accuracy, which are critical parameters in any probabilistic data stream management system.

Keywords: Probabilistic Stream, Sensor, Uncertainty, Continuous Query, Xtream, Orion.

1 Introduction

Many applications need to deal with enormous uncertain data streams. Some examples are sensor [1,2], RFID [3], GPS [4] and scientific [5] data streams. Values of attributes in such applications (e.g. sensor data) are error-prone, especially when we use outdated readings [6]. Scientific data including estimates, experimental measurements, and hypothetical data is also inherently uncertain. Most applications usually

E. Hüllermeier et al. (Eds.): SUM 2012, LNAI 7520, pp. 1–15, 2012.

ignore imprecision because handling uncertainty directly is inefficient. The problem would be even worse for real-time monitoring of high-rate uncertain input streams [7,8].

There are two different possibilities to deal with inevitable uncertain data: 1) data cleaning [9], and 2) handling uncertainty in data modeling and query processing. Cleaning data from uncertainties is ideal but not always feasible. Many applications prefer to deal with uncertainties and their inherent useful information as diamonds in the dirt [1]. Even if data cleaning makes sense for uncertain databases, it is irrelevant to real-time data stream processing. Thus, in this paper, we only focus on the second approach. Supporting uncertainties needs re-engineering of data model and query processing in data management systems. Data model should be designed in a way to support probabilistic distributions for attributes, correlations between tuples, and confidence about relevancy of a tuple in a stream. The most important change in query processing is supporting probabilistic threshold queries.

There are some challenges which make probabilistic data stream processing more complicated. The first one is correlation (temporal and spatial) between tuples. Supporting complex correlations needs graphical models [2]. However, sometimes it is easier for application to ignore these correlations and use a multi-purpose and efficient data management system instead of a graphical-based one. The other challenge is efficiency of query processing. Even if we ignore correlations between input tuples, it is not easy to always eliminate correlations between intermediate tuples. In this case, we have to use intentional semantics instead of extensional semantics [10] which vitiate complexity from PTIME to #P-complete [11].

In this paper, we present data model and query language of our system - Xtream. Then we focus on algorithms and approximations for basic operators (select, project, join and aggregate). After presenting the architecture of Xtream, we show how it outperforms Orion [12] regarding PLR benchmark [13]. Thus, our main contribution is presenting the first real *probabilistic data stream management system* and comparing it to a famous Probabilistic Data Base Management System (PDBMS) to show how changing the query processing into continuous will drastically improves the results.

Rest of the paper is organized as follows. We describe related work in section 2. First we focus on logical aspects of our system and introduce our new data model in section 3. Then we focus on physical aspects of our system and discuss about implementation of basic operators in section 4. In section 5, we evaluate and compare our system to Orion which is a PDBMS that supports continuous distributions. Finally, we conclude this paper in section 6.

2   Related Work

There is a broad range of related work on probabilistic databases [14,15] motivated by many applications need to manage large and uncertain data sets. The major challenge studied in probabilistic databases is the integration of query processing with probabilistic inference in an efficient way. Most of the *Probabilistic Databases (PDBs)* are using *Block Independent-Disjoint (BID)* model with lineages based on *Possible*

Worlds Semantics (PWS) [1]. These systems support limited correlations between uncertain values. The other variants are graphical-based systems [16] which support more correlations. However, they are more complex and less scalable.

Unlike other PDBs like *Trio* [17] and *MayBMS* [15], *Orion* [12] supports continuous distributions to cover sensor applications. Supporting continuous distributions raises new challenges about semantics as PWS is inherently discrete. Nevertheless, Orion states intuitively that it is based on PWS semantics. The main difference between Orion as a PDBMS and our system is real-time stream processing. Orion is good for querying over some historical sensor data but because of the database nature of it (in comparison to data streams), it is not a good choice for real-time monitoring of sensor data (as we show in our experimental evaluations in § 5).

Real-time probabilistic stream processing with continuous distributions first considered in *PODS*. Tran et al. [18] focused on two applications of object tracking in RFID networks and weather monitoring. PODS has a data model based on *Gaussian Mixture Model (GMM)* which supports arbitrary continuous distributions. PODS includes some join and aggregation algorithms but still it is not a complete system. In contrast, we have developed a real prototype for more investigation on different systemic aspects.

Event queries on indoor moving objects have been discussed in [19]. They used a graphical solution based on Hidden Markov Models and Cayuga query language which is completely different from SQL-based languages. These differences make their system very specific to some applications and somehow isolated from traditional databases. Our system is more general and suitable for combining uncertain data streams and relational databases and adaptable to wider range of applications.

To the best of our knowledge, *probabilistic data stream management* which is a new combination of *Data Stream Management Systems (DSMSs)* and *PDBs* were never considered as a complete system before and researchers only focused on different aspects like data model and query processing in isolation. Unlike them, we tried to develop a complete generic system for query evaluation on probabilistic data streams for the first time.

3   Probabilistic Data Model

Data model is the foundation of a data processing system and should be designed in a way that allows us to capture input data, process user defined queries and finally represent results for the end user. The most important difference between DSMS data models and Xtream's0 data model is supporting probabilistic domains. Although there are many useful recent models for supporting probabilistic values in PDBs [20,12,17,15], adapting those models to the stream-based nature of Xtream is not trivial. In this part, we define different parts of our data model.

Definition 1. P-Tuple: a P-tuple $T^{\mathcal{T}}$ of type $\mathcal{T}: (D_1, ..., D_n)$ is a pair (A, λ) in which A is $(a_1, ..., a_n)$ where a_k is an attribute in the form of a triple (l_k, u_k, p_k). l_k , u_k are uncertainty lower and upper bounds respectively. p_k is a bounded probability distribution function. λ is an event-based lineage.

Lineage is an expression based on the event model described in *PRA* and is a combination of atomic independent events using { ∧,∨, } operators [10]. Using intentional semantics based on lineage is strongly necessary in Xtream to compute correct probabilities. The anomaly of ignoring lineage is well described in [1,12,14]. However, using lineage poses more difficulties in probability evaluation and makes an abrupt change in complexity from PTIME to #P-complete [11].

Definition 2. P-Element: a P-element $E^\mathcal{T}$ is a pair$(\tau, T^\mathcal{T})$ where τ is an interval $[\tau_s, \tau_e)$ which determines tuple validity period and $T^\mathcal{T}$ is a P-tuple.

Definition 3. P-Stream: a P-stream $S^\mathcal{T}$ is an infinite list of ordered (non-decreasingly by start timestamp) P-elements with the same type $\mathcal{T}$.

Adding membership confidence to a P-tuple makes algorithms more complex. Thus, in this paper, we only use lineage in P-tuples and express membership as an additional event in lineage. The idea is similar to physical streams in PIPES [21]. We also define P-relations as a non-temporal counterpart of P-streams as follows:

Definition 4. P-Relation: a probabilistic relation $R^\mathcal{T}$ is a finite bag of P-tuples $T^\mathcal{T}$.

Time snapshot transforms a temporal probabilistic stream into a probabilistic relation (to leverage relational algebra). We define time snapshots formally as follows:

Definition 5. Time Snapshot (φ_τ): a time snapshot on P-stream $S^\mathcal{T}$ is $\varphi_\tau(S^\mathcal{T}) = \{T^\mathcal{T} | \exists([\tau_s, \tau_e), T^\mathcal{T}) \in S^\mathcal{T}, t_s \leq \tau < t_e\}$.

4   Probabilistic Queries

In this section we introduce our query language for Xtream. We focus on four basic operators (map, filter, join, aggregate) and present definitions, architectures, and algorithms for them.

4.1   The *Map* Operator

The *map* operator is a general form of project operator in relational algebra. It is a stateless operator which reforms a tuple based on some predefined functions and always consumes one tuple to generate another one. One of the most important challenges about map operator is eliminating some probabilistic attributes which are jointly distributed with other attributes [12]. However, in our system, we do not have this problem as we assumed that there is no intra-tuple correlation between probabilistic attributes. In the case of using map operator as a simple project operator, the implementation is straightforward (we should only eliminate some certain or uncertain attributes). In other cases, user is responsible for defining functions for *map* operator which can compute the distribution of new probabilistic variable based on some input certain/uncertain variables. We formally define the *map* operator below:

Definition 6. Map Operator (Λ): let $F :=< F_1, F_2, \ldots, F_n >$ be a list of functions and $T = (\tau, (A, \lambda))$ be a P-tuple where $A = (a_1, a_2, \ldots, a_n)$. The result of *map* operator on F and T $(\Lambda_F(T))$, is a new P-tuple $(\tau, (A', \lambda))$ with $A' = (F_1(a_1), F_2(a_2), \ldots, F_n(a_n))$.

Note that the schema of tuple will be changed after applying map operator. The only unchanged parameters are timestamp and the lineage.

4.2 The *Filter* Operator

The *filter* operator is similar to *select* operator in relational algebra but there are some tricky points about it. The *select* operator checks a condition on some certain attributes of a tuple and sends it to the output if condition is *true*. The first problem with *filter* operator is the semantics behind it. At least, there are two important possible alternatives for *filter* operator as illustrated in Fig. 1: 1) to pass the tuples untouched, or 2) to fit the tuple based on the condition.

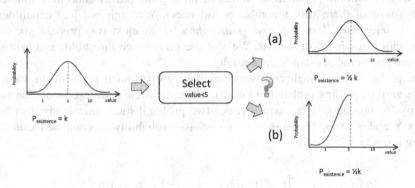

Fig. 1. Two alternatives for *filter* operator

We choose the second alternative (Fig. 1 (b)) based on WD-PWS semantics [22]. There is a noticeable difference between *filter* and *select* operators: unlike *select*, *filter* may change distribution of attributes of the result tuples.

The second challenge is about conditions which are evaluated to *true* or *false* in *relational databases*. Evaluating conditions on probabilistic attributes is more complex, especially when we have several probabilistic attributes. We formally define Filter operator below:

Definition 7. Tailor Function ($\mathfrak{T}$): the result of binary function $\mathfrak{T}$ on a set of attributes A and condition P, $\mathfrak{T}(A, P)$, is A' in which each probabilistic attribute $a_k: (l_k, u_k, p_k)$ on domain D_k is changed to $a_{k\prime}: (l_k, u_k, p'_k)$ where $\forall d \in D_k \; \neg P(d) \rightarrow p'_k(d) = 0 \wedge P(d) \rightarrow p'_k(d) = \alpha \, p_k(d)$ and $\alpha = 1/ \int_{x \in D_k \wedge p(x)} p_k(x)dx$.

Definition 8. Filter Operator (Φ): the unary filter operator on input P-stream S^T with condition P is: $\Phi_P(S^T) = \{(\tau, (\mathfrak{T}(A, P), \lambda')) | \exists \lambda \, (\tau, (A, \lambda)) \in S^T \wedge \lambda' = (\lambda \wedge (e_{new} \, as \, P(\mathfrak{T}(A, P)) = true)) \wedge \Pr(\lambda') > 0\}$.

4.3   The *Join* Operator

Join operation on uncertain streams is too costly, albeit very necessary in query processing. There are many different join operators in relational algebra (e.g. natural join, θ-join, equijoin, inner join, outer join, semi-join, Cartesian product). We define their counterparts for our model. In this paper we limit our join operators to the more general threshold-based probabilistic θ-join:

Definition 11. Threshold-based Probabilistic θ-Join ($\times_\theta^\rho$): threshold-based probabilistic θ-join of two P-streams, $S_1^T \times_\theta^\rho S_2^{T'}$, is a new P-stream $S_3^{T.T'} = \{(\tau, (A, \lambda)) | \exists\, t_1 \in S_1^T \exists t_2 \in S_2^{T'} (\tau = t_1.\tau \cap t_2.\tau \wedge A = t_1.A.t_2.A \wedge \lambda = t_1.\lambda \underline{\wedge} t_2.\lambda \underline{\wedge} e_{new}$ as $\theta(t_1, t_2) \wedge \Pr(\lambda) > \rho)\}$.

There are two main challenges in implementing join in Xtream: 1) evaluating conditions which combine probabilistic values in theta join, and 2) enormous number of results (most of them with ignorable confidences). We limit our join conditions to *equality, inequality, less-than and greater-than* for at most two probabilistic arguments with Gaussian distributions. We also use confidence thresholds and heuristics to ignore producing low-confidence results.

The probability of equality of two continuously distributed probabilistic values is always zero (regarding probability theory). Thus, in Xtream, we use resolution based equality [6] instead of exact equality. For two probabilistic continuously distributed values X and Y, $X =_c Y$ means $|X - Y| < c$. Its probability is computed using following formula [6]:

$$Pr(X =_c Y) = \int_{-\infty}^{+\infty} f_X(x)[F_Y(x + c) - F_Y(x - c)]dx \tag{1}$$

where f is the *pdf* function and F is the *CDF* function. However, it is costly to compute an integral. Thus, we use a trick to improve the efficiency of computing probability of equality. As we mentioned before, $X =_c Y$ is equal to $|X - Y| < c$; so instead of $Pr(X =_c Y)$ we compute $Pr(|X - Y| < c)$. Linear combination of two Gaussian variables leads to another Gaussian variable. For example, for $X = G(\mu_x, \sigma_x^2)$ and $Y = G(\mu_y, \sigma_y^2)$, $Z = X - Y = G(\mu_x - \mu_y, \sigma_x^2 + \sigma_y^2)$. As a result, $Pr(X =_c Y)$ is equal to $F_Z(c) - F_Z(-c)$ which is computable in a constant time using predefined standard normal tables.

For inequality, we compute the probability easily based on the idea of $Pr(X \neq_c Y) = 1 - Pr(X =_c Y)$. In addition, $Pr(X < Y)$ equals to $Pr(X - Y < 0)$ which is equal to $F_z(0)$.

Figure 2 illustrates the architecture of our binary join operator. Input tuples first go to the input queues and would be fetched based on their timestamps to preserve order of output tuples. After joining with partner tuples, each tuple should be stored temporarily in a state (aka synopsis). We use *SweepAreas* [21] for our join states. SweepAreas provide our join with a fast access to temporal tuples while a sweeping mechanism drops expired and useless tuples. We use a list-based implementation of SweepArea as probabilistic joins are a kind of similarity joins. We also use *symmetric nested-loop join (SNJ)*. The other alternative for our join algorithm is hash join which is irrelevant to similarity-based joins. The final box, *threshold filter*, drops low-confidence results based on the determined threshold. Changing threshold is a good method to handle overloads. We leave this idea for future.

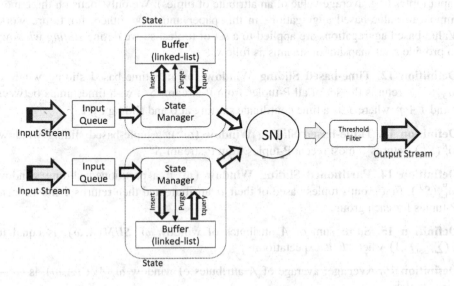

Fig. 2. The architecture of Xtream binary join operator

The algorithm of our threshold-based probabilistic θ-Join is presented in Algorithm 1.

Algorithm 1. Threshold-based Probabilistic -Join
Input: two P-streams $S_1^T, S_2^{T'}$, a predicate θ, and a threshold ρ.
Output: a P-stream $S_{out}^{T.T'}$
1   $S_{out}^{T.T'} \leftarrow \emptyset$;
2   let P_{remove} be a predicate which: $P_{remove}(T_1, T_2) = (T_1.\tau_E < T_2.\tau_S)?\ true:false$;
3   let SA_1, SA_2 be two SweepAreas($\leq_{\tau_E}, \theta, P_{remove}$) for $S_1^T, S_2^{T'}$ respectively;
4   **while** $S_1^T \neq \emptyset$ **and** $S_2^{T'} \neq \emptyset$
5   Let T_i be the next oldest P-tuple from $S_i\ i \in \{1,2\}$;
6      $j \leftarrow (i\ mod\ 2) + 1$;
7   SA_j.purgeElements(T_i, i); // drop elements T_j from SA_j where $T_j.\tau_E < T_i.\tau_S$
8   SA_i.insert(T_i);
9   **foreach** T_j **in** SA_j.tquery(T_i, i, ρ)   // $\Pr\left(\theta(T_i, T_j)\right) > \rho$
10   let $e_{new} \coloneqq \Pr\left(\theta(T_i, T_j)\right) > \rho$;
11   let $\lambda_{new} \coloneqq T_i.\lambda \wedge T_j.\lambda \wedge e_{new}$;
12   **if** $\Pr(\lambda_{new}) > \rho$
13          $\left(T_i.\tau \cap T_j.\tau, (T_i.T_j, \lambda_{new})\right) \hookrightarrow S_{out}^{T.T'}$

4.4   The *Aggregate* Operator

Aggregation operators are vital in Xtream because of the probabilistic nature of data. They could be either entity-based or value-based. Entity-based aggregations return tuples (or some parts of tuples) as their output (e.g. tuple with minimum value for a specific attribute) but value-based aggregations return a statistical value based of their input tuples (e.g. average value of an attribute of tuples). We only focus on three most important value-based aggregations in this paper and leave others for future work. Value-based aggregations are applied to a set of tuples, so we define *sliding windows* to provide a set snapshot of streams as follows:

Definition 12. Time-based Sliding Window (ω^T): a time-based sliding window, $\omega_{p,i}^T(S^T)$, returns the set of all P-tuples from P-stream, S^T, with timestamps between t and $t + p$ where t is a time coordinate starting at 0 and sliding by i.

Definition 13. Tuple-based Sliding Window (ω^R): a tuple-based sliding window, $\omega_i^R(S^T)$, returns i most recent P-tuples from P-stream S^T.

Definition 14. Partitioned Sliding Window (ω^P): a partitioned sliding window $\omega_{i,a}^P(S^T)$, first groups tuples based of their a attribute and then returns i most recent P-tuples for each group.

Definition 15. Sum: sum of A attributes of window ω, $SUM(A, \omega)$, is equal to $E(\sum_{t \in \omega} t.A)$ where E is expectation.

Definition 16. Average: average of A attributes of window ω, $AVG(A, \omega)$, is equal to $E(\frac{\sum_{t \in \omega} t.A}{|\omega|})$.

Definition 17. Count: count of P-tuples in window ω, $COUNT(\omega)$, is equal to $E(|\omega|)$.

Temporal validity of tuples and window size are very important in aggregation. For temporal tuples, we should remove the effect of expired tuples from aggregation result which forces us to keep a sketch of previously seen tuples. Aggregation functions are also usually required to be applied on a group of tuples. Thus, we should first apply a window operator on input stream, then group tuples, and finally apply aggregation function. Although we can apply window operator easily, grouping probabilistic tuples is a bit challenging, as tuples may belong to many groups with different probabilities. Thus, it is not possible to physically group tuples. Instead we do it in a logical way.

Figure 3 and Algorithm 2 illustrate our aggregation operator. Each time a new tuple comes to the operator, it computes the positive and negative changes on the aggregation result based on the new tuple and there is no need to refer to old tuples. The positive change (which is computed by f_+ function) will be applied to the current aggregation result immediately by f_{merge} function. The negative change, which wipes the effect of expired or dropped tuples out, should be stored temporally in a SweepArea. Whenever the operator decides to remove some tuples (because of the expiration or window specification), it extracts the negative changes and apply them on the aggregation result. Negative changes are computable using f_- function. The set of f_+, f_-, and f_{merge} functions can be easily defined for each of our aggregation functions. Note that we used *sliding windows* as embedded components in our operators as it is more efficient and also more compatible with pipeline architecture.

Our online data driven approach is much faster than partial aggregation approach presented in [21] as it needs a short constant time for each operation (including insert, remove, and evaluation). However, it limits us to a smaller number of aggregate functions.

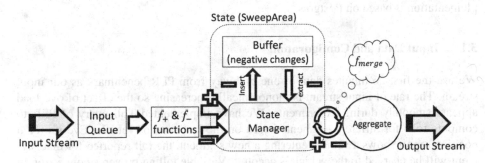

Fig. 3. The architecture of Xtream's aggregate operator

Algorithm 2. Aggregation

Input: a window $\omega(S^T)$ on input P-stream S^T, three predefined functions (f_+, f_-, f_{merge})

Output: a P-stream $S_{out}^{T'}$

1   $S_{out}^{T'} \leftarrow \emptyset$;
2   let AV be the current average value; $AV \leftarrow \emptyset$;
3   let P_{remove} be the validation predicate of ω;
4   let SA_1 be a SweepArea($\leq_{\tau_s}$, P_{remove});
5   **while** $S^T \neq \emptyset$
6   Let T be the next oldest P-tuple from S^T
7   Iterator $i \leftarrow SA_1$.extractElements(T, 1);
8   **while** i.hasNext() **do**
9       $nv \leftarrow i$.next();
10      $AV \leftarrow f_{merge}(AV, nv)$;
11   SA_1.insert($T.\tau, f_-(T)$);
12     $AV \leftarrow f_{merge}(AV, f_+(T))$;
13     $AV \hookrightarrow S_{out}^{T'}$

---

5   Empirical Study

We use PLR (Probabilistic Linear Road) benchmark [13] to evaluate our system. PLR simulates traffic characteristics of a simple expressway system based on variable tolling to handle traffic congestion of expressways. The main idea of PLR is to raise tolls on crowded expressways to encourage drivers toward alternative roads. In fact, it is a probabilistic version for it is ancestor, *Linear Road* benchmark [23]. In *linear road* benchmark, it is assumed that sensors are accurate which is unrealistic in the real-world. Therefore, PLR adds uncertainty to linear road and adapt it for evaluating probabilistic stream querying systems. In addition, PLR has changed queries in *linear road* from a historical nature into real-time suitable for stream processing systems. We compared Xtream with Orion (ver. 2.0). Orion is the only available probabilistic DBMS which supports continuous distributions as well as discrete ones. Orion's implementation is based on Postgres.

5.1   Input Data and Configuration

We use the first 26 minutes data of one highway from PLR benchmark as our input stream. The rate of input stream is monotonically increasing so the effect of overload appears smoothly during experiments. We have implemented Toll query of PLR to compute tolls for different segments based on the last state of vehicle. Every time a position report shows a vehicle entering a new segment, the toll reported for that segment will be charged to the vehicle's account. Variable tolling prevents congestion in different parts of expressways. Toll computation for a segment is based on three factors: 1) average number of vehicles, 2) average speed, and 3) nearest upstream accident. Because of the lack of space, we refer the reader to [13] for more details on input data types. Generally, the tuples indicate probabilistic positions of vehicles using Gaussian distributions.

The following experiments were conducted on a computer with 4 GB RAM, Core i7 CPU, running Ubuntu 11.10 and Orion 0.2. Although our system has 8 cores and Xtream supports multi-thread processing, we only use one thread to have a fare comparison between Xtream and Orion.

5.2   Evaluation Metrics

Evaluating probabilistic data stream processing systems is more tricky and complex than PDBs or DSMSs. The most important aspects are: 1) efficiency, and 2) accuracy. We focus on response time for efficiency and toll error for accuracy.

The most important point in evaluation is that ideal and real results do not have one-to-one relationship. Thus, we use the notion of variation distance between distributions [24] and compute sum of their differences for each time unit. In this way, both accuracy and response time affect the metric.

We also define precision as a probability threshold so that probability values less than it would be considered 0 (i.e. impossible). Each application can set its precision. We use this notion to ignore small enough probabilities in discretization. Here we show how changing precision will affect quality of results and performance.

In addition, it is a motivation for overload controlling (adaptation and load shedding) in Xtream which is left for future. We use four precisions 0, 0.1, 0.01, and 0.5 in our experiments.

5.3   Efficiency

We measure response time in Xtream as the difference between system time of *input tuples* and *output tuples*. In Orion implementation, a periodic ever-running function computes tolls for all of the segments. Thus, response times of the results are equal to corresponding function run-time. Experiments show that the continuous nature of Xtream makes it up to 10 times faster than Orion regarding selected precision (0 to 0.5). Figure 4 compares average response time for Orion and Xtream with four different precisions. To better distinguish between Xtream different versions, we removed Orion and depicted the graph in Fig. 5. As illustrated in the graphs, there is a noticeable difference between Orion and Xtream results as time goes by. Figure 6 also illustrates the global average response time for Orion and Xtream. Orion looks good enough in comparison to Xtream with zero precision in this graph. This is due to the number of output results. Although the rate of arriving input stream is monotonically increasing (Fig. 7), unlike Xtream (Fig. 8), Orion's output result rate (i.e. throughput) is decreasing (Fig. 9). In fact, fast exponentially increasing response time in Orion prevents generating complete results in subsequent periods. As a result, most of the output tuples, which are contributing in global average response time, belong to first periods (i.e. with short response times). The 2D pie chart in Fig. 10 illustrates the huge share of first periods in number of output tuples. Hence, Fig. 4 is a better clue to see the real difference between efficiency of Xtream and Orion. The difference between zero precision and others is also considerable.

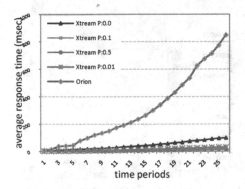

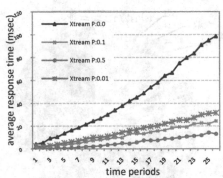

Fig. 4. Periodic average response time for Xtream and Orion

Fig. 5. Periodic average response time for Xtream with different precisions

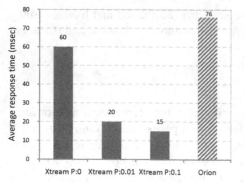

Fig. 6. Average response time for Xtream and Orion

Fig. 7. Rate of input data stream

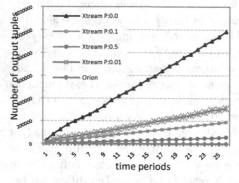

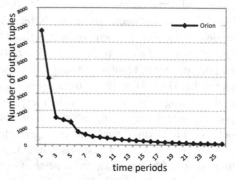

Fig. 8. Periodic number of output tuples (throughput) in Xtream and Orion

Fig. 9. Periodic number of output tuples (throughput) in Orion

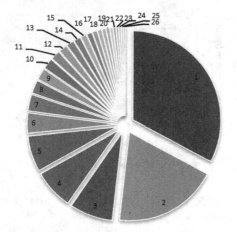

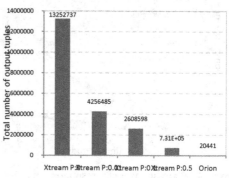

Fig. 10. Contribution of each period in total number of output tuples (i.e. and average response time) in Orion

Fig. 11. Total number of output tuples in Xtream and Orion

5.4   Accuracy

In addition to efficiency, accuracy of results is also important. We compute toll error as sum of differences between ideal and real toll functions. In this way, we do not need a one-to-one relationship between ideal and real results so we can easily compare two completely different systems (i.e. Xtream and Orion). Approximation error as well as response time are both effective in final toll errors. Thus, toll errors present accuracy and efficiency of systems in tandem. Figures 12 and 13 illustrate periodic and average toll errors for Xtream (with three different precisions) and Orion. Only for the case of 0.1 precision and in the first few periods, Orion generates less error. The average toll error of Orion is almost 160 times greater than Xtream with zero precision and about 20 times with 0.01 precision. The other difference between these two systems is that Orion's error is noticeably increasing but Xtream's errors are almost stable.

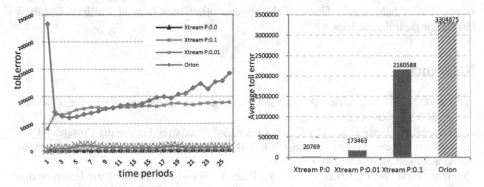

Fig. 12. Periodic toll error   **Fig. 13.** Average toll error

6   Conclusion

Continuous querying over probabilistic data streams is a new promising and growing field for database research community. Managing probabilistic data streams is necessary in developing complex and real-time monitoring systems. These systems should take uncertainties into account while dealing with bursty high-volume data streams.

Our approach to develop a probabilistic data stream management system was to start from a DSMS and extend it to support probabilistic data and queries. We started from data model and added lineage to tuples. Lineage is necessary to compute correct probabilities of final results. It also helps us to postpone confidence computation. Computing confidence based on complex lineages is another challenge. However, it is not a problem for short query plans as we used in our experiments. In addition, we redefined all of the operators to support uncertainty because even basic operators like filter (select) may perform on probabilistic streams in a different way. We used *WD-PWS semantics* to choose between different alternatives for operators. *WD-PWS* is an extension of *possible worlds semantics* which is well adapted to continuous distributions. Then, we implemented all operators in an efficient way to process probabilistic streams.

In the last part, we used *PLR* which is an extension of *linear road* benchmark for probabilistic streams and evaluated the idea of floating precisions for toll computation query. We showed how changing precision affect quality of results and quality of service. We leave automatic precision and expiration tuning as an open problem for future work. Moreover, we compared our system, Xtream, with Orion which is the only available PDBMS prototype that supports continuous distributions. The results show that Xtream is up to 10 times faster than Orion regarding selected precision. This noticeable difference in efficiency comes from the different architecture and operators in Xtream and Orion. In fact, Xtream is instinctively designed for long-time running continuous queries on probabilistic data streams while Orion periodically repeats queries based on the last snapshot of data.

Acknowledgments. We would like to thank Prof. Bernhard Seeger, Daniar Achakeyev, Philip Schmiegelt, Daniel Schäfer, and other members of Philipps Database Group for their discussions.

References

1. Dalvi, N., Ré, C., Suciu, D.: Probabilistic databases: diamonds in the dirt. Commun. ACM 52(7), 86–94 (2009)
2. Kanagal, B., Deshpande, A.: Efficient Query Evaluation over Temporally Correlated Probabilistic Streams. In: Proceedings of the 2009 IEEE International Conference on Data Engineering, ICDE 2009 (2009)
3. Tran, T., Sutton, C., Cocci, R., Nie, Y., Diao, Y., Shenoy, P.: Probabilistic Inference over RFID Streams in Mobile Environments. In: Proceedings of ICDE 2009 (2009)
4. Kanagal, B., Deshpande, A.: Online Filtering, Smoothing and Probabilistic Modeling of Streaming Data. In: ICDE, pp. 1160–1169 (2008)
5. Kurose, J., Lyons, E., McLaughlin, D., Pepyne, D., Philips, B., Westbrook, D., Zink, M.: An End-User-Responsive Sensor Network Architecture for Hazardous Weather Detection, Prediction and Response. In: Cho, K., Jacquet, P. (eds.) AINTEC 2006. LNCS, vol. 4311, pp. 1–15. Springer, Heidelberg (2006)
6. Cheng, R., Singh, S., Prabhakar, S., Shah, R., Vitter, J.S., Xia, Y.: Efficient join processing over uncertain data. In: Proceedings of the 15th ACM International Conference on Information and Knowledge Management, CIKM 2006 (2006)
7. Diao, Y., Li, B., Liu, A., Peng, L., Sutton, C., Tran, T., Zink, M.: Capturing Data Uncertainty in High-Volume Stream Processing. In: CIDR (2009)
8. Safaei, A.A., Haghjoo, M.S.: Parallel Processing of Data Stream Query Operators. Journal of Distributed and Parallel Databases 28(2-3) (2010)
9. Mayfield, C., Neville, J., Prabhakar, S.: ERACER: a database approach for statistical inference and data cleaning. In: Proceedings of the 2010 International Conference on Management of Data, SIGMOD 2010, pp. 75–86. ACM, New York (2010)
10. Fuhr, N., Rölleke, T.: A probabilistic relational algebra for the integration of information retrieval and database systems. ACM Trans. Inf. Syst. 15(1), 32–66 (1997)
11. Dalvi, N., Suciu, D.: Efficient query evaluation on probabilistic databases. The VLDB Journal 16(4), 523–544 (2007)

12. Singh, S., Mayfield, C., Shah, R., Prabhakar, S., Hambrusch, S., Neville, J., Cheng, R.: Database Support for Probabilistic Attributes and Tuples. In: Proceedings of the IEEE 24th International Conference on Data Engineering, April 07-12 (2008)
13. Karachi, A., Dezfuli, M.G., Haghjoo, M.S.: PLR: A Benchmark for Probabilistic Data Stream Management Systems. In: Pan, J.-S., Chen, S.-M., Nguyen, N.T. (eds.) ACIIDS 2012, Part III. LNCS, vol. 7198, pp. 405–415. Springer, Heidelberg (2012)
14. Benjelloun, O., Sarma, A.D., Halevy, A., Widom, J.: ULDBs: Databases with Uncertainty and Lineage. In: Proceedings of the 32nd International Conference on VLDB, pp. 953–964 (2006)
15. Antova, L., Koch, C., Olteanu, D.: From Complete to Incomplete Information and Back. In: Proceedings of the 2007 ACM SIGMOD International Conference on Management of Data, Beijing, China, June 11-14 (2007)
16. Sen, P., Deshpande, A., Getoor, L.: PrDB: managing and exploiting rich correlations in probabilistic databases. The VLDB Journal 18(5), 1065–1090 (2009)
17. Agrawal, P., Widom, J.: Continuous Uncertainty in Trio. In: MUD (2009)
18. Tran, T., Peng, L., Li, B., Diao, Y., Liu, A.: PODS: a new model and processing algorithms for uncertain data streams. In: Proceedings of the 2010 International Conference on Management of Data, SIGMOD 2010 (2010)
19. Ré, C., Letchner, J., Balazinksa, M., Suciu, D.: Event queries on correlated probabilistic streams. In: Proceedings of the 2008 ACM SIGMOD International Conference on Management of Data, SIGMOD 2008, pp. 715–728. ACM, New York (2008)
20. Ge, T., Zdonik, S.: Handling Uncertain Data in Array Database Systems. In: Proceedings of the IEEE 24th International Conference on Data Engineering, April 07-12 (2008)
21. Krämer, J., Seeger, B.: Semantics and implementation of continuous sliding window queries over data streams. ACM Trans. Database Syst. 34(1), Article 4 (April 2009)
22. Dezfuli, M.G., Haghjoo, M.S.: A Semantical Model for Probabilistic Data Stream Management Systems with Continuous Distributions. Submitted to Distributed and Parallel Databases Journal (2011)
23. Arasu, A., Cherniack, M., Galvez, E., Maier, D., Maskey, A.S., Ryvkina, E., Stonebraker, M., Tibbetts, R.: Linear road: a stream data management benchmark. In: Proceedings of the Thirtieth International Conference on Very Large Data Bases, VLDB 2004, vol. 30 (2004)
24. Mitzenmacher, M., Upfal, E.: Probability & Computing: Randomized Algorithms and Probabilistic Analysis. Cambridge U. Press (2005)

Applications of Ordinal Ranks to Flexible Query Answering*

Lucie Urbanova[1], Vilem Vychodil[1], and Lena Wiese[2]

[1] DAMOL (Data Analysis and Modeling Laboratory)
Dept. Computer Science, Palacky University, Olomouc
17. listopadu 12, CZ77146 Olomouc, Czech Republic
lucie.urbanova01@upol.cz, vychodil@acm.org
[2] Institute of Computer Science, University of Hildesheim
Samelsonplatz 1, 31141 Hildesheim, Germany
lena.wiese@uni-hildesheim.de

Abstract. Exact querying and retrieving relevant data from a database is a difficult task. We present an approach for flexibly answering algebraic queries using an extension of Codd's relational model with ordinal ranks based on residuated lattices and similarities on attribute domains.

Keywords: Flexible query answering, ranked data tables, complete residuated lattices, similarity, relational algebra.

1 Introduction

As nowadays massive amounts of data are stored in database systems, it becomes more and more difficult for a database user to exactly retrieve data that are relevant to him: it is not easy to formulate a database query such that:

1. on the one hand, the user retrieves all the answers that interest him; that is, false negatives and empty answers are avoided by also giving the user data that are closely related to his original query,
2. on the other hand, the user does not retrieve too much irrelevant data (that is, false positives in the form of "overabundant" and "unsatisfactory answers") – hence avoiding data not related to the user's original intention.

In this article, we assume the following setting: A user sends a query (expressed in relational algebra) to a relational database – that is, a set of data tables, where again each data table consists of a set of attributes and each attribute has been assigned a fixed domain of values (like strings or integers). However, based on the previous work in [3], we extend the relational database by ranks: each tuple in a data table has a rank value to denote how much a tuple matches the user's query. Ranks come from an ordinal scale bounded by 0 (no match) and 1

* Supported by grant no. P103/11/1456 of the Czech Science Foundation and internal grant of Palacky University no. PrF_2012_029. DAMOL is supported by project reg. no. CZ.1.07/2.3.00/20.0059 of the European Social Fund in the Czech Republic.

E. Hüllermeier et al. (Eds.): SUM 2012, LNAI 7520, pp. 16–29, 2012.

(full match) and we allow to have imperfect matches represented by intermediate ranks. These ordinal ranks have a *comparative* meaning: the higher the rank, the better it matches the user's query. Before the user starts querying the database, all ranks for tuples explicitly contained in the data tables are assumed to be 1, whereas implicitly all other tuples (not occurring in the data tuples but consisting of valid combinations of domain values) are ranked 0. By using ordinal ranks, query answering becomes more *flexible* in the following sense: The user will not only retrieve answer tuples with rank 1 (which might not even exist); instead the user will also retrieve answer tuples with ranks lower than 1 which still may contain relevant information for him. To obtain the ranks, we will employ a notion of similarity on each attribute domain: for any two values from a domain we assume a predefined value that denotes how similar the two domain values are. In particular, we show how to apply the notion of ranked data tables (RDTs) and complete residuated lattices

- to *rank* answer tuples according to their *relevance* for the user,
- to let a user specify *preferences* on equality conditions in his queries,
- and to suppress *irrelevant* answers by a global threshold.

2   Background on Ranked Data Tables

The flexible query answering approach we discuss in this paper is based on a similarity-based generalization of Codd's relational model of data [10]. As in the traditional Codd's model, we assume a set Y of attributes where each attribute $y \in Y$ has a specific fixed domain denoted D_y. A relation scheme consists of a finite subset of the attributes $R \subseteq Y$. Each data table (or relation instance) for a relation scheme R is a finite set of tuples where a tuple is a map $r \colon R \to \bigcup_{y \in R} D_y$ such that the value of the tuple for an attribute complies with the domain of the attribute, that is: $r(y) \in D_y$. The set of all tuples $r \colon R \to \bigcup_{y \in R} D_y$ will also be denoted by $\mathrm{Tupl}(R)$.

An important aspect of Codd's model is that data tables (relations) represent both the stored data and results of queries. In fact, in theory there is no distinction between the two roles as stored data can be seen as results of queries (show all stored data) and results of queries can again be stored. In both the cases, a data table $\mathcal{D}$ is a finite subset of $\mathrm{Tupl}(R)$, i.e., we may think of the tuples in $\mathcal{D}$ as tuples assigned a rank 1 indicating a match. Analogously, the tuples not present in $\mathcal{D}$ can be seen as tuples assigned a rank 0 indicating no match. Clearly, there is a one-to-one correspondence between data tables $\mathcal{D} \subseteq \mathrm{Tupl}(R)$ and such assignments where at most finitely many tuples from $\mathrm{Tupl}(R)$ are assigned rank 1. It is natural to interpret the ranks 0 and 1 as truth degrees which come from a two-element Boolean algebra with the usual interpretation (0 for falsity and 1 for truth) and ordering $0 < 1$. From this point of view, we argue that ordinal ranks which we use here in a more general setting are already present in the original Codd's model, only they do not appear explicitly. The primary role of ranks stems directly from the model – they indicate whether a tuple matches or does not match a given query (formulated in a particular query language, e.g. a relational algebra, and evaluated in a database instance).

Example 1. As a running example we assume a database of book stores and their stock. That is, we have a table of bookstores storing the name of a bookstore, the ZIP of the warehouse where the book is on stock, and the ISBN and price of each book sold in each store; and we have a table of books storing ISBN, title, and level of presentation. Hence a database instance may look as follows.

Stores	rank	name	zip	isbn	price
	1.0	Bookworm	38457	037-4-592-24599-7	59.95
	1.0	Bookworm	38457	834-3-945-25365-9	19.95
	1.0	Bookmarket	32784	834-3-945-25365-9	23.95
	1.0	BooksBooks	98765	945-7-392-66845-4	89.95

Books	rank	bookid	title	level
	1.0	037-4-592-24599-7	SQL	beginner
	1.0	834-3-945-25365-9	Databases	advanced
	1.0	945-7-392-66845-4	DB systems	professional

In this paper, to allow for flexible query answering, we utilize an extension of Codd's model which allows us to consider general ranks, not only 0 (false or no match) and 1 (true or match), coming from a general partially ordered set which is bounded by 0 and 1. Technically, the extension we use here results from Codd's model if we replace the two-valued Boolean algebra which serves as the structure of truth degrees by a more general structure. Hence, the approach we use in this paper builds upon Codd's model which is developed using a weaker metamathematics. There are several important practical consequences:

Clarity: The model stays purely relational. There is no "ad hoc" ranking module attached on top of the classic model. The classic model results from the model used here by a particular choice of the structure of ranks. Namely, if the structure is bivalent (the two-valued Boolean algebra), our model becomes the ordinary model with yes/no matches.

Ranked data tables are used instead of the classic data tables as the basic structures and represent both the results of queries and stored data. Each ranked data table $\mathcal{D}$ (an RDT) is a map assigning to each tuple $r \in \mathrm{Tupl}(R)$ a rank denoted $\mathcal{D}(r)$. The rank is interpreted as a degree to which r matches a query. The ranks have ordinal interpretation, $\mathcal{D}(r_1) > \mathcal{D}(r_2)$ means that r_1 is a better match than r_2, $\mathcal{D}(r) = 1$ means that r matches fully (a given query), $\mathcal{D}(s) = 0$ means that s does not match the query at all.

Support for imperfect matches: As in Codd's model, queries are represented by expressions which are evaluated in database instances (there are several equivalent query systems like relational algebra and domain relational calculus with range declarations). The fact that a general structure of ranks is used influences the rules how the queries are evaluated – there is a need to aggregate values of ranks that can be other than 0 and 1 and thus the operations of Boolean algebra can no longer do the job. In order to evaluate general queries, the structure of ranks shall be equipped by additional operations for aggregation of ranks. Nevertheless, the evaluation stays truth functional as in Codd's model. As a consequence, the results of queries are given only by the database instance and the structure of ranks.

Equalities replaced by similarities: Equalities on domains are an integral part of Codd's model and appear in restrictions (selections), natural joins, the semantics of functional dependencies, etc. In our model, equalities on domains (tacitly used in Codd's model) become explicit similarity relations, assigning to each two elements from a domain D_y of attribute y a degree to which they are similar. Similarity degrees come from the same scale as ranks and have the same ordinal interpretation: higher degrees mean higher similarity and thus higher preference if one is asked to choose between alternatives.

Other important aspects: The model is general and not limited just to data querying. There are results on various types of similarity-based dependencies in data [2] that can be exploited in the process of flexible query answering.

We now outline a fragment of the model which is sufficient to discuss flexible query answering with ordinal ranks. We use the unit interval on the reals $L = [0, 1]$ in all our examples which will both constitute the ranks of tuples in a data table as well as similarity degrees between any two values from a domain. In general, the scale can be an arbitrary set L bounded by 0 and 1. In order to be able to compare ranks and similarities, we equip L with a partial order $\leq$ so that $\langle L, \leq \rangle$ is a complete lattice. That means, for each subset of L there exists a supremum and an infimum with respect to $\leq$ and $\langle L, \leq \rangle$ can be alternatively denoted by $\langle L, \wedge, \vee, 0, 1 \rangle$ where $\wedge$ and $\vee$ denote the operations of infimum and supremum, respectively. In case of the real unit interval $L = [0, 1]$ and its natural ordering, the suprema and infima of finite nonempty subsets of $[0, 1]$ coincide with their maxima and minima. Moreover, we accompany the complete lattice with two binary operations that operate on any two elements of L, result in an element of L and play roles of truth functions of logical connectives "conjunction" and "implication" which are used in the process of evaluating queries as we shall see later. These operations are a multiplication $\otimes$ and a residuum $\rightarrow$. We postulate that $\langle L, \otimes, 1 \rangle$ is a commutative monoid and $\otimes$ and $\rightarrow$ satisfy a so-called adjointness property: $a \otimes b \leq c$ iff $a \leq b \rightarrow c$ (for all $a, b, c \in L$). Altogether, $\mathbf{L} = \langle L, \wedge, \vee, \otimes, \rightarrow, 0, 1 \rangle$ is called a *complete residuated lattice*.

Remark 1. The conditions for $\otimes$ and $\rightarrow$ were derived by Goguen [13] from a graded counterpart of *modus ponens* and were later employed in various multiple-valued logics with truth-functional semantics, most notably in BL [16], MTL [11] and their schematic extensions. Nowadays, adjointness is considered as a property which ensures that $\otimes$ and $\rightarrow$ are truth functions for multiple-valued conjunction and implication with reasonable properties. The properties are weaker than the properties of two-valued conjunction and implication but sufficient enough to have syntactico-semantically complete logics, see [16] and [9] for an overview of recent results. A particular case of a complete residuated lattice is a two-valued Boolean algebra if $L = [0, 1]$, $\wedge = \otimes$ is the truth function of ordinary conjunction, $\vee$ and $\rightarrow$ are truth functions of disjunction and implication, respectively.

Examples of complete residuated lattices include finite as well as infinite structures. For the real unit interval $L = [0, 1]$ and its natural ordering, all complete residuated lattices are given by a left-continuous t-norm $\otimes$, see [1,16]. Moreover, all complete residuated lattices with continuous $\otimes$ can be constructed by means of ordinal sums [8] from the following three pairs of adjoint operations:

Łukasiewicz	$a \otimes b = \max(a + b - 1, 0)$	$a \rightarrow b = \min(1 - a + b, 1)$
Gödel	$a \otimes b = \min(a, b)$	$a \rightarrow b = b$ if $a > b$; 1 otherwise
Goguen	$a \otimes b = a \cdot b$	$a \rightarrow b = \frac{b}{a}$ if $a > b$; 1 otherwise

Recall that an **L**-set (a fuzzy set) A in universe U is a map $A \colon U \rightarrow L$, $A(u)$ being interpreted as "the degree to which u belongs to A". A binary **L**-relation (a binary fuzzy relation) B on U is a map $B \colon U \times U \rightarrow L$, $B(u_1, u_2)$ interpreted as "the degree to which u_1 and u_2 are related according to B". See [1] for details.

Definition 1 (ranked data table). Let $R \subseteq Y$ be a relation scheme. A *ranked data table* on R (shortly, a RDT) is any map $\mathcal{D} \colon \mathrm{Tupl}(R) \rightarrow L$ such that there are at most finitely many tuples $r \in \mathrm{Tupl}(R)$ such that $\mathcal{D}(r) > 0$. The degree $\mathcal{D}(r)$ assigned to tuple r by $\mathcal{D}$ shall be called a *rank* of tuple r in $\mathcal{D}$.

Remark 2. The number of tuples which are assigned nonzero ranks in $\mathcal{D}$ is denoted by $|\mathcal{D}|$, i.e. $|\mathcal{D}|$ is the cardinality of $\{r \mid \mathcal{D}(r) > 0\}$. We call $\mathcal{D}$ *non-ranked* if $\mathcal{D}(r) \in \{0, 1\}$ for all tuples r. The non-ranked RDTs can be seen as initial data, i.e., relations in the usual sense representing stored data as in the classic model.

In this paper, we consider queries formulated using combinations of relational operations on RDTs. The operations extend the classic operations by considering general ranks. For our purposes, it suffices to introduce the following operations:

Intersection: An intersection (a $\otimes$-*intersection*) of RDTs $\mathcal{D}_1$ and $\mathcal{D}_2$ on relation scheme T is defined componentwise using $\otimes$: $(\mathcal{D}_1 \otimes \mathcal{D}_2)(t) = \mathcal{D}_1(t) \otimes \mathcal{D}_2(t)$ for all tuples t. If $\mathcal{D}_1$ and $\mathcal{D}_2$ are answers to queries Q_1 and Q_2, respectively, then $\mathcal{D}_1 \otimes \mathcal{D}_2$ is an answer to the conjunctive query "Q_1 and Q_2".

Projection: If $\mathcal{D}$ is an RDT on relation scheme T, the *projection* $\pi_R(\mathcal{D})$ of $\mathcal{D}$ onto $R \subseteq T$ is defined by $(\pi_R(\mathcal{D}))(r) = \bigvee_{s \in \mathrm{Tupl}(T \setminus R)} \mathcal{D}(rs)$ for each $r \in \mathrm{Tupl}(R)$. Note that rs denotes a usual concatenation of tuples r and s which is a set-theoretic union of r and s. Projection has the same meaning as in the Codd's model and the supremum $\bigvee$ aggregating the ranks $\mathcal{D}(rs)$ is used because of the existential interpretation of the projection.

Cross join (Cartesian product): For RDTs $\mathcal{D}_1$ and $\mathcal{D}_2$ on disjoint relation schemes S and T we define an RDT $\mathcal{D}_1 \bowtie \mathcal{D}_2$ on $S \cup T$, called a *cross join of $\mathcal{D}_1$ and $\mathcal{D}_2$* (or, a Cartesian product of $\mathcal{D}_1$ and $\mathcal{D}_2$), by $(\mathcal{D}_1 \bowtie \mathcal{D}_2)(st) = \mathcal{D}_1(s) \otimes \mathcal{D}_2(t)$. The cross join $\mathcal{D}_1 \bowtie \mathcal{D}_2$ contains tuples which consist of concatenations of tuples from $\mathcal{D}_1$ and $\mathcal{D}_2$. Note that in our model, we can have $|\mathcal{D}_1 \bowtie \mathcal{D}_2| < |\mathcal{D}_1| \cdot |\mathcal{D}_2|$ (e.g., if $\otimes$ is the Łukasiewicz conjunction).

Renaming attributes: The same operation as in the Codd's model [19].

If we apply these operations on non-ranked RDTs, we always obtain a nonranked RDT. RDTs with ranks other than 0 and 1 result from non-ranked RDTs by using similarity-based restrictions (selections): Given a nonranked RDT $\mathcal{D}$ and a formula like $y \approx d$ saying "(the value of the attribute) y is *similar to d*", we select from $\mathcal{D}$ only the tuples which match this similarity-based condition. Naturally, the condition shall be matched to degrees: each tuple from $\mathcal{D}$ is assigned a rank representing the degree to which the tuples matches the condition.

In order to formalize similarity-based restrictions, we equip each domain D_y with a binary **L**-relation $\approx_y$ on D_y which satisfies the following conditions: (i) for each $u \in D_y$: $u \approx_y u = 1$ (reflexivity), and (ii) for each $u, v \in D_y$: $u \approx_y v = v \approx_y u$ (symmetry). Such a relation shall be called a *similarity*. Taking into account similarities on domains, we introduce the following operation:

Restriction (selection): Let $\mathcal{D}$ be an RDT on T and let $y \in T$ and $d \in D_y$. A *similarity-based restriction* $\sigma_{y \approx d}(\mathcal{D})$ *of tuples in $\mathcal{D}$ matching $y \approx d$* is defined by $\big(\sigma_{y \approx d}(\mathcal{D})\big)(t) = \mathcal{D}(t) \otimes t(y) \approx_y d$. Considering $\mathcal{D}$ as a result of query Q, the rank of t in $\sigma_{y \approx d}(\mathcal{D})$ is interpreted as a degree to which "t matches the query Q and the y-value of t is similar to d". We can consider more general restrictions $\sigma_\varphi(\mathcal{D})$, where φ is a more complex formula than $y \approx d$.

Using cross joins and similarity-based restrictions, we can introduce *similarity-based equijoins* such as $\mathcal{D}_1 \bowtie_{p \approx q} \mathcal{D}_2 = \sigma_{p \approx q}(\mathcal{D}_1 \bowtie \mathcal{D}_2)$. Various other types of similarity-based joins can be introduced in our model. Details can be found in [3].

3   Similarity-Based Ranking of Database Answers

If a user specifies equality conditions in his query to select tuples from the database, these equality conditions may be too strict to give the user a satisfactory answer. The simplest setting to retrieve more relevant answers for the user is to replace equality $=$ in each condition with similarity $\approx$ and then use the similarity-based algebraic operators to obtain ranks for the answer tuples. The ranks will be based on the predetermined similarities between domain values. For simplicity, we concentrate in this paper on selection-projection-join (SPJ) queries. Other operators like division, union and intersection can be incorporated like in [3]. We may also introduce in our model a top_k operation in much the same sense as, e.g., in RankSQL [18] but in a truth-functional way. We do not discuss these issues here because of the limited scope of this paper.

Example 2. For example, we can ask for titles and prices of books sold by a bookstore with ZIP code 35455 by doing an equi-join over the ISBN:

$$\pi_{\text{title,price}}[\sigma_{\text{zip}=35455}(Books \bowtie_{\text{bookid}=\text{isbn}} Stores)]$$

Or ask for the name of a bookstore that sells "professional books" entitled "Databases":

$$\pi_{\text{name}}[\sigma_{\text{level}=\text{professional,title}=\text{Databases}}(Books \bowtie_{\text{bookid}=\text{isbn}} Stores)]$$

In general, we consider SPJ queries of the form:

$$\pi_A[\sigma_C(\rho_{\mathcal{R}}(\mathcal{D}_1) \bowtie_{\mathcal{E}_1} \cdots \bowtie_{\mathcal{E}_m} \rho_{\mathcal{R}}(\mathcal{D}_n))] \quad \text{where}$$

- $\mathcal{D}_i$ is a ranked data table
- C is a conjunction of selection conditions consisting of
    - equalities $y = d$ (where y is an attribute in the relation scheme and d is a constant from its domain D_y)
    - equalities $y_1 = y_2$ (where y_1 and y_2 are attributes in the relation scheme with a common domain $\mathcal{D}_i$)
- $\mathcal{E}_j$ is a set of join conditions as equalities between attributes $y_1 = y_2$ (where y_1 and y_2 are attributes in the relation scheme of some $\mathcal{D}_i$)
- $\mathcal{R}$ is a list of renaming conditions $y' \leftarrow y$ giving attribute y a new name y' (where y is an attribute in the relation scheme of the appropriate $\mathcal{D}_i$ but y' does not occur in any relation scheme)
- A is the set of projection attributes (after renaming according to $\mathcal{R}$)

Example 3. For the example SPJ queries there are no tuples in the example tables that satisfy the equality conditions. Now assume that similarity is defined on attribute zip as (35455≈38457)=0.9 as well (35455≈32784)=0.8 and (35455≈98765)=0.2; whereas for bookid and isbn we assume that similarity is defined by strict equality, that is it is 1 only for exactly the same ISBN and 0 otherwise: $(d \approx d) = 1$, but $(d \approx d') = 0$ for $d \neq d'$. To obtain a ranked data table as a flexible answer for the first query, we replace equality with similarity:

$$\pi_{\texttt{title,price}}[\sigma_{\texttt{zip}\approx 35455}(Books \bowtie_{\texttt{bookid}\approx\texttt{isbn}} Stores)]$$

results in the following ranked answer table:

rank	title	price
$0.9(= 0.9 \otimes 1)$	SQL	59.95
$0.9(= 0.9 \otimes 1)$	Databases	19.95
$0.8(= 0.8 \otimes 1)$	Databases	23.95
$0.2(= 0.2 \otimes 1)$	DB systems	89.95

By furthermore assuming similarity on the attributes level and zip we can also relax the second query. For example, let (professional≈advanced)=0.6 and (professional≈beginner)=0.1, as well as (Databases≈DB systems)=0.9 and (Databases≈SQL)=0.7. Then the query

$$\pi_{\texttt{name}}[\sigma_{\texttt{level}\approx\texttt{professional,title}\approx\texttt{Databases}}(Books \bowtie_{\texttt{bookid}\approx\texttt{isbn}} Stores)]$$

returns the table

rank	name
$0.9(= 1 \otimes 0.9 \otimes 1)$	BooksBooks
$0.6(= 0.1 \otimes 0.7 \otimes 1 \vee 0.6 \otimes 1 \otimes 1)$	Bookworm
$0.6(= 0.6 \otimes 1 \otimes 1)$	Bookmarket

4   Emphasizing Some Equality Conditions

Extending this simple setting, a user might want to express importance of some equality conditions in his queries. For one, he might want to express that he requires equality (that is full similarity to degree 1) in a selection condition. More generally, a user must be able to express that some conditions are more important for him – and hence to require a higher degree of satisfaction for these conditions; whereas for other conditions he is more willing to relax the equality requirement – and he will be content with a lower degree of satisfaction for these conditions. In our model, this kind of emphasis mechanism can be implemented using so-called residuated shifts as follows. Similarity-based restrictions $\sigma_{y\approx d}(\mathcal{D})$ that appear in our queries can be generalized so that we consider a more general formula of the form $a \Rightarrow y \approx d$ instead of $y \approx d$. In this setting, we introduce a *similarity-based restriction* $\sigma_{a\Rightarrow y\approx d}(\mathcal{D})$ *of tuples in* $\mathcal{D}$ *matching* $y \approx d$ *at least to degree* $a \in L$ defined with the help of the residuum operator $(\rightarrow)$ by

$$\left(\sigma_{a\Rightarrow y\approx d}(\mathcal{D})\right)(t) = \mathcal{D}(t) \otimes (a \rightarrow t(y) \approx_y d). \tag{1}$$

Using adjointness, we get that $1 \rightarrow a = a$ for all $a \in L$. Hence, $\sigma_{1\Rightarrow y\approx d}(\mathcal{D}) = \sigma_{y\approx d}(\mathcal{D})$. On the other hand, $0 \rightarrow a = 1$ for all $a \in L$, meaning $\sigma_{0\Rightarrow y\approx d}(\mathcal{D}) = \mathcal{D}$. Due to the monotony of $\otimes$ and antitony of $\rightarrow$ in the first argument, we get

$$\left(\sigma_{b\Rightarrow y\approx d}(\mathcal{D})\right)(t) \leq \left(\sigma_{a\Rightarrow y\approx d}(\mathcal{D})\right)(t)$$

whenever $a \leq b$. By a slight abuse of notation, the latter fact can be written as $\sigma_{b\Rightarrow y\approx d}(\mathcal{D}) \subseteq \sigma_{a\Rightarrow y\approx d}(\mathcal{D})$. Therefore, $a \in L$ in $\sigma_{a\Rightarrow y\approx d}(\mathcal{D})$ acts as a *threshold degree*, the lower the degree, the lower the emphasis on the condition $y \approx d$ and, in consequence, the larger the answer set. In the borderline cases, $\sigma_{0\Rightarrow y\approx d}(\mathcal{D})$ means no emphasis on the condition, $\sigma_{1\Rightarrow y\approx d}(\mathcal{D})$ means full emphasis, i.e., the original similarity-based selection.

Remark 3. Let us comment on the role of degrees as thresholds. Using adjointness, for all $a, b \in L$, we have $a \leq b$ iff $a \rightarrow b = 1$. Applied to (1), $a \rightarrow t(y) \approx_y d = 1$ iff the y-value of t is similar to d at least to degree a. Therefore, if $\mathcal{D}$ is a result of query Q, then the rank $\sigma_{a\Rightarrow y\approx d}(\mathcal{D})(t)$ shall be interpreted as a degree to which "t matches Q and the y-value of t is similar to d at least to degree a". This justifies the interpretation of $a \in L$ as a threshold degree. The benefit of using $a \rightarrow t(y) \approx_y d$ in (1) which is in fuzzy relational systems [1] called a residuated shift, is that the threshold is exceeded gradually and not just "exceeded in terms yes/no". For instance, if $a > t(y) \approx_y d$ and the similarity degree $t(y) \approx_y d$ is sufficiently close to a, the result of $a \rightarrow t(y) \approx_y d$ is not 1 but it shall be sufficiently close to 1, expressing the fact that the threshold has almost been exceeded, i.e., that the y-value of t is similar to d almost to degree a. This is illustrated by the following example.

Example 4. Assume we want to express that a similarity on the ZIP code of above 0.8 is perfectly fine for us, then the first query looks like this

$$\pi_{\texttt{title,price}}[\sigma_{0.8\Rightarrow(\texttt{zip}\approx35455)}(Books \bowtie_{\texttt{bookid}\approx\texttt{isbn}} Stores)]$$

and results in the following ranked answer table by evaluating $0.8 \to 0.9$ to 1.0, $0.8 \to 0.8$ to 1.0, and $0.8 \to 0.2$ to 0.2 in the Gödel algebra (this value would be 0.4 in Łukasiewicz and 0.25 in Goguen algebra):

rank	title	price
$1.0(= (0.8 \to 0.9) \otimes 1)$	SQL	59.95
$1.0(= (0.8 \to 0.9) \otimes 1)$	Databases	19.95
$1.0(= (0.8 \to 0.8) \otimes 1)$	Databases	23.95
$0.2(= (0.8 \to 0.2) \otimes 1)$	DB systems	89.95

In the second query, if we insist on books with professional level, but we are indeed interested in books with related titles up to a similarity of 0.5:

$$\pi_{\texttt{name}}\left[\sigma_{1\Rightarrow(\texttt{level}\approx\texttt{professional}),0.5\Rightarrow(\texttt{title}\approx\texttt{Databases})}\left(Books \bowtie_{\texttt{bookid}\approx\texttt{isbn}} Stores\right)\right]$$

then we get the answer table

rank	name
$1.0(= (1 \to 1) \otimes (0.5 \to 0.9) \otimes 1)$	BooksBooks
$0.6(= (1 \to 0.1) \otimes (0.5 \to 0.7) \otimes 1 \vee (1 \to 0.6) \otimes (0.5 \to 1) \otimes 1)$	Bookworm
$0.6(= (1 \to 0.6) \otimes (0.5 \to 1) \otimes 1)$	Bookmarket

5   Global Relevance Threshold for Subqueries

The results of queries in our model are influenced by the underlying structure of truth degrees **L** because the operations of **L** are used to aggregate ranks. Hence, different choices of **L** in general lead to different answer sets. By a careful choice of the structure of truth degrees, we can allow users to influence the size of the answer set so that the user can tune the structure to obtain an answer set of the most desirable size. In this section, we consider a situation where we want to emphasize answer tuples that satisfy at least some subqueries (similarity conditions) with a high degree. Opposed to this, answer tuples that satisfy all subqueries (similarity conditions) with lower degrees than desirable should be ranked considerably lower. The user might then specify a global threshold (as opposed to the local thresholds in Section 4) to denote that values above the threshold are relevant to him whereas values below the threshold are irrelevant. Interestingly, this idea can be implemented in our model by a choice of the structure **L** *without* altering the relational operations.

Remark 4. Note that the Goguen $\otimes$ (usual multiplication of real numbers) has the property that $a \otimes b > 0$ for all $a, b > 0$ ($\otimes$ is called strict), see Fig. 1 (right). The Łukasiewicz $\otimes$ does not have this property: for each $0 < a < 1$ there is $b > 0$ such that $a \otimes b = 0$ ($\otimes$ is called nilponent), see Fig. 1 (left). Thus, if one uses the Goguen operations on $[0, 1]$, each query considered in this paper has a nonempty answer set provided that all similarities have the property that $d_1 \approx_y d_2 > 0$ which can be technically ensured. In practice, the benefit of always having a nonempty answer set this way is foiled by having (typically) a large number of answers with very low ranks which match the initial query only to a

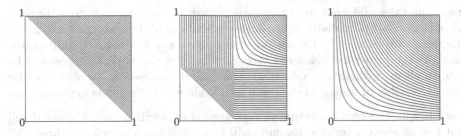

Fig. 1. Contour diagrams: Łukasiewicz $\otimes$ (left), $\otimes_{0.5}$ (middle), Goguen $\otimes$ (right)

very low degree and are (typically) not interesting for a user. On the contrary, using the Łukasiewicz operations instead of the Goguen ones, one can have a situation where there are subqueries satisfied to high degrees, say $a_1, \ldots, a_k$ but $a_1 \otimes \cdots \otimes a_k = 0$, i.e., the answer tuples will be lost since their rank will be zero.

Considering the behavior of the Goguen and Łukasiewicz operations mentioned in Remark 4 and our motivation to separate answers with sufficiently high degrees from the rest, we may consider complete residuated lattices $\mathbf{L}$ defined on the real unit interval that act as the Goguen structure on a subinterval $(\alpha, 1]$ and as the Łukasiewicz structure on the subinterval $[0, \alpha)$. The degree $\alpha \in [0, 1]$ is then a threshold that can be set by a user saying that "*if ranks of answer tuples exceed the threshold $\alpha \in L$, they shall not vanish from the answer set*". The combination of Goguen and Łukasiewicz structure we need for this particular purpose can be described as a result of algebraic operation called an ordinal sum [8]. For $\alpha \in [0, 1]$, we let

$$
a \otimes_\alpha b = \begin{cases} \alpha + \dfrac{(a - \alpha)(b - \alpha)}{1 - \alpha}, & \text{if } a, b \in (\alpha, 1), \\ \max(0, a + b - \alpha), & \text{if } a, b \in (0, \alpha), \\ \min(a, b), & \text{otherwise,} \end{cases}
$$

$$
a \rightarrow_\alpha b = \begin{cases} 1, & \text{if } a \le b, \\ \alpha + \dfrac{(1 - \alpha)(b - \alpha)}{a - \alpha}, & \text{if } 1 > a > b > \alpha, \\ \alpha - a + b, & \text{if } \alpha > a > b, \\ b, & \text{otherwise,} \end{cases}
$$

for all $a, b \in [0, 1]$. Then, $\mathbf{L}_\alpha = \langle L, \wedge, \vee, \otimes_\alpha, \rightarrow_\alpha, 0, 1 \rangle$ is a complete residuated lattice (an ordinal sum of an isomorphic copy of a Łukasiewicz structure and Goguen structure with the idempotent $\alpha \in L$, see [1,8,16] for details. Fig. 1 (middle) shows $\otimes_{0.5}$, i.e., the multiplication of $\mathbf{L}_{0.5}$ which behaves as the Łukasiewicz conjunction on $[0, 0.5]$ and the Goguen conjunction on $(0.5, 1]$.

Remark 5. Taking $\mathbf{L}_\alpha$ for the structure of degrees has the advantage that the threshold $\alpha \in L$ is *not* acting as a clear "cut" where answer tuples with a degree below the threshold are completely disregarded; instead, it leads to a stepwise degradation (but potentially with a degree still above 0) depending on how many subqueries are satisfied by the answer tuple only to a degree below the threshold: the more subqueries are below the threshold the closer the overall degree will be to 0. In effect, with this threshold we can flexibly increase or decrease the size (that is, number of tuples) in the result as it is shown by the following assertions.

The following theorem states that you get more results with a lower threshold and that results above a certain threshold will be ranked lower by a lower threshold.

Theorem 1. *Let $\mathcal{D}_{\mathbf{L}_\alpha}^Q$ be a result of an SPJ query Q when $\mathbf{L}_\alpha$ is used as the structure of truth degrees. If $\alpha < \beta$, then $|\mathcal{D}_{\mathbf{L}_\beta}^Q| \leq |\mathcal{D}_{\mathbf{L}_\alpha}^Q|$. If $\alpha < \beta$ and a tuple t satisfies all subqueries of Q to a degree greater than β, then $(\mathcal{D}_{\mathbf{L}_\alpha}^Q)(t) < (\mathcal{D}_{\mathbf{L}_\beta}^Q)(t)$.*

Proof (a sketch). The claim follows from fact that $a \otimes_\alpha b = 0$ implies $a \otimes_\beta b = 0$ whenever $\alpha \leq \beta$, i.e. α yields a greater (or equally sized) answer set than β. The second claim is a consequence of $a \otimes_\alpha b < a \otimes_\beta b$ for $a, b \in (\beta, 1)$. □

Example 5. Consider the following query with four similarity conditions

$$\sigma_{\texttt{title}\approx\texttt{SQL},\texttt{level}\approx\texttt{beginner},\texttt{zip}\approx\texttt{56571},\texttt{price}\approx\texttt{20.00}}(\textit{Books} \bowtie_{\texttt{bookid}\approx\texttt{isbn}} \textit{Stores})$$

In the table *Books* $\bowtie_{\texttt{bookid}\approx\texttt{isbn}}$ *Stores* we have the tuple $\langle$834-3-945-25365-9, Databases, advanced, Bookworm, 32784, 834-3-945-25365-9, 19.95$\rangle$. With the following given similarities (SQL≈Databases)=0.7, (beginner≈advanced)=0.4, (56571≈32784)=0.1, and (20.00≈19.95)=0.9, we see that two similarity conditions are satisfied at high degrees (0.7 and 0.9) while the other two are satisfied at low degrees (0.4 and 0.1). Comparing the different $\otimes$ operators we get:

Lukasiewicz	Gödel	Goguen	$\otimes_{0.5}$	$\otimes_{0.4}$	$\otimes_{0.3}$
0	0.1	0.0252	0	0.1	0.1

We see that for threshold $\alpha = 0.5$ the tuple is not included in the answer table (ranked 0); when decreasing the threshold to $\alpha = 0.4$ it is included (ranked 0.1) and stays in the answer for the lower threshold $\alpha = 0.3$.

Another potential tuple is $\langle$945-7-392-66845-4, DB systems, professional, BooksBooks, 98765, 945-7-392-66845-4, 89.95$\rangle$. With similarities (SQL≈DB systems)=0.3, (beginner≈professional)=0.1, (56571≈98765)=0.1, as well as (20.00≈89.95)=0.1, we see that all similarity conditions are satisfied at low degrees (0.3 and 0.1). Comparing the different $\otimes$ operators we get:

Lukasiewicz	Gödel	Goguen	$\otimes_{0.5}$	$\otimes_{0.4}$	$\otimes_{0.3}$
0	0.1	0.0003	0	0	0

Hence, even for low thresholds the tuple does not occur in the answer table.

6   Related Work

The notion of answering queries in a flexible and user-oriented manner has been investigated for a long time; see [12] for some recent examples and [21] for an extensive study on fuzzy querying (and a query language called SQLf). Other related work of Bosc et al considers extending the support of a fuzzy membership function in order to weaken fuzzy conditions in a query [4] where a tolerance relation can be used to flexibly model closeness in a domain. In [14], Hadjali and Pivert analyze the use of fuzzy views and study the influence of different fuzzy implication operators (like Gödel, Goguen and Łukasiewicz implication) on which views in a distributed database system are chosen to answer a fuzzy query. Pivert et al ([22]) handle the empty answer set (EAS) problem and the unsatisfactory answer set (UAS) problem, where there are either no answers at all or only answers with a low degree of satisfaction with respect to a user-defined threshold α. They efficiently determine minimal failing subqueries (MFS) of conjunctive queries defined by α-cuts; these MFS can give the user some explanation on why his query failed (that is, the answer set was empty or unsatisfactory). Although the notion of a threshold is common with our work, in [22] (sub-)queries below the threshold are totally disregarded, whereas our relevance threshold leads to more emphasis for answer tuples with some subqueries rated above the threshold (as opposed to lower ranks for answer tuples with all subqueries rated below it).

Vaneková and Vojtáš [23] provide an implementation of another fuzzy-set-based modeling of user preferences. They consider searching for data based on several attributes. A user expresses his search preferences as a membership function on the attribute domains.

In sum, most of the related works discussed above relies on gradual predicates or trapezoidal membership functions. In contrast, our approach relies on similarity defined on attribute domains. We argue that it might be equally difficult to define the membership degrees (see also the FITA/FATI problem discussed in [15]) for the fuzzy-set-based approaches as it is difficult to define the similarity values on attribute domains which our similarity-based approach relies on; hence no preference should be given to one or the other approach: while there are cases where user preferences can be easily modeled by fuzzy sets, in other cases it might be more natural to use similarities on domains. The paper [5] also provides a survey of related techniques (including similarity-based approaches) and argues how they are included in the fuzzy-set-based system. The similarity-based approaches surveyed there however do not take advantage of lattice-based fuzzy logic but rely more or less on metric calculations of distances. In this paper we showed how to put similarity-based flexible query answering on a sound logical foundation.

Preference queries – extensively studied by Chomicki (et al) [6,7,20] – are a further field of related work: a preference order $\prec$ must be predefined on attribute domains by a user; these attribute preference relations can then be combined into a tuple preference relation and this tuple preference relation can be used to return the most-preferred tuples upon a user query for example with the algebraic "winnow" operator. In particular, a special case of preference relations can be expressed by a scoring function f that is used to compare two constant values. The work in [24] extends preferences to work between sets of tuples. The advantage of preference orders is that they have some desirable properties (like transitivity: if A is preferred to B and B is preferred to C then A is also preferred to C). By iterating the winnow operator [6], a ranking of answer tuples (into best, second-best, ...) can be obtained; this is similar to our approach where the similarity degree of each answer tuple gives the user a fine-grained means of deciding if the tuple is relevant to him. A formal comparison between the properties of preference queries and the similarity-based approach we propose in this paper might be an interesting topic of future work.

Moreover, the user can specify dependencies between data – in particular, dependencies between different relations in the database. These dependencies can be given by logical rules and can be applied to the query to retrieve other answers that are relevant for the user under the background knowledge specified by the set of rules. This operation is known as "Goal Replacement" (e.g., [17]); it might be worthwhile to study its behavior in similarity-based query answering.

7   Conclusion

We applied similarities on attribute domains to rank tuples in answers to SPJ queries on relational databases. These ranks allow for flexible query answering as they have a comparative meaning and help users identify the answer tuples that match their intention best. Ranks are computed by i) replacing equality with similarities (Section 3), ii) emphasizing individual equality conditions with the residuated shift (Section 4), and iii) setting a global threshold to alter the size of the answer tables using ordinal sums of residuated lattices (Section 5). Our model has a strong theoretical background in the theory of ranked data tables where the classical bivalent structure is replaced by a structure with more than two truth values (that is, a lattice of ranks). Several hints towards future work have been given in Section 6. An efficient prototypical implementation of query answering in ranked data tables is under development at the Data Analysis and Modeling Laboratory of Palacky University; its performance will be closely analyzed in the future.

References

1. Bělohlávek, R.: Fuzzy Relational Systems: Foundations and Principles. Kluwer Academic Publishers, Norwell (2002)
2. Bělohlávek, R., Vychodil, V.: Data Tables with Similarity Relations: Functional Dependencies, Complete Rules and Non-redundant Bases. In: Li Lee, M., Tan, K.-L., Wuwongse, V. (eds.) DASFAA 2006. LNCS, vol. 3882, pp. 644–658. Springer, Heidelberg (2006)
3. Bělohlávek, R., Vychodil, V.: Query systems in similarity-based databases: logical foundations, expressive power, and completeness. In: ACM Symposium on Applied Computing (SAC), pp. 1648–1655. ACM (2010)
4. Bosc, P., HadjAli, A., Pivert, O.: Incremental controlled relaxation of failing flexible queries. Journal of Intelligent Information Systems (JIIS) 33(3), 261–283 (2009)
5. Bosc, P., Pivert, O.: Fuzzy queries and relational databases. In: 9th ACM Symposium on Applied Computing (SAC), pp. 170–174 (1994)

6. Chomicki, J.: Preference formulas in relational queries. ACM Transactions on Database Systems 28(4), 427–466 (2003)
7. Chomicki, J.: Logical foundations of preference queries. IEEE Data Engineering Bulletin 34(2), 3–10 (2011)
8. Cignoli, R., Esteva, F., Godo, L., Torrens, A.: Basic fuzzy logic is the logic of continuous t-norms and their residua. Soft Computing - A Fusion of Foundations, Methodologies and Applications 4, 106–112 (2000)
9. Cintula, P., Hájek, P.: Triangular norm based predicate fuzzy logics. Fuzzy Sets and Systems 161, 311–346 (2010)
10. Codd, E.F.: A relational model of data for large shared data banks. Communications of the ACM 26, 64–69 (1983)
11. Esteva, F., Godo, L.: Monoidal t-norm based logic: towards a logic for left-continuous t-norms. Fuzzy Sets and Systems 124(3), 271–288 (2001)
12. Galindo, J. (ed.): Handbook of Research on Fuzzy Information Processing in Databases. IGI Global (2008)
13. Goguen, J.A.: The logic of inexact concepts. Synthese 19, 325–373 (1979)
14. HadjAli, A., Pivert, O.: Towards Fuzzy Query Answering Using Fuzzy Views – A Graded-Subsumption-Based Approach. In: An, A., Matwin, S., Raś, Z.W., Ślęzak, D. (eds.) ISMIS 2008. LNCS (LNAI), vol. 4994, pp. 268–277. Springer, Heidelberg (2008)
15. HadjAli, A., Pivert, O.: A Fuzzy-Rule-Based Approach to the Handling of Inferred Fuzzy Predicates in Database Queries. In: Christiansen, H., De Tré, G., Yazici, A., Zadrozny, S., Andreasen, T., Larsen, H.L. (eds.) FQAS 2011. LNCS, vol. 7022, pp. 448–459. Springer, Heidelberg (2011)
16. Hájek, P.: Metamathematics of Fuzzy Logic. Kluwer Academic Publishers, Dordrecht (1998)
17. Inoue, K., Wiese, L.: Generalizing Conjunctive Queries for Informative Answers. In: Christiansen, H., De Tré, G., Yazici, A., Zadrozny, S., Andreasen, T., Larsen, H.L. (eds.) FQAS 2011. LNCS, vol. 7022, pp. 1–12. Springer, Heidelberg (2011)
18. Li, C., Chang, K.C.C., Ilyas, I.F., Song, S.: RankSQL: query algebra and optimization for relational top-k queries. In: Proc. 2005 ACM SIGMOD, pp. 131–142 (2005)
19. Maier, D.: The Theory of Relational Databases. Computer Science Press (1983)
20. Mindolin, D., Chomicki, J.: Contracting preference relations for database applications. Artificial Intelligence 175(7-8), 1092–1121 (2011)
21. Pivert, O., Bosc, P.: Fuzzy Preference Queries to Relational Databases. Imperial College Press (2012)
22. Pivert, O., Smits, G., HadjAli, A., Jaudoin, H.: Efficient Detection of Minimal Failing Subqueries in a Fuzzy Querying Context. In: Eder, J., Bielikova, M., Tjoa, A.M. (eds.) ADBIS 2011. LNCS, vol. 6909, pp. 243–256. Springer, Heidelberg (2011)
23. Vaneková, V., Vojtáš, P.: Fuzziness as a model of user preference in semantic web search. In: European Society of Fuzzy Logic and Technology Conference (EUSFLAT), pp. 998–1003 (2009)
24. Zhang, X., Chomicki, J.: Preference queries over sets. In: 27th International Conference on Data Engineering (ICDE), pp. 1019–1030. IEEE Computer Society (2011)

Managing Uncertainty in Databases
and Scaling It Up to Concurrent Transactions

Alfredo Cuzzocrea[1], Rubén de Juan Marín[2],
Hendrik Decker[2], and Francesc D. Muñoz-Escoí[2]

[1] ICAR-CNR and University of Calabria, I-87036 Cosenza, Italy
[2] Instituto Tecnológico de Informática, UPV, E-46022 Valencia, Spain

Abstract. Uncertainty and data integrity are closely related. The uncertainty of data can be modeled and maintained by existing database integrity technology. The usual requirement that integrity constraints must always be satisfied needs to be renounced for constraints that model uncertainty. That is possible due to the recently introduced inconsistency tolerance of methods for integrity checking, integrity repair and query answering. Inconsistency tolerance also enables the scaling up of uncertainty management to concurrent transactions.

1 Introduction

Uncertainty in databases is closely related to inconsistency, i.e., lack of integrity, in two ways. Firstly, the validity of answers in inconsistent databases obviously is uncertain. Secondly, conditions for stating properties of uncertainty of data can be modeled as integrity constraints. Thus, each constraint violation corresponds to some uncertainty in the database, no matter if the constraint models a regular integrity assertion or some specific uncertainty condition. This paper addresses both of the mentioned relations between uncertainty and inconsistency.

For instance, the denial $\leftarrow item(x, y), y < 75\%$ constrains entries x in the *item* table to have a probability (certainty) y of at least 75%. Similarly, the constraint $I = \leftarrow uncertain(x)$, where *uncertain* is defined by the database clause $uncertain(x) \leftarrow email(x, from(y)), \sim authenticated(y)$, bans each email message x that is uncertain because its sender y has not been authenticated. Likewise, *uncertain* could be defined, e.g., by $uncertain(x) \leftarrow item(x, null)$, indicating an uncertainty about each item x the attribute of which has a null value.

An advantage of modeling uncertainty by constraints is that the evolution of uncertainty across updates can then be monitored by inconsistency-tolerant methods for integrity checking, and uncertainty can then be eliminated by integrity repairing. For instance, each update U that tries to insert an email by a non-authenticated sender will be rejected by each method that checks U for integrity preservation, since U would violate I, in the preceding example. Ditto, stored email entries with unauthenticated senders or items with unknown attributes can be eliminated by repairing the violations of I in the database.

Conventional approaches to integrity management unrealistically require total constraint satisfaction before an update is checked and after a repair is done. However, methods for checking or repairing integrity or uncertainty must be

E. Hüllermeier et al. (Eds.): SUM 2012, LNAI 7520, pp. 30–43, 2012.

inconsistency-tolerant as soon as data that violate some constraint are admitted to persist across updates. In [20], we have shown that the total consistency requirement can be waived without further ado for most (though not all) known methods. Thus, they can be soundly applied in databases with persistent constraint violations, i.e. with extant inconsistency and uncertainty.

Rather than pretending that consistent databases certainly remain consistent across updates (as conventional methods do), inconsistency-tolerant methods just assure that inconsistency, i.e., uncertainty is not increased, neither by updates nor by repairs. Such increase or decrease is determined by violation measures [18] (called 'inconsistency metrics' in [17]). Some of these measures also serve to provide answers that have integrity in the presence of uncertainty, by adopting an inconsistency-tolerant approach proposed in [16], called AHI.

Inconsistency tolerance also enables uncertainty management for concurrent transactions. For making any guarantees of integrity preservation across concurrent transactions, the usual requirement is that each transaction maps each consistent state to a consistent successor state. Unfortunately, that excludes any prediction for what is going to happen in the presence of constraint violations, i.e., of uncertainty. However, we are going to see that the inconsistency tolerance of integrity management easily scales up to concurrent transactions, and concurrent query answering with AHI remains certain in the presence of uncertainty.

After some preliminaries in Section 2, we recapitulate inconsistency-tolerant integrity management (checking, repairing and query answering) in Section 3. In Section 4, we elaborate an example of how to manage uncertainty expressed by constraints. In Section 5, we outline how inconsistency-tolerant constraint management scales up to database systems with concurrent transactions. In Section 6, we address related work. In Section 7, we conclude.

2 Preliminaries

We use terminology and formalisms that are common for *datalog* [1]. Also, we assume some familiarity with transaction concurrency control [6].

Throughout the paper, we use symbols like D, I, IC, U for representing a database, an integrity constraint (in short, *constraint*), a finite set of constraints (also called *integrity theory*) and, resp., an update. We denote the result of executing an update U on D by D^U, and the truth value of a sentence (i.e., a closed formula) or a set of sentences S in D be denoted by $D(S)$. Any of the usual database semantics will do for large classes of databases and constraints.

Constraints often are asserted as denials, i.e., clauses with empty head of the form $\leftarrow B$, where the body B is a conjunction of literals that state what should not be true in any state of the database. For each constraint I that expresses what should be true, a denial form of I can be obtained by re-writing $\leftarrow \sim I$ in clausal form, as described, e.g., in [15]. Instead of leaving the head of denial constraints empty, a predicate that expresses some lack of consistency may be used in the head. For instance, $uncertain \leftarrow B$ explicitly states an uncertainty that is associated to each instance of B that is true in the database.

For each formula B, let $\forall B$ denote the universal closure of B.

3   Inconsistency-Tolerant Uncertainty Management

As argued in Section 1, violations of constraints, i.e., the inconsistency of given database states with their associated integrity theory, reflect uncertainty. Each update may violate or repair constraints, and thus increase or decrease the amount of uncertainty. Hence, checking updates for such increases, and decreasing uncertainty by repairing violated constraints, are essential for uncertainty management. Also mechanisms for providing answers that are certain in uncertain databases are needed. In 3.1, 3.2, and 3.3, we recapitulate and extend measure-based inconsistency-tolerance for integrity checking [17], repairing [18] and, resp., query answering [16], in terms of uncertainty.

3.1   Measure-Based Uncertainty-Tolerant Integrity Checking

In theory, an update is committed only if all integrity constraints remain totally satisfied. Since total integrity is rarely achieved, and in particular not in uncertain databases where uncertainty is modeled by constraints, integrity checking methods that are able to tolerate uncertainty are needed.

In [20], inconsistency-tolerant integrity checking has been formalized and discussed. In particular, it has been shown that many (but not all) existing integrity checking methods tolerate inconsistency and thus uncertainty, although most of them have been designed to be applied only if all constraints are totally satisfied before any update is checked. In [17], we have seen that integrity checking can be described by 'violation measures' [18], which are a form of inconsistency measures [25]. Such measures, called 'uncertainty measures' below, size the amount of violated constraints in pairs (D, IC). Thus, an update can be accepted if it does not increase the measured amount of constraint violations.

Definition 1. We say that $(\mu, \preccurlyeq)$ is an *uncertainty measure* (in short, a *measure*) if μ maps pairs (D, IC) to some metric space $(\mathbb{M}, \preccurlyeq)$ where $\preccurlyeq$ is a partial order, i.e. a binary relation on $\mathbb{M}$ that is antisymmetric, reflexive and transitive. For $E, E' \in \mathbb{M}$, let $E \prec E'$ denote that $E \preccurlyeq E'$ and $E \neq E'$.

In [17, 18, 23, 25], various axiomatic properties of uncertainty measures that go beyond Definition 1 are proposed. Here, we refrain from that, since the large variety of conceivable measures has been found to be "too elusive to be captured by a single definition" [23]. Moreover, several properties that are standard in measurement theory [4] and that are postulated also for inconsistency measures in [23, 25] do not hold for uncertainty measures, due to the non-monotonicity of database negation, as shown in [18]. To postulate a distance function for each measure, as in [18], is possible but not essential for the purpose of this paper.

Definition 2 captures each integrity checking method $\mathcal{M}$ (in short, method) as an I/O function that maps updates to $\{ok, ko\}$. The output ok means that the checked update is acceptable, and ko that it may not be acceptable. For deciding to ok or ko an update, $\mathcal{M}$ uses an uncertainty measure.

Definition 2. (*Uncertainty-tolerant Integrity Checking* (*abbr.: UTIC*))
An *integrity checking method* maps triples (D, IC, U) to $\{ok, ko\}$. Each such method $\mathcal{M}$ is called a *sound* (resp., *complete*) *UTIC method* if there is an uncertainty measure $(\mu, \preccurlyeq)$ such that, for each (D, IC, U), (1) (resp., (2)) holds.

$$\mathcal{M}(D, IC, U) = ok \;\Rightarrow\; \mu(D^U, IC) \preccurlyeq \mu(D, IC) \tag{1}$$

$$\mu(D^U, IC) \preccurlyeq \mu(D, IC) \;\Rightarrow\; \mathcal{M}(D, IC, U) = ok \tag{2}$$

If $\mathcal{M}$ is sound, it is also called a *μ-based UTIC method*.

The only real difference between conventional integrity checking and UTIC is that the former additionally requires total integrity before the update, i.e., that $D(IC) = true$ in the premise of Definition 2. The range of the measure μ used by conventional methods is the binary metric space $(\{true, false\}, \preccurlyeq)$ where $\mu(D, IC) = true$ means that IC is satisfied in D, $\mu(D, IC) = false$ that it is violated, and $true \prec false$, since, in each consistent pair (D, IC), there is a zero amount of uncertainty, which is of course less than the amount of uncertainty of each inconsistent pair (D, IC).

More differentiated uncertainty measures are given, e.g., by comparing or counting the sets of instances of violated constraints, or the sets of 'causes' of inconsistencies. Causes (characterized more precisely in 3.3.1) are defined in [16, 17] as the data whose presence or absence in the database is responsible for integrity violations. Other violation measures are addressed in [17].

As seen in [17], many conventional methods can be turned into measure-based uncertainty-tolerant ones, simply by waiving the premise $D(IC) = true$ and comparing violations in (D, IC) and (D^U, IC). If there are more violations in (D^U, IC) than in (D, IC), they output *ko*; otherwise, they may output *ok*. According to [20], the acceptance of U by an uncertainty-tolerant method guarantees that U does not increase the set of violated instances of constraints.

More generally, the following result states that uncertainty can be monitored and its increase across updates can be prevented by each UTIC method, in as far as uncertainty is modeled in the syntax of integrity constraints.

Theorem 1. Let D be a database and IC an integrity theory that models uncertainty in D. Then, the increase of uncertainty in D by any update U can be prevented by checking U with any sound UTIC method.

3.2   Uncertainty-Tolerant Integrity-Preserving Repairs

In essence, repairs consist of updates that eliminate constraint violations [28]. However, violations that are hidden or unknown to the application or the user may be missed when trying to repair a database. Moreover, as known from repairing by triggers [9], updates that eliminate some violation may inadvertently violate some other constraint. Hence, uncertainty-tolerant repairs are called for. Below, we recapitulate the definition of partial and total repairs in [20]. They are uncertainty-tolerant since some violations may persist after partial repairs. But they may not preserve integrity.

Definition 3. (*Repair*)
For a triple (D, IC, U), let S be a subset of IC such that $D(S) = false$. An update U is called a *repair* of S in D if $D^U(S) = true$. If $D^U(IC) = false$, U is also called a *partial repair* of IC in D. Otherwise, if $D^U(IC) = true$, U is called a *total repair* of IC in D.

Example 1. Let $D = \{p(1,2,3), p(2,2,3), p(3,2,3), q(1,3), q(3,2), q(3,3)\}$ and $IC = \{\leftarrow p(x,y,z) \wedge \sim q(x,z), \leftarrow q(x,x)\}$. Clearly, both constraints are violated. $U = \{delete\, q(3,3)\}$ is a repair of $\{\leftarrow q(3,3)\}$ in D and a partial repair of IC. It tolerates the uncertainty reflected by the violation of $\leftarrow p(2,2,3) \wedge \sim q(2,3)$ in D^U. However, U also causes the violation of $\leftarrow p(3,2,3) \wedge \sim q(3,3)$ in D^U. Thus, the partial repair $U' = \{delete\, q(3,3), delete\, p(3,2,3)\}$ is needed to eliminate the violation of $\leftarrow q(3,3)$ in D without causing any other violation.

Example 1 illustrates the need to check if a given update or partial repair is *integrity-preserving*, i.e., does not increase the amount of uncertainty. This problem is a generalization of what is known as *repair checking* [2]. The problem can be solved by UTIC, as stated in Theorem 2.

Theorem 2. Let $(\mu, \preccurlyeq)$ be an uncertainty measure, $\mathcal{M}$ a UTIC method based on $(\mu, \preccurlyeq)$, and U a partial repair of IC in D. For a tuple (D, IC), U preserves integrity w.r.t. μ, i.e., $\mu(D^U, IC) \preccurlyeq \mu(D, IC)$, if $\mathcal{M}(D, IC, U) = ok$.

For computing partial repairs, any off-the-shelve view update method can be used, as follows. Let $S = \{\leftarrow B_1, \ldots, \leftarrow B_n\}$ be a subset of constraints to be repaired in a database D. Candidate updates for satisfying the view update request can be obtained by running the view update request *delete violated* in $D \cup \{violated \leftarrow B_i \mid 0 \le i \le n\}$. For deciding if a candidate update U preserves integrity, U can be checked by UTIC, according to Theorem 2.

3.3 Certain Answers in Uncertain Databases

Violations of constraints that model uncertainty may impair the integrity of query answering, since the same data that cause the violations may also cause the computed answers. Hence, there is a need of an approach to provide answers that either have integrity and thus are certain, or that tolerate some uncertainty. An approach to provide answers that are certain in uncertain databases is outlined in 3.3.1, and generalized in 3.3.2 to provide answers that tolerate uncertainty.

3.3.1 Answers That Are Certain

Consistent query answering (abbr. *CQA*) [3] provides answers that are correct in each minimal total repair of IC in D. CQA uses semantic query optimization [10] which in turn uses integrity constraints for query answering. A similar approach is to infer consistent hypothetical answers abductively, together with a set of hypothetical updates that can be interpreted as integrity-preserving repairs [22].

A new approach to provide answers that have integrity (abbr. AHI) and thus certainty is proposed in [16]. AHI determines two sets of data: the causes by which an answer is deduced, and the causes that lead to constraint violations. For databases D and queries without negation in the body of clauses, causes are minimal subsets of ground instances of clauses in D by which positive answers or violations are deduced. For clauses with negation and negative answers, also minimal subsets of ground instances of the only-if halves of the completions of predicates in D [11] form part of causes. In general, causes are not unique.

An answer then is defined to have integrity if it has a cause that does not intersect with any of the causes of constraint violations, i.e., if it is deducible from data that are independent of those that violate constraints. Definition 4 below is a compact version of the definition of AHI in terms of certainty. Precise definitions of causes and details of computing AHI are in [16–18].

Definition 4. Let θ be an answer to a query $\leftarrow B$ in (D, IC), i.e., θ is either a substitution such that $D(\forall(B\theta)) = true$, where $\forall(B\theta)$ is the universal closure of $B\theta$, or $D(\leftarrow B) = true$, i.e., $\theta = no$.
a) Let B_θ stand for $\forall(B\theta)$ if θ is a substitution, or for $\leftarrow B$ if $\theta = no$.
b) θ *is certain* in (D, IC) if there is a cause C of B_θ in D such that $C \cap C_{IC} = \emptyset$, where C_{IC} is the union of all causes of constraint violations in (D, IC).

3.3.2 Answers That Tolerate Uncertainty

AHI is closely related to UTIC, since some convenient violation measures are defined by causes: cause-based methods accept an update U only if U does not increase the number or the set of causes of constraint violations [17]. Similar to UTIC, AHI is uncertainty-tolerant since it provides correct results in the presence of constraint violations. However, each answer accepted by AHI is independent of any inconsistent parts of the database, while UTIC may admit updates that violate constraints. For instance, U in Example 1 causes the violation of a constraint while eliminating some other violation. Now, suppose U is checked by some UTIC method based on a violation measure that assigns a greater weight to the eliminated violation than to the newly caused one. Thus, U can be *ok*-ed, since it decreases the measured amount of inconsistency.

In this sense, we are going to relax AHI to ATU: answers that tolerate uncertainty. ATU sanctions answers that are acceptable despite some amount of uncertainty involved in their derivation.

To quantify that amount, some 'tolerance measure' is needed. Unlike uncertainty measures which size the uncertainty in all of (D, IC), tolerance measures only size the uncertainty involved in the derivation of given answers or violations.

Definition 5. (*ATU*)
a) For answers θ to queries $\leftarrow B$ in (D, IC), a *tolerance measure* maps triples (D, IC, B_θ) to $(\mathbb{M}, \preceq)$, where $\mathbb{M}$ is a metric space partially ordered by $\preceq$.
b) Let τ be a tolerance measure and th a threshold value in $\mathbb{M}$ up to which uncertainty is tolerable. Then, an answer θ to some query $\leftarrow B$ in (D, IC) is said to *tolerate uncertainty up to th* if $\tau(D, IC, B_\theta) \preceq th$.

A first, coarse tolerance measure τ could be to count the elements of $C_\theta \cap C_{IC}$ where C_θ is the union of all causes of B_θ, and C_{IC} is as in Definition 4. Or, the application or its designer or the user may assign a specific weight to each element of each cause, similar to the tuple ranking in [5]. Then, τ can be defined by adding up the weights of elements in $C_\theta \cap C_{IC}$. Or, application-specific weights could be assigned to each ground instance I' of each $I \in IC$. Then, τ could sum up the weights of those I' that have a cause C' such that $C_\theta \cap C' \neq \emptyset$.

For example, $\tau(D, IC, B_\theta) = |C_{theta} \cap C_{IC}|$ counts elements in $C_\theta \cap C_{IC}$, where $|.|$ is the cardinality operator. Or, $\tau(D, IC, B_\theta) = \sum \{\omega(c) \mid c \in C_\theta \cap C_{IC}\}$ adds up the weights of elements in $C_\theta \cap C_{IC}$, where ω is a weight function.

4   Uncertainty Management – An Example

In this section, we illustrate the management of uncertainty by inconsistency-tolerant integrity management, and discuss some more conventional alternatives. In particular, we compare uncertainty-tolerant integrity management with brute-force constraint evaluation, conventional integrity checking that is not uncertainty-tolerant, total repairing, and CQA, in 4.1 – 4.6.

The predicates and their attributes below are open to interpretation. By assigning convenient meanings to predicates, it can be interpreted as a model of uncertainty in a decision support systems for, e.g., stock trading, or controlling operational hazards in a complex machine.

Let D be a database with the following definitions of view predicates ul, um, uh that model uncertainty of low, medium and, respectively, high degree:

$ul(x) \leftarrow p(x, x)$

$um(y) \leftarrow q(x, y), \sim p(y, x)$; $um(y) \leftarrow p(x, y), q(y, z), \sim p(y, z), \sim q(z, x)$

$uh(z) \leftarrow p(0, y), q(y, z), z > th$

where th be a threshold value greater than or equal to 0. Now, let uncertainty be denied by the following integrity theory:

$IC = \{\leftarrow ul(x), \quad \leftarrow um(x), \quad \leftarrow uh(x)\}$.

Note that IC is satisfiable, e.g., by $D = \{p(1, 2), p(2, 1), q(2, 1)\}$. Now, let the extensions of p and q in D be populated as follows.

$p(0, 0), p(0, 1), p(0, 2), p(0, 3), \ldots, p(0, 10000000),$

$p(1, 2), p(2, 4), p(3, 6), p(4, 8), \ldots, p(5000000, 10000000)$

$q(0, 0), q(1, 0), q(3, 0), q(5, 0), q(7, 0), \ldots, q(9999999, 0)$

It is easy to verify that the low-uncertainty denial $\leftarrow ul(x)$ is the only constraint that is violated in D, and that this violation is caused by $p(0, 0) \in D$.

Now, let us consider the update $U = insert\ q(0, 9999999)$.

4.1   Brute-Force Uncertainty Management

For later comparison, let us first analyse the general cost of brute-force evaluation of IC in D^U. Evaluating $\leftarrow ul(x)$ involves a full scan of p. Evaluating $\leftarrow um(x)$

involves access to the whole extension of q, a join of p with q, and possibly many lookups in p and q for testing the negative literals. Evaluating $\leftarrow uh(x)$ involves a join of p with q plus the evaluation of possibly many ground instances of $z > th$.

For large extensions of p and q, brute-force evaluation of IC clearly may last too long, in particular for safety-critical uncertainty monitoring in real time. In 4.2, we are going to see that it is far less costly to use an UTIC method that simplifies the evaluation of constraints by confining its focus on the data that are relevant for the update.

4.2   Uncertainty Management by UTIC

First of all, note that the use of customary methods that require the satisfaction of IC in D is not feasible in our example, since $D(IC) = false$. Thus, conventional integrity checking has to resort on brute-force constraint evaluation. We are going to see that checking U by an UTIC method is much less expensive than brute-force evaluation.

At update time, the following simplifications of medium and high uncertainty constraints are obtained from U. (No low uncertainty is caused by U since $q(0, 9999999)$ does not match $p(x, x)$.) These simplifications are obtained at hardly any cost, by simple pattern matching of U with pre-simplified constraints that can be compiled at constraint specification time.

$\leftarrow \sim p(9999999, 0)$; $\leftarrow p(x, 0)$, $\sim p(0, 9999999)$, $\sim q(9999999, x)$
$\leftarrow p(0, 0)$, $9999999 > th$

By a simple lookup of $p(9999999, 0)$ for evaluating the first of the three denials, it is inferred that $\leftarrow um$ is violated.

Now that a medium uncertainty has been spotted, there is no need to check the other two simplifications. Yet, let us do that, for later comparison in 4.3.

Evaluation of the second simplification from left to right essentially equals the cost of computing the answer $x = 0$ to the query $\leftarrow p(x, 0)$ and successfully looking up $q(9999999, 0)$. Hence, the second denial is $true$, i.e., there is no further medium uncertainty. Clearly, the third simplification is violated if $9999999 > th$ holds, since $p(0, 0)$ is true, i.e., U possibly causes high uncertainty.

Now, let us summarize this subsection. Validating U by UTIC according to Theorem 1 essentially costs a simple access to the p relation. Only one more look-up is needed for evaluating all constraints. And, apart from a significant cost reduction, UTIC prevents medium and high uncertainty constraint violations that would be caused by U if it were not rejected.

4.3   Methods That Are Uncertainty-Intolerant

UTIC is sound, but, in general, methods that are uncertainty-intolerant (i.e., not uncertainty-tolerant, e.g., those in [24, 26]) are unsound, as shown below.

Clearly, p is not affected by U. Thus, $D(ul(x)) = D^U(ul(x))$. Each integrity checking method that is uncertainty-intolerant, i.e., each conventional approach, assumes $D(IC) = true$. Thus, the method in [24] concludes that the unfolding

$\leftarrow p(x,x)$ of $\leftarrow ul(x)$ is satisfied in D and D^U. Hence, it concludes that also $\leftarrow p(0,0), 9999999 > th$ (the third of the simplifications in 4.2). is satisfied in D^U. However, that is wrong if $9999999 > th$ holds. Thus, uncertainty-intolerant integrity checking may wrongly infer that the high uncertainty constraint $\leftarrow uh(z)$ cannot be violated in D^U.

4.4   Uncertainty Management by Repairing (D, IC)

Conventional integrity checking requires $D(IC) = true$. To comply with that, all violations in (D, IC) must be repaired before each update. However, such repairs can be exceedingly costly, as argued below.

In fact, already the identification of all violations in (D, IC) may be prohibitively costly at update time. But there is only a single low uncertainty constraint violation in our example: $p(0,0)$ is the only cause of the violation $\leftarrow ul(0)$ in D. Thus, to begin with repairing D means to request $U = delete\ p(0,0)$, and to execute U if it preserves all constraints, according to Theorem 2.

To check U for integrity preservation means to evaluate the simplifications

$$\leftarrow q(0,0) \quad \text{and} \quad \leftarrow p(x,0), q(0,0), \sim q(0,x)$$

i.e., the two resolvents of $\sim p(0,0)$ and the clauses defining um, since U affects no other constraints. The second one is satisfied in D^U, since there is no fact matching $p(x,0)$ in D^U. However, the first one is violated, since $D^U(q(0,0)) = true$. Hence, also $q(0,0)$ must be deleted. That deletion affects the clause

$$um(y) \leftarrow p(x,y), q(y,z), \sim p(y,z), \sim q(z,x)$$

and yields the simplification $\leftarrow p(0,y), q(y,0), \sim p(y,0)$.

As is easily seen, this simplification is violated by each pair of facts of the form $p(0,o), q(o,0)$ in D, where o is an odd number in $[1, 9999999]$. Thus, deleting $q(0,0)$ for repairing the violation caused by deleting $p(0,0)$ causes the violation of each instance of the form $\leftarrow um(o)$, for each odd number o in $[1, 9999999]$.

Hence, repairing each of these instances would mean to request the deletion of many rows of p or q. We shall not further track those deletions, since it should be clear already that repairing D is complex and tends to be significantly more costly than UTIC. Another advantage of UTIC: since inconsistency can be temporarily tolerated, UTIC-based repairs do not have to be done at update time. Rather, they can be done off-line, at any convenient point of time.

4.5   Uncertainty Management by Repairing (D^U, IC)

Similar to repairing (D, IC), repairing (D^U, IC) also is more expensive than to tolerate extant constraint violations until they can be repaired at some more convenient time. That can be illustrated by the three violations in D^U, as identified in 4.1 and 4.2: the low uncertainty that already exists in D, and the medium and high uncertainties caused by U and detected by UTIC. To repair them obviously is even more intricate than to only repair the first of them as tracked in 4.4.

Moreover, for uncertainty management in safety-critical applications, it is no good idea to simply accept an update without checking for potential violations of constraints, and to attempt repairs only after the update is committed, since repairing takes time, during which an updated but unchecked state may contain possibly very dangerous uncertainty of any order.

4.6   AHI and ATU for Uncertainty Management

Checking and repairing uncertainty constraints involves their evaluation, by querying them. As already mentioned in 3.3.1, CQA is an approach to cope with constraint violations for query evaluation. However, the evaluation of constraints or simplifications thereof by CQA is unprofitable, since consistent query answers are defined to be those that are true in each minimally repaired database. Thus, for each queried denial constraint I, CQA will by definition return the empty answer, which indicates the satisfaction of I. Thus, answers to queried constraints computed by CQA have no meaningful interpretation.

For example, CQA computes the empty answer to the query $\leftarrow ul(x)$ and to $\leftarrow uh(z)$, for any extension of p and q. However, the only reasonable answers to $\leftarrow ul(x)$ and $\leftarrow uh(z)$ in D are $x = 0$ and, resp., $x = 9999999$, if $9999999 > th$. These answers correctly indicate low and high uncertainty in D and, resp., D^U.

For computing correct answers to queries (rather than to denials representing constraints), AHI and ATU are viable alternatives to CQA. A comparison which turned out to be advantageous for AHI has been presented in [16]. ATU goes beyond CQA and AHI by providing reasonable answers even if these answers depend on uncertain data that violate constraints, as we have seen in 3.3.2.

5   Scaling up Uncertainty Management to Concurrency

The number of concurrently issued transactions increases with the number of online users. So far, we have tacitly considered only serial executions of transactions. Such executions have many transactions wait for others to complete. Thus, the serialization of transactions severely limits the scalability of applications. Hence, to achieve high scalability, transactions should be executed concurrently, without compromising integrity, i.e., without increasing uncertainty.

Standard concurrency theory guarantees the preservation of integrity only if each transaction, when executed in isolation, translates a consistent state into a consistent successor state. More precisely, a standard result of concurrency theory says that, in a history H of concurrently executed transactions $T_1, \ldots, T_n$, each T_i preserves integrity if it preserves integrity when executed non-concurrently and H is serializable, i.e., the effects of the transactions in H are equivalent to the effects of a serial execution of $\{T_1, \ldots, T_n\}$. For convenience, let us capture this result by the following schematic rule:

$$isolated\ integrity\ +\ serializability\ \Rightarrow\ concurrent\ integrity \qquad (*)$$

Now, if uncertainty corresponds to integrity violation, and each transaction is supposed to operate on a consistent input state, then (*) does not guarantee that concurrently executed transactions on uncertain data would keep uncertainty at bay, even if they would not increase uncertainty when executed in isolation and the history of their execution was serializable.

Fortunately, however, the approaches and results in Section 3 straightforwardly scale up to concurrent transactions without further ado, as shown for inconsistency-tolerant integrity checking in [21], based on a measure that compares sets of violated instances of constraints before and after a transaction.

Theorem 3 below adapts Theorem 3 in [21] to measure-based UTIC in general. It asserts that a transaction T in a history H of concurrently executing transactions does not increase uncertainty if H is serializable and T preserves integrity whenever it is executed in isolation. On one hand, Theorem 3 weakens Theorem 3 [21] by assuming strict two-phase locking (abbr. S2PL) [6], rather than abstracting away from any implementation of serializability. On the other hand, Theorem 3 generalizes Theorem 3 [21] by using an arbitrary uncertainty measure μ, rather than the inconsistency measure mentioned above. A full-fledged generalization that would not assume any particular realization of serializability is possible along the lines of [21], but would be out of proportion in this paper.

Theorem 3. Let H be a S2PL history, μ an uncertainty measure and T a transaction in H that uses a μ-based UTIC method for checking the integrity preservation of its write operations. Further, let D be the committed state at which T begins in H, and D^T the committed state at which T ends in H. Then, $\mu(D^T, IC) \preccurlyeq \mu(D, IC)$.

The essential difference between (*) and Theorem 3 is that the latter is uncertainty-tolerant, the former is not. Thus, as opposed to (*), Theorem 3 identifies useful sufficient conditions for integrity preservation in the presence of uncertain data. Another important difference is that the guarantees of integrity preservation that (*) can make for T require the integrity preservation of all other transactions that may happen to be executed concurrently with T. As opposed to that, Theorem 3 does away with the standard premise of (*) that all transactions in H must preserve integrity in isolation; only T itself is required to have that property. Thus, the guarantees that Theorem 3 can make for individual transactions T are much better than those of (*).

To outline a proof of Theorem 3, we distinguish the cases that T either terminates by aborting or by committing its write operations. If T aborts, then Theorem 3 holds vacuously, since, by definition, no aborted transaction could have any effect whatsoever on any committed state. So, we can suppose that T commits. Let $\mathcal{M}$ be the μ-based method used by T. Since T commits, it follows that $\mathcal{M}(D, IC, W_T) = true$, where W_T is the write set of T, i.e., $D^T = D^{W_T}$, since otherwise, the writes of T would violate integrity and thus T would abort. Since H is S2PL, it follows that there is an equivalent serialization H' of H that preserves the order of committed states in H. Thus, D and D^T are also the committed states at beginning and end of T in H'. Hence, Theorem 3 follows from $\mathcal{M}(D, IC, W_T) = true$ and Definition 2 since H' is serial, i.e., non-concurrent. □

It follows from Theorem 2 that, similar to UTIC, also integrity repairing scales up to S2PL concurrency if realized as described in 3.2, i.e., if UTIC is used to check candidate repairs for integrity preservation.

Also AHI and ATU as defined in 3.3 scale up to concurrency, which can be seen as follows. Concurrent query answering is realized by read-only transactions. In S2PL histories, such transactions always read from committed states that are identical to states in equivalent serial histories, as described in the proof of Theorem 3. Hence, each answer can be checked for certainty or for being within the confines of tolerable uncertainty as described in 3.3.1 or, resp., 3.3.2.

6   Related Work

An early, not yet measure-based attempt to conceptualize some of the material in 3.1 has been made in [19]. Apart from that, it seems that integrity mainte-nance and query answering in the presence of uncertain data never have been approached in a uniform way, as in this paper. That is surprising since integrity, uncertainty and answering queries with certainty are obviously related.

Semantic similarities and differences between uncertainty and the lack of in-tegrity are observed in [27]. In that book, largely diverse proposals to handle data that suffer from uncertainty are discussed. In particular, approaches such as probabilistic and fuzzy set modeling, exception handling, repairing and para-consistent reasoning are discussed. However, no particular approach to integrity maintenance (checking or repairing) is considered. Also, no attention is paid to concurrency.

Several paraconsistent logics that tolerate inconsistency and thus uncertainty of data have been proposed, e.g., in [7, 8]. Each of them departs from classical first-order logic, by adopting some annotated, probabilistic, modal or multival-ued logic, or by replacing standard axioms and inference rules with non-standard axiomatizations. As opposed to that, UTIC fully conforms with standard datalog and does not need any extension of classical logic.

Work concerned with semantic inconsistencies in databases is also going on in the field of measuring inconsistency [25]. However, the violation measures on which UTIC is based have been conceived to work well also in databases with non-monotonic negation, whereas the inconsistency measures in the literature do not scale up to non-monotonicity, as argued in [18].

7   Conclusion

We have applied and extended recently developed concepts of logical inconsis-tency tolerance to problems of managing uncertainty in databases.

We have shown that certain forms of uncertainty of stored data can be mod-eled by integrity constraints and maintained by uncertainty-tolerant integrity management technology. In particular, updates can be monitored by UTIC, such that they do not increase uncertainty, and extant uncertainty can be partially repaired while tolerating remaining uncertainty. Also, we have outlined how databases can provide reasonable answers in the presence of uncertainty. More-over, we have highlighted that uncertainty tolerance is necessary and sufficient for scaling up uncertainty management in databases to concurrent transactions.

This result is significant since concurrency is a common and indeed indispensable feature of customary database management systems.

As illustrated in Section 4, the use of uncertainty-tolerant tools is essential, since wrong, possibly fatal conclusions can be inferred from deficient data by using a method that is uncertainty-intolerant. A lot of UTIC methods, and some intolerant ones, have been identified in [20, 17].

In ongoing research, we are elaborating a generalization of the results in Section 5 to arbitrarily serializable histories. Also, we are working on scaling up our results further to replicated databases and recoverable histories. Moreover, we envisage further applications of inconsistency-tolerant uncertainty management in the fields of OLAP, data mining, and data stream query processing, in order to complement our work in [12–14].

References

1. Abiteboul, S., Hull, R., Vianu, V.: Foundations of Databases. Addison-Wesley (1995)
2. Afrati, F., Kolaitis, P.: Repair Checking in Inconsistent Databases: Algorithms and Complexity. In: ICDT 2009. ACM Int. Conf. Proc. Series, vol. 361, pp. 31–41 (2009)
3. Arenas, M., Bertossi, L., Chomicki, J.: Consistent query answers in inconsistent databases. In: PODS 1999, pp. 68–79. ACM Press (1999)
4. Bauer, H.: Maß-und Integrationstheorie, 2nd edn. De Gruyter (1992)
5. Berlin, J., Motro, A.: TupleRank: Ranking Discovered Content in Virtual Databases. In: Etzion, O., Kuflik, T., Motro, A. (eds.) NGITS 2006. LNCS, vol. 4032, pp. 13–25. Springer, Heidelberg (2006)
6. Bernstein, P., Hadzilacos, V., Goodman, N.: Concurrency Control and Recovery in Database Systems. Addison-Wesley (1987)
7. Bertossi, L., Hunter, A., Schaub, T.: Introduction to Inconsistency Tolerance. In: Bertossi, L., Hunter, A., Schaub, T. (eds.) Inconsistency Tolerance. LNCS, vol. 3300, pp. 1–14. Springer, Heidelberg (2005)
8. Carnielli, W., Coniglio, M., D'Ottaviano, I. (eds.): The Many Sides of Logic. Studies in Logic, vol. 21. College Publications, London (2009)
9. Ceri, S., Cochrane, R., Widom, J.: Practical Applications of Triggers and Constraints: Success and Lingering Issues. In: Proc. 26th VLDB, pp. 254–262. Morgan Kaufmann (2000)
10. Chakravarthy, U., Grant, J., Minker, J.: Logic-based approach to semantic query optimization. Transactions on Database Systems 15(2), 162–207 (1990)
11. Clark, K.: Negation as Failure. In: Gallaire, H., Minker, J. (eds.) Logic and Data Bases, pp. 293–322. Plenum Press (1978)
12. Cuzzocrea, A.: OLAP Over Uncertain and Imprecise Data: Fundamental Issues and Novel Research Perspectives. In: Proc. XML and XQuery Workshop at 21st DEXA, pp. 331–336. IEEE CSP (2001)
13. Cuzzocrea, A.: Retrieving Accurate Estimates to OLAP Queries over Uncertain and Imprecise Multidimensional Data Streams. In: Bayard Cushing, J., French, J., Bowers, S. (eds.) SSDBM 2011. LNCS, vol. 6809, pp. 575–576. Springer, Heidelberg (2011)

14. Cuzzocrea, A., Decker, H.: Non-linear Data Stream Compression: Foundations and Theoretical Results. In: Corchado, E., Snášel, V., Abraham, A., Woźniak, M., Graña, M., Cho, S.-B. (eds.) HAIS 2012, Part III. LNCS, vol. 7208, pp. 622–634. Springer, Heidelberg (2012)

15. Decker, H.: The Range Form of Databases and Queries or: How to Avoid Floundering. In: Proc. 5th ÖGAI. Informatik-Fachberichte, vol. 208, pp. 114–123. Springer (1989)

16. Decker, H.: Answers That Have Integrity. In: Schewe, K.-D., Thalheim, B. (eds.) SDKB 2010. LNCS, vol. 6834, pp. 54–72. Springer, Heidelberg (2011)

17. Decker, H.: Inconsistency-Tolerant Integrity Checking Based on Inconsistency Metrics. In: König, A., Dengel, A., Hinkelmann, K., Kise, K., Howlett, R.J., Jain, L.C. (eds.) KES 2011, Part II. LNCS, vol. 6882, pp. 548–558. Springer, Heidelberg (2011)

18. Decker, H.: Measure-based Inconsistency-tolerant Maintenance of Database Integrity. In: 5th International Workshop on Semantics in Data and Knowledge Bases, SDKB. LNCS. Springer (to appear, 2012)

19. Decker, H., Martinenghi, D.: Integrity Checking for Uncertain Data. In: Proc. 2nd TDM Workshop on Uncertainty in Databases. CTIT Workshop Proceedings Series WP06-01, pp. 41–48. Univ. Twente, The Netherlands (2006)

20. Decker, H., Martinenghi, D.: Inconsistency-tolerant Integrity Checking. TKDE 23(2), 218–234 (2011)

21. Decker, H., Muñoz-Escoí, F.D.: Revisiting and Improving a Result on Integrity Preservation by Concurrent Transactions. In: Meersman, R., Dillon, T., Herrero, P. (eds.) OTM 2010. LNCS, vol. 6428, pp. 297–306. Springer, Heidelberg (2010)

22. Fung, T.H., Kowalski, R.: The IFF proof procedure for abductive logic programming. J. Logic Programming 33(2), 151–165 (1997)

23. Grant, J., Hunter, A.: Measuring the Good and the Bad in Inconsistent Information. In: Proc. 22nd IJCAI, pp. 2632–2637 (2011)

24. Gupta, A., Sagiv, Y., Ullman, J., Widom, J.: Constraint checking with partial information. In: Proc. 13th PODS, pp. 45–55. ACM Press (1994)

25. Hunter, A., Konieczny, S.: Approaches to Measuring Inconsistent Information. In: Bertossi, L., Hunter, A., Schaub, T. (eds.) Inconsistency Tolerance. LNCS, vol. 3300, pp. 191–236. Springer, Heidelberg (2005)

26. Lee, S.Y., Ling, T.W.: Further improvements on integrity constraint checking for stratifiable deductive databases. In: Proc. VLDB 1996, pp. 495–505. Morgan Kaufmann (1996)

27. Motro, A., Smets, P.: Uncertainty Management in Information Systems: From Needs to Solutions. Kluwer (1996)

28. Wijsen, J.: Database repairing using updates. ACM Transaction on Database Systems 30(3), 722–768 (2005)

Generalizing Naive and Stable Semantics in Argumentation Frameworks with Necessities and Preferences

Farid Nouioua

LSIS, Aix-Marseille Univ,
Avenue Escadrille Normandie Niemen, 13997, Marseille Cedex 20, France
farid.nouioua@lsis.org

Abstract. In [4] [5], the classical acceptability semantics are generalized by preferences. The extensions under a given semantics correspond to maximal elements of a relation encoding this semantics and defined on subsets of arguments. Furthermore, a set of postulates is proposed to provide a full characterization of any relation encoding the generalized stable semantics. In this paper, we adapt this approach to preference-based argumentation frameworks with necessities. We propose a full characterization of stable and naive semantics in this new context by new sets of adapted postulates and we present a practical method to compute them by using a classical Dung argumentation framework.

Keywords: abstract argumentation, acceptability semantics, necessities, preferences.

1 Introduction

Dung's abstract argumentation frameworks (AFs) [12] are a very influential model which has been widely studied and extended in different directions. Moreover, some works (see for example [2], [14]) started recently to bridge the gap between this model and the logical-based model [7] in which arguments are constructed from (possibly inconsistent) logical knowledge bases as couples of the form (support, claim). Among the various extensions of Dung model we are interested in this paper on two of them : adding information about preferences and representing support relations between arguments.

We need to handle preferences in argumentation theory because in real contexts, arguments may have different strengths. In Dung style systems, adding preferences may lead to the so-called "critical attacks" that arise when a less preferred argument attacks a more preferred one. Most of the first proposals on preference-based argumentation like [1] [6] [16] suggest to simply remove the critical attacks but the drawback of these approaches is to possibly tolerate non conflict-free extensions. A new approach proposed in [3] [4] [5] encodes acceptability semantics for preference-based argumentation frameworks (PAFs) as a relation on the powerset of the set of arguments and the extensions under a

E. Hüllermeier et al. (Eds.): SUM 2012, LNAI 7520, pp. 44–57, 2012.

given semantics as the maximal elements of this relation. This approach avoids the drawback mentioned above and ensures the recovering of the classical acceptability semantics when no critical attack is present. Moreover, [4] [5] give a set of postulates that must be verified by any relation encoding the stable semantics in preference-based frameworks.

At a different level, different approaches have been proposed to enrich Dung model by information expressing supports between arguments. The bipolar argumentation frameworks (BAFs) [10] [11] add an explicit support relation to Dung AFs and define new acceptability semantics. Their main drawback is that admissibility of extensions is no more guaranteed. [8] introduces the so-called deductive supports and proposes to use a meta Dung AF to obtain the extensions. Abstract dialectical frameworks [9] represent a powerful generalization of Dung AFs in which the acceptability conditions of an argument are more sophisticated. The acceptability semantics are redefined by adapting the Gelfond/Lifshitz reduct used in answer set programming (ASP). The Argumentation Frameworks with Necessities (AFNs) [17] are a kind of bipolar AFs where the support relation has the meaning of "necessity". The acceptability semantics are extended in a natural way that ensures admissibility without borrowing techniques from LPs or making use of a Meta Dung model. The aim of this paper is threefold:

- Adapting the approach proposed in [4] [5] for stable semantics to the case of preference-based AFNs (PAFNs). We show that by introducing some suitable notions, we obtain new postulates that are very similar to the original ones.
- Giving a full characterization of naive semantics in the context of PAFNs. Since the naive semantics is the counterpart of justified Lukaszewicz extensions [15] in default logic and of ι-answer sets [13] in logic programming (see [18]), this characterization allows a better understanding of these approaches.
- Computing a Dung AF whose naive and stable extensions correspond exactly to the generalized naive and stable extensions of the input PAFN.

The rest of the paper is organized as follows. In section 2 we recall some basics of AFs with necessities and/or preferences as well as the approach generalizing stable semantics by preferences. In section 3 we present some further notions that are useful to define the new postulates for PAFNs. Section 4 presents the stable and naive semantics in AFNs seen as dominance relations. Section 5 discusses the generalization of these semantics by preferences. In section 6 we characterize PAFNs by classical Dung AFs. Finally, section 7 concludes the paper.

2   Background

2.1   Argumentation Frameworks, Necessities and Preferences

A Dung AF [12] is a pair $F = \langle A, R \rangle$ where A is a set of arguments and R is a binary attack relation over A. A set $S \subseteq A$ attacks an argument b iff $(\exists a \in S)$ s.t $a \, R \, b$. S is *conflict-free* iff $(\not\exists a, b \in S)$ s.t. $a \, R \, b$. Many *acceptability semantics* have been proposed to define how sets of collectively acceptable arguments may be derived from an attack network. Among them we will focus in this paper on the *naive* and the *stable semantics*.

Definition 1. Let $F = \langle A, R \rangle$ be an AF and $S \subseteq A$. S is a naive extension of F iff S is a $\subseteq$-maximal conflict-free set. S is a stable extension of F iff S is conflict-free and $(\forall b \in A \setminus S)(\exists a \in S)$ s.t. $a\ R\ b$.

For an AF $F = \langle A, R \rangle$, we use the notation $Ext^N(F)$ (resp. $Ext^S(F)$) to denote the set of naive (resp. stable) extensions of F.

Argumentation frameworks with necessities (AFNs) (see [17] for details) represent a kind of bipolar extension of Dung AFs where the support relation is a *necessity* that captures situations in which one argument is necessary for another. Formally, an AFN is defined by $\Gamma = \langle A, R, N \rangle$ where A is a set of arguments, $R \subseteq A \times A$ is a binary *attack* relation over A and $N \subseteq A \times A$ is a *necessity* relation over A. For $a, b \in A$, $a\ N\ b$ means that the acceptance of a is necessary for the acceptance of b. We suppose that there are no cycles of necessities, i.e., $\nexists a_1, \ldots a_k$ for $k \geq 1$ such that $a_1 = a_k = a$ and $a_1\ N\ a_2 \ldots N\ a_k$.

To give the new definitions of acceptability semantics in AFNs, we need to introduce the key notions of coherence and strong coherence. The latter plays in some way the same role of conflict-freeness in Dung AFs.

Definition 2. Let $\Gamma = \langle A, R, N \rangle$ be an AFN and $S \subseteq A$. S is :

- *coherent* iff S is closed under N^{-1}, i.e. $(\forall a \in S)(\forall b \in A)$ if $b\ N\ a$ then $b \in S$.
- *strongly coherent* iff S is coherent and conflict-free (w.r.t. R).
- a *naive extension* of Γ iff S is a $\subseteq$-maximal strongly coherent set.
- a *stable extension* of Γ iff S is strongly coherent and $(\forall b \in A \setminus S)$ *either* $(\exists a \in S)$ *s.t.* $a\ R\ b$ *or* $(\exists a \in A \setminus S)$ *s.t.* $a\ N\ b$.

For an AFN $\Gamma = \langle A, R, N \rangle$, we use the notation $Ext^N(\Gamma)$ (resp. $Ext^S(\Gamma)$) to denote the set of naive (resp. stable) extensions of Γ.

The main properties of naive and stable extensions in AFs continue to hold for AFNs, namely : naive extensions always exist; naive extensions do not depend on the attacks directions; an AFN may have zero, one or several stable extensions and each stable extension is a naive extension but the inverse is not true. Besides, for an AFN $\Gamma = \langle A, R, N \rangle$ where $N = \emptyset$, strong coherence coincides with classical conflict-freeness and naive and stable extensions correspond to naive and stable extensions of the simple AF $\langle A, R \rangle$ respectively (in the sense of definition 1.).

Finally, a preference-based AF (PAF) (resp. a preference-based AFN (PAFN)) is defined by $\Lambda = \langle A, R, \geq \rangle$ (resp. $\Sigma = \langle A, R, N, \geq \rangle$) where $\langle A, R \rangle$ is an AF (resp. $\langle A, R, N \rangle$ is an AFN) and the additional element $\geq$ is a (partial or total) preorder on A. $a \geq b$ means that a it at least as strong as b.

2.2   Stable Semantics as a Dominance Relation in PAFs

The idea of representing acceptability semantics in a PAF as dominance relations has been first developed in [3] for grounded, stable and preferred semantics and a full characterization of stable semantics by a set of postulates has been proposed in [4] [5]. This approach encodes acceptability semantics, by taking into account the possible preferences, as a relation $\succeq$ on the powerset 2^A of the set of arguments : for $\mathcal{E}, \mathcal{E}' \in 2^A$, $\mathcal{E} \succeq \mathcal{E}'$ means that $\mathcal{E}$ is at least as good as $\mathcal{E}'$. $\succ$ denotes the strict version of $\succeq$. For a PAF $\Lambda = \langle A, R, \geq \rangle$, $\mathcal{E} \subseteq A$ is an extension under $\succeq$ iff $\mathcal{E}$ is a maximal element wrt $\succeq$, i.e., for each $\mathcal{E}' \subseteq A$, $\mathcal{E} \succeq \mathcal{E}'$. The set of extensions of Λ under $\succeq$ is denoted by $Ext_{\succeq}(\Lambda)$. The set of conflict-free sets of a PAF Λ (resp. an AF F) is denoted by $CF(\Lambda)$ (resp. $CF(F)$).

In [4] [5] the authors give a full characterisation of any dominance relation encoding stable semantics in PAFs (called pref-stable semantics) by means of the four postulates below. Let $\Lambda = \langle A, R, \geq \rangle$ be a PAF and $\mathcal{E}, \mathcal{E}' \in 2^A$:

Postulate 1. : for $\mathcal{E}, \mathcal{E}' \in 2^A$, $\dfrac{\mathcal{E} \in C\dot{F}(\Lambda) \quad \mathcal{E}' \notin CF(\Lambda)}{\mathcal{E} \succ \mathcal{E}'}$

Postulate 2. : for $\mathcal{E}, \mathcal{E}' \in CF(A)$, $\dfrac{\mathcal{E} \succeq \mathcal{E}'}{\mathcal{E}\setminus\mathcal{E}' \succeq \mathcal{E}'\setminus\mathcal{E}}$ $\qquad \dfrac{\mathcal{E}\setminus\mathcal{E}' \succeq \mathcal{E}'\setminus\mathcal{E}}{\mathcal{E} \succeq \mathcal{E}'}$

Postulate 3. : for $\mathcal{E}, \mathcal{E}' \in CF(A)$ s.t. $\mathcal{E} \cap \mathcal{E}' = \emptyset$,

$$\dfrac{(\exists x' \in \mathcal{E}')(\forall x \in \mathcal{E}) \; \neg(x \; R \; x' \; \wedge \; \neg(x' > x)) \; \wedge \; \neg(x > x')}{\neg(\mathcal{E} \succeq \mathcal{E}')}$$

Postulate 4. : for $\mathcal{E}, \mathcal{E}' \in CF(A)$ s.t. $\mathcal{E} \cap \mathcal{E}' = \emptyset$,

$$\dfrac{(\forall x' \in \mathcal{E}')(\exists x \in \mathcal{E}) \; s.t \; (x \; R \; x' \; \wedge \; \neg(x' > x)) \; \vee \; (x' \; R \; x \; \wedge \; x > x')}{\mathcal{E} \succeq \mathcal{E}'}$$

In other words, a dominance relation encodes a pref-stable semantics iff it satisfies the previous postulates. Such a relation is called a pref-stable relation. Postulate 1 ensures the conflict-freeness of extensions. Postulate 2 ensures that the comparison of two conflict-free sets depends entirely on their distinct elements. Postulates 3 and 4 compare distinct conflict-free sets and state when a set is considered as preferred or not to another one. It has been shown that the extensions under any pref-stable relation are the same and that pref-stable semantics generalizes classical stable semantics in the sense that for a PAF $\Lambda = \langle A, R, \geq \rangle$, if preferences do not conflict with attacks, then pref-stable extensions coincide with the stable extensions of the simple AF $\langle A, R \rangle$.

Different pref-stable relations exist. In [4] [5] the authors show the most general pref-stable relation ($\succeq_g$) and the most specific one ($\succeq_s$). The first (resp. the second) returns exactly the facts $\mathcal{E} \succeq_g \mathcal{E}'$ that can be proved by the postulates 1-4 (resp. whose negation cannot be proved by the postulates 1-4). For any pref-stable relation $\succeq$ we have : if $\mathcal{E} \succeq_g \mathcal{E}'$ then $\mathcal{E} \succeq \mathcal{E}'$ and if $\mathcal{E} \succeq \mathcal{E}'$ then $\mathcal{E} \succeq_s \mathcal{E}'$.

3   Emerging Necessities, Attacks and Preferences

In this section we introduce new notions representing *hidden* forms of necessities, attacks and preferences. These new notions are of great importance since they allow to extend in an easy and natural way the approach presented in [4] [5] for PAFs to the case of PAFNs.

The first notion is that of extended necessity relation : If a is necessary for b and b is necessary for c then we can deduce that a is indirectly necessary for c.

Definition 3. Let $\Sigma = \langle A, R, N, \geq \rangle$ be a PAFN. An extended necessity between a and b is denoted by $a\ N^+\ b$. It holds if there is a sequence $a_1, \ldots, a_k$ for $k \geq 2$ such that $a = a_1\ N\ a_2 \ldots N\ a_k = b$. $N^+(a)$ denotes the set of all the arguments that are related to a by an extended necessity, i.e., $N^+(a) = \{b \in A | b\ N^+\ a\}$. Moreover, we use the notation $a\ N^*\ b$ for any a such that $a\ N^+\ b$ or $a = b$ and we put $N^*(a) = N^+(a) \cup \{a\}$.

The interaction between attacks and necessities results in further implicit attacks that we call the extended attacks. In general, an extended attack between two arguments a and b holds whenever an element of $N^*(a)$ attacks (directly) an elements of $N^*(b)$. Indeed, if we accept a then we must accept a' which excludes b' (since a' attacks b') and this excludes in turn b.

Definition 4. Let $\Sigma = \langle A, R, N, \geq \rangle$ be a PAFN and $\Gamma = \langle A, R, N \rangle$ be its corresponding simple AFN. An extended attack from an argument a to an argument b is denoted by $a\ R^+\ b$. It holds iff $(\exists b' \in N^*(b))(\exists a' \in N^*(a))$ s.t $a'\ R\ b'$.

$CF^+(\Sigma)$ (and $CF^+(\Gamma)$) denotes the set of conflict-free subsets of A wrt R^+ and $SC(\Sigma)$ (and $SC(\Gamma)$) the set of strongly coherent subsets of A. It turns out that strong coherence is stronger than conflict freeness wrt R^+ :

Property 1. Let $\Sigma = \langle A, R, N, \geq \rangle$ be a PAFN and $S \subseteq A$. If S is strongly coherent then $S \in CF^+(\Sigma)$. The inverse is not true.

The different preference-based argumentation approaches more or less agree that the relevant problem to solve in handling preferences in AFs is that of critical attacks (a critical attack arises when an argument a attacks an argument b while b is better than a). It turns out that the interaction between preferences and necessities does not lead to a similar problem since a necessity between two arguments does not necessarily contradict the fact that one of these two arguments is better (or worse) than the second. To understand how preferences interact with necessities, we have to look at the very meaning of necessity. Indeed, accepting an argument requires the acceptance of all its (direct and indirect) necessary arguments. Thus, the input preference assigned to an argument a represents solely a rough preference. Its effective preference depends on that of all the elements of the set $N^*(a)$. To induce the effective preference of the arguments from their input (rough) preference we use the usual democratic relation .

Definition 5. Let $\Sigma = \langle A, R, N, \geq \rangle$ be a PAFN and $a, b \in A$. The *effective preference* relation between arguments is denoted by $\unrhd \subseteq A \times A$ and defined as follows : $(\forall a, b \in A)\ a \unrhd b$ iff $(\forall b' \in N^*(b) \setminus N^*(a))(\exists a' \in N^*(a) \setminus N^*(b))$ s.t $a' \geq b'$. $\rhd$ is the strict version of $\unrhd$, i.e., $a \rhd b$ iff $a \unrhd b$ and not $b \unrhd a$.

Finally we assume in the rest of the paper that the set A of arguments is finite and that there is no $a \in a$ s.t. $a\ R^+\ a$.

Example 1. Consider the PAFN $\Sigma = \langle A, R, N, \geq \rangle$ where $A = \{a, b, c, d\}$, $R = \{(c, a), (b, d)\}$, $N = \{(a, b), (c, d)\}$ and $a \geq c$, $a \geq d$, $d \geq b$. The corresponding AFN $\langle A, R, N \rangle$ is depicted in Fig. 1 :

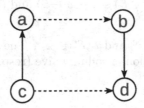

Fig. 1. The AFN corresponding to the PAFN Σ where $a \geq c$, $a \geq d$, $d \geq b$

By applying the previous definitions on this example we obtain :

$N^+ = N$ and $N^* = \{(a, a), (b, b), (c, c), (d, d), (a, b), (c, d)\}$.
$R^+ = \{(b, d), (c, a), (c, b), (d, a), (d, b)\}$.
$SC(\Sigma) = \{\emptyset, \{a\}, \{c\}, \{a, b\}, \{c, d\}\}$.
$CF^+(\Sigma) = \{\emptyset, \{a\}, \{b\}, \{c\}, \{d\}, \{a, b\}, \{c, d\}\}$.
$\unrhd = \{(a, a), (b, b), (c, c), (d, d), (a, c), (a, d), (b, a), (b, c), (b, d), (d, c)\}$.
$\rhd = \{(a, c), (a, d), (b, a), (b, c), (b, d), (d, c)\}$.

4 Generalizing Naive and Stable Semantics in AFNs

In this section we give a characterization of any relation $\succeq \subseteq A \times A$ that encodes naive or stable semantics in a simple AFN without preferences. The idea is that one must recover this characterization in the particular case of a PAFN where no conflict with attacks and necessities is caused by the presence of additional information about preferences. The characterization given here adapts the original one introduced in [4] [5] for the relations encoding stable semantics in simple AFs to the case where a necessity relation is present and give a new version dealing with naive semantics.

The following theorem[1] states the requirements that any relation encoding naive or stable semantics must fulfil :

Theorem 1. Let $\Gamma = \langle A, R, N \rangle$ be an AFN and $\succeq \subseteq 2^A \times 2^A$ then,

- $(\forall \mathcal{E} \in 2^A)$ $(\mathcal{E} \in Ext^N(\Gamma) \Leftrightarrow (\mathcal{E}$ is maximal wrt $\succeq^N))$ iff :
  1. $(\forall \mathcal{E} \in 2^A)$ if $\mathcal{E} \notin SC(\Gamma)$ then $(\exists \mathcal{E}' \in SC(\Gamma))$ s.t. $\neg(\mathcal{E} \succeq^N \mathcal{E}')$.
  2. if $\mathcal{E} \in SC(\Gamma)$ and $(\forall a' \notin \mathcal{E})(\exists a \in \mathcal{E})$ s.t. $(a \ R^+ \ a'$ or $a' \ R^+ \ a)$ then $(\forall \mathcal{E}' \in 2^A)$ $\mathcal{E} \succeq^N \mathcal{E}'$.
  3. if $\mathcal{E} \in SC(\Gamma)$ and $(\exists a' \notin \mathcal{E})$ s.t. $(\not\exists a \in \mathcal{E})$ and $(a \ R^+ \ a'$ or $a' \ R^+ \ a)$ then $(\exists \mathcal{E}' \in 2^A)$ s.t. $\neg(\mathcal{E} \succeq^N \mathcal{E}')$.

---

[1] Because of space limitation, the proofs of theorems and properties are not included in the paper.

– $(\forall \mathcal{E} \in 2^A) (\mathcal{E} \in Ext^S(\Gamma) \Leftrightarrow (\mathcal{E}$ is maximal wrt $\succeq^S))$ iff :
1. $(\forall \mathcal{E} \in 2^A)$ if $\mathcal{E} \notin SC(\Gamma)$ then $(\exists \mathcal{E}' \in SC(\Gamma))$ s.t. $\neg(\mathcal{E} \succeq^S \mathcal{E}')$.
2. if $\mathcal{E} \in SC(\Gamma)$ and $(\forall a' \notin \mathcal{E})(\exists a \in \mathcal{E})$ s.t. $(a \ R^+ \ a')$ then $(\forall \mathcal{E}' \in 2^A) \ \mathcal{E} \succeq^S \mathcal{E}'$.
3. if $\mathcal{E} \in SC(\Gamma)$ and $(\exists a' \notin \mathcal{E})$ s.t. $(\nexists a \in \mathcal{E})$ and $(a \ R^+ \ a')$ then $(\exists \mathcal{E}' \in 2^A)$ s.t. $\neg(\mathcal{E} \succeq^S \mathcal{E}')$.

Let $\Gamma = \langle A, R, N \rangle$ be an AFN and $\mathcal{E}, \mathcal{E}' \in 2^A$. The relation $\succeq_1^N$. (resp. $\succeq_1^S$) below is an example of a relation encoding naive (resp. stable) semantics in Γ :

$\mathcal{E} \succeq_1^N \mathcal{E}'$ (resp. $\mathcal{E} \succeq_1^S \mathcal{E}'$)iff:

– $\mathcal{E} \in SC(\Gamma)$ and $\mathcal{E}' \notin SC(\Gamma)$, or
– $\mathcal{E}, \mathcal{E}' \in SC(\Gamma)$ and $(\forall a' \in \mathcal{E}' \setminus \mathcal{E})(\exists a \in \mathcal{E} \setminus \mathcal{E}')$ s.t $(a \ R^+ \ a'$ or $a' \ R^+ \ a)$ (resp. $\mathcal{E}, \mathcal{E}' \in SC(\Gamma)$ and $(\forall a' \in \mathcal{E}' \setminus \mathcal{E})(\exists a \in \mathcal{E} \setminus \mathcal{E}')$ s.t $a \ R^+ \ a')$

Example 1 (Cont.). Let us consider again the PAFN of example 1 and take the corresponding simple AFN $\Gamma = \langle A, R, N \rangle$. It is easy to check that the maximal sets wrt $\succeq_1^N$ are $\{a, b\}$ and $\{c, d\}$ and only $\{c, d\}$ is a maximal set wrt $\succeq_1^S$. Notice that the same results are obtained by applying definition 2.

5   Generalizing Naive and Stable Semantics in PAFNs

The objective of this section is twofold. On the one hand we extend the full characterisation of Pref-Stable-Semantics (the generalization of stable semantics to PAFs) to the case of PAFNs. We call the resulting semantics the N-pref-stable semantics. On the other hand we give an additional full characterization of what we call the N-pref-naive extension which generalizes the naive extensions in PAFNs. We show that the relationship between stable and naive semantics in AFNs remains valid for the new generalized semantics and that the original semantics of simple AFNs are recovered if preferences are not in conflict with extended attacks. We call N-pref-naive (resp. N-pref-stable) relation any relation that encodes a N-pref-naive (resp. N-pref-stable) semantics. If there is no ambiguity we use indifferently the symbol $\succeq$ to denote a N-pref-naive or a N-pref-stable relation, otherwise we use $\succeq^N$ to denote a N-pref-naive relation and $\succeq^S$ to denote a N-pref-stable relation.

Consider the PAFN $\Sigma = \langle A, R, N, \geq \rangle$ and the relation $\succeq \subseteq 2^A \times 2^A$. We denote by $Ext_\succeq(\Sigma)$ the maximal subsets of arguments wrt $\succeq$ [2]. Let us first present and discuss the set of postulates we will use in characterizing N-pref-naive and N-pref-stable relations :

Postulate 1' : Let $\mathcal{E}, \mathcal{E}' \in 2^A$. Then : $\dfrac{\mathcal{E} \in SC(\Sigma) \qquad \mathcal{E}' \notin SC(\Sigma)}{\mathcal{E} \succ \mathcal{E}'}$

Postulate 2' : Let $\mathcal{E}, \mathcal{E}' \in SC(\Sigma)$. Then : $\dfrac{\mathcal{E} \succ \mathcal{E}'}{\mathcal{E} \setminus \mathcal{E}' \succeq \mathcal{E}' \setminus \mathcal{E}}$ $\qquad \dfrac{\mathcal{E} \setminus \mathcal{E}' \succeq \mathcal{E}' \setminus \mathcal{E}}{\mathcal{E} \succeq \mathcal{E}'}$

---

[2] We recall that $\mathcal{E} \in Ext_\succeq(\Sigma)$ iff $(\forall \mathcal{E}' \in 2^A) \ \mathcal{E} \succeq \mathcal{E}'$.

Postulate 3' : Let $\mathcal{E}, \mathcal{E}' \in CF^+(\Sigma)$ s.t. $\mathcal{E} \cap \mathcal{E}' = \emptyset$. Then :

$$\frac{(\exists x' \in \mathcal{E}')(\forall x \in \mathcal{E}) \, \neg(x \, R^+ \, x') \wedge \neg(x'R^+x) \wedge \neg(x \triangleright x') \wedge \neg(x' \triangleright x)}{\neg(\mathcal{E} \succeq \mathcal{E}')}$$

Postulate 3" : Let $\mathcal{E}, \mathcal{E}' \in CF^+(\Sigma)$ s.t. $\mathcal{E} \cap \mathcal{E}' = \emptyset$. Then :

$$\frac{(\exists x' \in \mathcal{E}')(\forall x \in \mathcal{E}) \, \neg(x \, R^+ \, x' \wedge \neg(x' \triangleright x)) \wedge \neg(x \triangleright x')}{\neg(\mathcal{E} \succeq \mathcal{E}')}$$

Postulate 4' : Let $\mathcal{E}, \mathcal{E}' \in CF^+(\Sigma)$ s.t. $\mathcal{E} \cap \mathcal{E}' = \emptyset$. Then :

$$\frac{(\forall x' \in \mathcal{E}')(\exists x \in \mathcal{E}) \, s.t \, (x \, R^+ \, x') \vee (x'R^+x)}{\mathcal{E} \succeq \mathcal{E}'}$$

Postulate 4" : Let $\mathcal{E}, \mathcal{E}' \in CF^+(\Sigma)$ s.t. $\mathcal{E} \cap \mathcal{E}' = \emptyset$. Then :

$$\frac{(\forall x' \in \mathcal{E}')(\exists x \in \mathcal{E}) \, s.t \, (x \, R^+ \, x' \wedge \neg(x' \triangleright x)) \vee (x' \, R^+ \, x \wedge x \triangleright x')}{\mathcal{E} \succeq \mathcal{E}'}$$

Postulates (1') and (2') are similar to postulates (1) and (2). They just replace conflict-freeness by strong coherence. Postulate (1) ensures that maximal elements of any relation satisfying it are strong coherent sets. Postulate (2') states that comparing two strongly coherent sets depends only on their distinct elements. Notice that if $\mathcal{E}, \mathcal{E}' \in SC(\Sigma)$ then it is obvious that $\mathcal{E} \setminus \mathcal{E}' \in CF^+(\Sigma)$ but it is not necessarily the case that $\mathcal{E} \setminus \mathcal{E}' \in SC(\Sigma)$. This is why the rest of postulates are defined on elements of $CF^+(\Sigma)$.

Postulate (3') and (3") are adapted forms of postulate (3). They capture the case where a set must not be better than another for any N-pref-naive and any N-pref-stable relation respectively. For N-pref-stable relations (postulate (3")) the adaptation consists just in replacing the direct attack by the extended one and the input preference relation by the effective one. The information about necessities is incorporated to these two notions in order to keep the original form of this postulate : a set must not be better than another whenever the second contains an argument that is neither successfully attacked nor strictly less preferred than any element of the first. For N-pref-naive relations (postulate(3')), a form of symmetry is introduced so that the orientation of attacks and preferences are no more important. Thus, a set must not be better than another whenever the second contains an element which is neither involved in an attack wrt R^+ (whatever its direction) nor compared by $\triangleright$ with any element of the first set.

Like postulate (4), postulates (4') and (4") capture the case where a set must be better than another for any N-pref-naive (resp. N-pref-stable) relation. For N-pref-stable relations (postulate 4") the adaptation consists again to just replacing the direct attack by the extended one and the input preference relation by the effective one. A set must be better than another whenever for each argument b of the second set there is an argument a in the first set such that either a attacks b (wrt. R^+) and b is not strictly better than a (wrt. $\triangleright$) or b attacks a but a is strictly better than b. For N-pref-naive relations (postulate(4')), a set must be considered as better than another whenever for each argument of the second set there is an argument in the first set which is in conflict with it (wrt. R^+). The N-pref-naive and N-pref-stable semantics are then defined as follows.

Definition 6. Let $\Sigma = \langle A, R, N, \geq \rangle$ be a PAFN and le us consider a relation $\succeq \subseteq 2^A \times 2^A$. $\succeq$ encodes N-pref-naive semantics iff it verifies the postulates 1',2',3',4' and $\succeq$ encodes N-pref-stable semantics iff it verifies the postulates 1',2',3",4".

All the results obtained for pref-stable semantics in PAFs (in absence of necessities) are easily generalized to the case of N-pref-stable semantics. Moreover, the relationship between naive and stable semantics is kept in the generalized semantics. First we have that N-pref-naive and N-pref-stable extensions are strongly coherent.

Property 2. Let $\Sigma = \langle A, R, N, \geq \rangle$ be a PAFN and $\succeq$ be a N-pref-naive or a N-pref-stable relation. If $\mathcal{E} \in Ext_\succeq(\Sigma)$ then $\mathcal{E} \in SC(\Sigma)$.

It turns out that postulate (3') is stronger than postulate (3") and postulate (4") is stronger than postulate (4'). This means that N-pref-naive relations derive more positive facts (of the form $\mathcal{E} \succeq \mathcal{E}'$) and allow less negative facts (of the form $\neg(\mathcal{E} \succeq \mathcal{E}')$) than N-pref-stable relations:

Theorem 2. Let $\Sigma = \langle A, R, N, \geq \rangle$ be a PAFN and $\succeq \subseteq A \times A$. if $\succeq$ satisfies postulate 3' then it satisfies postulate 3" and if it satisfies postulate 4" then it satisfies postulate 4'.

It is not difficult to check that one consequence of this result is that any N-pref-stable extension is also a N-pref-naive extension.

Corollary 1. Let $\Sigma = \langle A, R, N, \geq \rangle$ be a PAFN and $\succeq^N$ be a N-pref-naive relation and $\succeq^S$ a N-pref-stable relation. If $\mathcal{E} \in Ext_{\succeq^S}(\Sigma)$ then $\mathcal{E} \in Ext_{\succeq^N}(\Sigma)$.

As in the case of pref-stable semantics, all the N-pref-naive (resp. N-pref-stable) relations share the same maximal elements.

Theorem 3. Let $\Sigma = \langle A, R, N, \geq \rangle$ be a PAFN. For any pair of N-pref-naive (or N-pref-stable relations) $\succeq, \succeq' \subseteq A \times A$ we have : $Ext_\succeq(\Sigma) = Ext_{\succeq'}(\Sigma)$.

Although some N-pref-naive relations depend on the preferences (for example the most specific relation discussed later), the extensions themselves are independent from the preferences. This is true since there is always a N-pref-naive relation that is independent from the preferences (for example, the most general one discussed later in this section) and according to the previous theorem, any other N-pref-naive relation leads to the same extensions.

Theorem 4. Let $\Sigma = \langle A, R, N, \geq \rangle$ be a PAFN, $\Gamma = \langle A, R, N \rangle$ be its corresponding simple AFN and $\succeq^N$ be a N-pref-naive relation then : $Ext_{\succeq N}(\Sigma) = Ext^N(\Gamma)$ and $\succeq^N$ satisfies conditions 1,2,3 of theorem 1 for naive extensions.

The following theorem states that N-pref-stable semantics generalizes stable semantics for AFNs : if there is no conflict between extended attacks and effective preferences, the stable extensions of the corresponding AFN are recovered.

Theorem 5. Let $\Sigma = \langle A, R, N, \geq \rangle$ be a PAFN and $\Gamma = \langle A, R, N \rangle$ the corresponding AFN. If $\succeq^S$ is a N-Pref-Stable relation and ($\nexists a, b \in A$) s.t. $a \, R^+ b$ and $b \rhd a$, then : $Ext_{\succeq S}(\Sigma) = Ext^S(\Gamma)$ and $\succeq^S$ satisfies conditions 1,2,3 of theorem 1 for stable extensions.

Since N-pref-naive extensions are independent from preferences and each N-pref-stale extension is a N-pref-naive extension one can conclude the exact role of adding or updating preferences in an AFN :

Corollary 2. In a PAFN, preferences have no impact on naive extensions but they affect the selection function of stable extensions among naive extensions.

N-pref-stable semantics also generalizes pref-stable semantics. Indeed when the necessity relation is empty the N-pref-stable extensions coinside with the pref-stable extensions of the corresponding PAF.

Theorem 6. Let $\Sigma = \langle A, R, N, \geq \rangle$ be a PAFN with $N = \emptyset$, $\Lambda = \langle A, R, \geq \rangle$ be its corresponding simple PAF. If $\succeq^S$ is a N-pref-stable relation then : $Ext_{\succeq S}(\Sigma) = Ext_{\succeq S}(\Lambda)$ and $\succeq^S$ satisfies the postulates 1,2,3 and 4 characterizing pref-stable extensions of a PAF.

The most general N-pref-naive relation $\succeq_g^N$ of a PAFN $\Sigma = \langle A, R, N, \geq \rangle$ coincides with the relation $\succeq_1^N$ (see section 3) which does not depend on the preference relation. The most specific relation $\succeq_s^N$ is defined as follows[3]:

$$(\mathcal{E} \succeq_g^N \mathcal{E}') \text{ iff } (\mathcal{E}' \notin SC(\Sigma)) \text{ or } (\mathcal{E}, \mathcal{E}' \in SC(\Sigma) \text{ and}$$
$$(\forall a' \in \mathcal{E}' \setminus \mathcal{E})(\exists a \in \mathcal{E} \setminus \mathcal{E}') \text{ s.t. } (a \, R^+ a') \vee (a' \, R^+ a) \vee (a \rhd a') \vee (a' \rhd a))$$

The relation $\succeq_g^N$ and $\succeq_s^N$ verify the postulates of a N-pref-naive relation and any other N-pref-naive relation is stronger than $\succeq_g^N$ and weaker than $\succeq_s^N$.

Theorem 7. Let $\Sigma = \langle A, R, N, \geq \rangle$ be a PAFN. The relations $\succeq_g^N$ and $\succeq_s^N$ are N-pref-naive relations and for any N-pref-naive relation $\succeq^N$ we have : if $\mathcal{E} \succeq_g^N \mathcal{E}'$ then $\mathcal{E} \succeq^N \mathcal{E}'$ and if $\mathcal{E} \succeq^N \mathcal{E}'$ then $\mathcal{E} \succeq_s^N \mathcal{E}'$.

---

[3] We recall that the most general (resp. the most specific) N-pref-naive relation returns $\mathcal{E} \succeq_g^N \mathcal{E}'$ (resp. $\mathcal{E} \succeq_s^N \mathcal{E}'$) iff it can be proved (resp. it cannot be proved) with the postulates 1' to 4' that $\mathcal{E}$ is better than $\mathcal{E}'$(resp.$\mathcal{E}'$ is better than $\mathcal{E}$). Similar definitions are used for N-pref-stable relations.

For a PAFN $\Sigma = \langle A, R, N, \geq \rangle$, the most general and the most specific N-pref-stable relations $\succeq_g^S$ and $\succeq_s^S$ respectively, are defined as follows:

$$(\mathcal{E} \succeq_g^S \mathcal{E}') \text{ iff } (\mathcal{E} \in SC(\Sigma) \text{ and } (\mathcal{E}' \notin SC(\Sigma) \text{ or } \mathcal{E}, \mathcal{E}' \in SC(\Sigma) \text{ and }$$
$$(\forall a' \in \mathcal{E}' \setminus \mathcal{E})(\exists a \in \mathcal{E} \setminus \mathcal{E}') \text{ s.t. } ((a \ R^+ \ a') \wedge \neg(a' \rhd a)) \vee ((a' \ R^+ \ a) \wedge (a \rhd a')))$$

$$(\mathcal{E} \succeq_s^S \mathcal{E}') \text{ iff } (\mathcal{E}' \notin SC(\Sigma) \text{ or } (\mathcal{E}, \mathcal{E}' \in SC(\Sigma) \text{ and }$$
$$(\forall a' \in \mathcal{E}' \setminus \mathcal{E})(\exists a \in \mathcal{E} \setminus \mathcal{E}') \text{ s.t. } ((a \ R^+ \ a') \wedge \neg(a' \rhd a)) \vee (a \rhd a'))$$

A similar result of that given by theorem 7 is valid for N-pref-stable relations.

Theorem 8. Let $\Sigma = \langle A, R, N, \geq \rangle$ be a PAFN. The relations $\succeq_g^S$ and $\succeq_s^S$ are N-pref-stable relations and for any N-pref-stable relation $\succeq^S$ we have : if $\mathcal{E} \succeq_g^S \mathcal{E}'$ then $\mathcal{E} \succeq^S \mathcal{E}'$ and if $\mathcal{E} \succeq^S \mathcal{E}'$ then $\mathcal{E} \succeq_s^S \mathcal{E}'$.

Example 1 (cont.). Consider again the PAFN of example 1 and its corresponding simple AFN $\Gamma = \langle A, R, N \rangle$. We can easily check that $Ext_{\succeq N}(\Sigma) = Ext^N(\Gamma) = \{\{a, b\}, \{c, d\}\}$ and $Ext_{\succeq S}(\Sigma) = \{\{a, b\}\}$. We can remark that as expected, N-pref-naive extensions are not sensible to preferences contrarily to N-pref-stable extensions. Notice that for this particular example, the relations $\succeq_g^N$ and $\succeq_s^N$ (resp. $\succeq_g^S$ and $\succeq_s^S$) agree on the comparison between any two sets $\mathcal{E}, \mathcal{E}' \in CF(\Sigma)$ and between a set $\mathcal{E} \in CF(\Sigma)$ and a set $\mathcal{E}' \notin CF(\Sigma)$. But for $\succeq_s^N$ (resp. $\succeq_s^S$), we have also $\mathcal{E} \succeq_s^N \mathcal{E}'$ (resp. $\mathcal{E} \succeq_s^N \mathcal{E}'$) for any $\mathcal{E}, \mathcal{E}' \notin CF(\Sigma)$. In general, $\succeq_s^N$ (resp. $\succeq_s^S$) may have strictly more elements than $\succeq_g^N$ (resp. $\succeq_g^S$).

6   Characterisation in a Dung AF

In this section we give a characterization of N-pref-naive and N-pref-stable semantics in terms of classical naive and stable semantics of a Dung AF.

Under N-pref-naive semantics, $\mathcal{E}$ is an extension iff it is in conflict with each element outside it. Notice that this condition is independent from any preference.

Theorem 9. Let $\Sigma = \langle A, R, N, \geq \rangle$ be a PAFN, R^+ be the corresponding extended attack relation, $\rhd$ be the corresponding effective preference relation and $\succeq^N$ be a N-pref-naive relation, then $\mathcal{E} \in Ext_{\succeq N}(\Sigma)$ iff $\mathcal{E} \in SC(\Sigma)$ and $(\forall x' \in A \setminus \mathcal{E}')(\exists x \in \mathcal{E})$ s.t $(x \ R^+ \ x') \vee (x' \ R^+ \ x)$.

A set $\mathcal{E}$ is a N-pref-stable extension iff for each element b outside it there is an element a inside it such that either a attacks b (wrt R^+) and b is not strictly preferred to a (wrt. $\rhd$) or b attacks a but a is strictly preferred to b :

Theorem 10. Let $\Sigma = \langle A, R, N, \geq \rangle$ be a PAFN, R^+ be the corresponding extended attack relation, $\rhd$ be the corresponding effective preference relation and $\succeq^S$ be a N-pref-stable relation, then $\mathcal{E} \in Ext_{\succeq S}(\Sigma)$ iff $\mathcal{E} \in SC(\Sigma)$ and $(\forall x' \in A \setminus \mathcal{E}')(\exists x \in \mathcal{E})$ s.t $(x \ R^+ \ x' \wedge \neg(x' \rhd x)) \vee (x' \ R^+ \ x \wedge x \rhd x')$.

In Practice, The "structural" operations to perform on a PAF $\Sigma = \langle A, R, N, \geq \rangle$ to compute N-pref-naive and N-pref-stable extensions are the following :

– Compute the extended attack (R^+) and the effective preference ($\trianglerighteq$) relations.
– Compute the Dung AF $F = \langle A, Def \rangle$ where Def is a new attack relation obtained by inversing the direction of any attack of R^+ not in accordance with the preference relation $\trianglerighteq$. In other words, $Def = \{(a,b) \in A \times A | (a\ R^+\ b\ and\ \neg(b \triangleright a))\ or(b\ R^+\ a\ and\ a \triangleright b)\}$.
– Use the AF $F = \langle A, Def \rangle$ to compute as usual naive and stable extensions.

The N-pref-naive and the N-pref-stable extensions correspond respectively to the naive and stable extensions of the Dung AF whose set of arguments is that of the original PAFN and attack relation is R^+ after the inversion of any attack which is not in accordance with the effective preference relation $\trianglerighteq$.

Theorem 11. Let $\Sigma = \langle A, R, N, \geq \rangle$ be a PAFN, $\succeq^N$ be a N-pref-naive relation, $\succeq^S$ be a N-pref-stable relation, R^+ be the extended attack relation, $\trianglerighteq$ be the effective preference relation and F be the AF $\langle A, Def \rangle$ where Def is defined by $Def = \{(a,b) \in A \times A | (a\ R^+\ b\ and\ \neg(b \triangleright a))\ or(b\ R^+\ a\ and\ a \triangleright b)\}$ then: $Ext_{\succeq^N} = Ext^N(F)$ and $Ext_{\succeq^S} = Ext^S(F)$.

Example 1 (cont.). Consider again the PAFN of example 1. The Dung system with the extended attack R^+ is deicted in Fig. 2-(1). Inversing the directions of attacks wrt R^+ that are not compatible with the effective preference relation $\trianglerighteq$ (theses attacks are reperesented by thik arcs in Fig. 2-(1)) allows to compute the Dung AF $F = \langle A, Def \rangle$ with the new attack relation $Def = \{(a,c), (a,d), (b,c), (b,d)\}$ (see Fig. 2-(2)).

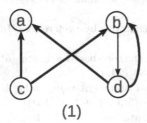 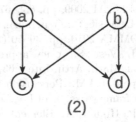

(1)   (2)

Fig. 2. (1) The Dung AF $\langle A, R^+ \rangle$ (before reparation), (2) The Dung AF $F = \langle A, Def \rangle$ (after reparation)

We can easily check that $Ext_{\succeq^N} = Ext^N(F) = \{\{a,b\}, \{c,d\}\}$ and $Ext_{\succeq^S} = Ext^S(F) = \{\{a,b\}\}$.

7   Conclusion

This paper builds upon the approach proposed in [4] [5] that introduces the interesting idea seeing acceptability semantics as a family of relations on the power set of the set of arguments. In our work we have extended this approach to the case of AFNs that represent a kind of bipolar argumentation frameworks where the support relation is a necessity. We have shown that a key point to perform this extension is to replace the input relations of necessities, attacks and preferences by the new relations of extended necessities, extended attacks and effective preferences. Thanks to these new notions, the obtained adapted form of stable semantics in preference-based AFNs is fully characterized by postualtes that are very similar to those proposed in [4] [5]. We have also extended the approach to the case of generalized naive semantics which represents the counterpart of justified extensions in default logics and ι-answer sets in logic programming. We have also shown how to represent any PAFN as a Dung AF so that a one to one correspondence is ensured between the (generalized) stable and naive extensions of the former and the (classical) stable and naive extensions of the second.

It has been shown in [17] that the argumentation frameworks we obtain by extending the necessity relation so that it can relate sets of arguments to single arguments allows to reach the same expressive power of arbitrary LPs. As a future work we want to use this idea to propose new argumentation-based approaches to handle preferences in logic programs.

References

1. Amgoud, L., Cayrol, C.: A reasoning model based on the production of acceptable arguments. Ann. Math. Artif. Intell. 34, 197–216 (2002)
2. Amgoud, L., Besnard, P.: Bridging the Gap between Abstract Argumentation Systems and Logic. In: Godo, L., Pugliese, A. (eds.) SUM 2009. LNCS, vol. 5785, pp. 12–27. Springer, Heidelberg (2009)
3. Amgoud, L., Vesic, S.: Repairing preference-based argumentation systems. In: Proceedings of IJCAI 2009, pp. 665–670 (2009)
4. Amgoud, L., Vesic, S.: Generalizing stable semantics by preferences. In: Proceedings of COMMA 2010, pp. 39–50 (2010)
5. Amgoud, L., Vesic, S.: A new approach for preference-based argumentation frameworks. Ann. Math. Artif. Intell. 63(2), 149–183 (2011)
6. Bench-Capon, T.J.M.: Persuasion in practical argument using value based argumentation frameworks. J. of Log. and Comp. 13(3), 429–448 (2003)
7. Besnard, P., Hunter, A.: Elements of Argumentation. The MIT Press (2008)
8. Boella, G., Gabbay, D.M., Van Der Torre, L., Villata, S.: Support in Abstract Argumentation. In: Proceedings of COMMA 2010, pp. 40–51 (2010)
9. Brewka, G., Woltran, S.: Abstract Dialectical Frameworks. In: Proceedings of KR 2010, pp. 102–111 (2010)
10. Cayrol, C., Lagasquie-Schiex, M.C.: On the Acceptability of Arguments in Bipolar Argumentation Frameworks. In: Godo, L. (ed.) ECSQARU 2005. LNCS (LNAI), vol. 3571, pp. 378–389. Springer, Heidelberg (2005)
11. Cayrol, C., Lagasquie-Schiex, M.C.: Coalitions of arguments: A tool for handling bipolar argumentation frameworks. Int. J. Intell. Syst. 25(1), 83–109 (2010)
12. Dung, P.M.: On the acceptability of arguments and its fundamental role in nonmonotonic reasoning, logic programming and n-person games. Artif. Intel. 77(2), 321–357 (1995)

13. Gebser, M., Gharib, M., Mercer, R., Schaub, T.: Monotonic Answer Set Programming. J. of Log. and Comp. 19(4), 539–564 (2009)
14. Gorogiannis, N., Hunter, A.: Instantiating abstract, argumentation with classical logic arguments: Postulates and properties. Artif. Intel. 175, 1479–1497 (2011)
15. Łukaszewicz, W.: Considerations on Default Logic: An Alternative Approach. Comput. Intel. 4, 1–16 (1988)
16. Modgil, S.: Reasoning about preferences in argumentation frameworks. Artif. Intel. 173, 901–934 (2009)
17. Nouioua, F., Risch, V.: Argumentation Frameworks with Necessities. In: Benferhat, S., Grant, J. (eds.) SUM 2011. LNCS, vol. 6929, pp. 163–176. Springer, Heidelberg (2011)
18. Nouioua, F., Risch, V.: A Reconstruction of Abstract Argumentation Admissible Semantics into Defaults and Answer Sets Programming. In: Proceedings of ICAART 2012, pp. 237–242 (2012)

Stable Semantics in Logic-Based Argumentation

Leila Amgoud

IRIT – CNRS, 118 route de Narbonne 31062, Toulouse – France

Abstract. This paper investigates the outputs of abstract logic-based argumentation systems under *stable* semantics. We delimit the number of stable extensions a system may have. We show that in the best case, an argumentation system infers exactly the common conclusions drawn from the maximal consistent subbases of the original knowledge base. This output corresponds to that returned by a system under the naive semantics. In the worst case, counter-intuitive results are returned. In the intermediary case, the system forgets intuitive conclusions. These two latter cases are due to the use of skewed attack relations. The results show that stable semantics is either useless or unsuitable in logic-based argumentation systems. Finally, we show that under this semantics, argumentation systems may inherit the problems of coherence-based approaches.

1 Introduction

An argumentation system for reasoning with inconsistent knowledge is built from a knowledge base using a monotonic logic. It consists of a set of *arguments*, *attacks* among them and a *semantics* for evaluating the arguments (see [4,10,7,12] for some examples of such systems).

Stable semantics is one of the prominent semantics proposed in [8]. A set of arguments is acceptable (or an *extension*) under this semantics, if it is free of conflicts and attacks any argument outside the set. Note that this semantics does not guarantee the existence of extensions for a system. In [6], the author studied the kind of outputs that may be returned under this semantics. However, the focus was on *one particular* argumentation system: it is grounded on propositional logic and uses 'assumption attack' [9]. The results show that each stable extension of the system is built from one maximal consistent subbase of the original knowledge base. However, it is not clear whether this is true for other attack relations or other logics. It is neither clear whether systems that have stable extensions return intuitive results. It is also unclear what is going wrong with systems that do not have stable extensions. Finally, the number of stable extensions that a system may have is unknown.

In this paper, we conduct an in-depth study on the outputs of argumentation systems under stable semantics. We consider *abstract* logic-based systems, i.e., systems that use *Tarskian logics* [15] and *any attack relation*. For the first time, the maximum number of stable extensions a system may have is delimited. It is the number of maximal (for set inclusion) consistent subbases of the knowledge base. Moreover, we show that stable semantics is either useless or unsuitable for these systems. Indeed, in the best case, such systems infer exactly the conclusions that are drawn from all the maximal consistent subbases. This corresponds exactly to the output of the same systems under naive

E. Hüllermeier et al. (Eds.): SUM 2012, LNAI 7520, pp. 58–71, 2012.

semantics. In the worst case, counter-intuitive results are returned. There is a third case where intuitive conclusions may be forgotten by the systems. These two last cases are due to the use of skewed attack relations. Finally, we show that argumentation systems that use stable semantics inherit the problems of coherence-based approaches [14].

The paper is organized as follows: Section 2 defines the logic-based argumentation systems we are interested in. Section 3 recalls three basic postulates that such systems should obey. Section 4 investigates the outcomes that are computed under stable semantics. Section 5 compares our work with existing ones and Section 6 concludes.

2 Logic-Based Argumentation Systems

Argumentation systems are built on an underlying *monotonic logic*. In this paper, we focus on Tarski's monotonic logics [15]. Indeed, we consider logics $(\mathcal{L}, \text{CN})$ where $\mathcal{L}$ is a set of well-formed *formulas* and CN is a *consequence operator*. It is a function from $2^{\mathcal{L}}$ to $2^{\mathcal{L}}$ which returns the set of formulas that are logical consequences of another set of formulas according to the logic in question. It satisfies the following basic properties:

1. $X \subseteq \text{CN}(X)$ (Expansion)
2. $\text{CN}(\text{CN}(X)) = \text{CN}(X)$ (Idempotence)
3. $\text{CN}(X) = \bigcup_{Y \subseteq_f X} \text{CN}(Y)^1$ (Finiteness)
4. $\text{CN}(\{x\}) = \mathcal{L}$ for some $x \in \mathcal{L}$ (Absurdity)
5. $\text{CN}(\emptyset) \neq \mathcal{L}$ (Coherence)

A CN that satisfies the above properties is *monotonic*. The associated notion of *consistency* is defined as follows:

Definition 1 (Consistency). *A set $X \subseteq \mathcal{L}$ is consistent wrt a logic $(\mathcal{L}, \text{CN})$ iff $\text{CN}(X) \neq \mathcal{L}$. It is inconsistent otherwise.*

Arguments are built from a *knowledge base* $\Sigma \subseteq \mathcal{L}$ as follows:

Definition 2 (Argument). *Let Σ be a knowledge base. An argument is a pair (X, x) s.t. $X \subseteq \Sigma$, X is consistent, and $x \in \text{CN}(X)^2$. An argument (X, x) is a sub-argument of (X', x') iff $X \subseteq X'$.*

Notations: Supp and Conc denote respectively the *support* X and the *conclusion* x of an argument (X, x). For all $\mathcal{S} \subseteq \Sigma$, $\text{Arg}(\mathcal{S})$ denotes the set of all arguments that can be built from $\mathcal{S}$ by means of Definition 2. Sub is a function that returns all the sub-arguments of a given argument. For all $\mathcal{E} \subseteq \text{Arg}(\Sigma)$, $\text{Concs}(\mathcal{E}) = \{\text{Conc}(a) \mid a \in \mathcal{E}\}$ and $\text{Base}(\mathcal{E}) = \bigcup_{a \in \mathcal{E}} \text{Supp}(a)$. $\text{Max}(\Sigma)$ is the set of all maximal (for set inclusion) consistent subbases of Σ. Finally, $\text{Free}(\Sigma) = \bigcap \mathcal{S}_i$ where $\mathcal{S}_i \in \text{Max}(\Sigma)$, and $\text{Inc}(\Sigma) = \Sigma \setminus \text{Free}(\Sigma)$.

An argumentation system is defined as follows.

---

[1] $Y \subseteq_f X$ means that Y is a finite subset of X.

[2] Generally, the support X is minimal (for set inclusion). In this paper, we do not need to make this assumption.

Definition 3 (Argumentation system). *An* argumentation system *(AS) over a knowledge base Σ is a pair $\mathcal{T} = (\text{Arg}(\Sigma), \mathcal{R})$ such that $\mathcal{R} \subseteq \text{Arg}(\Sigma) \times \text{Arg}(\Sigma)$ is an* attack relation. *For $a, b \in \text{Arg}(\Sigma)$, $(a, b) \in \mathcal{R}$ (or $a\mathcal{R}b$) means that a attacks b.*

The attack relation is left *unspecified* in order to keep the system very general. It is also worth mentioning that the set $\text{Arg}(\Sigma)$ may be infinite even when the base Σ is finite. This would mean that the argumentation system may be *infinite*[3]. Finally, arguments are evaluated using stable semantics.

Definition 4 (Stable semantics [8]). *Let $\mathcal{T} = (\text{Arg}(\Sigma), \mathcal{R})$ be an AS over a knowledge base Σ, and $\mathcal{E} \subseteq \text{Arg}(\Sigma)$ s.t. $\nexists\, a, b \in \mathcal{E}$ s.t. $a\mathcal{R}b$.*

- *$\mathcal{E}$ is a* naive *extension iff $\mathcal{E}$ is maximal (for set inclusion).*
- *$\mathcal{E}$ is a* stable *extension iff $\forall a \in \text{Arg}(\Sigma) \setminus \mathcal{E}$, $\exists b \in \mathcal{E}$ s.t. $b\mathcal{R}a$.*

It is worth noticing that each stable extension is a naive one but the converse is false. Let $\text{Ext}_x(\mathcal{T})$ denote the set of all extensions of $\mathcal{T}$ under semantics x (n and s will stand respectively for naive and stable semantics). When we do not need to specify the semantics, we use the notation $\text{Ext}(\mathcal{T})$ for short.

The extensions are used in order to define the conclusions to be drawn from Σ according to an argumentation system $\mathcal{T}$. The idea is to infer a formula x from Σ iff x is the conclusion of an argument in each extension. $\text{Output}(\mathcal{T})$ is the set of all such formulas.

Definition 5 (Output). *Let $\mathcal{T} = (\text{Arg}(\Sigma), \mathcal{R})$ be an AS over a knowledge base Σ. $\text{Output}(\mathcal{T}) = \{x \in \mathcal{L} \mid \forall \mathcal{E} \in \text{Ext}(\mathcal{T}), \exists a \in \mathcal{E}$ s.t. $\text{Conc}(a) = x\}$.*

$\text{Output}(\mathcal{T})$ coincides with the set of common conclusions of the extensions. Indeed, $\text{Output}(\mathcal{T}) = \bigcap \text{Concs}(\mathcal{E}_i)$, $\mathcal{E}_i \in \text{Ext}(\mathcal{T})$. Note also that when the base Σ contains only inconsistent formulas, then $\text{Arg}(\Sigma) = \emptyset$. Consequently, $\text{Ext}_s(\mathcal{T}) = \{\emptyset\}$ and $\text{Output}(\mathcal{T}) = \emptyset$. Without loss of generality, throughout the paper, we assume that Σ is *finite* and contains at least one consistent formula.

3   Postulates for Argumentation Systems

In [5], it was argued that logic-based argumentation systems should obey to some rationality postulates, i.e., desirable properties that any reasoning system should enjoy. The three postulates proposed in [5] are revisited and extended to any Tarskian logic in [1]. The first one concerns the closure of the system's output under the consequence operator CN. The idea is that the formalism should not forget conclusions.

Postulate 1 (Closure under CN). *Let $\mathcal{T} = (\text{Arg}(\Sigma), \mathcal{R})$ be an AS over a knowledge base Σ. $\mathcal{T}$ satisfies* closure *iff for all $\mathcal{E} \in \text{Ext}(\mathcal{T})$, $\text{Concs}(\mathcal{E}) = \text{CN}(\text{Concs}(\mathcal{E}))$.*

The second rationality postulate ensures that the acceptance of an argument should imply also the acceptance of all its sub-arguments.

---

[3] An AS is *finite* iff each argument is attacked by a finite number of arguments. It is *infinite* otherwise.

Postulate 2 (Closure under sub-arguments). *Let $\mathcal{T} = (\text{Args}(\Sigma), \mathcal{R})$ be an AS over a knowledge base Σ. $\mathcal{T}$ is closed under sub-arguments iff for all $\mathcal{E} \in \text{Ext}(\mathcal{T})$, if $a \in \mathcal{E}$, then $\text{Sub}(a) \subseteq \mathcal{E}$.*

The third rationality postulate ensures that the set of conclusions supported by each extension is consistent.

Postulate 3 (Consistency). *Let $\mathcal{T} = (\text{Arg}(\Sigma), \mathcal{R})$ be an AS over a knowledge base Σ. $\mathcal{T}$ satisfies consistency iff for all $\mathcal{E} \in \text{Ext}(\mathcal{T})$, $\text{Concs}(\mathcal{E})$ is consistent.*

In [1], the conditions under which these postulates are satisfies/violated are investigated. It is shown that the attack relation should be grounded on inconsistency. This is an obvious requirement especially for reasoning about inconsistent information.

Definition 6 (Conflict-dependent). *Let $\mathcal{T} = (\text{Arg}(\Sigma), \mathcal{R})$ be an AS. The attack relation $\mathcal{R}$ is conflict-dependent iff $\forall a, b \in \text{Arg}(\Sigma)$, if $a\mathcal{R}b$ then $\text{Supp}(a) \cup \text{Supp}(b)$ is inconsistent.*

4 The Outcomes of Argumentation Systems

As seen before, the acceptability of arguments is defined without considering neither their internal structure nor their origin. In this section, we fully characterize for the first time the 'concrete' outputs of an argumentation system under stable semantics. For that purpose, we consider only systems that enjoy the three rationality postulates introduced in the previous section. Recall that systems that violate them return undesirable outputs. Before presenting our study, we start first by analyzing the outputs of argumentation systems under the naive semantics. One may wonder why especially since this particular semantics is not used in the literature for evaluating arguments. The reason is that the only case where stable semantics ensures an intuitive output is where an argumentation system returns exactly the same output under stable and naive semantics.

4.1 Naive Semantics

Before characterizing the outputs of an AS under naive semantics, let us start by showing some useful properties. The next result shows that if each naive extension returns a consistent subbase of Σ, then the AS is certainly closed under sub-arguments.

Proposition 1. *Let $\mathcal{T} = (\text{Arg}(\Sigma), \mathcal{R})$ be an AS over a knowledge base Σ such that $\mathcal{R}$ is conflict-dependent. If $\forall \mathcal{E} \in \text{Ext}_n(\mathcal{T})$, $\text{Base}(\mathcal{E})$ is consistent, then $\mathcal{T}$ is closed under sub-arguments (under naive semantics).*

A consequence of the previous result is that under naive semantics, the satisfaction of both consistency and closure under sub-arguments is equivalent to the satisfaction of a stronger version of consistency.

Theorem 1. *Let $\mathcal{T} = (\text{Arg}(\Sigma), \mathcal{R})$ be an AS over a knowledge base Σ such that $\mathcal{R}$ is conflict-dependent. $\mathcal{T}$ satisfies consistency and closure under sub-arguments (under naive semantics) iff $\forall \mathcal{E} \in \text{Ext}_n(\mathcal{T})$, $\text{Base}(\mathcal{E})$ is consistent.*

In case of naive semantics, closure under the consequence operator CN is induced from the two other postulates: closure under sub-arguments and consistency.

Proposition 2. *Let* $\mathcal{T} = (\text{Arg}(\Sigma), \mathcal{R})$ *be an AS over a knowledge base* Σ *such that* $\mathcal{R}$ *is conflict-dependent. If* $\mathcal{T}$ *satisfies consistency and is closed under sub-arguments (under naive semantics), then it is also closed under* CN.

An important question now is: what is hidden behind naive semantics? We show that the naive extensions of *any* argumentation system that satisfies Postulates 2 and 3 *always* return maximal (for set inclusion) consistent subbases of Σ.

Theorem 2. *Let* $\mathcal{T} = (\text{Arg}(\Sigma), \mathcal{R})$ *be an AS over a knowledge base* Σ *such that* $\mathcal{R}$ *is conflict-dependent. If* $\mathcal{T}$ *satisfies consistency and is closed under sub-arguments (under naive semantics), then:*

– *For all* $\mathcal{E} \in \text{Ext}_n(\mathcal{T})$, $\text{Base}(\mathcal{E}) \in \text{Max}(\Sigma)$.
– *For all* $\mathcal{E}_i, \mathcal{E}_j \in \text{Ext}_n(\mathcal{T})$, *if* $\text{Base}(\mathcal{E}_i) = \text{Base}(\mathcal{E}_j)$ *then* $\mathcal{E}_i = \mathcal{E}_j$.

The previous result does not guarantee that all the maximal consistent subbases of Σ are captured. The next theorem confirms that *any* maximal consistent subbase of Σ defines a naive extension of an AS which satisfies consistency and closure under sub-arguments.

Theorem 3. *Let* $\mathcal{T} = (\text{Arg}(\Sigma), \mathcal{R})$ *be an AS over a knowledge base* Σ *such that* $\mathcal{R}$ *is conflict-dependent. If* $\mathcal{T}$ *satisfies consistency and is closed under sub-arguments (under naive semantics), then:*

– *For all* $\mathcal{S} \in \text{Max}(\Sigma)$, $\text{Arg}(\mathcal{S}) \in \text{Ext}_n(\mathcal{T})$.
– *For all* $\mathcal{S}_i, \mathcal{S}_j \in \text{Max}(\Sigma)$, *if* $\text{Arg}(\mathcal{S}_i) = \text{Arg}(\mathcal{S}_j)$ *then* $\mathcal{S}_i = \mathcal{S}_j$.

It follows that any argumentation system that satisfies the two postulates 2 and 3 enjoy a full correspondence between the maximal consistent subbases of Σ and the naive extensions of the system.

Theorem 4. *Let* $\mathcal{T} = (\text{Arg}(\Sigma), \mathcal{R})$ *be an AS over a knowledge base* Σ *such that* $\mathcal{R}$ *is conflict-dependent.* $\mathcal{T}$ *satisfies consistency and is closed under sub-arguments (under naive semantics) iff the naive extensions of* $\text{Ext}_n(\mathcal{T})$ *are exactly the* $\text{Arg}(\mathcal{S})$ *where* $\mathcal{S}$ *ranges over the elements of* $\text{Max}(\Sigma)$.

A direct consequence of the previous result is that the number of naive extensions is finite. This follows naturally from the finiteness of the knowledge base Σ.

Theorem 5. *Let* $\mathcal{T} = (\text{Arg}(\Sigma), \mathcal{R})$ *be an AS over a knowledge base* Σ *such that* $\mathcal{R}$ *is conflict-dependent and* $\mathcal{T}$ *satisfies consistency and is closed under sub-arguments (under naive semantics). If* Σ *is finite, then* $\mathcal{T}$ *has a finite number of naive extensions.*

Let us now characterize the set $\text{Output}(\mathcal{T})$ of inferences that may be drawn from a knowledge base Σ by an argumentation system $\mathcal{T}$ under naive semantics. It coincides with the set of formulas that are drawn by all the maximal consistent subbases of Σ.

Theorem 6. *Let* $\mathcal{T} = (\text{Arg}(\Sigma), \mathcal{R})$ *be an AS over a knowledge base* Σ *such that* $\mathcal{R}$ *is conflict-dependent,* $\mathcal{T}$ *satisfies consistency and is closed under sub-arguments (under naive semantics).* $\text{Output}(\mathcal{T}) = \bigcap \text{CN}(\mathcal{S}_i)$ *where* $\mathcal{S}_i \in \text{Max}(\Sigma)$.

In short, under naive semantics, *any* 'good' instantiation of Dung's abstract framework returns exactly the formulas that are drawn (with CN) by all the maximal consistent subbases of the base Σ. So whatever the attack relation that is chosen, the result will be the same. It is worth recalling that the output set contains exactly the so-called *universal conclusions* in the approach developed in [14] for reasoning about inconsistent propositional bases.

4.2 Stable Semantics

As for naive semantics, we show that under stable semantics, strong consistency induce closure under sub-arguments.

Proposition 3. *Let* $\mathcal{T} = (\text{Arg}(\Sigma), \mathcal{R})$ *be an AS over a knowledge base* Σ *such that* $\mathcal{R}$ *is conflict-dependent. If* $\forall \mathcal{E} \in \text{Ext}_s(\mathcal{T})$, $\text{Base}(\mathcal{E})$ *is consistent, then* $\mathcal{T}$ *is closed under sub-arguments (under stable semantics).*

The following theorem shows that satisfying consistency and closure under sub-arguments amounts exactly to satisfying the strong version of consistency.

Theorem 7. *Let* $\mathcal{T} = (\text{Arg}(\Sigma), \mathcal{R})$ *be an AS over a knowledge base* Σ *such that* $\mathcal{R}$ *is conflict-dependent.* $\mathcal{T}$ *satisfies consistency and closure under sub-arguments (under stable semantics) iff* $\forall \mathcal{E} \in \text{Ext}_s(\mathcal{T})$, $\text{Base}(\mathcal{E})$ *is consistent.*

Like for naive semantics, in case of stable semantics, closure under the consequence operator CN follows from closure under sub-arguments and consistency.

Proposition 4. *Let* $\mathcal{T} = (\text{Arg}(\Sigma), \mathcal{R})$ *be an AS over a knowledge base* Σ *such that* $\mathcal{R}$ *is conflict-dependent. If* $\mathcal{T}$ *satisfies consistency and is closed under sub-arguments (under stable semantics), then it is also closed under CN (under stable semantics).*

We now show that the stable extensions of *any* argumentation system, which satisfies Postulates 2 and 3, return maximal consistent subbases of Σ. This means that if one instantiates Dung's system and does not get maximal consistent subbases with stable extensions, then the instantiation certainly violates one or both of the two key postulates: consistency and closure under sub-arguments.

Theorem 8. *Let* $\mathcal{T} = (\text{Arg}(\Sigma), \mathcal{R})$ *be an AS over a knowledge base* Σ *such that* $\mathcal{R}$ *is conflict-dependent. If* $\mathcal{T}$ *satisfies consistency and closure under sub-arguments (under stable semantics), then:*

- *For all* $\mathcal{E} \in \text{Ext}_s(\mathcal{T})$, $\text{Base}(\mathcal{E}) \in \text{Max}(\Sigma)$.
- *For all* $\mathcal{E} \in \text{Ext}_s(\mathcal{T})$, $\mathcal{E} = \text{Arg}(\text{Base}(\mathcal{E}))$.
- *For all* $\mathcal{E}_i, \mathcal{E}_j \in \text{Ext}_s(\mathcal{T})$, *if* $\text{Base}(\mathcal{E}_i) = \text{Base}(\mathcal{E}_j)$ *then* $\mathcal{E}_i = \mathcal{E}_j$.

This result is strong as it characterizes the outputs under stable semantics of a large class of argumentation systems, namely, those grounded on Tarskian logics.

The previous result does not guarantee that each maximal consistent subbase of Σ has a corresponding stable extension in the argumentation system $\mathcal{T}$. To put it differently, it does not guarantee the equality $|\text{Ext}_s(\mathcal{T})| = |\text{Max}(\Sigma)|$. However, it shows that in case stable extensions exist, then their bases are certainly elements of $\text{Max}(\Sigma)$. This enables us to delimit the number of stable extensions that an AS may have.

Proposition 5. *Let* $\mathcal{T} = (\text{Arg}(\Sigma), \mathcal{R})$ *be an AS over a base* Σ *s.t.* $\mathcal{R}$ *is conflict-dependent and* $\mathcal{T}$ *satisfies consistency and closure under sub-arguments (under stable semantics). It holds that* $0 \leq |\text{Ext}_s(\mathcal{T})| \leq |\text{Max}(\Sigma)|$.

From this property, it follows that when the knowledge base is finite, the number of stable extensions is finite as well.

Property 1. *If* Σ *is finite, then* $\forall \mathcal{T} = (\text{Arg}(\Sigma), \mathcal{R})$ *s.t.* $\mathcal{T}$ *satisfies consistency and closure under sub-arguments (under stable semantics),* $|\text{Ext}_s(\mathcal{T})|$ *is finite.*

The fact that an argumentation system $\mathcal{T}$ verifies or not the equality $|\text{Ext}_s(\mathcal{T})| = |\text{Max}(\Sigma)|$ depends broadly on the attack relation that is chosen. Let $\Re_s$ be the set of *all* attack relations that ensure Postulates 2 and 3 under stable semantics ($\Re_s = \{\mathcal{R} \subseteq \text{Arg}(\Sigma) \times \text{Arg}(\Sigma) \mid \mathcal{R}$ is conflict-dependent and $(\text{Arg}(\Sigma), \mathcal{R})$ satisfies Postulates 2 and 3 under stable semantics} for all Σ). This set contains three *disjoints* subsets of attack relations: $\Re_s = \Re_{s1} \cup \Re_{s2} \cup \Re_{s3}$:

- $\Re_{s1}$: the relations which lead to $|\text{Ext}_s(\mathcal{T})| = 0$.
- $\Re_{s2}$: the relations which ensure $0 < |\text{Ext}_s(\mathcal{T})| < |\text{Max}(\Sigma)|$.
- $\Re_{s3}$: the relations which ensure $|\text{Ext}_s(\mathcal{T})| = |\text{Max}(\Sigma)|$.

Let us analyze separately each category of attack relations. We start with relations of the set $\Re_{s3}$. Those relations induce a one to one correspondence between the stable extensions of the argumentation system and the maximal consistent subbases of Σ.

Property 2. *Let* $\mathcal{T} = (\text{Arg}(\Sigma), \mathcal{R})$ *be an AS over a knowledge base* Σ *such that* $\mathcal{R} \in \Re_{s3}$. *For all* $\mathcal{S} \in \text{Max}(\Sigma)$, $\text{Arg}(\mathcal{S}) \in \text{Ext}_s(\mathcal{T})$.

An important question now is: do such attack relations exist? We are not interested in identifying all of them since they lead to the same result. It is sufficient to show whether they exist. Hopefully, such relations exist and *assumption attack* [9] is one of them.

Theorem 9. *The set* $\Re_{s3}$ *is not empty.*

We show now that argumentation systems based on this category of attack relations always have stable extensions.

Theorem 10. *For all* $\mathcal{T} = (\text{Arg}(\Sigma), \mathcal{R})$ *such that* $\mathcal{R} \in \Re_{s3}$, $\text{Ext}_s(\mathcal{T}) \neq \emptyset$.

Let us now characterize the output set of a system under stable semantics.

Theorem 11. *Let* $\mathcal{T} = (\text{Arg}(\Sigma), \mathcal{R})$ *be an AS over a knowledge base* Σ *such that* $\mathcal{R} \in \Re_{s3}$. $\text{Output}(\mathcal{T}) = \bigcap \text{CN}(\mathcal{S}_i)$ *where* $\mathcal{S}_i \in \text{Max}(\Sigma)$.

Note that this category of attack relations leads exactly to the same result as naive semantics. Thus, stable semantics does not play any particular role. Moreover, argumentation systems return the universal conclusions (of the coherence-based approach [14]) under any monotonic logic, not only under propositional logic as in [14].

Let us now analyze the first category ($\Re_{s1}$) of attack relations that guarantee the postulates. Recall that these relations prevent the existence of stable extensions.

Theorem 12. *Let* $\mathcal{T} = (\text{Arg}(\Sigma), \mathcal{R})$ *be an AS over a knowledge base* Σ *such that* $\mathcal{R} \in \Re_{s1}$. *It holds that* $\text{Output}(\mathcal{T}) = \emptyset$.

These attack relations are skewed, and may prevent intuitive conclusions from being drawn from a knowledge base. This is particularly the case of free formulas, i.e. in $\text{Free}(\Sigma)$, as shown next.

Property 3. *Let* $\mathcal{T} = (\text{Arg}(\Sigma), \mathcal{R})$ *be an AS over a knowledge base* Σ *such that* $\text{Free}(\Sigma) \neq \emptyset$. *If* $\mathcal{R} \in \Re_{s1}$, *then* $\forall x \in \text{Free}(\Sigma)$, $x \notin \text{Output}(\mathcal{T})$.

What about the remaining attack relations, i.e., those of $\Re_{s2}$ that ensure the existence of stable extensions? Systems that use these relations choose a proper subset of maximal consistent subbases of Σ and make inferences from them. Their output sets are defined as follows:

Theorem 13. *Let* $\mathcal{T} = (\text{Arg}(\Sigma), \mathcal{R})$ *be an AS over a knowledge base* Σ *such that* $\mathcal{R} \in \Re_{s2}$. $\text{Output}(\mathcal{T}) = \bigcap \text{CN}(\mathcal{S}_i)$ *where* $\mathcal{S}_i \in \text{Max}(\Sigma)$ *and* $\mathcal{S}_i = \text{Base}(\mathcal{E}_i)$ *with* $\mathcal{E}_i \in \text{Ext}_s(\mathcal{T})$.

These attack relations lead to an unjustified discrimination between the maximal consistent subbases of a knowledge base. Unfortunately, this is fatal for the argumentation systems which use them as they return counter-intuitive results (see Example 1).

Example 1. *Assume that* $(\mathcal{L}, \text{CN})$ *is propositional logic and let* Σ *contain three intuitively equally preferred formulas:* $\Sigma = \{x, x \to y, \neg y\}$. *This base has three maximal consistent subbases:*

- $\mathcal{S}_1 = \{x, x \to y\}$, $\mathcal{S}_2 = \{x, \neg y\}$, $\mathcal{S}_3 = \{x \to y, \neg y\}$.

The arguments that may be built from Σ *may have the following supports:* $\{x\}$, $\{x \to y\}$, $\{\neg y\}$, $\{x, x \to y\}$, $\{x, \neg y\}$, *and* $\{x \to y, \neg y\}$. *Assume now the attack relation shown in the figure below. For the sake of readability, we do not represent the conclusions of the arguments in the figure. An arrow from* X *towards* Y *is read as follows: any argument with support* X *attacks any argument with support* Y.

This argumentation system has two stable extensions:

- $\mathcal{E}_1 = \{a \in \text{Arg}(\Sigma) \mid \text{Supp}(a) = \{x, x \to y\}$ *or* $\text{Supp}(a) = \{x\}$ *or* $\text{Supp}(a) = \{x \to y\}\}$.
- $\mathcal{E}_2 = \{a \in \text{Arg}(\Sigma) \mid \text{Supp}(a) = \{x, \neg y\}$ *or* $\text{Supp}(a) = \{x\}$ *or* $\text{Supp}(a) = \{\neg y\}\}$.

It can be checked that this argumentation system satisfies consistency and closure under sub-arguments. The two extensions capture respectively the subbases S_1 and S_2.

It is worth noticing that the third subbase $S_3 = \{x \to y, \neg y\}$ is not captured by any stable extension. Indeed, the set $\text{Arg}(S_3) = \{a \in \text{Arg}(\Sigma) \mid \text{Supp}(a) = \{\neg y, x \to y\}$ or $\text{Supp}(a) = \{\neg y\}$ or $\text{Supp}(a) = \{x \to y\}\}$ is not a stable extension. S_3 is discarded due to the definition of the attack relation. Note that this leads to non-intuitive outputs. For instance, it can be checked that $x \in \text{Output}(\mathcal{T})$ whereas $\neg y \notin \text{Output}(\mathcal{T})$ and $x \to y \notin \text{Output}(\mathcal{T})$. Since the three formulas of Σ are assumed to be equally preferred, then there is no reason to privilege one compared to the others!

The example showed a skewed attack relation which led to 'artificial' priorities among the formulas of a base Σ: x is preferred to $\neg y$ and $x \to y$. The following result confirms this observation.

Theorem 14. *Let $\mathcal{T} = (\text{Arg}(\Sigma), \mathcal{R})$ be an AS over a knowledge base Σ. If $\mathcal{R} \in \Re_{s2}$, then $\exists x, x' \in \text{Inc}(\Sigma)$ such that $x \in \text{Output}(\mathcal{T})$ and $x' \notin \text{Output}(\mathcal{T})$.*

To sum up, there are three categories of attack relations that ensure the three rationality postulates. Two of them ($\Re_{s1}$ and $\Re_{s2}$) should be avoided as they lead to undesirable results. It is worth mentioning that we are not interested here in identifying those relations. Indeed, if they exist, they certainly lead to bad results and thus, should not be used. The third category of relations ($\Re_{s3}$) leads to "correct" results, but argumentation systems based on them return exactly the same results under naive semantics. Thus, stable semantics does not play any particular role in the logic-based argumentation systems we studied in the paper. Moreover, the outputs of the systems coincide with those of the coherence-based approach [14]. As a consequence, argumentation systems inherit the drawbacks of this approach. Let us illustrate this issue by the following example.

Example 2. *Assume that $(\mathcal{L}, \text{CN})$ is propositional logic and let $\Sigma = \{x, \neg x \wedge y\}$. This base has two maximal consistent subbases:*

- $S_1 = \{x\}$
- $S_2 = \{\neg x \wedge y\}$

According to the previous results, any instantiation of Dung's framework falls in one of the following cases:

- *Instantiations that use attack relations in $\Re_{s1}$ will lead to $\text{Output}(\mathcal{T}) = \emptyset$. This result is undesirable since y should be inferred from Σ since it is not part of the conflict.*
- *Instantiations that use attack relations in $\Re_{s2}$ will lead either to $\text{Output}(\mathcal{T}) = \text{CN}(\{x\})$ or to $\text{Output}(\mathcal{T}) = \text{CN}(\{\neg x \wedge y\})$. Both outputs are undesirable since they are unjustified. Why x and not $\neg x$ and vice versa?*
- *Instantiations that use attack relations in $\Re_{s3}$ will lead to $\text{Output}(\mathcal{T}) = \emptyset$. Like the first case, there is no reason to not conclude y.*

5 Related Work

This paper investigated the outputs of an argumentation system under stable semantics. There are some works in the literature which are somehow related to our. In [11,13], the authors studied whether some *particular* argumentation systems satisfy some of the rationality postulates presented in this paper. By particular system, we mean a system that is grounded on a particular logic and/or that uses a specific attack relation. In our paper the objective is different. We assumed abstract argumentation systems that satisfy the desirable postulates, and studied their outputs under stable semantics. Two other works, namely [6] and [3], share this objective. In [6], the author studied one particular system: the one that is grounded on propositional logic and uses the "assumption attack" relation [9]. The results got show that assumption attack belongs to our set $\Re_{s3}$. In [3], these results are generalized to argumentation systems that use the same attack relation but grounded on any Tarskian logic. Our work is more general since it completely abstracts from the attack relation. Moreover, it presents a *complete* view of the outputs under stable semantics.

6 Conclusion

This paper characterized for the first time the outputs (under *stable* semantics) of *any* argumentation system that is grounded on a Tarskian logic and that satisfies very basic rationality postulates. The study is very general since it keeps all the parameters of a system unspecified. Namely, Tarskian logics are abstract and no requirement is imposed on the attack relation except the property of conflict-dependency which is mandatory for ensuring the consistency postulate. We identified the maximum number of stable extensions a system may have. We discussed three possible categories of attack relations that may make a system satisfies the postulates. Two of them lead to counter-intuitive results. Indeed, either ad hoc choices are made or interesting conclusions are forgotten like the free formulas. Argumentation systems based on attack relations of the third category enjoy a one to one correspondence between the stable extensions and the maximal consistent subbases of the knowledge base. Consequently, their outputs are the common conclusions drawn from each maximal consistent subbase. This means that stable semantics does not play any particular role for reasoning with inconsistent information since the same result is returned by naive semantics. Moreover, the argumentation approach is equivalent to the coherence-based one. Consequently, it suffers from, the same drawbacks as this latter.

References

1. Amgoud, L.: Postulates for logic-based argumentation systems. In: WL4AI: ECAI Workshop on Weighted Logics for AI (2012)
2. Amgoud, L., Besnard, P.: Bridging the Gap between Abstract Argumentation Systems and Logic. In: Godo, L., Pugliese, A. (eds.) SUM 2009. LNCS, vol. 5785, pp. 12–27. Springer, Heidelberg (2009)

3. Amgoud, L., Besnard, P.: A Formal Analysis of Logic-Based Argumentation Systems. In: Deshpande, A., Hunter, A. (eds.) SUM 2010. LNCS, vol. 6379, pp. 42–55. Springer, Heidelberg (2010)
4. Besnard, P., Hunter, A.: A logic-based theory of deductive arguments. Artificial Intelligence 128(1-2), 203–235 (2001)
5. Caminada, M., Amgoud, L.: On the evaluation of argumentation formalisms. Artificial Intelligence Journal 171(5-6), 286–310 (2007)
6. Cayrol, C.: On the relation between argumentation and non-monotonic coherence-based entailment. In: IJCAI 1995, pp. 1443–1448 (1995)
7. D2.2. Towards a consensual formal model: inference part. Deliverable of ASPIC project (2004)
8. Dung, P.M.: On the acceptability of arguments and its fundamental role in nonmonotonic reasoning, logic programming and n-person games. Artificial Intelligence Journal 77, 321–357 (1995)
9. Elvang-Gøransson, M., Fox, J.P., Krause, P.: Acceptability of Arguments as Logical Uncertainty. In: Moral, S., Kruse, R., Clarke, E. (eds.) ECSQARU 1993. LNCS, vol. 747, pp. 85–90. Springer, Heidelberg (1993)
10. García, A., Simari, G.: Defeasible logic programming: an argumentative approach. Theory and Practice of Logic Programming 4, 95–138 (2004)
11. Gorogiannis, N., Hunter, A.: Instantiating abstract argumentation with classical logic arguments: Postulates and properties. Artificial Intelligence Journal 175(9-10), 1479–1497 (2011)
12. Pollock, J.L.: How to reason defeasibly. Artificial Intelligence Journal 57, 1–42 (1992)
13. Prakken, H.: An abstract framework for argumentation with structured arguments. Journal of Argument and Computation 1, 93–124 (2010)
14. Rescher, N., Manor, R.: On inference from inconsistent premises. Journal of Theory and Decision 1, 179–219 (1970)
15. Tarski, A.: On Some Fundamental Concepts of Metamathematics. In: Woodger, E.H. (ed.) Logic, Semantics, Metamathematics. Oxford Uni. Press (1956)

Appendix

Proof of Proposition 1. Let $\mathcal{T} = (\text{Arg}(\Sigma), \mathcal{R})$ be an AS over a knowledge base Σ such that $\mathcal{R}$ is conflict-dependent. Assume that $\mathcal{T}$ violates closure under sub-arguments. Thus, $\exists \mathcal{E} \in \text{Ext}_n(\mathcal{T})$ such that $\exists a \in \mathcal{E}$ and $\exists b \in \text{Sub}(a)$ with $b \notin \mathcal{E}$. This means that $\mathcal{E} \cup \{b\}$ is conflicting, i.e. $\exists c \in \mathcal{E}$ such that $b\mathcal{R}c$ or $c\mathcal{R}b$. Since $\mathcal{R}$ is conflict-dependent, then $\text{Supp}(b) \cup \text{Supp}(c)$ is inconsistent. However, $\text{Supp}(b) \subseteq \text{Supp}(a) \subseteq \text{Base}(\mathcal{E})$ and thus, $\text{Supp}(b) \cup \text{Supp}(c) \subseteq \text{Base}(\mathcal{E})$. This means that $\text{Base}(\mathcal{E})$ is inconsistent. This contradicts the assumption.   ∎

Proof of Theorem 1. Assume that an AS $\mathcal{T}$ satisfies Postulates 2 and 3, then from Proposition 5 (in [1]) it follows that $\forall \mathcal{E} \in \text{Ext}_n(\mathcal{T})$, $\text{Base}(\mathcal{E})$ is consistent.

Assume now that $\forall \mathcal{E} \in \text{Ext}_n(\mathcal{T})$, $\text{Base}(\mathcal{E})$ is consistent. Then, $\mathcal{T}$ satisfies consistency (Proposition 4, [1]). Moreover, from Proposition 1, $\mathcal{T}$ is closed under sub-arguments.   ∎

Proof of Proposition 2. Let $\mathcal{T} = (\text{Arg}(\Sigma), \mathcal{R})$ be an AS over a knowledge base Σ such that $\mathcal{R}$ is conflict-dependent. Assume that $\mathcal{T}$ is closed under sub-arguments and satisfies consistency. Assume also that $\mathcal{T}$ violates closure under CN. Thus, $\exists \mathcal{E} \in \text{Ext}_n(\mathcal{T})$

such that $\mathtt{Concs}(\mathcal{E}) \neq \mathtt{CN}(\mathtt{Concs}(\mathcal{E}))$. This means that $\exists x \in \mathtt{CN}(\mathtt{Concs}(\mathcal{E}))$ and $x \notin \mathtt{Concs}(\mathcal{E})$. Besides, $\mathtt{CN}(\mathtt{Concs}(\mathcal{E})) \subseteq \mathtt{CN}(\mathtt{Base}(\mathcal{E}))$. Thus, $x \in \mathtt{CN}(\mathtt{Base}(\mathcal{E}))$. Since CN verifies finiteness, then $\exists X \subseteq \mathtt{Base}(\mathcal{E})$ such that X is finite and $x \in \mathtt{CN}(X)$. Moreover, from Proposition 5 (in [1]), $\mathtt{Base}(\mathcal{E})$ is consistent. Then, X is consistent as well (from Property 2 in [2]). Consequently, the pair (X, x) is an argument. Besides, since $x \notin \mathtt{Concs}(\mathcal{E})$ then $(X, x) \notin \mathcal{E}$. This means that $\exists a \in \mathcal{E}$ such that $a\mathcal{R}(X, x)$ or $(X, x)\mathcal{R}a$. Finally, since $\mathcal{R}$ is conflict-dependent, then $\mathtt{Supp}(a) \cup X$ is inconsistent and consequently $\mathtt{Base}(\mathcal{E})$ is inconsistent. This contradicts the assumption.   ∎

Proof of Theorem 2. Let $\mathcal{T} = (\mathtt{Arg}(\Sigma), \mathcal{R})$ be an AS over a knowledge base Σ such that $\mathcal{R}$ is conflict-dependent. Assume that $\mathcal{T}$ satisfies consistency and is closed under sub-arguments. Let $\mathcal{E} \in \mathtt{Ext}_n(\mathcal{T})$. From Proposition 5 (in [1]), $\mathtt{Base}(\mathcal{E})$ is consistent.

Assume now that $\mathtt{Base}(\mathcal{E})$ is not maximal for set inclusion. Thus, $\exists x \in \Sigma \setminus \mathtt{Base}(\mathcal{E})$ such that $\mathtt{Base}(\mathcal{E}) \cup \{x\}$ is consistent. This means that $\{x\}$ is consistent. Thus, $\exists a \in \mathtt{Arg}(\Sigma)$ such that $\mathtt{Supp}(a) = \{x\}$. Since $x \notin \mathtt{Base}(\mathcal{E})$, then $a \notin \mathcal{E}$. Since $\mathcal{E}$ is a naive extension, then $\exists b \in \mathcal{E}$ such that $a\mathcal{R}b$ or $b\mathcal{R}a$. Since $\mathcal{R}$ is conflict-dependent, then $\mathtt{Supp}(a) \cup \mathtt{Supp}(b)$ is inconsistent. But, $\mathtt{Supp}(b) \subseteq \mathtt{Base}(\mathcal{E})$, this would mean that $\mathtt{Base}(\mathcal{E}) \cup \{x\}$ is inconsistent. Contradiction.

Let $\mathcal{E} \in \mathtt{Ext}_n(\mathcal{T})$. It is obvious that $\mathcal{E} \subseteq \mathtt{Arg}(\mathtt{Base}(\mathcal{E}))$ since the construction of arguments is monotonic. Let $a \in \mathtt{Arg}(\mathtt{Base}(\mathcal{E}))$. Thus, $\mathtt{Supp}(a) \subseteq \mathtt{Base}(\mathcal{E})$. Assume that $a \notin \mathcal{E}$, then $\exists b \in \mathcal{E}$ such that $a\mathcal{R}b$ or $b\mathcal{R}a$. Since $\mathcal{R}$ is conflict-dependent, then $\mathtt{Supp}(a) \cup \mathtt{Supp}(b)$ is inconsistent. Besides, $\mathtt{Supp}(a) \cup \mathtt{Supp}(b) \subseteq \mathtt{Base}(\mathcal{E})$. This means that $\mathtt{Base}(\mathcal{E})$ is inconsistent. Contradiction.

Let now $\mathcal{E}_i, \mathcal{E}_j \in \mathtt{Ext}_n(\mathcal{T})$. Assume that $\mathtt{Base}(\mathcal{E}_i) = \mathtt{Base}(\mathcal{E}_j)$. Then, $\mathtt{Arg}(\mathtt{Base}(\mathcal{E}_i)) = \mathtt{Arg}(\mathtt{Base}(\mathcal{E}_j))$. Besides, from the previous bullet, $\mathcal{E}_i = \mathtt{Arg}(\mathtt{Base}(\mathcal{E}_i))$ and $\mathcal{E}_j = \mathtt{Arg}(\mathtt{Base}(\mathcal{E}_j))$. Consequently, $\mathcal{E}_i = \mathcal{E}_j$.   ∎

Proof of Theorem 3. Let $\mathcal{T} = (\mathtt{Arg}(\Sigma), \mathcal{R})$ be an AS over a knowledge base Σ s.t. $\mathcal{R}$ is conflict-dependent. Assume that $\mathcal{T}$ satisfies Postulates 2 and 3.

Let $\mathcal{S} \in \mathtt{Max}(\Sigma)$, and assume that $\mathtt{Arg}(\mathcal{S}) \notin \mathtt{Ext}_n(\mathcal{T})$. Since $\mathcal{R}$ is conflict-dependent and $\mathcal{S}$ is consistent, then it follows from Proposition 5 in [2] that $\mathtt{Arg}(\mathcal{S})$ is conflict-free. Thus, $\mathtt{Arg}(\mathcal{S})$ is not maximal for set inclusion. So, $\exists a \in \mathtt{Arg}(\Sigma)$ such that $\mathtt{Arg}(\mathcal{S}) \cup \{a\}$ is conflict-free. There are two possibilities: i) $\mathcal{S} \cup \mathtt{Supp}(a)$ is consistent. But since $\mathcal{S} \in \mathtt{Max}(\Sigma)$, then $\mathtt{Supp}(a) \subseteq \mathcal{S}$, and this would mean that $a \in \mathtt{Arg}(\mathcal{S})$. ii) $\mathcal{S} \cup \mathtt{Supp}(a)$ is inconsistent. Thus, $\exists C \in \mathcal{C}_\Sigma$ such that $C \subseteq \mathcal{S} \cup \mathtt{Supp}(a)$. Let $X_1 = C \cap \mathcal{S}$ and $X_2 = C \cap \mathtt{Supp}(a)$. From Lemma 3 in [1], $\exists x_1 \in \mathtt{CN}(X_1)$ and $\exists x_2 \in \mathtt{CN}(X_2)$ such that the set $\{x_1, x_2\}$ is inconsistent. Note that (X_1, x_1) and (X_2, x_2) are arguments. Moreover, $(X_1, x_1) \in \mathtt{Arg}(\mathcal{S})$ and $(X_2, x_2) \in \mathtt{Sub}(a)$. Besides, since $\mathtt{Arg}(\mathcal{S}) \cup \{a\}$ is conflict-free, then $\exists \mathcal{E} \in \mathtt{Ext}(\mathcal{T})$ such that $\mathtt{Arg}(\mathcal{S}) \cup \{a\} \subseteq \mathcal{E}$. Thus, $(X_1, x_1) \in \mathcal{E}$. Since $\mathcal{T}$ is closed under sub-arguments then $(X_2, x_2) \in \mathcal{E}$. Thus, $\{x_1, x_2\} \subseteq \mathtt{Concs}(\mathcal{E})$. From Property 2 in [2], it follows that $\mathtt{Concs}(\mathcal{E})$ is inconsistent. This contradicts the fact that $\mathcal{T}$ satisfies consistency.

Let now $\mathcal{S}_i, \mathcal{S}_j \in \mathtt{Max}(\Sigma)$ be such that $\mathtt{Arg}(\mathcal{S}_i) = \mathtt{Arg}(\mathcal{S}_j)$. Assume that $\mathcal{S}_i \neq \mathcal{S}_j$, thus $\exists x \in \mathcal{S}_i$ and $x \notin \mathcal{S}_j$. Besides, $\mathcal{S}_i$ is consistent, then so is the set $\{x\}$. Consequently, $\exists a \in \mathtt{Arg}(\Sigma)$ such that $\mathtt{Supp}(a) = \{x\}$. It follows also that $a \in \mathtt{Arg}(\mathcal{S}_i)$ and thus $a \in \mathtt{Arg}(\mathcal{S}_j)$. By definition of an argument, $\mathtt{Supp}(a) \subseteq \mathcal{S}_j$. Contradiction.   ∎

Proof of Theorem 4. Let $\mathcal{T} = (\text{Arg}(\Sigma), \mathcal{R})$ be an AS over a knowledge base Σ such that $\mathcal{R}$ is conflict-dependent.

Assume that $\mathcal{T}$ satisfies Postulates 2 and 3. Them from Theorems 2 and 3, it follows that there is a full correspondence between $\text{Max}(\Sigma)$ and $\text{Ext}_n(\mathcal{T})$.

Assume now that there is a full correspondence between $\text{Max}(\Sigma)$ and $\text{Ext}_n(\mathcal{T})$. Then, $\forall \mathcal{E} \in \text{Ext}(\mathcal{T})$, $\text{Base}(\mathcal{E})$ is consistent. Consequently, $\mathcal{T}$ satisfies consistency. Moreover, from Proposition 1, $\mathcal{T}$ is closed under sub-arguments. ∎

Proof of Theorem 5. Let $\mathcal{T} = (\text{Arg}(\Sigma), \mathcal{R})$ be an AS over a knowledge base Σ such that $\mathcal{R}$ is conflict-dependent and $\mathcal{T}$ satisfies consistency and is closed under sub-arguments. From Theorem 4, it follows that $|\text{Ext}_n(\mathcal{T})| = |\text{Max}(\Sigma)|$. Since Σ is finite, then it has a finite number of maximal consistent subbases. Thus, the number of naive extensions is finite as well. ∎

Proof of Theorem 6. Let $\mathcal{T} = (\text{Arg}(\Sigma), \mathcal{R})$ be an AS over a knowledge base Σ such that $\mathcal{R}$ is conflict-dependent. Assume that $\mathcal{T}$ satisfies consistency and is closed under sub-arguments. Then, from Proposition 2, $\mathcal{T}$ enjoys closure under CN. Then, from Property 5 in [1], for all $\mathcal{E} \in \text{Ext}_n(\mathcal{T})$, $\text{Concs}(\mathcal{E}) = \text{CN}(\text{Base}(\mathcal{E}))$. Finally, from Theorem 4, there is a full correspondence between elements of $\text{Max}(\Sigma)$ and the naive extensions. Thus, for all $\mathcal{E}_i \in \text{Ext}_n(\mathcal{T})$, $\exists! \mathcal{S}_i \in \text{Max}(\Sigma)$ such that $\text{Base}(\mathcal{E}_i) = \mathcal{S}_i$. Thus, $\text{Concs}(\mathcal{E}_i) = \text{CN}(\mathcal{S}_i)$. By definition, $\text{Output}(\mathcal{T}) = \bigcap \text{Concs}(\mathcal{E}_i)$, thus $\text{Output}(\mathcal{T}) = \bigcap \text{CN}(\mathcal{S}_i)$. ∎

Proof of Proposition 3. Let $\mathcal{T} = (\text{Arg}(\Sigma), \mathcal{R})$ be an AS over a knowledge base Σ such that $\mathcal{R}$ is conflict-dependent. Assume that $\forall \mathcal{E} \in \text{Ext}_s(\mathcal{T})$, $\text{Base}(\mathcal{E})$ is consistent. Assume also that $\mathcal{T}$ violates closure under sub-arguments. Thus, $\exists \mathcal{E} \in \text{Ext}_s(\mathcal{T})$ such that $\exists a \in \mathcal{E}$ and $\exists b \in \text{Sub}(a)$ with $b \notin \mathcal{E}$. Since $\mathcal{E}$ is a stable extension, then $\exists c \in \mathcal{E}$ such that $c\mathcal{R}b$. Since $\mathcal{R}$ is conflict-dependent, then $\text{Supp}(b) \cup \text{Supp}(c)$ is inconsistent. However, $\text{Supp}(b) \subseteq \text{Supp}(a) \subseteq \text{Base}(\mathcal{E})$. Then, $\text{Supp}(b) \cup \text{Supp}(c) \subseteq \text{Base}(\mathcal{E})$. This means that $\text{Base}(\mathcal{E})$ is inconsistent. This contradicts the assumption. ∎

Proof of Theorem 7. Assume that an AS $\mathcal{T}$ satisfies Postulates 2 and 3, then from Proposition 5 (in [1]) it follows that $\forall \mathcal{E} \in \text{Ext}_s(\mathcal{T})$, $\text{Base}(\mathcal{E})$ is consistent.

Assume now that $\forall \mathcal{E} \in \text{Ext}_s(\mathcal{T})$, $\text{Base}(\mathcal{E})$ is consistent. Then, $\mathcal{T}$ satisfies consistency (Proposition 4, [1]). Moreover, from Proposition 3, $\mathcal{T}$ is closed under sub-arguments. ∎

Proof of Proposition 4. Let $\mathcal{T} = (\text{Arg}(\Sigma), \mathcal{R})$ be an AS over a knowledge base Σ such that $\mathcal{R}$ is conflict-dependent. Assume that $\mathcal{T}$ is closed under sub-arguments and satisfies consistency. Assume also that $\mathcal{T}$ violates closure under CN. Thus, $\exists \mathcal{E} \in \text{Ext}_s(\mathcal{T})$ such that $\text{Concs}(\mathcal{E}) \neq \text{CN}(\text{Concs}(\mathcal{E}))$. This means that $\exists x \in \text{CN}(\text{Concs}(\mathcal{E}))$ and $x \notin \text{Concs}(\mathcal{E})$. Besides, $\text{CN}(\text{Concs}(\mathcal{E})) \subseteq \text{CN}(\text{Base}(\mathcal{E}))$. Thus, $x \in \text{CN}(\text{Base}(\mathcal{E}))$. Since CN verifies finiteness, then $\exists X \subseteq \text{Base}(\mathcal{E})$ such that X is finite and $x \in \text{CN}(X)$. Moreover, from Proposition 5 (in [1]), $\text{Base}(\mathcal{E})$ is consistent. Then, X is consistent as well (from Property 2 in [2]). Consequently, the pair (X, x) is an argument. Besides, since $x \notin \text{Concs}(\mathcal{E})$ then $(X, x) \notin \mathcal{E}$. This means that $\exists a \in \mathcal{E}$ such that $a\mathcal{R}(X, x)$. Finally, since $\mathcal{R}$ is conflict-dependent, then $\text{Supp}(a) \cup X$ is inconsistent and consequently $\text{Base}(\mathcal{E})$ is inconsistent. This contradicts the assumption. ∎

Proof of Theorem 8. Let $\mathcal{T} = (\text{Arg}(\Sigma), \mathcal{R})$ be an AS over a knowledge base Σ such that $\mathcal{R}$ is conflict-dependent. Let $\mathcal{E} \in \text{Ext}_s(\mathcal{T})$. Since $\mathcal{T}$ satisfies Postulates 1, 2 and 3, then $\text{Base}(\mathcal{E})$ is consistent (from Proposition 3). Assume now that $\text{Base}(\mathcal{E})$ is not maximal for set inclusion. Thus, $\exists x \in \Sigma \setminus \text{Base}(\mathcal{E})$ such that $\text{Base}(\mathcal{E}) \cup \{x\}$ is consistent. This means that $\{x\}$ is consistent. Thus, $\exists a \in \text{Arg}(\Sigma)$ such that $\text{Supp}(a) = \{x\}$. Since $x \notin \text{Base}(\mathcal{E})$, then $a \notin \mathcal{E}$. Since $\mathcal{E}$ is a stable extension, then $\exists b \in \mathcal{E}$ such that $b\mathcal{R}a$. Since $\mathcal{R}$ is conflict-dependent, then $\text{Supp}(a) \cup \text{Supp}(b)$ is inconsistent. But, $\text{Supp}(b) \subseteq \text{Base}(\mathcal{E})$, this would mean that $\text{Base}(\mathcal{E}) \cup \{x\}$ is inconsistent. Contradiction.

Let $\mathcal{E} \in \text{Ext}_s(\mathcal{T})$. It is obvious that $\mathcal{E} \subseteq \text{Arg}(\text{Base}(\mathcal{E}))$ since the construction of arguments is monotonic. Let $a \in \text{Arg}(\text{Base}(\mathcal{E}))$. Thus, $\text{Supp}(a) \subseteq \text{Base}(\mathcal{E})$. Assume that $a \notin \mathcal{E}$, then $\exists b \in \mathcal{E}$ such that $b\mathcal{R}a$. Since $\mathcal{R}$ is conflict-dependent, then $\text{Supp}(a) \cup \text{Supp}(b)$ is inconsistent. Besides, $\text{Supp}(a) \cup \text{Supp}(b) \subseteq \text{Base}(\mathcal{E})$. This means that $\text{Base}(\mathcal{E})$ is inconsistent. Contradiction.

Let now $\mathcal{E}_i, \mathcal{E}_j \in \text{Ext}_s(\mathcal{T})$. Assume that $\text{Base}(\mathcal{E}_i) = \text{Base}(\mathcal{E}_j)$. Then, $\text{Arg}(\text{Base}(\mathcal{E}_i)) = \text{Arg}(\text{Base}(\mathcal{E}_j))$. Besides, from bullet 2 of this proof, $\mathcal{E}_i = \text{Arg}(\text{Base}(\mathcal{E}_i))$ and $\mathcal{E}_j = \text{Arg}(\text{Base}(\mathcal{E}_j))$. Consequently, $\mathcal{E}_i = \mathcal{E}_j$. $\blacksquare$

Proof of Proposition 5. Let $\mathcal{T} = (\text{Arg}(\Sigma), \mathcal{R})$ be an AS over a knowledge base Σ s.t. $\mathcal{R}$ is conflict-dependent and $\mathcal{T}$ satisfies consistency and closure under sub-arguments. If $\text{Ext}_s(\mathcal{T}) = \emptyset$, then $|\text{Ext}_s(\mathcal{T})| = 0$. If $\text{Ext}_s(\mathcal{T}) \neq \emptyset$, then $|\text{Ext}_s(\mathcal{T})| \leq |\text{Max}(\Sigma)|$ (from Theorem 8). $\blacksquare$

Proof of Property 2. Let $\mathcal{T} = (\text{Arg}(\Sigma), \mathcal{R})$ be an AS over a knowledge base Σ such that $\mathcal{R} \in \Re_{s3}$. Let $\mathcal{S} \in \text{Max}(\Sigma)$. Since $|\text{Ext}(\mathcal{T})| = |\text{Max}(\Sigma)|$, then from Theorem 8, $\exists \mathcal{E} \in \text{Ext}_s(\mathcal{T})$ such that $\text{Base}(\mathcal{E}) = \mathcal{S}$. Besides, from the same theorem, $\mathcal{E} = \text{Arg}(\text{Base}(\mathcal{E}))$, thus $\mathcal{E} = \text{Arg}(\mathcal{S})$. Consequently, $\text{Arg}(\mathcal{S}) \in \text{Ext}_s(\mathcal{T})$. $\blacksquare$

Proof of Theorem 9. It was shown under any Tarskian logic that the attack relation proposed in [9] and called *assumption attack* verifies the correspondence between stable extensions and maximal subbases. Thus, assumption attack belongs to $\Re_{s3}$. $\blacksquare$

Proof of Theorem 10. Let $\mathcal{T} = (\text{Arg}(\Sigma), \mathcal{R})$ be an AS over a knowledge base Σ such that $\mathcal{R} \in \Re_{s3}$. Then, $|\text{Ext}_s(\mathcal{T})| = |\text{Max}(\Sigma)|$. There are two cases: i) Σ contains only inconsistent formulas, thus $\text{Max}(\Sigma) = \{\emptyset\}$ and $\text{Ext}_s(\mathcal{T}) = \{\emptyset\}$ since $\text{Arg}(\Sigma) = \emptyset$. ii) Σ contains at lest one consistent formula x. Thus, $\exists \mathcal{S} \in \text{Max}(\Sigma)$ such $x \in \mathcal{S}$. Since $\mathcal{R} \in \Re_{s3}$, then $\text{Arg}(\mathcal{S}) \in \text{Ext}_s(\mathcal{T})$. $\blacksquare$

The Outcomes of Logic-Based Argumentation Systems under Preferred Semantics

Leila Amgoud

IRIT – CNRS, 118 route de Narbonne 31062, Toulouse – France

Abstract. Logic-based argumentation systems are developed for reasoning with inconsistent information. They consist of a set of arguments, attacks among them and a semantics for the evaluation of arguments. *Preferred* semantics is favored in the literature since it ensures the existence of *extensions* (i.e., acceptable sets of arguments), and it guarantees a kind of maximality, accepting thus arguments whenever possible.

This paper proposes the first study on the outcomes under preferred semantics of logic-based argumentation systems that satisfy basic rationality postulates. It focuses on systems that are grounded on Tarskian logics, and delimits the number of preferred extensions they may have. It also characterizes both their extensions and their sets of conclusions that are drawn from knowledge bases. The results are disappointing since they show that in the best case, the preferred extensions of a system are computed from the maximal consistent subbases of the knowledge base under study. In this case, the system is coherent, that is preferred extensions are stable ones. Moreover, we show that both semantics are useless in thic case since they ensure exactly the same result as naive semantics. Apart from this case, the outcomes of argumentation systems are counter-intuitive.

1 Introduction

An important problem in the management of knowledge-based systems is the handling of inconsistency. Inconsistency may be present because the knowledge base includes default rules (e.g. [16]) or because the knowledge comes from several sources of information (e.g. [9]).

Argumentation theory is an alternative approach for reasoning with inconsistent information. It is based on the key notion of *argument* which explains why a conclusion may be drawn from a given knowledge base. In fact, an argumentation system is a set of arguments, an attack relation and a semantics for evaluating the arguments (see [5,13,14,17] for some examples of such systems). Surprisingly enough, in most existing systems, there is no characterization of the kind of outputs that are drawn from a knowledge base. To say it differently, the properties of those outputs are unknown. These properties should broadly depend on the chosen semantics. It is worth mentioning that in all existing systems, Dung's semantics [11] or variants of them are used. The so-called *Preferred* semantics is the most favored one. It enjoys a kind of maximality which leads to the acceptance of arguments whenever possible. This semantics was mainly proposed as an alternative for stable semantics which does not guarantee the existence of stable extensions. While we may find some works that investigate the

E. Hüllermeier et al. (Eds.): SUM 2012, LNAI 7520, pp. 72–84, 2012.

outcomes of particular systems under stable semantics [8], there is no such work under preferred semantics. Thus, the outcomes under this semantics are still completely mysterious and unexplored.

This paper investigates for the first time the outcomes under preferred semantics of argumentation systems that are grounded on Tarskian logics [20] and that satisfy the basic rationality postulates proposed in [1]. We identify for the first time the maximum number of preferred extensions those systems may have, and characterize both their extensions and their sets of conclusions that are drawn from knowledge bases. The study completely abstracts from the logic and the attack relation. The results are disappointing. They show that in the best case, the preferred extensions of a system are computed from all the maximal consistent subbases of the knowledge base under study. In this case, the argumentation system is coherent, i.e., its preferred extensions coincide with its stable ones. In a companion paper [2], we have shown that stable semantics does not play any role in this case since the output of a system under this semantics is exactly what is returned by the same system under naive semantics (i.e., the maximal conflict-free sets of arguments). Consequently, preferred semantics is also useless in this case. In all the remaining cases we identified, the argumentation systems return counter-intuitive results. To sum up, preferred semantics is not commended for instantiations of Dung's framework with Tarskian logics.

The paper is organized as follows: we start by defining the logic-based argumentation systems we are interested in and by recalling the three basic postulates that such systems should obey. In a subsequent section, we investigate the properties of the preferred extensions of those systems. Next, we study the inferences that are drawn from a knowledge base by argumentation systems under preferred semantics. The last section is devoted to some concluding remarks.

2 Logic-Based Argumentation Systems and Rationality Postulates

In this paper, we consider abstract logic-based argumentation systems; that is systems that are grounded on *any* Tarskian logic [20] and that use *any* attack relation. Such abstraction makes our study very general.

According to Alfred Tarski, an abstract logic is a pair $(\mathcal{L}, \text{CN})$ where $\mathcal{L}$ is a set of well-formed formulas. Note that there is no particular requirement on the kind of connectors that may be used. CN is a consequence operator that returns the set of formulas that are logical consequences of another set of formulas according to the logic in question. It should satisfy the following basic properties:

1. $X \subseteq \text{CN}(X)$ (Expansion)
2. $\text{CN}(\text{CN}(X)) = \text{CN}(X)$ (Idempotence)
3. $\text{CN}(X) = \bigcup_{Y \subseteq_f X} \text{CN}(Y)$[1]   (Finiteness)
4. $\text{CN}(\{x\}) = \mathcal{L}$ for some $x \in \mathcal{L}$ (Absurdity)
5. $\text{CN}(\emptyset) \neq \mathcal{L}$ (Coherence)

The associated notion of *consistency* is defined as follows: A set $X \subseteq \mathcal{L}$ is *consistent* wrt a logic $(\mathcal{L}, \text{CN})$ iff $\text{CN}(X) \neq \mathcal{L}$. It is *inconsistent* otherwise. Besides, arguments are built from a *knowledge base* $\Sigma \subseteq \mathcal{L}$ as follows:

---

[1] $Y \subseteq_f X$ means that Y is a finite subset of X.

Definition 1 (Argument). *Let Σ be a knowledge base. An argument is a pair (X, x) s.t. $X \subseteq \Sigma$, X is consistent, and $x \in \text{CN}(X)^2$. An argument (X, x) is a sub-argument of another argument (X', x') iff $X \subseteq X'$.*

Notations: Supp and Conc denote respectively the *support* X and the *conclusion* x of an argument (X, x). For all $\mathcal{S} \subseteq \Sigma$, $\text{Arg}(\mathcal{S})$ denotes the set of all arguments that can be built from $\mathcal{S}$ by means of Definition 1. Sub is a function that returns all the sub-arguments of a given argument. For all $\mathcal{E} \subseteq \text{Arg}(\Sigma)$, $\text{Concs}(\mathcal{E}) = \{\text{Conc}(a) \mid a \in \mathcal{E}\}$ and $\text{Base}(\mathcal{E}) = \bigcup_{a \in \mathcal{E}} \text{Supp}(a)$. $\text{Max}(\Sigma)$ is the set of all maximal (for set inclusion) consistent subbases of Σ. $\text{Free}(\Sigma) = \bigcap \mathcal{S}_i$ where $\mathcal{S}_i \in \text{Max}(\Sigma)$, and $\text{Inc}(\Sigma) = \Sigma \setminus \text{Free}(\Sigma)$. Finally, $\mathcal{C}_\Sigma$ denote the set of all minimal conflicts[3] of Σ.

An argumentation system for reasoning over a knowledge base Σ is defined as follows.

Definition 2 (Argumentation system). *An argumentation system (AS) over a knowledge base Σ is a pair $\mathcal{T} = (\text{Arg}(\Sigma), \mathcal{R})$ such that $\mathcal{R} \subseteq \text{Arg}(\Sigma) \times \text{Arg}(\Sigma)$ is an attack relation. For $a, b \in \text{Arg}(\Sigma)$, $(a, b) \in \mathcal{R}$ (or $a\mathcal{R}b$) means that a attacks b.*

Throughout the paper, the attack relation is left *unspecified*.

Arguments are evaluated using preferred semantics [11]. For the purpose of this paper, we also need to recall the definition of stable semantics. Preferred semantics is based on two requirements: *conflict-freeness* and *defence*. Recall that a set $\mathcal{E}$ of arguments is *conflict-free* iff $\nexists a, b \in \mathcal{E}$ such that $a\mathcal{R}b$. It *defends* an argument a iff $\forall b \in \text{Arg}(\Sigma)$, if $b\mathcal{R}a$, then $\exists c \in \mathcal{E}$ such that $c\mathcal{R}b$.

Definition 3. *Let $\mathcal{T} = (\text{Arg}(\Sigma), \mathcal{R})$ be an AS and $\mathcal{E} \subseteq \text{Arg}(\Sigma)$.*

– *$\mathcal{E}$ is an admissible extension iff $\mathcal{E}$ is conflict-free and defends all its elements.*
– *$\mathcal{E}$ is a preferred extension iff it is a maximal (for set inclusion) admissible extension.*
– *$\mathcal{E}$ is a stable extension iff $\mathcal{E}$ is conflict-free and attacks any argument in $\text{Arg}(\Sigma) \setminus \mathcal{E}$.*

It is worth recalling that each stable extension is a preferred one but the converse is not true. Let $\text{Ext}_x(\mathcal{T})$ denote the set of all extensions of $\mathcal{T}$ under semantics x where p and s stand respectively for preferred and stable. When we do not need to specify the semantics, we use the notation $\text{Ext}(\mathcal{T})$ for short.

The set of conclusions drawn from a knowledge base Σ using an argumentation system $\mathcal{T} = (\text{Arg}(\Sigma), \mathcal{R})$ contains only the common conclusions of the extensions.

Definition 4 (Output). *Let $\mathcal{T} = (\text{Arg}(\Sigma), \mathcal{R})$ be an AS over a knowledge base Σ. $\text{Output}(\mathcal{T}) = \{x \in \mathcal{L} \mid \forall \mathcal{E} \in \text{Ext}(\mathcal{T}), \exists a \in \mathcal{E} \text{ s.t. } \text{Conc}(a) = x\}$.*

In [7], it was shown that not any instantiation of Dung's abstract argumentation framework is acceptable. Some instantiations like [15,18] may lead in some cases to undesirable outputs. Consequently, some rationality postulates that any system should obey

---

[2] Generally, the support X is minimal (for set inclusion). In this paper, we do not need to make this assumption.

[3] A set $C \subseteq \Sigma$ is a *minimal conflict* iff C is inconsistent and $\forall x \in C$, $C \setminus \{x\}$ is consistent.

were proposed. Postulates are desirable properties that any reasoning system should enjoy. In [1], those postulates were revisited and extended to any Tarskian logic. The first postulate concerns the closure of the system's output under the consequence operator CN. The idea is that the formalism should not forget conclusions.

Postulate 1 (Closure under CN). *Let* $\mathcal{T} = (\text{Arg}(\Sigma), \mathcal{R})$ *be an AS over a knowledge base.* $\mathcal{T}$ *satisfies* closure *iff for all* $\mathcal{E} \in \text{Ext}(\mathcal{T})$, $\text{Concs}(\mathcal{E}) = \text{CN}(\text{Concs}(\mathcal{E}))$.

The second rationality postulate ensures that the acceptance of an argument implies also the acceptance of all its sub-parts.

Postulate 2 (Closure under sub-arguments). *Let* $\mathcal{T} = (\text{Args}(\Sigma), \mathcal{R})$ *be an AS.* $\mathcal{T}$ *is* closed under sub-arguments *iff for all* $\mathcal{E} \in \text{Ext}(\mathcal{T})$, *if* $a \in \mathcal{E}$, *then* $\text{Sub}(a) \subseteq \mathcal{E}$.

The third postulate ensures that the set of conclusions supported by each extension is consistent.

Postulate 3 (Consistency). *Let* $\mathcal{T} = (\text{Arg}(\Sigma), \mathcal{R})$ *be an AS over a knowledge base* Σ. $\mathcal{T}$ *satisfies* consistency *iff for all* $\mathcal{E} \in \text{Ext}(\mathcal{T})$, $\text{Concs}(\mathcal{E})$ *is consistent.*

The following interesting result is shown in [1] under any acceptability semantics.

Proposition 1. *[1] Let* $\mathcal{T} = (\text{Arg}(\Sigma), \mathcal{R})$ *be an AS over a knowledge base* Σ. *If* $\mathcal{T}$ *satisfies consistency and closure under sub-arguments, then for all* $\mathcal{E} \in \text{Ext}(\mathcal{T})$, $\text{Base}(\mathcal{E})$ *is consistent.*

It was shown in [1] that in order to satisfy these postulates, the attack relation should capture inconsistency. This is an obvious requirement especially for reasoning about inconsistent information. Note also that *all* existing attack relations verify this property (see [14] for an overview of those relations defined under propositional logic).

Definition 5 (Conflict-dependent). *An attack relation* $\mathcal{R}$ *is* conflict-dependent *iff* $\forall a, b \in \text{Arg}(\Sigma)$, *if* $a\mathcal{R}b$ *then* $\text{Supp}(a) \cup \text{Supp}(b)$ *is inconsistent.*

3 Properties of Preferred Extensions

Throughout the paper, we assume argumentation systems $\mathcal{T} = (\text{Arg}(\Sigma), \mathcal{R})$ that are built over a knowledge base Σ. These systems are assumed to be sound in the sense that they enjoy the three rationality postulates described in the previous section. It is worth recalling that the attack relation is a crucial parameter in a system since the satisfaction of the postulates depends broadly on it. For instance, it was shown in [1] that argumentation systems that use symmetric relations may violate the consistency postulate. This is particularly the case when the knowledge base contains a ternary or a n-ary minimal conflict (with $n > 2$). Thus, such symmetric systems [10] should be avoided and are not concerned by our study.

Our aim in this section is to investigate the properties of preferred extensions of sound argumentation systems. We will answer the following interesting questions.

1. What is the number of preferred extensions an AS may have?
2. What is the link between each preferred extension and the knowledge base Σ?
3. Is the set of formulas underlying a preferred extension consistent?
4. What is the real added value of preferred semantics compared to stable semantics? To put it differently, does preferred semantics solve any problem encountered by stable one?

We start by showing that the argumentation systems that satisfy consistency and closure under sub-arguments, satisfy also the strong version of consistency. Indeed, the union of the supports of all arguments of each preferred extension is a consistent subbase of Σ. This result is interesting since it is in accordance with the idea that an extension represents a *coherent position/point of view*.

Proposition 2. *Let* $\mathcal{T} = (\text{Arg}(\Sigma), \mathcal{R})$ *be an AS s.t.* $\mathcal{R}$ *is conflict-dependent and* $\mathcal{T}$ *satisfies consistency and closure under sub-arguments. For all* $\mathcal{E} \in \text{Ext}_p(\mathcal{T})$, $\text{Base}(\mathcal{E})$ *is consistent.*

Proof. Let $\mathcal{T} = (\text{Arg}(\Sigma), \mathcal{R})$ be an AS s.t. $\mathcal{R}$ is conflict-dependent and $\mathcal{T}$ satisfies consistency and closure under sub-arguments. From Proposition 1, it follows immediately that for all $\mathcal{E} \in \text{Ext}_p(\mathcal{T})$, $\text{Base}(\mathcal{E})$ is consistent.

In addition to the fact that the subbase computed from a preferred extension is consistent, we show next that it is unique. Indeed, the subbase computed from one extension can never be a subset of a subbase computed from another extension. Thus, the preferred extensions of an argumentation system return completely different subbases of Σ.

Proposition 3. *Let* $\mathcal{T} = (\text{Arg}(\Sigma), \mathcal{R})$ *be an AS s.t.* $\mathcal{R}$ *is conflict-dependent and* $\mathcal{T}$ *satisfies Postulates 2 and 3. For all* $\mathcal{E}_i, \mathcal{E}_j \in \text{Ext}_p(\mathcal{T})$, *if* $\text{Base}(\mathcal{E}_i) \subseteq \text{Base}(\mathcal{E}_j)$ *then* $\mathcal{E}_i = \mathcal{E}_j$.

Proof. Let $\mathcal{T} = (\text{Arg}(\Sigma), \mathcal{R})$ be an AS s.t. $\mathcal{R}$ is conflict-dependent and $\mathcal{T}$ satisfies consistency and closure under sub-arguments. Assume that $\mathcal{E}_i, \mathcal{E}_j \in \text{Ext}_p(\mathcal{T})$ and $\text{Base}(\mathcal{E}_i) \subseteq \text{Base}(\mathcal{E}_j)$. We first show that $\forall a \in \text{Arg}(\text{Base}(\mathcal{E}_i))$, $\mathcal{E}_j \cup \{a\}$ is conflict-free. Let $a \in \text{Arg}(\text{Base}(\mathcal{E}_i))$. Assume that $\mathcal{E}_j \cup \{a\}$ is not conflict-free. Thus, $\exists b \in \mathcal{E}_j$ such that $a\mathcal{R}b$ or $b\mathcal{R}a$. Since $\mathcal{R}$ is conflict-dependent, then $\text{Supp}(a) \cup \text{Supp}(b)$ is inconsistent. Besides, $\text{Supp}(a) \subseteq \text{Base}(\mathcal{E}_j)$. Thus, $\text{Base}(\mathcal{E}_j)$ is inconsistent. This contradicts Proposition 2.

Let $\mathcal{E} = \mathcal{E}_j \cup (\mathcal{E}_i \setminus \mathcal{E}_j)$. From above, it follows that $\mathcal{E}$ is conflict-free. Moreover, $\mathcal{E}$ defends any element in $\mathcal{E}_j$ (since $\mathcal{E}_j \in \text{Ext}_p(\mathcal{T})$) and any element in $\mathcal{E}_i \setminus \mathcal{E}_j$ (since $\mathcal{E}_i \in \text{Ext}_p(\mathcal{T})$). Thus, $\mathcal{E}$ is an admissible set. This contradicts the fact that $\mathcal{E}_j \in \text{Ext}_p(\mathcal{T})$.

In [2], we have shown that the subbases computed from the stable extensions of any argumentation system that satisfies the postulates are maximal (for set inclusion) consistent subbases of Σ. In what follows, we show that this is not necessarily the case for preferred extensions. Note that this does not mean that a preferred extension can never return a maximal consistent subbase. The previous result guarantees that the maximal consistent subbases containing a non-maximal subbase computed from a given extension will never be returned by any other extension of the same system.

Proposition 4. *Let* $\mathcal{T} = (\mathrm{Arg}(\Sigma), \mathcal{R})$ *be an AS s.t.* $\mathcal{R}$ *is conflict-dependent and* $\mathcal{T}$ *satisfies consistency and closure under sub-arguments. Let* $\mathcal{E} \in \mathrm{Ext}_p(\mathcal{T})$. *If* $\mathrm{Base}(\mathcal{E}) \notin$ $\mathrm{Max}(\Sigma)$, *then* $\forall \mathcal{S} \in \mathrm{Max}(\Sigma)$ *s.t.* $\mathrm{Base}(\mathcal{E}) \subset \mathcal{S}$, $\nexists \mathcal{E}' \in \mathrm{Ext}_p(\mathcal{T})$ *s.t.* $\mathrm{Base}(\mathcal{E}') = \mathcal{S}$.

Proof. Let $\mathcal{T} = (\mathrm{Arg}(\Sigma), \mathcal{R})$ be an AS s.t. $\mathcal{R}$ is conflict-dependent and $\mathcal{T}$ satisfies consistency and closure under sub-arguments. Let $\mathcal{E} \in \mathrm{Ext}_p(\mathcal{T})$. Assume that $\mathrm{Base}(\mathcal{E}) \notin \mathrm{Max}(\Sigma)$ and that $\exists \mathcal{S} \in \mathrm{Max}(\Sigma)$ s.t. $\mathrm{Base}(\mathcal{E}) \subset \mathcal{S}$ and $\exists \mathcal{E}' \in \mathrm{Ext}_p(\mathcal{T})$ s.t. $\mathrm{Base}(\mathcal{E}') = \mathcal{S}$. Thus, $\exists x \in \mathcal{S} \setminus \mathrm{Base}(\mathcal{E}')$. Moreover, $\exists a \in \mathcal{E}'$ such that $x \in \mathrm{Supp}(a)$ and $a \notin \mathcal{E}$. Besides, from Proposition 2, it holds that $\mathcal{E} = \mathcal{E}'$. Contradiction.

The non-maximality of the subbases that are computed from preferred extensions is due to the existence of *undecided arguments* under preferred semantics. Indeed, in [6] another way of defining Dung's semantics was provided. It consists of labeling the nodes of the graph corresponding to the argumentation system with three possibles values: {in, out, undec}. The value *in* means that the argument is accepted, the value *out* means that the argument is attacked by an accepted arguments, and finally the value *undec* means that the argument is neither accepted nor attacked by an accepted argument. It is thus possible that some formulas appear only in undecided arguments.

Another particular property of preferred extensions is the fact that they may not be closed in terms of arguments. Indeed, they may not contain all the arguments that may be built from their bases. Indeed, it is possible that $\mathcal{E}$ is a preferred extension of a system and $\mathcal{E} \neq \mathrm{Arg}(\mathrm{Base}(\mathcal{E}))$. Surprisingly enough, the supports and conclusions of the missed arguments are conclusions of the extensions. Thus, even if an argument of $\mathrm{Arg}(\mathrm{Base}(\mathcal{E}))$ does not belong to the extension $\mathcal{E}$, all the formulas of its supports are conclusions of arguments in the extension, and the same holds for its conclusion.

Proposition 5. *Let* $\mathcal{T} = (\mathrm{Arg}(\Sigma), \mathcal{R})$ *be an AS s.t.* $\mathcal{T}$ *is closed under* CN *and under sub-arguments. Let* $\mathcal{E} \in \mathrm{Ext}_p(\mathcal{T})$. *For all* $a \in \mathrm{Arg}(\mathrm{Base}(\mathcal{E}))$, $\mathrm{Supp}(a) \subseteq \mathrm{Concs}(\mathcal{E})$ *and* $\mathrm{Conc}(a) \in \mathrm{Concs}(\mathcal{E})$.

Proof. Let $\mathcal{T} = (\mathrm{Arg}(\Sigma), \mathcal{R})$ be an AS s.t. $\mathcal{T}$ is closed under CN and under sub-arguments. Let $\mathcal{E} \in \mathrm{Ext}_p(\mathcal{T})$ and $a \in \mathrm{Arg}(\mathrm{Base}(\mathcal{E}))$. Thus, $\mathrm{Supp}(a) \subseteq \mathrm{Base}(\mathcal{E})$. Since $\mathcal{T}$ is closed under sub-arguments, then $\mathrm{Base}(\mathcal{E}) \subseteq \mathrm{Concs}(\mathcal{E})$ (proved in [1]). Thus, $\mathrm{Supp}(a) \subseteq \mathrm{Concs}(\mathcal{E})$. Besides, by monotonicity of CN, $\mathrm{CN}(\mathrm{Supp}(a)) \subseteq$ $\mathrm{CN}(\mathrm{Base}(\mathcal{E}))$. Since $\mathcal{T}$ is also closed under CN, then $\mathrm{Concs}(\mathcal{E}) = \mathrm{CN}(\mathrm{Base}(\mathcal{E}))$ (proved in [1]). Thus, $\mathrm{CN}(\mathrm{Supp}(a)) \subseteq \mathrm{Concs}(\mathcal{E})$ and $\mathrm{Conc}(a) \in \mathrm{Concs}(\mathcal{E})$.

We show next that the free part of Σ (i.e., the formulas that are not involved in any conflict) is inferred by *any* argumentation system under preferred semantics. The reason is that the set of arguments built from $\mathrm{Free}(\Sigma)$ is an admissible extension of any argumentation systems whose attack relations are conflict-dependent. Thus, this is true even for systems that do not satisfy the postulates.

Proposition 6. *Let* $\mathcal{T} = (\mathrm{Arg}(\Sigma), \mathcal{R})$ *be s.t.* $\mathcal{R}$ *is conflict-dependent. The set* $\mathrm{Arg}(\mathrm{Free}(\Sigma))$ *is an admissible extension of* $\mathcal{T}$.

Proof. Let $(\mathrm{Arg}(\Sigma), \mathcal{R})$ be s.t. $\mathcal{R}$ is conflict-dependent. Let $a \in \mathrm{Arg}(\mathrm{Free}(\Sigma))$. Assume that $\exists b \in \mathrm{Arg}(\Sigma)$ s.t. $a\mathcal{R}b$ or $b\mathcal{R}a$. Since $\mathcal{R}$ is conflict-dependent, then $\exists C \in$

$\mathcal{C}_{\Sigma}$ such that $C \subseteq \text{Supp}(a) \cup \text{Supp}(b)$. By definition of an argument, both $\text{Supp}(a)$ and $\text{Supp}(b)$ are consistent. Then, $C \cap \text{Supp}(a) \neq \emptyset$. This contradicts the fact that $\text{Supp}(a) \subseteq \text{Free}(\Sigma)$. Thus, $\text{Arg}(\text{Free}(\Sigma))$ is *conflict-free and can never be attacked*.

We show next that the set $\text{Arg}(\text{Free}(\Sigma))$ is contained in every preferred extension. This is true for any argumentation system that uses a conflict-dependent attack relation. That is, it is always true.

Proposition 7. *Let* $\mathcal{T} = (\text{Arg}(\Sigma), \mathcal{R})$ *be s.t.* $\mathcal{R}$ *is conflict-dependent. For all* $\mathcal{E} \in \text{Ext}_p(\mathcal{T})$, $\text{Arg}(\text{Free}(\Sigma)) \subseteq \mathcal{E}$.

Proof. Let $(\text{Arg}(\Sigma), \mathcal{R})$ be s.t. $\mathcal{R}$ is conflict-dependent. Assume that $\exists \mathcal{E} \in \text{Ext}_p(\mathcal{T})$ such that $\text{Arg}(\text{Free}(\Sigma)) \not\subseteq \mathcal{E}$. Thus, either $\mathcal{E} \cup \text{Arg}(\text{Free}(\Sigma))$ is conflicting or $\mathcal{E}$ does not defend elements of $\text{Arg}(\text{Free}(\Sigma))$. Both cases are impossible since arguments of $\text{Arg}(\text{Free}(\Sigma))$ neither attack nor are attacked by any argument.

The next result shows that formulas of $\text{Free}(\Sigma)$ are always drawn from Σ under preferred semantics.

Proposition 8. *Let* $\mathcal{T} = (\text{Arg}(\Sigma), \mathcal{R})$ *be s.t.* $\mathcal{R}$ *is conflict-dependent. It holds that* $\text{Free}(\Sigma) \subseteq \text{Output}(\mathcal{T})$.

Proof. Let $\mathcal{T} = (\text{Arg}(\Sigma), \mathcal{R})$ be s.t. $\mathcal{R}$ is conflict-dependent. From Proposition 7, $\forall \mathcal{E} \in \text{Ext}_p(\mathcal{T})$, $\text{Arg}(\text{Free}(\Sigma)) \subseteq \mathcal{E}$. Besides, $\forall x \in \text{Free}(\Sigma)$, $(\{x\}, x) \in \text{Arg}(\text{Free}(\Sigma))$, thus $(\{x\}, x) \in \mathcal{E}$. Consequently, $\forall \mathcal{E} \in \text{Ext}_p(\mathcal{T})$, $\text{Free}(\Sigma) \subseteq \text{Concs}(\mathcal{E})$. It follows that $\text{Free}(\Sigma) \subseteq \bigcap \text{Concs}(\mathcal{E})$ where $\mathcal{E} \in \text{Ext}_p(\mathcal{T})$.

In [2], we have shown that formulas of the set $\text{Free}(\Sigma)$ may be missed by argumentation systems under stable semantics. This is particularly the case when the systems do not have stable extensions. We have also seen that this problem is due to the use of skewed attack relations. Even if those relations ensure the rationality postulates, the corresponding systems do not return satisfactory results since they may miss intuitive conclusions like $\text{Free}(\Sigma)$. The previous results show that since preferred semantics guarantees the existence of preferred extensions, then it guarantees also the inference of elements of $\text{Free}(\Sigma)$.

The previous results make it possible to delimit the maximum number of preferred extensions a system may have. It is the number of consistent subbases of Σ that contain the free part of Σ and which are pairwise different. Note that this number is less than the number of consistent subbases of Σ.

Proposition 9. *Let* $\mathcal{T} = (\text{Arg}(\Sigma), \mathcal{R})$ *be an AS s.t.* $\mathcal{R}$ *is conflict-dependent and* $\mathcal{T}$ *satisfies consistency and closure under sub-arguments. It holds that* $1 \leq |\text{Ext}_p(\mathcal{T})| \leq |\text{Cons}(\Sigma)|$ *where* $\text{Cons}(\Sigma) = \{\mathcal{S} \mid \mathcal{S} \subseteq \Sigma,\ \mathcal{S} \text{ is consistent and } \text{Free}(\Sigma) \subseteq \mathcal{S}\}$.

Proof. From Proposition 2, each preferred extension returns a consistent subbase of Σ. From Proposition 3, it is not possible to have the same subbase several times. Finally, from Proposition 7, each preferred extension contains $\text{Arg}(\text{Free}(\Sigma))$.

Until now, we showed that preferred extensions reflect coherent points of view since they rely on consistent subbases of Σ. We also showed that when Σ is finite, each argumentation system that enjoy the rationality postulates has a *finite number of preferred extensions*. Proposition 9 provides the maximum number of such extensions. We thus answered all our questions.

4 Inferences under Preferred Semantics

In this section, we investigate the characteristics of the set $\texttt{Output}(\mathcal{T})$ of any argumentation system $\mathcal{T}$ that satisfies the postulates. Indeed, we study the kind of inferences that are made by an argumentation system under preferred semantics. From the results of the previous section, this set is defined as follows.

Proposition 10. *Let* $\mathcal{T} = (\texttt{Arg}(\Sigma), \mathcal{R})$ *be an AS s.t.* $\mathcal{R}$ *is conflict-dependent and* $\mathcal{T}$ *satisfies the three postulates. It holds that* $\texttt{Output}(\mathcal{T}) = \bigcap \texttt{CN}(\mathcal{S}_i)$ *s.t.* $\mathcal{S}_i \in \texttt{Cons}(\Sigma)$ *and* $\mathcal{S}_i = \texttt{Base}(\mathcal{E}_i)$ *where* $\mathcal{E}_i \in \texttt{Ext}_p(\mathcal{T})$.

Proof. Let $\mathcal{T} = (\texttt{Arg}(\Sigma), \mathcal{R})$ be an AS s.t. $\mathcal{R}$ is conflict-dependent and $\mathcal{T}$ satisfies the three postulates. Let $\mathcal{E} \in \texttt{Ext}_p(\mathcal{T})$. Since $\mathcal{T}$ is closed both under $\texttt{CN}$ and under subarguments, then $\texttt{Concs}(\mathcal{E}) = \texttt{CN}(\texttt{Base}(\mathcal{E}))$ (result shown in [1]). From Proposition 7, $\texttt{Free}(\Sigma) \subseteq \texttt{Base}(\mathcal{E})$. Moreover, from Proposition 2, $\texttt{Base}(\mathcal{E})$ is consistent. Thus, $\texttt{Base}(\mathcal{E}) \in \texttt{Cons}(\Sigma)$. From Definition 4, $\texttt{Output}(\mathcal{T}) = \bigcap \texttt{Concs}(\mathcal{E}_i)$, $\mathcal{E}_i \in \texttt{Ext}_p(\mathcal{T})$. Thus, $\texttt{Output}(\mathcal{T}) = \bigcap \texttt{Base}(\mathcal{E}_i)$, $\mathcal{E}_i \in \texttt{Ext}_p(\mathcal{T})$.

It is worth noticing that preferred semantics is more powerful than stable semantics *only* in case stable extensions do not exist and $\texttt{Free}(\Sigma) \neq \emptyset$. Indeed, in this case the output set of any argumentation system is empty $((\texttt{Output}(\mathcal{T})) = \emptyset)$ under stable semantics. Thus, the free formulas of Σ will not be inferred while they are guaranteed under preferred semantics. However, this does not mean that outputs under preferred semantics are "complete" and "intuitive". We show that some argumentation systems may miss in some cases some interesting conclusions. Worse yet, they may even return counter-intuitive ones. Let us illustrate our ideas on the following example.

Example 1. *Let us consider the following propositional knowledge base* $\Sigma = \{x, \neg x \land y\}$. *The two formulas are equally preferred. From Proposition 10, it follows that any reasonable argumentation that may be built over* Σ *will have one of the three following outputs:*

- $\texttt{Output}_1(\mathcal{T}) = \emptyset$. *This is the case of systems that have a unique and empty extension, or those which have two extensions* $\mathcal{E}_1$ *and* $\mathcal{E}_2$ *where* $\texttt{Base}(\mathcal{E}_1) = \{x\}$ *and* $\texttt{Base}(\mathcal{E}_2) = \{\neg x \land y\}$.
- $\texttt{Output}_2(\mathcal{T}) = \texttt{CN}(\{x\})$. *This is the case of systems that have* $\mathcal{E}_1$ *as their unique extension.*
- $\texttt{Output}_3(\mathcal{T}) = \texttt{CN}(\{\neg x \land y\})$. *This is the case of systems that have* $\mathcal{E}_2$ *as their unique extension.*

Let us analyze the three cases. In the first one, the result is not satisfactory. Indeed, one may expect to have y *as a conclusion since it is not part of the conflict in* Σ. *Assume that*

x stands for "sunny day" and y for "My dog is sick". It is clear that the two information x and y are independent. This shows that argumentation systems are syntax-dependent. The two other outputs ($\text{Output}_2(\mathcal{T})$ and $\text{Output}_3(\mathcal{T})$) are not satisfactory neither. The reason in these cases is different. For instance, in $\text{Output}_2(\mathcal{T})$, the formula x is inferred from Σ while $\neg x$ is not deduced. This discrimination between the two formulas is not justified since the two formulas of Σ are assumed to be equally preferred.

Let us now analyze in detail all the possible situations that *may* occur. Throughout this sub-section, $\Re_p$ denotes the set of all attack relations that ensure the three postulates for any argumentation system, that is for any Σ. Indeed, $\Re_p = \{ \mathcal{R} \subseteq \text{Arg}(\Sigma) \times \text{Arg}(\Sigma) \mid \mathcal{R}$ is conflict-dependent and $(\text{Arg}(\Sigma), \mathcal{R})$ satisfies Postulates 1, 2 and 3 under preferred semantics$\}$ for all Σ. We distinguish two categories of attack relations: those that always lead to a unique extension ($\Re_u$) and those that may lead to multiple extensions ($\Re_m$), where $\Re_p = \Re_u \cup \Re_m$.

Unique Extension. Let us focus on argumentation systems $\mathcal{T} = (\text{Arg}(\Sigma), \mathcal{R})$ that satisfy the three postulates and that use attack relations of the set $\Re_u$. Thus, $\text{Ext}_p(\mathcal{T}) = \{\mathcal{E}\}$. Three possible situations may occur:

$\text{Ext}_p(\mathcal{T}) = \{\emptyset\}$: In this case, the output set is empty, and consequently, there is no free formula, i.e. all the formulas of Σ are involved in at least one conflict.

Property 1. *Let* $\mathcal{T} = (\text{Arg}(\Sigma), \mathcal{R})$ *be an AS s.t.* $\mathcal{R} \in \Re_u$. *If* $\text{Ext}_p(\mathcal{T}) = \{\emptyset\}$, *then* $\text{Output}(\mathcal{T}) = \emptyset$ *and* $\text{Free}(\Sigma) = \emptyset$.

Proof. Since $\text{Ext}_p(\mathcal{T}) = \{\emptyset\}$, then from Definition 4 $\text{Output}(\mathcal{T}) = \emptyset$. From Proposition 8, it follows that $\text{Free}(\Sigma) = \emptyset$.

Note that the fact that $\text{Free}(\Sigma) = \emptyset$ does not imply that $\text{Ext}_p(\mathcal{T}) = \{\emptyset\}$. At a first glance, the previous result may seem reasonable since all the formulas in Σ are conflicting and are all equally preferred. However, Example 1 shows that this is not the case since there are some interesting formulas that may be missed.

$\text{Ext}_p(\mathcal{T}) = \{\text{Arg}(\text{Free}(\Sigma))\}$: Argumentation systems that have the unique preferred extension $\text{Arg}(\text{Free}(\Sigma))$ return as conclusions all the formulas that follow under CN from $\text{Free}(\Sigma)$.

Property 2. *Let* $\mathcal{T} = (\text{Arg}(\Sigma), \mathcal{R})$ *be an AS s.t.* $\mathcal{R} \in \Re_u$. *If* $\text{Ext}_p(\mathcal{T}) = \{\text{Arg}(\text{Free}(\Sigma))\}$, *then* $\text{Output}(\mathcal{T}) = \text{CN}(\text{Free}(\Sigma))$.

Proof. Let $\mathcal{T} = (\text{Arg}(\Sigma), \mathcal{R})$ be an AS s.t. $\mathcal{R} \in \Re_u$. Assume that $\text{Ext}_p(\mathcal{T}) = \{\text{Arg}(\text{Free}(\Sigma))\}$. Let $\mathcal{E} = \text{Arg}(\text{Free}(\Sigma))$. Since $\mathcal{T}$ is closed both under sub-arguments and CN, then $\text{Concs}(\mathcal{E}) = \text{CN}(\text{Base}(\mathcal{E}))$ ([1]). Thus, $\text{Concs}(\mathcal{E}) = \text{CN}(\text{Free}(\Sigma))$. Besides, $\text{Output}(\mathcal{T}) = \text{Concs}(\mathcal{E})$, thus $\text{Output}(\mathcal{T}) = \text{CN}(\text{Free}(\Sigma))$.

It is worth mentioning that such outputs correspond exactly to the so-called *free consequences* developed in [4] for handling inconsistency in propositional knowledge bases. The authors in [4] argue that this approach is very conservative. Indeed, if $\text{Free}(\Sigma)$ is empty, then nothing can be drawn from Σ. This may lead to miss intuitive formulas as shown in the next example.

Example 2. *Let us consider the following propositional knowledge base* $\Sigma = \{x, \neg x \wedge y, z\}$. *It can be checked that* $\text{Free}(\Sigma) = \{z\}$. *Thus, any reasonable argumentation system that may be built over* Σ *and that uses an attack relation of category* $\Re_u$ *will have the set* $\text{CN}(\{z\})$ *as output. However, y should also be inferred from Σ.*

$\text{Ext}_p(\mathcal{T}) = \{\mathcal{E}\}$ where $\text{Arg}(\text{Free}(\Sigma)) \subset \mathcal{E}$: In this case, there is at least one argument in the extension $\mathcal{E}$ whose support contains at least one formula which is involved in at least one conflict in Σ. However, since $\text{Base}(\mathcal{E})$ is consistent, then there are some formulas involved in the same conflict which are not considered. Then, there is a discrimination between elements of Σ which leads to ad hoc results as shown by the following result.

Proposition 11. *Let* $\mathcal{T} = (\text{Arg}(\Sigma), \mathcal{R})$ *be an AS s.t.* $\mathcal{R} \in \Re_u$. *If* $\text{Ext}_p(\mathcal{T}) = \{\mathcal{E}\}$ *and* $\text{Arg}(\text{Free}(\Sigma)) \subset \mathcal{E}$, *then* $\exists x \in \text{Inc}(\Sigma)$ *s.t.* $x \in \text{Output}(\mathcal{T})$ *and* $\exists x' \in \text{Inc}(\Sigma)$ *s.t.* $x' \notin \text{Output}(\mathcal{T})$.

Proof. Let $\mathcal{T} = (\text{Arg}(\Sigma), \mathcal{R})$ be an AS s.t. $\mathcal{R} \in \Re_u$. Assume that $\text{Ext}_p(\mathcal{T}) = \{\mathcal{E}\}$ and $\text{Arg}(\text{Free}(\Sigma)) \subset \mathcal{E}$. Thus, $\exists a \in \mathcal{E}$ and $a \notin \text{Arg}(\text{Free}(\Sigma))$. Consequently, $\text{Supp}(a) \nsubseteq \text{Free}(\Sigma)$. Thus, $\exists x \in \text{Supp}(a)$ and $x \notin \text{Free}(\Sigma)$. Thus, $x \in \text{Inc}(\Sigma)$. Moreover, since $\mathcal{T}$ is closed under sub-arguments, then from [1], $\text{Base}(\mathcal{E}) \subseteq \text{Output}(\mathcal{T})$, then $x \in \text{Output}(\mathcal{T})$. Besides, $\{x\}$ is consistent since $\text{Supp}(a)$ is consistent. Thus, $\exists C \in \mathcal{C}_\Sigma$ such that $|C| > 1$ and $x \in C$. Since $\mathcal{T}$ satisfies consistency, then $C \nsubseteq \text{Output}(\mathcal{T})$. Thus, $\exists x' \in C$ such that $x' \notin \text{Output}(\mathcal{T})$.

Let us illustrate this result on the following critical example.

Example 2 (Cont): Assume again the propositional knowledge base $\Sigma = \{x, \neg x \wedge y, z\}$. Argumentation systems of the previous category may return either $\mathcal{E}_1$ or $\mathcal{E}_2$ (not both) such that $\text{Base}(\mathcal{E}_1) = \{x, z\}$ and $\text{Base}(\mathcal{E}_2) = \{\neg x \wedge y, z\}$. In the first case, $\text{Output}(\mathcal{T}) = \text{CN}(\{x, z\})$. Thus, $x \in \text{Output}(\mathcal{T})$ while $\neg x \notin \text{Output}(\mathcal{T})$. In the second case, $\text{Output}(\mathcal{T}) = \text{CN}(\{\neg x \wedge y, z\})$, thus $\neg x \in \text{Output}(\mathcal{T})$ while $x \notin \text{Output}(\mathcal{T})$. Both cases are undesirable since there is no reason to privilege x over $\neg x$ and vice versa. Remember the case where x stands for "sunny day" and y for "my dog is sick".

Multiple Extensions. Let us now tackle the second category of attack relations: the ones that may lead to multiple preferred semantics. Let $\mathcal{T} = (\text{Arg}(\Sigma), \mathcal{R})$ be an AS s.t. $\mathcal{R} \in \Re_m$ and let $\text{Ext}_p(\mathcal{T}) = \{\mathcal{E}_1, \ldots, \mathcal{E}_n\}$ such that $n \geq 1$. We have seen previously that each preferred extension gives birth to a consistent subbase of Σ. This subbase may be either maximal (for set inclusion) or not. Moreover, the subbases of some extensions of the same system may be maximal while those of the remaining extensions not. In what follows, we study three possible cases.

Case 1: In this case, an argumentation system has at least one preferred extension whose corresponding base is not maximal (i.e. $\exists \mathcal{E} \in \text{Ext}_p(\mathcal{T})$ such that $\text{Base}(\mathcal{E}) \notin \text{Max}(\Sigma)$). The output set may be counter-intuitive since some priority will be given to some formula. Let us consider the following example.

Example 3. *Let us consider the following propositional knowledge base that contains four equally preferred formulas:* $\Sigma = \{x, x \rightarrow y, z, z \rightarrow \neg y\}$. *It can be checked that* $\text{Free}(\Sigma) = \emptyset$. *An argumentation system that fits in Case 1 would have for instance, two preferred extensions* $\mathcal{E}_1$ *and* $\mathcal{E}_2$ *such that* $\text{Base}(\mathcal{E}_1) = \{x, x \rightarrow y, z\}$ *and* $\text{Base}(\mathcal{E}_2) = \{z, z \rightarrow \neg y\}$. *Note that* $\text{Base}(\mathcal{E}_1) \in \text{Max}(\Sigma)$ *while* $\text{Base}(\mathcal{E}_2) \notin \text{Max}(\Sigma)$. *The output of this system is* $\text{Output}(\mathcal{T}) = \text{CN}(\{z\})$. *Thus,* $z \in \text{Output}(\mathcal{T})$ *while the three other formulas of* Σ *are not elements of* $\text{Output}(\mathcal{T})$. *This result is unjustified since all the formulas of* Σ *are involved in the conflict and are equally preferred.*

Case 2: The bases of all the preferred extensions of an argumentation system are maximal (for set inclusion). However, not all the maximal consistent subsets of Σ have a corresponding preferred extension (i.e. $\forall \mathcal{E} \in \text{Ext}_p(\mathcal{T})$, $\text{Base}(\mathcal{E}) \in \text{Max}(\Sigma)$ and $|\text{Ext}_p(\mathcal{T})| < |\text{Max}(\Sigma)|$). The same problem described in Case 1 is encountered here. Indeed, some formulas are privileged over others in an ad hoc way. Let us consider the following example.

Example 3 (Cont): Let us consider again the knowledge base of Example 3. An argumentation system that fits in Case 2 would have for instance, two preferred extensions $\mathcal{E}_1$ and $\mathcal{E}_2$ such that $\text{Base}(\mathcal{E}_1) = \{x, x \rightarrow y, z\}$ and $\text{Base}(\mathcal{E}_2) = \{x, x \rightarrow y, z \rightarrow \neg y\}$. Note that both subsets are maximal. It is easy to check that $x, x \rightarrow y \in \text{Output}(\mathcal{T})$ while $z, z \rightarrow \neg y \notin \text{Output}(\mathcal{T})$. This result is again unjustified.

Case 3: The bases of all the preferred extensions of an argumentation system are maximal (for set inclusion). Moreover, any maximal consistent subsets of Σ has a corresponding preferred extension in the argumentation system (i.e. $\forall \mathcal{E} \in \text{Ext}_p(\mathcal{T})$, $\text{Base}(\mathcal{E}) \in \text{Max}(\Sigma)$ and $|\text{Ext}_p(\mathcal{T})| = |\text{Max}(\Sigma)|$). The outputs of such systems are exactly the common conclusions that are drawn under CN from the maximal consistent subsets of Σ.

Property 3. *Let* $\mathcal{T} = (\text{Arg}(\Sigma), \mathcal{R})$ *be an AS s.t.* $\mathcal{R} \in \Re_m$. *If* $\forall \mathcal{E} \in \text{Ext}_p(\mathcal{T})$, $\text{Base}(\mathcal{E}) \in \text{Max}(\Sigma)$ *and* $|\text{Ext}_p(\mathcal{T})| = |\text{Max}(\Sigma)|$, *then* $\text{Output}(\mathcal{T}) = \bigcap \mathcal{S}_i$ *where* $\mathcal{S}_i \in \text{Max}(\Sigma)$.

Proof. This follows from the definition of $\text{Output}(\mathcal{T})$ and the fact that for each preferred extension $\mathcal{E}$, $\text{Concs}(\mathcal{E}) = \text{CN}(\text{Base}(\mathcal{E}))$.

It is worth noticing that this output corresponds to the *universal consequences* developed in [19] for handling inconsistency in propositional knowledge bases. Thus, argumentation systems of this category generalize the coherence-based approach to any logic. Consequently, they inherit its problems, namely the one described in Example 1 (missing intuitive conclusions). It is also worth recalling that there exist attack relations that lead to this result. Assumption attack developed in [12] is one of them. Indeed, any argumentation system that use this relation will have the output described in Property 3. Finally, we have shown in another paper that the stable extensions of any argumentation system (that satisfies consistency and closure under sub-arguments) return maximal consistent subsets of Σ. We show next that when all the maximal subsets of Σ have a corresponding stable extension in a system, then this latter is certainly coherent, i.e., its preferred extensions are stable ones.

Attack relation	Cases	Output	Problem
$\mathcal{R} \in \Re_u$	$\text{Ext}_p(\mathcal{T}) = \{\emptyset\}$	$\emptyset$	M
	$\text{Ext}_p(\mathcal{T}) = \{\text{Arg}(\text{Free}(\Sigma))\}$	$\text{CN}(\text{Free}(\Sigma))$	M
	$\text{Ext}_p(\mathcal{T}) = \{\mathcal{E}\}$ and $\text{Arg}(\text{Free}(\Sigma)) \subseteq \mathcal{E}$	$\text{CN}(S), S \in \text{Cons}(\Sigma)$	U
$\mathcal{R} \in \Re_m$	$\exists \mathcal{E}_i$ s.t. $\text{Base}(\mathcal{E}_i) \notin \text{Max}(\Sigma)$	$\bigcap_{i=1.k} \text{CN}(S_i), \{S_1, \dots, S_k\} \subseteq \text{Cons}(\Sigma)$	U
	$\forall \mathcal{E}_i, \text{Base}(\mathcal{E}_i) \in \text{Max}(\Sigma)$ and $|\text{Ext}_p(\mathcal{T})| < |\text{Max}(\Sigma)|$	$\bigcap_{i=1.k} \text{CN}(S_i), \{S_1, \dots, S_k\} \subset \text{Max}(\Sigma)$	U
	$\forall \mathcal{E}_i, \text{Base}(\mathcal{E}_i) \in \text{Max}(\Sigma)$ and $|\text{Ext}_p(\mathcal{T})| = |\text{Max}(\Sigma)|$	$\bigcap \text{CN}(S_i), S_i \in \text{Max}(\Sigma)$	M

Fig. 1. Outcomes under preferred semantics (M stands for missing conclusions and U for undesirable ones)

Proposition 12. *Let* $\mathcal{T} = (\text{Arg}(\Sigma), \mathcal{R})$ *be an AS over a knowledge base* Σ *s.t.* $\mathcal{R} \in \Re_p$. *If* $|\text{Ext}_s(\mathcal{T})| = |\text{Max}(\Sigma)|$, *then* $\mathcal{T}$ *is coherent.*

Proof. Let $\mathcal{T} = (\text{Arg}(\Sigma), \mathcal{R})$ be an AS over a knowledge base Σ such that $\mathcal{R} \in \Re_p$. By definition of stable semantics, $\text{Ext}_s(\mathcal{T}) \subseteq \text{Ext}_p(\mathcal{T})$. Assume now that $\mathcal{E} \in \text{Ext}_p(\mathcal{T})$. From Proposition 2, $\text{Base}(\mathcal{E})$ is consistent. Thus, $\exists S \in \text{Max}(\Sigma)$ such that $\text{Base}(\mathcal{E}) \subseteq S$. From Property 4 in [3], it holds that $\text{Arg}(\text{Base}(\mathcal{E})) \subseteq \text{Arg}(S)$. However, $\mathcal{E} \subseteq \text{Arg}(\text{Base}(\mathcal{E}))$ and $\text{Arg}(S) \in \text{Ext}_s(\mathcal{T})$. Thus, $\mathcal{E} \subseteq \text{Arg}(S)$ where $\text{Arg}(S) \in \text{Ext}_p(\mathcal{T})$. This contradicts the fact that $\mathcal{E}$ is a preferred extension, thus maximal.

The results of this section show that reasoning under preferred semantics is not recommended. Figure 1 summaries the different outputs that may be encountered under this semantics.

5   Conclusion

In this paper, we characterized for the first time the outcomes of argumentation systems under preferred semantics. To the best of our knowledge there is no work that tackled this issue. In [8], the author studied the outcomes of a very particular system under stable semantics. In [14], the authors focused on argumentation systems that are grounded on propositional logic and studied the properties of various systems using specified attack relations. The focus was mainly on the satisfaction of rationality postulates. In this paper, we assume abstract logic-based argumentations in which neither the underlying logic nor the attack relations are specified. This abstraction makes our results more general and powerful. Moreover, among all the possible instantiations of this setting, we considered those that satisfy some basic rationality postulates. Indeed, systems that violate those postulates should be avoided as they certainly lead to undesirable results. A first important result consists of delimiting the maximum number of preferred extensions a system may have. We have shown that if the knowledge base under study is finite, then any argumentation system buit over it has a finite number of preferred extensions. Then, we have shown that from each preferred extension, a (maybe maximal) consistent subbase of the knowledge base is computed. This subbase contains all the formulas that are not involved in any conflict. We have then shown that in the best case, reasoning under preferred semantics may lead to missing some interesting conclusions. This is mainly due to the fact that argumentation systems are syntax-dependent. Moreover, they coincide with the coherence-based approach [4,19], thus inherit all its weaknesses. In the worst case, preferred semantics will lead to undesirable conclusions. The main problem here is that the attack relation defines some "artificial" priorities between the formulas of the knowledge base leading to ad hoc outputs. The only good news is that preferred semantics performs better than stable one in the sense that it guarantees the inference of the free formulas (i.e., the formulas that are not involved in any conflict).

To sum up, we have shown (for the large class of argumentation systems we discussed) that preferred semantics should be avoided.

References

1. Amgoud, L.: Postulates for logic-based argumentation systems. page WL4AI: ECAI Workshop on Weighted Logics for AI (2012)
2. Amgoud, L.: Stable Semantics in Logic-Based Argumentation. In: Hüllermeier, E., et al. (eds.) SUM 2012. LNCS (LNAI), vol. 7520, pp. 58–71. Springer, Heidelberg (2012)
3. Amgoud, L., Besnard, P.: Bridging the Gap between Abstract Argumentation Systems and Logic. In: Godo, L., Pugliese, A. (eds.) SUM 2009. LNCS, vol. 5785, pp. 12–27. Springer, Heidelberg (2009)
4. Benferhat, S., Dubois, D., Prade, H.: How to infer from inconsistent beliefs without revising? In: IJCAI 1995, pp. 1449–1457 (1995)
5. Bondarenko, A., Dung, P., Kowalski, R., Toni, F.: An abstract, argumentation-theoretic approach to default reasoning. Artificial Intelligence 93, 63–101 (1997)
6. Caminada, M.: On the Issue of Reinstatement in Argumentation. In: Fisher, M., van der Hoek, W., Konev, B., Lisitsa, A. (eds.) JELIA 2006. LNCS (LNAI), vol. 4160, pp. 111–123. Springer, Heidelberg (2006)
7. Caminada, M., Amgoud, L.: An axiomatic account of formal argumentation. In: AAAI 2005, pp. 608–613 (2005)
8. Cayrol, C.: On the relation between argumentation and non-monotonic coherence-based entailment. In: Proceedings of the 14th International Joint Conference on Artificial Intelligence, pp. 1443–1448 (1995)
9. Cholvy, L.: Automated Reasoning with Merged Contradictory Information Whose Reliability Depends on Topics. In: Froidevaux, C., Kohlas, J. (eds.) ECSQARU 1995. LNCS, vol. 946, pp. 125–132. Springer, Heidelberg (1995)
10. Coste-Marquis, S., Devred, C., Marquis, P.: Symmetric Argumentation Frameworks. In: Godo, L. (ed.) ECSQARU 2005. LNCS (LNAI), vol. 3571, pp. 317–328. Springer, Heidelberg (2005)
11. Dung, P.M.: On the acceptability of arguments and its fundamental role in nonmonotonic reasoning, logic programming and n-person games. Artificial Intelligence Journal 77, 321–357 (1995)
12. Elvang-Gøransson, M., Fox, J.P., Krause, P.: Acceptability of Arguments as Logical Uncertainty. In: Moral, S., Kruse, R., Clarke, E. (eds.) ECSQARU 1993. LNCS, vol. 747, pp. 85–90. Springer, Heidelberg (1993)
13. García, A., Simari, G.: Defeasible logic programming: an argumentative approach. Theory and Practice of Logic Programming 4, 95–138 (2004)
14. Gorogiannis, N., Hunter, A.: Instantiating abstract argumentation with classical logic arguments: Postulates and properties. Artificial Intelligence Journal 175(9-10), 1479–1497 (2011)
15. Governatori, G., Maher, M., Antoniou, G., Billington, D.: Argumentation semantics for defeasible logic. Journal of Logic and Computation 14(5), 675–702 (2004)
16. Pearl, J.: System z: A natural ordering of defaults with tractable applications to default reasoning. In: TARK 1990, pp. 121–135 (1990)
17. Pollock, J.L.: How to reason defeasibly. Artificial Intelligence Journal 57, 1–42 (1992)
18. Prakken, H., Sartor, G.: Argument-based extended logic programming with defeasible priorities. Journal of Applied Non-Classical Logics 7, 25–75 (1997)
19. Rescher, N., Manor, R.: On inference from inconsistent premises. Journal of Theory and Decision 1, 179–219 (1970)
20. Tarski, A.: On Some Fundamental Concepts of Metamathematics. In: Woodger, E.H. (ed.) Logic, Semantics, Metamathematics. Oxford Uni. Press (1956)

Abstract Argumentation
via Monadic Second Order Logic*

Wolfgang Dvořák[1], Stefan Szeider[2], and Stefan Woltran[2]

[1] Theory and Applications of Algorithms Group, University of Vienna
[2] Institute of Information Systems, Vienna University of Technology

Abstract. We propose the formalism of Monadic Second Order Logic (MSO) as a unifying framework for representing and reasoning with various semantics of abstract argumentation. We express a wide range of semantics within the proposed framework, including the standard semantics due to Dung, semi-stable, stage, cf2, and resolution-based semantics. We provide building blocks which make it easy and straight-forward to express further semantics and reasoning tasks. Our results show that MSO can serve as a *lingua franca* for abstract argumentation that directly yields to complexity results. In particular, we obtain that for argumentation frameworks with certain structural properties the main computational problems with respect to MSO-expressible semantics can all be solved in linear time. Furthermore, we provide a novel characterization of resolution-based grounded semantics.

1 Introduction

Starting with the seminal work by Dung [18] the area of argumentation has evolved to one of the most active research branches within Artificial Intelligence (see, e.g., [6]). Dung's abstract argumentation frameworks, where arguments are seen as abstract entities which are just investigated with respect to how they relate to each other, in terms of "attacks", are nowadays well understood and different semantics (i.e., the selection of sets of arguments which are jointly acceptable) have been proposed. In fact, there seems to be no single "one suits all" semantics, but it turned out that studying a particular setting within various semantics and to compare the results is a central research issue within the field. Different semantics give rise to different computational problems, such as deciding whether an argument is acceptable with respect to the semantics under consideration, that require different approaches for solving these problems.

This broad range of semantics for abstract argumentation demands for a *unifying framework* for representing and reasoning with the various semantics. Such a unifying framework would allow us to see what the various semantics have in common, in what they differ, and ideally, it would offer generic methods for solving the computational problems that arise within the various semantics. Such a unifying framework should be general enough to accommodate most of the significant semantics, but simple enough to be decidable and computationally feasible.

* Dvořák's and Woltran's work has been funded by the Vienna Science and Technology Fund (WWTF) through project ICT08-028 and Szeider's work has been funded by the European Research Council (ERC), grant reference 239962 (COMPLEX REASON).

E. Hüllermeier et al. (Eds.): SUM 2012, LNAI 7520, pp. 85–98, 2012.

In this paper we propose such a unifying framework. We express several semantics within the framework, and we study its properties. The proposed unifying framework is based on the formalism of Monadic Second Order Logic (MSO), which is a fragment of Second Order logic with relational variables restricted to unary. MSO provides higher expressiveness than First Order Logic while it has more appealing algorithmic properties than full Second Order logic. Furthermore, MSO plays an important role in various parts of Computer Science. For instance, by Büchi's Theorem, a formal language is regular if and only if it can be expressed by MSO (this also provides a link between MSO and finite automata); furthermore, by Courcelle's Theorem, MSO expressible properties can be checked in linear time on structures of bounded treewidth.

Main Contributions. The results in this paper can be summarized as follows:

(1) We express a wide range of semantics within our proposed framework, including the standard semantics due to Dung, semi-stable, stage, cf2, and resolution-based semantics. For the latter, we present a new characterization that admits an MSO-encoding without quantification over sets of attacks and thus provides additional algorithmic implications.
(2) We provide MSO-building blocks which make it easy and straight-forward to express other semantics or to create new ones or variants.
(3) We also illustrate that any labeling-based semantics can be canonically expressed within our framework. We show that the main computational problems can be solved in linear time for all semantics expressible in our framework when restricted to argumentation frameworks of certain structures. This includes decision problems such as skeptical and brave acceptance, but also counting problems, for instance, determining how many extensions contain a given argument.

Our results show that MSO is indeed a suitable unifying framework for abstract argumentation and can serve as a *lingua franca* for further investigations. Furthermore, recent systems [28,29] showed quite impressive performance for evaluating MSO formulas over graphs, thus the proposed framework can be exploited as a rapid-prototyping approach to experiment with established and novel argumentation semantics.

Finally, we want to emphasise that in contrast to existing work [19,20,21,22] our goal is not to provide new complexity results for a particular argumentation semantics, but we propose MSO as a general logical framework for specifying argumentation semantics, having fixed-parameter tractability results as a neat side effect (compared to other approaches discussed below). Thus, in contrast to previous work, where MSO techniques were used as an auxiliary tool for achieving tractability results for particular semantics, our approach intends to raise MSO to a new conceptual level.

Related Work. Using MSO as a tool to express AI formalisms has been advocated in [26,27]. In terms of abstract argumentation MSO-encodings were given in [19,22] and implications in terms of parameterized complexity also appeared in [20,21].

Finding a uniform logical representation for abstract argumentation has been subject of several papers. While [7] used propositional logic for this purpose, [24] showed that quantified propositional logic admits complexity-adequate representations. Another branch of research focuses on logic programming as common grounds for different argumentation semantics, see [32] for a survey. Finally also the use of constraint satisfaction techniques was suggested [1,8]. All this research was mainly motivated by implementation issues and led to systems such as ASPARTIX [23]. As mentioned above, also MSO can serve this purpose, but in addition yields further results "for free", in particular in terms of complexity.

2 Background

We start this section by introducing (abstract) argumentation frameworks [18] and recalling the semantics we study in this paper (see also [4]).

Definition 1. *An* argumentation framework (AF) *is a pair $F = (A, R)$ where A is a set of arguments and $R \subseteq A \times A$ is the attack relation. The pair $(a, b) \in R$ means that a attacks b. We say that an argument $a \in A$ is* defended *(in F) by a set $S \subseteq A$ if, for each $b \in A$ such that $(b, a) \in R$, there exists a $c \in S$ such that $(c, b) \in R$.*

Example 1. In the following we use the AF $F = (\{a, b, c, d, e\}, \{(a, b), (b, a), (b, c), (c, d), (d, e), (e, c)\})$ as running example. The graph representation is given as follows:

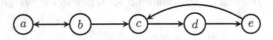

Semantics for argumentation frameworks are given via a function σ which assigns to each AF $F = (A, R)$ a set $\sigma(F) \subseteq 2^A$ of extensions. We first consider for σ the functions *naive*, *stb*, *adm*, *com*, *prf*, *grd*, *stg*, and *sem* which stand for naive, stable, admissible, complete, preferred, grounded, stage, and semi-stable semantics, respectively. Towards the definition of these semantics we introduce two more formal concepts.

Definition 2. *Given an AF $F = (A, R)$, the* characteristic function $\mathcal{F}_F : 2^A \Rightarrow 2^A$ *of F is defined as $\mathcal{F}_F(S) = \{ x \in A \mid x \text{ is defended by } S \}$. For a set $S \subseteq A$ and an argument $a \in A$, we write $S \rightarrowtail^R a$ (resp. $a \rightarrowtail^R S$) in case there is an argument $b \in S$, such that $(b, a) \in R$ (resp. $(a, b) \in R$). Moreover, for a set $S \subseteq A$, we denote the set of arguments attacked by S as $S_R^\oplus = \{ x \mid S \rightarrowtail^R x \}$, and resp. $S_R^\ominus = \{ x \mid x \rightarrowtail^R S \}$, and define the* range *of S as $S_R^+ = S \cup S_R^\oplus$.*

Example 2. In our running example $\mathcal{F}_F(\{a\}) = \{a\}$, $\mathcal{F}_F(\{b\}) = \{b, d\}$, $\{a\}_R^\oplus = \{b\}$, $\{b\}_R^\oplus = \{a, c\}$ and $\{a\}_R^+ = \{a, b\}$, $\{b\}_R^+ = \{a, b, c\}$.

Definition 3. *Let $F = (A, R)$ be an AF. A set $S \subseteq A$ is* conflict-free *(in F), if there are no $a, b \in S$, such that $(a, b) \in R$. $cf(F)$ denotes the collection of conflict-free sets of F. For a conflict-free set $S \in cf(F)$, it holds that*

- $S \in naive(F)$, *if there is no $T \in cf(F)$ with $T \supset S$;*
- $S \in stb(F)$, *if $S_R^+ = A$;*
- $S \in adm(F)$, *if $S \subseteq \mathcal{F}_F(S)$;*
- $S \in com(F)$, *if $S = \mathcal{F}_F(S)$;*
- $S \in grd(F)$, *if $S \in com(F)$ and there is no $T \in com(F)$ with $T \subset S$;*
- $S \in prf(F)$, *if $S \in adm(F)$ and there is no $T \in adm(F)$ with $S \subset T$;*

- $S \in sem(F)$, if $S \in adm(F)$ and there is no $T \in adm(F)$ with $S_R^+ \subset T_R^+$;
- $S \in stg(F)$, if there is no $T \in cf(F)$, with $S_R^+ \subset T_R^+$.

We recall that for each AF F, $stb(F) \subseteq sem(F) \subseteq prf(F) \subseteq com(F) \subseteq adm(F)$ holds, and that each of the considered semantics σ except stb satisfies $\sigma(F) \neq \emptyset$. Moreover grd yields a unique extension for each AF F (in what follows identified by $grd(F)$), which is the least fix-point of the characteristic function $\mathcal{F}_F$.

Example 3. Our running example F has four admissible sets, i.e. $adm(F) = \{\emptyset, \{a\}, \{b\}, \{b, d\}\}$, with $\{a\}$ and $\{b, d\}$ being the preferred extensions. The grounded extension is the empty set, moreover $com(F) = \{\emptyset, \{a\}, \{b, d\}\}$ and $stb(F) = sem(F) = stg(F) = \{\{b, d\}\}$.

On the base of these semantics one can define the family of resolution-based semantics [3], with the resolution-based grounded semantics being its most popular instance.

Definition 4. *Given AF $F = (A, R)$, a resolution $\beta \subset R$ of F is a $\subseteq$-minimal set of attacks such that for each pair $\{(a, b), (b, a)\} \subseteq R$ ($a \neq b$) either $(a, b) \in \beta$ or $(b, a) \in \beta$. We denote the set of all resolutions of an AF F by $\gamma(F)$. Given a semantics σ, the corresponding resolution-based semantics σ^* is given by $\sigma^*(F) =$*
$$\min_{\subseteq} \bigcup_{\beta \in \gamma(F)} \{\sigma((A, R \setminus \beta))\}.$$

Example 4. For our example AF F we get the two resolutions $\{(a, b)\}$ and $\{(b, a)\}$. In the case of resolution based grounded semantics this yields two candidates for extensions $grd((A, R \setminus \{(a, b)\})) = \{b, d\}$ and $grd((A, R \setminus \{(b, a)\})) = \{a\}$. As they are not in $\subseteq$-relation both are resolution-based grounded extensions and thus $grd^*(F) = \{\{a\}, \{b, d\}\}$.

Finally, let us consider the semantics cf2, which was introduced in [5] as part of a general schema for argumentation semantics. cf2 semantics gained some interest as it handles even and odd length cycles of attacks in a similar way. Towards a definition of cf2 semantics we need the following concepts.

Definition 5. *Given an AF $F = (A, R)$ and a set $S \subseteq A$. By $SCC(F)$ we denote the set of all strongly connected components of F. $D_F(S)$ denotes the set of arguments $a \in A$ attacked by an argument $b \in S$ occurring in a different component. Finally, for $F = (A, R)$ and a set S of arguments, $F|_S := (A \cap S, R \cap (S \times S))$ and $F - S := F|_{A \setminus S}$.*

Example 5. To illustrate D_F consider our example AF and the set $\{b, d\}$. We have two components $\{a, b\}$ and $\{c, d, e\}$ and that b attacks c. Hence $c \in D_F(\{b, d\})$. Also b attacks a and d attacks e but as both conflicts are within one component they do not add to the set $D_F(\{b, d\})$ and we have $D_F(\{b, d\}) = \{c\}$.

Definition 6. *Given an AF $F = (A, R)$, for $S \subseteq A$ we have that $S \in cf2(F)$ if one of the following conditions holds: (i) $|SCC(F)| = 1$ and $S \in naive(F)$; (ii) $\forall C \in SCC(F) : C \cap S \in cf2(F|_C - D_F(S))$.*

Example 6. For our example AF we obtain $cf2(F) = \{\{a, e\}, \{a, d\}, \{a, c\}, \{b, d\}\}$.

Labeling-based semantics. So far we have considered so-called extension-based semantics. However, there are several approaches defining argumentation semantics via certain kind of labelings instead of extensions. As an example we consider the complete labelings from [11].

Definition 7. *Given an AF* $F = (A, R)$, *a function* $\mathcal{L} : A \to \{\text{in, out, undec}\}$ *is a* complete labeling *iff the following conditions hold: (i)* $\mathcal{L}(a) = $ in *iff for each* b *with* $(b, a) \in R$, $\mathcal{L}(b) = $ out; *(ii)* $\mathcal{L}(b) = $ out *iff there exists* b *with* $(b, a) \in R$, $\mathcal{L}(b) = $ in.

There is a one-to-one mapping between complete extensions and complete labelings, such that the set of arguments labeled with "in" corresponds to a complete extension.

Example 7. The example AF has three complete labelings corresponding to the three complete extensions: the labeling $\mathcal{L}_1$ corresponding to $\emptyset$ with $\mathcal{L}_1(a) = \mathcal{L}_1(b) = \mathcal{L}_1(c) = \mathcal{L}_1(d) = \mathcal{L}_1(e) = $ undec; the labeling $\mathcal{L}_2$ corresponding to $\{a\}$ with $\mathcal{L}_2(a) = $ in, $\mathcal{L}_2(b) = $ out, and $\mathcal{L}_2(c) = \mathcal{L}_2(d) = \mathcal{L}_2(e) = $ undec; and the labeling $\mathcal{L}_3$ corresponding to $\{b, d\}$ with $\mathcal{L}_1(b) = \mathcal{L}_1(d) = $ in and $\mathcal{L}_3(a) = \mathcal{L}_3(c) = \mathcal{L}_3(e) = $ out.

Monadic Second Order Logic. Informally, Monadic Second Order Logic can be seen as an extension of First Order Logic that admits quantification over sets. First Order Logic is built from variables $x, y, z, \ldots$ referring to elements of the universe, atomic formulas $R(t_1, \ldots, t_k)$, $t_1 = t_2$, with t_i being variables or constants, the usual Boolean connectives, and quantification $\exists x$, $\forall x$. MSO$_1$ extends the language of First Order Logic by set variables $X, Y, Z, \ldots$, atomic formulas $t \in X$ with t a variable or constant, and quantification over set variables. We further consider MSO$_2$ an extension of MSO$_1$ which is only defined on graphs (which is perfectly fine for our purposes). MSO$_2$ adds variables $X^E, Y^E, Z^E, \ldots$ ranging over sets of edges of the graph and quantification over such variables. In the following when talking about MSO we refer to MSO$_2$.

For an MSO formula ϕ we usually write $\phi(x_1, \ldots, x_p, X_1, \ldots X_q)$ to denote that the free variables of ϕ are $x_1, \ldots, x_p, X_1, \ldots X_q$. For a graph $G = (V, E)$, $v_1, \ldots, v_p \in V$, and $A_1, \ldots A_q \subseteq V$, we write $G \models \phi(v_1, \ldots, v_p, A_1, \ldots A_q)$ to denote that the formula ϕ holds true for G if x_i is instantiated with v_i and X_j is instantiated with A_j, $1 \leq i \leq p$, $1 \leq j \leq q$.

3 Encoding Argumentation Semantics in MSO

Building Blocks. We first introduce some shorthands simplifying notation when dealing with subset relations and the range of extensions.

$$X \subseteq Y = \forall x \, (x \in X \to x \in Y) \qquad x \notin X = \neg(x \in X)$$
$$X \subset Y = X \subseteq Y \land \neg(Y \subseteq X) \qquad x \in X_R^+ = x \in X \lor \exists y(y \in X \land (y, x) \in R)$$
$$X \not\subseteq Y = \neg(X \subseteq Y) \qquad X \subseteq_R^+ Y = \forall x \, (x \in X_R^+ \to x \in Y_R^+)$$
$$X \not\subset Y = \neg(X \subset Y) \qquad X \subset_R^+ Y = X \subseteq_R^+ Y \land \neg(Y \subseteq_R^+ X)$$

Another important notion that underlies argumentation semantics is the notion of a set being conflict-free. The following MSO formula encodes that a set X is conflict-free w.r.t. the attack relation R:

$$cf_R(X) = \forall x, y \, ((x, y) \in R \to (\neg x \in X \lor \neg y \in X))$$

Next we give a building block for maximizing extensions using an (MSO expressible) order $\sqsubseteq$:

$$\max{}_{A,P(.),\sqsubseteq}(X) = P(X) \wedge \neg\exists Y\left(Y \subseteq A \wedge P(Y) \wedge X \sqsubset Y\right)$$

Clearly we can also implement minimization by inverting the order, i.e., $\min{}_{A,P(.),\sqsubseteq}(X) = \max{}_{A,P(.),\sqsupseteq}(X)$.

Standard Encodings. In the following we provide MSO-characterizations for the different argumentation semantics. The characterizations for adm, stb, prf are borrowed from [19] while those for sem, stg are borrowed from [22].

$$naive_{A,R}(X) = \max{}_{A,cf_R(.),\subseteq}(X)$$
$$adm_R(X) = cf_R(X) \wedge \forall x,y \left(((x,y) \in R \wedge y \in X) \to\right.$$
$$\exists z(z \in X \wedge (z,x) \in R))$$
$$com_{A,R}(X) = adm_R(X) \wedge \forall x((x \in A \wedge x \notin X) \to$$
$$\exists y((y,x) \in R \wedge \neg\exists z(z \in X \wedge (z,y) \in R)))$$
$$grd_{A,R}(X) = \min{}_{A,com_{A,R}(.),\subseteq}(X)$$
$$stb_{A,R}(X) = cf_R(X) \wedge \forall x(x \in A \to x \in X_R^+)$$
$$prf_{A,R}(X) = \max{}_{A,adm_R(.),\subseteq}(X)$$
$$sem_{A,R}(X) = \max{}_{A,adm_R(.),\subseteq_R^+}(X)$$
$$stg_{A,R}(X) = \max{}_{A,cf_R(.),\subseteq_R^+}(X)$$

These characterisations are straight-forward translations of the definitions and thus can be easily checked to be correct.

Based on the above characterizations, we proceed with encodings for the resolution-based semantics as follows. Via $res_R(X^E)$, given as

$$\forall x,y\left(X^E \subseteq R \wedge (x,x) \in R \to (x,x) \in X^E \wedge\right.$$
$$(x \neq y \wedge (x,y) \in R) \to ((x,y) \in X^E \leftrightarrow (y,x) \notin X^E)),$$

we express modified frameworks $(A, R \setminus \beta)$ where β is a resolution according to Definition 4. Now resolution-based semantics are characterised by

$$\sigma_{A,R}^*(X) = \exists X^E(res_R(X^E) \wedge \sigma_{A,X^E}(X) \wedge \tag{1}$$
$$\forall Y\forall Y^E(res_R(Y^E) \wedge \sigma_{A,Y^E}(Y) \to Y \not\subset X)).$$

Labeling-based semantics. There are several approaches to define argument semantics via different kind of argumentation labelings and almost all argumentation semantics admit a characterization via argument labelings. The general concept behind labelings is to use a fixed set of labels and assign to each argument a subset of them, or just a single label. Such labelings are valid if for each argument the assigned labels satisfy certain (qualitative) conditions concerning the labels of attacking arguments and the labels of the attacked arguments. Additionally one might demand that the set of arguments labeled by a specific label is maximal or minimal. All these properties can be easily expressed in MSO, which we illustrate for complete labelings. We encode an in, out, undec labeling $\mathcal{L}$ as a triple $(\mathcal{L}_{\text{in}}, \mathcal{L}_{\text{out}}, \mathcal{L}_{\text{undec}})$ where $\mathcal{L}_l := \{\, a \in A \mid \mathcal{L}(a) = l \,\}$. To have these three sets disjoint, one uses the formula $\varphi = \forall x \in A((x \in \mathcal{L}_{\text{in}} \vee x \in \mathcal{L}_{\text{out}} \vee x \in \mathcal{L}_{\text{undec}}) \wedge (x \notin \mathcal{L}_{\text{in}} \vee x \notin \mathcal{L}_{\text{out}}) \wedge (x \notin \mathcal{L}_{\text{in}} \vee x \notin \mathcal{L}_{\text{undec}}) \wedge (x \notin \mathcal{L}_{\text{undec}} \vee x \notin \mathcal{L}_{\text{out}}))$. Now we can give an MSO formula $com_{A,R}(\mathcal{L}_{\text{in}}, \mathcal{L}_{\text{out}}, \mathcal{L}_{\text{undec}})$ expressing whether such a triple is a complete labeling:

$$\varphi \wedge \forall x \in X(x \in \mathcal{L}_{\text{in}} \leftrightarrow (\forall y \in X((y,x) \in R \rightarrow y \in \mathcal{L}_{\text{out}})))$$
$$\wedge \forall x \in X(x \in \mathcal{L}_{\text{out}} \leftrightarrow (\exists y \in X((y,x) \in R \wedge y \in \mathcal{L}_{\text{in}})))$$

Further, one can directly encode preferred labelings, which are defined as complete labelings with maximal $\mathcal{L}_{\text{in}}$.

$$prf_{A,R}(\mathcal{L}_{\text{in}}, \mathcal{L}_{\text{out}}, \mathcal{L}_{\text{undec}}) = com_{A,R}(\mathcal{L}_{\text{in}}, \mathcal{L}_{\text{out}}, \mathcal{L}_{\text{undec}}) \wedge \neg \exists \mathcal{L}'_{\text{in}}, \mathcal{L}'_{\text{out}}, \mathcal{L}'_{\text{undec}}$$
$$(\mathcal{L}_{\text{in}} \subset \mathcal{L}'_{\text{in}} \wedge com_{A,R}(\mathcal{L}'_{\text{in}}, \mathcal{L}'_{\text{out}}, \mathcal{L}'_{\text{undec}}))$$

MSO-characterization for cf2. The original definition of *cf2* semantics is of recursive nature and thus not well suitable for a direct MSO-encoding. Hence we use an alternative characterisation of *cf2* [25]. For this purpose we need the following definitions.

Definition 8. *Given an AF* $F = (A, R)$, $B \subseteq A$, *and* $a, b \in A$, *we define* $a \Rightarrow^B_F b$ *if and only if there exists a sequence* $(b_i)_{1 \leq i \leq n}$ *with* $b_i \in B$, $b_1 = a$, $b_n = b$ *and* $(b_i, b_{i+1}) \in R$.

The relation $\Rightarrow^B_F$ can be encoded in MSO by first defining a relation $\hat{R}_{R,B}(u,v) = (u,v) \in R \wedge u \in B \wedge v \in B$ capturing the allowed attacks and borrowing the following MSO-encoding for reachability [12]: $reach_R(x,y) = \forall X(x \in X \wedge [\forall u, v(u \in X \wedge R(u,v) \rightarrow v \in X)] \rightarrow y \in X)$. Finally we obtain $\Rightarrow^B_R (x,y) = reach_{\hat{R}_{R,B}}(x,y)$.

Definition 9. *For AF* $F = (A, R)$ *and sets* $D, S \subseteq A$ *we define:* $\Delta_{F,S}(D) = \{\, a \in A \mid \exists b \in S : b \neq a, (b,a) \in R, a \not\Rightarrow^{A \backslash D}_F b \,\}$. $\Delta_{F,S}$ *denotes the least fixed-point of* $\Delta_{F,S}(\cdot)$.

Example 8. Consider our example AF F and the set $\{b, d\}$. Towards the least fixed-point $\Delta_{F,\{b,d\}}$ first consider $\Delta_{F,\{b,d\}}(\emptyset)$. The arguments attacked by $\{b, d\}$ are a, c, e, but a and e having paths back to their attackers and thus $\Delta_{F,\{b,d\}}(\emptyset) = \{c\}$. Next consider $\Delta_{F,\{b,d\}}(\{c\})$. Still d attacks e but $e \not\Rightarrow^{A \backslash \{c\}}_F d$. Thus $\Delta_{F,\{b,d\}}(\{c\}) = \{c, e\}$ which is also the least fixed-point $\Delta_{F,\{b,d\}}$ of $\Delta_{F,\{b,d\}}(\cdot)$.

One can directly encode whether an argument x is in the operator $\Delta_{F,S}(D)$ by $\Delta_{A,R,S,D}(x) = x \in A \wedge \exists b \in S(b \neq x \wedge (b,x) \in R \wedge \neg \Rightarrow^{A \backslash D}_F (x,b))\}$ and thus also whether x is in the least fixed-point $\Delta_{F,S}$, by $\Delta_{A,R,S}(x) = \exists X \subseteq A(x \in X \wedge \forall a(a \in X \leftrightarrow \Delta_{A,R,S,X}(a)) \wedge \neg \exists Y \subset X(\forall b(b \in Y \leftrightarrow \Delta_{A,R,S,Y}(b))))$.

Definition 10. *For AF* F *we define the separation of* F *as* $[[F]] = \bigcup_{C \in SCCs(F)} F|_C$.

Example 9. To obtain the separation of our example AF we have to delete all attacks that are not within an single SCC. That is we simple remove the attack (b, c) and obtain the AF $(\{a, b, c, d, e\}, \{(a, b), (b, a), (c, d), (d, e), (e, c)\})$ as the separation of our example AF.

The attack relation of the separation of an AF (A, R) is given by $R_{[[(A,R)]]}(x, y) = x \in A \wedge y \in A \wedge (x, y) \in R \wedge \Rightarrow_R^A (y, x)$.

The following result provides an alternative characterization for *cf2* semantics.

Proposition 1 ([25]). *For any AF F, $S \in cf2(F)$ iff $S \in cf(F) \cap naive([[F - \Delta_{F,S}]])$.*

Example 10. For example consider the *cf2* extension $\{a, d\}$ of our running example. Clearly $\{a, d\} \in cf(F)$ and as illustrated before $\Delta_{F,\{b,d\}} = \{c, e\}$. We obtain $[[F - \Delta_{F,S}]] = (\{a, b, d\}, \{(a, b), (b, a)\})$ and thus also $\{a, d\} \in naive([[F - \Delta_{F,S}]])$.

Using the above Proposition we obtain the following MSO characterisation of *cf2*.

$$cf2(X) = cf_R(X) \wedge naive_{\hat{A}, R_{[[(\hat{A},R)]]}}(X) \quad \text{where} \quad \hat{A}(x) = x \in A \wedge \neg \Delta_{A,R,X}(x)$$

4   Algorithmic Implications

Most computational problems studied for AFs are computationally intractable (see, e.g., [19]), while the importance of efficient algorithms is evident. An approach to deal with intractable problems comes from parameterized complexity theory and is based on the fact, that many hard problems become polynomial-time tractable if some problem parameter is bounded by a fixed constant. In case the order of the polynomial bound is independent of the parameter one speaks of *fixed-parameter tractability* (FPT).

One popular parameter for graph-based problems is *treewidth* [9] which intuitively measures how tree-like a graph is. One weakness of treewidth is that it only captures sparse graphs. The parameter *clique-width* [17] generalizes treewidth, in the sense that each graph class of bounded treewidth has also bounded clique-width, but clique-width also captures a wide range of dense graphs.[1]

Both parameters have already been considered for abstract argumentation [19,20,21] and are closely related to MSO by means of meta-theorems. One such meta-theorem is due to [13] and shows that one can solve any graph problem that can be expressed in MSO_1 in linear time for graphs of clique-width bounded by some fixed constant k, when given together with a certain algebraic representation of the graph, a so called k-expression. A similar result is Courcelle's seminal meta-theorem [15,16] for MSO_2 and treewidth (which is also based on a certain structural decomposition of the graph, a so called tree-decomposition). Together with results from [10,30] stating that also k-expressions and tree-decompositions can be computed in linear time if k is bounded by a constant we get the following meta-theorem.

---

[1] As we do not make direct use of them, we omit the formal definitions of treewidth and clique-width here; the interested reader is referred to other sources [19,20]. We just note that these parameters are originally defined for undirected graphs, but can directly be used for AFs, as well.

Theorem 1. *For every fixed MSO formula $\phi(x_1, \ldots, x_i, X_1, \ldots X_j, X_1^E, \ldots X_l^E)$ and integer c, there is a linear-time algorithm that, given a graph (V, E) of treewidth $\leq c$, $v_k \in V$, $A_k \subseteq V$, and $B_k \subseteq E$ decides whether $(V, E) \models \phi(v_1, \ldots, v_i, A_1, \ldots A_j, B_1, \ldots B_l)$. If ϕ is in MSO_1, then this also holds for graphs of clique-width $\leq c$.*

The theorem can be extended to capture also counting and enumeration problems [2,14].

In the next theorem we give fixed-parameter tractability results w.r.t. the parameters treewidth and clique-width for the main reasoning problems in abstract argumentation.

Theorem 2. *For each argumentation semantics σ that is expressible in MSO, the following tasks are fixed-parameter tractable w.r.t. the treewidth of the given AF:*

- *Deciding whether an argument $a \in A$ is in at least one σ-extension (Credulous acceptance).*
- *Deciding whether an argument $a \in A$ is in each σ-extension (Skeptical acceptance).*
- *Verifying that a set $E \subseteq A$ is a σ-extension (Verification).*
- *Deciding whether there exists a σ-extension (Existence).*
- *Deciding whether there exists a non-empty σ-extension (Nonempty).*
- *Deciding whether there is a unique σ-extension (Unique).*

If σ is expressible in MSO_1, then the above tasks are also fixed-parameter tractable w.r.t. the clique-width of the AF.

Proof. The result follows by Theorem 1 and the following MSO-encodings: Credulous acceptance: $\phi_{\mathsf{Cred}}^\sigma(x) = \exists X \ (x \in X \wedge \sigma_R(X))$; Skeptical acceptance: $\phi_{\mathsf{Skept}}^\sigma(x) = \forall X \ (\sigma_R(X) \to x \in X)$; Verification: $\phi_{\mathsf{Ver}}^\sigma(X) = \sigma_R(X)$; Existence: $\phi_{\mathsf{Exists}}^\sigma = \exists X \sigma_R(X)$; Nonempty: $\phi_{\mathsf{Exists}\neg\emptyset}^\sigma = \exists X \exists x (\sigma_R(X) \wedge x \in X)$; and Unique: $\phi_U^\sigma = \exists X \sigma_R(X)) \wedge \neg \exists Y (Y \neq X \wedge \sigma_R(Y))$. We would like to note that these encodings do not use quantification over edge sets whenever σ is free of such a quantification. $\square$

MSO is also a gentle tool for studying the relation between different semantics, as illustrated by Theorem 3.

Theorem 3. *For any argumentation semantics σ, σ' expressible in MSO, the following tasks are fixed-parameter tractable w.r.t. the treewidth of the given AF.*

- *Deciding whether $\sigma(F) = \sigma'(F)$ (Coincidence).*
- *Deciding whether arguments skeptically accepted w.r.t. σ are also skeptically accepted w.r.t. σ' (Skepticism 1).*
- *Deciding whether arguments credulously accepted w.r.t. σ are also credulously accepted w.r.t. σ' (Skepticism 2).*
- *Deciding whether $\sigma(F) \subseteq \sigma'(F)$ (Skepticism 3).*

If σ is expressible in MSO_1 the above tasks are also fixed-parameter tractable w.r.t. the clique-width of the AF.

Proof. The result follows by Theorem 1 and the following MSO-encodings: Coincidence: $\phi^\sigma_{\text{Coin}}(x) = \forall X\,(\sigma_R(X) \leftrightarrow \sigma'_R(X))$; Skepticism 1: $\phi^\sigma_{\text{sk1}}(x) = \forall x(\phi^\sigma_{\text{Skept}}(x) \to \phi^{\sigma'}_{\text{Skept}}(x))$; Skepticism 2: $\phi^\sigma_{\text{sk1}}(x) = \forall x(\phi^\sigma_{\text{Cred}}(x) \to \phi^{\sigma'}_{\text{Cred}}(x))$; Skepticism 3: $\phi^\sigma_{\text{sk1}}(x) = \forall X(\sigma_{A,R}(X) \to \sigma'_{A,R}(X))$. $\qquad\qquad\square$

One prominent instantiation of the first problem mentioned in Theorem 3 is deciding whether an AF is coherent, i.e., whether stable and preferred extensions coincide.

Most of the characterizations we have provided so far are actually in MSO$_1$ and by the above results we obtain fixed-parameter tractability for treewidth and clique-width. The notable exception is the schema (1) we provided for the resolution-based semantics. There is no straight forward way to reduce this MSO$_2$ formula into MSO$_1$ (and thus providing complexity results in terms of clique-width) and in general it is unclear whether this is possible at all. Surprisingly, in the case of resolution-based grounded semantics one can get rid off the explicit quantification over sets of attacks as we show next.

5 An MSO$_1$-Characterization for grd^*

We provide a novel characterisation of resolution-based grounded semantics that avoids the quantification over sets of attacks in schema, as in (1), and thus yields an MSO$_1$-encoding. To this end we first restrict the class of resolutions we have to consider when showing that a given set is a complete extension of some resolved AF.

Lemma 1. *For each AF $F = (A, R)$ and $E \in grd^*(F)$, there exists a resolution β with $\{ (b, a) \mid a \in E, b \notin E, \{(a, b), (b, a)\} \subseteq R \} \subseteq \beta$ such that $E \in com(A, R \setminus \beta)$.*

Proof. As $E \in grd^*(F)$ we have that there exists a resolution β' such that $E \in grd(A, R \setminus \beta')$. Now let us define β as $\{ (b, a) \mid a \in E, \{(a, b), (b, a)\} \subseteq R \} \cup (\beta' \cap (A \setminus E \times A \setminus E))$. Clearly E is conflict-free in $(A, R \setminus \beta)$. Next we show that (i) $E^\oplus_{R\setminus\beta'} = E^\oplus_{R\setminus\beta}$ and (ii) $E^\ominus_{R\setminus\beta'} \supseteq E^\ominus_{R\setminus\beta}$.

For (i), let us first consider $b \in E^\oplus_{R\setminus\beta'}$. Then there exists $(a, b) \in R \setminus \beta'$ with $a \in E$ and by construction also $(a, b) \in R \setminus \beta$ and thus $b \in E^\oplus_{R\setminus\beta}$. Now let us consider $b \in E^\oplus_{R\setminus\beta}$. Then there exists $(a, b) \in R \setminus \beta$ with $a \in E$ and by construction either $(a, b) \in R \setminus \beta'$ or $(b, a) \in R \setminus \beta'$. In the first case clearly $b \in E^\oplus_{R\setminus\beta'}$. In the latter case b attacks E and as E is admissible in $(A, R \setminus \beta')$ there exists $c \in E$ such that $(c, b) \in R \setminus \beta'$, hence $b \in E^\oplus_{R\setminus\beta'}$. For (ii) consider $b \in E^\ominus_{R\setminus\beta}$, i.e., exists $a \in E$ such that $(b, a) \in R \setminus \beta$. By the construction of β we have that $(a, b) \notin R$ and therefore $(b, a) \in R \setminus \beta'$. Hence also $b \in E^\ominus_{R\setminus\beta'}$.

As $E \in adm(A, R \setminus \beta')$ we have that $E^\ominus_{R\setminus\beta'} \subseteq E^\oplus_{R\setminus\beta'}$ and by the above observations then also $E^\ominus_{R\setminus\beta} \subseteq E^\oplus_{R\setminus\beta}$. Thus E is an admissible set. Finally let us consider an argument $a \in A \setminus E^\oplus_{R\setminus\beta}$. In the construction of β the incident attacks of a are not effected and hence $\{a\}^\ominus_{R\setminus\beta'} = \{a\}^\ominus_{R\setminus\beta}$. That is E defends a in $(A, R \setminus \beta)$ iff E defends a in $(A, R \setminus \beta')$. Now as $E \in com(A, R \setminus \beta')$ we have that a is not defended and hence $E \in com(A, R \setminus \beta)$. $\qquad\qquad\square$

With this result at hand, we can give an alternative characterization for grd^*.

Lemma 2. *For each AF $F = (A, R)$ and $E \subseteq A$, $E \in grd^*(F)$ if and only if the following conditions hold:*

1. *there exists a resolution β with $\{ (b, a) \mid a \in E, \{(a, b), (b, a)\} \subseteq R \} \subseteq \beta$ and $E \in com(A, R \setminus \beta)$*
2. *E is $\subseteq$-minimal w.r.t. (1).*

Proof. Let us first recall that, by definition, the grounded extension is the $\subseteq$-minimal complete extension and hence $grd^* = com^*$.

$\Rightarrow$: Let $E \in grd^*(F)$. Then by Lemma 1, E fulfills condition (1). Further we have that each set E satisfying (1) is a complete extension of a resolved AF. As by definition E is $\subseteq$-minimal in the set of all complete extensions of all resolved AFs it is also minimal for those satisfying (1).

$\Leftarrow$: As E satisfies (1) it is a complete extension of a resolved AF. Now towards a contradiction let us assume it is not a resolution-based grounded extension. Then there exists $G \in grd^*(F)$ with $G \subset E$. But by Lemma 1 G fulfills condition (1) and thus $G \subset E$ contradicts (2). □

In the next step we look for an easier characterization of condition (1).

Lemma 3. *For each AF $F = (A, R)$ and $E \subseteq A$ the following statements are equivalent:*

1. *There exists a resolution β with $\{ (b, a) \mid a \in E, \{(a, b), (b, a)\} \subseteq R \} \subseteq \beta$ and $E \in com(A, R \setminus \beta)$.*
2. *$E \in com(A, R \setminus \{ (b, a) \mid a \in E, \{(a, b), (b, a)\} \subseteq R \})$ and $grd^*(A \setminus E_R^+, R \cap ((A \setminus E_R^+) \times (A \setminus E_R^+))) = \{\emptyset\}$.*

Proof. In the following we will use the shorthands $R^* = R \setminus \{ (b, a) \mid a \in E, \{(a, b), (b, a)\} \subseteq R \}$ and $(A', R') = (A \setminus E_R^+, R \cap ((A \setminus E_R^+) \times (A \setminus E_R^+)))$.

$(1) \Rightarrow (2)$: Consider a resolution β such that $E \in com(A, R \setminus \beta)$. We first show that then also $E \in com(A, R^*)$. By construction we have that for arbitrary $b \in A$ that (a) $E \rightarrowtail^R b$ iff $E \rightarrowtail^{R \setminus \beta} b$ iff $E \rightarrowtail^{R^*} b$, and (b) $b \rightarrowtail^{R \setminus \beta} E$ iff $b \rightarrowtail^{R^*} E$. Hence we have that (i) $E \in adm(A, R \setminus \beta)$ iff $E \in adm(A, R^*)$ and (ii) $E_R^+ = E_{R \setminus \beta}^+ = E_{R^*}^+$. By definition of complete semantics, $E \in com(A, R \setminus \beta)$ is equivalent to for each argument $b \in A \setminus E$ there exists an argument $c \in A$ such that $c \rightarrowtail^{R \setminus \beta} b$ and $E \not\rightarrowtail^{R \setminus \beta} c$. As $R^* \supseteq R \setminus \beta$ we obtain that $(c, b) \in R \setminus \beta$ implies $(c, b) \in R^*$. Using (a) we obtain that $E \in com(A, R \setminus \beta)$ implies for each argument $b \in A \setminus E$ existence of an argument $c \in A$ such that $(c, b) \in R^*$ and $E \not\rightarrowtail^{R^*} c$, i.e., $E \in com(A, R^*)$.

Now addressing $grd^*(A', R') = \{\emptyset\}$ we again use the assumption $E \in com(A, R \setminus \beta)$, i.e., each argument which is defended by E is already contained in E, we have that $grd(A \setminus E_{R \setminus \beta}^+, R \setminus \beta \cap ((A \setminus E_R^+) \times (A \setminus E_R^+))) = grd(A', R' \setminus \beta) = \{\emptyset\}$. Note that $\beta' = \beta \cap R'$ is a resolution of (A', R') and that $grd(A', R' \beta) = grd(A', R' \setminus \beta') = \{\emptyset\}$. We can conclude that $grd^*(A', R') = \{\emptyset\}$.

$(1) \Leftarrow (2)$: Consider $\beta' \in \gamma(F)$ s.t. $grd(A', R' \setminus \beta') = \{\emptyset\}$; such a β' exists since $grd^*(A', R') = \{\emptyset\}$. Now consider the resolution $\beta = \{ (b, a) \mid a \in E, \{(a, b), (b, a)\} \subseteq R \} \cup \beta'$. Again, by construction of β we have that for arbitrary $b \in A$: (a) $E \rightarrowtail^R b$ iff $E \rightarrowtail^{R \setminus \beta} b$ iff $E \rightarrowtail^{R^*} b$, and (b) $b \rightarrowtail^{R \setminus \beta} E$ iff

$b \rightarrowtail^{R^*} E$. Hence we obtain that $E \in adm(A, R \setminus \beta)$. Using $R = E_{R \setminus \beta}^+ = E_{R^*}^+$ we have $grd(A \setminus E_{R \setminus \beta}^+, (R \setminus \beta) \cap ((A \setminus E_R^+) \times (A \setminus E_R^+))) = grd(A', R' \setminus \beta') = \{\emptyset\}$. Thus, $E \in com(A, R \setminus \beta)$. $\qquad \square$

Finally we exploit a result from [3].

Proposition 2 ([3]). *For every AF $F = (A, R)$, $grd^*(F) = \{\emptyset\}$ iff for each minimal SCC S of F at least one one of the following conditions holds: (i) S contains a self-attacking argument; (ii) S contains a non-symmetric attack; and (iii) S contains an undirected cycle.*

Based on the above observations we obtain the following characterization of resolution-based grounded semantics.

Theorem 4. *For each AF $F = (A, R)$, the grd^*-extensions are the $\subseteq$-minimal sets $E \subseteq A$ such that:*

1. *$E \in com(A, R')$ with $R' = R \setminus \{ (b, a) \mid a \in E, \{(a, b), (b, a)\} \subseteq R \})$.*
2. *Each minimal SCC S of $\hat{F} = (A \setminus E_R^+, R \cap A \setminus E_R^+ \times A \setminus E_R^+)$ satisfies one of the following conditions: S contains a self-attacking argument; S contains a non-symmetric attack; or S contains an undirected cycle*

Proof. By Lemma 3, condition (1) in Lemma 2 is equivalent to $E \in com(A, R \setminus \{ (b, a) \mid a \in E, \{(a, b), (b, a)\} \subseteq R \})$ and $grd^*(A \setminus E_R^+, R \cap ((A \setminus E_R^+) \times (A \setminus E_R^+))) = \{\emptyset\}$. The former being condition (1) of the theorem. The latter, due to Proposition 2, is equivalent to condition (2) of the theorem. $\qquad \square$

Having Theorem 4 at hand we can build an $\mathrm{MSO_1}$-encoding as follows. First we encode the attack relation R' as $R'_E(x, y) = (x, y) \in R \wedge \neg(x \in E \wedge y \notin E \wedge (x, y) \in R \wedge (y, x) \in R)$. Then the AF $\hat{F} = (\hat{A}, \hat{R})$ is given by:

$$\hat{A}_{A,R,E}(x) = x \in A \wedge x \notin E \wedge \neg \exists y \in E : R'_E(y, x)$$

$$\hat{R}_{E,R}(x, y) = (x, y) \in R \wedge A_{A,R,E}^*(x) \wedge A_{A,R,E}^*(y)$$

Based on reachability we can easily specify whether arguments are strongly connected $SC_R(x, y) = reach_R(x, y) \wedge reach_R(y, x)$, and a predicate that captures all arguments in minimal SCCs $minSCC_{A,R}(x) = A(x) \wedge \neg \exists y (A(y) \wedge reach_R(y, x) \wedge \neg reach_R(x, y))$. It remains to encode the check for each SCC.

$$C1_R(x) = \exists y(SC_R(x, y) \wedge (y, y) \in R)$$
$$C2_R(x) = \exists y, z(SC_R(x, y) \wedge SC_R(x, z) \wedge (y, z) \in R \wedge (z, y) \notin R)$$
$$C3_R(x) = \exists X(\exists y \in X \wedge \forall y \in X[SC_R(x, y) \wedge$$
$$\exists u, v \in X : u \neq v \wedge (u, y) \in R \wedge (y, v) \in R])$$
$$C_R(x) = C1_R(x) \vee C2_R(x) \vee C3_R(x)$$

Finally using Theorem 4 we obtain an $\mathrm{MSO_1}$-encoding for resolution-based grounded semantics:

$$grd^*_{A,R}(X) = cand_{A,R}(X) \wedge \neg \exists Y (cand_{A,R}(Y) \wedge Y \subset X)$$

where $cand_{A,R}(X)$ stands for

$$com_{A,R'_X}(X) \wedge \forall x(minSCC_{\hat{A}_{A,R,E}, \hat{R}_{E,R}}(x) \rightarrow C_{\hat{R}_{E,R}}(x)).$$

6 Conclusion

In this paper we have shown that Monadic Second Order Logic (MSO) provides a suitable unifying framework for abstract argumentation. We encoded the most popular semantics within MSO and gave building blocks illustrating that MSO can naturally capture several concepts that are used for specifying semantics. This shows that MSO can be used as rapid prototyping tool for the development of new semantics.

Moreover, we gave a new characterisation of resolution-based grounded semantics that admits an MSO_1-encoding. This shows that reasoning in this semantics is tractable for frameworks of bounded clique-width. In fact, the collection of encodings we provided here shows that acceptance as well as other reasoning tasks are fixed-parameter tractable for several semantics w.r.t. the clique-width (hence also for treewidth).

For future work we suggest to study whether also other instantiations of the resolution-based semantics can be expressed in MSO_1 (recall that we provided already a schema for MSO_2-encodings). Moreover, it might be interesting to compare the performance of MSO tools with dedicated argumentation systems. Finally, we want to advocate the use of MSO for automated theorem discovery [31]. In fact, our encodings allow us to express meta-statements like "does it hold for AFs F that each σ-extension is also a σ'-extension." Although we have to face undecidability for such formulas, there is the possibility that MSO-theorem provers come up with a counter-model. Thus, in a somewhat similar way as Weydert [33], who used a First Order Logic encoding of complete semantics to show certain properties for semi-stable semantics of infinite AFs, MSO can possibly be used to support the argumentation researcher in obtaining new insights concerning the wide range of different argumentation semantics.

References

1. Amgoud, L., Devred, C.: Argumentation Frameworks as Constraint Satisfaction Problems. In: Benferhat, S., Grant, J. (eds.) SUM 2011. LNCS, vol. 6929, pp. 110–122. Springer, Heidelberg (2011)
2. Arnborg, S., Lagergren, J., Seese, D.: Easy problems for tree-decomposable graphs. J. Algorithms 12(2), 308–340 (1991)
3. Baroni, P., Dunne, P.E., Giacomin, M.: On the resolution-based family of abstract argumentation semantics and its grounded instance. Artif. Intell. 175(3-4), 791–813 (2011)
4. Baroni, P., Giacomin, M.: Semantics of abstract argument systems. In: Rahwan, I., Simari, G. (eds.) Argumentation in Artificial Intelligence, pp. 25–44. Springer (2009)
5. Baroni, P., Giacomin, M., Guida, G.: SCC-recursiveness: A general schema for argumentation semantics. Artif. Intell. 168(1-2), 162–210 (2005)
6. Bench-Capon, T.J.M., Dunne, P.E.: Argumentation in artificial intelligence. Artif. Intell. 171(10-15), 619–641 (2007)
7. Besnard, P., Doutre, S.: Checking the acceptability of a set of arguments. In: Delgrande, J.P., Schaub, T. (eds.) Proc. NMR, pp. 59–64 (2004)
8. Bistarelli, S., Campli, P., Santini, F.: Finding partitions of arguments with Dung's properties via SCSPs. In: Chu, W.C., Wong, W.E., Palakal, M.J., Hung, C.C. (eds.) Proc. SAC, pp. 913–919. ACM (2011)
9. Bodlaender, H.L.: A tourist guide through treewidth. Acta Cybernetica 11, 1–21 (1993)
10. Bodlaender, H.L.: A linear-time algorithm for finding tree-decompositions of small treewidth. SIAM J. Comput. 25(6), 1305–1317 (1996)

11. Caminada, M., Gabbay, D.M.: A logical account of formal argumentation. Studia Logica 93(2), 109–145 (2009)
12. Courcelle, B., Engelfriet, J.: Graph structure and monadic second-order logic, a language theoretic approach, vol. 2012. Cambridge University Press (2011)
13. Courcelle, B., Makowsky, J.A., Rotics, U.: Linear time solvable optimization problems on graphs of bounded clique-width. Theory Comput. Syst. 33(2), 125–150 (2000)
14. Courcelle, B., Makowsky, J.A., Rotics, U.: On the fixed parameter complexity of graph enumeration problems definable in monadic second-order logic. Discr. Appl. Math. 108(1-2), 23–52 (2001)
15. Courcelle, B.: Recognizability and second-order definability for sets of finite graphs. Tech. Rep. I-8634, Université de Bordeaux (1987)
16. Courcelle, B.: Graph rewriting: an algebraic and logic approach. In: Handbook of Theoretical Computer Science, vol. B, pp. 193–242. Elsevier, Amsterdam (1990)
17. Courcelle, B., Engelfriet, J., Rozenberg, G.: Context-free Handle-rewriting Hypergraph Grammars. In: Ehrig, H., Kreowski, H.-J., Rozenberg, G. (eds.) Graph Grammars 1990. LNCS, vol. 532, pp. 253–268. Springer, Heidelberg (1991)
18. Dung, P.M.: On the acceptability of arguments and its fundamental role in nonmonotonic reasoning, logic programming and n-person games. Artif. Intell. 77(2), 321–358 (1995)
19. Dunne, P.E.: Computational properties of argument systems satisfying graph-theoretic constraints. Artif. Intell. 171(10-15), 701–729 (2007)
20. Dvořák, W., Szeider, S., Woltran, S.: Reasoning in argumentation frameworks of bounded clique-width. In: Baroni, P., Cerutti, F., Giacomin, M., Simari, G.R. (eds.) Proc. COMMA. FAIA, vol. 216, pp. 219–230. IOS Press (2010)
21. Dvořák, W., Pichler, R., Woltran, S.: Towards fixed-parameter tractable algorithms for abstract argumentation. Artif. Intell. 186, 1–37 (2012)
22. Dvořák, W., Woltran, S.: Complexity of semi-stable and stage semantics in argumentation frameworks. Inf. Process. Lett. 110(11), 425–430 (2010)
23. Egly, U., Gaggl, S.A., Woltran, S.: Answer-set programming encodings for argumentation frameworks. Argument and Computation 1(2), 147–177 (2010)
24. Egly, U., Woltran, S.: Reasoning in argumentation frameworks using quantified boolean formulas. In: Dunne, P.E., Bench-Capon, T.J.M. (eds.) Proc. COMMA. FAIA, vol. 144, pp. 133–144. IOS Press (2006)
25. Gaggl, S.A., Woltran, S.: cf2 semantics revisited. In: Baroni, P., Cerutti, F., Giacomin, M., Simari, G.R. (eds.) Proc. COMMA. FAIA, vol. 216, pp. 243–254. IOS Press (2010)
26. Gottlob, G., Pichler, R., Wei, F.: Bounded treewidth as a key to tractability of knowledge representation and reasoning. Artif. Intell. 174(1), 105–132 (2010)
27. Gottlob, G., Szeider, S.: Fixed-parameter algorithms for artificial intelligence, constraint satisfaction, and database problems. The Computer Journal 51(3), 303–325 (2006)
28. Kneis, J., Langer, A., Rossmanith, P.: Courcelle's theorem - a game-theoretic approach. Discrete Optimization 8(4), 568–594 (2011)
29. Langer, A., Reidl, F., Rossmanith, P., Sikdar, S.: Evaluation of an MSO-solver. In: Proc. of ALENEX 2012, pp. 55–63. SIAM (2012)
30. Oum, S., Seymour, P.D.: Approximating clique-width and branch-width. J. Comb. Theory, Ser. B 96(4), 514–528 (2006)
31. Tang, P., Lin, F.: Discovering theorems in game theory: Two-person games with unique pure nash equilibrium payoffs. Artif. Intell. 175(14-15), 2010–2020 (2011)
32. Toni, F., Sergot, M.: Argumentation and Answer Set Programming. In: Balduccini, M., Son, T.C. (eds.) Gelfond Feschrift. LNCS (LNAI), vol. 6565, pp. 164–180. Springer, Heidelberg (2011)
33. Weydert, E.: Semi-stable extensions for infinite frameworks. In: Proc. BNAIC, pp. 336–343 (2011)

An Approach to Argumentation Considering Attacks through Time

Maximiliano C.D. Budán[1,2,3], Mauro Gómez Lucero[1,2],
Carlos I. Chesñevar[1,2], and Guillermo R. Simari[2]

[1] Argentine National Council of Scientific and Technical Research (CONICET)
[2] AI R&D Lab (LIDIA) – Universidad Nacional del Sur in Bahía Blanca
[3] Universidad Nacional de Santiago del Estero
{mcdb,mjg,cic,grs}@cs.uns.edu.ar

Abstract. In the last decade, several argument-based formalisms have emerged, with application in many areas, such as legal reasoning, autonomous agents and multi-agent systems; many are based on Dung's seminal work characterizing Abstract Argumentation Frameworks (AF). Recent research in the area has led to Temporal Argumentation Frameworks (TAF), that extend AF by considering the temporal availability of arguments. A new framework was introduced in subsequent research, called Extended Temporal Argumentation Framework (E-TAF), extending TAF with the capability of modeling the availability of attacks among arguments. E-TAF is powerful enough to model different time-dependent properties associated with arguments; moreover, we will present an instantiation of the abstract framework E-TAF on an extension of Defeasible Logic Programming (DeLP) incorporating the representation of temporal availability and strength factors of arguments varying over time, associating these characteristics with the language of DeLP. The strength factors are used to model different more concrete measures such as reliability, priorities, etc.; the information is propagated to the level of arguments, then the E-TAF definitions are applied establishing their temporal acceptability.

Keyword: Argumentation, Temporal Argumentation, Defeasible Logic Programming, Argument and Computation.

1 Introduction

Argumentation represents a powerful paradigm to formalize commonsense reasoning. In a general sense, argumentation can be defined as the study of the interaction of arguments for and against conclusions, with the purpose of determining which conclusions are acceptable [6,21]. Several argument-based formalisms have emerged finding application in building autonomous agents and multi-agent systems. An agent may use argumentation to perform individual reasoning to resolve conflicting evidence or to decide between conflicting goals [2,5]; Multiple agents may also use dialectical argumentation to identify and reconcile differences between themselves, through interactions such as negotiation, persuasion, and joint deliberation [18,22,20].

E. Hüllermeier et al. (Eds.): SUM 2012, LNAI 7520, pp. 99–112, 2012.
© Springer-Verlag Berlin Heidelberg 2012

Reasoning about time is a central issue in commonsense reasoning, thus becoming a valuable feature when modeling argumentation capabilities for intelligent agents [3,15]. Recent research has introduced Temporal Argumentation Frameworks (TAF) extending Dung's AF with the consideration of argument's temporal availability [10,11]. In TAF, arguments are valid only during specific time intervals (called *availability intervals*). Thus, the set of acceptable arguments associated with a TAF may vary over time. Even though arguments in TAF are only available on certain time intervals, their attacks are assumed to be static and permanent over these intervals.

Recently, in [9] a novel framework, called Extended Temporal Argumentation Framework (E-TAF) was introduced, enriching a TAF with the capability of modeling the availability of attacks among arguments. This additional feature of E-TAF permits to model strength of arguments varying over time, *i.e.*, an attack can be only available in a given time interval signifying that the attacking argument is stronger than the attacked one on this attack interval. The notion of argument strength is a generalization of different possible measures for comparing arguments, such as reliability, priorities, etc.

In this work, to provide a concrete, fully specified (non-abstract) knowledge representation and reasoning formalism. We present an instantiation of the abstract framework E-TAF based on the argumentation formalism *Defeasible Logic Programming (DeLP)*, a logic programming approach to argumentation that has proven to be successful for real-world applications (*e.g.*, [13,5,7]). This instantiation, called ST-DeLP, incorporates the representation of temporal availability and strength factors varying over time associated with the elements of the language of DeLP, following a different intuition from the one presented in [17]. It also specifies how arguments are built, and how availability and strength of arguments are obtained from the corresponding information attached to the language elements from which are built. After determining the availability of attacks by comparing strength of conflicting arguments over time, E-TAF definitions are applied to establish temporal acceptability of arguments. Thus, the main contribution of this paper lies on the integration of time and strength in the context of argumentation systems.

2   Abstract Argumentation

We will summarize the abstract argumentation framework introduced in Dung's seminal work [12]; the reader is directed to that reference for a complete presentation. To simplify the representation and analysis of pieces of knowledge, Dung introduced the notion of *Argumentation Framework (AF)* as a convenient abstraction of a defeasible argumentation system. In the *AF*, an argument is considered as an abstract entity with unspecified internal structure, and its role in the framework is completely determined by the relation of attack it maintains with other arguments.

Definition 1 (Argumentation Framework [12]). *An argumentation framework (AF) is a pair $\langle AR, Attacks \rangle$, where AR is a set of arguments, and Attacks is a binary relation on AR, i.e., $Attacks \subseteq AR \times AR$.*

Given an *AF*, an argument A is considered *acceptable* if it can be defended by arguments in AR of all the arguments in AR that attack it (attackers). This intuition is formalized in the following definitions, originally presented in [12].

Definition 2 (Acceptability). *Let $AF = \langle AR, Attacks \rangle$ be an argumentation framework.*

- *A set $S \subseteq AR$ is called* conflict-free *if there are no arguments $A, B \in S$ such that $(A, B) \in Attacks$.*
- *An argument $A \in AR$ is* acceptable *with respect to a set $S \subseteq AR$ iff for each $B \in AR$, if B attacks A then there is $C \in S$ such that $(C, B) \in Attacks$; in such case it is said that B is attacked by S.*
- *A conflict-free set $S \subseteq AR$ is* admissible *iff each argument in S is acceptable with respect to S.*
- *An admissible set $E \subseteq AR$ is a* complete extension *of AF iff E contains each argument that is acceptable with respect to E.*
- *A set $E \subseteq AR$ is the* grounded extension *of AF iff E is a complete extension that is minimal with respect to set inclusion.*

Dung [12] also presented a fixed-point characterization of the grounded semantics based on the characteristic function F defined below.

Definition 3. *Let $\langle AR, Attacks \rangle$ be an AF. The associated characteristic function $F : 2^{AR} \to 2^{AR}$, is $F(S) =_{def} \{A \in AR \mid A$ is acceptable w.r.t. $S\}$.*

The following proposition suggests how to compute the grounded extension associated with a *finitary AF* (*i.e.*, such that each argument is attacked by at most a finite number of arguments) by iteratively applying the characteristic function starting from $\emptyset$. See [4,16] for details on semantics of *AF*s.

Proposition 1 ([12]). *Let $\langle AR, Attacks \rangle$ be a finitary AF. Let $i \in \mathbb{N} \cup \{0\}$ such that $F^i(\emptyset) = F^{i+1}(\emptyset)$. Then $F^i(\emptyset)$ is the least fixed point of F, and corresponds to the grounded extension associated with the AF.*

3   Defeasible Logic Programming

Defeasible Logic Programming (DeLP), is a formalism that combines results of *Logic Programming* and *Defeasible Argumentation*. DeLP provides representational elements able to represent information in the form of strict and weak rules in a declarative way, from which arguments supporting conclusions can be constructed, providing a defeasible argumentation inference mechanism for obtaining the *warranted* conclusions. The defeasible argumentation characteristics of DeLP supplies means for building applications dealing with incomplete and contradictory information in real world, dynamic domains. Thus, the resulting approach is suitable for representing agents' knowledge and for providing an argumentation based reasoning mechanism to these agents.

Below we present the essential definitions of DeLP, see [14] for full details.

Definition 4 (DeLP program). *A DeLP program $\mathcal{P}$ is a pair (Π, Δ) where (1) Δ is a set of defeasible rules of the form $L \prec P_1, \ldots, P_n$, with $n > 0$, and (2) Π is a set of strict rules of the form $L \leftarrow P_1, \ldots, P_n$, with $n \geq 0$. In both cases L and each P_i are literals, i.e., a ground atom A or a negated ground atom $\sim A$, where '$\sim$' represents the strong negation.*

Pragmatically, strict rules can be used to represent strict (non defeasible) information, while defeasible rules are used to represent tentative or weak information. It is important to remark that the set Π must be consistent as it represents strict (undisputed) information.

Definition 5 (Defeasible derivation). *Let $\mathcal{P}$ be a DeLP program and L a ground literal. A defeasible derivation of L from $\mathcal{P}$ consists of a finite sequence $L_1, \ldots, L_n = L$ of ground literals, such that for each i, $1 \leq i \leq n$, L_i is a fact or there exists a rule R_i in $\mathcal{P}$ (strict or defeasible) with head L_i and body $B_1, \ldots, B_m$, such that each literal on the body of the rule is an element L_j of the sequence appearing before L_i ($j \leq i$). We will use $\mathcal{P} \mid\!\sim L$ to denote that there exists a defeasible derivation of L from $\mathcal{P}$.*

We say that a given set S of DeLP clauses is contradictory if and only if $S \mid\!\sim L$ and $S \mid\!\sim \sim L$ for some literal L.

Definition 6 (Argument). *Let L be a literal and $\mathcal{P} = (\Pi, \Delta)$ be a DeLP program. An argument for L is a pair $\langle A, L \rangle$, where A is a minimal (w.r.t. set inclusion), non contradictory set of defeasible rules of Δ, such that $A \mid\!\sim L$. We say that an argument $\langle B, L \rangle$ is a sub-argument of $\langle A, L \rangle$ iff $B \subseteq A$.*

DeLP provides an argumentation based mechanism to determine *warranted* conclusions. This procedure involves constructing arguments from programs, identifying conflicts or *attacks* among arguments, evaluating pairs of arguments in conflict to determine if the attack is successful, becoming a *defeat*, and finally analyzing defeat interaction among all relevant arguments to determine warrant [14].

Definition 7 (Disagreement). *Let $\mathcal{P} = (\Pi, \Delta)$ be a DeLP program. Two literals L and L' are in disagreement if and only if the set $\Pi \cup \{L, L'\}$ is contradictory.*

Definition 8 (Attack). *Let $\mathcal{P} = (\Pi, \Delta)$ be a DeLP program. Let $\langle A_1, L_1 \rangle$ and $\langle A_2, L_2 \rangle$ be two arguments in $\mathcal{P}$. We say that $\langle A_1, L_1 \rangle$ counter-argues, rebuts, or attacks $\langle A_2, L_2 \rangle$ at the literal L if and only if there is a sub-argument $\langle A, L \rangle$ of $\langle A_2, L_2 \rangle$ such that L and L_1 are in disagreement. The argument $\langle A, L \rangle$ is called disagreement sub-argument, and the literal L will be the counter-argument point.*

In this work, a complete presentation of the inference mechanism of DeLP is not necessary since our formalization will be based on an extension of Dung's approach to argumentation semantics.

4   Modeling Temporal Argumentation with TAF

A *Timed Abstract Framework* (TAF) [10,11] is a recent extension of Dung's formalism where arguments are active only during specific intervals of time; this intervals are called availability intervals. Attacks between arguments are considered *only* when both the attacker and the attacked arguments are available. Thus, when identifying the set of acceptable arguments the outcome associated with a TAF may vary in time.

To represent time we assume that a correspondence was defined between the time line and the set $\mathbb{R}$ of real numbers. A *time interval*, representing a period of time without interruptions, will be represented as a real interval $[a - b]$ (we use '−' instead of ',' as a separator for readability reasons). To indicate that one of the endpoints (extremes) of the interval is to be excluded, following the notation for real intervals, the corresponding square bracket will be replaced with a parenthesis, *e.g.*, $(a - b]$ to exclude the endpoint a.

To model discontinuous periods of time we introduce the notion of *time intervals set*. Although a time intervals set suggests a representation as a set of sets (set of intervals), we chose a flattened representation as a set of reals (the set of all real numbers contained in any of the individual time intervals). In this way, we can directly apply traditional set operations and relations on time intervals sets.

Definition 9 (Time Intervals Set). *A* time intervals set *is a subset $S \subseteq \mathbb{R}$.*

When convenient we will use the set of sets notation for time intervals sets; that is, a time interval set $S \subseteq \mathbb{R}$ will be denoted as the set of all disjoint and $\subseteq$-maximal individual intervals included in the set. For instance, we will use $\{(1 - 3], [4.5 - 8)\}$ to denote the time interval set $(1 - 3] \cup [4.5 - 8)$

Now we formally introduce the notion of *Timed Argumentation Framework*, which extends the *AF* of Dung by adding the availability function. This additional component will be used to capture those time intervals where arguments are available.

Definition 10 (Timed Argumentation Framework). *A timed argumentation framework (or TAF) is a 3-tuple $\langle AR, Attacks, Av \rangle$ where AR is a set of arguments, Attacks is a binary relation defined over AR and Av is an availability function for timed arguments, defined as $Av : AR \longrightarrow \wp(\mathbb{R})$, such that $Av(A)$ is the set of availability intervals of an argument A.*

Example 1. *Consider the TAF $\Phi = \langle AR, Attacks, Av \rangle$ where:*
$AR = \{A, B, C, D, E, F, G\}$
$Attacks = \{(B, A), (C, B), (E, A), (G, E), (F, G), (G, D)\}$
$Av = \{(A, \{[10 - 50], [80 - 120]\}); (B, \{[55 - 100]\}); (C, \{[40 - 90]\}); (D, \{[10 - 30]\}); (E, \{[20 - 75]\}); (F, \{[5 - 30]\}); (G, \{[10 - 40]\})\}$ *(See Fig. 1)*

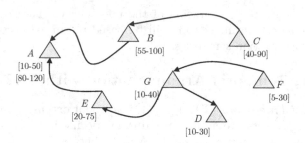

Fig. 1. TAF corresponding to example 1

The following definitions formalize argument acceptability in TAF, and are extensions of the acceptability notions presented in section 2 for AF. Firstly, we present the notion of timed argument profile, t-profile, that binds an argument to a set of time intervals; these profiles constitute a fundamental component for the formalization of time-based acceptability.

Definition 11 (T-Profile). *Let $\Phi = \langle AR, Attacks, Av \rangle$ be a TAF. A timed argument profile in Φ, or just t-profile, is a pair $\rho = (A, \tau)$ where $A \in AR$ and τ is a set of time intervals; $(A, Av(A))$ is called the* basic *t-profile of A.*

Since the availability of arguments varies in time, the acceptability of a given argument A will also vary in time. The following definitions extend Dung's original formalization for abstract argumentation by considering t-profiles instead of arguments.

Definition 12 (Defense of A from B w.r.t. S). *Let A and B be arguments. Let S be a set of t-profiles. The defense t-profile of A from B w.r.t. S is $\rho_A = (A, \tau_A^B)$, where: $\tau_A^B =_{def} (Av(A) - Av(B)) \bigcup_{\{(C,\tau_C) \in S \mid C \ Attacks \ B\}} (Av(A) \cap Av(B) \cap \tau_C)$.*

Intuitively, A is defended from the attack of B when B is not available $(Av(A) - Av(B))$, but also in those intervals where, although the attacker B is available, B is in turn attacked by an argument C in the base set S of t-profiles. The following definition captures the defense profile of A, but considering all its attacking arguments.

Definition 13 (Acceptable t-profile of A w.r.t. S). *Consider a set S of t-profiles. The acceptable t-profile for A w.r.t. a set S is $\rho_A = (A, \tau_A)$, where $\tau_A =_{def} \bigcap_{\{B \ Attacks \ A\}} \tau_A^B$ and (A, τ_A^B) is the defense t-profile of A from B w.r.t. S.*

Since an argument must be defended against all its attackers that are considered acceptable, we have to intersect the set of time intervals in which it is defended of each of its attackers.

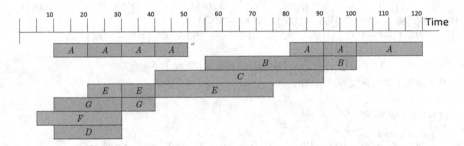

Fig. 2. Representation of the arguments associated with Ex. 2 in a time line

Definition 14 (Acceptability). *Let* $AF = \langle AR, Attacks, Av \rangle$ *be a temporal argumentation framework.*

- *A set S of t-profiles is called* t-conflict-free *if there are no t-profiles (A, τ_A), $(B, \tau_B) \in S$ such that $(A, B) \in Attacks$ and $\tau_A \cap \tau_B \neq \emptyset$.*
- *A t-conflict-free set S of t-profiles is a* t-admissible *set iff for all $(A, \tau_A) \in S$ it holds that (A, τ_A) is the acceptable t-profile of A w.r.t. S.*
- *A t-admissible set S is a* t-complete *extension of TAF iff S contains all the t-profiles that are acceptable with respect to S.*
- *A set S is the* t-grounded *extension of TAF iff S is t-complete and minimal with respect to set inclusion.*

In particular, the fixed point characterization for grounded semantics proposed by Dung can be directly applied to TAF by considering the following modified version of the characteristic function.

Definition 15. *Let $\langle AR, Attacks, Av \rangle$ be a TAF. Let S be a set of t-profiles. The associated characteristic function is defined as follows: $F(S) =_{def} \{(A, \tau) \mid A \in AR$ and (A, τ) is the acceptable t-profile of A w.r.t. $S\}$.*

Example 2. *Suppose we want to establish the acceptability of A in the TAF Φ presented in example 1. Let us obtain the t-grounded extension of Φ by applying the fixed point characterization.*

$F^0(\emptyset) = \emptyset$
$F^1(\emptyset) = \{(A, \{[10 - 20), (100 - 120]\}); (C, \{[40 - 90]\}); (F, \{[5 - 30]\}); (B, \{(90 - 100]\});$
$(E, \{(40 - 75]\}); (G, \{(30 - 40]\})\}$
$F^2(\emptyset) = \{(A, \{[10 - 40], [80 - 90], (100 - 120]\}); (C, \{[40 - 90]\}); (F, \{[5 - 30]\});$
$(B, \{(90 - 100]\}); (E, \{[20 - 30], (40 - 75]\}); (G, \{(30 - 40]\})\}$
$F^3(\emptyset) = \{(A, \{[10 - 20), (30 - 40), [80 - 90], (100 - 120]\}); (C, \{[40 - 90]\});$
$(F, \{[5 - 30]\}); (B, \{(90 - 100]\}); (E, \{[20 - 30], (40 - 75]\}); (G, \{(30 - 40]\})\}$
$F^4(\emptyset) = F^3(\emptyset)$
Consequently, $F^3(\emptyset)$ is the t-grounded extension of Φ. Next we describe how the temporal availability of A in $F^3(\emptyset)$ was obtained from $F^2(\emptyset)$. By applying definition 12:
$\tau_A^B = (Av(A) - Av(B)) \bigcup_{\{(C, \tau_C)\}} (Av(A) \cap Av(B) \cap \tau_C) =$

$$= \qquad (\{[10 \quad - \quad 50], [80 \quad - \quad 120]\} \quad - \quad \{[55 \quad - \quad 100]\}) \cup$$
$$(\{[10-50],[80-120]\} \cap \{[55-100]\} \cap \{[40-90]\}) =$$
$$= \{[10-50],(100-120]\} \cup [80-90] = \{[10-50],[80-90],(100-120]\}$$

$$\tau_A^E = (Av(A) - Av(E)) \bigcup\nolimits_{\{(G,\tau_G)\}} (Av(A) \cap Av(B) \cap \tau_G) =$$
$$= \{[10-20),(30-40],[80-120]\}$$

By applying definition 13:
$$\tau_A = \cap_{\{X \ Attacks \ A\}} \tau_A^X = \tau_A^B \cap \tau_A^E =$$
$$= \{[10-50],[80-90],(100-120]\} \cap \{[10-20),(30-40],[80-120]\} =$$
$$= \{[10-20),(30-40],[80-90],(100-120]\}$$

5   E-TAF: A TAF Extension with Time Intervals for Attacks

In this section we present E-TAF [9], an extension of TAF that takes in consideration not only the availability of the arguments but also looks into the availability of attacks. Adding time intervals to attacks is a meaningful extension for several domains; consider for example the notion of *statute of limitations* common in the law of many countries. A statute of limitations is an enactment in a common law legal system that sets the maximum time after an event that legal proceedings based on that event may be initiated. One reason for having a statute of limitations is that over time evidence can be corrupted or disappear; thus, the best time to bring a lawsuit is while the evidence is still acceptable and as close as possible to the alleged illegal behavior. Consider the following situation: (1) *John has left debts unpaid in Alabama, US, during 2008,* (2) *He has canceled them in 2009, but paying with counterfeited US dollars, committing fraud,* (3) *This fraud was detected on Jan 1, 2010.* A possible argument exchange for prosecuting John could be as follows:

- Arg_1: (Plaintiff) John left debts unpaid in Alabama in 2008 [Jan 1, 2008-+∞)
- Arg_2: (Defendant) John paid all his debts in Alabama for 2008 [Jan 1,2009-+∞)
- Arg_3: (Plaintiff) John did not cancel his debts in Alabama for 2008, as he paid them with counterfeited US dollars, committing fraud [Jan 1,2010-+∞)

According to the statute of limitations for Alabama,[1] the attack from Arg_3 to Arg_2 would be valid only until Jan 1, 2012 (for 2 years from the moment it was discovered). Note that Arg_3 is valid by itself (as the fraud was committed anyway), but the statute of limitations imposes a time-out on the attack relationship between arguments Arg_3 and Arg_2. Thus, John would be not guilty of committing fraud if the dialogue would have taken place in 2012, as the attack from Arg_3 to Arg_2 would not apply.

Next we formalize the definition of *extended* TAF, which provides the elements required to capture timed attacks between timed arguments.

---

[1] The statute of limitations may vary in different countries; for the case of the U.S. see e.g. www.statuteoflimitations.net/fraud.html

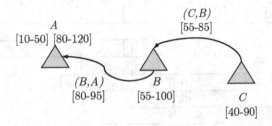

Fig. 3. E-TAF: example

Definition 16 (Extended TAF). *An extended timed abstract argumentation framework (or simply E-TAF) is a 4-tuple* $\langle AR, Attacks, ARAv, ATAv \rangle$ *where:*

– *AR is a set of arguments,*
– *Attacks is a binary relation defined over AR,*
– $ARAv : AR \longrightarrow \wp(\mathbb{R})$ *is the availability function for timed arguments, and*
– $ATAv : Attacks \longrightarrow \wp(\mathbb{R})$ *is the availability function for timed attacks, where* $ATAv((A, B)) \subseteq ARAv(A) \cap ARAv(B)$.

The condition $ATAv((A, B)) \subseteq ARAv(A) \cap ARAv(B)$ ensures that the availability of the attack cannot exceed the availability of the arguments involved.

Example 3. *E-TAF*= $\langle AR, Attacks, ARAv, ATAv \rangle$
$AR = \{A, B, C\}$
$Attacks = \{(B, A); (C, B)\}$
$ARAv = \{(A, \{[10 - 50], [80 - 120]\}); (B, \{[55 - 100]\}); (C, \{[40 - 90]\})\}$
$ATAv = \{((B, A), [80 - 95]); ((C, B), [55 - 85])\}$ *(See Fig. 3)*

The following definitions are extensions of the definitions 12 and 13, taking into account the availability of attacks.

Definition 17 (Defense t-profile of A from B). *Let S be a set of t-profiles. Let A and B be arguments. The defense t-profile of A from B w.r.t. S is* $\rho_A = (A, \tau_A^B)$, *where* $\tau_A^B = [ARAv(A) - ATAv((B, A))] \cup$
$\bigcup_{\{(C, \tau_C) \in S \ | \ C \ Attacks \ B\}} (ARAv(A) \cap ATAv((B, A)) \cap ATAv((C, B)) \cap \tau_C)$.

The notion of acceptable t-profile of A w.r.t. S remains unchanged in E-TAF with respect to the corresponding definition in TAF.

Definition 18 (Acceptable t-profile of A). *Let* $\langle AR, Attacks, ARAv, ATAv \rangle$ *be an E-TAF. Let S be a set of t-profiles. The acceptable t-profile for A w.r.t. S is* $\rho_A = (A, \tau_A)$, *where* $\tau_A = \cap_{\{B \ Attacks \ A\}} \tau_A^B$ *and* (A, τ_A^B) *is the defense t-profile of A from B w.r.t. S.*

The formalization of acceptability for TAF directly applies to E-TAF, except for the conflict-free notion which has to be recast as shown in the next definition.

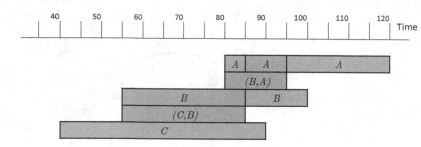

Fig. 4. Representation of the temporal attacks relations

Definition 19 (Conflict-freeness.   Characteristic   function).   *Let* $\langle AR, Attacks, ARAv, ATAv \rangle$ *be an E-TAF. A set S of t-profiles is called conflict-free if there are no t-profiles* (A, τ_A), $(B, \tau_B) \in S$ *such that* $(A, B) \in Attacks$ *and* $\tau_A \cap \tau_B \cap ATAv((B, A)) \neq \emptyset$. *The associated characteristic function for* $\langle AR, Attacks, ARAv, ATAv \rangle$ *is defined as follows:*
$$F(S) =_{def} \{(A, \tau) \mid A \in AR \text{ and } (A, \tau) \text{ is the acceptable t-profile of } A \text{ w.r.t. } S\}.$$

Example 4. *Suppose we want to establish the acceptability of A in the E-TAF in example 3. In this case, for simplicity, we will restrict the temporal availability of A to the interval* [80 − 120]. *Let us obtain the t-grounded extension of E-TAF by applying the fixed point characterization:*

$$F^0(\emptyset) = \emptyset$$
$$F^1(\emptyset) = \{(A, \{(95 - 120]\}); (C, \{[40 - 90]\}); (B, \{(85 - 100]\})\}$$
$$F^2(\emptyset) = \{(A, \{[80 - 85], (95 - 120]\}); (C, \{[40 - 90]\}); (B, \{(85 - 100]\})\}$$
$$F^3(\emptyset) = F^2(\emptyset)$$

Consequently, $F^3(\emptyset)$ *is the t-grounded extension of the E-TAF. Next we describe how the temporal availability of A was obtained in* $F^3(\emptyset)$ *by applying the definitions 17 and 18 from* $F^2(\emptyset)$. *By applying definition 17, we get:*
$$\tau_A^B \quad = \quad (ARAv(A) \quad - \quad ATAv(B, A)) \bigcup_{\{(C, \tau_C)\}} (ARAv(A)$$
$$\cap ATAv((B, A)) \cap ATAv((C, B)) \cap \tau_C) =$$
$$= (\{[80 - 120]\} - \{[80 - 95]\}) -_{\{(C, \tau_C)\}} (\{[80 - 120]\} \cap \{[80 - 95]\} \cap \{[55 - 85]\} \cap$$
$$\{[40 - 90]\}) = \{(95 - 120]\} \cup \{[80 - 85]\} = \{[80 - 85], (95 - 120]\}$$

By applying definition 18: $\tau_A = \cap_{\{X \text{ Attacks } A\}} \tau_A^X$, *where* $\tau_A^B = \{[80 - 85], (95 - 120]\}$

6   Temporal Availability and Strength variation on DeLP

In this section we present ST-DeLP, an instantiation of the abstract framework E-TAF based on the rule-based argumentation framework DeLP. This instantiation incorporates the ability to represent temporal availability and strength factors varying over time, associated with rules composing arguments. This information is then propagated to the level of arguments, and will be used to define temporal availability of attacks in E-TAF.

This association of temporal and strength information to DeLP clauses is formalized through the definition of ST-program, presented below.

Definition 20 (ST-program). *A ST-program $\mathcal{P}$ is a set of clauses of the form (γ, τ, υ), called ST-clauses, where: (1) γ is a DeLP clause, (2) τ is a set of time intervals for a ST-clause, and (3) $\upsilon : \mathbb{R} \longrightarrow [0, 1]$ is a function that determines the strength factor for a ST-clause.*

We will say that (γ, τ, υ) is a strict (defeasible) ST-clause iff γ is a strict (defeasible) DeLP clause. Then, given a ST-program $\mathcal{P}$ we will distinguish the subset Π of strict ST-clauses, and the subset Δ of defeasible ST-clauses.

Next we will introduce the notion of argument and sub-argument in ST-DeLP. Informally, an argument A is a tentative proof (as it relies on information with different strength) from a consistent set of clauses, supporting a given conclusion Q, and specifying its strength varying on time. Given a set S of ST-clauses, we will use $Clauses(S)$ to denote the set of all DeLP clauses involved in ST-clauses of S. Formally, $Clauses(S) = \{\gamma \mid (\gamma, \tau, \upsilon) \in A\}$.

Definition 21 (ST-argument). *Let Q be a literal, and $\mathcal{P}$ be a ST-program. We say that $\langle A, Q, \tau, \upsilon \rangle$ is an ST-argument for a goal Q from $\mathcal{P}$, if $A \subseteq \Delta$, where:*

(1) *$Clauses(\Pi \cup A) \mathrel{\vdash\!\!\!\sim} Q$*
(2) *$Clauses(\Pi \cup A)$ is non contradictory.*
(3) *$Clauses(A)$ is such that there is no $A_1 \subsetneq A$ such that A_1 satisfies conditions (1) and (2) above.*
(4) *$\tau = \tau_1 \cap, ..., \cap \tau_n$ for each ST-clause $(\gamma_i, \tau_i, \upsilon_i) \in A$.*
(5) *$\upsilon : \mathbb{R} \longrightarrow [0, 1]$, such that $\upsilon(\alpha) = MIN(\upsilon_1(\alpha), ..., \upsilon_n(\alpha))$, for each $(\gamma_i, \tau_i, \upsilon_i) \in A$, where $\alpha \in \mathbb{R}$.*

Definition 22 (ST-subargument). *Let $\langle A, L, \tau_1, \upsilon_1 \rangle$ and $\langle B, Q, \tau_2, \upsilon_2 \rangle$ be two arguments. We will say that $\langle B, Q, \tau_2, \upsilon_2 \rangle$ is a ST-subargument of $\langle A, L, \tau_1, \upsilon_1 \rangle$ if and only if $B \subseteq A$. Notice that the goal Q may be a sub-goal associated with the proof of goal L from A.*

As in DeLP, ST-arguments may be in conflict. However in ST-DeLP we must also take into account the availability of conflicting arguments. Then, two ST-arguments involving contradictory information will be in conflict only when their temporal availability intersects.

Definition 23 (Counter-arguments). *Let $\mathcal{P}$ be an ST-program, and let $\langle A_1, L_1, \tau_1, \upsilon_1 \rangle$ and $\langle A_2, L_2, \tau_2, \upsilon_2 \rangle$ be two ST-arguments w.r.t. $\mathcal{P}$. We will say that $\langle A_1, L_1, \tau_1, \upsilon_1 \rangle$ counter-argues $\langle A_2, L_2, \tau_2, \upsilon_2 \rangle$ if and only if there exists a ST-subargument $\langle A, L, \tau, \upsilon \rangle$ (called disagreement ST-subargument) of $\langle A_2, L_2, \tau_2, \upsilon_2 \rangle$ such that L_1 and L disagree, provided that $\tau_1 \cap \tau_2 \neq \emptyset$.*

To define the acceptability of arguments in ST-DeLP we will just construct an E-TAF based on the available ST-arguments. The E-TAF defined will capture the temporal availability and the strength of ST-arguments. Let $\mathcal{P}$ be a ST-program. The E-TAF obtained from $\mathcal{P}$ is $\Psi = \langle AR, Attacks, ARA\upsilon, ATA\upsilon \rangle$ where:

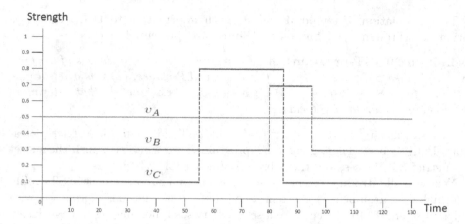

Fig. 5. Representation of the strength functions v_A, v_B, v_C

- AR represents the set of all the ST-arguments from $\mathcal{P}$.
- *Attacks* represents the counter-argument relation among ST-arguments.
- $ARAv(A) =_{def} \tau_A$, where $A \in AR$ and τ_A is the time-intervals set associated with A in $\mathcal{P}$.
- $ATAv((B, A)) =_{def} \{\alpha \in \mathbb{R} \mid \alpha \in \tau_A \cap \tau_B \text{ and } v_B(\alpha) \geq v_A(\alpha)\}$

Notice that an attack from B to A is available only in the time intervals where the strength of B is grater or equal than the strength of A.

Example 5. *Let us consider the following ST-program:*

$$\mathcal{P} = \begin{cases} (a \prec s, k, \ \{[75 - 140]\}, \ v_1) & (t, \ \{[0 - 150]\}, \ v_6) \\ (k \prec m, \ \{[0 - 70], [80 - 120]\}, \ v_2) & (j, \ \{[0 - 150]\}, \ v_7) \\ (\sim k \leftarrow p, \ \{[55 - 120]\}, \ v_3) & (l, \ \{[0 - 150]\}, \ v_8) \\ (p \prec t, l, \ \{[30 - 100]\}, \ v_4) & (\sim p \prec j, \ \{[40 - 90]\}, \ v_9) \\ (s, \ \{[0 - 150]\}, \ v_5) & (m, \ \{[0 - 150]\}, \ v_10) \end{cases}$$

where the strength functions are defined below:

$v_1(\alpha) = 0.5$	$v_6(\alpha) = 1$
$v_2(\alpha) = 0.9$	$v_7(\alpha) = 1$
$v_3(\alpha) = 1$	$v_8(\alpha) = 1$
$v_4(\alpha) = \begin{cases} 0.3 \ \alpha < 80 \\ 0.7 \ 80 \leq \alpha \leq 95 \\ 0.3 \ \alpha > 95 \end{cases}$	$v_9(\alpha) = \begin{cases} 0.1 \ \alpha < 55 \\ 0.8 \ 55 \leq \alpha \leq 85 \\ 0.1 \ \alpha > 85 \end{cases}$
$v_5(\alpha) = 1$	$v_3(\alpha) = 1$

Now we construct the E-TAF corresponding to the previous ST-program.

$AR= \{A, \ B, \ C\}$, *where* A *stands for* $\langle A, a, [80, 120], v_A \rangle$, B *stands for* $\langle B, \sim k, [55, 100], v_B \rangle$ *and* C *stands for* $\langle C, \sim p, [40, 90], v_C \rangle$ *and where the strength functions* v_A, v_B *and* v_C *are depicted in Fig. 5.*

$Attacks = \{(B,A), (C,B)\}$
$ARAv(A) = \{[80, 120]\}$ $ARAv(B) = \{[55, 100]\}$ $ARAv(C) = \{[40, 90]\}$
$ATAv((B,A)) = \{\alpha \in \mathbb{R} \mid \alpha \in \{[80, 95]\} \text{ and } v_B(\alpha) \geq v_A(\alpha)\} = \{[80, 95]\}$
$ATAv((C,B)) = \{[55, 85]\}$

This framework coincides with the E-TAF presented in example 3, Fig. 4, for which argument acceptability was already analyzed.

7 Conclusions – Related and Future Work

Argumentation based formalisms has been successfully applied for reasoning in a single agent, and in multi-agent domains. Dung's AF has been proven fruitful for developing several extensions with application in different contexts (*e.g.*, [8,1]). Reasoning about time is a main concern in many areas, *e.g.*, automated agent deliberation, and recently, an abstract argument based formalization has been defined, called *Temporal Abstract Framework* (TAF), that extends the Dung's formalism by introducing the temporal availability of arguments into account.

In a recent work, a novel extension of TAF called *Extended Temporal Argumentation Framework* (E-TAF [9]) was introduced; this extension takes into account not only the availability of arguments but also the availability of attacks, allowing to model strength of arguments varying over time (strength understood as a generalization of different possible measures for comparing arguments, such as reliability, priorities, etc.).

In this paper we proposed an instantiation of E-TAF with the argumentation formalism *Defeasible Logic Programming* (DeLP), obtaining a formalization we called ST-DeLP. This instantiation provides a concrete knowledge representation and reasoning formalism that allows to specify temporal availability and strength of knowledge at the object language level. This information is propagated to the level of arguments, and acceptability is analyzed as formalized in E-TAF.

As future work we will develop an implementation of ST-DeLP by using the existing DeLP system [2] as a basis. The resulting implementation will be exercised in different domains requiring to model strength varying over time. We are also interested in analyzing the salient features of our formalization in the context of other argumentation frameworks, such as the ASPIC+ framework [19], where rationality postulates for argumentation are explicitly considered.

References

1. Amgoud, L., Devred, C.: Argumentation Frameworks as Constraint Satisfaction Problems. In: Benferhat, S., Grant, J. (eds.) SUM 2011. LNCS, vol. 6929, pp. 110–122. Springer, Heidelberg (2011)
2. Amgoud, L., Prade, H.: Using arguments for making and explaining decisions. Artificial Intelligence 173(3-4), 413–436 (2009)
3. Augusto, J.C., Simari, G.R.: Temporal defeasible reasoning. Knowledge and Information Systems 3(3), 287–318 (2001)

---

[2] See http://lidia.cs.uns.edu.ar/delp

4. Baroni, P., Giacomin, M.: Semantics of abstract argument systems. In: Rahwan, I., Simari, G.R. (eds.) Argumentation in Artificial Intelligence, pp. 24–44. Springer (2009)
5. Beierle, C., Freund, B., Kern-Isberner, G., Thimm, M.: Using defeasible logic programming for argumentation-based decision support in private law. In: COMMA. Frontiers in Artificial Intelligence and Applications, vol. 216, pp. 87–98. IOS Press (2010)
6. Besnard, P., Hunter, A.: Elements of Argumentation. MIT Press (2008)
7. Brena, R., Chesñevar, C.I.: Information distribution decisions supported by argumentation. In: Encyclopedia of Decision Making and Decision Support Technologies, pp. 309–315. Information Science Reference, USA (2008)
8. Brewka, G., Dunne, P.E., Woltran, S.: Relating the semantics of abstract dialectical frameworks and standard AFs. In: Walsh, T. (ed.) IJCAI, pp. 780–785. IJCAI/AAAI (2011)
9. Budán, M.C.D., Lucero, M.G., Chesñevar, C.I., Simari, G.R.: Modeling time and reliability in structured argumentation frameworks. In: KR 2012, pp. 578–582 (2012)
10. Cobo, M.L., Martínez, D.C., Simari, G.R.: On admissibility in timed abstract argumentation frameworks. In: ECAI. Frontiers in Artificial Intelligence and Applications, vol. 215, pp. 1007–1008. IOS Press (2010)
11. Cobo, M.L., Martinez, D.C., Simari, G.R.: Acceptability in Timed Frameworks with Intermittent Arguments. In: Iliadis, L., Maglogiannis, I., Papadopoulos, H. (eds.) Artificial Intelligence Applications and Innovations, Part II. IFIP AICT, vol. 364, pp. 202–211. Springer, Heidelberg (2011)
12. Dung, P.M.: On the acceptability of arguments and its fundamental role in nonmonotonic reasoning and logic programming and n-person games. Artificial Intelligence 77, 321–357 (1995)
13. Galitsky, B., McKenna, E.W.: Sentiment extraction from consumer reviews for providing product recommendations. Patent Application, US 2009/0282019 A1 (November 2009)
14. García, A.J., Simari, G.R.: Defeasible logic programming: An argumentative approach. Theory Practice of Logic Programming 4(1), 95–138 (2004)
15. Mann, N., Hunter, A.: Argumentation using temporal knowledge. In: COMMA 2008, pp. 204–215 (2008)
16. Modgil, S., Caminada, M.: Proof theories and algorithms for abstract argumentation frameworks. In: Rahwan, I., Simari, G.R. (eds.) Argumentation in Artificial Intelligence, pp. 105–132. Springer (2009)
17. Pardo, P., Godo, L.: t-DeLP: A Temporal Extension of the Defeasible Logic Programming Argumentative Framework. In: Benferhat, S., Grant, J. (eds.) SUM 2011. LNCS, vol. 6929, pp. 489–503. Springer, Heidelberg (2011)
18. Pasquier, P., Hollands, R., Rahwan, I., Dignum, F., Sonenberg, L.: An empirical study of interest-based negotiation. Autonomous Agents and Multi-Agent Systems 22(2), 249–288 (2011)
19. Prakken, H.: An abstract framework for argumentation with structured arguments. Argument and Computation 1, 93–124 (2010)
20. Rahwan, I., Ramchurn, S.D., Jennings, N.R., Mcburney, P., Parsons, S., Sonenberg, L.: Argumentation-based negotiation. Knowl. Eng. Rev. 18, 343–375 (2003)
21. Rahwan, I., Simari, G.R.: Argumentation in Artificial Intelligence. Springer (2009)
22. van der Weide, T.L., Dignum, F., Meyer, J.-J.C., Prakken, H., Vreeswijk, G.A.W.: Multi-criteria argument selection in persuasion dialogues. In: AAMAS 2011, Richland, SC, vol. 3, pp. 921–928 (2011)

Drift Detection and Characterization for Fault Diagnosis and Prognosis of Dynamical Systems

Antoine Chammas, Moamar Sayed-Mouchaweh, Eric Duviella,
and Stéphane Lecoeuche

Univ Lille Nord de France, F-59000 Lille, France
EMDouai, IA, F-59500 Douai, France
{antoine.chammas,moamar.sayed-mouchaweh,eric.duviella,
stephane.lecoeuche}@mines-douai.fr

Abstract. In this paper, we present a methodology for drift detection and characterization. Our methodology is based on extracting indicators that reflect the health state of a system. It is situated in an architecture of fault diagnosis/prognosis of dynamical system that we present in this paper. A dynamical clustering algorithm is used as a major tool. The feature vectors are clustered and then the parameters of these clusters are updated as each feature vector arrives. The cluster parameters serve to compute indicators for drift detection and characterization. Then, a prognosis block uses these drift indicators to estimate the remaining useful life. The architecture is tested on a case study of a tank system with different scenarios of single and multiple faults, and with different dynamics of drift.

Keywords: Fault Diagnosis, Prognosis, Drift, Dynamical Clustering.

1 Introduction

Incipent faults are undesirable changes in a process behavior. The affected process state passes from normal to failure through an intermediate state where it is in degraded mode. This degraded mode is also called faulty mode [6]. Condition-Based Maintenance (CBM) was introduced to try to maintain the correct equipment at the right time. It enables the pre-emptive maintenance of systems that are subject to incipient faults. For this purpose, CBM uses a supervision system (diagnosis/prognosis) in order to determine the equipments health using real-time data [16,26], and act only when maintenance is actually necessary. Prognosis is the ability to predict accurately the Remaining Useful Life (RUL) of a failing component or subsystem. Prognostics and Health Management (PHM) is the tool used by a CBM system. The reason is that a PHM provides, via a long-term prediction of the fault evolution, information on the time where a subsystem or a component will no longer perform its intended function. In order to estimate the RUL, it is important to accurately determine the fault conditions: detection of fault and isolation of fault. Moreover, it is important to continuously determine the current condition state of a process when operating online. These requirements are achieved by a diagnosis module.

Approaches for diagnosis have been largely studied in the literature. Globally, we can cite model-based approaches and data-driven based approaches. The model-based

E. Hüllermeier et al. (Eds.): SUM 2012, LNAI 7520, pp. 113–126, 2012.

approach [6] needs an analytical model of the studied process, which is not often trivial to obtain. Data-driven approaches use directly historical data extracted from the system [27,13,7]. Amongst them appears the Pattern Recognition (PR) approach in which operating modes of the system are represented by classes or clusters in a feature space. Let d be the dimensionality of the feature space. Any d-feature vector in this space is called a pattern [22]. Clusters are similar patterns in restricted regions in the feature space corresponding to different functioning modes. In many applications, available historical training sets of data corresponding to normal and failure functioning modes provide an initial feature space on which a classification decision model is built.

A drift is a gradual change in the distribution of patterns that is mainly caused by an incipient fault. For this reason, dynamical clustering appears to be an appropriate way to follow the evolution of a drift because it allows continuous updates of the distribution of patterns, thus of the decision model [24,2,12]. The idea behind this approach is to iteratively update the parameters of clusters in the feature space as new data arrives. This allows to follow the temporal evolution of a drift, and thus the computation of Condition Monitoring (CM) data on which a prognosis module relies.

In fact, prognosis approaches rely on the CM data that reflects the defect evolution in order to estimate the RUL. In general, prognosis approaches can be divided into three categories [26,4,16,21] according to the defect evolution model; Model-based prognosis, data-driven prognosis and experience-based prognosis. In the model-based prognosis, the defect evolution law is characterised by a deterministic model or a stochastic model [7]. It could be directly related to time, such as the crack grow length law (the Paris-rule) and the pneumatic erosion linear model [14]. It could also be analytically related to observable variables in a process, such as, in [19], a physical relationship between fault severity and machine vibration signals was used. In experience-based approach, prognosis is based on the evaluation of a fiability function (like the bathtub curve which is modeled by Weibull distributions) or on stochastic deterioration function like the gamma function [18]. Finally, data-driven based approaches are used when no physical model of the defect evolution is available. Data-driven approaches can be roughly divided into statistical based methods and neural network based methods [21,3]. Statistical methods for prognosis are based on CM measurements related to the health condition or state of the physical asset. There are two types of CM measurements: Direct CM (DCM) measurements and Indirect CM (ICM) measurements [23,7]. The difference is that in the former, the degradation state is directly observable, but in the latter, the degradation state must be reconstructed. In the case of DCM, the estimation of the RUL is the estimation of the CM measurements to reach a pre-defined threshold level. In the case of ICM, it is possible to extract some meaningful indicators that are able to represent the health state of a system. Then, the estimation of the RUL is done by exploiting these indicators. The idea is to evaluate their trajectory into a future horizon [21] until it reaches a failure situation. In order to project the health indicator into future horizons, time series prediction models can be used, such as exponential smoothing and regression based models. In [20], an exponential smoothing was used on sensor measurements to estimate degradation trends. In [28], a logistic regression was used to asses the machine condition and an ARMA (auto regressive moving average) model was used for the prediction of the trend of the degraded performance of the machine. In [11], fault was reconstructed by Principal Component Analysis (PCA) on sensor measurements followed by a denoising step using wavelet networks. Then, the fault trend was recursively identified by a recursive least square auto-regressive (AR) model.

Our contribution in this paper is to derive indicators for detection and characterization of a drift in a process. The computation of these indicators takes place inside an architecture of diagnosis and prognosis. This architecture is depicted and explained in the next section. The health state is reconstructed using metrics applied on iteratively updated cluster parameters. The indicators we compute allow us to detect drift, pinpoint the fault behind causing the drift, and estimate the RUL.

In section 2, the proposed approach for diagnosis/prognosis is shown and depicted. To do so, in a first phase, the essential characteristics of a drift are described. Then, in a second phase, the main functionalities of the dynamical clustering algorithm are described. Then, the scheme of our architecture is shown. In section 3, we describe the drift detection method we use and the computation of the indicators characterizing the drift. In section 4, we show how we exploit these indicators for prognosis purposes. We conclude in section 5 our work with some perspectives.

2   Architecture for Diagnosis and Prognosis

In this section, our diagnosis/prognosis architecture is shown. In a first place, the main characteristics of a drift are detailed. Then, the dynamical clustering algorithm we use is explained. After these two subsections, we show in figure 2 how we relate all these aspects together in the goal of detecting and characterizing drifts.

2.1   Drift Characteriztion

A drift is the case when the process passes gradually from a normal operating mode to failure passing by faulty situation. This is reflected in the feature space by a change in the parameters of the distribution of the patterns [9]. In the literature, many papers deal with the characterization of drifts [25,29,15]. Concerning diqgnosis and condition monitoring, the criterias that are interesting to fully characterize a drift are:

1. Severity: it is an indicator that assesses the amplitude of a drift.
2. Speed: the speed is defined as the velocity of change of the severity indicator.
3. Predictability/Direction: when a drift occurs, the distribution of the patterns starts to change. The patterns gradually start appearing in new regions. When a known fault is causing the drift, the patterns tend to move towards the failure region correspond-ing to this fault. In the opposite case, when an unknown fault is causing the drift, the patterns tend to move towards an unknown region in space. Predictability or di-rection criteria is the ability to say towards which region patterns are apporaching in the feature space.

2.2   Dynamical Clustering Algorithm

Our architecture is based on monitoring parameters of clusters who are dynamically updated at each instant t. The dynamical clustering algorithm we use is AuDyC and it stands for Auto-Adaptive Dynamical Clustering. It uses a technique that is inspired from the mixed Gaussian model [24,2]. The patterns are modeled using Gaussian prototypes P^j characterized by a center $M_{P^j} \in \mathbf{R}^{d \times 1}$ and a covariance matrix $\Sigma_{P^j} \in \mathbf{R}^{d \times d}$ where d is the dimensionality of the feature space. Each gaussian prototype forms a cluster. The parameters of each cluster are its mean and covariance matrix. A minimum number of N_{win} patterns are necessary to define one cluster, where N_{win} is a user defined threshold. Each subsequent input feature vector (pattern) is compared to a cluster database and assigned to a cluster that is most similar to it. After assigment to a cluster, the parameters of the latter are updated. For more details on the functionalities; the rules of recursive adaptation and the similarity criteria in AuDyC, please refer to [24,10].

The initial construction of the knowledge base is done using AuDyC. Initially, we are given:

- S_n: the set of data corresponding to normal operating mode. let $N_{S_n} = card(S_n)$ be the cardinality of S_n.
- $S_{f_i}, i = 1, ..., n_f$: sets of data corresponding to different failure modes. n_f is the number of sets given.

We start the training process with an empty cluster database. The first training set used for learning is S_n. The first feature vector is inserted in the database as the initial cluster, denoted $C_n(t = 1)$. Each subsequent input feature vector is assigned to $C_n(t)$ and the parameters (center and covariance matrix) of the latter are updated. The process iteratively continues on each feature vector of the training set S_n. After a certain time, the cluster $C_n(t)$ *converge* to a region in the feature space. At this point, we define $C_N = C_n(N_{S_n})$ the last cluster that is covering the region corresponding to normal operating mode. The same is done to the training sets $S_{f_i}, i = 1, ..., n_f$ and the result is the obtainment of $C_{f_i}, i = 1, ..., n_f$, the last clusters covering failure modes (see figure (1(a))). Finally, we define, for drift detection purpose (as we will see later), $S_1 = \{\mu_{C_n}(j), j = N_{S_n} - n_1 + 1, ..., N_{S_n}\}$, where n_1 is user defined, a sample of the last n_1 means of clusters corresponding to normal operating mode.

Once this offline model is built, it can be plugged to the online system that is providing a data stream of patterns. As each pattern is available, it is compared to the knowledge base model constructed on the training set. Under normal operating conditions, features will be assigned to C_N. After each assignment, the parameters (mean and covariance) of C_N are updated, *i.e.* the cluster is updated and will be denoted $C_e(t)$ as to say the evolving cluster at time t. We have $C_e(t = 0) = C_N$ and under normal operating mode (without drift), $C_e(t) \approx C_N$. After the occurence of a drift, there are three possible cases (see figure 1(b,c,d)) for the trajectory of the mean of $C_e(t)$:

- Towards a known region in case fault is known,
- Towards an unknown region in case of unknown fault,
- Possible change in direction due to multiple faults.

2.3   General Scheme of the Architecture

At this point, we can give an overview of our architecture in figure 2.

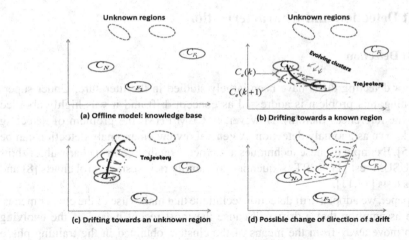

Fig. 1. Different possible drifting scenarios (resulting from AuDyC adaptation mechanism)

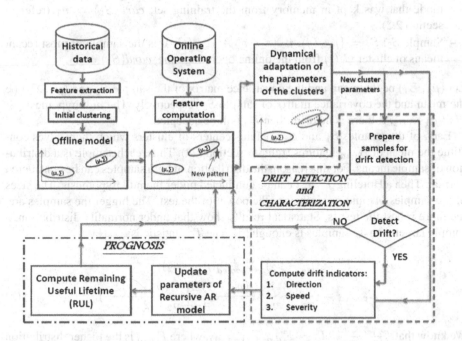

Fig. 2. Diagnosis and prognosis architecture

3 Drift Detection and Characterization

3.1 Drift Detection

Methods for detecting drifts have been largely studied in the literature. Under supervised learning, this problem is addressed as concept drift and it was highly aborded [29] (and the references therein). However, in CBM/PHM, the problem of detecting change is known as anomaly detection. A general review on anomaly detection can be found in [5]. Example of these techniques are process quality control charts algorithms such as CUSUM test [1], SPRT (sequential probability ratio test) control charts [8] and hypothesis tests [17,11].

In this paper, we adopt a drift detection technique that makes use of the cluster means. Under the assumption that a drift will cause the most recent means of the evolving clusters to move away from the means of the clusters obtained in the training phase under normal operating conditions, and since we use a parametric technique (Gaussian model) to model the data in the feature space, a hypothesis test on the mean can be used to detect drifts. The samples used for this test are:

- Sample S_1: it is the sample of cluster means corresponding to normal operating mode that was kept in memory from the training set; $card(S_1) = n_1$ (refer to section 2.2).
- Sample S_2: $S_2 = \{\mu_{C_e}(j), j = t - n_2 + 1, ..., t\}$. It is the sample of most recent means of cluster $C_e(t)$ from the online operating time. $card(S_2) = n_2$.

Let (μ_1, Σ_1) be the mean and the covariance matrix of the sample S_1 and (μ_2, Σ_2) be the mean and the covariance matrix of sample S_2 respectively. The null hypothesis is: $H_o : \mu_1 = \mu_2$. If there is no drift, then H_o will be accepted.

Each of the samples S_1 and S_2 contains centers of clusters which themselves constitue the mean of N_{win} features (refer to section 2.2). Thus, each sample is a distribution of sample means. For this reason, the distribution of both samples can be considered normal. Then a Hoteling T^2 test can be conducted under the null hypothesis. The sizes of the samples (n_1 and n_2) influence the power of the test. The bigger the samples are, the more robust the test is. Statistical results show that under normality distribution, a sample larger than 30 samples is enough [5]. The T^2 statistic is:

$$T^2 = (\mu_1 - \mu_2)^T \Sigma_{pooled}^{-1}(\mu_1 - \mu_2), \tag{1}$$

$$\Sigma_{pooled} = (\frac{1}{n_1} + \frac{1}{n_2}).(\frac{(n_1 - 1)\Sigma_1 + (n_2 - 1)\Sigma_2}{n_1 + n_2 - 2}). \tag{2}$$

We know that $\frac{n_2+n_1-d-1}{d.(n_1+n_2-2)}T^2 \sim F_{(d,n_1+n_2-d-1)}$, where $F_{(p,q)}$ is the Fisher distribution with p and q degrees of freedom. We remind that d is the dimensionality of the feature space. Thus, for a given confidence level α, drift is confirmed if:

$$T^2 \leq \frac{d.(n_1 + n_2 - 2)}{n_2 + n_1 - d - 1} F_{(d,n_1+n_2-d-1)|\alpha}, \tag{3}$$

3.2 Drift Characteristization Indicators

Once a drift is detected, computation of drift indicators at each current time step is necessary. In order to fully characterize a drift, three drift indicators are required:

- Direction indicator: based on the colinearity of the trajectory of the evolving cluster $C_e(t)$ with known failure clusters.
- Severity indicator: based on the distance of $C_e(t)$ compared to C_N and C_{f_i}.
- Speed indicator: based on the derivative of the severity indicator, the speed indicator reflects the speed of evolution of the fault.

Direction Indicator: It is used to pinpoint the cause of the fault (isolation) by studying the direction of the movement of the evolving cluster. For this reason, let $P_{C_e}(t) = (\mu_e(t), \Sigma_e(t))$ and $P_{C_e}(t-1) = (\mu_e(t-1), \Sigma_e(t-1))$ the parameters of the evolving clusters at time t and $(t-1)$ respectively, and $P_{C_{f_i}}, i = 1, ..., n_f$ the parameters of the failure clusters corresponding to failure number i, we will define the following vectors:

- $D_e(t) = \frac{\mu_e(t) - \mu_e(t-1)}{||\mu_e(t) - \mu_e(t-1)||}$, the unitary vector relating the centers of two consecutive clusters $C_e(t)$.
- $D_i(t) = \frac{\mu_{f_i}(t) - \mu_e(t)}{||\mu_{f_i}(t) - \mu_e(t)||}$, the unitary vector relating the centers of $C_e(t)$ at time t, and the center of the failure cluster C_{f_i}.

At each step, let $p(t) = \max\limits_{1 \leq i \leq n_f}(D_e^T(t)D_i(t))$, and $\Gamma(t) = arg(\max\limits_{1 \leq i \leq n_f}(D_e^T(t)D_i(t)))$. $p(t)$ is the maximum of the scalar product between $D_e(t)$ and all the vectors $D_i(t)$. The closer the value of $p(t)$ is to 1, the more $D_e(t)$ and $D_i(t)$ are closer to be colinear. If $p(t) = 1$, then the drift is linear, *i.e.* the trajectory of the centers of the cluster $C_e(t)$ is a line. The values of the direction indicator and the rules of assignment at each step follow this algorithm:

1. First calculate $p(t)$ and $\Gamma(t)$,
2. If $p(t) \geq p_M$, where p_M is a user-defined threshold, then it is safe to say that the movement is towards the failure cluster whose number is Γ In this case, we give a value for direction such that $direction = \Gamma$. In the case where $p(t) < p_M$, the movement of the drift is considered to be towards an unknown region. The value of direction in this case is 0. The threshold p_M is user defined and depends on the application. The larger its value is, the more the user is assuming a linear drift.

Severity Indicator: The severity indicator must reflect how far the evolving class is from normal class and how close it is getting to the faulty class. Because the clusters are Gaussians, the Kullback-Leibler divergence metric is used to compare the clusters. The equation of this divergence between two multivariate Gaussian prototypes $P_1 = (\mu_1, \Sigma_1)$ and $P_2 = (\mu_2, \Sigma_2)$ is:

$$d_{KL}(P_1, P_2) = 0.5[ln(\frac{|\Sigma_{P_2}|}{|\Sigma_{P_1}|}) - d + tr(\Sigma_{P_2}^{-1} \Sigma_{P_1}) + (\mu_{P_1} - \mu_{P_2})^T \Sigma_{P_2}^{-1}(\mu_{P_1} - \mu_{P_2})],$$

$$(4)$$

where $|\Sigma|$ is the determinant of the matrix $|\Sigma|$. Two cases are possible:

- The direction of drift is towards a known failure cluster C_{f_i} : in this case the severity indicator is denoted $sv_i(t)$ and is given by:

$$sv_i(t) = \frac{d_{KL}(C_n, C_e(t))}{d_{KL}(C_n, C_e(t)) + d_{KL}(C_e(t), C_{f_i})}. \tag{5}$$

- The direction of the drift is towards an unknown region in space: in this case, the severity indicator will be the divergence of the evolving class from the normal class and will be denoted $sv(t) = d_{KL}(C_n, C_e(t))$.

From equation (5), it is clear that a severity indicator $sv_i(t)$ can take values ranging from 0 to 1. Under normal operating conditions, $C_e(t) \approx C_N$ thus $sv_i(t) \to 0$. Under failure operating conditions, $C_e(t) \approx C_{f_i}$ thus $sv_i(t) \to 1$.

Speed Indicator: The speed indicator is the derivative of the severity indicator. It reflects the speed of the evolution of clusters. In case of drift isolated, *i.e.* cluster $C_e(t)$ is evolving from normal to a known faulty cluster, the speed indicator will be:

$$sp_i(t) = sv_i(t) - sv_i(t-1), \tag{6}$$

whereas in case of drifting towards an unknown region, the speed indicator is not computed.

4 Prognosis

Given the drift indicators, the prognosis module must be able to compute the remaining useful lifetime (RUL). The severity indicators calculated on the earlier section reflects the state of health of the system. A recursive auto regressive (RAR) model will be fit on the severity indicator in order to project it's path to futur horizon. Two cases are possible:

- Drifting towards a known failure cluster C_i: in this case, the RAR model for the corresponding severity indicator sv_i is used to compute the RUL. The computation of the RUL is described in the subsection below.
- Drifting towards a unknown failure cluster: in this case, the RUL cannot be computed because the region towards the cluster is evolving is unknown.

4.1 Dynamics of Drift and RUL Computation

Given a severity indicator $sv_i(t), i = 1, ..., n_f$, the RUL is the time value for which the equation (7) is satisfied:

$$sv_i(t + RUL) = 1. \tag{7}$$

A RAR model of order m can be written as:

$$sv_i(t) = \sum_{j=1}^{m} (a_j.sv_i(t-j)) + e(t), \tag{8}$$

where $a_j, j = 1...m$ are model parameters and $e(t)$ is the model noise, which is assumed to be zero mean and i.i.d. Let

$$\Theta_i(t) = [a_1(t), a_2(t), ..., a_m(t)]^T,$$

be the parameter vector at time t, and

$$\phi_i(t) = [sv_i(t-1) \,...\, sv_i(t-m)]^T$$

the vector of the last m values of the severity indicator.

The parameter vector $\Theta_i(t)$ is recursively updated using the recursive least squares algorithm:

$$\hat{\Theta}_i(t) = \hat{\Theta}_i(t-1) + K_i(t).(sv_i(t) - \hat{sv}_i(t)), \tag{9}$$

$$\hat{sv}_i(t) = \phi_i^T(t).\hat{\Theta}_i(t-1), \tag{10}$$

$$K_i(t) = Q_i(t).\phi_i(t), \tag{11}$$

$$Q_i(t) = \frac{P_i(t-1)}{1 + \phi_i^T(t).P_i(t-1).\phi_i(t)}, \tag{12}$$

$$P_i(t) = P_i(t-1) - P_i(t-1).\phi_i(t).\phi_i^T(t).Q_i(t), \tag{13}$$

where $\hat{\Theta}_i(0) = 0$ and $P_i(0) = \gamma I, \gamma >> 0$ an arbitrary positive number.

At a step t, multi step prediction beginning at $sv_i(t)$ can be obtained in an iterative way:

$$\hat{sv}_i(k + H) = \hat{\Theta}_i^T(t)\hat{\phi}_i(t + H), \tag{14}$$

where $\hat{\Theta}_i(t)$ is the estimated parametor vector at time t and $\hat{\phi}_i^T(t + H) = [\hat{sv}_i(t + H - 1), ..., \hat{sv}_i(t + H - m)]$. In case of non detected drift, RUL has no sense and will be given the value -1 in our algorithm. In case of drift detected and fault isolated, the calculation of the RUL follows this algorithm:

H=0,
while $\hat{sv}_i(t + H) < 0.99$
H++; update $\hat{\phi}_i^T(t + H)$; calculate $\hat{sv}_i(t + H)$ (eq (14)); end while
$RUL = H.$

5   Case Study: Tank System

The database used for the testing our diagnosis/prognosis architecture was simulated using the benchmark of a tank system. Different scenarios including known faults and unknown faults were simulated and the results are shown. The tank system is shown in figure 3. Under normal operating mode, the level of water is kept between two thresholds, h_1^{HIGH} and h_1^{LOW}. When the level of water reaches h_1^{HIGH}, P_1 is closed and V_2 is opened. When the level of water reaches h_1^{LOW}, P_1 is opened and V_2 is closed. The valve V_1 is used to simulate leak in the tank. The surface of the valves V_1 and V_2 is the same: $S_{V_1} = S_{V_2} = S_V$ and the surface of the pump pipe is S_P. The instrumentation used consists of only one sensor for the level of water in the tank. It is denoted by h_1.

5.1 Considered Faults

In order to test our architecture, three faults are considered; two known faults and one unknown:

1. Fault1 (known): gradual increase of the surface of the valve V_1 leading to a gradual increase of the flow of water leaking from the tank. This surface increases from $0\% \times S_V$ to $30\% \times S_V$ considered as the maximum intensity of leakage. At this stage, the system is considered in failure. When the surface is between $0\% \times S_V$ and $30\% \times S_V$, the system is faulty (degraded operation).
2. Fault2 (known): clogging of the pump P1 meaning that the flow of water that the pump is delivering is decreasing with time. The same principle for the simulation of V_1 is used. A clogging of $30\% \times S_P$ corresponds to failure.
3. Fault3 (unknown): clogging of the valve V_2. The same principle for the simulation of V_1 is used. A clogging of $30\% \times S_V$ means failure.

5.2 Feature Extraction

At the beginning, three data sets are given. One corresponding to normal operating mode, and two corresponding to fault1 and fault2 operating modes respectively. In figure 4, we show the sensor measurements under normal operating mode. We can clearly see that a cycle is a sequence of a filling period followed by a draining period. The features extracted from this signal are the time required for the filling T_1 and the time required for draining T_2. At the end of each cycle, one feature is extracted thus the time unit considered will be the cycles denoted by cy.

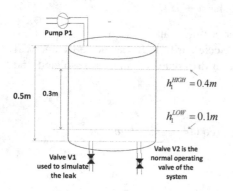

Fig. 3. The tank system

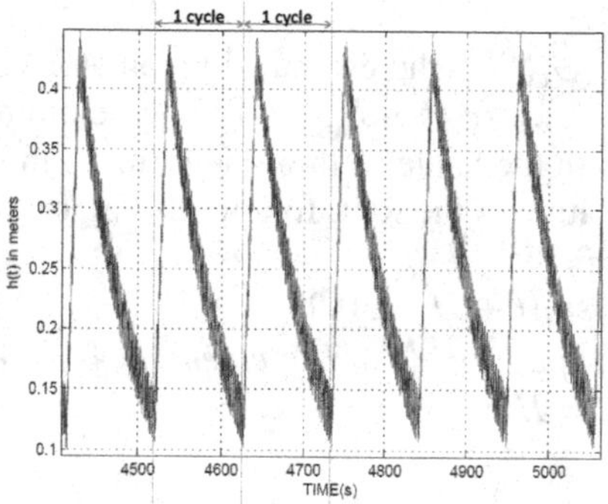

Fig. 4. Six filling/draining cycles under normal operating conditions

5.3 Scenarios

Three scenarios were considered:

- Scenario1: only fault1 is considered. The drift is simulated linearly from 0% to 30% in $26cy$.
- Scenario2: only fault2 is considered. The drift is simulated linearly from 0% to 30% in $60cy$.

– Scenario3: In this scenario, fault1 and fault3 are considered. After the occurence of fault1, fault3 is activated with some delay and both faults are thus activated together. This results to a change in the direction of the drift and the idea behind is to test the ability of our algorithm to detect this case.

5.4   Results

The parameters of the overall algorithm are: $\lambda_m = 0.02$ and $N_{win} = 30$ (section 2), $\alpha = 99.7\%$, $n_1 = 60$, $n_2 = 30$, $F_{(d,n_1+n_2-d-1|\alpha)} = 3.6$ and $p_M = 0.85$ (section 3). In all scenarios, drift was started at $t_{drift} = 20cy$.

The results for the scenarios are depicted in the figures below.

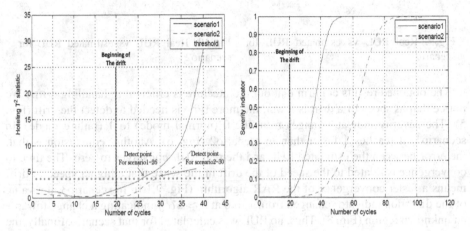

Fig. 5. Drift detection results for scenarios 1 and 2

Fig. 6. Severity indicator results for scenarios 1 and 2

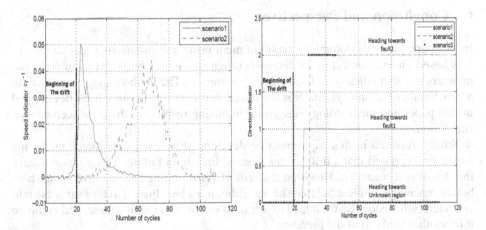

Fig. 7. Speed indicator results for scenarios 1 and 2

Fig. 8. Direction indicator results for scenarios 1, 2 and 3

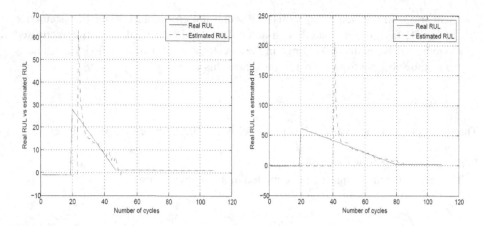

Fig. 9. Real RUL vs estimated RUL for scenario1

Fig. 10. Real RUL vs estimated RUL for scenario2

The obtained results confirm that our architecture is efficient in handling drifts. We notice that as speed of a drift is lower, the more time is needed to detect the drift (Fig. 5). The severity indicator ranges from the 0 (normal mode) to 1 (failure mode). In scenario1, it reaches 1 faster than in scenario2 (Fig. 6). Another conclusion is that, the error between the estimated RUL and the real RUL converges to zero. The time to convergence is related to the speed of the drift in the sense that a higher speed of drift means a faster convergence of the RAR algorithm (Fig. 9,10). In scenario3, the value of the direction indicator changed from 1 to 0 at $t \approx 47cy$ indicating a change towards an unknown region (Fig. 8). Thus, no RUL was calculated for that scenario. Finally, the speed indicator in scenario1 shows that the speed increases very fast to its maximum value, whereas in scenario2, the increase of speed is slower (Fig. 7).

6 Conclusion and Perspectives

In this paper, a methodology of condition monitoring and prognosis was established. It is based on an architecture of diagnosis/prognosis. It was based on monitoring parameters of a dynamically updated cluster parameters. The methodology was tested on a benchmark of a tank system. It was showed that under the assumptions developped in this paper, the methodology has given promising results for different scenarios of simulation.

Surely, research in this paper must be developped in a stepwise direction. In this paper, we supposed that a drift led the system directly to failure. This a possible case in a lot of real scenarios. However, the drift of functionning can stabilize somewhere before reaching complete failure. The algorithm must be adjusted to take this case into consideration. Secondly, we intend to test our system on a real database, and compare it to another methods of the literature.

References

1. Basseville, M., Nikiforov, I.: Detection of Abrupt Changes: Theory and Application. Prentice Hall, Inc. (1993)
2. Boubacar, H.A., Lecoueuche, S., Maouche, S.: Audyc neural network using a new gaussian densities merge mechanism. In: 7th Conference on Adaptive and Neural Computing Algorithms, pp. 155–158 (2004)
3. Brotherton, T., Jahns, G., Jacobs, J., Wroblewski, D.: Prognosis of faults in gas turbine engines. In: IEEE Aerospace Conference (2000)
4. Byington, C., Roemer, M., Kacprzynski, G., Galie, T.: Prognostic enhancements to diagnostic systems for improved condition-based maintenance. In: IEEE Aerospace Conference (2002)
5. Chandola, V., Banerjee, A., Kumar, V.: Anomaly detection: A survey. Technical report, ACM Computing Surveys (2009)
6. Isermann, R.: Model-based fault detection and diagnosis - status and applications. Annual Reviews in Control 29, 71–85 (2005)
7. Jardine, A.K.S., Lin, D., Banjevic, D.: A review on machinery diagnostics and prognostics implementing condition-based maintenance. Mechanical Systems and Signal Processing 20, 1483–1510 (2006)
8. Kuncheva, L.: Using control charts for detecting concept change in streaming data. Technical report, Technical Report BCS-TR-001 (2009)
9. Kuncheva, L.I.: Classifier Ensembles for Changing Environments. In: Roli, F., Kittler, J., Windeatt, T. (eds.) MCS 2004. LNCS, vol. 3077, pp. 1–15. Springer, Heidelberg (2004)
10. Lecoueuche, S., Lurette, C.: Auto-Adaptive and Dynamical Clustering Neural Network. In: Kaynak, O., Alpaydın, E., Oja, E., Xu, L. (eds.) ICANN 2003 and ICONIP 2003. LNCS, vol. 2714, pp. 350–358. Springer, Heidelberg (2003)
11. Li, G., Qin, J., Ji, Y., Zhou, D.-H.: Reconstruction based fault prognosis for continuous processes. Control Engineering Practice 18, 1211–1219 (2010)
12. Iverson, D.L.: Inductive system health monitoring. In: Proceedings of The 2004 International Conference on Artificial Intelligence (IC-AI 2004). CSREA Press, Las Vegas (2004)
13. Markou, M., Singh, S.: Novelty detection: a review–part i: statistical approaches. Signal Processing 83, 2481–2497 (2003)
14. Meeker, W.Q., Escobar, L.A.: Statistical Methods for Reliability Data. Wiley series in probability and statistics, applied probability and statistics section. Jhon Wiley and Sons, New York (1998)
15. Minku, L.L., Yao, X.: Ddd: A new ensemble approach for dealing with concept drift. IEEE Transactions on Knowledge and Data Engineering (2011)
16. Muller, A., Marquez, A.C., Iunga, B.: On the concept of e-maintenance: Review and current research. Reliability Engineering and System Safety 93, 1165–1187 (2008)
17. MvBain, J., Timusk, M.: Fault detection in variable speed machinery: Statistical parametrization. Journal of Sound and Vibration 327, 623–646 (2009)
18. Noortwijk, V.: A survey of the application of gamma process in maintenance. Reliability Engineering & System Safety 94, 2–21 (2009)
19. Oppenheimer, C., Loparo, K.: Physically based diagnosis and prognosis of cracked rotor shafts. In: Preceeding of SPIE (2002)
20. Peysson, F., Boubezoul, A., Oulasdine, M., Outbib, R.: A data driven prognostic methodology without a priori knowledge. In: Proceedings of the 7th IFAC Symposium on Fault Detection, Supervision and Safety of Technical Processes (2009)
21. Peysson, F., Oulasdine, M., Outbib, R., Leger, J.-B., Myx, O., Allemand, C.: Damage trajectory analysis based prognostic. In: International Conference on Prognostics and Health Management, PHM 2008 (2008)

22. Sayed-Mouchaweh, M.: Semi-supervised classification method for dynamic applications. Fuzzy Sets and Systems 161, 544–563 (2012)
23. Si, X.-S., Wang, W., Hu, C.-H., Zhou, D.-H.: Remaining useful life estimation - a review on the statistical data driven approaches. European Journal of Operational Research 213, 1–14 (2011)
24. Traore, M., Duviella, E., Lecoeuche, S.: Comparison of two prognosis methods based on neuro fuzzy inference system and clustering neural network. In: SAFE PROCESS (2009)
25. Tsymbal, A.: The problem of concept drift: definitions and related work. Trinity College, Dublin, Ireland, TCD-CS-2004-15 (2004)
26. Vachtsevanos, G., Lewis, F., Roemer, M., Hess, A., Wu, B.: Intelligent Fault Diagnosis and Prognosis for Engineering Systems. Jhon Wiley and Sons, Inc. (2006)
27. Venkatasubramanian, V., Rengaswamy, R., Yin, K., Kavuri, S.: A review of process fault detection and diagnosis part iii: Process history based methods. Computers and Chemical Engineering 27, 327–346 (2003)
28. Yan, J., Lee, J., Koc, M.: Predictive algorithm for machine degradation using logistic regression. In: MIM 2002 (2002)
29. Indre zliobate. Learning under concept drift: an overview. Technical report, Vilnius University (2009)

An Attempt to Employ Genetic Fuzzy Systems to Predict from a Data Stream of Premises Transactions

Bogdan Trawiński[1], Tadeusz Lasota[2], Magdalena Smętek[1], and Grzegorz Trawiński[3]

[1] Wrocław University of Technology, Institute of Informatics,
Wybrzeże Wyspiańskiego 27, 50-370 Wrocław, Poland
[2] Wrocław University of Environmental and Life Sciences, Dept. of Spatial Management
ul. Norwida 25/27, 50-375 Wrocław, Poland
[3] Wrocław University of Technology, Faculty of Electronics,
Wybrzeże S. Wyspiańskiego 27, 50-370 Wrocław, Poland
{magdalena.smetek,bogdan.trawinski}@pwr.wroc.pl,
tadeusz.lasota@up.wroc.pl, grzegorztrawinski@wp.pl

Abstract. An approach to apply ensembles of genetic fuzzy systems, built over the chunks of a data stream, to aid in residential premises valuation was proposed. The approach consists in incremental expanding an ensemble by systematically generated models in the course of time. The output of aged component models produced for current data is updated according to a trend function reflecting the changes of premises prices since the moment of individual model generation. An experimental evaluation of the proposed method using real-world data taken from a dynamically changing real estate market revealed its advantage in terms of predictive accuracy.

Keywords: genetic fuzzy systems, data stream, sliding windows, ensembles, predictive models, trend functions, property valuation.

1 Introduction

The area of data stream mining has attracted the attention of many researchers during the last fifteen years. Processing data streams represents a novel challenge because it requires taking into account memory limitations, short processing times, and single scans of arriving data. Many strategies and techniques for mining data streams have been devised. Gaber in his recent overview paper categorizes them into four main groups: two-phase techniques, Hoeffding bound-based, symbolic approximation-based, and granularity-based ones [11]. Much effort is devoted to the issue of concept drift which occurs when data distributions and definitions of target classes change over time [9], [22], [27], [29]. Among the instantly growing methods of handling concept drift in data streams Tsymbal distinguishes three basic approaches, namely instance selection, instance weighting, and ensemble learning [25], the latter has been systematically overviewed in [16],[23]. In adaptive ensembles, component models are generated from sequential blocks of training instances. When a new block arrives, models are examined and then discarded or modified based on the results of the

E. Hüllermeier et al. (Eds.): SUM 2012, LNAI 7520, pp. 127–140, 2012.

evaluation. Several methods have been proposed for that, e.g. accuracy weighted ensembles [26] and accuracy updated ensembles [4].

One of the most developed recently learning technologies devoted to dynamic environments have been evolving fuzzy systems [20]. Data-driven fuzzy rule based systems (FRBS) are characterized by three important features. Firstly, they are able of approximating any real continuous function on a compact set with an arbitrary accuracy [6], [13]. Secondly, they have the capability of knowledge extraction and representation when modeling complex systems in a way that they could be understood by humans [1]. Thirdly, they can be permanently updated on demand based on new incoming samples as is the case for on-line measurements or data streams. The technologies provide such updates with high performance, both in computational times and predictive accuracy. The major representatives of evolving fuzzy approaches are FLEXFIS [21] and eTS [2] methods. The former incrementally evolves clusters (associated with rules) and performs a recursive adaptation of consequent parameters by using local learning approach. The latter is also an incremental evolving approach based on recursive potential estimation in order to extract the most dense regions in the feature space as cluster centers (rule representatives).

Ensemble models have been drawing the attention of machine learning community due to its ability to reduce bias and/or variance compared with their single model counterparts. The ensemble learning methods combine the output of machine learning algorithms to obtain better prediction accuracy in the case of regression problems or lower error rates in classification. The individual estimator must provide different patterns of generalization, thus the diversity plays a crucial role in the training process. To the most popular methods belong bagging [3], boosting [27], and stacking [28]. Bagging, which stands for bootstrap aggregating, devised by Breiman [3] is one of the most intuitive and simplest ensemble algorithms providing good performance. Diversity of learners is obtained by using bootstrapped replicas of the training data. That is, different training data subsets are randomly drawn with replacement from the original training set. So obtained training data subsets, called also bags, are used then to train different classification and regression models. Theoretical analyses and experimental results proved benefits of bagging, especially in terms of stability improvement and variance reduction of learners for both classification and regression problems [5], [10]. However, the aforementioned approaches are devoted to static environments, i.e. they assume that all training data are available before the learning phase is conducted. Once this phase is completed, the learning system is no more capable of updating the generated model. It means the system is not adaptive and the model cannot evolve over time.When processing a data stream using nonincremental method, we cannot apply any resampling technique as bootstrap, holdout or cross-valitation. Instead, we should try to select a chunk of data coming in within the shortest time period possible, and use it to train a model, and validate the model using the data coming in the next period.

The goal of the study presented in this paper was to make an attempt to apply a nonincremental genetic fuzzy systems to build reliable predictive models from a data stream. The approach was inspired by the observation of a real estate market of in one big Polish city in recent years when it experienced a violent growth of residential premises prices. Our method consists in the utilization of aged models to compose ensembles and correction of the output provided by component models was updated with trend functions reflecting the changes of prices in the market over time.

2   Motivation and GFS Ensemble Approach

Property and real estate appraisals play the crucial role in many areas of social life and economic activity as well as for private persons, especially for asset valuation, sales transactions, property taxation, insurance estimations, and economic and spatial planning. The values of properties change with market conditions in the course of time and must be periodically updated, and the value estimation is based on the current indicators of real estate market, first of all on recent real estate sales transactions. The accuracy of real estate valuation models depends on proper identifying the relevant attributes of properties and finding out the actual interrelationship between prices and attributes. In current sales comparison approaches for residential premises, it is necessary to have transaction prices of the properties sold whose attributes are similar to the one being appraised. If good comparable transactions are available, then it is possible to obtain reliable estimates for the prices of the residential premises.

The approach based on fuzzy logic is especially suitable for property valuation because professional appraisers are forced to use many, very often inconsistent and imprecise sources of information, and their familiarity with a real estate market and the land where properties are located is frequently incomplete. Moreover, they have to consider various rice drivers and complex interrelation among them. An appraiser should make on-site inspection to estimate qualitative attributes of a given property as well as its neighbourhood. They have also to assess such subjective factors as location attractiveness and current trend and vogue. So, their estimations are to a great extent subjective and are on uncertain knowledge, experience, and intuition rather than on objective data.

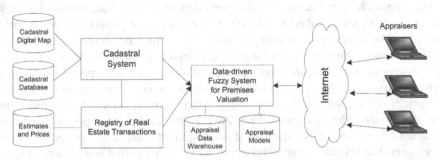

Fig. 1. Information systems to assist with real estate appraisals

So, the appraisers should be supported by automated valuation systems which incorporate data driven models for premises valuation developed employing sales comparison method. The data driven models, considered in the paper, were generated using real-world data on sales transactions taken from a cadastral system and a public registry of real estate transactions. The architecture of the proposed system is shown in Fig. 1. The appraiser accesses the system through the Internet and input the values of the attributes of the premises being evaluated into the system, which calculates the output using a given model. The final result as a suggested value of the property is sent back to the appraiser. We explore data-driven fuzzy rule-based systems (FRBS) as a specific data-driven model architecture used in the framework shown in Fig. 1, which were recognized to be able of approximating any real continuous function on a compact set with an arbitrary accuracy. Moreover, FRBSs allow for knowledge extraction and representation by modeling complex systems in a way understandable by humans. So, the interpretability of fuzzy systems is a characteristic that favors this type of models because it is often required to explain the behavior of a given real appraisal model.

So far, we have investigated several methods to construct regression models to assist with real estate appraisal based on fuzzy approach: i.e. genetic fuzzy systems as both single models [14] and ensembles built using various resampling techniques [12], [18], but in this case the whole datasets had to be available before the process of training models started. All property prices were updated to be uniform in a given point of time. An especially good performance revealed evolving fuzzy models applied to cadastral data [17], [19]. Evolving fuzzy systems are appropriate for modeling the dynamics of real estate market because they can be systematically updated on demand based on new incoming samples and the data of property sales ordered by the transaction date can be treated as a data stream. In this paper we present our first attempt to employ evolutionary fuzzy approach to explore data streams to model dynamic real estate market. The problem is not trivial because on the one hand a genetic fuzzy system needs a number of samples to be trained and on the other hand the time window to determine a chunk of training data should be as small as possible to retain the model accuracy at an acceptable level. The processing time in this case is not a very important issue because property valuation models need not to be updated and/or generated from scratch in an on-line mode.

Our approach is grounded on the observation of a real estate market in one big Polish city with the population of 640 000. The residential premises prices in Poland depend on the form of the ownership of the land on which buildings were erected. For historical reasons the majority of the land in Poland is council-owned or state-owned. The owners of flats lease the land on terms of the so-called perpetual usufruct, and consequently, most flat sales transactions refer to the perpetual usufruct of the land. The prices of flats with the land ownership differ from the ones of flats with the land lease. Moreover, the apartments built after 1996 attain higher prices due to new construction technologies, quality standards, and amenities provided by the developers. Furthermore, apartments constructed in this period were intended mainly for investments and trades which also led to the higher prices. To our study we selected sales transaction data of apartments built before 1997 and where the land was leased on terms of perpetual usufruct. Therefore the dynamics of real estate market concerns more the prices of residential premises rather than other basic attributes of properties such as usable area, number of rooms, floor, number of storeys in a building, etc.

Having a real-world dataset referring to residential premises transactions accomplished in the city, which after cleansing counted 5212 samples, we were able to determine the trend of price changes within 11 years from 1998 to 2008. It was modelled by the polynomial of degree three. The chart illustrating the change trend of average transactional prices per square metre is given in Fig. 2.

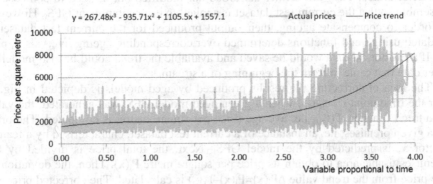

Fig. 2. Change trend of average transactional prices per square metre over time

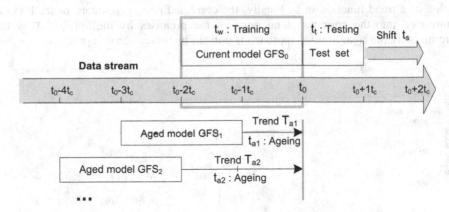

Fig. 3. GFS ensemble approach to predict from a data stream

The idea of the GFS ensemble approach to predict from a data stream is illustrated in Fig. 3. The data stream is partitioned into data chunks according to the periods of a constant length t_c. In the case of transactional data t_c can be equal to one, two, or more months. The sliding time window of the length t_w is equal to the multiple of t_c so that $t_w = jt_c$, where $j=1,2,3,...$. The window determines the scope of training data to generate from scratch a property valuation model, in our case GFS. It is assumed that the models generated over a given training dataset is valid for the next interval which specifies the scope for a test dataset. Similarly, the interval t_t which delineates a test dataset is equal to the multiple of t_c so that $t_t = kt_c$, where $k=1,2,3,...$. The sliding window is shifted step by step of a period t_s in the course of time, and likewise, the interval t_s is equal to the multiple of t_c so that $t_s=lt_c$, where $l=1,2,3,...$.

Let us consider a point of time t_0 at which the current model GFS_0 was generated from scratch over data that came in between time t_0-t_w and t_0. In the experiments reported in the next section as t_0 we took January 1, 2006, which corresponded the value of 2.92 on the x axis in the chart in Fig. 2. In this time the prices of residential premises were growing quickly due to the run on real estate. The models created earlier, i.e. GFS_1, GFS_2, etc. have aged gradually and in consequence their accuracy has deteriorated. They are neither discarded nor modified but utilized to compose an ensemble so that the current test dataset is applied to each component GFS_i. However, in order to compensate ageing, their output produced for the current test dataset is updated using trend functions determined over corresponding ageing intervals t_{ai} plus t_w. If all historical data would be saved and available the trend could be also modelled over data that came in from the beginning of a stream.

The idea of correcting the results produced by aged models is depicted in Fig. 4. For the time point $t_{gi}=t_0-t_{ai}$, when a given aged model GFS_i was generated, the value of a trend function $T(t_{gi})$, i.e. average price per square metre, is computed. The price of a given premises, i.e. an instance of a current test dataset, characterised by a feature vector $\mathbf{x}$, is predicted by the model GFS_i. Next, the total price is divided by the premises usable area to obtain its price per square metre $P_i(\mathbf{x})$. Then, the deviation of the price from the trend value $\Delta P_i(\mathbf{x})=P_i(\mathbf{x})-T(t_{gi})$ is calculated. The corrected price per square metre of the premises $P_i'(\mathbf{x})$ is worked out by adding this deviation to the trend value in the time point t_0 using the formula $P_i'(\mathbf{x})=\Delta P_i(\mathbf{x})+T(t_0)$, where $T(t_0)$ is the value of a trend function in t_0. Finally, the corrected price per square metre $P'(\mathbf{x})$ is converted into the corrected total price of the premises by multiplying it by the premises usable area. Similar approach is utilized by professional appraisers.

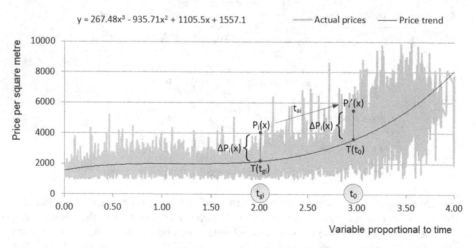

Fig. 4. The idea of correcting the output of aged models

The resulting output of the ensemble for a given instance of the test dataset is computed as the arithmetic mean of the results produced by the component models and corrected by corresponding trend functions. Moreover, weighting component models in an ensemble can be applied according to time past or estimated model accuracy. In the study we applied a simple method proposed in our work in 2006 [15], where the weights assigned to a component model are inversely proportional to its ageing time: $w_i=1-i/N$, where i is the index of a model, $i=0,1,2,..,N-1$, and N denotes the number of component models encompassed by the ensemble.

In this paper we present our first attempt to employ evolutionary fuzzy approach to explore data streams to model dynamic real estate market. The problem is not trivial because on the one hand a genetic fuzzy system needs a number of samples to be trained without overfitting and on the other hand the time window to determine a chunk of training data should be as small as possible to diminish the ageing impact and retain the model accuracy at an acceptable level. The processing time in this case is not a decisive factor because property valuation models need not to be updated and/or generated from scratch in an on-line mode. Thus, our approach, outlined above, raises a challenge to find the trade-off between the length of a sliding window delimiting the training dataset and the deteriorating impact of ageing models, overfitting, and computational efficiency. Moreover, the issue of how different trend functions modeled over different time intervals affect the accuracy of single and ensemble fuzzy models could be explored. In addition to this, different weighing techniques of component aged models in an ensemble could be investigated.

3   Experimental Setup and Results

The investigation was conducted with our experimental system implemented in Matlab environment using Fuzzy Logic, Global Optimization, Neural Network, and Statistics toolboxes. The system was designed to carry out research into machine learning algorithms using various resampling methods and constructing and evaluating ensemble models for regression problems.

Real-world dataset used in experiments was drawn from an unrefined dataset containing above 50 000 records referring to residential premises transactions accomplished in the Polish big city, mentioned in the previous section, within 11 years from 1998 to 2008. In this period most transactions were made with non-market prices when the council was selling flats to their current tenants on preferential terms. First of all, transactional records referring to residential premises sold at market prices were selected. Then the dataset was confined to sales transaction data of apartments built before 1997 and where the land was leased on terms of perpetual usufruct.

The final dataset counted the 5213 samples. Five following attributes were pointed out as main price drivers by professional appraisers: usable area of a flat (*Area*), age of a building construction (*Age*), number of storeys in the building (Storeys), number of rooms in the flat including a kitchen (*Rooms*), the distance of the building from the city centre (*Centre*), in turn, price of premises (*Price*) was the output variable.

The property valuation models were built by genetic fuzzy systems over chunks of data stream determined by the time span of 3, 6, 9, and 12 months, as described in the previous section. The parameters of the architecture of fuzzy systems as well as genetic algorithms are listed in Table 1. Similar designs are described in [7], [8], [14]. As test dataset the chunks of data stream specified by the time intervals of one and three months were employed. These intervals followed a time point t_0, which was set to January 1, 2006, i.e. the point of 2.92 on the x axis in the chart in Fig. 2. As a performance function the mean absolute error (MSE) was used, and as aggregation functions of ensembles arithmetic averages were employed.

Table 1. Parameters of GFS used in experiments

Fuzzy system	Genetic Algorithm
Type of fuzzy system: Mamdani	Chromosome: rule base and mf, real-coded
No. of input variables: 5	Population size: 100
Type of membership functions (mf): triangular	Fitness function: MSE
No. of input mf: 3	Selection function: tournament
No. of output mf: 5	Tournament size: 4
No. of rules: 15	Elite count: 2
AND operator: prod	Crossover fraction: 0.8
Implication operator: prod	Crossover function: two point
Aggregation operator: probor	Mutation function: custom
Defuzzyfication method: centroid	No. of generations: 100

The trends were modelled using Matlab function polyfit(x,y,n), which finds the coefficients of a polynomial p(x) of degree n that fits the y data by minimizing the sum of the squares of the deviations of the data from the model (least-squares fit). Due to the relatively short ageing periods, we used in our study linear trend functions to model the changes of premises prices.

The resulting output of the ensemble for a given instance of the test dataset is computed as the arithmetic mean of the results produced by the component models and corrected by corresponding trend functions. Moreover, weighting component models in an ensemble was applied, using the weights inversely proportional to ageing time. The weights were determined according to the method described in the preceding section and are listed in Table 2, where the number of component models is given in the heading and the model indices in the first column. As can be seen the weights assigned to respective component GFSs are distinct for different size of ensembles.

Table 2. Weight values decreasing with the age of models

#	2	3	4	5	6	7	8	9	10	11	12
0	1.00	1.00	1.00	1.00	1.00	1.00	1.00	1.00	1.00	1.00	1.00
1	0.50	0.67	0.75	0.80	0.83	0.86	0.88	0.89	0.90	0.91	0.92
2		0.33	0.50	0.60	0.67	0.71	0.75	0.78	0.80	0.82	0.83
3			0.25	0.40	0.50	0.57	0.63	0.67	0.70	0.73	0.75
4				0.20	0.33	0.43	0.50	0.56	0.60	0.64	0.67
5					0.17	0.29	0.38	0.44	0.50	0.55	0.58
6						0.14	0.25	0.33	0.40	0.45	0.50
7							0.13	0.22	0.30	0.36	0.42
8								0.11	0.20	0.27	0.33
9									0.10	0.18	0.25
10										0.09	0.17
11											0.08

The performance of ageing single models created by genetic fuzzy systems (GFS) in terms of MAE is illustrated graphically in Figures 5-9. The x axis shows the age of models, i.e. the how many months passed from time when respective models were created to the time point t_0. The models were generated with training datasets determined by sliding windows of t_l=3, 6, 9, and 12 months respectively. The windows were shifted of t_s=3 months. Two the same test datasets, current for t_0, determined by the interval of 1 and 3 months were employed to all current and aged models. Following denotation was used in the legend of Figures 5-9: noT and withT mean that output produced by respective models for the current test dataset was not updated and updated, respectively, using linear trend functions determined over corresponding ageing intervals t_{ai} plus t_w. In turn, the numbers in round brackets, i.e. (i/j), denote a time span for training and test datasets respectively.

In the charts it is clearly seen that the MAE values for models whose output was not corrected by trend functions grow as ageing time increases. The reverse relation can be noticed when the results produced by models were updated with trend functions. Furthermore, the older models with trend correction reveal better performance than the less aged ones. This can be explained by less fluctuations of premises prices in earlier time intervals (see Fig. 2). Moreover, the shorter time span (starting from t_0) of test data the lower MAE value may indicate that data included in test sets also undergo ageing. In turn, no significant differences in accuracy can be observed between models generated for 3 and 12 month time intervals (see Fig. 9).

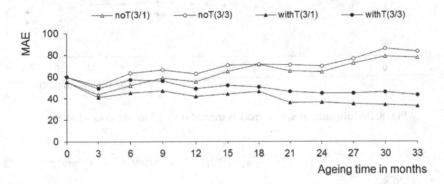

Fig. 5. Performance of ageing models trained over 3 month data windows

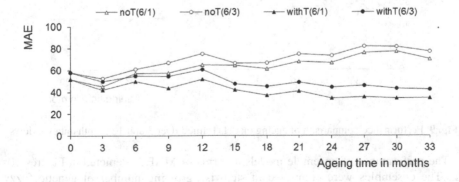

Fig. 6. Performance of ageing models trained over 6 month data windows

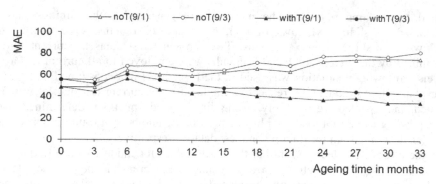

Fig. 7. Performance of ageing models trained over 9 month data windows

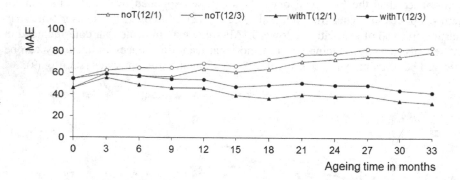

Fig. 8. Performance of ageing models trained over 12 month data windows

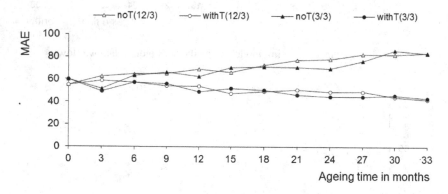

Fig. 9. Performance comparison of ageing models trained over 3 and 12 month data windows

The performance of ensemble models in terms of MAE is depicted in Figures 10-12. The ensembles were composed of stepwise growing number of genetic fuzzy systems (GFS), presented in Fig. 5 and 8. To a single model, current for t_0, more and more aged models, built over training data of the same time span, were added. The same test datasets, current for t_0, determined by the interval of 1 and 3 months were applied to each ensemble. However, in the paper we present only the results for the longer period. Following denotation was used in the legend of Figures 10-12: noT and withT analogously to Fig. 5-9 indicate whether the output provided by component models was updated with trend functions or not. The meaning of the numbers in round brackets remains the same. In turn the letter w denotes that component models were weighted using the weights inversely proportional to ageing time.

In the charts can be seen that the MAE values for ensembles where the output of component models was corrected by trend functions decrease as the number of GFSs grows. The reverse relation can be noticed when the results produced by component models were not updated with trend functions. Moreover, adding weights to models without trend correction leads to better performance but has no advantageous effect on ensembles with trend correction. This can be explained by better accuracy of older models with updated output. Finally, the ensembles encompassing models generated for 3 month time intervals reveal better accuracy the ones built over 12 month windows (see Fig. 12).

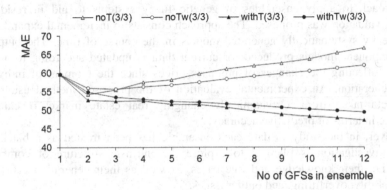

Fig. 10. Performance of ensembles comprising GFSs trained over 3 month data windows

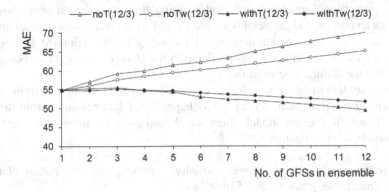

Fig. 11. Performance of ensembles comprising GFSs trained over 12 month data windows

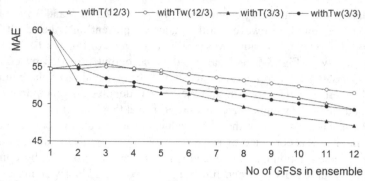

Fig. 12. Performance comparison of ensembles comprising GFSs trained over 3 and 12 month data windows

4   Conclusions and Future Work

An approach to apply ensembles of genetic fuzzy systems to aid in residential premises valuation was proposed. The approach consists in incremental expanding an ensemble by systematically generated models in the course of time. The output of aged component models produced for current data is updated according to a trend function reflecting the changes of premises prices since the moment of individual model generation. An experimental evaluation of the proposed method using real-world data taken from a dynamically changing real estate market revealed its advantage in terms of predictive accuracy.

However, in the study to date each ensemble has been treated as a black box. Further investigation is planned to explore the intrinsic structure of component models, i.e. their knowledge and rule bases, as well as their generation efficiency, interpretability, overfitting, and outlier issues.

Moreover, the weighting component models requires more thorough investigation. The technique we applied in our study consisted in assigning weights that were inversely proportional to component model ageing time. A question arises how big differences among individual weights should be and what should they depend on. It is planned to explore the weights determined according to the estimated accuracy of component models as well as set proportional to the change rate of prices per square metre within the sliding time window.

Another problem to tackle is as follows. When we try to build models from scratch over relatively small amount of data it may happen that data coming within the next period will not fit a given model. Then we should consider to employ clustering, random oracle, or stratification.

Acknowledgments. This paper was partially supported by the Polish National Science Centre under grant no. N N516 483840.

References

1. Alonso, J.M., Magdalena, L., González-Rodríguez, G.: Looking for a good fuzzy system interpretability index: An experimental approach. International Journal of Approximate Reasoning 51, 115–134 (2009)
2. Angelov, P.P., Filev, D.: An approach to online identification of Takagi-Sugeno fuzzy models. IEEE Transactions on Systems, Man and Cybernetics, part B 34(1), 484–498 (2004)
3. Breiman, L.: Bagging Predictors. Machine Learning 24(2), 123–140 (1996)
4. Brzeziński, D., Stefanowski, J.: Accuracy Updated Ensemble for Data Streams with Concept Drift. In: Corchado, E., Kurzyński, M., Woźniak, M. (eds.) HAIS 2011, Part II. LNCS (LNAI), vol. 6679, pp. 155–163. Springer, Heidelberg (2011)
5. Bühlmann, P., Yu, B.: Analyzing bagging. Annals of Statistics 30, 927–961 (2002)
6. Castro, J.L., Delgado, M.: Fuzzy systems with defuzzification are universal approximators. IEEE Transactions on System, Man and Cybernetics 26, 149–152 (1996)
7. Cordón, O., Gomide, F., Herrera, F., Hoffmann, F., Magdalena, L.: Ten years of genetic fuzzy systems: current framework and new trends. Fuzzy Sets and Systems 141, 5–31 (2004)
8. Cordón, O., Herrera, F.: A Two-Stage Evolutionary Process for Designing TSK Fuzzy Rule-Based Systems. IEEE Tr. on Sys., Man, and Cyb.-Part B 29(6), 703–715 (1999)
9. Elwell, R., Polikar, R.: Incremental Learning of Concept Drift in Nonstationary Environments. IEEE Transactions on Neural Networks 22(10), 1517–1531 (2011)
10. Fumera, G., Roli, F., Serrau, A.: A theoretical analysis of bagging as a linear combination of classifiers. IEEE Transactions on Pattern Analysis and Machine Intelligence 30(7), 1293–1299 (2008)
11. Gaber, M.M.: Advances in data stream mining. Wiley Interdisciplinary Reviews: Data Mining and Knowledge Discovery 2(1), 79–85 (2012)
12. Kempa, O., Lasota, T., Telec, Z., Trawiński, B.: Investigation of Bagging Ensembles of Genetic Neural Networks and Fuzzy Systems for Real Estate Appraisal. In: Nguyen, N.T., Kim, C.-G., Janiak, A. (eds.) ACIIDS 2011, Part II. LNCS (LNAI), vol. 6592, pp. 323–332. Springer, Heidelberg (2011)
13. Kosko, B.: Fuzzy systems as universal approximators. IEEE Transactions on Computers 43(11), 1329–1333 (1994)
14. Król, D., Lasota, T., Trawiński, B., Trawiński, K.: Investigation of Evolutionary Optimization Methods of TSK Fuzzy Model for Real Estate Appraisal. International Journal of Hybrid Intelligent Systems 5(3), 111–128 (2008)
15. Król, D., Szymański, M., Trawiński, B.: The recommendation mechanism in an internet information system with time impact coefficient. International Journal of Computer Science Applications 3(2), 65–80 (2006)
16. Kuncheva, L.I.: Classifier Ensembles for Changing Environments. In: Roli, F., Kittler, J., Windeatt, T. (eds.) MCS 2004. LNCS, vol. 3077, pp. 1–15. Springer, Heidelberg (2004)
17. Lasota, T., Telec, Z., Trawiński, B., Trawiński, K.: Investigation of the eTS Evolving Fuzzy Systems Applied to Real Estate Appraisal. Journal of Multiple-Valued Logic and Soft Computing 17(2-3), 229–253 (2011)
18. Lasota, T., Telec, Z., Trawiński, G., Trawiński, B.: Empirical Comparison of Resampling Methods Using Genetic Fuzzy Systems for a Regression Problem. In: Yin, H., Wang, W., Rayward-Smith, V. (eds.) IDEAL 2011. LNCS, vol. 6936, pp. 17–24. Springer, Heidelberg (2011)

19. Lughofer, E., Trawiński, B., Trawiński, K., Kempa, O., Lasota, T.: On Employing Fuzzy Modeling Algorithms for the Valuation of Residential Premises. Information Sciences 181, 5123–5142 (2011)
20. Lughofer, E.: Evolving Fuzzy Systems – Methodologies, Advanced Concepts and Applications. STUDFUZZ, vol. 266. Springer, Heidelberg (2011)
21. Lughofer, E.: FLEXFIS: A robust incremental learning approach for evolving TS fuzzy models. IEEE Transactions on Fuzzy Systems 16(6), 1393–1410 (2008)
22. Maloof, M.A., Michalski, R.S.: Incremental learning with partial instance memory. Artificial Intelligence 154(1-2), 95–126 (2004)
23. Minku, L.L., White, A.P., Yao, X.: The Impact of Diversity on Online Ensemble Learning in the Presence of Concept Drift. IEEE Transactions on Knowledge and Data Engineering 22(5), 730–742 (2010)
24. Schapire, R.E.: The strength of weak learnability. Mach. Learning 5(2), 197–227 (1990)
25. Tsymbal, A.: The problem of concept drift: Definitions and related work. Technical Report. Department of Computer Science, Trinity College, Dublin (2004)
26. Wang, H., Fan, W., Yu, P.S., Han, J.: Mining concept-drifting data streams using ensemble classifiers. In: Getoor, L., et al. (eds.) KDD 2003, pp. 226–235. ACM Press (2003)
27. Widmer, G., Kubat, M.: Learning in the presence of concept drift and hidden contexts. Machine Learning 23, 69–101 (1996)
28. Wolpert, D.H.: Stacked Generalization. Neural Networks 5(2), 241–259 (1992)
29. Zliobaite, I.: Learning under Concept Drift: an Overview. Technical Report. Faculty of Mathematics and Informatics. Vilnius University, Vilnius (2009)

Navigating Interpretability Issues in Evolving Fuzzy Systems*

Edwin Lughofer

Johannes Kepler University of Linz, Austria
edwin.lughofer@jku.at

Abstract. In this position paper, we are investigating interpretability issues in the context of evolving fuzzy systems (EFS). Current EFS approaches, developed during the last years, are basically providing methodologies for precise modeling tasks, i.e. relations and system dependencies implicitly contained in on-line data streams are modeled as accurately as possible. This is achieved by permanent dynamic updates and evolution of structural components. Little attention has been paid to the interpretable power of these evolved systems, which, however, originally was one fundamental strength of fuzzy models over other (data-driven) model architectures. This paper will present the (little) achievements already made in this direction, discuss new concepts and point out open issues for future research. Various well-known and important interpretability criteria will serve as basis for our investigations.

Keywords: evolving fuzzy systems, interpretability criteria, on-line assurance.

1 Introduction

1.1 Motivation and State-of-the-Art

In today's industrial systems, the automatic adaptation of system models with new incoming data samples plays more and more a major role. This is due to an increasing complexity of the systems, as changing environments, upcoming new systems states or new operation modes not contained in pre-collected training sets are often not covered by an initial model setup in an off-line stage: in fact, collecting data samples for all possible system states and environmental behaviors would require a huge amount of operator's time, causing significant expenses for companies. Evolving models as part of the evolving intelligent system community [2] are providing methodologies for permanent model updates, mostly in single-pass and incremental manner, thus being able to dynamically update their structures in changing environments [28] — in fact, learning in non-stationary environments can be seen as a comprehensive field of research in this direction as including evolving as well as incremental data stream learning strategies for various machine learning and soft computing models/concepts.

---

* This work was funded by the Austrian fund for promoting scientific research (FWF, contract number I328-N23, acronym IREFS). This publication reflects only the authors' views.

E. Hüllermeier et al. (Eds.): SUM 2012, LNAI 7520, pp. 141–153, 2012.

Improved transparency and interpretability of the evolved models may be useful in several real-world applications where the operators and experts intend to gain a deeper understanding which interrelations, dependencies exist in the system in order to enrich their own knowledge and to be able to interpret the characteristics of the system on a more deeper level. For instance, in quality control applications, it could be very useful to know which interrelations of system variables/features were violated in case of faults; This may serve as valuable input to fault diagnosis and reasoning approaches [13] for tracing back the origin of the faults and providing appropriate automatic feedback control reactions [21]. Another example are decision support [10] systems, where it is sometimes necessary to know why certain decisions where made by classification models, especially in medical applications, see for instance [31].

Evolving fuzzy systems (EFS) are a powerful tool to tackle these demands as they are offering model architectures from the field of fuzzy logic and systems including components with a clear linguistically interpretable meaning [7]. Indeed, when trained from data, instead of designed manually by an expert, they are loosing some of their interpretable power, but still maintain some interpretability, especially when specific improvement techniques are applied, see e.g. [33] [11]. In the evolving sense, so far the main focus has been laid on precise modeling techniques, which are aiming for high accuracy on new unseen data and in a mathematical sense for coming close to optimality in an incremental optimization context — see [18] for a survey of precise EFS approaches. Little attention has been paid how to assure or at least how to improve interpretability of these types of models (most of these are only covering techniques for reducing complexity of models). The upper part in the framework visualized in Figure 1 shows the components which are integrated in current EFS approaches, often leading to weird fuzzy partitions and confusing rule bases. The lower part demonstrates fuzzy modeling from the viewpoint of human beings. Their expert knowledge, usually acquired within a long-term experience by working at the corresponding system, is used as basis for designing fuzzy rule bases directly by encoding this knowledge as a combination of linguistic terms and rules.

1.2 Aim of the Paper

Thus, in this paper, we are discussing the (little) achievements already made in this direction, discuss new concepts and point out open issues for future research. We are aiming for laying a bridge between linguistic and precise fuzzy modeling in an incremental learning context (indicated by the dashed arrows in Figure 1). Within this context, well-known interpretability criteria from the world of batch trained resp. expert-based fuzzy systems will serve as basis for our investigations, in particular 1) *distinguishability and simplicity*, 2) *consistency*, 3) *coverage and completeness*, 4) *feature importance levels*, 5) *rule importance levels*, and 6) *interpretation of consequents*, covering both, high-level (rule-based level) and low-level (fuzzy-set level) interpretation spirits according to [33] (there investigated within the scope of completely off-line fuzzy models). One important point in our investigations will be the on-line capability of concepts for

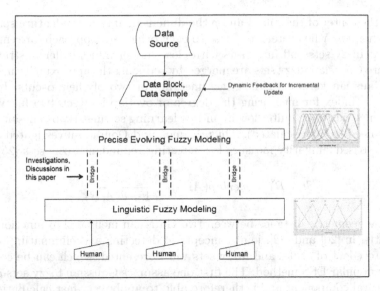

Fig. 1. Framework showing precise evolving and linguistic fuzzy modeling components (based on data and human input)

improving/ensuring the above criteria, in order to be applicable in fast processes along the incremental learning methods of EFS.

2 Distinguishability and Simplicity

While *distinguishability* expects the usage of structural components (rules, fuzzy sets) which are clearly separable as non-overlapping and non-redundant, *simplicity* goes a step further and expects models with a low complexity, at least as low as possible in order to still achieve a reasonable accuracy.

2.1 Distinguishability

In current evolving fuzzy systems guided by precise incremental and recursive modeling techniques, trying to position the structural components following the natural distribution of the data, a low distinguishability may appear whenever rules and fuzzy sets are moved, adapted, re-set due to new incoming data samples from an infinite stream. For instance, consider two data clouds representing two local distributions of the data which are adequately modeled by two rules in the fuzzy systems; after some time, the gap between the two clouds is filled up with new data, which makes the two clouds indistinguishable, and so also the two rules.

Thus, during the last years, several attempts were made to address this issue. In [16], redundant rules are detected by a similarity measure S expressed by the sum of the absolute deviations between the normalized coordinates of two rule (cluster) centers:

$$S(A, B) = 1 - \frac{\|c_A - c_B\|}{p} \tag{1}$$

with c_A the center of the rule A and p the dimensionality of the feature space; S is high when two rule centers are close to each other. An approach for removing redundant fuzzy sets and fuzzy sets with close model values is demonstrated in [23], where Gaussian fuzzy sets are merged by reducing the approximate merging of two Gaussian kernels to the exact merging of two of their α-cuts. It uses the Jaccard index for measuring the degree of overlap between two fuzzy sets, which turned out to be quite slow in on-line learning settings (two time-intensive integrals need to be calculated). Thus, Ramos and Dourado investigated a fast geometric-based similarity measure for detecting redundant fuzzy sets [25]:

$$S(A, B) = overlap(A, B) = \frac{1}{1 + d(A, B)} \qquad (2)$$

with d the geometric distance between two Gaussian membership functions.

Recently, in [20] and [22], joint concepts of detecting and eliminating redundant (antecedent of) rules and fuzzy sets were presented, which can be coupled with any popular EFS method. The first one assumes Gaussian fuzzy sets as basic structural component and is therefore able to apply very fast calculations of the overlap degree between two rules as well as fuzzy sets. This is achieved by using virtual projections of the antecedent space onto the one-dimensional axis and aggregate over the maximal component-wise membership degree of intersection points between two Gaussians:

$$S(A, B) = overlap(A, B) = Agg_{j=1}^{p} overlap_{A,B}(j) \qquad (3)$$

with

$$overlap_{A,B}(j) = \max(\mu(inter_x(1)), \mu(inter_x(2))) \qquad (4)$$

$\mu(x)$ the membership degree to the univariate Gaussian and $inter_x(1, 2)$ the two intersection points. In [22], the generalization to arbitrary fuzzy sets is conducted by defining the overlap degree in form of a two-sided fuzzy inclusion measure.

In all approaches, the overlap resp. similarity degree is normalized to $[0, 1]$, where 0 means no overlap/similarity and 1 full overlap/similarity. Thus, if $S(A, B)$ is greater than a threshold (e.g. 0.8), merging of rules (defined by centers c^A and c^B, spreads σ^A and σ^B) can be conducted in $O(p)$ computation time with p the dimensionality of the feature space (for details on derviation see [20]):

$$c_j^{new} = \frac{c_j^A k_A + c_j^B k_B}{k_A + k_B}$$

$$\sigma_j^{new} = \sqrt{\frac{k_A(\sigma_j^A)^2}{k_A + k_B} + (c_j^A - c_j^{new})^2 + \frac{(c_j^{new} - c_j^B)^2}{k_A + k_B} + \frac{k_B \sigma_j^B}{k_A + k_B}}$$

$$k_{new} = k_A + k_B \qquad (5)$$

where k_A denotes the number of samples falling into rule A and k_B the number of samples falling into rule B, $k_A > k_B$. This is usually applied to rules updated during the last incremental learning cycle.

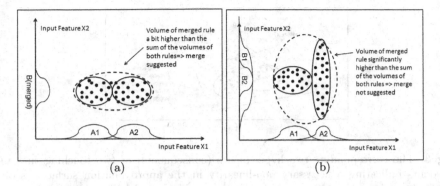

Fig. 2. (a): two rules (solid ellipsoids) which are touching each other and are homogeneous in the sense that the (volume, orientation of the) merged rule (dotted ellipsoid) is in conformity with the original two rules; (b): two rules (solid ellipsoids) which are touching each other and are not homogeneous (merged rule too extremely blown up)

2.2   Simplicity

Simplicity assurance can be seen as an extension, generalization of distinguishability assurance, as it deals with complexity reduction in a wider sense. In the well-known *SAFIS* approach [26], rule pruning is implicitly integrated in the incremental learning algorithm: rules are pruned whenever the statistical influence of a fuzzy rule measured in terms of its relative membership activation level (relative to the activation levels of all rules so far) is getting lower than a pre-defined threshold. In this sense, the rule can be seen as unimportant for the final model output over past data cycles and therefore deleted. Another attempt for deleting weakly activated rules is demonstrated in *simple_eTS* [1], where rules with very low support are deleted.

Here, we want to present an additional possibility by extending the considerations made in the previous paragraph to the case where rules are moving together, but are not significantly overlapping. Following the similarity-based criteria in the previous paragraph including geometric properties, we consider a new geometric-based touching criterion for ellipsoids. The condition whether two rules A and B overlap or touch can be deduced as follows (proof is left to the reader):

$$d(c_A, c_B) \leq \frac{\sum_{k=1}^{p} |c_{A;k} - c_{B;k}|(fac * \sigma_{A;k} + fac * \sigma_{B;k})}{\sum_{k=1}^{p} |c_{A;k} - c_{B;k}|} + \epsilon \qquad (6)$$

The second part of the merge condition is driven by the fact that the two rules should form a homogenous region when joined together. In order to illustrate this, the two rules shown in Figure 2 (a) can be safely merged together, wherever the two rules shown in (b) are representing data clouds with different orientation, shape characteristics, thus merging has to be handled with care. In fact, merging these two rules to one (shown by a dotted ellipsoid) would unnaturally

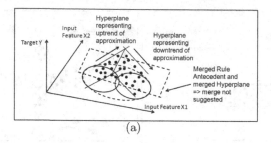

(a)

Fig. 3. The corresponding two hyper-planes (consequents) of two touching rule antecedents indicating a necessary non-linearity in the approximation surface, as one hyper-plane shows an up-trend and the other a downtrend in the functional dependency between input features X1, X2 and target Y; thus, a merged rule (indicated by dotted ellipsoids and hyper-plane) would yield an imprecise presentation and cause a low model quality in this region

blow up the range of influence of the new rule, yielding an imprecise presentation of the two data clouds, affecting the predictive quality of the model in that part. Therefore, we suggest to add a smooth homogeneity criterion based on the volumes of the original and (hypothetically) merged rules antecedents, characterizing a significant blow up of the merged rules compared to the original rules.

$$V_{merged} \leq p(V_A + V_B) \tag{7}$$

with p the dimensionality of the feature space and V the volume of an ellipsoid in main position.

The third part of the merge condition takes into account the consequents of the rules. In the sense of rules grown together, but not necessarily overlapping, consequents of the rules may point into different directions, accounting for a necessary non-linearity aspect implicitly contained in the data. In a classification context, this can be simply verified by comparing the majority class of two neighboring rules. In a regression context, we have to catch the degree of continuation of the local trend over the two nearby lying rules. This is essential as a different trend indicates a necessary non-linearity contained in the functional relation/approximation between inputs and outputs — see Figure 3. Thus, we propose a similarity measure based on the degree of deviation in the hyper-planes' gradient information, i.e. consequent parameter vectors without the intercept. This can be measured in terms of the angle ϕ between the two normal vectors defining the hyper-planes:

$$S_{cons}(A, B) = \frac{\phi}{\pi} \tag{8}$$

being maximal similar when ϕ is close to 180 degrees, indicating that the two hyper-planes the same trend of the approximation surface. Rules fulfilling all three conditions (only updated rules need to be checked in each incremental learning cycle) can be merged using (5).

3   Consistency

The inconsistency of the rule base can be measured in terms of the contradiction levels of two or more fuzzy rules contained in the rule base. For the sake of simplicity, we only consider the contradiction between two fuzzy rules, which can be easily generalized to multiple rule contradictions.

Definition 1. *Rule A is contradictory to Rule B if and only if* $S_{ante}(A, B) \geq S_{cons}(A, B)$ *with* $S_{ante}(A, B) \geq thresh$ *with* $thresh \in [0.5, 1.0]$.

We suggest the following three-way function for the consistency degree:

$$Cons(A, B) = \begin{cases} 0 & S_{ante} \geq thresh \wedge S_{ante} \geq S_{cons} \\ 1 & S_{ante} \geq thresh \wedge S_{ante} < S_{cons} \\ e^{-\frac{(\frac{S_{ante}(A,B)}{S_{cons}(A,B)} - 1)^2}{(\frac{1}{S_{ante}})^2}} & S_{ante} < thresh \end{cases} \quad (9)$$

The last part assures that in case of a low antecedent similarity the consistency is usually close to 1. The consistency of the overall rule base is then given by using the aggregation of the consistencies of the single rules:

$$Cons_{all} = Agg^C_{i,j=1; i \neq j} Cons(A_i, B_j) \quad (10)$$

A practicable operator for Agg could be the mean as lying in $[0, 1]$ and indicating an average expected consistency.

A high consistency level in the rule base in EFS within an incremental learning context can be achieved by applying the merging strategy as described in the previous section, using the overlap degree (3) as similarity for the antecedents and (8) as similarity for the consequents in case of Takagi-Sugeno type models and (3) as similarity on the two consequent fuzzy sets of the two rules in case of Mamdani fuzzy systems. Thus, only weak inconsistencies may be left (with $S_{ante} < thresh$ and $S_{cons} << S_{ante}$).

4   Coverage and Completeness

Coverage is guaranteed whenever fuzzy sets with infinite support are used (such as for instance Gaussian membership functions). As long as coverage can be seen as a special case of ϵ-completeness (with ϵ a very small number) in the fuzzy partition space, we are providing the definition of ϵ-completeness.

Definition 2. *A fuzzy partition containing the fuzzy sets* $\mu_1, ..., \mu_C$ *is said to be* ϵ-complete *whenever for two adjacent fuzzy sets* μ_m *and* μ_{m+1} *the overlap degree is at least* ϵ, *with* $\epsilon > 0$ *a small positive number.*

The number of approaches which are taking this criterion into account and assuring always significantly covered input states, are quite rare: one of the most well-known approaches supporting this concept internally during the incremental learning scheme, is the *SOFNN* approach [14]. In fact, the rule evolution machine itself relies on the concepts of ϵ-completeness. Another approach which assures a minimum of overlap between adjacent fuzzy sets is the *SAFIS* approach, [26]. There, an overlap factor κ is defined and integrated into the "influence of a fuzzy rule" concept, which is responsible whether new rules are evolved or not. Here, we are additionally investigating an enhanced concept directly integrated in the incremental optimization procedure of non-linear antecedent parameters.

Minimizing the non-linear least squares error measure $J = \sum_{k=1}^{N}(\hat{y}(k)-y(k))^2$ fits the non-linear parameters as close as possible to the data streams samples, thus yielding a precise model. In order to investigate ϵ-completeness of fuzzy partitions, the idea is to relax this hard fitting criterion by an additional terms which punishes those partitions in which fuzzy sets have a lower degree of overlap. Aggregating this on rule level, we suggest to include the following term in the optimization problem:

$$\sum_{k=1}^{N} \frac{\prod_{i=1}^{C} \max(0, \epsilon - \mu_i(\boldsymbol{x}_k))}{\epsilon^C} \tag{11}$$

with N the number of data samples seen so far, C the number of rules in the system and $\mu_i(\boldsymbol{x})$ the membership degree of the current sample to rule i; taking the minimum over the degrees to the fuzzy sets, reflects the minimum completeness over all antecedent parts (fuzzy partitions). If $\mu_i(\boldsymbol{x})$ is close to 0 for all rules, the sample is not sufficiently covered and therefore the (11) approaches 1 due to the normalization term ϵ^C. Assuming $[0,1]$ normalized target values y, the combined optimization criterion becomes:

$$J_{ext} = \alpha(\sum_{k=1}^{N}(y_k - \sum_{i=1}^{C} l_i(\boldsymbol{x}_k)\Psi_i(\Phi_{nonlin})(\boldsymbol{x}_k))^2) + \beta \sum_{k=1}^{N}(\frac{\prod_{i=1}^{C} \max(0, \epsilon - \mu_i(\boldsymbol{x}_k))}{\epsilon^C}) \tag{12}$$

with α and β denoting the importance level of precision versus completeness, thus being able to tradeoff accuracy and interpretability in a continuous manner, and Ψ_i the ith basis functions (normalized membership degree to the ith rule). By including the derivatives of both terms w.r.t. all non-linear parameters in the Jacobian, incremental optimization formulas such as the incremental Levenberg-Marquardt update formulas can be applied in the same manner as in [30]. The derivatives of the centers and spreads in Gaussian fuzzy sets w.r.t to the first function are presented in [18], Section 2.7, Chapter 2, the derivatives of the second term Pun w.r.t. to non-linear parameters Φ_n are (proof left to the reader):

$$\frac{\partial Pun}{\partial \Phi_n} = \begin{cases} \sum_{k=1}^{N} \frac{1}{\epsilon^C} \frac{\partial \mu_n(\boldsymbol{x}_k)}{\partial \Phi_n} \prod_{i=1, i\neq n}^{C} \max(0, \epsilon - \mu_i(\boldsymbol{x}_k)) & \epsilon - \mu_n(\boldsymbol{x}_k) > 0 \\ 0 & else \end{cases} \tag{13}$$

5   Feature Weights

Feature weights can be seen as important from two viewpoints of interpretability:

1. They may give rise to the importance of features either in a global or even in a local sense, thus making the features themselves interpretable for the process, i.e. telling an expert or operator which features have a higher and which ones a lower influence on the final model output.
2. They may also lead to a reduction of the rule lengths, as features with weights smaller than ϵ (also called *out-weighted features*) have a very low impact on the final model output and therefore can be eliminated from the rule antecedent and consequent parts.

An approach which is addressing incremental feature weighting strategies within the context of evolving fuzzy classifiers is presented in [19]. It operates on a global basis, hence features are either seen as important or unimportant for the whole model, and it is focussed on dynamic classification problems. The basic idea is that feature weights $\lambda_1, ..., \lambda_p$ for the p features included in the learning problem are calculated based on a stable separability criterion [9]:

$$J = trace(S_w^{-1} S_b) \tag{14}$$

with S_w the within scatter matrix modeled by the sum of the covariance matrices for each class, and S_b the between scatter matrix, modeled by the sum of the degree of mean shift between classes. For normalization purposes to $[0, 1]$, finally the feature weights are defined by:

$$\lambda_j = 1 - \frac{J_j - min_{1,...,p}(J_j)}{max_{j=1,...,p}(J_j) - min_{1,...,p}(J_j)} \tag{15}$$

hence the feature with the weakest discriminatory power (and therefore maximal J_j) is assigned a weight value of 0 and the feature with strongest discriminatory power a weight of 1. The idea is that p criteria $J_1, ..., J_p$ are incrementally learned, where J_i represents (14) by excluding the ith feature, thus a high J_i close to J using all features points to an unnecessary feature, as the features were dropped without much affecting the separation power of the feature space.

Another attempt for automatic feature weighting and reduction is demonstrated in [3] for on-line regression problems, exploiting the Takagi-Sugeno fuzzy systems architecture with local linear consequent functions. The basic idea is to track the relative contribution of each feature, compared with the contributions of all features, to the model output over time.

6   Rule Importance Levels

Rule weights may serve as important corner stones for a smooth rule reduction during learning procedures, as rules with low weights can be seen as unimportant and may be pruned or even re-activated at a later stage. Furthermore, rule weights can be used to handle inconsistent rules in a rule base, see e.g. [24], thus serving for another possibility to tackle the problem of consistency, compare with Section 3.

The usage of rule weights and their updates during incremental learning phases, was, to our best knowledge, not studied so far in the evolving fuzzy community (open issue). A possibility would be to integrate the weights as multiplication factors of the rule membership degrees μ into the fuzzy model's inference scheme. In this sense, the rule weights are appearing as non-linear parameters, which may be optimized within incremental procedures such as recursive gradient descent (RGD), recursive Levenberg-Marquardt (RLM) [30] or recursive Gauss-Newton as applied in [12], according to the least-squares error problem or also using the punished measure as in (12).

Within an on-line modeling context, the interpretation of updating rule weights is the following: some rules which seem to be important at the stage when they are evolved may turn out to be less important at a later stage, without necessarily become overlapping to other rules (this issue is handled by the distinguishability considerations in Section 2). Current approaches such as $simpl_eTS$ [1] or $eTS+$ [3] tackle this issue by simply deleting the rules from the rule base (based on some criteria like utility or ages) in order to reduce complexity and enhance transparency as well as computation time. However, such rules may become important again at a later stage. Updating the rule weights may help here to find a smooth transition between out-weighting and reactivation of such rules; also, some rules may be a bit important and therefore should contribute a little to the final model output (e.g. with weight 0.3) and also within the optimization and learning procedures. The importance of rules can be directly linked with the approximation quality of the evolved fuzzy system through rule weights (as part of error minimization). This direct link 'rule importance↔system error' is also taken into account in the recently published approach by Leng et al. [15].

7   Interpretation of Consequents

Here, we are further investigating the consequents of Takagi-Sugeno type fuzzy systems [29], which are singletons, hyper-planes, higher order of polynomials or recently introduced as a linear combination of kernel functions [8]. The interpretation of singletons is quite clear, as they are representing single values, whose meanings the experts are usually aware of. Higher order polynomials are rarely used in the fuzzy community, and the interpretation of mixture models with kernels is from linguistic point of view almost impossible. Thus, we are focussing on the most convenient and widely used hyper-planes.

Indeed, in literature, see for instance [7] [11] [33], the hyper-plane type consequents are often seen as non-interpretable and therefore excluded from any linguistic modeling point of view. However, in practical applications they may offer some important insights regarding several topics:

– Trend analysis in certain local regions: this may be important for on-line control applications to gain knowledge about the control behavior of the signal: a constant behavior or rapidly changing behavior can be simply recognized by inspecting the consequent parameters (weights of the single variables) in the corresponding local regions.

- Feature importance levels: as already discussed in Section 5, a kind of local feature weighting effect can be achieved by interpreting the local gradients as sensitivity factors w.r.t. to changes of single variables in the corresponding regions.
- A physical interpretation of the consequent parameters can be achieved when providing a rule centering approach [6] and interpreting them as Taylor series expansion.

A pre-requisite to assure all the interpretation capabilities listed above are well-organized hyper-planes snuggling along the basic trend of the approximation surface and hence really providing partial functional tendencies; otherwise, they may point to any direction. The data-driven technique which is able to fulfill this requirement is the so-called *local learning* concept, which tries to estimate the consequent function of each rule separately: for the batch modeling case, the nice behavior of the local learning of consequent functions was verified in [32]; for the incremental evolving case, a deeper investigation of this issue was conducted in [23]. Local learning in incremental on-line mode relies on the concept of recursive fuzzily weighted least squares [4], whereas global learning relies on the conventional recursive least squares approach [17]. In [5], local learning was compared with global learning (learning all consequent parameters in one sweep) within the context of multi-model classifiers: it turned out that local learning was not only able to extract consequent functions with a higher interpretable capability, but also more robust fuzzy models in terms of accuracy and numerical stability during training. Due to these investigations, the older global approach can be seen as obsolete, more or less.

Smoothening with incremental regularization [27] techniques may further help to improve the interpretation of consequents, as preventing over-fitting according to significant noise levels.

8 Conclusion

This paper addresses several issues for improving the (linguistic) interpretability of evolving fuzzy systems, and does this in a balanced scheme of demonstrating already well-known techniques, extending the state-of-the-art, providing novel concepts and finally pointing out some important future topics. New concepts were presented mainly in the fields of 1.) *simplicity* by highlighting possible fast homogenous geometric-based criteria for rule reduction, 2.) by investigating techniques for assuring ϵ-completeness of fuzzy partitions and rule bases, and 3.) by introducing rule weights for tracking rule importance levels. Future important topics to be addressed are 1.) model-based reliability concepts ensuring enhanced interpretation of model predictions, 2.) steps in the direction of ensuring local property of EFS (= firing of only 2^p rules at the same time, with p the dimensionality of the feature space), and 3.) analysis and improvement of interpretability in multi model EFC architectures.

References

1. Angelov, P., Filev, D.: SimpLeTS: A simplified method for learning evolving Takagi-Sugeno fuzzy models. In: Proceedings of FUZZ-IEEE 2005, Reno, Nevada, U.S.A., pp. 1068–1073 (2005)
2. Angelov, P., Filev, D., Kasabov, N.: Evolving Intelligent Systems — Methodology and Applications. John Wiley & Sons, New York (2010)
3. Angelov, P.P.: Evolving Takagi-Sugeno fuzzy systems from streaming data, eTS+. In: Angelov, P., Filev, D., Kasabov, N. (eds.) Evolving Intelligent Systems: Methodology and Applications, pp. 21–50. John Wiley & Sons, New York (2010)
4. Angelov, P.P., Filev, D.: An approach to online identification of Takagi-Sugeno fuzzy models. IEEE Transactions on Systems, Man and Cybernetics, Part B: Cybernetics 34(1), 484–498 (2004)
5. Angelov, P.P., Lughofer, E., Zhou, X.: Evolving fuzzy classifiers using different model architectures. Fuzzy Sets and Systems 159(23), 3160–3182 (2008)
6. Bikdash, M.: A highly interpretable form of Sugeno inference systems. IEEE Transactions on Fuzzy Systems 7(6), 686–696 (1999)
7. Casillas, J., Cordon, O., Herrera, F., Magdalena, L.: Interpretability Issues in Fuzzy Modeling. Springer, Heidelberg (2003)
8. Cheng, W.Y., Juang, C.F.: An incremental support vector machine-trained ts-type fuzzy system for online classification problems. Fuzzy Sets and Systems 163(1), 24–44 (2011)
9. Dy, J.G., Brodley, C.E.: Feature selection for unsupervised learning. Journal of Machine Learning Research 5, 845–889 (2004)
10. Feng, J.: An intelligent decision support system based on machine learning and dynamic track of psychological evaluation criterion. In: Kacpzryk, J. (ed.) Intelligent Decision and Policy Making Support Systems. Springer, Heidelberg (2008)
11. Gacto, M.J., Alcala, R., Herrera, F.: Interpretability of linguistic fuzzy rule-based systems: An overview of interpretability measures. Information Sciences 181(20), 4340–4360 (2011)
12. Kalhor, A., Araabi, B.N., Lucas, C.: An online predictor model as adaptive habitually linear and transiently nonlinear model. Evolving Systems 1(1), 29–41 (2010)
13. Korbicz, J., Koscielny, J.M., Kowalczuk, Z., Cholewa, W.: Fault Diagnosis - Models, Artificial Intelligence and Applications. Springer, Heidelberg (2004)
14. Leng, G., McGinnity, T.M., Prasad, G.: An approach for on-line extraction of fuzzy rules using a self-organising fuzzy neural network. Fuzzy Sets and Systems 150(2), 211–243 (2005)
15. Leng, G., Zeng, X.-J., Keane, J.A.: An improved approach of self-organising fuzzy neural network based on similarity measures. Evolving Systems 3(1), 19–30 (2012)
16. Lima, E., Hell, M., Ballini, R., Gomide, F.: Evolving fuzzy modeling using participatory learning. In: Angelov, P., Filev, D., Kasabov, N. (eds.) Evolving Intelligent Systems: Methodology and Applications, pp. 67–86. John Wiley & Sons, New York (2010)
17. Ljung, L.: System Identification: Theory for the User. Prentice Hall PTR, Prentic Hall Inc., Upper Saddle River, New Jersey (1999)
18. Lughofer, E.: Evolving Fuzzy Systems — Methodologies, Advanced Concepts and Applications. Springer, Heidelberg (2011)
19. Lughofer, E.: On-line incremental feature weighting in evolving fuzzy classifiers. Fuzzy Sets and Systems 163(1), 1–23 (2011)

20. Lughofer, E., Bouchot, J.-L., Shaker, A.: On-line elimination of local redundancies in evolving fuzzy systems. Evolving Systems 2(3), 165–187 (2011)
21. Lughofer, E., Eitzinger, C., Guardiola, C.: On-line quality control with flexible evolving fuzzy systems. In: Sayed-Mouchaweh, M., Lughofer, E. (eds.) Learning in Non-Stationary Environments: Methods and Applications. Springer, New York (2012)
22. Lughofer, E., Hüllermeier, E.: On-line redundancy elimination in evolving fuzzy regression models using a fuzzy inclusion measure. In: Proceedings of the EUSFLAT 2011 Conference, Aix-Les-Bains, France, pp. 380–387. Elsevier (2011)
23. Lughofer, E., Hüllermeier, E., Klement, E.P.: Improving the interpretability of data-driven evolving fuzzy systems. In: Proceedings of EUSFLAT 2005, Barcelona, Spain, pp. 28–33 (2005)
24. Pal, N.R., Pal, K.: Handling of inconsistent rules with an extended model of fuzzy reasoning. Journal of Intelligent and Fuzzy Systems 7, 55–73 (1999)
25. Ramos, J.V., Dourado, A.: Pruning for interpretability of large spanned eTS. In: Proceedings of the 2006 International Symposium on Evolving Fuzzy Systems (EFS 2006), Lake District, UK, pp. 55–60 (2006)
26. Rong, H.-J.: Sequential adaptive fuzzy inference system for function approximation problems. In: Sayed-Mouchaweh, M., Lughofer, E. (eds.) Learning in Non-Stationary Environments: Methods and Applications. Springer, New York (2012)
27. Rosemann, N., Brockmann, W., Neumann, B.: Enforcing local properties in online learning first order TS-fuzzy systems by incremental regularization. In: Proceedings of IFSA-EUSFLAT 2009, Lisbon, Portugal, pp. 466–471 (2009)
28. Sayed-Mouchaweh, M., Lughofer, E.: Learning in Non-Stationary Environments: Methods and Applications. Springer, New York (2012)
29. Takagi, T., Sugeno, M.: Fuzzy identification of systems and its applications to modeling and control. IEEE Transactions on Systems, Man and Cybernetics 15(1), 116–132 (1985)
30. Wang, W., Vrbanek, J.: An evolving fuzzy predictor for industrial applications. IEEE Transactions on Fuzzy Systems 16(6), 1439–1449 (2008)
31. Wetter, T.: Medical Decision Support Systems. In: Brause, R., Hanisch, E. (eds.) ISMDA 2000. LNCS, vol. 1933, pp. 1–3. Springer, Heidelberg (2000)
32. Yen, J., Wang, L., Gillespie, C.W.: Improving the interpretability of TSK fuzzy models by combining global learning and local learning. IEEE Transactions on Fuzzy Systems 6(4), 530–537 (1998)
33. Zhou, S.M., Gan, J.Q.: Low-level interpretability and high-level interpretability: a unified view of data-driven interpretable fuzzy systems modelling. Fuzzy Sets and Systems 159(23), 3091–3131 (2008)

Certain Conjunctive Query Answering in SQL

Alexandre Decan, Fabian Pijcke, and Jef Wijsen

Université de Mons, Mons, Belgium,
{alexandre.decan,jef.wijsen}@umons.ac.be, fabian.pijcke@gmail.com

Abstract. An uncertain database **db** is defined as a database in which distinct tuples of the same relation can agree on their primary key. A repair (or possible world) of **db** is then obtained by selecting a maximal number of tuples without ever selecting two distinct tuples of the same relation that agree on their primary key. Given a query Q on **db**, the *certain answer* is the intersection of the answers to Q on all repairs. Recently, a syntactic characterization was obtained of the class of acyclic self-join-free conjunctive queries for which certain answers are definable by a first-order formula, called certain first-order rewriting [15]. In this article, we investigate the nesting and alternation of quantifiers in certain first-order rewritings, and propose two syntactic simplification techniques. We then experimentally verify whether these syntactic simplifications result in lower execution times on real-life SQL databases.

1 Introduction

Uncertainty can be modeled in the relational model by allowing primary key violations. Primary keys are underlined in the conference planning database $\mathbf{db_0}$ in Fig. 1. There are still two candidate cities for organizing SUM 2016 and SUM 2017. The table S shows controversy about the attractiveness of Mons, while information about the attractiveness of Gent is missing. A *repair* (or

R	Conf	Year	Town
	SUM	2012	Marburg
	SUM	2016	Mons
	SUM	2016	Gent
	SUM	2017	Rome
	SUM	2017	Paris

S	Town	Attractiveness
	Charleroi	C
	Marburg	A
	Mons	A
	Mons	B
	Paris	A
	Rome	A

Fig. 1. Uncertain database $\mathbf{db_0}$

possible world) is obtained by selecting a maximal number of tuples, without selecting two tuples with the same primary key value. Database $\mathbf{db_0}$ has 8 repairs, because there are two choices for SUM 2016, two choices for SUM 2017, and two choices for Mons' attractiveness. The following conjunctive query asks in which years SUM took place (or will take place) in a city with A attractiveness:

$$Q_0 = \{y \mid \exists z \big(R(\text{`SUM'}, y, z) \wedge S(\underline{z}, \text{`A'}) \big) \}.$$

E. Hüllermeier et al. (Eds.): SUM 2012, LNAI 7520, pp. 154–167, 2012.

The *certain (query) answer* is the intersection of the query answers on all repairs, which in this example is $\{2012, 2017\}$. Notice incidentally that $Q_0(\mathbf{db}_0)$ also contains 2016, but that answer is not certain, because in some repairs, the organizing city of SUM 2016 does not have A attractiveness.

For every database, the certain answer to Q_0 is obtained by the following first-order query:

$$\varphi_0 = \{y \mid \exists z R(\text{`SUM'}, \underline{y}, z) \wedge \forall z \Big( R(\text{`SUM'}, \underline{y}, z) \to \begin{bmatrix} S(\underline{z}, \text{`A'}) \wedge \\ \forall v \big( S(\underline{z}, v) \to v = \text{`A'} \big) \end{bmatrix} \Big) \}.$$

We call φ_0 a certain first-order rewriting for Q_0. Certain first-order rewritings are of practical importance, because they can be encoded in SQL, which allows to obtain certain answers using standard database technology. However, it is well-known that not all conjunctive queries have a certain first-order rewriting [4,16], and it remains an open problem to syntactically characterize the conjunctive queries that have one. Nevertheless, for conjunctive queries that are acyclic and self-join-free, such characterization has recently been found [15,17].

In this article, we focus on the class of acyclic self-join free conjunctive queries that have a certain first-order rewriting. We first provide algorithm NaiveFo which takes such query as input, and constructs its certain first-order rewriting. We then provide two theorems indicating that rewritings produced by algorithm NaiveFo can generally be "simplified" by (i) reducing the number of (alternations of) quantifier blocks and/or by (ii) reducing the quantifier nesting depth. Finally, the implementation of our theory shows that certain SQL rewriting is an effective and efficient technique for computing certain answers.

This article is organized as follows. Section 2 provides notations and definitions. In particular, we provide measures for describing the syntactic complexity of a first-order formula. Section 3 discusses related work. Section 4 introduces the construct of *attack graph* which is essential for the purpose of certain first-order rewriting. Section 5 gives the code of algorithm NaiveFo. Section 6 shows how rewritings can be simplified with respect to the complexity measures of Section 2. Section 7 reports on our experiments conducted on real-life SQL databases. Section 8 concludes the article.

2 Notations and Terminology

We assume a set of *variables* disjoint from a set **dom** of *constants*. We will assume some fixed total order on the set of variables, which will only serve to "serialize" sets of variables into sequences in a unique way. If x is a sequence of variables and constants, then $\text{vars}(x)$ is the set of variables that occur in x.

Let U be a set of variables. A *valuation over U* is a total mapping θ from U to **dom**. Such valuation θ is often extended to be the identity on constants and on variables not in U.

Key-Equal Atoms. Every *relation name* R has a fixed *signature*, which is a pair $[n, k]$ with $n \geq k \geq 1$: the integer n is the *arity* of the relation name and

$\{1, 2, \ldots, k\}$ is the *primary key*. If R is a relation name with signature $[n, k]$, then $R(s_1, \ldots, s_n)$ is an R-*atom* (or simply atom), where each s_i is a constant or a variable $(1 \leq i \leq n)$. Such atom is commonly written as $R(\boldsymbol{x}, \boldsymbol{y})$ where the primary key value $\boldsymbol{x} = s_1, \ldots, s_k$ is underlined and $\boldsymbol{y} = s_{k+1}, \ldots, s_n$. An atom is *ground* if it contains no variables. Two ground atoms $R_1(\underline{\boldsymbol{a}_1}, \boldsymbol{b}_1), R_2(\underline{\boldsymbol{a}_2}, \boldsymbol{b}_2)$ are *key-equal* if $R_1 = R_2$ and $\boldsymbol{a}_1 = \boldsymbol{a}_2$. The arity of an atom F, denoted $\mathsf{arity}(F)$, is the arity of its relation name.

Database and Repair. A *database schema* is a finite set of *relation names*. All constructs that follow are defined relative to a fixed database schema.

A *database* is a finite set $\mathbf{db}$ of ground atoms using only the relation names of the schema. Importantly, a database can contain distinct, key-equal atoms. Intuitively, if a database contains distinct, key-equal atoms A and B, then only one of A or B can be true, but we do not know which one. In this respect, the database contains uncertainty. A database $\mathbf{db}$ is *consistent* if it does not contain two distinct atoms that are key-equal. A *repair* of a database $\mathbf{db}$ is a maximal (under set inclusion) consistent subset of $\mathbf{db}$.

Conjunctive Queries. A *conjunctive query* is a pair (q, V) where $q = \{R_1(\underline{\boldsymbol{x}_1}, \boldsymbol{y}_1), \ldots, R_n(\underline{\boldsymbol{x}_n}, \boldsymbol{y}_n)\}$ is a finite set of atoms and V is a subset of the variables occurring in q. Every variable of V is *free*; the other variables are *bound*. This query represents the first-order formula $\exists u_1 \ldots \exists u_k \left(R_1(\underline{\boldsymbol{x}_1}, \boldsymbol{y}_1) \wedge \cdots \wedge R_n(\underline{\boldsymbol{x}_n}, \boldsymbol{y}_n) \right)$, in which $u_1, \ldots, u_k$ are all the variables of $\mathsf{vars}(\boldsymbol{x}_1 \boldsymbol{y}_1 \ldots \boldsymbol{x}_n \boldsymbol{y}_n) \backslash V$. If $V = \emptyset$, then Q is called *Boolean*. We write $\mathsf{vars}(q)$ for the set of variables that occur in q. We denote by $|q|$ the number of atoms in q, and we define $\mathsf{aritysum}(q) = \sum_{F \in q} \mathsf{arity}(F)$.

Let $\mathbf{db}$ be a database. Let $Q = (q, V)$ be a conjunctive query, and let $U = \mathsf{vars}(q)$. Let $\boldsymbol{x} = \langle x_1, \ldots, x_m \rangle$ be the variables of V ordered according to the total order on the set of variables. The *answer* to Q on $\mathbf{db}$, denoted $Q(\mathbf{db})$, is defined as follows:

$$Q(\mathbf{db}) = \{\theta(\boldsymbol{x}) \mid \theta \text{ is a valuation over } U \text{ such that } \theta(q) \subseteq \mathbf{db}\}.$$

In particular, if Q is Boolean, then either $Q(\mathbf{db}) = \{\langle\rangle\}$ (representing **true**) or $Q(\mathbf{db}) = \{\}$ (representing **false**).

We say that Q has a *self-join* if some relation name occurs more than once in q; if Q has no self-join, then it is called *self-join-free*. We write SJFCQ for the class of self-join-free conjunctive queries.

Certain Conjunctive Query Answering. Let $Q = (q, V)$ be a conjunctive query. The *certain answer* to Q on $\mathbf{db}$, denoted $Q_{\mathsf{sure}}(\mathbf{db})$, is defined as follows:

$$Q_{\mathsf{sure}}(\mathbf{db}) = \bigcap \{Q(\mathbf{rep}) \mid \mathbf{rep} \text{ is a repair of } \mathbf{db}\}.$$

Let $\langle x_1, \ldots, x_m \rangle$ be the ordered sequence of variables in V. We say that the certain answer to Q is *first-order computable* if there exists a first-order formula $\varphi(x_1, \ldots, x_m)$, with free variables $x_1, \ldots, x_m$, such that for every database $\mathbf{db}$, for every $a \in \mathbf{dom}^m$, $a \in Q_{\mathsf{sure}}(\mathbf{db}) \iff \mathbf{db} \models \varphi(a)$. The formula φ, if it exists, is called a *certain first-order rewriting* for Q; its encoding in SQL is a *certain SQL rewriting*.

Notational Conventions. We use letters F, G, H, I for atoms appearing in a query. For $F = R(\underline{x}, y)$, we denote by $\mathsf{KeyVars}(F)$ the set of variables that occur in x, and by $\mathsf{Vars}(F)$ the set of variables that occur in F, that is, $\mathsf{KeyVars}(F) = \mathsf{vars}(x)$ and $\mathsf{Vars}(F) = \mathsf{vars}(x) \cup \mathsf{vars}(y)$.

Acyclic Conjunctive Queries. A *join tree* τ for a conjunctive query $Q = (q, V)$ is an undirected tree whose vertices are the atoms of q such that whenever the same variable x occurs in two atoms F and G, then either $x \in V$ (i.e., x is free) or x occurs in each atom on the unique path linking F and G. We will assume that join trees are edge-labeled, such that an edge between F and G is labeled with the set $(\mathsf{Vars}(F) \cap \mathsf{Vars}(G)) \setminus V$. A conjunctive query Q is called *acyclic* if it has a join tree [2].

Quantifier Rank and Quantifier Alternation Depth. The *quantifier rank* of a first-order formula φ, denoted by $\mathsf{qr}(\varphi)$, is the depth of the quantifier nesting in φ and is defined as usual (see, for example, [10, page 32]):

- If φ is quantifier-free, then $\mathsf{qr}(\varphi) = 0$.
- $\mathsf{qr}(\varphi_1 \wedge \varphi_2) = \mathsf{qr}(\varphi_1 \vee \varphi_2) = \max\left(\mathsf{qr}(\varphi_1), \mathsf{qr}(\varphi_2)\right)$;
- $\mathsf{qr}(\neg\varphi) = \mathsf{qr}(\varphi)$;
- $\mathsf{qr}(\exists x\varphi) = \mathsf{qr}(\forall x\varphi) = 1 + \mathsf{qr}(\varphi)$.

A first-order formula φ is said to be in *prenex normal form* if it has the form $Q_1x_1 \ldots Q_nx_n\psi$, where Q_i's are either $\exists$ or $\forall$ and ψ is quantifier-free. We say that φ has *quantifier alternation depth* m if $Q_1x_1 \ldots Q_nx_n$ can be divided into m blocks such that all quantifiers in a block are of the same type and quantifiers in two consecutive blocks are different.

For formulas not in prenex normal form, the number of quantifier blocks is counted as follows. A universally quantified formula is a formula whose main connective is $\forall$. An existentially quantified formula is a formula whose main connective is $\exists$. The number of quantifier blocks in a first-order formula φ, denoted $\mathsf{qbn}(\varphi)$, is defined as follows.

- If φ is quantifier-free, then $\mathsf{qbn}(\varphi) = 0$.
- $\mathsf{qbn}(\varphi_1 \wedge \varphi_2) = \mathsf{qbn}(\varphi_1 \vee \varphi_2) = \mathsf{qbn}(\varphi_1) + \mathsf{qbn}(\varphi_2)$;
- $\mathsf{qbn}(\neg\varphi) = \mathsf{qbn}(\varphi)$;
- if φ is not universally quantified and $n \geq 1$, then $\mathsf{qbn}(\forall x_1 \ldots \forall x_n\varphi) = 1 + \mathsf{qbn}(\varphi)$; and
- if φ is not existentially quantified and $n \geq 1$, then $\mathsf{qbn}(\exists x_1 \ldots \exists x_n\varphi) = 1 + \mathsf{qbn}(\varphi)$.

For example, if φ is $\exists x \exists y(\exists u\varphi_1 \wedge \exists v\varphi_2)$ and φ_1, φ_2 are both quantifier-free, then $\mathsf{qbn}(\varphi) = 3$. Notice that φ has a prenex normal form with quantifier alternation depth equal to 1. Clearly, if φ is in prenex normal form, then the quantifier alternation depth of φ is equal to $\mathsf{qbn}(\varphi)$.

Proposition 1. *Every first-order formula φ has an equivalent one in prenex normal form with quantifier alternation depth less than or equal to* $\mathsf{qbn}(\varphi)$.

3 Related Work

Certain (or consistent) query answering was founded in the seminal work by Arenas, Bertossi, and Chomicki [1]. The current state of the art can be found in [3]. Fuxman and Miller [8] were the first ones to focus on certain first-order rewriting of SJFCQ queries under primary key constraints, with applications in the ConQuer system [7]. Their results have been generalized by Wijsen [15,17], who obtained a syntactic characterization of those acyclic SJFCQ queries that allow certain first-order rewriting. It follows from the proof of Corollary 5 in [14] that acyclicity is also implicit in the work of Fuxman and Miller.

   Given a Boolean query Q, $\mathsf{CERTAINTY}(Q)$ is the decision problem that takes as input an (uncertain) database **db** and asks whether Q evaluates to true on every repair. Wijsen [16] showed that the class SJFCQ contains Boolean queries Q such that $\mathsf{CERTAINTY}(Q)$ is in **P** but not first-order expressible. It is an open conjecture that for every Boolean SJFCQ query Q, it is the case that $\mathsf{CERTAINTY}(Q)$ is in **P** or **coNP**-complete. For queries with exactly two atoms, such dichotomy was recently shown true [9].

   Maslowski and Wijsen [12,11] have studied the complexity of the counting variant of $\mathsf{CERTAINTY}(Q)$, denoted $\natural\mathsf{CERTAINTY}(Q)$. Given a database **db**, the problem $\natural\mathsf{CERTAINTY}(Q)$ asks to determine the exact number of repairs of **db** that satisfy some Boolean query Q. They showed that that for every Boolean SJFCQ query Q, it is the case that $\natural\mathsf{CERTAINTY}(Q)$ is in **P** or $\natural$**P**-complete. The problem $\natural\mathsf{CERTAINTY}(Q)$ is closely related to query answering in probabilistic data models [5]. From the probabilistic database angle, our uncertain databases are a restricted case of *block-independent-disjoint* probabilistic databases [5,6]. A *block* in a database **db** is a maximal subset of key-equal atoms. If $\{R(\underline{a}, b_1), \ldots, R(\underline{a}, b_n)\}$ is a block of size n, then every atom of the block has a probability of $1/n$ to be selected in a repair of **db**. Every repair is a possible world, and all these worlds have the same probability.

4 Attack Graph

Let $Q = (q, V)$ be an acyclic SJFCQ query and $U = \mathsf{vars}(q)$. For every $F \in q$, we denote by $\mathsf{FD}(Q, F)$ the set of functional dependencies that contains $X \to Y$ whenever q contains some atom G with $G \neq F$ such that $X = \mathsf{KeyVars}(G) \setminus V$ and $Y = \mathsf{Vars}(G) \setminus V$. For every $F \in q$, we define:

$$F^{+,Q} = \{x \in U \mid \mathsf{FD}(Q, F) \models (\mathsf{KeyVars}(F) \setminus V) \to x\}.$$

Example 1. For the Boolean query Q in Fig. 2, we have $\mathsf{FD}(Q, F) \equiv \{x \to y, x \to z\}$, $\mathsf{FD}(Q, G) \equiv \{u \to x, x \to y, x \to z\}$, $\mathsf{FD}(Q, H) \equiv \{u \to x, x \to z\}$, and $\mathsf{FD}(Q, I) \equiv \{u \to x, x \to y\}$. It follows $F^{+,Q} = \{u\}$, $G^{+,Q} = \{x, y, z\}$, $H^{+,Q} = \{x, z\}$, and $I^{+,Q} = \{x, y\}$. More elaborated examples can be found in [15].

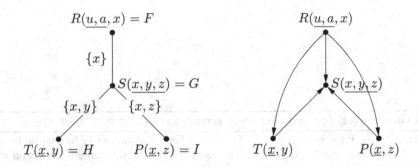

Fig. 2. Join tree (left) and attack graph (right) for Boolean query $Q = (q, \emptyset)$ with $q = \{R(\underline{u}, \underline{a}, x), S(\underline{x}, \underline{y}, z), T(\underline{x}, y), P(\underline{x}, z)\}$. It is understood that a is a constant.

Let τ be a join tree for Q. The *attack graph* of τ is a directed graph whose vertices are the atoms of q. There is a directed edge from F to some atom G if for every edge e on the unique path in τ that links F and G, there exists a variable x in e's edge label such that $x \notin F^{+,Q}$.

Example 2. See Fig. 2 (right). There is a directed edge from F to H because the edge labels $\{x\}$ and $\{x, y\}$ on the path between F and H are not contained in $F^{+,Q} = \{u\}$. The attack graph is acyclic.

It is known [17] that if τ_1 and τ_2 are two join trees for Q, then the attack graphs of τ_1 and τ_2 are identical. The attack graph of Q is then defined as the attack graph of any join tree for Q. If the attack graph contains a directed edge from atom F to atom G, then we say that F *attacks* G. An atom F of Q is said to be *unattacked* if the attack graph of Q contains no directed edge that points to F (i.e., F has zero indegree).

The following result was shown in [15] for Boolean acyclic SJFCQ queries. The extension to queries with free variables is easy and already indicated in [15].

Theorem 1. *Let $Q = (q, V)$ be an acyclic SJFCQ query. Then, Q has a certain first-order rewriting if and only if the attack graph of Q is acyclic.*

5 Naive Algorithm

Algorithm NaiveFo implements definitions given in [15]. The algorithm takes two inputs: an acyclic SJFCQ query $Q = (q, V)$ and a directed acyclic graph that contains Q's attack graph. The algorithm computes a certain first-order rewriting φ for Q. For example, the formula φ_0 in Section 1 was obtained by applying the algorithm on Q_0.

Algorithm NaiveFo is called "naive" because it does not attempt to mini-mize the alternations or nesting depth of quantifiers. This may be problematic when the rewritings are translated into SQL for execution, as illustrated by the following example.

Function NaiveFo(q,V,E) Construct certain first-order rewriting

---

Input: $Q = (q, V)$ is an acyclic SJFCQ query, where V is the set of free variables of q.

$E \subseteq q \times q$ is an acyclic set of directed edges containing the attack graph of Q.

Result: certain first-order rewriting φ for Q.

begin

    if $q = \emptyset$ **then**

        $\varphi \leftarrow$ **true**;

    else

        choose an atom $F = R(\underline{x_1, \ldots, x_k}, y_1, \ldots, y_\ell)$ that is unattacked in E;

        $V' \leftarrow V$;

        $X \leftarrow \emptyset$;

        foreach $i \leftarrow 1$ **to** k **do**

            if x_i *is a variable* **and** $x_i \notin V'$ **then**

                $V' \leftarrow V' \cup \{x_i\}$;

                $X \leftarrow X \cup \{x_i\}$;

        $Y \leftarrow \emptyset$;

        NEW $\leftarrow \emptyset$;

        foreach $i \leftarrow 1$ **to** ℓ **do**

            if y_i *is a constant* **or** $y_i \in V'$ **then**

                let z_i be a new variable;

                NEW $\leftarrow$ NEW $\cup \{i\}$;

            else   `/* `y_i` is a variable not in `V'` */`

                let z_i be the same variable as y_i;

                $V' \leftarrow V' \cup \{y_i\}$;

                $Y \leftarrow Y \cup \{y_i\}$;

        $q' \leftarrow q \setminus \{F\}$;

        $E' \leftarrow E \setminus (\{F\} \times q)$;

        $V' \leftarrow V' \cap \mathsf{vars}(q')$;

$$\varphi \leftarrow \exists X \left[\begin{array}{l} \exists Y\, R(\underline{x_1, \ldots, x_k}, y_1, \ldots, y_\ell) \wedge \\ \forall z_1 \ldots \forall z_\ell \Big(R(\underline{x_1, \ldots, x_k}, z_1, \ldots, z_\ell) \to \left[\begin{array}{l} \bigwedge_{i \in \mathrm{NEW}} z_i = y_i \wedge \\ \mathrm{NaiveFo}(q', V', E') \end{array} \right] \Big) \end{array} \right];$$

    return φ;

Example 3. For each $m \geq 1$, assume relation name R_i with signature $[2, 1]$, and let $\lfloor m \rfloor = \{R_1(\underline{x_1}, b), \ldots, R_m(\underline{x_m}, b)\}$, where b is a constant. For $m \geq 1$, let $\llbracket m \rrbracket = (\lfloor m \rfloor, \emptyset)$, a Boolean query whose attack graph has no edges. Formulas φ_1, φ_2, and φ_3 are three possible certain first-order rewritings for $\llbracket m \rrbracket$. The formula φ_1 is returned by algorithm NaiveFo, while φ_2 and φ_3 result from some syntactic simplification techniques described in Section 6. In particular, φ_2 minimizes the number of quantifier blocks, and φ_3 minimizes the nesting depth of quantifiers.

$$\varphi_1 = \exists x_1 \Big(R_1(\underline{x_1}, b) \wedge \forall z_1 \Big(R_1(\underline{x_1}, z_1) \to z_1 = b \wedge$$

$$\exists x_2 \Big(R_2(\underline{x_2}, b) \wedge \forall z_2 \Big(R_2(\underline{x_2}, z_2) \to z_2 = b \wedge$$

$$\ddots$$

$$\exists x_m \big(R_m(\underline{x_m}, b) \wedge \forall z_m \big(R_m(\underline{x_m}, z_m) \to z_m = b \big) \big) \cdots \Big) \Big) \Big) \Big)$$

$$\varphi_2 = \exists x_1 \ldots \exists x_m \Big(\bigwedge_{i=1}^{m} R_i(\underline{x_i}, b) \wedge \forall z_1 \ldots \forall z_m \big(\bigwedge_{i=1}^{m} R_i(\underline{x_i}, z_i) \to \bigwedge_{i=1}^{m} z_i = b \big) \Big)$$

$$\varphi_3 = \bigwedge_{i=1}^{m} \exists x_i \Big(R_i(\underline{x_i}, b) \wedge \forall z_i \big(R_i(\underline{x_i}, z_i) \to z_i = b \big) \Big)$$

Notice that φ_1, φ_2, and φ_3 each contain m existential and m universal quantifiers. The following table gives the quantifier rank and the number of quantifier blocks for these formulas; recall that these measures were defined in Section 2.

i	$\mathsf{qr}(\varphi_i)$	$\mathsf{qbn}(\varphi_i)$
1	$2m$	$2m$
2	$2m$	2
3	2	$2m$

The differences in syntactic complexity persist in SQL. Assume that for each $i \in \{1, \ldots, m\}$, the first and the second attribute of each R_i are named A and B respectively. Thus, A is the primary key attribute. For $m = 2$, the queries Q1, Q2, and Q3 in Fig. 3 are direct translations into SQL of φ_1, φ_2, and φ_3.[1] The fact that φ_2 only has one $\forall$ quantifier block results in Q2 having only one NOT EXISTS. Notice further that Q2 requires m tables in each FROM clause, whereas Q3 takes the intersection of m SQL queries, each with a single table in the FROM clause.

The foregoing example shows that formulas returned by algorithm NaiveFo can be "optimized" so as to have lower quantifier rank and/or less (alternations of) quantifiers blocks. The theoretical details will be given in the next section.

---

[1] In practice, we construct rewritings in tuple relational calculus (TRC), and then translate TRC into SQL. Such translations are well known (see, e.g., Chapter 3 of [13]). We omit the details of these translations in this article because of space limitations.

```
Q1 = SELECT 'true' FROM R1 AS r11
     WHERE  NOT EXISTS (SELECT * FROM R1 AS r12
                        WHERE  r12.A = r11.A
                        AND    (r12.B <> 'b'
                        OR     NOT EXISTS (SELECT * FROM R2 AS r21
                                           WHERE  NOT EXISTS (SELECT * FROM R2 AS r22
                                                              WHERE  r22.A = r21.A
                                                              AND    r22.B <> 'b'))))

Q2 = SELECT 'true' FROM R1 AS r11, R2 AS r21
     WHERE  NOT EXISTS (SELECT * FROM R1 AS r12, R2 AS r22
                        WHERE  r12.A = r11.A AND r22.A = r21.A
                        AND    (r12.B <> 'b' OR r22.B <> 'b'))

Q3 = SELECT 'true' FROM R1 AS r11
     WHERE  NOT EXISTS (SELECT * FROM R1 AS r12
                        WHERE  r12.A = r11.A AND r12.B <> 'b')
     INTERSECT
     SELECT 'true' FROM R2 AS r21
     WHERE  NOT EXISTS (SELECT * FROM R2 AS r22
                        WHERE  r22.A = r21.A AND r22.B <> 'b')
```

Fig. 3. Q1, Q2, Q3 are certain SQL rewritings for $\|2\|$

6   Syntactic Simplifications

Consider the second last line of algorithm NaiveFo, which specifies the first-order formula φ returned by a call NaiveFo(q, V, E) with $q \neq \emptyset$. Since NaiveFo is called recursively once for each atom of q, the algorithm can return a formula with $2|q|$ quantifier blocks and with quantifier rank as high as aritysum(q).

In this section, we present some theoretical results that can be used for constructing "simpler" certain first-order rewritings. Section 6.1 implies that NaiveFo can be easily modified so as to return formulas with less (alternations of) quantifier blocks. Section 6.2 presents a method for decreasing the quantifier rank. Importantly, our simplifications do not decrease (nor increase) the number of $\exists$ or $\forall$ quantifiers in a formula; they merely group quantifiers of the same type in blocks and/or decrease the nesting depth of quantifiers.

6.1   Reducing the Number of Quantifier Blocks

Algorithm NaiveFo constructs a certain first-order rewriting by treating one unattacked atom at a time. The next theorem implies that multiple unattacked atoms can be "rewritten" together, which generally results in less (alternations of) quantifier blocks, as expressed by Corollary 1.

Theorem 2. *Let $Q = (q, V)$ be an acyclic SJFCQ query. Let $S \subseteq q$ be a set of unattacked atoms in Q's attack graph. Let $X = \left(\bigcup_{F \in S} \mathsf{KeyVars}(F) \right) \setminus V$. If φ is a certain first-order rewriting for $(q, V \cup X)$, then $\exists X \varphi$ is a certain first-order rewriting for Q.*

Corollary 1. *Let (Q, V) be an acyclic SJFCQ query whose attack graph is acyclic. Let p be the number of atoms on the longest directed path in the attack graph of Q. There exists a certain first-order rewriting φ for Q such that $\mathsf{qbn}(\varphi) \leq 2p$.*

Example 4. The longest path in the attack graph of Fig. 2 contains 3 atoms. From Corollary 1 and Proposition 1, it follows that the query of Fig. 2 has a certain first-order rewriting with quantifier alternation depth less than or equal to 6.

6.2   Reducing the Quantifier Rank

Consider query $\lVert m \rVert$ with $m \geq 1$ in Example 3. Algorithm NaiveFo will "rewrite" the atoms $R_i(\underline{x_i}, b)$ sequentially ($1 \leq i \leq m$). However, since these atoms have no bound variables in common, it is correct to rewrite them "in parallel" and then join the resulting formulas. This idea is generalized in the following theorem.

Definition 1. *Let $Q = (q, V)$ be an SJFCQ query with $q \neq \emptyset$. An independent partition of Q is a (complete disjoint) partition $\{q_1, \ldots, q_k\}$ of q such that for $1 \leq i < j \leq k$, $\mathsf{vars}(q_i) \cap \mathsf{vars}(q_j) \subseteq V$.*

Theorem 3. *Let $Q = (q, V)$ be an acyclic SJFCQ query. Let $\{q_1, \ldots, q_k\}$ be an independent partition of Q. For each $1 \leq i \leq k$, let φ_i be a certain first-order rewriting for $Q_i = (q_i, V_i)$, where $V_i = V \cap \mathsf{vars}(q_i)$. Then, $\bigwedge_{i=1}^{k} \varphi_i$ is a certain first-order rewriting for Q.*

We show next that Theorem 3 gives us an upper bound on the quantifier rank of certain first-order rewritings. Intuitively, given a join tree τ, we define $\mathsf{diameter}(\tau)$ as the maximal sum of arities found on any path in τ. A formal definition follows.

Definition 2. *Let $Q = (q, V)$ be an acyclic SJFCQ query. Let τ be a join tree for Q. A chain in τ is a subset $q' \subseteq q$ such that the subgraph of τ induced by q' is a path graph. We define $\mathsf{diameter}(\tau)$ as the largest integer n such that $n = \mathsf{aritysum}(q')$ for some chain q' in τ.*

Example 5. The join tree τ in Fig. 2 (left) contains three maximal (under set inclusion) chains: $\{F, G, H\}$, $\{F, G, I\}$, $\{G, H, I\}$. The chains containing F have the greatest sum of arities; we have $\mathsf{diameter}(\tau) = 3 + 3 + 2 = 8$.

Corollary 2. *Let $Q = (q, V)$ be an acyclic SJFCQ query whose attack graph is acyclic. Let τ be a join tree for Q. There exists a certain first-order rewriting φ for Q such that $\mathsf{qr}(\varphi) \leq \mathsf{diameter}(\tau)$.*

Corollaries 1 and 2 show upper bounds on the number of quantifier blocks or the quantifier rank of certain first-order rewritings. Algorithm NaiveFo can be easily modified so as to diminish either of those measures. It is not generally possible to minimize both measures simultaneously. For example, there seems to be no certain first-order rewriting φ for $\lVert m \rVert$ such that $\mathsf{qbn}(\varphi) = \mathsf{qr}(\varphi) = 2$ (cf. the table in Section 5).

Schema	Arity	Number of tuples	Size
PAGE[id, namespace, title, ...]	12	4,862,082	417 MB
CATEGORY[id, title, ...]	6	296,002	13 MB
CATEGORYLINKS[from, to, ...]	7	14,101,121	1.8 GB
INTERWIKI[prefix, url, ...]	6	662	40 KB
EXTERNALLINKS[from, to, ...]	3	6,933,703	1.3 GB
IMAGELINKS[from, to]	2	12,419,720	428 MB

$$Q_{4,\text{fr}} = \exists^* \begin{bmatrix} \text{PAGE}(\underline{x}, n, l, \ldots) \wedge \text{CATEGORYLINKS}(\underline{x}, t, \ldots) \wedge \\ \text{CATEGORY}(\underline{y}, t, \ldots) \wedge \text{INTERWIKI}(\underline{\text{'fr'}}, u, \ldots) \end{bmatrix}$$

$$Q_{5,\text{fr}} = \exists^* \begin{bmatrix} \text{PAGE}(\underline{x}, n, l, \ldots) \wedge \text{CATEGORYLINKS}(\underline{x}, t, \ldots) \wedge \\ \text{CATEGORY}(\underline{y}, t, \ldots) \wedge \text{EXTERNALLINKS}(\underline{x}, u, \ldots) \wedge \\ \text{INTERWIKI}(\underline{\text{'fr'}}, u, \ldots) \end{bmatrix}$$

$$Q_{6,\text{fr}} = \exists^* \begin{bmatrix} \text{PAGE}(\underline{x}, n, l, \ldots) \wedge \text{CATEGORYLINKS}(\underline{x}, t, \ldots) \wedge \\ \text{CATEGORY}(\underline{y}, t, \ldots) \wedge \text{EXTERNALLINKS}(\underline{x}, u, \ldots) \wedge \\ \text{INTERWIKI}(\underline{\text{'fr'}}, u, \ldots) \wedge \text{IMAGELINKS}(\underline{x}, i) \end{bmatrix}$$

$$Q_{nB,\text{fr}}(u) = \exists x \exists n \exists l \exists t \exists y \begin{bmatrix} \text{PAGE}(\underline{x}, n, l, \ldots) \wedge \text{CATEGORYLINKS}(\underline{x}, t, \ldots) \wedge \\ \text{CATEGORY}(\underline{y}, t, \ldots) \wedge \text{EXTERNALLINKS}(\underline{x}, \underline{u}, \ldots) \wedge \\ \text{INTERWIKI}(\underline{\text{'fr'}}, u, \ldots) \end{bmatrix}$$

Fig. 4. Database schema and queries

7   Experiments

Databases and Queries We used a snapshot of the relational database containing Wikipedia's meta-data.[2] All experiments were conducted using MySQL version 5.1.61 on a machine with Intel core i7 2.9GHz CPU and 4GB RAM, running Gentoo Linux.

The database schema and the database size are shown in Fig. 4 (top). Attributes not shown are not relevant for our queries. To allow primary key violations, all primary key constraints and unique indexes were dropped and replaced by nonunique indexes. An inconsistent database was obtained by adding $\lceil N/10000 \rceil$ conflicting tuples to each relation with N tuples. Newly added tuples are *crossovers* of existing tuples: if $R(\underline{a_1}, b_1)$ and $R(\underline{a_2}, b_2)$ are distinct tuples, then $R(\underline{a_1}, b_2)$ is called a crossover.

In our experiments, we used eight acyclic SJFCQ queries which are in accordance with the intended semantics of the database schema. All queries are variations of $Q_{5,\text{fr}}$, which asks: "Is there some page with a category link to some category and with an external link to the wiki identified by 'fr'?". To help understanding, we point out that attribute CATEGORYLINKS.to refers to CATEGORY.title, not to CATEGORY.id. Four queries are shown in Fig. 4 (bottom). Each position that is not shown contains a new distinct variable not occurring elsewhere. Four other queries are obtained by replacing 'fr' with 'zu'.

---

[2] The dataset is publicly available at
http://dumps.wikimedia.org/frwiki/20120117/.

Table 1. Execution time in seconds for eight conjunctive queries and their certain SQL rewritings according to three different rewriting algorithms: ν=naive, β reduces $\mathsf{qbn}(\cdot)$, and ρ reduces $\mathsf{qr}(\cdot)$

	$\mathsf{qbn}(\cdot)$	$\mathsf{qr}(\cdot)$	'fr' variant on consistent/inconsistent database	'zu' variant on consistent/inconsistent database
Q_4	1	4	0.0004/0.0004	0.0004/0.0004
ν_4	7	7	0.2106/0.2114	0.2125/0.2117
β_4	3	5	> 1 hour	> 1 hour
ρ_4	5	4	0.2101/0.2108	0.2098/0.2106
Q_5	1	5	0.0004/0.0004	0.0004/0.0004
ν_5	9	9	0.0019/0.0020	2.1482/2.1472
β_5	5	7	0.0014/0.0015	2.1154/2.1152
ρ_5	7	6	0.0014/0.0015	2.1304/2.1298
Q_6	1	6	0.0004/0.0004	0.0004/0.0004
ν_6	11	11	0.0020/0.0021	2.1730/2.1725
β_6	5	8	0.0014/0.0015	2.1209/2.1211
ρ_6	8	6	0.0015/0.0016	2.1450/2.1439
Q_{nB}	1	4	0.0112/0.0095	0.0120/0.0099
ν_{nB}	8	8	0.0018/0.0019	2.1379/2.1374
β_{nB}	4	6	0.0015/0.0015	2.1098/2.1103
ρ_{nB}	6	5	0.0015/0.0016	2.1272/2.1274

That is, $Q_{4,\mathrm{zu}}$ is obtained from $Q_{4,\mathrm{fr}}$ by replacing 'fr' with 'zu'. Likewise for $Q_{5,\mathrm{zu}}$, $Q_{6,\mathrm{zu}}$, and $Q_{nB,\mathrm{zu}}$. All queries are Boolean except for $Q_{nB,\mathrm{fr}}$ and $Q_{nB,\mathrm{zu}}$. For the Boolean queries, all 'fr' variants evaluate to true on the consistent database, while the 'zu' variants evaluate to false.

Measurements. To measure the execution time $t_{Q,\mathrm{db}}$ of query Q on database **db**, we execute Q ten times on **db** and take the second highest execution time. The average and standard deviation for $t_{Q,\mathrm{db}}$ are computed over 100 such measurements. Since the coefficient of variability (i.e., the ratio of standard deviation to average) was consistently less than 10^{-4} (except for the problematic queries $\beta_{4,\mathrm{fr}}$ and $\beta_{4,\mathrm{zu}}$), standard deviations have been omitted in Table 1.

In Table 1, the symbol ν refers to naive certain SQL rewritings, β refers to rewritings that decrease the number of quantifier blocks, and ρ refers to rewritings that decrease the quantifier rank. We have developed software that takes as input an acyclic SJFCQ query, tests whether it has a certain first-order rewriting, and if so, returns three certain SQL rewritings (the naive one and two simplifications). The generated SQL code is obviously awkward; for example, the SQL query $\nu_{6,\mathrm{fr}}$ contains 10 nested NOT EXISTS.

The values for $\mathsf{qbn}(\cdot)$ and $\mathsf{qr}(\cdot)$ were obtained by applying the definitions of Section 2 on formulas in tuple relational calculus (TRC). The motivation for this is that SQL uses aliases ranging over tables, just like TRC. The quantifier rank of a formula generally decreases when it is translated from first-order logic into TRC. For example, $\varphi = \exists x \exists y \exists z \big(R(\underline{x}, y) \wedge S(\underline{y}, z)\big)$ is translated into $\psi = \exists r \exists s \big(R(r) \wedge S(s) \wedge r.2 = s.1\big)$, where $\mathsf{qr}(\varphi) = 3$ and $\mathsf{qr}(\psi) = 2$. The quantifier rank of a Boolean SJFCQ query in TRC is equal to its number of atoms.

Observations. The certain SQL rewritings of $Q_{4,\text{fr}}$ (and $Q_{4,\text{zu}}$) show some anomalous behavior: their execution time is much higher than rewritings of other queries involving more atoms. Since we have not been able to find the causes of this behavior in MySQL, we leave these queries out from the discussion that follows.

The execution times of the certain SQL rewritings on large databases seem acceptable in practice. A further analysis reveals the following:

- As can be expected, most conjunctive queries (indicated by Q) execute faster than their certain SQL rewritings (indicated by ν, β, ρ). The nonBoolean query $Q_{nB,\text{fr}}$ is the only exception.
- In most rewritings, replacing 'fr' with 'zu' results in a considerable increase of execution time. Recall that the 'fr' variants of the Boolean queries evaluate to true on the consistent database, while the 'zu' variants evaluate to false. A possible explanation for the observed time differences is that 'fr' variants of Boolean queries can terminate as soon as some valuation for the tuple variables in the query makes the query true, whereas 'zu' variants have to range over all possible valuations.
- Execution times on consistent and inconsistent databases are almost identical. Recall that both databases only have nonunique indexes.
- Syntactic simplifications have a fairly low effect on execution times. Define the *speedup* of some simplified rewriting as the ratio of the execution time of the naive rewriting to that of the simplified one. For example, the speedup of $\rho_{5,\text{fr}}$ on the consistent database is equal to $\frac{0.0019}{0.0014} \approx 1.36$. All speedups are between 1 and 1.5.

8   Conclusion

We focused on the class of acyclic SJFCQ queries that have a certain first-order rewriting. A syntactic characterization of this class is given in [15,17]. We first implemented this earlier theory in a simple algorithm NaiveFo for constructing certain first-order rewritings. We then proposed two syntactic simplifications for such rewritings, which consist in reducing the number of quantifier blocks and reducing the quantifier rank. Our implementation indicates that certain SQL rewriting is an effective and efficient technique for obtaining certain answers. Also, it seems that naive rewritings perform well on existing database technology, and that syntactic simplifications do not result in important speedups.

References

1. Arenas, M., Bertossi, L.E., Chomicki, J.: Consistent query answers in inconsistent databases. In: PODS, pp. 68–79. ACM Press (1999)
2. Beeri, C., Fagin, R., Maier, D., Yannakakis, M.: On the desirability of acyclic database schemes. J. ACM 30(3), 479–513 (1983)
3. Bertossi, L.E.: Database Repairing and Consistent Query Answering. Synthesis Lectures on Data Management. Morgan & Claypool Publishers (2011)
4. Chomicki, J., Marcinkowski, J.: Minimal-change integrity maintenance using tuple deletions. Inf. Comput. 197(1-2), 90–121 (2005)
5. Dalvi, N.N., Ré, C., Suciu, D.: Probabilistic databases: diamonds in the dirt. Commun. ACM 52(7), 86–94 (2009)
6. Dalvi, N.N., Re, C., Suciu, D.: Queries and materialized views on probabilistic databases. J. Comput. Syst. Sci. 77(3), 473–490 (2011)
7. Fuxman, A., Fazli, E., Miller, R.J.: Conquer: Efficient management of inconsistent databases. In: Özcan, F. (ed.) SIGMOD Conference, pp. 155–166. ACM (2005)
8. Fuxman, A., Miller, R.J.: First-order query rewriting for inconsistent databases. J. Comput. Syst. Sci. 73(4), 610–635 (2007)
9. Kolaitis, P.G., Pema, E.: A dichotomy in the complexity of consistent query answering for queries with two atoms. Inf. Process. Lett. 112(3), 77–85 (2012)
10. Libkin, L.: Elements of Finite Model Theory. Springer (2004)
11. Maslowski, D., Wijsen, J.: A dichotomy in the complexity of counting database repairs. J. Comput. Syst. Sci. (in press)
12. Maslowski, D., Wijsen, J.: On counting database repairs. In: Fletcher, G.H.L., Staworko, S. (eds.) LID, pp. 15–22. ACM (2011)
13. Ullman, J.D.: Principles of Database and Knowledge-Base Systems, vol. I. Computer Science Press (1988)
14. Wijsen, J.: On the consistent rewriting of conjunctive queries under primary key constraints. Inf. Syst. 34(7), 578–601 (2009)
15. Wijsen, J.: On the first-order expressibility of computing certain answers to conjunctive queries over uncertain databases. In: Paredaens, J., Gucht, D.V. (eds.) PODS, pp. 179–190. ACM (2010)
16. Wijsen, J.: A remark on the complexity of consistent conjunctive query answering under primary key violations. Inf. Process. Lett. 110(21), 950–955 (2010)
17. Wijsen, J.: Certain conjunctive query answering in first-order logic. ACM Trans. Database Syst. 37(2) (2012)

Restoring Consistency
in P2P Deductive Databases

Luciano Caroprese and Ester Zumpano

DEIS, Univ. della Calabria, 87030 Rende, Italy
{lcaroprese,zumpano}@deis.unical.it

Abstract. This paper proposes a logic framework for modeling the interaction among deductive databases and computing consistent answers to logic queries in a P2P environment. As usual, data are exchanged among peers by using logical rules, called mapping rules. The declarative semantics of a P2P system is defined in terms of weak models. Under this semantics each peer uses its mapping rules to import minimal sets of mapping atoms allowing to satisfy its local integrity constraints. An equivalent and alternative characterization of minimal weak model semantics, in terms of prioritized logic programs, is also introduced and the computational complexity of P2P logic queries is investigated.

1 Introduction

Peer-to-peer (P2P) systems have recently become very popular in the social, academic and commercial communities. A large amount of traffic in the Internet is due to P2P applications.

The possibility for the users for sharing knowledge from a large number of informative sources, have enabled the development of new methods for data integration easily usable for processing distributed and autonomous data.

Each peer, joining a P2P systems relies on the peers belonging to the same environment and can both *provide or import data*. More specifically, each peer joining a P2P system exhibits a set of *mapping rules*, i.e. a set of semantic correspondences to a set of peers which are already part of the system (neighbors). Thus, in a P2P system the entry of a new source, *peer*, is extremely simple as it just requires the definition of the mapping rules. By using mapping rules, as soon as it enters the system, a peer can participate and access all data available in its neighborhood, and through its neighborhood it becomes accessible to all the other peers in the system.

Recently, there have been several proposals which consider the integration of information and the computation of queries in an open ended network of distributed peers [2, 7, 15] as well as the problem of schema mediation and query optimization in P2P environments [18, 19, 21, 25].

However, many serious theoretical and practical challenges need an answer.

Previously proposed approaches investigate the data integration problem in a P2P system by considering each peer as initially consistent, therefore the introduction of inconsistency is just relied to the operation of importing data from

E. Hüllermeier et al. (Eds.): SUM 2012, LNAI 7520, pp. 168–179, 2012.
© Springer-Verlag Berlin Heidelberg 2012

other peers. These approaches assume that for each peer it is preferable to import as much knowledge as possible.

Our previous works [8, 10–13] follow this same direction. The interaction among deductive databases in a P2P system has been modeled by importing, by means of mapping rules, maximal sets of atoms not violating integrity constraints, that is *maximal sets of atoms that allow the peer to enrich its knowledge while preventing inconsistency anomalies.*

This paper stems from a different perspective. A peer can be initially inconsistent. In this case, the P2P system it joins has to provide support to restore consistency. The basic idea, yet very simple, is the following: in the case of inconsistent database the information provided by the neighbors can be used in order to restore consistency, that is to only integrate the missing portion of a correct, but incomplete database. Then, an inconsistent peer, in the interaction with different peers, just imports the information allowing to restore consistency, that is *minimal sets of atoms allowing the peer to enrich its knowledge so that restoring inconsistency anomalies.*

The framework here presented deals with inconsistencies that can be Šrepairedš by adding more information. Inconsistency with respect to functional dependencies or denial constraints are outside the scope of this work as violations of these constraints can only be repaired by removing information.

The declarative semantics presented in this paper stems from the observations that in real world P2P systems peers often use the available import mechanisms to extract knowledge from the rest of the system only if this knowledge is strictly needed to repair an inconsistent local database.

The following example will intuitively clarify our perspective.

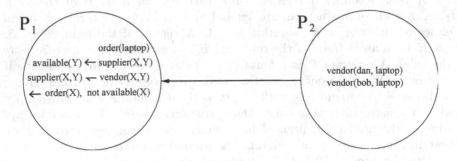

Fig. 1. A P2P system

Example 1. Consider the P2P system depicted in Figure 1. The peer $\mathcal{P}_2$ stores information about vendors of devices and contains the facts *vendor(dan, laptop)*, whose meaning is *'Dan is a vendor of laptops'*, and *vendor(bob, laptop)*, whose meaning is *'Bob is a vendor of laptops'*. The peer $\mathcal{P}_1$ contains the fact *order(laptop)*, stating that there exists the order of a laptop, the standard rule *available(Y) ← supplier(X,Y)*, stating that a device Y is available if there is a supplier X of Y, and the constraint *← order(X), not available(X)*, stating that cannot exist the order of a device which is not available. Moreover, it also exhibits the mapping rule (whose syntax will be formally defined in the following section) $supplier(X, Y) \leftarrow vendor(X, Y)$, used to import tuples from the relation $vendor$ of P_2 into the relation $supplier$ of P_1.

The local database of P_1 is inconsistent because the ordered device $laptop$ is not available (there is no supplier of laptops). P_1 'needs' to import the minimal set of atoms in order to restore consistency from its neighbors.   □

The paper presents a logic-based framework for modeling the interaction among peers. It is assumed that each peer consists of a database, a set of standard logic rules, a set of mapping rules and a set of integrity constraints. In such a context, a query can be posed to any peer in the system and the answer is provided by using locally stored data and all the information that can be consistently imported from its neighbors. In synthesis, the main contributions are:

- A formal declarative semantics for P2P systems, called *Minimal Weak Model* semantics, which uses the mapping rules among peers to import only minimal sets of atoms that make local databases consistent.
- An alternative equivalent semantics, called *Preferred Stable Model* semantics, based on the rewriting of mapping rules into standard logic rules with priorities.
- Preliminary results on the complexity of answering queries.

2   Background

We assume that there are finite sets of predicate symbols, constants and variables. A *term* is either a constant or a variable. An *atom* is of the form $p(t_1, \ldots, t_n)$ where p is a predicate symbol and $t_1, \ldots, t_n$ are terms. A *literal* is either an atom A or its negation *not A*. A *rule* is of the form $H \leftarrow \mathcal{B}$, where H is an atom (*head* of the rule) and $\mathcal{B}$ is a conjunction of literals (*body* of the rule). A program $\mathcal{P}$ is a finite set of rules. $\mathcal{P}$ is said to be positive if it is negation free. The definition of a predicate p consists of all rules having p in the head. A ground rule with empty body is a *fact*. A rule with empty head is a *constraint*. It is assumed that programs are *safe*, i.e. variables appearing in the head or in negated body literals are range restricted as they appear in some positive body literal. The ground instantiation of a program $\mathcal{P}$, denoted by $ground(\mathcal{P})$ is built by replacing variables with constants in all possible ways. An interpretation is a set of ground atoms. The truth value of ground atoms, literals and rules with respect to an interpretation M is as follows: $val_M(A) = A \in M$, $val_M(not\ A) = not\ val_M(A)$, $val_M(L_1, \ldots, L_n) = min\{val_M(L_1), \ldots, val_M(L_n)\}$ and $val_M(A \leftarrow L_1, \ldots, L_n) = val_M(A) \geq val_M(L_1, \ldots, L_n)$, where A is an atom, $L_1, \ldots, L_n$ are literals and $true > false$. An interpretation M is a model for a program $\mathcal{P}$, if all rules in $ground(\mathcal{P})$ are $true$ w.r.t. M. A model M is said to be minimal if there is no model N such that $N \subset M$. We denote the set of minimal models of a program $\mathcal{P}$ with $\mathcal{MM}(\mathcal{P})$. Given an interpretation M and a predicate symbol g, $M[g]$ denotes the set of

g-tuples in M. The semantics of a positive program $\mathcal{P}$ is given by its unique minimal model which can be computed by applying the *immediate consequence operator* $\mathbf{T}_{\mathcal{P}}$ until the fixpoint is reached ($\mathbf{T}_{\mathcal{P}}^{\infty}(\emptyset)$). The semantics of a program with negation $\mathcal{P}$ is given by the set of its stable models, denoted as $\mathcal{SM}(\mathcal{P})$. An interpretation M is a *stable model* (or *answer set*) of $\mathcal{P}$ if M is the unique minimal model of the positive program $\mathcal{P}^M$, where $\mathcal{P}^M$ is obtained from $ground(\mathcal{P})$ by (i) removing all rules r such that there exists a negative literal $not\ A$ in the body of r and A is in M and (ii) removing all negative literals from the remaining rules [16]. It is well known that stable models are minimal models (i.e. $\mathcal{SM}(\mathcal{P}) \subseteq \mathcal{MM}(\mathcal{P})$) and that for negation free programs, minimal and stable model semantics coincide (i.e. $\mathcal{SM}(\mathcal{P}) = \mathcal{MM}(\mathcal{P})$).

Prioritized Logic Programs

Several works have investigated various forms of priorities into logic languages [5, 6, 14, 22]. In this paper we refer to the extension proposed in [22]. A *(partial) preference relation* $\succeq$ among atoms is defined as follows. Given two atoms e_1 and e_2, the statement $e_2 \succeq e_1$ is a *priority* stating that for each a_2 instance of e_2 and for each a_1 instance of e_1, a_2 has higher priority than a_1. If $e_2 \succeq e_1$ and $e_1 \not\succeq e_2$ we write $e_2 \succ e_1$. If $e_2 \succ e_1$ the sets of ground instantiations of e_1 and e_2 have empty intersection. This property is evident. Indeed, assuming that there is a ground atom a which is an instance of e_1 and e_2, the statements $a \succeq a$ and $a \not\succeq a$ would hold at the same time (a contradiction). The relation $\succeq$ is transitive and reflexive. A *prioritized logic program* (PLP) is a pair $(\mathcal{P}, \Phi)$ where $\mathcal{P}$ is a program and Φ is a set of priorities. Φ^* denotes the set of priorities which can be reflexively or transitively derived from Φ. Given a prioritized logic program $(\mathcal{P}, \Phi)$, the relation $\sqsupseteq$ is defined over the stable models of $\mathcal{P}$ as follows. For any stable models M_1, M_2 and M_3 of $\mathcal{P}$: (i) $M_1 \sqsupseteq M_1$; (ii)$M_2 \sqsupseteq M_1$ if a) $\exists e_2 \in M_2 - M_1$, $\exists e_1 \in M_1 - M_2$ such that $(e_2 \succeq e_1) \in \Phi^*$ and b) $\not\exists e_3 \in M_1 - M_2$ such that $(e_3 \succ e_2) \in \Phi^*$; (iii) if $M_2 \sqsupseteq M_1$ and $M_1 \sqsupseteq M_0$, then $M_2 \sqsupseteq M_0$.

If $M_2 \sqsupseteq M_1$ holds, then we say that M_2 is *preferable* to M_1 w.r.t. Φ. Moreover, we write $M_2 \sqsupset M_1$ if $M_2 \sqsupseteq M_1$ and $M_1 \not\sqsupseteq M_2$. An interpretation M is a *preferred stable model* of $(\mathcal{P}, \Phi)$ if M is a stable model of $\mathcal{P}$ and there is no stable model N such that $N \sqsupset M$. The set of preferred stable models of $(\mathcal{P}, \Phi)$ will be denoted by $\mathcal{PSM}(\mathcal{P}, \Phi)$.

3   P2P Systems: Syntax and FOL Semantics

A *(peer) predicate* symbol is a pair $i:p$ where i is a *peer identifier* and p is a predicate symbol. A *(peer) atom* is of the form $i:A$ where A is a standard atom. A *(peer) literal* L is of the form A or $not\ A$ where A is a peer atom. A *(peer) rule* is of the form $A \leftarrow A_1, \ldots, A_n$ where A is a peer atom and $A_1, \ldots, A_n$ are peer atoms or built-in atoms. A *(peer) integrity constraint* is of the form $\leftarrow L_1, \ldots, L_m$ where $L_1, \ldots, L_m$ are peer literals or built-in atoms. Whenever the peer is understood, the peer identifier can be omitted. The definition of a predicate $i:p$ consists of all rules having as head predicate symbol $i:p$. In the following, we assume that for each peer $\mathcal{P}_i$ there are three distinct sets of predicates called, respectively, *base*, *derived* and *mapping predicates*. A base predicates is defined by ground facts; a derived predicate $i : p$ is defined by *standard rules*, i.e. peer rules using in the body only predicates defined in the peer $\mathcal{P}_i$; a mapping predicate $i:p$ is defined by *mapping rules*, i.e peer rules using in the body only predicates defined in other peers. Without loss of generality, we assume that every mapping predicate is defined by only one rule of the form $i : p(X) \leftarrow j : q(X)$ with $j \neq i$. The definition of a mapping predicate $i : p$ consisting of n rules of the form $i : p(X_k) \leftarrow \mathcal{B}_k$ with $1 \leq k \leq n$, can be rewritten into $2n$ rules of the form $i : p_k(X_k) \leftarrow \mathcal{B}_k$ and $i : p(X) \leftarrow i : p_k(X)$, with $1 \leq k \leq n$. Observe that standard and mapping rules are always positive whereas negation is allowed in integrity constraints.

Definition 1. A peer $\mathcal{P}_i$ is a tuple $\langle \mathcal{D}_i, \mathcal{LP}_i, \mathcal{MP}_i, \mathcal{IC}_i \rangle$ where (i) $\mathcal{D}_i$ is a (local) database consisting of a set of facts; (ii) $\mathcal{LP}_i$ is a set of standard rules; (iii) $\mathcal{MP}_i$ is a set of mapping rules and (iv) $\mathcal{IC}_i$ is a set of integrity constraints over predicates defined in $\mathcal{D}_i$, $\mathcal{LP}_i$ and $\mathcal{MP}_i$. A P2P system $\mathcal{PS}$ is a set of peers $\{\mathcal{P}_1, \ldots, \mathcal{P}_n\}$. □

Given a P2P system $\mathcal{PS} = \{\mathcal{P}_1, \ldots, \mathcal{P}_n\}$ where $\mathcal{P}_i = \langle \mathcal{D}_i, \mathcal{LP}_i, \mathcal{MP}_i, \mathcal{IC}_i \rangle$, we denote as $\mathcal{D}, \mathcal{LP}, \mathcal{MP}$ and $\mathcal{IC}$ respectively the global sets of ground facts, standard rules, mapping rules and integrity constraints: $\mathcal{D} = \mathcal{D}_1 \cup \cdots \cup \mathcal{D}_n$, $\mathcal{LP} = \mathcal{LP}_1 \cup \cdots \cup \mathcal{LP}_n$, $\mathcal{MP} = \mathcal{MP}_1 \cup \cdots \cup \mathcal{MP}_n$ and $\mathcal{IC} = \mathcal{IC}_1 \cup \cdots \cup \mathcal{IC}_n$. With a little abuse of notation we shall also denote with $\mathcal{PS}$ both the tuple $\langle \mathcal{D}, \mathcal{LP}, \mathcal{MP}, \mathcal{IC} \rangle$ and the set $\mathcal{D} \cup \mathcal{LP} \cup \mathcal{MP} \cup \mathcal{IC}$. Given a peer $\mathcal{P}_i$, $MPred(\mathcal{P}_i)$, $DPred(\mathcal{P}_i)$ and $BPred(\mathcal{P}_i)$ denote, respectively, the sets of mapping, derived and base predicates defined in $\mathcal{P}_i$. Analogously, $MPred(\mathcal{PS})$, $DPred(\mathcal{PS})$ and $BPred(\mathcal{PS})$ define the sets of mapping, derived and base predicates in $\mathcal{PS}$.

A peer $\mathcal{P}_i$ is *locally consistent* if the database $\mathcal{D}_i$ and the standard rules in $\mathcal{LP}_i$ are consistent w.r.t. $\mathcal{IC}_i$, $(\mathcal{D}_i \cup \mathcal{LP}_i \models \mathcal{IC}_i)$.

FOL Semantics. The FOL semantics of a P2P system $\mathcal{PS} = \{\mathcal{P}_1, \ldots, \mathcal{P}_n\}$ is given by the minimal model semantics of $\mathcal{PS} = \mathcal{D} \cup \mathcal{LP} \cup \mathcal{MP} \cup \mathcal{IC}$. For a given P2P system $\mathcal{PS}$, $\mathcal{MM}(\mathcal{PS})$ denotes the set of minimal models of $\mathcal{PS}$. As $\mathcal{D} \cup \mathcal{LP} \cup \mathcal{MP}$ is a positive program, $\mathcal{PS}$ may admit zero or one minimal model. In particular, if $\mathcal{MM}(\mathcal{D} \cup \mathcal{LP} \cup \mathcal{MP}) = \{M\}$ then $\mathcal{MM}(\mathcal{PS}) = \{M\}$ if $M \models \mathcal{IC}$, otherwise $\mathcal{MM}(\mathcal{PS}) = \emptyset$. The problem with such a semantics is that local inconsistencies make the global system inconsistent.

Preferred Weak Model Semantics. In [8] the authors introduced the *preferred weak model semantics*. This semantics is based on the idea to use the mapping rules to import in each peer *as much knowledge as possible* without violating local integrity constraints. Under this semantics only facts not making the local databases inconsistent are imported, and the preferred weak models are those in which peers import maximal sets of facts not violating integrity constraints. This approach assumes that the local databases are originally consistent.

4   Minimal Weak Model Semantics

The semantics presented in this paper stems from the observations that in real world P2P systems often the peers use the available import mechanisms to extract knowledge from the rest of the system only if this knowledge is strictly needed to repair an inconsistent local database. In more formal terms, each peer uses its mapping rules to import minimal sets of mapping atoms allowing to satisfy local integrity constraints.

We introduce a new interpretation of mapping rules, which will now be denoted with a different syntax of the form $H \leftharpoondown B$. Intuitively, $H \leftharpoondown B$ means that if the body conjunction B is *true* in the source peer the atom H can be imported in the target peer, that is H is *true* in the target peer only if it implies (directly or even indirectly) the satisfaction of some constraints that otherwise would be violated. The following example should make the meaning of mapping rules crystal clear.

Example 2. Consider again the P2P system presented in Example 1.

As we observed, the local database of P_1 is inconsistent because the ordered device *laptop* is not available. The peer P_1 has to import some supplier of laptops in order to make its database consistent. Then, the mapping rule $supplier(X, Y) \leftharpoondown vendor(X, Y)$ will be used to import one supplier from the corresponding facts of P_2: $supplier(dan, laptop)$ or $supplier(bob, laptop)$. P_1 will not import both facts because just one of them is sufficient to satisfy the local integrity constraint $\leftarrow order(X), not\ available(X)$.

We observe that if P_1 does not contain any fact its database is consistent and no fact is imported from P_2. $\square$

Before formally presenting the minimal weak model semantics we introduce some notation. Given a mapping rule $r = A \leftharpoondown B$, with $St(r)$ we denote the corresponding logic rule $A \leftarrow B$. Analogously, given a set of mapping rules $\mathcal{MP}$, $St(\mathcal{MP}) = \{St(r) \mid r \in \mathcal{MP}\}$ and given a P2P system $\mathcal{PS} = \mathcal{D} \cup \mathcal{LP} \cup \mathcal{MP} \cup \mathcal{IC}$, $St(\mathcal{PS}) = \mathcal{D} \cup \mathcal{LP} \cup St(\mathcal{MP}) \cup \mathcal{IC}$.

In the next two subsections we present two alternative and equivalent characterizations of the minimal weak model semantics. The first semantics is based on a different satisfaction of mapping rules, whereas the second one is based on the rewriting of mapping rules into prioritized rules [6, 22].

4.1   Minimal Weak Models

Informally, the idea is that for a ground mapping rule $A \leftharpoondown B$, the atom A *could be inferred* only if the body B is *true*. Formally, given an interpretation M, a ground standard rule $D \leftarrow C$ and a ground mapping rule $A \leftharpoondown B$, $val_M(C \leftarrow D) = val_M(C) \geq val_M(D)$ whereas $val_M(A \leftharpoondown B) = val_M(A) \leq val_M(B)$.

Definition 2. Given a P2P system $\mathcal{PS} = \mathcal{D} \cup \mathcal{LP} \cup \mathcal{MP} \cup \mathcal{IC}$, an interpretation M is a *weak model* for $\mathcal{PS}$ if $\{M\} = \mathcal{MM}(St(\mathcal{PS}^M))$, where $\mathcal{PS}^M$ is the program obtained from $ground(\mathcal{PS})$ by removing all mapping rules whose head is *false* w.r.t. M. $\square$

We shall denote with $M[\mathcal{D}]$ (resp. $M[\mathcal{LP}]$, $M[\mathcal{MP}]$) the set of ground atoms of M which are defined in $\mathcal{D}$ (resp. $\mathcal{LP}$, $\mathcal{MP}$).

Definition 3. Given two weak models M and N, we say that M is *preferable* to N, and we write $M \sqsupseteq N$, if $M[\mathcal{MP}] \subseteq N[\mathcal{MP}]$. Moreover, if $M \sqsupseteq N$ and $N \not\sqsupseteq M$ we write $M \sqsupset N$. A weak model M is said to be *minimal* if there is no weak model N such that $N \sqsubset M$.   □

The set of weak models for a P2P $\mathcal{PS}$ system will be denoted by $\mathcal{WM}(\mathcal{PS})$, whereas the set of minimal weak models will be denoted by $\mathcal{MWM}(\mathcal{PS})$. We say that a P2P system $\mathcal{PS}$ is *consistent* if $\mathcal{MWM}(\mathcal{PS}) \neq \emptyset$; otherwise it is *inconsistent*.

Proposition 1. *For any P2P system $\mathcal{PS}$, $\sqsupseteq$ defines a partial order on the set of weak models of $\mathcal{PS}$.*   □

We observe that if each peer of a P2P system are locally consistent then no mapping atom is inferred. Clearly not always a minimal weak model exists. It happens when there is at least a peer which is locally inconsistent and there is no way to import mapping atoms that could repair its local database so that its consistency can be restored.

Example 3. Consider the P2P system $\mathcal{PS}$ presented in Example 2. The weak models of the system are:

$M_1 = \{vendor(dan, laptop), vendor(bob, laptop), order(laptop),$
$supplier(dan, laptop), available(laptop)\}$,

$M_2 = \{vendor(dan, laptop), vendor(bob, laptop), order(laptop),$
$supplier(bob, laptop), available(laptop)\}$, and

$M_3 = \{vendor(dan, laptop), vendor(bob, laptop), order(laptop),$
$supplier(dan, laptop), supplier(bob, laptop), available(laptop)\}$,

whereas the minimal weak models are M_1 and M_2 because they contain minimal subsets of mapping atoms (resp. $\{supplier(dan, laptop)\}$ and $\{supplier(bob, laptop)\}$).   □

4.2   Prioritized Programs and Preferred Stable Models

Now we present an alternative characterization of the minimal weak model semantics based on the rewriting of mapping rules into prioritized rules [6, 22]. For the sake of notation we consider exclusive disjunctive rules of the form $A \oplus A' \leftarrow \mathcal{B}$ whose meaning is that if $\mathcal{B}$ is *true* then exactly one of A or A' must be *true*. Note that the rule $A \oplus A' \leftarrow \mathcal{B}$ is just shorthand for the rules $A \leftarrow \mathcal{B}, not\ A'$ and $A' \leftarrow \mathcal{B}, not\ A$ and the integrity constraint $\leftarrow A, A'$.

   Given a pair $P = (A, B)$, where A and B are generic objects, $P[1]$ (resp. $P[2]$) denotes the object A (resp. B).

Definition 4. Given a P2P system $\mathcal{PS} = \mathcal{D} \cup \mathcal{LP} \cup \mathcal{MP} \cup \mathcal{IC}$ and a mapping rule $r = i : p(x) \leftarrow \mathcal{B}$, then

- $Rew(r)$ denotes the pair $(i : p(x) \oplus i : p'(x) \leftarrow \mathcal{B},\ i : p'(x) \succeq i : p(x))$, consisting of a disjunctive mapping rule and a priority statement,
- $Rew(\mathcal{MP}) = (\{Rew(r)[1] \| \ r \in \mathcal{MP}\}, \{Rew(r)[2] \| \ r \in \mathcal{MP}\})$ and
- $Rew(\mathcal{PS}) = (\mathcal{D} \cup \mathcal{LP} \cup Rew(\mathcal{MP})[1] \cup \mathcal{IC},\ Rew(\mathcal{MP})[2]).$ □

In the above definition the atom $i : p(x)$ (resp. $i : p'(x)$) means that the fact $p(x)$ is imported (resp. not imported) in the peer $\mathcal{P}_i$. For a given mapping rule r, $Rew(r)[1]$ (resp. $Rew(r)[2]$) denotes the first (resp. second) component of $Rew(r)$.

Intuitively, the rewriting $Rew(r) = (A \oplus A' \leftarrow \mathcal{B}, A' \succeq A)$ of a mapping rule $r = A \leftarrow \mathcal{B}$ means that if $\mathcal{B}$ is *true* in the *source peer* then two alternative actions can be performed in the *target peer*: A can be either imported or not imported; but the action of *not importing* A is preferable over the action of *importing* A.

Example 4. Consider again the system analyzed in Example 3. The rewriting of the system is: $Rew(\mathcal{PS}) = \{vendor(dan, laptop), vendor(bob, laptop),$ $order(laptop),$

$\qquad supplier(X, Y) \oplus supplier'(X, Y) \leftarrow vendor(X, Y),$
$\qquad available(Y) \leftarrow supplier(X, Y),$
$\qquad \leftarrow order(X), not\ available(X)\},$
$\qquad \{supplier'(X, Y) \succeq supplier(X, Y)\}).$

$Rew(\mathcal{PS})[1]$ has three stable models:

$\quad M_1 = \{vendor(dan, laptop), vendor(bob, laptop), order(laptop),$
$supplier(dan, laptop), supplier'(bob, laptop), available(laptop)\},$

$\quad M_2 = \{vendor(dan, laptop), vendor(bob, laptop), order(laptop),$
$supplier'(dan, laptop), supplier(bob, laptop), available(laptop)\},$ and

$\quad M_3 = \{vendor(dan, laptop), vendor(bob, laptop), order(laptop),$
$supplier(dan, laptop), supplier(bob, laptop), available(laptop)\},$

The preferred stable models are M_1 and M_2. □

Given a P2P system $\mathcal{PS}$ and a preferred stable model M for $Rew(\mathcal{PS})$ we denote with $St(M)$ the subset of non-primed atoms of M and we say that $St(M)$ is a preferred stable model of $\mathcal{PS}$. We denote the set of preferred stable models of $\mathcal{PS}$ as $\mathcal{PSM}(\mathcal{PS})$. The following theorem shows the equivalence of preferred stable models and minimal weak models.

Theorem 1. *For every P2P system* $\mathcal{PS}$, $\mathcal{PSM}(\mathcal{PS}) = \mathcal{MWM}(\mathcal{PS})$. □

5   Discussion

Complexity Results. We consider now the computational complexity of calculating minimal weak models and answers to queries.

Proposition 2. *Given a P2P system* $\mathcal{PS}$, *checking if there exists a minimal weak model for* $\mathcal{PS}$ *is a NP-complete problem.*   □

As a P2P system may admit more than one preferred weak model, the answer to a query is given by considering *brave* or *cautious* reasoning (also known as *possible* and *certain* semantics).

Definition 5. *Given a P2P system* $\mathcal{PS} = \{\mathcal{P}_1, \ldots, \mathcal{P}_n\}$ *and a ground peer atom A, then A is* true *under*

- *brave reasoning if* $A \in \bigcup_{M \in \mathcal{MWM}(\mathcal{PS})} M$,
- *cautious reasoning if* $A \in \bigcap_{M \in \mathcal{MWM}(\mathcal{PS})} M$.   □

We assume here a simplified framework not considering the distributed complexity as we suppose that the complexity of communications depends on the number of computed atoms which are the only elements exported by peers. The upper bound results can be immediately found by considering analogous results on stable model semantics for prioritized logic programs. For disjunction-free $(\vee - free)$[1] prioritized programs deciding whether an atom is *true* in some preferred model is Σ_2^p-complete, whereas deciding whether an atom is *true* in every preferred model is Π_2^p-complete [22].

Theorem 2. *Let* $\mathcal{PS}$ *be a consistent P2P system, then*

1. *Deciding whether an atom A is* true *in some preferred weak model of* $\mathcal{PS}$ *is in* Σ_2^p.
2. *Deciding whether an atom A is* true *in every preferred weak model of* $\mathcal{PS}$ *is in* Π_2^p *and* $co\mathcal{NP}$-*hard.*   □

Related Works. The problem of integrating and querying databases in a P2P system has been investigated in [3, 7, 15].

In [7] a new semantics for a P2P system, based on epistemic logic, is proposed. The paper also shows that the semantics is more suitable than traditional semantics based on FOL (First Order Logic) and proposes a sound, complete and terminating procedure that returns the certain answers to a query submitted to a peer.

In [15] a characterization of P2P database systems and a model-theoretic semantics dealing with inconsistent peers is proposed. The basic idea is that if a peer does not have models all (ground) queries submitted to the peer are *true* (i.e. are *true* with respect to all models). Thus, if some databases are inconsistent it does not mean that the entire system is inconsistent.

The semantics in [15] coincides with the epistemic semantics in [7].

An interesting approach for answering queries in a Peer to Peer data exchange system has been recently proposed in [3]. Given a peer P in a P2P system a

---

[1] The symbol $\vee$ denotes inclusive disjunction and is different from $\oplus$ as the latter denotes exclusive disjunction. It should be recalled that inclusive disjunction allows more than one atom to be true while exclusive disjunction allows only one atom to be true.

solution for P is a database instance that respects the exchange constraints and trust relationship P has with its 'immediate neighbors' and stays as close as possible to the available data in the system.

In [19] the problem of schema mediation in a Peer Data Management System (PDMS) is investigated. A flexible formalism, PPL, for mediating peer schemas, which uses the GAV and LAV formalism to specify mappings, is proposed. The semantics of query answering for a PDMS is defined by extending the notion of certain answer.

In [25] several techniques for optimizing the reformulation of queries in a PDMS are presented. In particular the paper presents techniques for pruning semantic paths of mappings in the reformulation process and for minimizing the reformulated queries.

The design of optimization methods for query processing over a network of semantically related data is investigated in [21].

None of the previous proposals take into account the possibility of modeling some preference criteria while performing the data integration process. New interesting semantics for data exchange systems that goes in this direction has been recently proposed in [3, 4, 13].

In [3, 4] it is proposed a new semantics that allows for a cooperation among pairwise peers that related each other by means of data exchange constraints (i.e. mapping rules) and trust relationships. The decision by a peer on what other data to consider (besides its local data) does not depend only on its data exchange constraints, but also on the trust relationship that it has with other peers.

The data exchange problem among distributed independent sources has been investigated in [8, 9, 12]. In [8] the authors define a declarative semantics for P2P systems that allows to import in each peer maximal subsets of atoms which do not violate the local integrity constraints. The framework has been extended in [12] where a mechanism to set different degrees of reliability for neighbor peers has been provided and in [9] in which 'dynamic' preferences allow to import, in the case of conflicting sets of atoms, looking at the properties of data provided by the peers.

6   Concluding Remarks and Directions for Further Research

In this paper we have introduced a logic programming based framework for P2P deductive databases. It is based on the assumption, coherent with many real world P2P systems, that each peer uses its mapping rules to import minimal sets of mapping atoms allowing to satisfy its local integrity constraints. We have presented a different characterization of the semantics based on preferred stable models for prioritized logic programs. Moreover, we have also provided some preliminary results on the complexity of answering queries in different contexts.

The approach we follow in this paper does not contrast the previous one in [8, 10–13]. It is not alternative, but complementary.

In a more general framework two different forms of mapping rules can be managed, one allowing to import maximal sets of atoms not violating integrity constraints (say *Max-Mapping rules*) and one allowing to import minimal sets of atoms (say *Min-Mapping rules*). This second kind of mapping rules also allow to restore consistency.

Note that, at the present, the logical framework equally trusts its neighbor peers. No preference is given to the information sources providing data. Our proposal will be enriched by allowing, in the presence of multiple alternatives, the selection of data from the most reliable sources.

References

1. Arenas, M., Bertossi, L., Chomicki, J.: Consistent Query Answers in Inconsistent Databases. In: PODS Conf., pp. 68–79 (1999)
2. Bernstein, P.A., Giunchiglia, F., Kementsietsidis, A., Mylopulos, J., Serafini, L., Zaihrayen, I.: Data Management for Peer-to-Peer Computing: A Vision. In: WebDB, pp. 89–94 (2002)
3. Bertossi, L., Bravo, L.: Query Answering in Peer-to-Peer Data Exchange Systems. In: EDBT Workshops, pp. 476–485 (2004)
4. Bertossi, L., Bravo, L.: Query The semantics of consistency and trust in peer data exchange systems. In: LPAR, pp. 107–122 (2007)
5. Brewka, G., Eiter, T.: Preferred Answer Sets for Extended Logic Programs. Artificial Intelligence 109(1-2), 297–356 (1999)
6. Brewka, G., Niemela, I., Truszczynski, M.: Answer Set Optimization. In: Int. Joint Conf. on Artificial Intelligence, pp. 867–872 (2003)
7. Calvanese, D., De Giacomo, G., Lenzerini, M., Rosati, R.: Logical foundations of peer-to-peer data integration. In: PODS Conf., pp. 241–251 (2004)
8. Caroprese, L., Greco, S., Zumpano, E.: A Logic Programming Approach to Querying and Integrating P2P Deductive Databases. In: FLAIRS, pp. 31–36 (2006)
9. Caroprese, L., Zumpano, E.: Aggregates and priorities in P2P data management systems. In: IDEAS, pp. 1–7 (2011)
10. Caroprese, L., Molinaro, C., Zumpano, E.: Integrating and Querying P2P Deductive Databases. In: IDEAS, pp. 285–290 (2006)
11. Caroprese, L., Zumpano, E.: Consistent Data Integration in P2P Deductive Databases. In: Prade, H., Subrahmanian, V.S. (eds.) SUM 2007. LNCS (LNAI), vol. 4772, pp. 230–243. Springer, Heidelberg (2007)
12. Caroprese, L., Zumpano, E.: Modeling Cooperation in P2P Data Management Systems. In: An, A., Matwin, S., Raś, Z.W., Ślęzak, D. (eds.) Foundations of Intelligent Systems. LNCS (LNAI), vol. 4994, pp. 225–235. Springer, Heidelberg (2008)
13. Caroprese, L., Zumpano, E.: Handling Preferences in P2P Systems. In: Lukasiewicz, T., Sali, A. (eds.) FoIKS 2012. LNCS, vol. 7153, pp. 91–106. Springer, Heidelberg (2012)
14. Delgrande, J.P., Schaub, T., Tompits, H.: A Framework for Compiling Preferences in Logic Programs. TPLP 3(2), 129–187 (2003)
15. Franconi, E., Kuper, G.M., Lopatenko, A., Zaihrayeu, I.: A Robust Logical and Computational Characterisation of Peer-to-Peer Database Systems. In: DBISP2P, pp. 64–76 (2003)
16. Gelfond, M., Lifschitz, V.: The Stable Model Semantics for Logic Programming. In: Proc. Fifth Conf. on Logic Programming, pp. 1070–1080 (1998)

17. Greco, G., Greco, S., Zumpano, E.: Repairing and Querying Inconsistent Databases. IEEE Transaction on Knowledge and Data Engineering (2003)
18. Gribble, S., Halevy, A., Ives, Z., Rodrig, M., Suciu, D.: What can databases do for peer-to-peer? In: WebDB, pp. 31–36 (2006)
19. Halevy, A., Ives, Z., Suciu, D., Tatarinov, I.: Schema mediation in peer data management systems. In: Int. Conf. on Database Theory, pp. 505–516 (2003)
20. Lenzerini, M.: Data integration: A theoretical perspective. In: Proc. ACM Symposium on Principles of Database Systems, pp. 233–246 (2002)
21. Madhavan, J., Halevy, A.Y.: Composing mappings among data sources. In: Int. Conf. on Very Large Data Bases, pp. 572–583 (2003)
22. Sakama, C., Inoue, K.: Prioritized logic programming and its application to commonsense reasoning. Artificial Intelligence (123), 185–222 (2000)
23. Schaub, T., Wang, K.: A Comparative Study of Logic Programs with Preference. In: IJCAI, pp. 597–602 (2001)
24. Simons, S., Niemela, I., Soininen, T.: Extending and implementing the stable model semantics. Artif. Intell. 138(1-2), 181–234 (2002)
25. Tatarinov, I., Halevy, A.: Efficient Query reformulation in Peer Data Management Systems. In: SIGMOD, pp. 539–550 (2004)
26. Ullman, J.D.: Principles of Database and Knowledge Base Systems, vol. 1. Computer Science Press, Potomac (1988)

Tractable Cases of Clean Query Answering under Entity Resolution via Matching Dependencies*

Jaffer Gardezi[1] and Leopoldo Bertossi[2]

[1] University of Ottawa, SITE.
Ottawa, Canada
jgard082@uottawa.ca
[2] Carleton University, SCS
Ottawa, Canada
bertossi@scs.carleton.ca

Abstract. Matching Dependencies (MDs) are a recent proposal for declarative entity resolution. They are rules that specify, given the similarities satisfied by values in a database, what values should be considered duplicates, and have to be matched. On the basis of a chase-like procedure for MD enforcement, we can obtain clean (duplicate-free) instances; possibly several of them. The clean answers to queries (which we call the resolved answers) are invariant under the resulting class of instances. Identifying the clean versions of a given instance is generally an intractable problem. In this paper, we show that for a certain class of MDs, the characterization of the clean instances is straightforward. This is an important result, because it leads to tractable cases of resolved query answering. Further tractable cases are derived by making connections with tractable cases of CQA.

1 Introduction

For various reasons, such as errors or variations in format, integration of data from different sources, etc., databases may contain different coexisting representations of the same external, real world entity. Those "duplicates" can be entire tuples or values within them. To obtain accurate information, in particular, query answers from the data, those tuples or values should be merged into a single representation.

Identifying and merging duplicates is a process called *entity resolution* (ER) [13, 16]. Matching dependencies (MDs) are a recent proposal for declarative duplicate resolution [17, 18]. An MD expresses, in the form of a rule, that if the values of certain attributes in a pair of tuples are similar, then the values of other attributes in those tuples should be matched (or merged) into a common value.

For example, the MD $R_1[X_1] \approx R_2[X_2] \rightarrow R_1[Y_1] \doteq R_2[Y_2]$ is a symbolic expression saying that, if an R_1-tuple and R_2-tuple have similar values for their attributes X_1, X_2, then their values for attributes Y_1, Y_2 should be made equal. This is a *dynamic* dependency, in the sense that its satisfaction is checked against a pair of instances: the first one where the antecedent holds, and the second one where the identification of values takes place. This semantics of MDs was sketched in [18].

In this paper we use a refinement of that original semantics that was put forth in [23] (cf. also [24]). It improves wrt the latter in that it disallows changes that are irrelevant to

---

* Research supported by the NSERC Strategic Network on Business Intelligence (BIN ADC05) and NSERC/IBM CRDPJ/371084-2008.

E. Hüllermeier et al. (Eds.): SUM 2012, LNAI 7520, pp. 180–193, 2012.

the duplicate resolution process. Actually, [23] goes on to define the clean versions of the original database instance D_0 that contains duplicates. They are called the *resolved instances* (RIs) of D_0 wrt the given set M of matching dependencies. A resolved instance is obtained as the fixed point of a chase-like procedure that starts from D_0 and iteratively applies or enforces the MDs in M. Each step of this chase generates a new instance by making equal the values that are identified as duplicates by the MDs.

In [23] it was shown that resolved instances always exist, and that they have certain desirable properties. For example, the set of allowed changes is just restrictive enough to prevent irrelevant changes, while still guaranteeing existence of resolved instances. The resolved instances that minimize the *overall number* of attribute value changes wrt the original instance are called *minimally resolved instances* (MRIs). On this basis, given a query Q posed to a database instance D_0 that may contain duplicates, we defined the *resolved answers* wrt Σ as the query answers that are true of all the minimally resolved instances [23].

The concept of resolved query answer has similarities to that of *consistent query answer* (CQA) in a database that fails to satisfy a set of integrity constraints [3, 7, 8]. The consistent answers are invariant under the *repairs* of the original instance. However, data cleaning and CQA are different problems. For the former, we want to compute a clean instance, determined by MDs; for the latter, the goal is obtaining semantically correct query answers. MDs are not (static) ICs. In principle, we could see clean instances as repairs, treating MDs similarly to static FDs. However, the existing repair semantics do not capture the matchings as dictated by MDs (cf. [23, 24] for a more detailed discussion).

The motivation for defining the concept of resolved answers to a query is that even in a database instance containing duplicates, much or most of the data may be duplicate-free. One can therefore obtain useful information from the instance without having to perform data cleaning on the instance. This would be convenient if the user does not want, or cannot afford, to go through a data cleaning process. In other situations the user may not have write access to the data being queried, or any access to the data sources, as in virtual data integration systems [25, 9].

In this paper we show that for a certain sets of MDs whose members depend cyclically on each other, it is possible to characterize the form of the minimally resolved instances for any given instance. In particular, we introduce a recursively defined predicate for identifying the sets of duplicate values within a database instance. This predicate can be combined with a query, opening the ground for tractability via a query rewriting approach to the problem of retrieving the resolved answers to the query.

We also establish connections between the current problem and consistent query answering (CQA) to obtain further tractable cases. When the form of a set of MDs is such that application of one MD cannot affect the application of another MD in the set, the resolved instances of a given database instance are similar to repairs of the instance wrt a set of functional dependencies (FDs). This allows us to apply results on CQA under FDs [15, 21, 30].

This paper is organized as follows. In Section 2 we introduce basic concepts and notation of MDs. In Section 3, we define the important concepts used in this paper, in particular, (minimally) resolved instances and resolved answers to queries. Section

4 contains the main result of this paper, which is a characterization of the minimally resolved instances for certain sets of MDs with cyclic dependency graphs. Section 5 makes some connections with CQA. Section 6 concludes the paper and discusses related and future work. The missing and not already formally published proofs can be found all in [22].

2 Preliminaries

We consider a relational schema $\mathcal{S}$ that includes an enumerable, possibly infinite domain U, and a finite set $\mathcal{R}$ of database predicates. Elements of U are represented by lower case letters near the beginning of the alphabet. $\mathcal{S}$ determines a first-order (FO) language $L(\mathcal{S})$. An instance D for $\mathcal{S}$ is a finite set of ground atoms of the form $R(\bar{a})$, with $R \in \mathcal{R}$, say of arity n, and $\bar{a} \in U^n$. $R(D)$ denotes the extension of R in D. Every predicate $R \in \mathcal{S}$ has a set of attribute, denoted $attr(R)$. As usual, we sometimes refer to attribute A of R by $R[A]$. We assume that all the attributes of a predicate are different, and that we can identify attributes with *positions* in predicates, e.g. $R[i]$, with $1 \leq i \leq n$. If the ith attribute of predicate R is A, for a tuple $t = (c_1, \ldots, c_n) \in R(D)$, $t_R^D[A]$ (usually, simply $t_R[A]$ or $t[A]$ if the instance is understood) denotes the value c_i. For a sequence $\bar{A}$ of attributes in $attr(R)$, $t[\bar{A}]$ denotes the tuple whose entries are the values of the attributes in $\bar{A}$. For a tuple t in a relation instance with attribute A, the pair (t, A) is called a *value position* (usually simply, *position*). In that case, $t[A]$ is the value taken by that position (for the given instance). Attributes have and may share subdomains of U.

In the rest of this section, we summarize some of the assumptions, definitions, notation, and results from [23], that we will need.

We will assume that every relation in an instance has an auxiliary attribute, a surrogate key, holding values that act as *tuple identifiers*. Tuple identifiers are never created, destroyed or changed. They do not appear in MDs, and are used to identify different versions of the same original tuple that result from the matching process. We usually leave them implicit; and "tuple identifier attributes" are commonly left out when specifying a database schema. However, when explicitly represented, they will be the "first" attribute of the relation. For example, if $R \in \mathcal{R}$ is n-ary, $R(t, c_1, \ldots, c_n)$ is a tuple with id t, and is usually written as $R(t, \bar{c})$. We usually use the same symbol for a tuple's identifier as for the tuple itself. Tuple identifiers are unique over the entire instance.[1]

Two instances over the same schema that share the same tuple identifiers are said to be *correlated*. In this case it is possible to unambiguously compare their tuples, and as a result, also the instances.

As expected, some of the attribute domains, say A, have a built-in binary similarity relation $\approx_A$. That is, $\approx_A \subseteq Dom(A) \times Dom(A)$. It is assumed to be reflexive and symmetric. Such a relation can be extended to finite lists of attributes (or domains therefor), componentwise. For single attributes or lists of them, the similarity relation is generically denoted with $\approx$.

A *matching dependency* (MD) [17], involving predicate R, is an expression (or rule), m, of the form

$$m : R[\bar{A}] \approx R[\bar{A}] \rightarrow R[\bar{B}] \doteq R[\bar{B}], \tag{1}$$

---

[1] An alternative to the use of tuple ids could be the *dynamic mappings* introduced in [28].

with $\bar{A} = (A_1, ..., A_k)$ and $\bar{B} = (B_1, ..., B_{k'})$ lists of attributes from $attr(R)$.[2] We assume the attributes in $\bar{A}$ are all different, similarly for $\bar{B}$. The set of attributes on the left-hand-side (LHS) of the arrow in m is denoted with $LHS(m)$. Similarly for the right-hand-side (RHS). The condition on the LHS of (1) means that, for a pair of tuples t_1, t_2 in (an instance of) R, $t_1[A_i] \approx_i t_2[A_i]$, $1 \leq i \leq k$. Similarly, the expression on the RHS means $t_1[B_i] \doteq t_2[B_i]$, $1 \leq i \leq k'$. Here, $\doteq$ means that the values should be updated to the same value. Accordingly, the intended semantics of (1) is that, for an instance D, if any pair of tuples $t_1, t_2 \in R(D)$ satisfy the similarity conditions on the LHS, then for the same tuples (or tuple ids), the attributes on the RHS have to take the same values [18], possibly through updates that may lead to a new version of D.

Attributes that appear to the right of the arrow in an element of a set M of MDs are called *changeable attributes*. We assume that all sets M of MDs are in *standard form*, i.e. for no two different MDs $m_1, m_2 \in M$, $LHS(m_1) = LHS(m_2)$. All sets of MDs can be put in this form. MDs in a set M can interact in the sense that a matching enforced by one of them may create new similarities that lead to the enforcement of another MD in M. This intuition is captured through the *MD-graph*.

Definition 1. [24] Let M be a set of MDs in standard form. The *MD-graph* of M, denoted $MDG(M)$, is a directed graph with a vertex m for each $m \in M$, and an edge from m to m' iff $RHS(m) \cap LHS(m') \neq \emptyset$.[3] If $MDG(M)$ contains edges, M is called *interacting*. Otherwise, it is called *non-interacting* (NI). □

3   Matching Dependencies and Resolved Answers

Updates as prescribed by an MD m are not arbitrary. The updates based on m have to be justified by m, as captured through the notion of *modifiable value position* in an instance. Values in modifiable positions are the only ones that are allowed to change under a legal update. The notion of modifiable position depends on the syntax of the MDs, but also on the instance at hand on which updates that identify values are to be applied, because the tuple t in a position (t, A) belongs to that instance. We give an example illustrating some issues involved in the definition of modifiability (cf. Definition 2 below).

Example 1. Consider $m\colon R[A] \approx R[A] \to R[B] \doteq R[B]$ on schema $R[A, B]$, and the instance $R(D)$ shown below.

$R(D)$	A	B
t_1	a_1	c_1
t_2	a_2	c_1
t_3	a_3	c_3
t_4	b_1	c_3
t_5	b_2	c_3

Assume the only non-trivial similarities are $a_1 \approx a_2 \approx a_3$ and $b_1 \approx b_2$. One might be tempted to declare positions (t_i, B) and (t_j, B) as modifiable whenever $t_i[A] \approx t_j[A]$ holds in D. In this case, (t_4, B) and (t_5, B) would be classified as modifiable.

---

[2] We consider this class to simplify the presentation. However, the results in this paper also apply to the more general case of MDs of the form $R_1[\bar{A}] \approx R_2[\bar{B}] \to R_1[\bar{C}] \doteq R_2[\bar{D}]$, with the corresponding attributes in $\bar{A}, \bar{B}$ (and in $\bar{C}, \bar{D}$) sharing domains, in particular, similarity relations [22].

[3] That is, they share at least one corresponding pair of attributes.

However, the values in those positions should not be allowed to change, since $t_4[B] = t_5[B]$ and no duplicate resolution is needed. Consequently, we might consider adding the requirement that $t_i[B] \neq t_j[B]$, which would make (t_2, B) and (t_3, B) modifiable, but (t_1, B) non-modifiable.

This is problematic, because a legal update would in general lead to $t_1[B] \neq t_2[B]$ in the new instance (unless the update value for (t_2, B) and (t_3, B) is chosen to be c_1). This would go against the intended meaning of the MD, which tells us that (t_1, B), (t_2, B), and (t_3, B) represent the same entity. As a consequence, $t_1[B] = t_2[B] = t_3[B]$ should hold in the updated instance.   □

Example 1 shows that defining modifiability of a value position[4] in terms of just a pair of tuples does not lead to an appropriate restriction on updates. The definition below uses recursion to take larger groups of positions into account.

Definition 2. Let D be an instance, M a set of MDs, and $\mathcal{P}$ be a set of positions (t, G), where t is a tuple of D and G is an attribute of t. (a) For a tuple $t_1 \in R(D)$ and C an attribute of R, the position (t_1, C) is *modifiable wrt* $\mathcal{P}$ if there exist $t_2 \in R(D)$, an $m \in M$ of the form $R[\bar{A}] \approx R[\bar{A}] \rightarrow R[\bar{B}] \doteq R[\bar{B}]$, and an attribute B of $\bar{B}$, such that $(t_2, B) \in \mathcal{P}$ and one of the following holds:

1. $t_1[\bar{A}] \approx t_2[\bar{A}]$, but $t_1[B] \neq t_2[B]$.
2. $t_1[\bar{A}] \approx t_2[\bar{A}]$ and (t_2, B) is modifiable wrt $\mathcal{P} \setminus \{(t_2, B)\}$.

(b) The position (t_1, B) is *modifiable* if it is modifiable wrt $\mathcal{V} \setminus \{(t_1, B)\}$, where $\mathcal{V}$ is the set of all positions (t, G) with t a tuple of D and G an attribute of t.   □

This definition 2 is recursive. The base case occurs when either case 1. applies (with any $\mathcal{P}$) or when there is no tuple/attribute pair in $\mathcal{P}$ that can satisfy part (a). Notice that the recursion must eventually terminate, since the latter condition must be satisfied when $\mathcal{P}$ is empty, and each recursive call reduces the size of $\mathcal{P}$.

Example 2. (example 1 continued) Since $a_2 \approx a_3$ and $c_1 \neq c_3$, (t_2, B) and (t_3, B) are modifiable (base case). Since $a_1 \approx a_2$ and (t_2, B) is modifiable, with case 2. of Definition 2, we obtain that (t_1, B) is also modifiable.

For (t_5, B) to be modifiable, it must be modifiable wrt $\{(t_i, B) \mid 1 \leq i \leq 4\}$, and via t_4. According to case 2. of Definition 2, this requires (t_4, B) to be modifiable wrt $\{(t_i, B) \mid 1 \leq i \leq 3\}$. However, this is not the case since there is no t_i, $1 \leq i \leq 3$, such that $t_4[A] \approx t_i[A]$. Therefore (t_5, B) is not modifiable. A symmetric argument shows that (t_4, B) is not modifiable either.

Notice that the recursive nature of Definition 2 requires defining modifiability in terms of a set of value positions (the set $\mathcal{P}$ in the definition). This set allows us to keep track of positions that have already been "tried". For example, to determine the modifiability of (t_5, B), we must determine whether or not (t_4, B) is modifiable. However, since we have already eliminated (t_5, B) from consideration when deciding about (t_4, B), we avoid an infinite loop.   □

---

[4] Not to be confused with the notion of changeable attribute.

Definition 3. [23] Let D, D' be correlated instances, and M a set of MDs. (D, D') *satisfies* M, denoted $(D, D') \vDash M$, iff: 1. For any pair of tuples $t_1, t_2 \in R(D)$, if there exists an $m \in M$ of the form $R[\bar{A}] \approx R[\bar{A}] \to R[\bar{B}] \doteq R[\bar{B}]$ and $t_1[\bar{A}] \approx t_2[\bar{A}]$, then for the corresponding tuples (i.e. with same ids) $t'_1, t'_2 \in R(D')$, it holds $t'_1[\bar{B}] = t'_2[\bar{B}]$. 2. For any tuple $t \in R(D)$ and any attribute G of R, if (t, G) is a non-modifiable position, then $t'_R[G] = t_R[G]$. □

Intuitively, D' in Definition 3 is a new version of D that is produced after a single update. Since the update involves matching values (i.e. making them equal), it may produce "duplicate" tuples, i.e. that only differ in their tuple ids. They could possibly be merged into a single tuple in the a data cleaning process. However, we keep the two versions. In particular, D and D' have the same number of tuples. Keeping or eliminating duplicates will not make any important difference in the sense that, given that tuple ids are never updated, two duplicates will evolve in exactly the same way as subsequent updates are performed. Duplicate tuples will never be subsequently "unmerged".

This definition of MD satisfaction departs from [18], which requires that updates preserve similarities. Similarity preservation may force undesirable changes [23]. The existence of the updated instance D' for D is guaranteed [23]. Furthermore, wrt [18], our definition does not allow unnecessary changes from D to D'. Definitions 2 and 3 imply that only values of changeable attributes are subject to updates.

Definition 3 allows us to define a *clean instance* wrt M as the result of a chase-like procedure, each step being satisfaction preserving.

Definition 4. [23] (a) A *resolved instance* (RI) for D wrt M is an instance D', such that there are instances $D_1, D_2, ... D_n$ with: $(D, D_1) \vDash M$, $(D_1, D_2) \vDash M$,..., $(D_{n-1}, D_n) \vDash M$, and $(D_n, D') \vDash M$, and $(D', D') \vDash M$. We say D' is *stable*. (b) D' is a *minimally resolved instance* (MRI) for D wrt M if it is a resolved instance and it minimizes the overall number of attribute value changes wrt D. (c) $MRI(D, M)$ denotes the class of MRIs of D wrt M. □

Example 3. Consider the MD $R[A] \approx R[A] \to R[B] \doteq R[B]$ on predicate R, and the instance D. It has several resolved instances, among them, four that minimize the number of value changes. One of them is D_1. A resolved instance that is not minimal in this sense is D_2.

$R(D)$	A	B		$R(D_1)$	A	B		$R(D_2)$	A	B
t_1	a_1	c_1		t_1	a_1	c_1		t_1	a_1	c_1
t_2	a_1	c_2		t_2	a_1	c_1		t_2	a_1	c_1
t_3	b_1	c_3		t_3	b_1	c_3		t_3	b_1	c_1
t_4	b_1	c_4		t_4	b_1	c_3		t_4	b_1	c_1

□

In this work, as in [23, 24], we are investigating what we could call "the pure case" of MD-based entity resolution. It adheres to the original semantics outlined in [18], which does not specify how the matchings are to be done, but only which values must be made equal. That is, the MDs have implicit existential quantifiers (for the values in common). The semantics we just introduced formally captures this pure case. We find situations like this in other areas of data management, e.g. with referential integrity constraints, tuple-generating dependencies in general [1], schema mappings in data exchange [5], etc. A "non-pure" case, that uses matching functions to realize the matchings as prescribed by MDs, is investigated in [11, 12, 4]. Since there is always an RI [23], there is always an MRI for an instance D wrt M.

The *resolved* answers to a query are *certain* for the class of MRIs for D wrt M.

Definition 5. [23] Let $\mathcal{Q}(\bar{x})$ be a query expressed in the first-order language $L(\mathcal{S})$ associated to schema $\mathcal{S}$ of an instance D. A tuple of constants $\bar{a}$ from U is a *resolved answer* to $\mathcal{Q}(\bar{x})$ wrt the set M of MDs, denoted $D \models_M \mathcal{Q}[\bar{a}]$, iff $D' \models \mathcal{Q}[\bar{a}]$, for every $D' \in MRI(D, M)$. We denote with $ResAn(D, \mathcal{Q}, M)$ the set of resolved answers to $\mathcal{Q}$ from D wrt M. □

Example 4. (example 1 cont.) Since the only MRI for the original instance D is $R(D')=\{\langle t_1, a_1, c_1 \rangle, \langle t_2, a_2, c_1 \rangle, \langle t_3, a_3, c_1 \rangle, \langle t_4, b_1, c_3 \rangle, \langle t_5, b_2, c_3 \rangle\}$, the resolved answers to the query $\mathcal{Q}(x, y)$: $R(x, y)$ are $\{\langle a_1, c_1 \rangle, \langle a_2, c_1 \rangle, \langle a_3, c_1 \rangle, \langle b_1, c_3 \rangle, \langle b_2, c_3 \rangle\}$. □

For a query $\mathcal{Q}$ and set of MDs M, the *resolved answer problem* is the problem of deciding, given a tuple $\bar{a}$ and instance D, whether or not $\bar{a} \in ResAn(D, \mathcal{Q}, M)$. More precisely, it is defined by

$$RA_{\mathcal{Q},M} := \{(D, \bar{a}) \mid \bar{a} \in ResAn(D, \mathcal{Q}, M)\}. \tag{2}$$

4   Hit-Simple-Cyclic Sets of MDs

In general, the resolved answer decision problem is NP-hard.

Theorem 1. [23] The resolved answer decision problem can be intractable for join-free conjunctive queries and pairs of interacting MDs. More precisely, for the the query $\mathcal{Q}(x, z)$: $\exists y R(x, y, z)$, and the following set M of MDs

$$m_1 : R[A] \approx R[A] \rightarrow R[B] \doteq R[B]$$
$$m_2 : R[B] \approx R[B] \rightarrow R[C] \doteq R[C]$$

the resolved answer (decision) problem is *NP*-hard (in data). □

Generally, intractability of the resolved answer problem arises when choices of update values made during one update in the chase sequence can affect subsequent updates. For the case in Theorem 1, when the instance is updated according to m_1, the choice of update values for values in the B column affects subsequent updates made to the values in the C column according to m_2. The resolved answer problem is tractable for non-interacting sets of MDs, because there is no dependence of updates on previous updates.

In this section, we define a class of sets of MDs, called *hit-simple-cyclic* (HSC) sets, for which the resolved answer problem is tractable for an important class of conjunctive queries. Specifically, we introduce a recursively-defined predicate that can be used to identify the sets of values that must be updated to obtain an MRI and the possible values to which they can be updated. For HSC sets, the interaction of the MDs does not lead to intractability, as it is the case for the set of MDs in Theorem 1. This is because the stability requirement of Definition 4 imposes a simple form on MRIs, making it unnecessary to consider the many possible chase sequences.

Definition 6. A set M of MDs is *simple-cycle* (SC) if its MD graph $MDG(M)$ is (just) a cycle, and in all MDs $m \in M$, at most one attribute in $LHS(m)$ is changeable.   □

Example 5. For schema $R[A, C, F, G]$, consider the following set M of MDs:

$$m_1: \ R[A] \approx R[A] \rightarrow R[C, F, G] \doteq R[C, F, G],$$
$$m_2: \ R[C] \approx R[C] \rightarrow R[A, F, G] \doteq R[A, F, G].$$

$MDG(M)$ is a cycle, because attributes in $RHS(m_2)$ appear in $LHS(m_1)$, and viceversa. Furthermore, M is SC, because $LHS(m_1)$ and $LHS(m_2)$ are singletons.   □

SC sets of MDs can be easily found in practical applications. For them, it is easy to characterize the form taken by an MRI.

Example 6. Consider the instance D and a SC set of MDs below, where the only similarities are: $a_i \approx a_j$, $b_i \approx b_j$, $d_i \approx d_j$, $e_i \approx e_j$, with $i, j \in \{1, 2\}$.

$R(D)$	A	B
1	a_1	d_1
2	a_2	e_2
3	b_1	e_1
4	b_2	d_2

$m_1: \ R[A] \approx R[A] \rightarrow R[B] \doteq R[B],$
$m_2: \ R[B] \approx R[B] \rightarrow R[A] \doteq R[A].$

If the MDs are applied twice, successively, starting from D, a possible result is:

$R(D)$	A	B		$R(D_1)$	A	B		$R(D_2)$	A	B
1	a_1	d_1		1	b_2	d_1		1	a_2	e_1
2	a_2	e_2	$\rightarrow$	2	a_2	d_1	$\rightarrow$	2	a_2	d_1
3	b_1	e_1		3	a_2	e_1		3	b_2	d_1
4	b_2	d_2		4	b_2	e_1		4	b_2	e_1

It should be clear that, in any sequence of instances $D_1, D_2, \ldots$, obtained from D by applying the MDs, the updated instances must have the following pairs of equal values or matchings (shown through the tuple ids) in Table 1. In any stable instance, the pairs of values in the above tables must be equal. Given the alternating behavior, this can only be the case if all values in A are equal, and similarly for B, which can be achieved with a single update, choosing any value as the common value for each of A and B.

Table 1. Table of matchings

D_i i odd	A	B		D_i i even	A	B
tuple (id) pairs	$(1, 4), (2, 3)$	$(1, 2), (3, 4)$		tuple (id) pairs	$(1, 2), (3, 4)$	$(1, 4), (2, 3)$

In particular, an MRI requires the common value for each attribute to be set to a most common value in the original instance. For D there are 16 MRIs.   □

Example 7. (example 5 cont.) The relation R subject to the given M, has two "keys", $R[A]$ and $R[C]$. A relation like this may appear in a database about people: $R[A]$ could be used for the person's name, $R[C]$ the address, and $R[F]$ and $R[G]$ for non-distinguishing information, e.g. gender and age.   □

We now define an extension of the class of SC sets of MDs.

Definition 7. A set M of MDs is *hit-simple-cyclic* iff the following hold: (a) In all MDs $m \in M$, at most one attribute in $LHS(m)$ is changeable. (b) Each vertex v_1 in $MDG(M)$ is on at least one cycle, or there is a vertex v_2 on a cycle with at least two vertices such that there is an edge from v_1 to v_2. □

Notice that SC sets are also HSC sets. An example of the MD graph of an HSC set of MDs is shown in Figure 1.

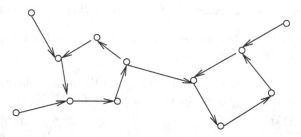

Fig. 1. The MD-graph of an HSC set of MDs

As the previous examples suggest, it is possible to provide a full characterization of the MRIs for an instance subject to an HSC set of MDs, which we do next. For this result, we need a few definitions and notations.

For an SC set M and $m \in M$, if a pair of tuples satisfies the similarity condition of *any* MD in M, then the values of the attributes in $RHS(m)$ must be merged for these tuples. Thus, in Example 6, a pair of tuples satisfying *either* $R[A] \approx R[A]$ *or* $R[B] \approx R[B]$ have *both* their $R[A]$ and $R[B]$ attributes updated to the same value. More generally, for an HSC set M of MDs, and $m \in M$, there is only a *subset* of the MDs such that, if a pair of tuples satisfies the similarity condition of an MD in the subset, then the values of the attributes in $RHS(m)$ must be merged for the pair of tuples. We now formally define this subset.

Definition 8. Let M be a set of MDs, and $m \in M$. (a) The *previous set* of m, denoted $PS(m)$, is the set of all MDs $m' \in M$ with a path in $MDG(M)$ from m' to m. (b) The previous set $PS(M)$ of a set M of MDs is $\bigcup_{m \in M} PS(m)$. □

When applying a set of MDs to an instance, consistency among updates must be enforced. This generally requires computing the transitive closure of similarity relations. For example, suppose both m_1 and m_2 have the conjunct $R[A] \doteq R[A]$. If t_1 and t_2 satisfy the condition of m_1, and t_2 and t_3 satisfy the condition of m_2, then $t_1[A]$ and $t_3[A]$ must be updated to the same value, since updating them to different values would require $t_2[A]$ to be updated to two different values at once. We formally define this transitive closure relation.

Definition 9. Consider an instance D, and set of MDs $M = \{m_1, m_2, \ldots, m_n\}$. (a) For MD $m : R[\bar{A}] \approx R[\bar{A}] \rightarrow R[\bar{B}] \doteq R[\bar{B}]$, T_m is the reflexive, symmetric, transitive closure of the binary relation that relates pairs of tuples t_1 and t_2 in D satisfying $t_1[A] \approx t_2[A]$. (b) For a subset M' of M, $T_{M'}$ is the reflexive, symmetric, transitive closure of $\{T_m \mid m \in M'\}$. □

In the case of HSC sets of MDs, the MRIs for a given instance can be characterized simply using the T relation. This result is stated formally below.

Proposition 1. [22] For a set of MDs M and attribute A, let M_A be the set of all $m \in M$ such that $A \in RHS(m)$. For M HSC and D an instance, each MRI for D wrt M is obtained by setting, for each attribute A and each equivalence class E of $T_{PS(M_A)}$, the value of all $t[A], t \in E$, to one of the most frequent values for $t[A], t \in E$. □

Example 8. (example 6 cont.) We represent tuples by their ids. We have:

$$
\begin{aligned}
T_{m_1} &= \{(1,2),(3,4),(2,1),(4,3)\} \cup \{(i,i) \mid 1 \le i \le 4\}, \\
T_{m_2} &= \{(1,4),(2,3),(4,1),(3,2)\} \cup \{(i,i) \mid 1 \le i \le 4\}, \\
T_{\{m_1,m_2\}} &= \{(i,j) \mid 1 \le i,j \le 4\}, \\
T_{PS(M_A)} &= T_{PS(m_2)} = T_{\{m_1,m_2\}}, \quad T_{PS(M_B)} = T_{PS(m_1)} = T_{\{m_1,m_2\}}.
\end{aligned}
$$

The (single) equivalence class of $T_{PS(M_A)}$ $(T_{PS(M_B)})$ is $\{(i,A) \mid 1 \le i \le 4\}$ $(\{(i,B) \mid 1 \le i \le 4\})$. From Proposition 1, the 16 MRIs are obtained by setting all $R[A]$ and $R[B]$ attribute values to one of the four existing (and, actually, equally frequent) values for them. □

It is possible to prove using Proposition 1 that, for HSC sets, the resolved answer problem is efficiently solvable for join-free conjunctive queries like the one in Theorem 1. In fact, it is shown in [22] that for HSC sets and a significant class of conjunctive queries with restricted joins, the resolved answer problem is solvable in polynomial time using a query rewriting technique.

Definition 10. Let Q be a conjunctive query without built-ins, and M a set of MDs. Q is an *unchangeable join* conjunctive query if there are no existentially quantified variables in a join in Q in the position of a changeable attribute. $UJCQ$ denotes this class of queries. □

Example 9. For schema $S = \{R[A,B]\}$, let M consist of the single MD $R[A] \approx R[A] \to R[B] \doteq R[B]$. Attribute B is changeable, and A is unchangeable. The query $Q_1(x,z) : \exists y(R(x,y) \wedge R(z,y))$ is not in $UJCQ$, because the bound and repeated variable y is for the changeable attribute B. However, the query $Q_2(y) : \exists x \exists z(R(x,y) \wedge R(x,z))$ is in $UJCQ$: the only bound, repeated variable is x which is for the unchangeable attribute A. If variables x and y are swapped in the first atom of Q_2, the query is not $UJCQ$. □

Theorem 2. [22] For a HSC (or non-interacting) set of MDs M and a $UJCQ$ query Q, there is an effective rewriting Q' that is efficiently evaluable and returns the resolved answers to Q. □

The rewritten queries of the theorem are expressed in FO logic with an embedded recursively defined predicate that expresses the transitive closure of Definition 9 (e.g. in Datalog) plus a the *count* aggregation operator (of number of different attribute values). They can be evaluated in quadratic time in data [22].

5   A CQA Connection

MDs can be seen as a new form of integrity constraint (IC), with a dynamic semantics. An instance D violates an MD m if there are unresolved duplicates, i.e. tuples t_1 and t_2 in D that satisfy the similarity conditions of m, but differ in value on some pairs of attributes that are expected to be matched according to m. The instances that are consistent with a set of MDs M (or *self-consistent* from the point of view of the dynamic semantics) are resolved instances of themselves with respect to M. Among classical ICs, the closest analogues of MDs are functional dependencies (FDs).

Now, given a database instance D and a set of ICs Σ, possibly not satisfied by D, *consistent query answering* (CQA) is the problem of characterizing and computing the answers to queries Q that are true in all *repairs* of D, i.e. the instances D' that are consistent with Σ and minimally differ from D [3]. Minimal difference between instances can be defined in different ways. Most of the research in CQA has concentrated on the case of the set-theoretic symmetric difference of instances, as sets of tuples, which in the case of repairs is made minimal under set inclusion, as originally introduced in [3]. Also the minimization of the *cardinality* of this set-difference has been investigated [27, 2]. Other forms of minimization measure the differences in terms of changes of attribute values between D and D' (as opposed to entire tuples) [20, 29, 19, 10], e.g. the number of attribute updates can be used for comparison. Cf. [7, 14, 8] for CQA.

Because of their practical importance, much work on CQA has been done for the case where Σ is a set of functional dependencies (FDs), and in particular for sets, $\mathcal{K}$, of key constraints (KCs) [15, 21, 30], with the distance being the set-theoretic symmetric difference under set inclusion. In this case, on which we concentrate in the rest of this section, a *repair* D' of an instance D becomes a maximal subset of D that satisfies $\mathcal{K}$, i.e. $D' \subseteq D$, $D' \models \mathcal{K}$, and there is no D'' with $D' \subsetneq D'' \subseteq D$, with $D'' \models \mathcal{K}$ [15].

Accordingly, for a FO query $Q(\bar{x})$ and a set of KCs $\mathcal{K}$, $\bar{a}$ is a *consistent answer* from D to $Q(\bar{x})$ wrt $\mathcal{K}$ when $D' \models Q[\bar{a}]$, for every repair D' of D. For fixed $Q(\bar{x})$ and $\mathcal{K}$, the *consistent query answering problem* is about deciding membership in the set $CQA_{Q,\mathcal{K}} = \{(D, \bar{a}) \mid \bar{a} \text{ is a consistent answer from } D \text{ to } Q \text{ wrt } \mathcal{K}\}$.

Notice that this notion of minimality involved in repairs wrt FDs is tuple and set-inclusion oriented, whereas the one that is implicitly related to MDs and MRIs via the matchings (cf. Definition 4) is attribute and cardinality oriented.[5] However, the connection can still be established.

For certain classes of conjunctive queries and ICs consisting of a single KC per relation, *CQA* is tractable. This is the case for the $\mathcal{C}_{forest}$ class of conjunctive queries [21], for which there is a FO rewriting methodology for computing the consistent answers. $\mathcal{C}_{forest}$ excludes repeated relations (self-joins), and allows joins only between non-key and key attributes. Similar results were subsequently proved for a larger class of queries that includes some queries with repeated relations and joins between non-key attributes [30]. The following result allows us to take advantage of tractability results for CQA in our MD setting.

---

[5] Cf. [23] for a discussion of the differences between FDs and MDs seen as ICs, and their repair processes.

Proposition 2. [22] Let D be a database instance for a single predicate R whose set of attributes is $\bar{A} \cup \bar{B}$, with $\bar{A} \cap \bar{B} = \emptyset$; and m the MD $R[\bar{A}] = R[\bar{A}] \to R[\bar{B}] \doteq R[\bar{B}]$. There is a polynomial time reduction from $RA_{\mathcal{Q},\{m\}}$ (cf. (2)) to $CQA_{\mathcal{Q},\{\kappa\}}$, where κ is the key constraint $R \colon \bar{A} \to \bar{B}$. $\square$

This result can be easily generalized to several relations with one such MD defined on each. The reduction takes an instance D for $RA_{\mathcal{Q},\{m\}}$ and produces an instance D' for $CQA_{\mathcal{Q},\{\kappa\}}$. The schema of D' is the same as for D, but the extension of the relation is changed wrt D via counting. Definitions for those aggregations can be inserted into query $\mathcal{Q}$, producing a rewriting $\mathcal{Q}'$. Thus, we obtain:

Theorem 3. [22] Let $\mathcal{S}$ be a schema with $\mathcal{R} = \{R_1[\bar{A}_1, \bar{B}_1], \ldots, R_n[\bar{A}_n, \bar{B}_n]\}$ and $\mathcal{K}$ the set of KCs $\kappa_i \colon R_i[\bar{A}_i] \to R_i[\bar{B}_i]$. Let $\mathcal{Q}$ be a FO query for which there is a polynomial-time computable FO rewriting $\mathcal{Q}'$ for computing the consistent answers to $\mathcal{Q}$. Then there is a polynomial-time computable FO query $\mathcal{Q}''$ extended with aggregation[6] for computing the resolved answers to $\mathcal{Q}$ from D wrt the set of MDs $m_i \colon R_i[\bar{A}_i] = R_i[\bar{A}_i] \to R_i[\bar{B}_i] \doteq R_i[\bar{B}_i]$. $\square$

The aggregation in $\mathcal{Q}''$ in Theorem 3 arises from the *generic* transformation of the instance that is used in the reduction involved in Proposition 2, but here becomes implicit in the query.

   This theorem can be applied to decide/compute resolved answers in those cases where a FO rewriting for CQA (aka. consistent rewriting) has been identified.

Example 10. The query $\mathcal{Q} \colon \exists x \exists y \exists z \exists w (R(x,y,w) \land S(y,w,z))$ is in the class $\mathcal{C}_{forest}$ for relational predicates $R[A, B, C]$ and $S[C, E, F]$ and KCs $A \to BC$ and $CE \to F$. By Theorem 3 and the results in [21], there is a polynomial-time computable FO query with counting that returns the resolved answers to $\mathcal{Q}$ wrt the MDs $R[A] = R[A] \to R[B,C] \doteq R[B,C]$ and $S[C,E] = S[C,E] \to S[F] \doteq S[F]$, namely,

$$\mathcal{Q}'' \colon \exists x \exists y \exists z \exists w [R'(x,y,w) \land S(y,w,z) \land \forall y' \forall w'(R'(x,y',w') \to \exists z' S(y',w',z')),$$

where $R'(x,y,w) := \exists w'\{R(x,y,w') \land \forall y'[Count\{w'' \mid R(x,y',w'')\} \le$
$$Count\{w'' \mid R(x,y,w'')\}]\} \land$$
$$\exists y'\{R(x,y',w) \land \forall v[Count\{y'' \mid R(x,y'',v)\} \le Count\{y'' \mid R(x,y'',w)\}]\}.$$

Here, $Count\{x \mid E(x)\}$, for E a first-order expression and x a variable, denotes the number of distinct values of x that satisfy E in the database instance at hand. This "resolved rewriting" wrt the MDs is obtained from the consistent rewriting $\mathcal{Q}'$ for $\mathcal{Q}$ wrt the FDs, by replacing R by R' as indicted above, i.e. using

$$\mathcal{Q}' \colon \exists x \exists y \exists z \exists w [R(x,y,w) \land S(y,w,z) \land \forall y' \forall w'(R(x,y',w') \to \exists z' S(y',w',z')),$$

S in $\mathcal{Q}'$ could also be replaced by a similar S' to obtain $\mathcal{Q}''$. However, in this example it is not necessary, because the key values are not changed when the MDs are applied, so the join condition is not affected (the join is on the key values for S, but on the non-key values for R).

---

[6] This is a proper extension of FO query languages [26, Chapter 8].

Notice that $\mathcal{Q}$ is not in $UJCQ$ because variable y is existentially quantified, participates in a join, and occurs at the position of the changeable attribute $R[B]$ (cf. Definition 10). Therefore, Theorem 2 cannot be used to obtain a query rewriting in this case.   □

The example shows that via the CQA connection we obtain rewritable and tractable cases of resolved answering that are different from those provided by Theorem 2.

In this paper we have concentrated on tractable cases of resolved query answering. However, the CQA connection can also be exploited to obtain intractability results, which we briefly illustrate.

Theorem 4. [22] Consider the relational predicate $R[A, B, C]$, the MD m: $R[A] = R[A] \rightarrow R[B, C] \doteq R[B, C]$, and the query $\mathcal{Q}$: $\exists x \exists y \exists y' \exists z (R(x, y, c) \wedge R(z, y', d) \wedge y = y')$. $RA_{\mathcal{Q}, \{m\}}$ is $coNP$-complete (in data).   □

This result can be obtained through a reduction and a result in [15, Thm. 3.3]. Notice that the query in Theorem 4 is not $UJCQ$.

6   Conclusions

Matching dependencies first appeared in [17], and their semantics is given in [18]. The original semantics was refined in [11, 12], including the use of *matching functions* for matching two attribute values. An alternative refinement of the semantics, which is the one used in this paper, is given in [23, 24]. Cf. [22] for a thorough complexity analysis, as well as the derivation of a query rewriting algorithm for the resolved answer problem.

This paper builds on the MD-based approach to duplicate resolution introduced in [23]. The latter paper introduced the framework used in this paper, and proved that the resolved answer problem is intractable in some cases. In this paper, we presented a case for which minimally resolved instances can be identified in polynomial time. From this, it follows that the resolved answers can be efficiently retrieved from a dirty database in this case [22]. We also derived other tractable cases using results from CQA.

We used minimal resolved instances (MRIs) as our model of a clean database. Another possibility is to use arbitrary, not necessarily minimal, resolved instances (RIs). This has the advantage of being more flexible in that it takes into account all possible ways of repairing the database. In some cases, the RIs can be characterized by expressing the chase rules in Datalog. Such a direct approach is difficult in the case of MRIs, because of the global nature of the minimality constraint.

References

[1] Abiteboul, S., Hull, R., Vianu, V.: Foundations of Databases. Addison-Wesley (1995)
[2] Afrati, F., Kolaitis, P.: Repair checking in inconsistent databases: Algorithms and complexity. In: Proc. ICDT, pp. 31–41. ACM Press (2009)
[3] Arenas, M., Bertossi, L., Chomicki, J.: Consistent query answers in inconsistent databases. In: Proc. PODS, pp. 68–79. ACM Press (1999)
[4] Bahmani, Z., Bertossi, L., Kolahi, S., Lakshmanan, L.: Declarative entity resolution via matching dependencies and answer set programs. In: Proc. KR, pp. 380–390. AAAI Press (2012)
[5] Barcelo, P.: Logical foundations of relational data exchange. SIGMOD Record 38(1), 49–58 (2009)
[6] Benjelloun, O., Garcia-Molina, H., Menestrina, D., Su, Q., Euijong Whang, S., Widom, J.: Swoosh: A generic approach to entity resolution. VLDB Journal 18(1), 255–276 (2009)

[7] Bertossi, L.: Consistent query answering in databases. ACM Sigmod Record 35(2), 68–76 (2006)

[8] Bertossi, L.: Database Repairing and Consistent Query Answering. Synthesis Lectures on Data Management. Morgan & Claypool (2011)

[9] Bertossi, L., Bravo, L.: Consistent Query Answers in Virtual Data Integration Systems. In: Bertossi, L., Hunter, A., Schaub, T. (eds.) Inconsistency Tolerance. LNCS, vol. 3300, pp. 42–83. Springer, Heidelberg (2005)

[10] Bertossi, L., Bravo, L., Franconi, E., Lopatenko, A.: The complexity and approximation of fixing numerical attributes in databases under integrity constraints. Information Systems 33(4), 407–434 (2008)

[11] Bertossi, L., Kolahi, S., Lakshmanan, L.: Data cleaning and query answering with matching dependencies and matching functions. In: Proc. ICDT. ACM Press (2011)

[12] Bertossi, L., Kolahi, S., Lakshmanan, L.: Data cleaning and query answering with matching dependencies and matching functions. Theory of Computing Systems (2012), doi: 10.1007/s00224-012-9402-7

[13] Bleiholder, J., Naumann, F.: Data fusion. ACM Computing Surveys 41(1), 1–41 (2008)

[14] Chomicki, J.: Consistent Query Answering: Five Easy Pieces. In: Schwentick, T., Suciu, D. (eds.) ICDT 2007. LNCS, vol. 4353, pp. 1–17. Springer, Heidelberg (2006)

[15] Chomicki, J., Marcinkowski, J.: Minimal-change integrity maintenance using tuple deletions. Information and Computation 197(1/2), 90–121 (2005)

[16] Elmagarmid, A., Ipeirotis, P., Verykios, V.: Duplicate record detection: A survey. IEEE Trans. Knowledge and Data Eng. 19(1), 1–16 (2007)

[17] Fan, W.: Dependencies revisited for improving data quality. In: Proc. PODS, pp. 159–170. ACM Press (2008)

[18] Fan, W., Jia, X., Li, J., Ma, S.: Reasoning about record matching rules. In: Proc. VLDB, pp. 407–418 (2009)

[19] Flesca, S., Furfaro, F., Parisi, F.: Querying and repairing inconsistent numerical databases. ACM Trans. Database Syst. 35(2) (2010)

[20] Franconi, E., Palma, A.L., Leone, N., Perri, S., Scarcello, F.: Census Data Repair: A Challenging Application of Disjunctive Logic Programming. In: Nieuwenhuis, R., Voronkov, A. (eds.) LPAR 2001. LNCS (LNAI), vol. 2250, pp. 561–578. Springer, Heidelberg (2001)

[21] Fuxman, A., Miller, R.: First-order query rewriting for inconsistent databases. J. Computer and System Sciences 73(4), 610–635 (2007)

[22] Gardezi, J., Bertossi, L.: Query answering under matching dependencies for data cleaning: Complexity and algorithms. arXiv:1112.5908v1

[23] Gardezi, J., Bertossi, L., Kiringa, I.: Matching dependencies with arbitrary attribute values: semantics, query answering and integrity constraints. In: Proc. Int. WS on Logic in Databases (LID 2011), pp. 23–30. ACM Press (2011)

[24] Gardezi, J., Bertossi, L., Kiringa, I.: Matching dependencies: semantics, query answering and integrity constraints. Frontiers of Computer Science 6(3), 278–292 (2012)

[25] Lenzerini, M.: Data integration: a theoretical perspective. In: Proc. PODS 2002, pp. 233–246 (2002)

[26] Libkin, L.: Elements of Finite Model Theory. Springer (2004)

[27] Lopatenko, A., Bertossi, L.: Complexity of Consistent Query Answering in Databases Under Cardinality-Based and Incremental Repair Semantics. In: Schwentick, T., Suciu, D. (eds.) ICDT 2007. LNCS, vol. 4353, pp. 179–193. Springer, Heidelberg (2006)

[28] Vianu, V.: Dynamic functional dependencies and database aging. J. ACM 34(1), 28–59 (1987)

[29] Wijsen, J.: Database repairing using updates. ACM Trans. Database Systems 30(3), 722–768 (2005)

[30] Wijsen, J.: On the first-order expressibility of computing certain answers to conjunctive queries over uncertain databases. In: Proc. PODS, pp. 179–190. ACM Press (2010)

A Distance between Continuous Belief Functions

Dorra Attiaoui[1], Pierre-Emmanuel Doré, Arnaud Martin[2],
and Boutheina Ben Yaghlane[1]

[1] LARODEC, ISG Tunis, 41 Rue de la Liberté,
Cité Bouchoucha 2000 Le Bardo, Tunisie
[2] UMR 6074 IRISA, Université de Rennes1, BP 3021, 22302 Lannion cedex France
`attiaoui.dorra@gmail.com,`
`{pierre-emmanuel.dore,Arnaud.Martin}@univ-rennes1.fr,`
`boutheina.yaghlane@ihec.rnu.tn`

Abstract. In the theory of belief functions, distances between basic belief assignments are very important in many applications like clustering, conflict measuring, reliability estimation. In the discrete domain, many measures have been proposed, however, distance between continuous belief functions have been marginalized due to the nature of these functions. In this paper, we propose an adaptation inspired from the Jousselme's distance for continuous belief functions.

Keywords: Theory of belief functions, continuous belief functions, distance, scalar product.

1 Introduction

The theory of belief function (also referred to as the mathematical theory of evidence) is one of the most popular quantitative approach. It is known for its ability to represent uncertain and imprecise information.

It is a strong formalism widely used in many research areas: medical, image processing,... It has been used thanks to its ability to manage imperfect information.

In the discrete case, information fusion, has known a large success, focusing on the study of conflict between belief functions. Among these researches, it has been considered that a distance between two bodies of evidence can be interpreted as a conflict measure used during combination as presented in [7]. Smets, in [15], extended the theory of belief functions on real numbers thinking over continuous belief functions by presenting a complete description.

In the discrete case, distances between probability distributions can be considered as a definition of dissimilarity in the theory of belief functions. Ristic and Smets in [10], defined a distance based on Dempster's conflict factor, others proposed geometrical ones like Jousselme *et al.* in [5].

In this paper we are interested to adapt the notion of distance to the continuous belief functions, and study the behavior according to two different types of distributions.

E. Hüllermeier et al. (Eds.): SUM 2012, LNAI 7520, pp. 194–205, 2012.
© Springer-Verlag Berlin Heidelberg 2012

This paper is organized as follow: in Section 2 we recall some basic concepts of the theory of belief functions in the discrete case, and present the notion of distance. Therefore, in Section 3, we introduce the continuous belief functions, some of their properties and characterization. After this, we propose in Section 4 an adaptation of the Jousselem's distance using Smet's formalism. Finally, in Section 5, the proposed distance is used to measure the distance to the continuous belief functions where the probability density functions are following in the first place all the normal then the exponential distribution, and the contribution of using this distance instead of a classical scalar product.

2 Belief Function Theory Background

This section recalls the necessary background notions related to the theory of belief functions. It has been developed by Dempster in his work on upper and lower probabilities [1]. Afterwards, it was formalized in a mathematical framework by Shafer in [11]. This theory is able to deal and represent imperfect (uncertain, imprecise and /or incomplete) information.

2.1 Discrete Belief Functions

Let us consider a variable x taking values in a finite set $\Omega = \{\omega_1, \cdots, \omega_n\}$ called *the frame of discernment*.

A *basic belief assignment* (*bba*) is defined on the set of all subsets of Ω, named power set and noted 2^Ω. It affects a real value from $[0, 1]$ to every subset of 2^Ω reflecting sources amount of belief on this subset. A bba m verifies:

$$\sum_{X \subseteq \Omega} m(X) = 1. \tag{1}$$

Given a bba m we can associate some other functions defined as:

- Credibility: measures the strength of the evidence in favor of a set of propositions for all $X \in 2^\Omega \setminus \emptyset$:

$$bel(X) = \sum_{Y \subseteq X, Y \neq \emptyset} m(Y). \tag{2}$$

- Plausibility: quantifies the maximum amount of belief that could be given to a X of the frame of discernment for all $X \in 2^\Omega \setminus \emptyset$:

$$pl(X) = \sum_{Y \in 2^\Omega, Y \cap X \neq \emptyset} m(Y). \tag{3}$$

- Commonality: measures the set of *bbas* affected to the focal elements included in the studied set, for all $X \in 2^\Omega$:

$$q(X) = \sum_{Y \supseteq X} m(Y). \tag{4}$$

– Pignistic probability transformation: proposed in[12], transforms a *bba* m into a probability measure for all $X \in 2^\Omega$:

$$betP(X) = \sum_{Y \neq \emptyset} \frac{|X \cap Y|}{|Y|} \frac{m(Y)}{1 - m(\emptyset)}. \tag{5}$$

The bba (m) and the functions *bel*, *pl* are different expressions of the same information.

2.2   Combination Rules

In the belief function theory, Dempster in [1] proposed the first combination rule. It is defined for two bbas $m_1, m_2, \forall X \in 2^\Omega$ with $X \neq \emptyset$ by:

$$m_{DS}(X) = \frac{1}{1-k} \sum_{A \cap B = X} m_1(A)m_2(B), \tag{6}$$

where k is generally called the global conflict of the combination or its inconsistency, defined by $k = \sum_{A \cap B = \emptyset} m_1(A)m_2(B)$ and $1 - k$ is a normalization constant.

The Dempster combination rule is not the only one used to combine, Dubois and Prade proposed a disjunctive one. Smets [14] proposed to consider an open world, therefore the conjunctive rule is a non-normalized one and for two basic belief assignments m_1, m_2 for all $X \in 2^\Omega$ by:

$$m_{conj}(X) = \sum_{A \cap B = X} m_1(A)m_2(B) := (m_1 \oplus m_2)(X). \tag{7}$$

and $k = m_{conj}(\emptyset)$ is considered as a non expected solution.

This combination rule is very useful because, in one hand it decreases the vagueness and, on the other hand, it increases the belief of the observed focal elements.

3   A Distance between Two Discrete Belief Functions

The aim of this paper is to define a distance between continuous belief functions, let us begin by introducing some basic concepts relative to distances.

3.1   Properties of the Distance

The distance defined between two elements A and B in a set I satisfies the following requirement:

– Nonnegativity: $d(A, B) \geq 0$.
– Nondegeneracy: $d(A, B) = 0 \Leftrightarrow A = B$.
– Symmetry: $d(A, B) = d(B, A)$.
– Triangle inequality: $d(A, B) \leq d(A, C) + d(C, B)$.

3.2   Scalar Product

On a defined set, a distance can be measured with a scalar product between two vectors f and g. A dot product has a symmetric and a bilinar form, and is defined positive.

- $\langle f, g \rangle = \langle g, f \rangle$
- $\langle f, f \rangle > 0$
- $\langle f, f \rangle = 0 \Rightarrow f = \vec{0}$: the zero vector.

We have $\langle f, f \rangle = \|f\|^2$ which is the square norm of f.

   To measure a distance using a scalar product, we use the following expression:

$$d_{SP}(f, g) = \sqrt{\frac{1}{2}(\|f\|^2 + \|g\|^2 - 2\langle f, g \rangle))} \tag{8}$$

In the next section, we will present a particular measure based on a scalar product able to define a distance between two discrete belief functions in the theory of belief functions.

3.3   A Distance between Two Discrete Belief Functions

The distance introduced in [5] is the most appropriate distance to measure the dissimilarity between two *bbas* m_1, m_2 according to [6] after making a comparison of distances in belief functions theory. for two *bbas*: m_1, m_2 on 2^Ω:

$$d(m_1, m_2) = \sqrt{\frac{1}{2}(\|m_1\|^2 + \|m_2\|^2 - 2\langle m_1, m_2 \rangle))} \tag{9}$$

and $\langle m_1, m_2 \rangle$ is the scalar product defined by:

$$\langle m_1, m_2 \rangle = \sum_{i=1}^{n} \sum_{j=1}^{n} m_1(A_i) m_2(A_j) \frac{|A_i \cap A_j|}{|A_t \cup A_j|} \tag{10}$$

where $n = |2^\Omega|$.

   Therefore, $d(m_1, m_2)$ is considered as an illustration of the scalar product where the factor $\frac{1}{2}$ is needed to normalize d and guarantee that $0 \leq d(m_1, m_2) \leq 1$.

   According to [6], this metric distance respects all the properties expected by a distance and it can be considered as an appropriate measure of the difference or the lack of similarity between two *bbas*.

   This distance is based on the dissimilarity of Jaccard defined as:

$$Jac(A, B) = \frac{|A \cap B|}{|A \cup B|}. \tag{11}$$

Moreover, $d(m_1, m_2)$ can be called a total conflict measure, which is an interesting property to compute the total conflict based on a measurable distance like presented in [7].

4   Continuous Belief Functions

Strat in [17], and Smets in [15] proposed a definition of continuous belief functions. However, using the belief function framework to model information in a continuous frame is not an easy task due to the complex nature of the focal elements. Hence Strat decided to assign mass only on the intervals of $\mathbb{R}$. This representation reduces the number of the propositions and simplifies the computation of the intersections by focusing only on the contiguous intervals.

Smets extented the Transferable Belief Model (TBM) to continuous realm and defined belief functions over the extended set of reals noted $\overline{\mathbb{R}} = \mathbb{R} \cup \{-\infty, +\infty\}$. Comparing to the discrete domain, on real numbers, in [15] *bba* became *basic belief densities* (*bbd*) defined on an interval $[a, b]$ of $\mathbb{R}$.

4.1   Belief Functions on $\overline{\mathbb{R}}$

Smets generalized the classical *bba* into a basic belief density (*bbd*) noted m^I on the interval I. In the definition of the *bbd*, all focal elements are closed intervals or $\emptyset$. Given a normalized *bbd* m^I, he defined an other function f on $\mathbb{R}^2$, where $f(a, b) = m^I([a, b])$ for $a \leq b$ and $f(a, b) = 0$ whenever $a > b$.

f is called a probability density function (*pdf*) on $\mathbb{R}^2$.

- Credibility:

$$bel^{\overline{\mathbb{R}}}([a, b]) = \int_{x=a}^{x=b} \int_{y=x}^{y=b} m^{\overline{\mathbb{R}}}([x, y]) dy dx. \tag{12}$$

- Plausibility:

$$pl^{\overline{\mathbb{R}}}([a, b]) = \int_{x=-\infty}^{x=b} \int_{y=max(a,x)}^{y=+\infty} m^{\overline{\mathbb{R}}}([x, y]) dy dx. \tag{13}$$

- Commonality:

$$q^{\overline{\mathbb{R}}}([a, b]) = \int_{x=-\infty}^{x=a} \int_{y=b}^{y=+\infty} m^{\overline{\mathbb{R}}}([x, y]) dy dx. \tag{14}$$

Smets' approach is based on the description of focal elements from a continuous function, where the frame of discernment is built having on connected sets of $\overline{\mathbb{R}}$.

Nguyen shown in [8] that a random set can define a belief function. He introduced the notion of a "source" made of a probability space and a multivalued mapping Γ able to define the lower probability (and consequently a basic belief assignment).

Doré *et al.* in [2] proposed a similar approach.. They proposed to use Γ to describe the set of focal elements of a continuous belief function. In this case, they used probability space to assign a mass to focal sets.

4.2   Belief Functions Associated to a Probability Density

A probability density function is an expression of an expert's belief. This probability can be expressed according to a basic belief density which is described using a normal (Gaussian), exponential distribution.

In order to choose the most appropriate one among all the belief functions, we apply the principle of least commitment for evidential reasoning proposed by Dubois and Prade in [3], and Hsia in [4] this principle can be considered as a natural approach for selecting the less specific *bba* from a subset , It consists in selecting the least committed *bba*, and supports the idea that one should never give more support than justified to any subset of Ω.

Among all the belief functions, the consonant, where focal elements are nested (there is a use of a relation of total ordering). The consonant are considered as the most appropriate functions because they express the principle of least commitment.

We are applying this principle to the consonant *bbd* in our illustration in section 6, to represent a normal distribution that is having a bell shape.

5   A Distance between Continuous Belief Functions

Traditional distances known for the discrete case cannot be used due to the nature of the continuous belief functions like described in section4. First of all, we have shown in section 3.2, a scalar product is able to compute a distance between two *bbd*. To handle the problem of the nature of these functions, a scalar product is defined on $\overline{\mathbb{R}}$ by:

$$\langle f, g \rangle = \int_{x=-\infty}^{+\infty} \int_{y=-\infty}^{+\infty} f([x,y]) g([x,y]) dx dy \tag{15}$$

In this section, we will introduce a new method to evaluate the similarity based on Jousselme's distance with Smets' formalism on continuous belief functions. A similarity /dissimilarity measure quantifies how much two distributions are different. Using the properties of belief functions on real numbers, we are now able to define a distance between two densities in a interval I.

$$\langle f_1, f_2 \rangle = \tag{16}$$

$$\int_{-\infty}^{+\infty} \int_{y_i=x_i}^{+\infty} \int_{-\infty}^{+\infty} \int_{y_j=x_j}^{y_j=+\infty} f_1(x_i, y_i) f_2(x_j, y_j) \delta(x_i, x_j, y_i, y_j) dy_j dx_j dy_i dx_i$$

The scalar product of the two continuous *pdfs* is noted: $\langle f_1, f_2 \rangle$ satisfying all proprieties stated in section 3.2, with a function δ defined as $\delta : \mathbb{R} \longrightarrow [0,1]$

$$\delta(x_i, x_j, y_i, y_j) = \frac{\lambda(\llbracket max(x_i, x_j), min(y_i, y_j) \rrbracket)}{\lambda(\llbracket max(y_i, y_j), min(x_i, x_j) \rrbracket)} \tag{17}$$

where λ represents the Lebesgue measure used for the interval's length, and $\delta(x_i, x_j, y_i, y_j)$ is an extension of the measure of Jaccard applied for the intervals in the case of continuous belief functions where $[\![a,b]\!]$ refers to (17).

$$[\![a,b]\!] = \begin{cases} \emptyset, & \text{if } a > b \\ [a,b], & \text{otherwise.} \end{cases} \tag{18}$$

Therefore, the distance is defined by:

$$d(f_1, f_2) = \sqrt{\frac{1}{2}(\|f_1\|^2 + \|f_2\|^2 - 2\langle f_1, f_2 \rangle)} \tag{19}$$

This distance can be used between two or more belief functions.

Let's consider σf a set of $bbds$. We measure the distance between one bbd and the $n-1$ other one by:

$$d(f_i, \sigma f) = \frac{1}{n-1} \sum_{j=1, i \neq j}^{n} d(f_i, f_j) \tag{20}$$

6   Illustrations

As mentioned in the previous section, the result of our work is a distance between two or several $bbds$.

In this section we consider the cases of two different kinds of distributions: the first one is a normal representation and the second is an exponential one. We use these distributions to deduce the different $bbds$ and then we are able to measure the distance between two or several continuous belief functions.

The aim here is to have a probability distribution that includes uncertainty, so we model it using the basic belief densities.

6.1   Basic Belief Densities Induced by Normal Distributions

In this analysis we will focus on a normal probability density function, like presented [9]. The focal sets of the belief functions are the intervals $[\mu - x, \mu + x]$ of $\mathbb{R}$, with μ: the mean of the normal distribution and $x \in \mathbb{R}^+$ and σ: the standard deviation. We consider a normal distribution $\mathcal{N}(x; \mu; \sigma)$, with $x \geq \mu$:

$$\varphi(x) = 2(x - \mu)^2 \frac{1}{\sigma\sqrt{2\pi^3}} e^{\frac{(x-\mu)^2}{2\sigma}}, \tag{21}$$

where $\varphi(x)$ is the basic belief density associated to the Gaussian, when we apply the principle of least commitment where x is the representation of the intervals previously mentioned.

This function is null at $x = \mu$, increases with x and reaches a maximum of $4/(\sigma e\sqrt{2\pi})$ at $x = \mu + \sqrt{2}\sigma$, then decreases to 0 at x goes to infinity like presented by Smets in [15].

6.2 Analysis of a Distance between Belief Densities Induced by Normal Distributions

In this part, we will make a comparison between two distances one is measured using a classical scalar product and the other is our adaptation of Jousselme's distance for two and several continuous belief functions.

In Figure 1, we consider four *pdfs* having a normal distribution.

Table 1. Probability density distributions

pdf	1	2	3	4
μ	0	0	4	4
σ	1	0.5	1	0.5

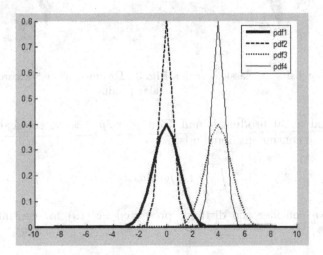

Fig. 1. Four *pdfs* following normal distribution

First of all we measure the distance between f_1 and the rest of the *pdfs* using (19). After that measure the average of the distance between all the *bbds* according to (20).

The average of the distances is $d(f_1, \sigma f) = 0.634$.

According to the Figure 1, the smaller is the distance, the more similar are the distributions. The distance between f_1 and f_4 is the biggest one, this means that those two distributions are the farest from each other comparing to the distance between f_1, f_2, and f_1, f_3.

The distance between f_1 and f_2 is the smaller one, this can be explained in Figure 1 by the fact that f_1 is the nearest one tho f_2.

In Figure 2, we fix $\mu_1 = 0$, $\sigma_1 = 0.5$, and for the second *pdf*, $0 \le \mu_2 \le 10$ with a step $0, 5$ and $0.1 \le \sigma_1 \le 3$ with a step $=0.1$. For the normal distribution, we only use focal elements where $y = 2\mu - x$ if $\mu_1 \ne \mu_2$, then the distance based

Table 2. Distances measured

distance	value
$d(f_1, f_2)$	0.3873
$d(f_1, f_3)$	0.7897
$d(f_1, f_4)$	0.8247

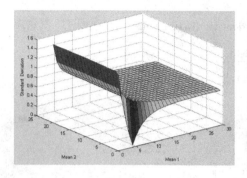 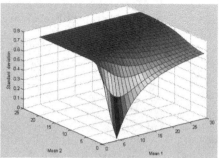

Fig. 2. Distance using a classical scalar product

Fig. 3. Distance using Jaccard based on scalar product

on the classical scalar product is null. Else, $y = 2\mu = x$, so the scalar product presented in the continuous domain is:

$$\int_{y=\mu}^{+\infty} f(y)g(y)dxdy. \tag{22}$$

After that, we consider the distance presented in (19) for continuous belief functions.

Classical Scalar Product for Belief Densities Induced by Normal Distribution: In Figure 2, we notice that there is a remarkable drop in the value of the distance followed by a discontinuity in the 3D representation. When $\mu_1 = \mu_2$, the distance based on classical scalar product is not null, we can say that the standard deviation has an impact on the distance. Moreover, when σ_2 decreases, in this case the distance based on the classical scalar product has a rising values. However, reaching a certain point, especially when $\sigma_1 = \sigma_2$, we observe that the distance is null, so here we are in presence of similar distributions.

Moreover, the distance based on classical scalar product does not have a normalized value $d > 1, 4$. Unfortunately, the mean does not really have an outstanding impact on the distance using the classical scalar product as a measure. Based on that, the distance based on classical scalar product is almost useless in this case, which can be considered as a bad representation of the distance. All these elements created a trays' phenomenon in the obtained Figure 2.

Continuous Distance for Belief Densities Induced by Normal Distribution: By varying the values of μ_2 between $[0, 10]$ with a step of 0.5 we obtain Figure 3. When $\mu_1 = \mu_2$, and $\sigma_1 = \sigma_2$, we are dealing with similar distributions, we obtain a null distance. Otherwise, wa also notice that when the difference between σ_1 and σ_2 increases, the distance rises. We have a similar behavior also when the difference between the means μ_1 and μ_2, every time, the similarity between the two distributions drops which is explained by a growth of the value of the distance.

Comparing to the results obtained with the classical scalar product, we observe that our distance takes in consideration the standard deviation, that does have a real impact when computing this measure of similarity.

The difference between them, is the function δ presented in (17). δ, that allows us to have a more specific distance between two belief functions and use wisely the mean to have a better measure of the distance.

It is more useful, then and accurate to use our distance proposed in (19).

6.3   Basic Belief Densities Induced by Exponential Distributions

In this section, we will suppose that the probability distribution is following an exponential density. The specific expression is for the probability density.

$$f(y) = \frac{y}{\theta^2} e^{\frac{-y}{\theta}} \tag{23}$$

It is obtained when we use the Least Commitment Principle presented in Section 4.2, on a set of basic belief densities associated to the exponential distribution, where θ is the mean and the focal elements are in the intervals $[0, x]$.

6.4   Analysis of a Distance between Belief Densities Induced by Exponential Distributions

We measure the distance between 2 exponential distributions according to a classical scalar product definition, and our adaptation of Jousselme's distance for continuous belief functions.

Figures 4 and 5 show respectively the results obtained after computing the two distances. We are dealing with two exponential distributions f_1, f_2, where $\theta_1 = 1$ and $\theta_2 \in [0.1, 10]$ with a discretization step $= 0.1$.

Classical Scalar Product for Belief Densities Induced by Exponential Distributions: At the begging of Figure 4, when $\theta_1 = 1$ and $\theta_2 = 0.1$, the distance based on classical scalar product has the highest value. When the value of θ_2 increases, the distance between the distributions decreases continually until it makes a discontinuity when $\theta_1 = \theta_2$ where it has a null value, this means that $f_1 = f_2$.

After that, the value of θ_2 gets far from θ_1, the probability distribution f_2 becomes different from f_1 and that generates a non null distance that increases suddenly just when θ_2 gets bigger, then the distance based on classical scalar

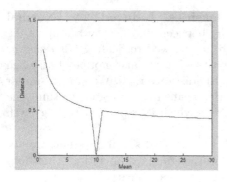
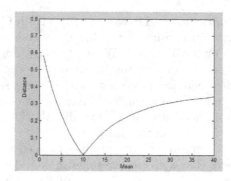

Fig. 4. Distance using a classical scalar product   **Fig. 5.** Distance using Jaccard based scalar product

product decreases and gets stable and reaches a value near to 0.4. Based on these observations, the distance using the classical scalar product is not really adapted for continuous belief functions because it creates a discontinuity, and except the point where $\theta_1 = \theta_2$, the variation of this distance does not reflect the difference between θ_1 and θ_2.

Continuous Distance for Belief Densities Induced by Exponential Distributions: The behavior of the distance in Figure 5 is starting at the highest value $d(f_1, f_2) = 0.6$ with the similar exponential distribution presented previously. When the value of θ_2 gets closer to θ_1, $d(f_1, f_2)$ decreases and reaches a null value where $f_1 = f_2$ (the distributions are similar to each other). Unlike the the distance based on classical scalar product, the increase of $d(f_1, f_2)$ is gradual as θ_2 gets bigger values, the distance grows continually.

7   Conclusion

In this paper we have introduced an adaptation of the Jousselme's distance for the continuous belief functions according to Smets' formalism. This measure is able to define a distance between two or several basic belief density functions.

This distance is based on the function δ, which take into account the imprecision of the focal elements, whereas the classical scalar product. This distance has all the properties of a classical distance.

To illustrate the behavior of the proposed distance in different situations, we used different probability distributions: a normal and an exponential one, from which we deduced basic belief densities. Afterwards, we compared our distance to the results obtained when using a distance based on a classical scalar product.

In future work, we will use the proposed distance in order to define a conflict measure for continuous belief functions.

References

1. Dempster, A.P.: Upper and Lower probabilities induced by a multivalued mapping. Annals of Mathematical Statistics 38, 325–339 (1967)
2. Doré, P.E., Martin, A., Abi-Zied, I., Jousselme, A.L., Maupin, P.: Belief functions induced by multimodal probability density functions, an application to search and rescue problem. RAIRO - Operation Research 44(4), 323–343 (2010)
3. Dubois, D., Prade, H.: The Principle of Minimum Specificity as a basis for Evidential Reasoning. In: Bouchon, B., Yager, R.R. (eds.) Uncertainty in Knowledge-Based Systems. LNCS, vol. 286, pp. 75–84. Springer, Heidelberg (1987)
4. Hsia, Y.T.: Characterizing belief with minimum commitment. In: International Joint Conference on Artificial Intelligence, IJCAI 1991, pp. 1184–1189. Morgan Kaufman, San Mateo (1991)
5. Jousselme, A.L., Grenier, D., Bossé, E.: A new distance between two bodies of evidence. Information Fusion 2, 91–101 (2001)
6. Jousselme, A.L., Maupin, P.: On some properties of distances in evidence theory, Belief, Brest, France (2010)
7. Martin, A., Jousselme, A.L., Osswald, C.: Conflict measure for the discounting operation on belief functions. In: International Conference on Information Fusion, Cologne, Germany (2008)
8. Nguyen, H.T.: On random sets and belief functions. Journal of Mathematical Analysis and Applications 65, 531–542 (1978)
9. Ristic, R., Smets, P.: Belief function theory on the continuous space with an application to model based classification, pp. 4–9 (2004)
10. Ristic, R., Smets, P.: The TBM global distance measure for the association of uncertain combat ID declarations. Information Fusion, 276–284 (2006)
11. Shafer, G.: A mathematical theory of evidence. Princeton University Press (1976)
12. Smets, P.: Constructing the pignistic probability function in a context of uncertainty. Uncertainty in Artificial Intelligence 5, 29–39 (1990)
13. Smets, P.: The Combination of Evidence in the Transferable Belief Model. IEEE Transactions on Pattern Analysis and Machine Intelligence 12(5), 447–458 (1990)
14. Smets, P.: The combination of evidence in the Transferable belief Model. IEEE Transactions on Pattern Analysis and Machine Intelligence 12(5), 447–458 (1990)
15. Smets, P.: Belief Functions on real numbers. International Journal of Approximate Reasoning 40(3), 181–223 (2005)
16. Smets, P., Kennes, R.: The transferable belief model. Artificial Intelligence 66, 191–234 (1994)
17. Strat, T.: Continuous belief function for evidential reasoning. In: Proceedings of the 4th National Conference on Artificial Intelligence, University of Texas at Austin (1984)

On the Complexity of the Graphical Representation and the Belief Inference in the Dynamic Directed Evidential Networks with Conditional Belief Functions

Wafa Laâmari[1], Boutheina Ben Yaghlane[2], and Christophe Simon[3]

[1] LARODEC Laboratory - Institut Supérieur de Gestion de Tunis, Tunisia
[2] Institut des Hautes Études
Commerciales de Carthage, Université de Tunis, Tunisia
[3] Université de Lorraine, Centre de Recherche en Automatique de Nancy, UMR 7039,
Vandoeuvre-lès-Nancy, F-54506, France
CNRS, Centre de Recherche en Automatique de Nancy, UMR 7039,
Vandoeuvre-lès-Nancy, F-54506, France

Abstract. Directed evidential graphical models are important tools for handling uncertain information in the framework of evidence theory. They obtain their efficiency by compactly representing (in)dependencies between variables in the network and efficiently reasoning under uncertainty. This paper presents a new dynamic evidential network for representing uncertainty and managing temporal changes in data. This proposed model offers an alternative framework for dynamic probabilistic and dynamic possibilistic networks. A complexity study of representation and reasoning in the proposed model is also presented in this paper.

1 Introduction

A wide variety of network-based approaches have been developed for modeling available knowledge in real-world applications such as probabilistic, possibilistic and evidential graphical models. Graphical models [4], [7] provide representational and computational aspects making them a powerful tool allowing efficiently representing knowledge and reasoning under uncertainty which may be either aleatory or epistemic.

The temporal dimension is another very important aspect which must be taken into account when reasoning under uncertainty. Several methods have been developed to address this problem, including those relying on network-based approaches. Murphy [5] developed dynamic Bayesian networks (DBNs) that aimed to represent the temporal dimension under the probabilistic formalism. This formalism provides different tools to handle efficiently the aleatory uncertainty, but not the epistemic uncertainty.

E. Hüllermeier et al. (Eds.): SUM 2012, LNAI 7520, pp. 206–218, 2012.

Recently, Heni et al. [2] proposed dynamic possibilistic networks (DPNs) for modeling uncertain sequential data.

Weber et al. [12] also introduced dynamic evidential networks (DENs) to model the temporal evolution of uncertain knowledge. Based on an extension of the Bayes' theorem to the representation of the Dempster-Shafer's belief functions theory [8], DENs do not fully exploit the abilities of the evidential formalism.

The aim of the paper is to propose a new network-based approach considering the temporal dimension: the dynamic directed evidential network with conditional belief functions (DDEVN) and to present a computational complexity of the graphical representation and the reasoning in the proposed model.

The remainder of this paper is organised as follows. In section 2, we sketch the directed evidential networks with conditional belief functions (DEVN). We briefly recall in section 3 the necessary background leading up to the formalism of dynamic graphical models. In section 4, we introduce the DDEVN which extends static DEVN to enable modeling changes over time and we give the propagation algorithm in the DDEVN. Section 5 is devoted to a short illustration of the new framework in the reliability area. Section 6 presents a complexity study of DDEVNs.

2 Directed Evidential Networks with Conditional Belief Functions (DEVN)

In the evidential networks with conditional belief functions (ENCs), first proposed by Smets [10] and studied later by Xu in [13], relations between variables are represented using conditional belief functions. These models allow to model only binary relations between variables. To make it possible to represent the relations for any number of nodes, Ben Yaghlane proposed in [1] the directed evidential network with conditional belief functions (DEVN).

A DEVN is a directed graphical model formalizing the uncertainty in the knowledge by the means of the evidence theory framework. For each root node X in the DEVN, having a frame of discernment Ω_X constituted by q mutually exhaustive and exclusive hypotheses, an a priori belief function $M(X)$ has to be defined over the 2^q focal sets A_i^X by the following equation:

$$M(X) = [m(\emptyset) \quad m(A_1^X)....m(A_i^X)....m(A_{2q-1}^X)] \ . \tag{1}$$

with

$$m(A_i^X) \geq 0 \quad and \quad \sum_{A_i^X, A_i^X \in 2^{\Omega_X}} m(A_i^X) = 1 \ . \tag{2}$$

where $m(A_i^X)$ is the belief that X verifies the hypotheses of the focal element A_i^X.

For other nodes, a conditional belief function $M[Pa(X)](X)$ is specified for each possible hypothesis A_i^X knowing the focal sets of the parents of X.

The Belief propagation in the DEVN relies on the use of a secondary computational data structure called the modified binary join tree [1].

2.1   Modified Binary Join Tree (MBJT)

The modified binary join tree [1], called MBJT, is a refinement of the binary join tree (BJT) [9]. Unlike the BJT, the MBJT emphasizes explicitly the conditional relations in the DEVN for using them when performing the inference process [1].

The MBJT construction process is based on the fusion algorithm [9] with some signifiant modifications described in the following algorithm:

Algorithm 1: Construction of the Static MBJT

Input: a DEVN, an elimination Sequence
Output: a MBJT
1.   Determine the subsets that form the hypergraph H from the DEVN
2.   Arrange the subsets of H in a binary join tree using the elimination sequence
3.   Attach singleton subsets to the binary join tree
4.   Make the join tree binary again if it becomes non-binary when attaching a singleton subset to it
5.   Draw rectangles containing the conditional relations between variables instead of circles containing just the list of these variables (to obtain the MBJT)

Note that if we just perform steps 1, 2, 3 and 4, we obtain a BJT. These steps are more detailed in [9] and more details for the MBJT are given in [1].

3   Basic Backgrounds on Dynamic Directed Networks

A dynamic directed network [2], [6], [12] is a directed graphical model representing the temporal dimension. Each time step k ($k \geq 0$) is represented in the dynamic directed network by $G_k = (N_k, E_k)$, where N_k is a non empty finite set of nodes representing all the variables at the time slice k, and E_k is a set of directed edges representing the conditional independencies between variables of time slice k.

The transition from time k to time $k + 1$ is represented by the set $E(k, k + 1)$ which includes all the directed edges linking two nodes X_k and Y_{k+1}[1] belonging to the two consecutive time slices k and $k+1$. Each edge in $E(k, k+1)$ represents a qualitative dependency between the two nodes it links and denotes a transition function defined by a conditional value table as follows: .

---

[1] Nodes X_k and Y_{k+1} may represent the same variable or two different variables

$$F[X_k](Y_{k+1}) =$$

$$\begin{bmatrix} f[A_1^{X_k}](A_1^{Y_{k+1}}) & \cdots & f[A_1^{X_k}](A_{Q_Y}^{Y_{k+1}}) \\ \cdots & \cdots & \cdots \\ f[A_{Q_X}^{X_k}](A_1^{Y_{k+1}}) & \cdots & f[A_{Q_X}^{X_k}](A_{Q_Y}^{Y_{k+1}}) \end{bmatrix}. \tag{3}$$

where $A_i^{X_k}$ is the i-th state of X_k and $A_j^{Y_{k+1}}$ is the j-th state of Y_{k+1}.

Dynamic directed graphical models used in [6], [2], [12] are supposed to be:

- Stationary: that means that the transition function defined between the current time slice k and the future time slice $k + 1$ does not depend on k. Therefore, only one conditional distribution $F[X_k](Y_{k+1})$ is required to illustrate the temporal dependencies between a node Y_{k+1} and its parent node X_k at any two consecutive time slices.
- Markovian: that means that the distribution $F(Y_{k+1})$ of the variable Y at the future time step $k + 1$ depends only on the distributions of its parent nodes in this time slice and also on the distribution of its parent nodes in the immediately preceding time step which is the present time step k. Thus, the future time slice $k + 1$ is conditionally independent of the past given the present time slice k [5].

The dynamic directed networks' representation presents a drawback because of the new sets of nodes N_k and edges E_k which are introduced in the graphical model to represent each new time step k. With a large number of time slices, the graphical structure becomes huge, and as a result the inference process becomes cumbersome and time consuming. To overcome this problem, the proposed representation of dynamic directed models keeps the network in a compact form with only two consecutive time slices [5].

Thanks to the new representation, a dynamic directed network is simply defined as a couple (GS_0, GS_k), where GS_0 denotes the graphical structure corresponding to the initial time slice $k = 0$ and GS_k denotes a 2-time slices directed graphical model (2-TDGM) in which only two nodes are introduced to represent the same variable at successive time steps: the first node is used to model a variable in the time slice k and the second one is used to represent it at the time slice $k + 1$.

The concept of the outgoing interface I_k has been defined in [5] as:

$$I_k = \{X_k \in N_k \ / \ (X_k, Y_{k+1}) \in E(k, k + 1) \ and \ Y_{k+1} \in N_{k+1}\}$$

where N_k is the set of nodes modeling the time slice k, N_{k+1} is the set of nodes modeling the time slice $k+1$ and $E(k, k+1)$ is the set of edges linking two nodes X_k and Y_{k+1} belonging to the successive time slices k and $k + 1$.

A 1.5 DDGM is a graph obtained by the elimination of nodes not belonging to the outgoing interface I_k from the 2-TDGM .

4   Dynamic Directed Evidential Networks with Conditional Belief Functions (DDEVNs)

A dynamic directed evidential network with conditional belief functions (DDEVN) is a DEVN taking into account the temporal dimension. DDEVNs are introduced in a way similar to other dynamic directed networks, in the sense that the considered problems are those whose dynamics can be modeled as stochastic processes which are stationary and Markovian.

4.1   Graphical Representaion of Dynamic Directed Evidential Networks with Conditional Belief Functions

Dynamic directed evidential network with conditional belief functions is defined as a couple (D_0, D_k), where D_0 is the DEVN representing the time slice $k = 0$ and D_k is the triplet $(G_k, G_{k+1}, E(k, k + 1))$ denoting a 2-time slice DEVN (2-TDEVN) with only two successive time slices k and $k + 1$. The temporal dependencies between variables are represented in DDEVN by transition-belief mass using equation 3.

4.2   Inference in Dynamic Directed Evidential Networks with Conditional Belief Functions

Reasoning is made in the DDEVN through an extension of the exact Interface algorithm of Murphy [5], [6] to this network.

Starting from an observed situation at the initial time step $k = 0$ and transition distributions illustrating the temporal dependencies between two consecutive time slices, a 2-TDEVN allows to compute the belief mass distribution of a variable X at any time slice $k = T$ thanks to its two slices k and $k + 1$.

For belief propagation in the DDEVN, two MBJTs are created. The first one, denoted M_0, represents the initial time step $k = 0$, while the second, denoted M_k, corresponds to each time slice $k > 0$.

Depending on the time slice k, the propagation process is performed for computing node marginals, either in MBJT M_0 or in MBJT M_k. M_0 is used for computing the marginals of nodes at time slice $k = 0$, while M_k is used when computing marginals in a time step $k \geq 1$.

As when reasoning in the junction tree [6], [2], the key idea is that when advancing from the past time step $k - 1$ to the current time step k, we need to store the marginals of variables in the outgoing interface I_k that will be useful as observations introduced in the corresponding nodes in the next inference. By recursively performing the bidirectional message-passing scheme in M_k, we can compute the marginals of variables at any time step.

The construction of the MBJTs are detailed by the algorithms 2 and 3:

Algorithm 2: Construction and initialization of M_0

Input: a 2-TDEVN
Output: a MBJT M_0, an outgoing interface I_0
1. Identify I_0 by selecting nodes in the outgoing interface of time slice k = 0
$I_0 \leftarrow \{x \in N_0 \ / \ (x,y) \in E(0,1) \ and \ y \in N_1\}$
2. Eliminate each node belonging to N_1 from the 2-TDEVN
3. Construct the MBJT M_0 from the resulting structure using Algorithm 1
4. Let P be the set of the given potentials
   For i=1 to length(P)
     If P(i) is an a priori belief function distribution
       assign P(i) to the corresponding singleton node in
       M_0.
     Else
       assign P(i) to the corresponding conditional node in
       M_0.
     End if
   End for

Algorithm 3: Construction and initialization of M_k

Input: a 2-TDEVN
Output: a MBJT M_k, an outgoing interface I_k
1. $I_k \leftarrow \{x \in N_k \ / \ (x,y) \in E(k,k+1) \ and \ y \in N_{k+1}\}$
2. Eliminate from the 2-TDEVN each node belonging to N_k and not to I_k (to obtain a 1.5 DDEVN)
3. Construct the MBJT M_k from the 1.5 DDEVN using Algorithm 1
4. Let P be the set of potentials relative only to variables of time slice k+1
   For i=1 to length(P)
     If P(i) is an a priori belief function distribution
       assign P(i) to the corresponding singleton node in
       M_k.
     Else
       assign P(i) to the corresponding conditional node in
       M_k.
     End if
   End for

Since DDEVNs are a subclass of DEVNs, the inference algorithm developed for DEVNs can be adapted and applied for reasoning with DDEVNs.

The propagation process in the DDEVN is given by the algorithm 4:

Algorithm 4: Propagation in the DDEVN

Input: M_0 and M_k
Output: marginal distributions
1. Performing the propagation process in $M_0{}^2$
    Let n be the number of singleton nodes in I_0
    For $i=1$ to n
      node$=I_0(i)$;
      Marg_distr= Marginal_distribution(node);
      Add Marg_Distr to the interface potentials IP_0;
    End for
2. **If** $k > 0$
      For $i=1$ to k
        For $j=1$ to length(IP_{i-1})
          Current_potential$=IP_{i-1}(j)$;
          Associate Current_potential to the corresponding
          singleton node in M_k;
        End for
        Performing the propagation process in M_k;
        Let n be the number of singleton nodes in I_i;
        For $nb=1$ to n
          node$=I_i(nb)$;
          Marg_Distr= Marginal_Distribution(node);
          Add Marg_Distr to IP_i;
        End for
      End for
    End if
3. **If** $k = 0$
      Compute_Marginals(M_0)^3
    Else
      Compute_Marginals(M_k)
    End if

---

5   Illustrative Case Study

For the sake of illustration, let us apply the DDEVN to the reliability analysis of the well known bridge system [11]. The bridge system consists of five components. Each component Ci has two disjoint states ($\{Up\},\{Down\}$): Up, shortly written U, is the working state and $Down$, shortly written D, is the fail state.

---

² The propagation process is performed as in the static MBJT. For details, the reader is referred to [1].

³ To compute the marginal for a node, we combine its own initial potential with the messages received from all the neighbors during the propagation process [1].

Thus, the frame of discernment of each component is $\Omega = \{Up, Down\}$ and its corresponding power set is $2^{\Omega} = \{\emptyset, \{Up\}, \{Down\}, \{Up, Down\}\}$.

In the classical reliability analysis, the reliability of the bridge system is its probability to be in the state $\{Up\}$ during a mission time.

Fig.1 (a) shows the DAG corresponding to the bridge system in which Ci represents the i-th component of the system, Aj represents the j-th *And* gate, Oz represents the z-th *Or* gate and node R represents the bridge system reliability.

5.1   Application of the DDEVN to the Bridge System Reliability

The bridge system is modeled by the DDEVN shown in Fig.1 (b). Each node Ci_k represents the i-th component in the time slice k, node Oz_k represents the z-th *Or* gate in the time slice k, Aj_k represents the j-th *And* gate in the time slice k and node R_k represents the state of the system in the k-th time slice.

Fig.1 (c) shows the 1.5 DDEVN created from the 2-TDEVN given in Fig.1 (b) by removing all nodes in the time slice k not belonging to I_k.

The belief mass distributions of the five components at the time step $k+1$ which depend on their distributions at the time step k are represented in tables 1,2,3,4 and 5. Table 7 represents the a priori mass distributions of the components at the time step 0. The conditional mass distribution relative to node R_k is given in table 6. The conditional belief mass distributions relative to nodes Aj_k and Oz_k are defined equivalent to the logical *And* and *Or* gates.

Tables 1, 2, 3 and 4. Conditional Mass Tables $M[C1_k](C1_{k+1})$, $M[C2_k](C2_{k+1})$, $M[C3_k](C3_{k+1})$ and $M[C4_k](C4_{k+1})$

$C1_{k+1} \backslash C1_k$	U	D	$U \cup D$
U	0.994	0	0.000
D	0.004	1	0.001
$U \cup D$	0.002	0	0.999

$C2_{k+1} \backslash C2_k$	U	D	$U \cup D$
U	0.995	0	0.000
D	0.003	1	0.003
$U \cup D$	0.002	0	0.997

$C3_{k+1} \backslash C3_k$	U	D	$U \cup D$
U	0.996	0	0.000
D	0.002	1	0.001
$U \cup D$	0.002	0	0.999

$C4_{k+1} \backslash C4_k$	U	D	$U \cup D$
U	0.997	0	0.000
D	0.002	1	0.002
$U \cup D$	0.001	0	0.998

Tables 5 and 6. Conditional Mass Tables $M[C5_k](C5_{k+1})$ and $M[O2_k](R_k)$

$C5_{k+1} \backslash C5_k$	U	D	$U \cup D$
U	0.996	0	0.000
D	0.003	1	0.003
$U \cup D$	0.001	0	0.997

$R_k \backslash O2_k$	U	D	$U \cup D$
U	1	0	0
D	0	1	0
$U \cup D$	0	0	1

Table 7. The a Priori Mass Tables $M(C1_0)$, $M(C2_0)$, $M(C3_0)$, $M(C4_0)$ and $M(C5_0)$

Ci_0	U	D	$U \cup D$
$m(Ci_0)$	1.0	0.0	0.0

Construction and Initialization of M_0 and M_k: Using the 2-TDEVN in Fig.1, the MBJTs M_0 and M_k shown in Fig.2 and Fig.3 are constructed by applying the first three steps of algorithms 2 and 3.

Using the a priori and the conditional mass tables, M_0 and M_k are initialized by assigning each belief function distribution to the corresponding node (the fourth step of algorithms 2 and 3).

Performing the Propagation Process in the DDEVN: Suppose now that we wish to compute the reliability of the bridge system at time step $k = 1400$. We first perform the inference process in M_0 and we compute the marginals of nodes in the outgoing interface $I_0 = \{C1_0, C2_0, C3_0, C4_0, C5_0\}$ (the first step of algorithm 4).

The marginal distributions of nodes in the outgoing interface I_0 will be used when performing the propagation in the MBJT M_k in the next time slice ($k = 1$). They will be repectively introduced in nodes $C1_0$, $C2_0$, $C3_0$, $C4_0$ and $C5_0$ of M_k. After performing the inference algorithm, M_k yields the marginals of nodes $C1_1$, $C2_1$, $C3_1$, $C4_1$ and $C5_1$ (forming the outgoing interface I_1) which are the sufficient information needed to continue the propagation in the following time slice $k = 2$.

After carrying out the inference process in the MBJT M_k recursively for 1400 time slices, we obtain the following distribution for node R_{1400} corresponding to the reliability of the bridge system $M(R_{1400}) = [m(\emptyset) = 0 \ m(\{Up\}) = 0.000014 \ m(\{Down\}) = 0.982673 \ m(\{Up, Down\}) = 0.017313]$.

6 Complexity Analysis

6.1 Complexity of the MBJTs Construction Process

The complexity of algorithms 2 and 3 used for constructing and initializing M_0 and M_k depends on the run-time complexity of the MBJT construction process described by algorithm 1.

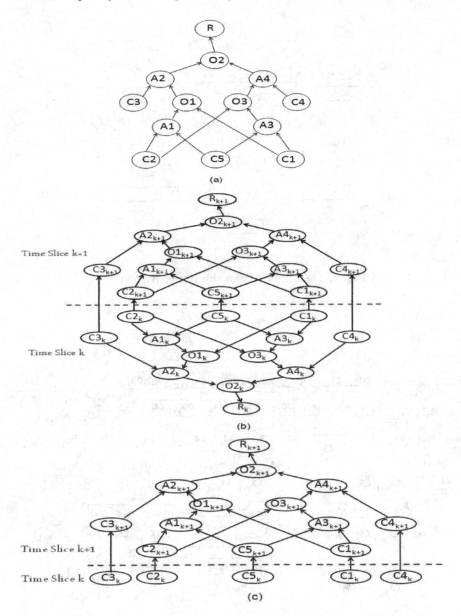

Fig. 1. The DAG (a), the DDEVN (2-TDEVN) (b) and the 1.5 DDEVN (c) for the Bridge System

Evaluation of the run-time complexity of the MBJT construction process for the worst-case scenario is made by examining the structure of algorithm 1. This algorithm is composed of five steps. Therefore, its running time is the total amount of time needed to run the five steps.

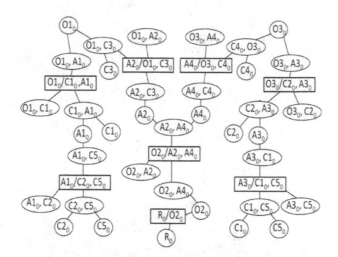

Fig. 2. The MBJT M_0 for the Bridge System

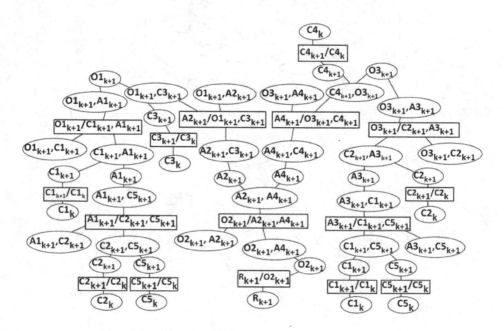

Fig. 3. The MBJT M_k for the Bridge System

The first step in the MBJT construction process takes $O(N^2)$ time to determine all the subsets of the hypergraph H, where N is the number of nodes in the network [4].

---

[4] The network used to construct the MBJT is the DEVN in the static case, while it is the 2-TDEVN in the dynamic case.

Step 2 is composed of two imbricated loops. The outer loop iterates from S to 2, where S is the number of the hypergraph's subsets. Starting with S subsets forming a set named $Sets$, the algorithm eliminate for a worst case evaluation two subsets from $Sets$ in each iteration and adds one new subset before passing to the next iteration, and as a result the number of subsets in $Sets$ decreases by one from one iteration to another. This outer loop iterates and stops when having just one subset in $Sets$, so running the outer loop body consumes $O(S)$ times. When passing through the outer loop, an instruction allowing to extract from $Sets$ the subsets containing the variable to be eliminated, is carried out before passing to run the inner loop. This instruction is run once in each iteration of the outer loop and consumes S times for the worst-case since on the first pass through the outer loop, $Sets$ contains S subsets. The inner loop iterates from E to 2, where E is the number of subsets containing the current variable to be eliminated. In fact, starting initially with E subsets which form a set called $Varsets$, the algorithm eliminates in each iteration two subsets from $Varsets$ and adds another subset resulting from their union. So $|Varsets|$ decreases by 1 in each iteration. The inner loop ends when having just one subset in $Varsets$ and this will occur after $E - 1$ iterations. The union operation of the two selected subsets used in each iteration consumes V^2 times, where V is the number of variables in the largest subset in $Sets$. As a result, the inner loop takes $O(E * V^2)$. Step 2 takes $O(S * [S + E * V^2])$ times.

Step 3 of the MBJT construction process aims at attaching the singleton nodes not added yet to the BJT constructed in step 2. To attach a singleton node to the obtained BJT, all the BJT's nodes are visited in the worst case. Thus, the third step consumes $O(s * N_1)$ times, where s is the number of singleton nodes not added yet to the BJT constructed in step2 and N_1 is the number of nodes in this BJT.

The computational complexity of making the join tree binary again which is the fourth step is relative to $O(N_1^2)$, where N_1 is the number of nodes in the BJT.

The last step carried out to obtain the MBJT takes $O(N_1 * N)$ times.

6.2   The Computational Complexity of the Inference Process

The MBJT is a structure for local computation of marginals. In each time slice k, the number of messages exchanged between nodes during the bidirectional message-passing scheme is $2 * (l - c) = 2 * l - 2 * c$, where l is the number of undirected edges in the MBJT, and c is the number of conditional nodes. In fact, when having two joint nodes separated by a conditional node denoted by C, the edge connecting the first joint node and C and the edge that connects C and the second joint node are regarded as one edge because the conditional node C will neither send nor receive messages, and only two messages are at most exchanged in this case between the two joint nodes through the two edges [3]. So the total number of the exchanged messages when inferring beliefs in DDEVNs is $T * 2 * (l - c)$, where T is the number of time slices.

The complexity of belief propagation in MBJT is exponential in the maximum node size and takes per time slice $O([j * 2^{dz}] + [c * 2^d * d^{z-1}])$ times, where j is the number of joint nodes in the MBJT, d is the maximum number of states of variables in the initial network and z is the maximum node size in the MBJT. Inferring beliefs in DDEVNs is relative to $O([J * 2^{dz}] + [C * 2^d * d^{z-1}])$, where J=j*T, C=c*T and T is the number of time slices.

7   Conclusion

Inspired by the same aims of DBNs and DPNs and also with the intent to make the DEVN able to dynamically model a problem under uncertainty, we have proposed in this paper the dynamic DEVN that we called DDEVN. DDEVN extends the static DEVN to deal with temporal changes in data. DDEVN was applied to the reliability study of a well known system, the bridge system, to show that this new framework is a powerful tool for efficiently modeling the temporal evolution of this system in uncertain reasoning. We have also presented a computational analysis of the graphical representation and the belief propagation in DDEVNs. In future work, the development of new algorithms to perform the propagation process in the DDEVN will be of a great interest.

References

[1] Ben Yaghlane, B., Mellouli, K.: Inference in directed evidential networks based on the transferable belief model. Int. J. Approx. Reasoning 48(2), 399–418 (2008)
[2] Heni, A., Ben Amor, N., Benferhat, S., Alimi, A.: Dynamic Possibilistic Networks: Representation and Exact Inference. In: IEEE International Conference on Computational Intelligence for Measurement Systems and Applications (CIMSA 2007), Ostuni, Italy, pp. 1–8 (2007)
[3] Laâmari, W., Ben Yaghlane, B., Simon, C.: Comparing Evidential Graphical Models for Imprecise Reliability. In: Deshpande, A., Hunter, A. (eds.) SUM 2010. LNCS, vol. 6379, pp. 191–204. Springer, Heidelberg (2010)
[4] Lauritzen, S.L.: Graphical Models. Clarendon Press, Oxford (1996)
[5] Murphy, K.: Dynamic Bayesian Networks: Representation, Inference and Learning. PhD thesis, Dept. Computer Science, UC, Berkeley (2002)
[6] Murphy, K.: Probabilistic Graphical Models. Michael Jordan (2002)
[7] Pearl, J.: Probabilistic Reasoning in Intelligent Systems: Networks of Plausible Inference. Morgan Kaufmann, San Mateo (1988)
[8] Shafer, G.: A Mathematical Theory of Evidence. Princeton University Press, Princeton (1976)
[9] Shenoy, P.P.: Binary join trees fo computing marginals in the Shenoy-Shafer architecture. Int. J. Approx. Reasoning 17(2-3), 239–263 (1997)
[10] Smets, P.: Belief function: the disjunctive rule of combination and the generalized Bayesian theorem. Int. J. Approx. Reasoning 9, 1–35 (1993)
[11] Torres-Toledano, J., Sucar, L.: Bayesian Networks for Reliability Analysis of Complex Systems. In: Coelho, H. (ed.) IBERAMIA 1998. LNCS (LNAI), vol. 1484, pp. 195–206. Springer, Heidelberg (1998)
[12] Weber, P., Simon, C.: Dynamic evidential networks in system reliability analysis: A Dempster Shafer approach. In: 16th Mediterranean Conference on Control and Automation, France, vol. 93, pp. 262–267 (2008)
[13] Xu, H., Smets, P.: Evidential Reasoning with Conditional Belief Functions. In: Heckerman, D., Poole, D., Lopez De Mantaras, R. (eds.) UAI 1994, pp. 598–606. Morgan Kaufmann, San Mateo (1994)

Revision over Partial Pre-orders: A Postulational Study

Jianbing Ma[1], Salem Benferhat[2], and Weiru Liu[1]

[1] School of Electronics, Electrical Engineering and Computer Science,
Queen's University Belfast, Belfast BT7 1NN, UK
{jma03,w.liu}@qub.ac.uk
[2] CRIL-CNRS, UMR 8188, Faculté Jean Perrin, Université d'Artois,
rue Jean Souvraz, 62307 Lens, France
benferhat@cril.fr

Abstract. Belief revision is the process that incorporates, in a consistent way, a new piece of information, called input, into a belief base. When both belief bases and inputs are propositional formulas, a set of natural and rational properties, known as AGM postulates, have been proposed to define genuine revision operations. This paper addresses the following important issue : How to revise a partially pre-ordered information (representing initial beliefs) with a new partially pre-ordered information (representing inputs) while preserving AGM postulates? We first provide a particular representation of partial pre-orders (called units) using the concept of closed sets of units. Then we restate AGM postulates in this framework by defining counterparts of the notions of logical entailment and logical consistency. In the second part of the paper, we provide some examples of revision operations that respect our set of postulates. We also prove that our revision methods extend well-known lexicographic revision and natural revision for both cases where the input is either a single propositional formula or a total pre-order.

1 Introduction

The problem of belief revision is a major issue in several Artificial Intelligence applications to manage the dynamics of information systems. Roughly speaking, belief revision results from the effect of inserting new piece of information while preserving some consistency conditions. In the logical setting, a simple form of a belief revision assumes that both initial beliefs, denoted by K, and input information, denoted by μ, are represented by propositional formulas. In this framework, the revision of K by μ consists in producing a new formula denoted by $K * \mu$, where $*$ represents a revision operation. Extensive works have studied and characterized the revision operation $*$ from semantics, syntactic, computational and axiomatics points of views. In particular, Alchourron, Gärdenfors and Makinson [1] proposed an elegant set of rationality postulates [14], known as AGM postulates, that any revision operation $*$ should satisfy. These postulates are mainly based on two important principles: success principle and minimal change principle. The success principle states that the input μ is a sure piece of information and hence should be entailed from $K * \mu$. The minimal change principle states that the revised base $K * \mu$ should be as close as possible to initial beliefs K.

E. Hüllermeier et al. (Eds.): SUM 2012, LNAI 7520, pp. 219–232, 2012.

In particular if K and μ are consistent then $K * \mu$ should be simply equivalent to the propositional conjunction of K and μ.

Since AGM proposal, many extensions [10] have been proposed to take into account complex belief and inputs. For instance, in the context of uncertain information, a so-called Jeffrey's rule [12,7], has been proposed for revising probability distributions. In evidence theory, revision of mass functions are proposed [26,27]. Similarly, in possibility theory and ordinal conditional functions framework, the so-called transmutation [33] have been proposed. It modifies the ranking or possibility degrees of interpretations so as to give priority to the input information. Various forms of ranking revisions have been suggested in (e.g., [6,13,8,22,9,23,24]).

In the logical setting, belief revision has also been extensively studied. In [8] four postulates have been added to AGM postulates in order to characterize iterated belief revision operators which transform a given ordering on interpretations, in presence of new information, into a new ordering. In [29,3] a lexicographic strategy, associated with a set of three rationality postulates, have been defined to revise a total pre-order by a new total pre-order. In [5,4], different strategies have been proposed to revise an epistemic state represented by a partial pre-order on the possible worlds.

This paper deals with a flexible representation of information where both initial beliefs and input are represented by partial pre-orders. Despite its importance in many applications, there are very few works that address revision methods of a partial pre-order by a partial pre-order. In [32], revision of partial orders is studied in a standard expansion and contraction way. But it does not provide concrete revision results because of the use of certain kinds of selection functions. In a short paper [21], a framework that studies revision on partial pre-orders is developed and two main revision operators are proposed. However, there are no postulates addressed in that short paper, neither did the paper discussed the relationship between the revision operators and various revision strategies proposed in the literature.

A natural question addressed in this paper is whether it is possible to reuse AGM postulates while both initial beliefs and inputs are partial pre-orders. The answer is Yes. The idea is not to change rationality postulates, but to modify the representation of beliefs and adapt main logical concepts such as logical entailment and consistency. More precisely, we will follow the representation of partial pre-orders proposed in [21]. A partial pre-oder over a set of symbols is viewed as a closed set of *units*. Each unit represents an individual constraint between symbols (a pair of symbols with an ordering connective). The revision of partial pre-order by another partial pre-order is then viewed as revision of closed set of units by another closed set of units. In this paper, we show that AGM postulates have natural counterparts when initial beliefs and input are represented by sets of units. We also provide the counterpart of success postulates and minimal change principle in our frameworks. The reformulation of AGM postulates is possible once logical entailment is interpreted as set inclusion between sets of units, and logical consistency is interpreted as the absence of cycles between sets of units associated with partial pre-orders of inputs and initial beliefs. Additional postulates are also studied in this paper.

In the second part of the paper, we prove that the two revision operators proposed in [21] satisfy the proposed postulates for partial pre-order revision as well as some rational properties, such as the iteration property. Furthermore, we also prove that our revision operators are extensions of the well-known lexicographic revision [30] and natural belief revision [6].

To summarize, this paper makes the following main contributions:

- We propose a set of rationality postulates that the unit-based revision shall follow.
- We prove some important properties among these revision strategies. We also prove that these revision strategies satisfy certain rationality postulates.
- When reducing to classical belief revision, we prove that our revision strategies extend some existing revision strategies, such as, lexicographic and natural revisions.

The remainder of the paper is organized as follows. We introduce some necessary notations and definitions in Section 2. In Section 3, we discuss the principles a revision rule on partial pre-orders shall satisfy and propose AGM-style postulates for partial pre-orders revision. We then discuss, in Sections 4, two examples of belief revision strategies, proposed in [21], that satisfy AGM-style postulates and analyse their properties. Section 5 shows that the well-known lexicographic and natural revision can be recovered in our frameworks. We then review some related works in Section 6. Section 7 concludes the paper.

2 Notations and Definitions

We use W to denote a finite set of symbols. Let $\preceq$ be a pre-order over W where $w \preceq w'$ means that w is at least as preferred as w'. Two operators, $\prec$ and $\approx$, are defined from $\preceq$ in a usual sense. Note that as $\preceq$ implies $\prec$ or $\approx$ while $\prec$ (or $\approx$) is a simple relation, in this paper, we only focus on simple relations $\prec$ and $\approx$. Each $w \prec w'$ or $w \approx w'$ is called a *unit* for $w \neq w'$. A partial pre-order is represented by a finite set of units denoted by S, we use $Sym(S)$ to denote the set of symbols from W appearing in S.

Definition 1. *A set of units S is closed iff*

- $w \approx w' \in S$ *implies* $w' \approx w \in S$;
- *for any different w_1, w_2, w_3 in W, if $w_1 \ R_1 \ w_2 \in S$ and $w_2 \ R_2 \ w_3 \in S$ and $w_1 \ R_1 \ w_2 \wedge w_2 \ R_2 \ w_3$ implies $w_1 \ R_3 \ w_3$, then $w_1 \ R_3 \ w_3 \in S$, (where R_i is either $\approx$ or $\prec$).*

We can see that a closed set of units corresponds to a partial pre-order in the usual sense (i.e., a transitive, reflexive binary relation). A set S can be extended to a unique minimal closed set based on transitivity and symmetry of $\approx$ and transitivity of $\prec$. We use $Cm(S)$ to denote this unique minimal closed set extended from S.

Example 1. *Let $S = \{w_1 \approx w_2, w_2 \prec w_3\}$, then $Cm(S) = \{w_1 \approx w_2, w_2 \approx w_1, w_2 \prec w_3, w_1 \prec w_3\}$.*

S is closed when it cannot be extended further. This is the counterpart of the deductive closure of a belief base K in classical logics.

Definition 2. *A subset C of S is a cycle if $C = \{w_1 \ R_1 \ w_2, w_2 \ R_2 \ w_3, \cdots, w_n \ R_n \ w_1\}$ s.t. $\exists R_i, R_i$ is $\prec$ for $1 \leq i \leq n$. C is minimal if there does not exist a cycle C' s.t. $Cm(C') \subset Cm(C)$.*

If S has a cycle, then S is said to be inconsistent. Otherwise it is said to be consistent or free of cycles. If S is closed and contains cycles, then all minimal cycles are of the form $\{a \prec b, b \prec a\}$ or $\{a \prec b, b \approx a\}$, i.e., only two units.

Any unit $w \ R \ w'$ is called a *free unit* if $w \ R \ w'$ is not involved in any cycle in S. The concept of free unit is the counterpart of free formula concept in logic-based inconsistency handling [2,11].

Example 2. *Let $S = \{w_1 \prec w_2, w_2 \approx w_3, w_3 \approx w_4, w_4 \prec w_1, w_3 \prec w_1\}$, then $C_1 = \{w_1 \prec w_2, w_2 \approx w_3, w_3 \prec w_1\}$ is a minimal cycle whilst $C_2 = \{w_1 \prec w_2, w_2 \approx w_3, w_3 \approx w_4, w_4 \prec w_1\}$ is a cycle but not minimal, since the sub sequence $w_3 \approx w_4, w_4 \prec w_1$ in C_2 can be replaced by $w_3 \prec w_1$ and hence forms C_1.*

For any set of units S, we use $[S]$ to count the number of semantically distinct units in S such that $w \approx w'$ and $w' \approx w$ are counted as one instead of two. So for $S = \{w_1 \approx w_2, w_2 \approx w_1, w_3 \prec w_1, w_3 \prec w_2\}$, we have $[S] = 3$.

Without loss of generality, subsequently, if without other specifications, we assume that a set of units S and any new input S_I are both closed and free of cycles. For convenience, we use $\mathcal{S}_{CC}$ to denote the set of all closed and consistent sets of units (free of cycles) w.r.t. a given W and $\{\approx, \prec\}$.

3 Principles and Postulates of Unit-Based Revision

3.1 Motivations

Let $\odot$ be a revision operator which associates a resultant set of units $\hat{S} = S \odot S_I$ with two given sets, one represents the prior state (S) and the other new evidence (S_I). This section provides natural properties, for the unit-based revision operation $\odot$, which restate the AGM postulates [14] in our context. As in standard belief revision (an input and initial are sets of propositional formulas), we also consider the two following principles as fundamental:

Success Postulate: It states that information conveyed by the input evidence should be retained after revision. In our context, this also means that an input partial pre-order (or its associated set of units) must be preserved, namely $S_I \subseteq S \odot S_I$. In particular if two possible worlds have the same possibility conveyed by the input, then they should still be equally possible after revision, regardless their ordering in the prior state. This clearly departs from the work reported in [3] (where a tie in the input could be broken by the prior state) for instance. In the degenerate case, where the input is fully specified by a total pre-oder then the result of revision should be simply equal to the input. This situation is similar to the case where in standard AGM postulates where the input is a propositional formula having exactly one model (and hence the result of revision is that formula), or in frameworks of probabilistic revision where applying Jeffrey's rule of conditioning to the situation where the input if specified by a probability distribution simply gives that probability distribution.

Minimal Change Principle: It states that the prior information should be altered as little as possible while complying with the Success postulate. This means in our context that as few of units (individual binary ordering relations) as possible to be removed from the prior state after revision. Specifying minimal change principle needs to define the concept of conflicts in our framework. Roughly speaking, two sets of units (here associated with input and initial beliefs) are conflicting if the union of their underlying partial pre-orders contains cycles. Some units (from the set of units representing initial beliefs) should be removed to get rid of cycles, and minimal change requires that this set of removed units should be as small as possible.

3.2 AGM-Style Postulates for Unit-Based Revision

We now rephrase basic rationality postulates when initial epistemic state and input are no longer propositional formulas but sets of units. More precisely, we adapt the well-known AGM postulates (reformulated in [14] (KM)) to obtain a set of revised postulates in which an agent's original beliefs and an input are represented as sets of units. These revised postulates, dubbed UR0-UR6, are as follows:

UR0 $S \odot S_I$ is a closed set of units.
UR1 $Cm(S_I) \subseteq S \odot S_I$.
UR2 If $S_I \cup S$ is consistent,
    then $S \odot S_I = Cm(S_I \cup S)$.
UR3 If S_I is consistent, then $S \odot S_I$ is also consistent.
UR4 If $Cm(S_1) = Cm(S_2)$ and $Cm(S_{I1}) = Cm(S_{I2})$,
    then $S_1 \odot S_{I1} = S_2 \odot S_{I2}$.
UR5 $S \odot (S_{I1} \cup S_{I2}) \subseteq Cm((S \odot S_{I1}) \cup S_{I2})$.
UR6 If $(S \odot S_{I1}) \cup S_{I2}$ is consistent,
    then $Cm((S \odot S_{I1}) \cup S_{I2}) \subseteq S \odot (S_{I1} \cup S_{I2})$.

UR0 simply states that $S \odot S_I$ is closed based on the five inference rules on units (Def. 1).

UR1 formalizes the success postulate. Note that the counterpart of logical entailment here is represented by set inclusion. However it departs from the standard propositional logic definition of entailment where $\phi \models \mu$ means that the set of models of ϕ is included in the set of models of μ. In our context, a set of units S is said to entail another set of units S_1 if $Cm(S_1) \subseteq Cm(S)$.

UR2 indicates that if the prior state and the input are consistent, then the revision result is simply the minimal closure of the disjunction of the prior state and the input. Here we need to point out that consistency here is represented by the absence of cycles. Our definition of consistency (between two sets of units) is stronger than that in logic-based revision (generally defined between two propositional formulae representing the initial beliefs and an input). That is, in belief revision, consistency is required between two formulas (initial state vs. input). In this paper, consistency is required for pre-orders (initial state vs. input) which contain more information than formulas. This is similar to asking for the consistency between two epistemic states not the consistency between their belief sets. Therefore, **UR2** is stronger than its counterpart R2 in logic-based revision [8]

R2: if $\Phi \wedge \mu$ is consistent, then $Bel(\Phi \circ \mu) \equiv Bel(\Phi) \wedge \mu$, or its weaker version defined when initial epistemic state is represented by a partial pre-order [4,5]: if $Bel(\Phi) \models \mu$, then $Bel(\Phi \circ \mu) \equiv Bel(\Phi) \wedge \mu$. This is not surprising, since in our framework, all units play an equal role in the belief change, while in logic-based revision, only units that determine belief sets play crucial roles[1].

Also, here we should point out that we do not define Bel in our framework because in general, elements (w_1, w_2) can be more than possible worlds. However, when reducing to belief revision scenarios, Bel cab be defined in the standard way.

UR3 ensures consistency of the revision result given the consistency of the input.

UR4 is a kind of *syntactic irrelevance* postulate, in which we use minimal closure equivalence to replace logical equivalence used in usual syntactic irrelevance postulates.

UR5 and **UR6** depict the conditions that the order of revision and disjunction operations is exchangeable.

In summary, this set of postulates are natural counterparts of KM postulates in our framework.

Remarks: Most of existing approaches modify AGM postulates to cope with new revision procedures (e.g. revising total pre-orders). Our approach brings a fresh perspective to the problem of representing/revising pre-orders in which AGM can be used. Our results may retrospectively appear to be obvious but note that it has not been considered before. Actually, it was not obvious that in order to revise a partial pre-order by another (where a new representation of partial pre-order is needed), one can still use AGM postulates.

3.3 Additional Postulates

In [25], a postulate proposed for iterated belief revision on epistemic states is defined as follows.

ER4* $\Phi \circ_E \Psi^{\mathscr{F}} \circ_E \Theta^{\mathscr{F}'} = \Phi \circ_E \Theta^{\mathscr{F}'}$ where partition $\mathscr{F}'$ of W is a refinement of partition $\mathscr{F}$.

In which $\circ_E$ is an epistemic state revision operator, and Φ, $\Psi^{\mathscr{F}}$, and $\Theta^{\mathscr{F}'}$ are all epistemic states which generalize formula-based belief representations.

In our framework **ER4*** is rewritten as follows:

UER4* For any $S, S_I, S_I' \in \mathcal{S}_{CC}$ such that $S_I \subseteq S_I'$, then $S \odot S_I \odot S_I' = S \odot S_I'$.

UER4* states that given a prior set and two new inputs, if the latter input has a finer structure than the former input, then the latter totally shadows the former in iterated revision. This is clearly the counterpart of **ER4*** and represents an important result for iterated revision.

We also propose additional properties inspired from Darwiche and Pearl's iterated belief revision postulates :

---

[1] This is an essential difference between our framework with the framework in [4]. As an instance of this difference, R2 is questioned in [4] and hence is not valid in the framework of [4] whilst it is valid (after translation) in our framework.

UCR1 If $w_1 \prec w_2 \in Cm(S_I)$, then $w_1 \prec w_2 \in Cm(S \odot S_I)$.

UCR2 If $w_1 \approx w_2 \in Cm(S_I)$, then $w_1 \approx w_2 \in Cm(S \odot S_I)$.

UCR3 If w_1 and w_2 are incomparable in $Cm(S_I)$, then $w_1 \; R \; w_2 \in Cm(S)$ iff $w_1 \; R \; w_2 \in Cm(S \odot S_I)$ (R is $\prec$ or $\approx$).

Here UCR1-3 are inspired from the semantic expression of DP postulates CR1-CR4. UCR3 is its counterpart of CR1-CR2 in our framework, and it can be seen as a counterpart of the relevance criterion [19]. It says that in case two elements are incomparable in the input, then the ordering between these two elements should be the same in both initial and revised state. This is inspired from CR1 and CR2 where the ordering of models (resp. countermodels) of input is the same in both initial and revised states.

UCR1-2 are inspired from CR3 and CR2, respectively. Also note that UCR1-2 are implied by UR1.

Note that here any ties (e.g., $w_1 \approx w_2$) in the input are preserved while in the original work of [8], they are broken based on the information given by initial beliefs.

4   Examples of Revision Operators for Partial-Preorders

In this section, we give two examples of revision operators, introduced in [21], that satisfy all postulates.

Match Revision. The key idea of match revision is to remove any units in S which join at least one minimal cycle in $S \cup S_I$. Therefore, these units are potentially conflicting with S_I.

Definition 3. *(Match Revision Operator, [21]) For any $S, S_I \in \mathcal{S}_{CC}$, let $S' = Cm(S \cup S_I)$ and let C be the set of all minimal cycles of S', then the match revision operator $\odot_{match}$ is defined as:*
$S \odot_{match} S_I = Cm(S' \setminus (\bigcup_{C \in \mathcal{C}} C \setminus S_I))$.

Example 3. *Let $S = \{w_3 \prec w_2, w_2 \prec w_4, w_4 \prec w_1, w_3 \prec w_4, w_3 \prec w_1, w_2 \prec w_1\}$ and $S_I = \{w_1 \prec w_2, w_4 \prec w_3\}$, then we have six minimal cycles in $Cm(S \cup S_I)$. That is*

$C_1 : w_1 \prec w_2, w_2 \prec w_1, \qquad C_2 : w_1 \prec w_4, w_4 \prec w_1,$
$C_3 : w_2 \prec w_4, w_4 \prec w_2, \qquad C_4 : w_3 \prec w_2, w_2 \prec w_3,$
$C_5 : w_3 \prec w_4, w_4 \prec w_3, \qquad C_6 : w_1 \prec w_3, w_3 \prec w_1.$

Hence we have: $S \odot_{match} S_I = Cm(\{w_1 \prec w_2, w_4 \prec w_3\}) = \{w_1 \prec w_2, w_4 \prec w_3\}$.

However, the match revision operator removes too many units from the prior state S, as we can see from Example 3. In fact, if certain units are removed from S, then there will be no cycles in $S \cup S_I$, hence some other units subsequently could have been retained. That is, there is no need to remove all the conflicting units at once, but one after the other. This idea leads to the following inner and outer revision operators.

Inner Revision. The basic idea of inner revision is to insert each unit of S_I one by one into S, and in the meantime, remove any unit in S that are inconsistent with the inserted unit. Of course, the revision result depends on the order in which these units from S_I are inserted to S. Hence, only the units that exist in all revision results for any insertion order should be considered credible for the final, consistent revision result.

For a set of units S, let $\mathsf{PMT}(S)$ denote the set of all permutations of the units in S. For example, if $S = \{w_1 \prec w_3, w_2 \prec w_3\}$, then $\mathsf{PMT}(S) = \{(w_1 \prec w_3, w_2 \prec w_3), (w_2 \prec w_3, w_1 \prec w_3)\}$.

Definition 4. *For any $S, S_I \in \mathcal{S}_{CC}$, let $\vec{t} = (t_1, \cdots, t_n)$ be a permutation in $\mathsf{PMT}(S_I)$, then the result of sequentially inserting $\vec{t}$ into S one by one[2], denoted as $S_{\vec{t}}$, is defined as follows:*

- *Let S_i be the resulted set by sequentially inserting $t_1, \cdots, t_i$ one by one. Let $S' = Cm(S_i \cup \{t_{i+1}\})$ and $\mathcal{C}$ be the set of all minimal cycles of S', then $S_{i+1} = S' \setminus (\bigcup_{C \in \mathcal{C}} C \setminus S_I)$.*
- *$S_{\vec{t}} = S_n$.*

The inner revision operator is defined as follows.

Definition 5. *(Inner Revision Operator, [21]) For any $S, S_I \in \mathcal{S}_{CC}$, the inner revision operator is defined as:*

$$S \odot_{in} S_I = Cm(\bigcap_{\vec{t} \in \mathsf{PMT}(S_I)} S_{\vec{t}}). \tag{1}$$

Example 4. *Let $S = \{w_3 \prec w_2, w_2 \prec w_4, w_4 \prec w_1, w_3 \prec w_4, w_3 \prec w_1, w_2 \prec w_1\}$ and $S_I = \{w_1 \prec w_2, w_4 \prec w_3\}$, then we have $S_{(w_4 \prec w_3, w_1 \prec w_2)} = \{w_1 \prec w_2, w_4 \prec w_3, w_3 \prec w_1, w_4 \prec w_1, w_4 \prec w_2, w_3 \prec w_2\}$ and $S_{(w_1 \prec w_2, w_4 \prec w_3)} = \{w_1 \prec w_2, w_4 \prec w_3, w_3 \prec w_1, w_3 \prec w_2, w_4 \prec w_2, w_4 \prec w_1\}$. Hence $S \odot_{in} S_I = \{w_1 \prec w_2, w_4 \prec w_3, w_3 \prec w_1, w_3 \prec w_2, w_4 \prec w_2, w_4 \prec w_1\}$.*

Examining Unit-Based Postulates. For operators $\odot_{in}$ and $\odot_{match}$, we have the following results.

Proposition 1. *The revision operators $\odot_{in}$ and $\odot_{match}$ satisfy UR0-UR6.*

This proposition shows that our revision operators satisfy all the counterparts of AGM-style postulates proposed in Section 3.

Next we show that our revision operations satisfy **UER4*** and iteration properties:

Proposition 2.   – *For any $S, S_I, S_I' \in \mathcal{S}_{CC}$ such that $S_I \subseteq S_I'$, then $S \odot_{in} S_I \odot_{in} S_I' = S \odot_{in} S_I'$. We also have $S \odot_{match} S_I \odot_{match} S_I' = S \odot_{match} S_I'$.*
- *The revision operators $\odot_{in}$ and $\odot_{match}$ satisfy UCR1-UCR3.*

To summarize, from the Proposition 2, we can conclude that match and inner revision operator satisfies many of the well-known postulates for belief revision and iterated belief revision, which demonstrates that these strategies are rational and sound.

---

[2] As two units like $w \approx w'$ and $w' \approx w$ are in fact the same, in this and the next section, this type of units are considered as one unit and be inserted together.

5   Recovering Lexicographic and Natural Belief Revision

Now we come to show that the two well-known belief revision strategies (lexicographic and natural revision) can be encoded as stated in our framework.

5.1   Recovering Lexicographic Revision

We first assume that an input, representing a propositional formula μ (a typical input in the belief revision situation), is described by a *modular order* where:

- each model of μ is preferred to each model of $\neg\mu$,
- models (resp. counter-models) of μ are incomparable.

That is, given μ, the corresponding input representing μ is denoted as $S_I^\mu = \{w \prec w' : \forall w \models \mu, w' \models \neg\mu\}$. Clearly for any consistent μ, $S_I^\mu \in \mathcal{S}_{CC}$.
Then we have the following result.

Proposition 3. *For any $S \in \mathcal{S}_{CC}$ and any consistent μ, we have $S \odot_{in} S_I^\mu = S_I^\mu \cup \{wRw' : w, w' \models \mu \text{ and } wRw' \in S\} \cup \{wRw' : w, w' \models \neg\mu \text{ and } wRw' \in S\}$.*

Proof of Proposition 3: A sketch of the proof can be shown as follows. Let $\hat{S} = S \odot_{in} S_I^\mu$, then it is easy to show the following steps:

- $S_I^\mu \subseteq \hat{S}$.
  The input is reserved in the revision result.
- $\{wRw' : w, w' \models \mu \text{ and } wRw' \in S\} \in \hat{S}$, $\{wRw' : w, w' \models \neg\mu \text{ and } wRw' \in S\} \in \hat{S}$.
  It is easy to see that any unit $wRw' \in S$ such that w, w' are both models of μ or both counter-models of μ is consistent with S_I^μ. So any such unit is in $\hat{S}$.
- $\hat{S}$ only contains units which can be induced from the above two steps. In fact, we can find that $S_I^\mu \cup \{wRw' : w, w' \models \mu \wedge wRw' \in S\} \cup \{wRw' : w, w' \models \neg\mu \wedge wRw' \in S\}$ already forms a total pre-order over W (and hence is complete) and obviously it is consistent.   □

That is, given the input S_I^μ representing μ, inner revision operator reduces to a lexicographic revision [30] in the belief revision case. Hence obviously it follows the spirit of AGM postulates [1], Darwiche and Pearl's iterated belief revision postulates [8], and the Recalcitrance postulate [30].

In [3] an extension of lexicographic revision of an epistemic state $\triangleright_{initial}$ (viewed as a total pre-order), by an input in the form of another total pre-order, denoted here by $\triangleright_{input}$, is defined. The obtained result is a new epistemic state, denoted by $\triangleright_{lex}$ (*lex* for lexicographic ordering), and defined as follows:

- $\forall w_1, w_2 \in W$, if $w_1 \triangleright_{input} w_2$ then $w_1 \triangleright_{lex} w_2$.
- $\forall w_1, w_2 \in W$, if $w_1 =_{input} w_2$ then $w_1 \triangleright_{lex} w_2$ if and only if $w_1 \triangleright_{initial} w_2$.

Namely, $\triangleright_{lex}$ is obtained by refining $\triangleright_{input}$ by means of the initial ordering $\triangleright_{initial}$ for breaking ties in $\triangleright_{input}$.

To recover this type of revision, it is enough to interpret ties in $\triangleright_{input}$ as incomparable relations, namely:

Proposition 4. *Let $\triangleright_{initial}$ and $\triangleright_{input}$ be two total pre-orders and $\triangleright_{lex}$ be the result of refining $\triangleright_{input}$ by $\triangleright_{initial}$ as defined above. Let $S_{initial}$ be the set of all units in $\triangleright_{initial}$, and S_{input} be the set of strict relations in $\triangleright_{input}$, namely, $S_{input} = \{\omega_1 \prec \omega_2$ such that $\omega_1 \triangleright_{input} \omega_2\}$ (i.e., ties in $\triangleright_{input}$ are not included in S_{input}). Then we have : $S_{initial} \odot_{in} S_{input} = Cm(\{\omega_1 \prec \omega_2$ such that $\omega_1 \triangleright_{lex} \omega_2\})$.*

The above proposition shows that the lexicographic inference can be recovered when both initial beliefs and input are total pre-ordered. Of course, our approach goes beyond lexicographic belief revision since inputs can be partially pre-ordered.

5.2 Recovering Natural Belief Revision

This section shows that inner revision allows to recover the well-known natural belief revision proposed in [6] and hinted by Spohn [31]

Let $\triangleright_{initial}$ be a total pre-order on the set of interpretations representing initial epistemic state. Let μ be a new piece of information. We denote by $\triangleright_N$ he result of applying natural belief revision of $\triangleright_{initial}$ by ϕ. $\omega =_N \omega'$ denotes that ω and ω' are equally plausible in the result of revision. Natural belief revision of $\triangleright_{initial}$ by μ consists in considering the most plausible models of μ in $\triangleright_{initial}$ as the most plausible interpretations in $\triangleright_N$. Namely, $\triangleright_N$ is defined as follows:

– $\forall \omega \in \min(\phi, \triangleright_{initial}), \forall \omega' \in \min(\phi, \triangleright_{initial}), \omega =_N \omega'$
– $\forall \omega \in \min(\phi, \triangleright_{initial}), \forall \omega' \notin \min(\phi, \triangleright_{initial}), \omega \triangleright_N \omega'$
– $\forall \omega \notin \min(\phi, \triangleright_{initial}), \forall \omega' \notin \min(\phi, \triangleright_{initial}), \omega \triangleright_N \omega'$ iff $\omega \triangleright_{initial} \omega'$.

In order to recover natural belief revision we will again apply inner revision. Let us describe the input. We will denote by ϕ as a propositional formula whose models are those of μ which are minimal in $\triangleright_{initial}$. The input is described by the following *modular order* $\triangleright_{input}$ where :

– each model of ϕ is preferred to each model of $\neg\phi$, namely : $\forall \omega, \omega'$, if $\omega \models \phi$ and $\omega' \models \neg\phi$ then $\omega \triangleright_{input} \omega'$
– models of ϕ are equally plausible, namely : $\forall \omega, \omega'$, if $\omega \models \phi$ and $\omega' \models \phi$ then $\omega =_{input} \omega'$
– models of $\neg\phi$ are incomparable.

Then we have :

Proposition 5. *Let $\triangleright_{initial}$ be a total pre-order associated with initial beliefs. Let μ be a proposition formula and $\triangleright_N$ be the result of revising $\triangleright_{initial}$ by μ as defined above. Let $\triangleright_{input}$ be a partial pre-order defined above. Let $S_{initial}$ (resp. $S_{initial}, S_N$) be the set of all units in $\triangleright_{initial}$ (resp. $\triangleright_{input}, \triangleright_N$). Then we have : $S_{initial} \odot_{in} S_{input} = S_N$.*

The above proposition shows that the natural belief revision can be recovered using inner revision. And again, our framework goes beyond natural belief revision since initial beliefs can be partially pre-ordered (while it is defined as totally pre-ordered in [6]).

6 Discussions and Related Works

In this section, first we briefly review some related works and then we present some discussions on the difference between our approach and other revision approaches.

[15] and [17] in fact focus on revising with conditionals. Our unit $w \prec w'$ can be translated into $(w|w \lor w')$ in the framework proposed in [17]. However, in [17], they only consider initial and input epistemic states as ordinal conditional functions which cannot encode a partial pre-order in general.

In [32], revision of partial orders is studied in a standard expansion and contraction way, in which contraction uses a cut function which can be seen as a selection function, and hence the result is not deterministic, whilst all revision operators proposed in this paper provide deterministic results.

[28] only considers merging of partial pre-orders (which follows a different definition from this paper) instead of revision, so it departs from the work we investigated in this paper.

Furthermore, we already showed that our approach can recover, as a particular case, the lexicographic inference ([29], etc) when both the initial state and the input are either a propositional formula or a total pre-order.

Lastly, note that our revision operations are totally different from Lang's works on preference (e.g. [20]) and Kern-Isberner's revision with conditionals (e.g., [16]).

Remarks: A key question should be answered is: what is the difference between revision in this paper and in other papers?

In existing approaches, when a revision strategy is extended to deal with some complex task, two steps are commonly followed:

- generalizing the concept of "theory" (initial state) and input. For instance, in [25], when the task is to revise an epistemic state, the representation of initial epistemic state and the input was generalized that can recover almost all common uncertainty representations. However, in most works, these representations extend the concept of "propositional formulas" directly.
- extending or modifying AGM postulates, and several of them "get rid" of some postulates. For instance, in [4], postulate R2 is removed and replaced by other postulates. [18] has suggested new postulates to deal with the idea *improvements* and drop some of the AGM postulates.

In our approach, however, we propose revision operators from a different perspective.

- First, we keep all the AGM postulates even if our aim is to generalize the revision process to deal with a very flexible structure which is a partial pre-order.
- Second, we consider very different components of the revision operation. Initial epistemic state is no longer a propositional formula but a set of units. Similarly for the input. With this change, some standard concepts need to be adapted, in particular the concepts of consistency and entailment. We can illustrate this by the following example taken from [4].

Example 5. *Let* $W = \{w_1, w_2, w_3, w_4\}$. *Consider the following partial pre-order representing the agent's initial epistemic state* Φ: $w_3 \prec_B w_2 \prec_B w_1$ *which means*

w_3 is the most preferred, w_1 is the least preferred, and w_4 is incomparable to others.
Assume that the input is propositional formula μ with models w_2 and w_4, denoted as $[\mu] = \{w_2, w_4\}$ (where $[\phi]$ represents the set of models of any propositional formula ϕ). The aim is to revise the partial pre-order with μ.

Let us see how the two approaches behave (from representational point of view and from axiomatic point of view).
In [4], from a representational point of view,
Epistemic state = Partial pre-order Φ
Input   = Propositional formula μ
Output   = $\Phi \circ \mu$ which is a propositional formula
From an axiomatic point of view, the example shows that R2 postulate (see R2 in Section 3) maybe questionable. That is, in the example, the former approach only keeps the possible world w_4 that is incomparable to others while the latter also keeps the minimal possible world w_2 in the set of comparable possible worlds, which is more reasonable.
Indeed, in this example:

$$[Bel(\Phi)] = \{w : w \in W \wedge \nexists w', w' \prec_B w\} = \{w_3, w_4\},$$

$$[\mu] = \{w_2, w_4\}.$$

So $Bel(\Phi)$ and μ are consistent and hence according to R2, $Bel(\Phi \circ \mu) = \{w_4\}$. This is questionable as a result.
Therefore, in [4], they consider R2 is no longer valid when epistemic states are partial pre-orders and they propose different alternatives for this postulate.
In our approach, from a representational point of view,
Epistemic state=Partial pre-orders represented by a set of units, i.e., $S = Cm(\{w_3 \prec w_2, w_2 \prec w_1\})$
Input   = Partial pre-orders represented by a set of units, i.e., $S_I = \{w_2 \prec w_1, w_2 \prec w_3, w_4 \prec w_1, w_4 \prec w_3\})$

From an axiomatic point of view, we keep all AGM (or KM) postulates but coherence and entailment do not have the same meaning. In our approach, S and S_I are not consistent since in S we have $w_3 \prec w_2$ while in S_I we have $w_2 \prec w_3$. Therefore, the result of revision is not $S \cup S_I$.
What can we obtain with our approach in this example? We get $\hat{S} = Cm\{w_3 \prec w_1, w_2 \prec w_3, w_4 \prec w_3\}$ in which the expected result is obtained with minimal models: $\{w_2, w_4\}$.
In fact, in the definition of our revision operation, we do not focus on $Bel(\Phi)$, instead, we focus on *small components* that compose the partial pre-orders.

7   Conclusion

Although logic-based belief revision is fully studied, revision strategies for ordering information have seldom been addressed. In this paper, we investigated the issue of revising a partial pre-order by another partial pre-order. We proposed a set of rationality postulates to regulate this kind of revision. We also proved several revision operators satisfy these postulates as well as some rational properties.

The fact that our revision operators satisfy the counterparts of the AGM style postulates and the iterated revision postulate shows that our revision strategies provide rational and sound approaches to handling revision of partial pre-orders. In addition, when reducing to classical belief revision situation, our revision strategies become the lexicographic revision. This is another indication that our revision strategies have a solid foundation rooted from the classic belief revision field.

For future work, we will study the relationship between our revision framework and revision strategies proposed for preferences.

References

1. Alchourrón, C.E., Gärdenfors, P., Makinson, D.: On the logic of theory change: Partial meet functions for contraction and revision. Symb. Log. 50, 510–530 (1985)
2. Benferhat, S., Dubois, D., Prade, H.: Representing default rules in possibilistic logic. In: Procs. of KR, pp. 673–684 (1992)
3. Benferhat, S., Konieczny, S., Papini, O., Pérez, R.P.: Iterated revision by epistemic states: Axioms, semantics and syntax. In: Proc. of ECAI 2000, pp. 13–17 (2000)
4. Benferhat, S., Lagrue, S., Papini, O.: Revision of partially ordered information: Axiomatization, semantics and iteration. In: IJCAI 2005, pp. 376–381 (2005)
5. Bochman, A.: A logical theory of nonmonotonic inference and belief change. Springer (2001)
6. Boutilier, C.: Revision sequences and nested conditionals. In: Procs. of IJCAI, pp. 519–525. Morgan Kaufmann Publishers Inc., San Francisco (1993)
7. Chan, H., Darwiche, A.: On the revision of probabilistic beliefs using uncertain evidence. Artif. Intell. 163(1), 67–90 (2005)
8. Darwiche, A., Pearl, J.: On the logic of iterated belief revision. Artificial Intelligence 89, 1–29 (1997)
9. Dubois, D., Moral, S., Prade, H.: Belief change rules in ordinal and numerical uncertainty theories. Handbook of Defeasible Reasoning and Uncertainty Management Systems 3, 311–392 (1998)
10. Fermé, E., Hansson, S.O. (eds.): Journal of Philosophical Logic. Special Issue on 25 Years of AGM Theory, vol. 40(2). Springer, Netherlands (2011)
11. Hunter, A., Konieczny, S.: Shapley inconsistency values. In: Proc. of KR 2006, pp. 249–259 (2006)
12. Jeffrey, R.: The logic of decision, 2nd edn. Chicago University Press (1983)
13. Jin, Y., Thielscher, M.: Iterated belief revision, revised. Artificial Intelligence 171, 1–18 (2007)
14. Katsuno, H., Mendelzon, A.O.: Propositional knowledge base revision and minimal change. Artificial Intelligence 52, 263–294 (1991)
15. Kern-Isberner, G.: Conditionals in Nonmonotonic Reasoning and Belief Revision. LNCS (LNAI), vol. 2087. Springer, Heidelberg (2001)
16. Kern-Isberner, G.: Handling conditionals adequately in uncertain reasoning and belief revision. Journal of Applied Non-Classical Logics 12(2), 215–237 (2002)
17. Kern-Isberner, G., Krümpelmann, P.: A constructive approach to independent and evidence retaining belief revision by general information sets. In: Procs. of IJCAI, pp. 937–942 (2011)

18. Konieczny, S., Pérez, R.P.: Improvement operators. In: Procs. of KR, pp. 177–187 (2008)
19. Kourousias, G., Makinson, D.: Parallel interpolation, splitting, and relevance in belief change. J. Symb. Log. 72(3), 994–1002 (2007)
20. Lang, J., van der Torre, L.: From belief change to preference change. In: Procs. of ECAI, pp. 351–355 (2008)
21. Ma, J., Benferhat, S., Liu, W.: Revising partial pre-orders with partial pre-orders: A unit-based revision framework. In: (Short paper) Procs. of KR (2012)
22. Ma, J., Liu, W.: A general model for epistemic state revision using plausibility measures. In: Procs. of ECAI, pp. 356–360 (2008)
23. Ma, J., Liu, W.: Modeling belief change on epistemic states. In: Proc. of 22th Flairs, pp. 553–558. AAAI Press (2009)
24. Ma, J., Liu, W.: A framework for managing uncertain inputs: An axiomization of rewarding. Inter. Journ. of Approx. Reasoning 52(7), 917–934 (2011)
25. Ma, J., Liu, W., Benferhat, S.: A belief revision framework for revising epistemic states with partial epistemic states. In: Procs. of AAAI 2010, pp. 333–338 (2010)
26. Ma, J., Liu, W., Dubois, D., Prade, H.: Revision rules in the theory of evidence. In: Procs. of ICTAI, pp. 295–302 (2010)
27. Ma, J., Liu, W., Dubois, D., Prade, H.: Bridging jeffrey's rule, agm revision and dempster conditioning in the theory of evidence. International Journal on Artificial Intelligence Tools 20(4), 691–720 (2011)
28. Ma, J., Liu, W., Hunter, A.: Modeling and reasoning with qualitative comparative clinical knowledge. International Journal of Intelligent Systems 26(1), 25–46 (2011)
29. Nayak, A.C.: Iterated belief change based on epistemic entrenchment. Erkenntnis 41, 353–390 (1994)
30. Nayak, A.C., Pagnucco, M., Peppas, P.: Dynamic belief revision operators. Artificial Intelligence 146, 193–228 (2003)
31. Spohn, W.: Ordinal conditional functions: A dynamic theory of epistemic states. Causation in Decision, Belief Change, and Statistics 2, 105–134 (1988)
32. Tamargo, L.H., Falappa, M.A., García, A.J., Simari, G.R.: A Change Model for Credibility Partial Order. In: Benferhat, S., Grant, J. (eds.) SUM 2011. LNCS, vol. 6929, pp. 317–330. Springer, Heidelberg (2011)
33. Williams, M.A.: Transmutations of knowledge systems. In: Proc. of KR 1994, pp. 619–629 (1994)

Representing Uncertainty by Possibility Distributions Encoding Confidence Bands, Tolerance and Prediction Intervals

Mohammad Ghasemi Hamed[1,2], Mathieu Serrurier[1], and Nicolas Durand[1,2]

[1] IRIT - 118 route de Narbonne 31062, Toulouse Cedex 9, France
[2] ENAC/MAIAA - 7 avenue Edouard Belin 31055 Toulouse, France

Abstract. For a given sample set, there are already different methods for building possibility distributions encoding the family of probability distributions that may have generated the sample set. Almost all the existing methods are based on parametric and distribution free confidence bands. In this work, we introduce some new possibility distributions which encode different kinds of uncertainties not treated before. Our possibility distributions encode statistical tolerance and prediction intervals (regions). We also propose a possibility distribution encoding the confidence band of the normal distribution which improves the existing one for all sample sizes. In this work we keep the idea of building possibility distributions based on intervals which are among the smallest intervals for small sample sizes. We also discuss the properties of the mentioned possibility distributions.

Keywords: imprecise probabilities, possibility distribution, confidence band, confidence region, tolerance interval, prediction interval, normal distribution, distribution free.

1 Introduction

In 1978, Zadeh introduced the possibility theory [31] as an extension of his theory on fuzzy sets. Possibility theory offers an alternative to the probability theory when dealing with some kinds of uncertainty. Possibility distribution can be viewed as a family of probability distributions. Then, the possibility distribution contains all the probability distributions that are respectively upper and lower bounded by the possibility and the necessity measure [10]. For a given sample set, there are already different methods for building possibility distribution which encodes the family of probability distributions that may have generated the sample set[3,25,2]. The mentioned methods are almost all based on parametric and distribution free confidence bands.

In this paper we review some methods for constructing confidence bands for the normal distribution and for constructing distribution free confidence bands (γ-C distribution). Then we propose a possibility distribution for a sample set drawn from an unknown normal distribution based on Frey [15] confidence band which improves the existing possibility distribution proposed by Aregui et al. [2]

E. Hüllermeier et al. (Eds.): SUM 2012, LNAI 7520, pp. 233–246, 2012.

for all sample sizes. We also introduce possibility distribution which encodes tolerance intervals, named as γ-Confidence Tolerance Possibility distribution (γ-CTP distribution). The proposed possibility distribution uses tolerance intervals to build the maximal specific possibility distribution that bounds each population quantile of the true distribution (with a fixed confidence level) that might have generated our sample set. The distribution obtained will bound each confidence interval of inter-quantiles independently. This latter is different from a possibility distribution encoding a confidence band, because a possibility distribution encoding a confidence band will simultaneously bounds all population quantiles of the true distribution (with a fixed confidence level) that might have generated our sample set. Finally, we consider possibility distributions encoding prediction intervals (prediction possibility distribution). In this case, each α-cut will contain the next observation with a confidence level equal to $1 - \alpha$. Each of the proposed possibility distributions encodes a different kind of uncertainty that is not expressed by the other ones. We show that γ-C distribution is always less specific than γ-CTP distribution which is itself less specific than the prediction possibility distribution. This is due to the fact that the distributions properties are less and less strong. Note that the confidence level is usually chosen by the domain expert.

This paper is structured as follows: we begin with a background on possibility theory. Then we review possibility distribution encoding confidence bands and their relationship with confidence regions. In this section we introduce a method which improves existing possibility distributions. Next we see how to encode tolerance intervals and prediction intervals by possibility distributions. Finally, we end with a discussion on the mentioned possibility distributions and some illustrations.

2 Background

2.1 Possibility Theory

Possibility theory, introduced by Zadeh [31,12], was initially created in order to deal with imprecision and uncertainty due to incomplete information. This kind of uncertainty may not be handled by probability theory, especially when a priori knowledge about the nature of the probability distribution is lacking. In possibility theory, we use a membership function π to associate a distribution on the universe of discourse Ω. In this paper, we only consider the case $\Omega = \mathbb{R}$.

Definition 1. *A possibility distribution π is a function from Ω to $(\mathbf{R} \rightarrow [0,1])$.*

The definition of the possibility measure Π is based on the possibility distribution π such that:

$$\Pi(A) = sup(\pi(x), \forall x \in A). \tag{1}$$

The necessity measure is defined by the possibility measure

$$\forall A \subseteq \Omega, N(A) = 1 - \Pi(A^C) \tag{2}$$

where A^C is the complement of the set A. A distribution is normalized if : $\exists x \in \Omega$ such that $\pi(x) = 1$. When the distribution π is normalized, we have : $\Pi(\emptyset) = 0, \Pi(\Omega) = 1$.

Definition 2. *The α-cut A_α of a possibility distribution $\pi(\cdot)$ is the interval for which all the point located inside have a possibility membership $\pi(x)$ greater or equal to α.*

$$A_\alpha = \{x | \pi(x) \geq \alpha, x \in \Omega\}, \tag{3}$$

2.2 Possibility Distribution Encoding a Family of Probability Distribution

In fact, one interpretation of possibility theory, based on Zadeh's [31] consistency principle of possibility "what is probable should be possible", is to consider a possibility distribution as a family of probability distributions (see [10] for an overview). In the following, we denote f as density function, F as its Cumulative Distribution Function (CDF) and P as its probability measure. Thus, a possibility distribution π will represent the family of the probability distributions Θ for which the measure of each subset of Ω's will be bounded by its possibility measures :

Definition 3. *A possibility measure Π is equivalent to the family Θ of probability distribution F such that*

$$\Theta = \{F | \forall A \in \Omega, P(A) \leq \Pi(A)\}, A \subseteq \Omega. \tag{4}$$

Definition 4. *Given a probability density function $f(\cdot)$ with finite number of modes, we define the interval I_β^* defined below as "smallest β-content interval" of f.*

$$I_\beta^* = \{x | x \in f^{-1}[d], \forall d \in [c, +\inf)\} \tag{5}$$

where $\int_{\{x | f(x) \geq c\}} f(x)dx = \beta$ and $c > 0$.

We know that $Pr(I_\beta^*) = \beta$ and this interval is unique only if f has a finite number of modes. Now let θ be a set of Cumulative Distribution Function (CDF) F defined by a possibility distribution function $\pi(\cdot)$. Thus, an alternative to equations (4) is:

$$\forall \alpha \in [0, 1], \forall F \in \Theta, I_{F,\beta}^* \subseteq A_{\pi,\alpha} \tag{6}$$

where $\beta = 1 - \alpha$ and $A_{\pi,\alpha}$ is the α-cut of possibility distribution $\pi(\cdot)$. Thus, a possibility distribution encodes a family of probability distributions for which each quantile is bounded by a possibility α-cut. By considering the definition of necessity, we obtain the following inequalities:

$$N(A) \leq P(A) \leq \Pi(A), A \subset \Omega. \tag{7}$$

Thus by using the possibility and necessity measures, like in the Dempster-Shafer theory, we can define upper and lower values to describe how an event is likely to occur.

2.3   Probability-Possibility Transformation

In many cases it is desirable to move from the probability framework to the possibility framework. This is why several transformations based on various principles such as consistency (what is probable is possible) or information invariance have already been proposed [8,9,22,13,11]. Dubois et al.[14] suggest that when moving from probability to possibility framework we should use the "maximum specificity" principle which aims at finding the most informative possibility distribution. Formally the maximum specificity principle is defined as follow:

Definition 5. *Given the maximum specific possibility distribution (m.s.p.d)* π^* *that encodes the probability distribution function F (i.e. $\forall A \subseteq \Omega, N^*(A) \leq P(A) \leq \Pi^*(A)$) we have, for all π such as $\forall A \subseteq \Omega, N(A) \leq P(A) \leq \Pi(A)$, $\pi^*(x) \leq \pi(x), \forall x \in \Omega$.*

Because the possibility distribution explicitly handles the imprecision and is also based on an ordinal structure rather than an additive one, it has a weaker representation than the probability one. This kind of transformation (probability to possibility) may be desirable when we are in presence of weak source of knowledge or when it is computationally harder to work with the probability measure than with the possibility measure. The "most specific" possibility distribution is defined for a probability distribution having a finite number of modes and the equation is as below [11] :

$$\pi_t(x) = sup(1 - P(I_\beta^*), x \in I_\beta^*) \tag{8}$$

where π_t is the "most specific" possibility distribution, I_β^* is the smallest β-content interval [11]. Then, in the spirit of equation 6, given f and its transformation π^* we have :

$$A_\alpha^* = I_\beta^* \quad where \ \alpha = 1 - \beta.$$

Figure (1) presents the maximum specific transformation (in blue) of a normal probability distribution (in green) with mean and variance respectively equal to $0, 1$ ($\mathcal{N}(0,1)$).

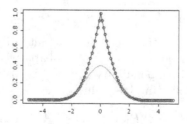

Fig. 1. The m.s.p.d for $\mathcal{N}(0,1)$

Definition 6. *The interval between the lower and upper quantiles of the same level are called inter-quantiles. The inter-quantile at level p is defined by*

$$[F^{-1}(p), F^{-1}(1-p)], 0 < p < 0.5 \tag{9}$$

where $F^{-1}(\cdot)$ is the inverse function of the continuous strictly-monotone CDF $F(\cdot)$.

An inter-quantile at level p contains β proportion of the distribution where $\beta = 1 - 2p$. We will call a β-content inter-quantile I_β, the interval that contains β proportion of the underlying distribution, we have $Pr(I_\beta) = \beta$.

Proposition 1. *The maximum specific possibility distribution (m.s.p.d) $\pi^*(\cdot)$ of unimodal symmetric probability density function $f(\cdot)$ can be built just by calculating the β-content inter-quantile I_β of $f(\cdot)$ for all the values of β, where $\beta \in [0, 1]$.*

It is evident that for any unimodal symmetric p.d.f $f(\cdot)$, *the smallest β-content interval I_β^* of f is also its inter-quantile at level $\frac{1-\beta}{2}$*. Therefore the proposition 1 can be used.

3   Possibility Distribution Encoding Confidence Bands

3.1   Confidence Band

Definition 7. *The confidence band for a CDF F is a function which associates to each x an interval $[L(x), U(x)]$ such as :*

$$P(\forall x, L(x) \leq F(x) \leq U(x)) \geq \gamma \text{ where } \forall x, 0 \leq L(x) \leq U(x) \leq 1 \tag{10}$$

In frequentist statistics, a confidence band is an interval defined for each value x of the random variable X such that for a repeated sampling, the frequency of $F(x)$ located inside the interval $[L(x), U(x)]$ for all the values of X tends to the confidence coefficient γ.

Note that given any γ level confidence band, we can use it to infer confidence intervals of the quantile function $Q(\beta) = F^{-1}(\beta) = \inf\{x \in R : \beta \leq F(x)\}$, and that for all $\beta \in (0, 1)$. In other word the confidence band gives simultaneously confidence intervals for all $F^{-1}(\beta), \beta \in (0, 1)$. Therefore such confidence intervals derived from confidence bands are *Simultaneous confidence Intervals (SMI) for all population quantiles.* We can take advantage of this property to derive simultaneous γ-confidence intervals for β-content inter-quantiles of the unknown CDF $F(\cdot)$ and we will denote them by I_β^C.

By using proposition [1] and tables of confidence band stated in the statistic literature [21,15,7,5,20,1], we can encode simultaneous γ-confidence intervals for β-content inter-quantiles I_β^C, of an unknown CDF $F(\cdot)$ by a possibility distribution represented by π_γ^C:

$$\pi_\gamma^C(x) = 1 - \max_{x \in I_{1-\alpha}^C} (\alpha) \text{ where } A_\alpha = I_\beta^C, \beta = 1 - \alpha \tag{11}$$

By construction, the obtained distribution has the following property:

Proposition 2. *Let π_γ^C be a possibility distribution obtained by equation (11) we have:*

$$P(\forall x, \forall \alpha \in (0,1), P(x \in A_\alpha) \geq 1 - \alpha) \geq \gamma$$

Suppose that K is a set of estimated CDF $\hat{F}$ for F, and that in a repeated sampling the frequency of the function $\hat{F}$ being equal to the true CDF F ($\forall x, \hat{F}(x) = F(x)$), tends to $1 - \alpha$. The function set K will be a confidence band for $F(x)$. This is expressed formally below:

$$P(\exists \hat{F} \in K, F = \hat{F}) = 1 - \alpha \tag{12}$$

Equations (10) and (12) are two different views representing confidence band. In equations (10), the goal is to find a set K composed of estimated cumulative distribution functions being one "choice" for estimating the true CDF F. When F belongs to a parametric family, we can use the confidence region of its parameter vector to construct its confidence band [21,7,15]. Therefore, confidence bands build by considering confidence regions are described by equation (12).

3.2 Possibility Distribution Encoding Normal Confidence Bands

As we saw above, we can use the confidence region of parameter of a probability distribution to infer its confidence band. Cheng and Iles [7] and kanofsky [21] used this approach to infer the confidence band of the normal distribution. Aregui et al. [3], proposed to construct possibility distributions for sample set drawn from a known parametric family with an unknown parameter vector. Their possibility distribution encoded the Cheng et al. [7] confidence band. In another paper, Aregui et al. [2] used confidence region for the parameters of the normal distribution to infer a possibility distribution. This possibility distribution encodes all normal distributions having their parameters inside the desired confidence region of the "true parameters" belonging to the "real distribution" that has generated this sample set. As we saw previously, encoding the confidence region of parameters results in a possibility distribution which encodes the whole confidence band of the normal distribution. The band encoded by their method was built by the "Smallest Mood exact" confidence region [4]. The "Smallest Mood exact" region contains exactly the desired confidence level and it was the the second smallest confidence region (after the "likelihood-ratio test") in [4]. This region is easy to obtain and is particularly useful for small sample sizes. (Note that if we want to construct the m.s.p.d encoding confidence band of a normal distribution, we have to find the smallest possible SMIs which leads us to use the tightest confidence band.) In [15], Frey proposed the minimum-area confidence region and the minimum area based confidence band for the normal distribution. She showed that her minimum area confidence band improves other bands for all sample sizes. In the same way we propose a possibility distribution which encodes the Frey confidence band. In figures (2,3) we compared our possibility distribution named "0.95 Frey C.P.D." (0.95 Frey Confidence Possibility Distribution) which is encoded in blue with the Mood based and Smallest Mood based confidence possibility distribution.

Proposition 3. *The Frey normal confidence band improves the confidence band resulted by the "Smallest Mood exact" region, in area sense, and this for all sample sizes.*

Sketch of proof : The "smallest Mood confidence region" is found by the optimal confidence levels ψ_1 and ψ_2 for the mean and variance confidence intervals and also for each sample size such that $(1 - \psi_1)(1 - \psi_1) = \gamma = 0.95$ [4]. We can observe that the Frey confidence band for $\gamma = 0.95$ gives confidence band with smaller area (integral over all confidence intervals of inter-quantiles) than the band issued from a Mood confidence region $(1-\psi)^2 = 0.95^2$ where $(1-\psi) = 0.95$ and this for all sample sizes. We know that we the confidence band built from Mood confidence region $(1 - \psi)^2$ where $(1 - \psi) = 0.95$ is everywhere smaller than the confidence band built from the "smallest Mood confidence region" for $\gamma = 0.95$. The statement above holds for every value of $\gamma \in (0.01, 1)$ ■

3.3 Possibility Distribution Encoding Distribution Free Confidence Bands

For distribution free confidence bands, the most known method is the Kolmogorof [5] statistic for small sample sizes and the Kolmogorof-Smirnof test for large sample sizes. Some other methods have also been suggested based on the weighted version of the Kolmogorof-Smirnof test [1]. Owen also proposed a nonparametric likelihood confidence band for a distribution function. Remark that Owen's nonparametric likelihood bands are narrower in the tails and wider in the center than Kolmogorov-Smirnov bands and are asymmetric on the empirical cumulative distribution function. Frey [20] suggested another approach in which the upper and lower bounds of the confidence band are chosen to minimize a narrowness criterion and she compared her results to other methods. The optimal bands have a nice property : by choosing appropriate weights, you may obtain bands that are narrow in whatever region of the distribution is of interest. Masson et al. [25], suggested simultaneous confidence intervals of the the the multinomial distribution to build possibility distributions and in another paper, Aregui et al. [3] proposed the Kolmogorof confidence band [5] to construct predictive belief functions[26] for sample set drawn from an unknown distribution. Thus, we propose to use the Frey band to construct the possibility distribution since it allows to have narrower α-cuts for the α's of interest.

4   Possibility Distribution Encoding Tolerance Interval

A tolerance interval, is an interval which guarantees with a specified confidence level γ, to contain a specified proportion β of the population. Confidence bounds or limits are endpoints within which we expect to find a stated proportion of the population. As the sample set grows, a parameter's confidence interval downsizes toward zero. In the same way, increasing the sample size leads the tolerance interval bounds to converge toward a fixed value. We name a $100\%\beta$ tolerance interval(region) with confidence level $100\%\gamma$, a β-content γ-coverage tolerance interval (region) and we represent it by $I^T_{\gamma,\beta}$.

Having a sample set which come from a CDF $F(\cdot)$ with unknown parameters and for a given confidence level γ, we encode all the β-content γ-coverage tolerance intervals of $F(\cdot)$, $\forall \beta \in (0,1)$, by a possibility distribution and we name it "γ-confidence tolerance possibility distribution" (γ-CTP distribution represented by π^{CTP}_γ). When we do not know the distribution of the sample set, we can use β-content γ-coverage distribution free tolerance intervals ,$\forall \beta \in (0,1)$, of the unknown probability distribution in order to build Distribution Free γ-Confidence Tolerance Possibility (γ-DFCTP distribution represented by π^{DFCTP}_γ) distribution. The possibility distributions π^{CTP}_γ and π^{DFCTP}_γ will have by construction, the following property:

Proposition 4. Let π^{CTP}_γ (or π^{DFCTP}_γ) be a possibility distribution that encodes tolerance intervals, we have:

$$\forall \alpha \in (0,1),\ P(\forall x,\ P(x \in A_\alpha) \geq 1-\alpha) \geq \gamma \ where \ A_\alpha = I^T_{\gamma,\beta}, \beta = 1-\alpha$$

Note that, it may also be interesting to fix the proportion β and make the confidence coefficient vary , $\gamma \in (0,1)$, to have a β-content tolerance possibility distribution.

Possibility Distribution Encoding Tolerance Interval for the Normal Distribution. When our sample set comes from a univariate normal distribution, the lower and upper tolerance bounds (x_l and x_u ,respectively) are calculated by formulas (13) and (14) where, $\bar{X}$ is the sample mean, S the sample standard deviation, $\chi^2_{1-\gamma,n-1}$ represents the p-value of the chi-square distribution with $n-1$ degree of freedom and $Z^2_{1-\frac{1-\beta}{2}}$ is the squared of the critical value of the standard normal distribution with probability $(1 - \frac{1-\beta}{2})$ [19]. Hence, the boundaries of a β-content γ-coverage tolerance interval for a random sample of size n drawn from an unknown normal distribution are defined as follows:

$$x_l = \bar{X} - \mathbf{k}S,\ x_u = \bar{X} + \mathbf{k}S \tag{13}$$

$$\mathbf{k} = \sqrt{\frac{(n-1)(1+\frac{1}{n})Z^2_{1-\frac{1-\beta}{2}}}{\chi^2_{1-\gamma,n-1}}} \tag{14}$$

For more details on tolerance intervals see [16].

By using proposition (1), we can find the boundaries of the $(1-\alpha)$-cut $A_{1-\alpha} = [x_l, x_u]$ of the possibility distribution which are calculated by (13), then we obtain the possibility distribution π^{CTP}_γ as computed below, where $\Phi(\cdot)$ is the CDF of the standard normal distribution.

$$\pi^{CTP}_\gamma(x) = 2(1 - \Phi\left(\sqrt{\frac{\chi^2_{(1-\gamma,n-1)}(\frac{x-\bar{X}}{S})^2}{(n-1)(1+\frac{1}{n})}}\right)) \tag{15}$$

Possibility   Distribution   Encoding   Distribution   Free   Tolerance Interval. Let $\{x_1, x_2, \cdots, x_n\}$ be n independent observations from the random variable X and let $f(x)$ be its continuous probability density function. A distribution free tolerance region is the region between two tolerance limits where the probability that this region contains β proportion of the unknown probability distribution function is equal to γ. The mentioned tolerance limits are functions $L_1(x_1, x_2, \cdots, x_n) = x_r$ and $L_2(x_1, x_2, \cdots, x_n) = x_s$ constructed based on the order statistics of the observations.

$$\int_{x_s}^{x_r} f(x)dx \geq \beta, \tag{16}$$

In order to find the distribution free β-content γ-coverage tolerance interval (region) of continuous random variable x, we have to find the smallest n and the order statistics x_r and x_s for which the probability that equation (16) holds is greater or equal to γ. Equation (16) has a sampling distribution which was first defined by Wilks [30] for a univariate random variable with symmetrical values of r and s. Wilks [30] definition puts the following constraints on r and s: $s = n - r + 1$ and $0 < r < s \leq n$. Wald [27] proposed a non-symmetrical multivariate generalization of the Wilks method. Note that because distribution free tolerance intervals are based on order statistics, the sample size required for a given distribution free tolerance interval may increase with the interval's confidence level (γ) or the interval's proportion β. For example, in order to have 95% 0.99-content tolerance interval between the first and last element of a sample set, using formula in [18], we need $n = 473$. For the calculation of the sample size requirement for tolerance intervals. The reader can refer to [16] and [18].

The construction of possibility distribution based on distribution free tolerance intervals (region) raises some problems, because for a given sample set there are many ways to choose the r and s order statistics. If we choose them symmetrically such that $r = n - s + 1$, then the possibility distribution which encodes these intervals does not guarantee that its α-cuts include the mode and the α-cuts are neither the smallest ones. In fact, for any symmetric unimodal distribution, if we choose r and s order statistics in a symmetrical way, we will have tolerance intervals which are also the smallest possible ones and also include the mode of the distribution (see proposition (1)). Thus the Distribution Free γ-Confidence Tolerance Possibility (π_γ^{DFCTP}) distribution is constructed by equation below where x_r and x_s are the limits for the distribution free $I_{\gamma,\beta}^T$ of our sample set.

$$\pi_\gamma^{DFCTP}(x) = 1 - \max_{x \in I_{\gamma,1-\alpha}^T} (\alpha) \; where \; A_\alpha = I_{\gamma,\beta}^T = [x_r, x_s], \beta = 1 - \alpha$$

5   Possibility Distribution Encoding Prediction Intervals

Let us now define a prediction interval and its associated possibility distribution. A prediction interval uses past observations to estimate an interval for what the future values will be, however other confidence intervals and credible intervals of parameters give an estimate for the unknown value of a true population parameter.

Definition 8. *Let $x_1, x_2, \cdots, x_n$ be a random sample drawn from an arbitrary distribution, then the interval $[x_l, x_u]$ is a $100\%(1 - \alpha)$ prediction interval such that:*

$$P\left(x_l \leq x \leq x_u\right) = 1 - \alpha.$$

The prediction interval for the next future observation from a normal distribution is given by [17]:

$$\frac{x_{n+1} - \overline{X}_n}{S\sqrt{1 + 1/n}} \sim t_{n-1} \tag{17}$$

$$I_\beta^{Prev} = [\overline{X}_n - t_{(\frac{\alpha}{2},n-1)}S\sqrt{1 + \frac{1}{n}}, \overline{X}_n + t_{(1-\frac{\alpha}{2},n-1)}S\sqrt{1 + \frac{1}{n}}] \tag{18}$$

So the equation (18) gives a two tailed $1 - \alpha$ prediction interval for the next observation x_{n+1} , where $\overline{X}_n$ represents the estimated mean from the n past observations , $t_{(1-\frac{\alpha}{2},n-1)}$ is the $100(\frac{1+p}{2})th$ quantile of Student's t-distribution with $n - 1$ degrees of freedom and $\beta = 1 - \alpha$.

By using proposition (1) and equation (18), we can infer a prediction possibility (π^{Prev}) distributions for a sample set which comes from a normal distribution with an unknown mean and variance. π^{Prev} is computed as below where $T_{n-1}(\cdot)$ is the CDF of the Student distribution with $n - 1$ degree of freedom. Equation (5) shows the properties of the α-cuts of π^{Prev}.

$$\pi^{Prev}(x) = 2(1 - T_{n-1}\left(\left|\frac{x_{n+1} - \overline{X}_n}{S\sqrt{1 + 1/n}}\right|\right)) \tag{19}$$

By construction, the obtained distribution has the following property:

Proposition 5. *Let π^{prev} be a possibility distribution that encodes prediction intervals using equation (19) build from a random sample set $X = \{x_1, \ldots, x_n\}$ we have:*

$$\forall \alpha \in (0,1), P(x_{n+1} \in A_\alpha) \geq 1 - \alpha \ where \ A_\alpha = I_\beta^{Prev}$$

For distribution free prediction intervals, the reader can find more information in [16],[23] and [6].

6   Discussion and Illustrations

We have seen three different types of intervals and their encoding possibility distributions. The most known approach is to choose the possibility distribution which is encoded by confidence bands. However, depending on the application, we might be interested to infer other possibility distributions than the one that encodes conventional SMIs. We can deduce, from the propositions 2, 4 and 5, that:

$$\forall x, \ \pi_\gamma^C(x) \geq \pi_\gamma^{CTP}(x) \geq \pi^{Prev}(x).$$

The choice of the distributions depends on the application. In this work, because of the lack of space, we just focused on two-sided intervals. It might be useful to construct one-sided possibility distributions encoding one-sided intervals. One can also be interested to encode percentiles of a distribution with a possibility distribution, but note that instead of using possibility distribution encoding two-sided percentiles we can use π_γ^{CTP} (two-sided). For more information the reader can refer to [16]. The reviewed distributions can be used for different purpose in uncertainty management. Wallis [29] used the Wald et al.[28] normal tolerance limits to find tolerance intervals for linear regression. In the same way, we can use our γ-CTP distribution to build probabilistic regression which encodes tolerance bounds of the response variable. Note that we are not restricted to linear possibilistic linear regression with homoscedastic and normal errors. We can also apply our γ-CTP and γ-DFCTP distributions to do possibilistic non-parametric and parametric regression with homoscedastic or heteroscedastic errors.

Figure (2) shows the $\pi_{0.95}^C$ for a sample set of size 10 with sample mean and sample variance respectively equal to 0 and 1, figure (2) represents the same concept for $n = 25$. This figure illustrates the proposition 3. Indeed, we can see that our possibility distribution is more informative than the Aregui et al. possibility distribution.

In figure (5) the blue color is used to represent π^{Prev} for different sample sets drawn from the normal distribution, all having the same sample parameters, $(\overline{X}, S) = (0, 1)$ but different sample sizes. The green distribution represents the probability-possibility transformation of $\mathcal{N}(0, 1)$.

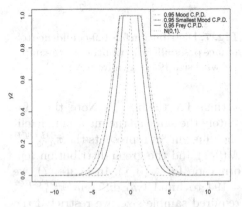 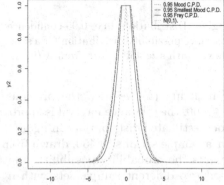

Fig. 2. Possibility distribution encoding normal confidence band for a sample set of size 10 having $(\overline{X}, S) = (0, 1)$

Fig. 3. Possibility distribution encoding normal confidence band for a sample set of size 25 having $(\overline{X}, S) = (0, 1)$

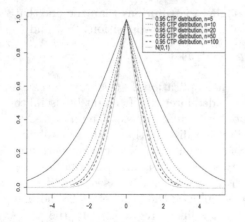

Fig. 4. 0.95-confidence tolerance possibility distribution for different sample sizes having $(\overline{X}, S) = (0, 1)$

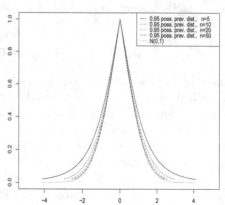

Fig. 5. 0.95-confidence prevision possibility distribution for different sample sizes having $(\overline{X}, S) = (0, 1)$

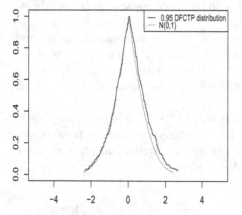

Fig. 6. Distribution free 0.95-confidence tolerance possibility distribution for a sample set with size 450 drawn from $\mathcal{N}(0, 1)$.

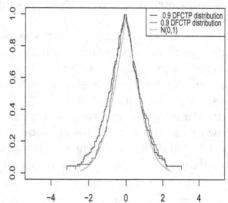

Fig. 7. Distribution free 0.9-confidence tolerance possibility distributions for a sample set with size 194 drawn from $\mathcal{N}(0, 1)$

In figure (4) we used the previous settings for the $\pi_{0.95}^{CTP}$. Note that, for $n \geq 100$, the tolerance interval is approximately the same as the maximum likelihood estimated distribution. In figure 6, the blue curves represents the $\pi_{0.95}^{DFCTP}$ for a sample set of size 450, drawn from $\mathcal{N}(0, 1)$ and the green distribution represents the probability-possibility transformation for $\mathcal{N}(0, 1)$. In figure (7), we used two different sample sets with $n = 194$ to build two different $\pi_{0.9}^{DFCTP}$. In this example, in order to reduce the required sample size, we restricted the biggest β to 0.98.

7 Conclusion

In this work, we proposed different possibility distributions encoding different kind of uncertainties. We also proposed a possibility distribution encoding confidence band of the normal distribution which improves the existing one for all sample sizes. Building possibility distributions which encode tolerance intervals and prediction intervals are also new concepts that we introduced in this work. For future works, we propose to build in the same way the possibility distributions encoding distribution free tolerance regions [27] and tolerance regions for the multivariate normal distribution [24]. We also propose to use our distributions for possibilistic regression.

Acknowledgement. We would like to thank Prof. Jesse Frey for her help and guideline during this work.

References

1. Anderson, T.W., Darling, D.A.: Asymptotic theory of certain "goodness of fit" criteria based on stochastic processes. Ann. of Math. Stat. 23(2), 193–212 (1952)
2. Aregui, A., Denœux, T.: Consonant Belief Function Induced by a Confidence Set of Pignistic Probabilities. In: Mellouli, K. (ed.) ECSQARU 2007. LNCS (LNAI), vol. 4724, pp. 344–355. Springer, Heidelberg (2007)
3. Aregui, A., Denœux, T.: Constructing predictive belief functions from continuous sample data using confidence bands. In: ISIPTA, pp. 11–20 (July 2007)
4. Arnold, B.C., Shavelle, R.M.: Joint confidence sets for the mean and variance of a normal distribution. Am. Stat. 52(2), 133–140 (1998)
5. Birnbaum, Z.W.: Numerical tabulation of the distribution of kolmogorov's statistic for finite sample size. J. Amer. Statistical Assoc. 47(259), 425–441 (1952)
6. Chakraborti, S., van der Laan, P.: Precedence probability and prediction intervals. J. Roy. Stat.Society. Series D (The Statistician) 49(2), 219–228 (2000)
7. Cheng, R.C.H., Iles, T.C.: Confidence bands for cumulative distribution functions of continuous random variables. Technometrics 25(1), 77–86 (1983)
8. Civanlar, M.R., Trussell, H.J.: Constructing membership functions using statistical data. Fuzzy Sets Syst. 18, 1–13 (1986)
9. Delgado, M., Moral, S.: On the concept of possibility-probability consistency. Fuzzy Set. Syst. 21(3), 311–318 (1987)
10. Didier, D.: Possibility theory and statistical reasoning. Compu. Stat. Data An. 51, 47–69 (2006)
11. Dubois, D., Foulloy, L., Mauris, G., Prade, H.: Probability-possibility transformations, triangular fuzzy sets and probabilistic inequalities. Rel. Comp. (2004)
12. Dubois, D., Prade, H.: Fuzzy sets and systems - Theory and applications. Academic Press, New York (1980)
13. Dubois, D., Prade, H.: Fuzzy sets and probability: Misunderstandings, bridges and gaps. In: IEEE Fuzzy Sys., pp. 1059–1068 (1993)
14. Dubois, D., Prade, H., Sandri, S.: On possibility/probability transformations. In: IFSA, pp. 103–112 (1993)
15. Frey, J., Marrero, O., Norton, D.: Minimum-area confidence sets for a normal distribution. J. Stat. Plan. Inf. 139, 1023–1032 (2009)

16. Hahn, G.J., Meeker, W.Q.: Statistical Intervals: A Guide for Practitioners. John Wiley and Sons (1991)
17. Hahn, G.J.: Factors for calculating two-sided prediction intervals for samples from a normal distribution. JASA 64(327), 878–888 (1969)
18. Hanson, D.L., Owen, D.B.: Distribution-free tolerance limits elimination of the requirement that cumulative distribution functions be continuous. Technometrics 5(4), 518–522 (1963)
19. Howe, W.G.: Two-sided tolerance limits for normal populations, some improvements. J. Amer. Statistical Assoc. 64(326), 610–620 (1969)
20. Jesse, Frey: Optimal distribution-free confidence bands for a distribution function. J. Stat. Plan. Inf. 138(10), 3086–3098 (2008)
21. Kanofsky, P., Srinivasan, R.: An approach to the construction of parametric confidence bands on cumulative distribution functions. Biometrika 59(3), 623–631 (1972)
22. Klir, G.J.: A principle of uncertainty and information invariance. Internat. J. General Systems 17(23), 249–275 (1990)
23. Konijn, H.S.: Distribution-free and other prediction intervals. Am. Stat. 41(1), 11–15 (1987)
24. Krishnamoorthy, K., Mathew, T.: Comparison of approximation methods for computing tolerance factors for a multivariate normal population. Technometrics 41(3), 234–249 (1999)
25. Masson, M., Denoeux, T.: Inferring a possibility distribution from empirical data. Fuzzy Sets Syst. 157, 319–340 (2006)
26. Smets, P.: The transferable belief model and other interpretations of dempster-shafer's model, pp. 326–333 (1990)
27. Wald, A.: An extension of wilks' method for setting tolerance limits. Ann. of Math. Stat. 14(1), 45–55 (1943)
28. Wald, A., Wolfowitz, J.: Tolerance limits for a normal distribution. The Annals of Mathematical Statistics 17(2), 208–215 (1946)
29. Wallis, W.A.: Tolerance intervals for linear regression. In: Proc. Second Berkeley Symp. on Math. Statist. and Prob., pp. 43–51 (1951)
30. Wilks, S.S.: Determination of sample sizes for setting tolerance limits. Ann. of Math. Stat. 12(1), 91–96 (1941)
31. Zadeh, L.A.: Fuzzy sets as a basis for a theory of possibility. Fuzzy Set. Syst. 1(1), 3–28 (1978)

Harmonic Wavelets Based Identification
of Nonlinear and Time-Variant Systems

Ioannis A. Kougioumtzoglou[1] and Pol D. Spanos[2]

[1] Institute for Risk and Uncertainty and Centre for Engineering Sustainability, University of
Liverpool, Liverpool, L69 3GH, UK
kougioum@liverpool.ac.uk
[2] Department of Mechanical Engineering and Materials Science, Rice University, Houston,
Texas 77005, USA
spanos@rice.edu

Abstract. Structural systems subject to non-stationary excitations can often
exhibit time-varying nonlinear behavior. In such cases, a reliable identification approach is critical for successful damage detection and for designing an
effective structural health monitoring (SHM) framework. In this regard, an
identification approach for nonlinear time-variant systems based on the localization properties of the harmonic wavelet transform is developed herein.
The developed approach can be viewed as a generalization of the well established reverse MISO spectral identification approach to account for non-
stationary inputs and time-varying system parameters. Several linear and
nonlinear time-variant systems are used to demonstrate the reliability of the
approach. The approach is found to perform satisfactorily even in the case of
noise-corrupted data.

Keywords: nonlinear system, harmonic wavelet, system identification,
multiple-input-single-output model.

1 Introduction

Most structural systems are likely to exhibit nonlinear and time-varying behavior
when subjected to severe earthquake, wind and sea wave excitations, or to extreme
events due to climate change such as hurricanes, storms and floods (e.g.
Kougioumtzoglou and Spanos, 2009). The need to identify the damage in an early
stage and to ensure the safety and functionality of the structure has made health
monitoring an important research field (e.g. Farrar and Worden, 2007). It can be
readily seen that for the purpose of structural health monitoring (SHM) reliable
identification approaches are necessary to quantify the nonlinear time-varying behavior of the structures.

Regarding nonlinear system identification various interesting approaches have
been developed such as the ones based on the Volterra-Wiener (VW) representation
theory (e.g. Bedrosian and Rice, 1971; Schetzen, 1980). Nevertheless, the VW

E. Hüllermeier et al. (Eds.): SUM 2012, LNAI 7520, pp. 247–260, 2012.
© Springer-Verlag Berlin Heidelberg 2012

approach requires significant computational effort and appears unattractive for multi-degree-of-freedom (MDOF) systems.

An alternative approach utilizes a representation of the nonlinear restoring forces as a set of parallel linear sub-systems. As a result, the identification of a nonlinear system can be achieved by adopting a multiple-input/single-output (MISO) linear system analysis approach (e.g. Rice and Fitzpatrick, 1988; Bendat et al., 1992; Bendat et al., 1995; Bendat, 1998) and by utilizing measured stationary excitation-response data. Several marine/offshore system identification applications have been presented based on the aforementioned reverse MISO approach. For instance, Spanos and Lu (1995) addressed the nonlinearity induced by the structure-environment interaction in marine applications, whereas identification of MDOF moored structures was performed by Raman et al. (2005) and by Panneer Selvam et al. (2006). In Zeldin and Spanos (1998) the MISO approach was examined from a novel perspective and it was shown that it can be interpreted as a Gram-Schmidt kind of orthogonal decomposition. A more detailed presentation of nonlinear system identification techniques can be found in review papers such as the one by Kerschen et al. (2006).

On the other hand, wavelet-based identification approaches (e.g. Kijeswki and Kareem, 2003; Spanos and Failla, 2005) appear promising in detecting the evolutionary features of structures (e.g. degradation) and can result in effective SHM (e.g. Hou et al., 2006). Indicatively, early attempts on wavelet-based identification approaches include the work by Staszewski (1998) based on the ridges and skeletons of the wavelet transform. Further, Lamarque et al. (2000) introduced a wavelet-based formula similar to the logarithmic decrement formula to estimate damping. Ghanem and Romeo (2000) presented a wavelet-Galerkin approach for identification of linear time-variant (LTV) systems. The approach was also applied to nonlinear system identification (Ghanem and Romeo, 2001). In Chakraborty et al. (2006) the modal parameters of a linear MDOF system were identified using a modified Littlewood-Paley (LP) wavelet basis function.

In this paper, a generalization of the reverse MISO identification approach is developed based on the harmonic wavelet transform. To this aim, the nonlinear system is expressed as a combination of linear sub-systems in the wavelet domain. Further, time and frequency dependent generalized harmonic wavelet based frequency response functions (GHW-FRFs) are defined and a conditioning procedure is used to de-correlate the inputs of the equivalent MISO system. In this regard, the approach can address cases of nonlinear systems with time-varying parameters by utilizing non-stationary excitation-response measured data. It is noted that one significant advantage of the approach is that the non-stationary processes involved can be non-Gaussian in general with an arbitrary evolutionary power spectrum (EPS). Various examples of linear and nonlinear time-variant structural systems are used to demonstrate the accuracy of the approach. Cases of signals corrupted with noise are also included to further highlight the reliability and robustness of the approach.

2   Harmonic Wavelets

The family of generalized harmonic wavelets, proposed by Newland (1994), utilizes two parameters (m,n) for the definition of the bandwidth at each scale level. One of its main advantages relates to the fact that these two parameters, in essence, decouple the time-frequency resolution achieved at each scale from the value of the central frequency; this is not the case with other wavelet bases such as the Morlet.

Generalized harmonic wavelets have a band-limited, box-shaped frequency spectrum. A wavelet of (m, n) scale and (k) position in time attains a representation in the frequency domain of the form

$$\Psi^{G}_{(m,n),k}(\omega) = \begin{cases} \dfrac{1}{(n-m)\Delta\omega}e^{-i\frac{\omega k T_{o}}{n-m}}, & m\Delta\omega \leq \omega < n\Delta\omega \\ 0, & otherwise \end{cases} \tag{1}$$

where m, n an k are considered to be positive integers, and

$$\Delta\omega = \frac{2\pi}{T_{o}}, \tag{2}$$

where T_0 is the total duration of the signal under consideration. The inverse Fourier transform of Eq.(1) gives the time-domain representation of the wavelet which is equal to

$$\psi^{G}_{(m,n),k}(t) = \frac{e^{in\Delta\omega\left(t-\frac{kT_0}{n-m}\right)} - e^{im\Delta\omega\left(t-\frac{kT_0}{n-m}\right)}}{i(n-m)\Delta\omega\left(t-\dfrac{kT_0}{n-m}\right)}, \tag{3}$$

and is in general complex-valued, with magnitude (e.g. Spanos et al., 2005)

$$\left|\psi^{G}_{(m,n),k}(t)\right| = \frac{\sin\left(\pi(n-m)\left(\dfrac{t}{T_o} - \dfrac{k}{n-m}\right)\right)}{\pi(n-m)\left(\dfrac{t}{T_o} - \dfrac{k}{n-m}\right)}, \tag{4}$$

and phase

$$\phi^{G}_{(m,n),k}(t) = \pi(n+m)\left(\frac{t}{T_o} - \frac{k}{n-m}\right). \tag{5}$$

Newland (1994) showed that a collection of harmonic wavelets spanning adjacent non-overlapping intervals at different scales forms an orthogonal basis. The continuous generalized harmonic wavelet transform (GHWT) is defined as

$$W^G_{(m,n),k}[f] = \frac{n-m}{T_o} \int_{-\infty}^{\infty} f(t)\overline{\psi^G_{(m,n),k}(t)}dt,$$ (6)

and projects any finite energy signal $f(t)$ on this wavelet basis. In Eq.(6) the bar over a symbol denotes complex conjugation. Further, the orthogonality properties of such a basis allow for perfect reconstruction of the original signal $f(t)$ according to the equation (e.g. Newland, 1994; Spanos et al., 2005)

$$f(t) = 2\,\mathrm{Re}\left[\sum_{m,n}\sum_k \left(W^G_{(m,n),k}[f]\psi^G_{(m,n),k}(t)\right)\right],$$ (7)

where in Eq.(7) $f(t)$ is assumed to be a zero-mean signal. Considering, next, Parseval's theorem and the non-overlapping character of the different energy bands it was shown in Spanos and Kougioumtzoglou (2012) that the EPS can be estimated by the equation

$$S_{ff}(\omega,t) = S^{ff}_{(m,n),k} = \frac{E\left[\left|W^G_{(m,n),k}[f]\right|^2\right]}{(n-m)\Delta\omega}, \quad m\Delta\omega \le \omega < n\Delta\omega, \quad \frac{kT_o}{n-m} \le t < \frac{(k+1)T_o}{n-m}$$ (8)

In Eq.(8) the EPS of the process f is assumed to have a constant value in the associated time and frequency intervals. Similarly, the cross EPS of two random processes f and g can be estimated as (see also Spanos and Kougioumtzoglou, 2011)

$$S^{fg}_{(m,n),k} = \frac{E\left[W^G_{(m,n),k}[f]\overline{W^G_{(m,n),k}}[g]\right]}{(n-m)\Delta\omega}, \quad m\Delta\omega \le \omega < n\Delta\omega, \quad \frac{kT_o}{n-m} \le t < \frac{(k+1)T_o}{n-m}$$ (9)

In fact, in Spanos and Kougioumtzoglou (2012) Eq.(8) was used to obtain estimates not only for separable but non-separable in time and frequency EPS as well. Further, in Spanos et al. (2005) and in Huang and Chen (2009) EPS estimation applications were presented based on expressions similar to Eqs.(8) and (9).

3   Identification Approach

3.1   Harmonic Wavelet Based Input-Output Relationships

Consider a single-degree-of-freedom (SDOF) linear time-variant (LTV) system whose motion is governed by the differential equation.

$$\ddot{x} + C(t)\dot{x} + K(t)x = w(t) \tag{10}$$

where $C(t)$ represents the time-varying damping; $K(t)$ represents the time-varying stiffness; and $w(t)$ denotes a zero-mean non-stationary excitation. Applying the GHWT on both sides of Eq.(10) yields

$$W^G_{(m,n),k}[\ddot{x}] + W^G_{(m,n),k}[C(t)\dot{x}] + W^G_{(m,n),k}[K(t)x] = W^G_{(m,n),k}[w] \tag{11}$$

Further, assuming that the damping and stiffness elements are slowly-varying in time, and thus, approximately constant over the compact support of the harmonic wavelet in the time domain, Eq.(11) becomes

$$W^G_{(m,n),k}[\ddot{x}] + C_k W^G_{(m,n),k}[\dot{x}] + K_k W^G_{(m,n),k}[x] = W^G_{(m,n),k}[w] \tag{12}$$

where (C_k, K_k) represent the values of the damping and stiffness elements in the $k - th$ translation (time) interval. Focusing next on the GHWT of the velocity and applying integration by parts yields

$$W^G_{(m,n),k}[\dot{x}] = \frac{n-m}{T_o}\left[x(t)\overline{\psi^G_{(m,n),k}}(t)\right]_{-\infty}^{\infty} - \frac{n-m}{T_o}\int_{-\infty}^{\infty} x(t)\frac{d\overline{\psi^G_{(m,n),k}}(t)}{dt}dt \tag{13}$$

Taking into account the compact support of the harmonic wavelet in the time domain Eq.(13) becomes

$$W^G_{(m,n),k}[\dot{x}] = -\frac{n-m}{T_o}\int_{-\infty}^{\infty} x(t)\frac{d\overline{\psi^G_{(m,n),k}}(t)}{dt}dt \tag{14}$$

Further, considering Eq.(3) and noticing that

$$\int_{m\Delta\omega}^{n\Delta\omega} e^{i\omega\left(t-\frac{kT_0}{n-m}\right)}d\omega = \frac{e^{in\Delta\omega\left(t-\frac{kT_0}{n-m}\right)} - e^{im\Delta\omega\left(t-\frac{kT_0}{n-m}\right)}}{i(n-m)\Delta\omega\left(t-\frac{kT_0}{n-m}\right)}(n-m)\Delta\omega \tag{15}$$

yields

$$\int_{m\Delta\omega}^{n\Delta\omega} e^{i\omega\left(t-\frac{kT_0}{n-m}\right)}d\omega = \psi^G_{(m,n),k}(t)\left((n-m)\Delta\omega\right) \tag{16}$$

Assuming next that the frequency band is small enough Eq.(16) can be recast in the form

$$\psi^G_{(m,n),k}(t) = \exp\left(i\omega_{c,(m,n),k}\left(t - \frac{kT_0}{n-m}\right)\right),\qquad(17)$$

where

$$\omega_{c,(m,n),k} = \frac{(n+m)\Delta\omega}{2}.\qquad(18)$$

Combining Eqs.(14) and (17) and manipulating yields

$$W^G_{(m,n),k}[\dot{x}] = i\omega_{c,(m,n),k}\frac{n-m}{T_o}\int_{-\infty}^{\infty} x(t)\overline{\psi^G_{(m,n),k}(t)}dt,\qquad(19)$$

or, equivalently

$$W^G_{(m,n),k}[\dot{x}] = i\omega_{c,(m,n),k}W^G_{(m,n),k}[x].\qquad(20)$$

Applying a similar analysis for the GHWT of the acceleration yields

$$W^G_{(m,n),k}[\ddot{x}] = -\omega^2_{c,(m,n),k}W^G_{(m,n),k}[x].\qquad(21)$$

Taking into account Eqs.(20-21) Eq.(12) becomes

$$W^G_{(m,n),k}[x]\left(-\omega^2_{c,(m,n),k} + i\omega_{c,(m,n),k}C_k + K_k\right) = W^G_{(m,n),k}[w].\qquad(22)$$

Applying next complex conjugation to Eq.(22) yields

$$\overline{W^G_{(m,n),k}}[x]\left(-\omega^2_{c,(m,n),k} - i\omega_{c,(m,n),k}C_k + K_k\right) = \overline{W^G_{(m,n),k}}[w].\qquad(23)$$

Combining Eqs.(22-23) and taking expectations results in

$$E\left[\left|W^G_{(m,n),k}[x]\right|^2\right] = \frac{1}{\left(K_k - \omega^2_{c,(m,n),k}\right)^2 + \left(\omega_{c,(m,n),k}C_k\right)^2}E\left[\left|W^G_{(m,n),k}[w]\right|^2\right].\qquad(24)$$

Considering next Eq.(24), and taking into account Eq.(8) yields

$$S^{xx}_{(m,n),k} = \frac{1}{\left(K_k - \omega^2_{c,(m,n),k}\right)^2 + \left(\omega_{c,(m,n),k}C_k\right)^2}S^{ww}_{(m,n),k}.\qquad(25)$$

It can be readily seen that Eq.(25) resembles the celebrated spectral input-output relationship of the linear stationary random vibration theory (e.g. Roberts and Spanos, 2003). Eq.(25) can be viewed as an equivalent relationship in the harmonic wavelet domain. This leads to defining the GHW-FRF as

$$H^G_{(m,n),k} = \left(-\omega^2_{c,(m,n),k} + i\omega_{c,(m,n),k}C_k + K_k\right)^{-1}$$
(26)

Note that the GHW-FRF is defined in the associated time and frequency intervals, or, in other words, it is frequency and time dependent. Utilizing next Eq.(26), Eq.(25) can take the equivalent form

$$S^{xx}_{(m,n),k} = \left|H^G_{(m,n),k}\right|^2 S^{ww}_{(m,n),k}$$
(27)

Further, manipulating Eq.(22) and taking expectations yields

$$E\left[W^G_{(m,n),k}[x]\overline{W^G_{(m,n),k}}[w]\right] = H^G_{(m,n),k}E\left[\left|W^G_{(m,n),k}[w]\right|^2\right]$$
(28)

or, equivalently

$$S^{xw}_{(m,n),k} = H^G_{(m,n),k}S^{ww}_{(m,n),k}$$
(29)

where Eq.(9) has been considered.

To recapitulate, harmonic wavelet based auto- and cross-spectral input-output relationships have been derived and will be used in the ensuing analysis to develop the harmonic wavelet based reverse MISO identification approach.

3.2   Harmonic Wavelet Based Reverse MISO Identification Approach

Consider a nonlinear SDOF system whose motion is governed by the differential equation

$$\ddot{u} + C(t)\dot{u} + K(t)u + h[u,\dot{u}] = w(t)$$
(30)

where $h[.]$ is an arbitrary nonlinear function which depends on the response displacement and velocity. In the following, it is assumed that the nonlinear function $h[.]$ can be expressed as a superposition of zero-memory nonlinear transformations and linear sub-systems (e.g. Zeldin and Spanos, 1998). Specifically,

$$h[u,\dot{u}] = \sum_{l=1}^{M} A_l\left(\frac{d}{dt}\right)p_l(u)$$
(31)

where A represents polynomial functions; p represents zero-memory nonlinear transformations; and M accounts for the total number of base functions used in the representation of the nonlinear restoring force $h[.]$. The terms u and p associated with the unknown structural parameters are interpreted as the inputs x of the MISO system. Thus, the composed MISO system can be described by the equation

$$\sum_{l=1}^{s} A_l\left(\frac{d}{dt}\right) x_l(t) + n(t) = w(t),$$ (32)

where s denotes the total number of the input variables used in the equivalent MISO representation of the nonlinear structural system; and $n(t)$ accounts for possible extraneous noise. Note that in the developed MISO formulation the traditional input/output roles of the excitation/response quantities have been reversed. Applying next the GHWT on both sides of Eq.(32) yields

$$\sum_{l=1}^{s} A^G_{l,(m,n),k} W^G_{(m,n),k}[x_l] + W^G_{(m,n),k}[n] = W^G_{(m,n),k}[w].$$ (33)

It can be readily seen that the unknowns in the system identification problem consist of the different GHW-FRFs. In the ensuing analysis, stationary random data MISO analysis techniques for identification of linear FRFs (e.g. Bendat, 1998) are properly adapted and generalized for the case of GHW-FRFs.

Obviously, the number of multiple inputs (s) depends on the nature of the nonlinearity term $h[.]$). Without loss of generality, assume that the nonlinear system under consideration can be described by the reverse two-input/single output model of Fig.(1). Then, Eq.(33) becomes

$$A^G_{1,(m,n),k} W^G_{(m,n),k}[x_1] + A^G_{2,(m,n),k} W^G_{(m,n),k}[x_2] + W^G_{(m,n),k}[n] = W^G_{(m,n),k}[w],$$ (34)

where (A) is simply the reciprocal of the linear GHW-FRF of Eq.(26), namely

$$A^G_{1,(m,n),k} = \left(H^G_{(m,n),k}\right)^{-1}.$$ (35)

Note that in the MISO model $W^G_{(m,n),k}[x_1]$ and $W^G_{(m,n),k}[x_2]$ represent, in general, correlated inputs. In Bendat (1998) techniques are described to replace the original set of correlated inputs with a new set of uncorrelated inputs determined by using conditioned power spectra (PS) (see also Rice and Fitzpatrick, 1988; Spanos and Lu, 1995). The generalization of the therein described techniques to the case of time and frequency dependent harmonic wavelet based EPS is rather straightforward. Specifically, the original model can be replaced by an equivalent MISO model with new uncorrelated inputs $W^G_{(m,n),k}[z_1]$ and $W^G_{(m,n),k}[z_2]$ passing through new linear systems L_1 and L_2. Their relationship is given by the equations

$$A^G_{2,(m,n),k} = L^G_{2,(m,n),k} \tag{36}$$

and

$$A^G_{1,(m,n),k} = L^G_{1,(m,n),k} - A^G_{2,(m,n),k} \frac{S^{x_1 x_2}_{(m,n),k}}{S^{x_1 x_1}_{(m,n),k}} \tag{37}$$

where

$$L^G_{2,(m,n),k} = \frac{S^{x_2 w}_{(m,n),k} - \dfrac{S^{x_1 x_2}_{(m,n),k}}{S^{x_1 x_1}_{(m,n),k}} S^{x_2 x_1}_{(m,n),k}}{S^{x_2 x_2}_{(m,n),k} - \dfrac{\left| S^{x_1 x_2}_{(m,n),k} \right|^2}{S^{x_1 x_1}_{(m,n),k}}} \tag{38}$$

and

$$L^G_{1,(m,n),k} = \frac{S^{x_1 w}_{(m,n),k}}{S^{x_1 x_1}_{(m,n),k}} \tag{39}$$

4   Numerical Examples

In the numerical applications following, the non-separable spectrum of the form

$$S(\omega,t) = S_1 \left(\frac{\omega}{5\pi} \right)^2 e^{-0.15t} t^2 e^{-\left(\frac{\omega}{5\pi} \right)^2 t}, \quad t \geq 0, \quad -\infty < \omega < \infty \tag{40}$$

where $S_1 = 50$, is considered to produce the excitation realizations. This spectrum comprises some of the main characteristics of seismic shaking, such as decreasing of the dominant frequency with time (e.g. Liu, 1970; Spanos and Solomos, 1983). Sample paths compatible with Eq.(40) are generated using the concept of spectral representation of a stochastic process. In this regard, a representation of the non-stationary process takes the form (e.g. Spanos and Zeldin, 1998; Liang et al., 2007)

$$z_{non}(t) = \sum_{n=0}^{N-1} \sqrt{4 S_{z_{non}}(n\Delta\omega, t)\Delta\omega} \cos\left((n\Delta\omega)t + \phi_n\right) \tag{41}$$

where the independent random phases following a uniform distribution over the interval $[0, 2\pi]$. As far as the calculation of the wavelet coefficients is concerned, a computationally efficient algorithm, which takes advantage of the fast Fourier transform

(FFT) scheme, can be found in Newland (1997, 1999). Further, a standard 4th order Runge-Kutta numerical integration scheme is utilized for the solution of the governing nonlinear equations of motion. Also, the effect of noise is taken into consideration. Specifically, Gaussian white noise is added to simulate the measurement noise corresponding to signal-to-noise ratio equal to 40 dB; that is the standard deviation of the added white noise is equal to 10% of the standard deviation of the signal.

4.1   Duffing Oscillator with Time-Varying Parameters

A Duffing oscillator with smoothly time-varying parameters is considered next to demonstrate the efficiency of the approach to address nonlinear systems. The parameters of the system of Eq.(30) become

$$C(t) = 2 + 0.01t^2 , \tag{42}$$

$$K(t) = 100 - 0.1t^2 , \tag{43}$$

and

$$h[u, \dot{u}] = \varepsilon K(t) u^3 , \tag{44}$$

where the nonlinearity magnitude has a value equal to unity.

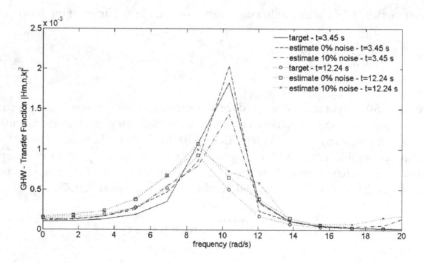

Fig. 1. Comparison between the target value of the squared modulus of the GHW-FRF and estimates derived from noiseless and noise corrupted data at different time instants

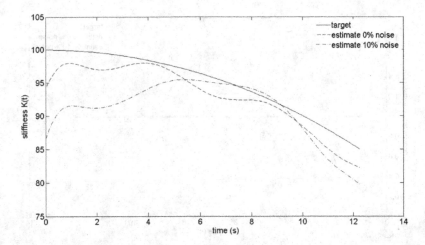

Fig. 2. Comparison between the target value of the stiffness and estimates derived from noise-less and noise corrupted data

The squared modulus of the estimated GHW-FRF is plotted in Fig.(1) and compared with the theoretical values for different time instants. Further, the time-varying parameters have been estimated by utilizing the real and imaginary parts of the GHW-FRF of Eq.(35). The stiffness, damping and nonlinearity magnitude estimates are plotted in Figs.(2), (3) and (4), respectively. The developed approach captures successfully the time-varying character of the parameters values even in the presence of noise.

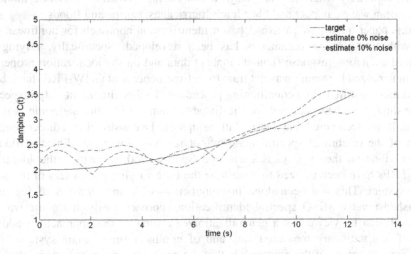

Fig. 3. Comparison between the target value of the damping and estimates derived from noise-less and noise corrupted data

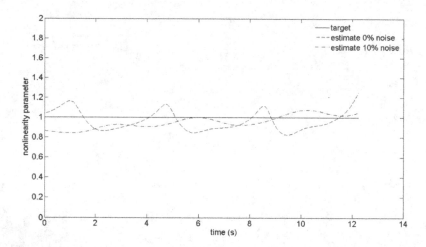

Fig. 4. Comparison between the target value of the nonlinearity magnitude and estimates derived from noiseless and noise corrupted data

5   Concluding Remarks

Structural systems subject to natural or man-made hazards, and to extreme events due to climate change, most often exhibit a time-varying and nonlinear behavior. Thus, reliable identification approaches are needed to identify and quantify efficiently the time-dependent values of the nonlinear system parameters. The importance of such approaches is paramount for further development of health monitoring procedures of systems, especially when they are subjected to excitations such as seismic motions, winds, ocean waves, impact and blast loads, hurricanes, storms and floods.

In this paper a harmonic wavelet based identification approach for nonlinear systems with time-varying parameters has been developed. Specifically, relying on measured excitation-response (non-stationary) data and on the localization properties of the generalized harmonic wavelet transform time-dependent GHW-FRFs have been defined and identified via a conditioning procedure. To this aim, input-output spectral relationships have been derived for the linear system case in the harmonic wavelet domain. It has been noted that these relationships can be construed as a direct generalization of the celebrated spectral input-output relationships of the linear stationary random vibration theory (e.g. Roberts and Spanos, 2003). Further, the identified GHW-FRFs have been utilized to determine the time-varying parameters of the associated system. This has been done in conjunction with an extension of the well-established reverse MISO spectral identification approach to the harmonic wavelet domain. It can be viewed as a generalization/extension of the approach to address cases of non-stationary measured data and of nonlinear time-variant systems. One significant advantage of the approach is that the non-stationary processes involved in the identification approach can be non-Gaussian, in general, with arbitrary evolutionary power spectra (EPS). The reliability of the identification approach, a prerequisite for effective SHM frameworks, has been demonstrated for a number of linear and nonlinear time-variant structural systems including cases of noise-corrupted data by several numerical simulations.

References

Bedrosian, E., Rice, S.O.: The output properties of Volterra systems (Nonlinear systems with memory) driven by harmonic and Gaussian Inputs. Proceedings of the IEEE 59, 1688–1707 (1971)

Bendat, J.S., Palo, P.A., Coppolini, R.N.: A general identification technique for nonlinear differential equations of motion. Probabilistic Engineering Mechanics 7, 43–61 (1992)

Bendat, J.S., Palo, P.A., Coppolini, R.N.: Identification of physical parameters with memory in nonlinear systems. International Journal of Non-Linear Mechanics 30, 841–860 (1995)

Bendat, J.S.: Nonlinear systems techniques and applications. John Wiley and Sons (1998)

Chakraborty, A., Basu, B., Mitra, M.: Identification of modal parameters of a mdof system by modified L-P wavelet packets. Journal of Sound and Vibration 295, 827–837 (2006)

Farrar, C.R., Worden, K.: An introduction to structural health monitoring. Phil. Trans. R. Soc. A 365, 303–315 (2007)

Ghanem, R., Romeo, F.: A wavelet-based approach for the identification of linear time-varying dynamical systems. Journal of Sound and Vibration 234, 555–576 (2000)

Ghanem, R., Romeo, F.: A wavelet-based approach for model and parameter identification of nonlinear systems. International Journal of Non-Linear Mechanics 36, 835–859 (2001)

Hou, Z., Hera, A., Shinde, A.: Wavelet-based structural health monitoring of earthquake excited structures. Computer-Aided Civil and Infrastructure Engineering 21, 268–279 (2006)

Huang, G., Chen, X.: Wavelets-based estimation of multivariate evolutionary spectra and its application to nonstationary downburst winds. Engineering Structures 31, 976–989 (2009)

Kerschen, G., Worden, K., Vakakis, A., Golinval, J.-C.: Past, present and future of nonlinear system identification in structural dynamics. Mechanical Systems and Signal Processing 20, 505–592 (2006)

Kijewski, T., Kareem, A.: Wavelet transforms for system identification in civil engineering. Computer-Aided Civil and Infrastructure Engineering 18, 339–355 (2003)

Kougioumtzoglou, I.A., Spanos, P.D.: An approximate approach for nonlinear system response determination under evolutionary excitation. Current Science, Indian Academy of Sciences 97, 1203–1211 (2009)

Lamarque, C.-H., Pernot, S., Cuer, A.: Damping identification in multi-degree-of-freedom systems via a wavelet-logarithmic decrement - Part 1: Theory. Journal of Sound and Vibration 235, 361–374 (2000)

Liang, J., Chaudhuri, S.R., Shinozuka, M.: Simulation of non-stationary stochastic processes by spectral representation. Journal of Engineering Mechanics 133, 616–627 (2007)

Liu, S.C.: Evolutionary power spectral density of strong-motion earthquakes. Bulletin of the Seismological Society of America 60, 891–900 (1970)

Newland, D.E.: Harmonic and musical wavelets. Proceedings of the Royal Society London A 444, 605–620 (1994)

Newland, D.E.: Practical signal analysis: Do wavelets make any difference? In: Proceedings of 16th ASME Biennial Conference on Vibration and Noise, Sacramento (1997)

Newland, D.E.: Ridge and phase identification in the frequency analysis of transient signals by harmonic wavelets. Journal of Vibration and Acoustics 121, 149–155 (1999)

Paneer Selvam, R., Bhattacharyya, S.K.: System identification of a coupled two DOF moored floating body in random ocean waves. Journal of Offshore Mechanics and Arctic Engineering 128, 191–202 (2006)

Raman, S., Yim, S.C.S., Palo, P.A.: Nonlinear model for sub- and super-harmonic motions of a MDOF moored structure, Part 1 – System identification. Journal of Offshore Mechanics and Arctic Engineering 127, 283–290 (2005)

Rice, H.J., Fitzpatrick, J.A.: A generalized technique for spectral analysis of nonlinear systems. Mechanical Systems and Signal Processing 2, 195–207 (1988)

Roberts, J.B., Spanos, P.D.: Random Vibration and Statistical Linearization. Dover Publications, New York (2003)

Schetzen, M.: The Volterra and Wiener theories of nonlinear systems. John Wiley and Sons (1980)

Spanos, P.D., Solomos, G.P.: Markov approximation to transient vibration. Journal of Engineering Mechanics 109, 1134–1150 (1983)

Spanos, P.D., Lu, R.: Nonlinear system identification in offshore structural reliability. Journal of Offshore Mechanics and Arctic Engineering 117, 171–177 (1995)

Spanos, P.D., Zeldin, B.A.: Monte Carlo treatment of random fields: A broad perspective. Applied Mechanics Reviews 51, 219–237 (1998)

Spanos, P.D., Tezcan, J., Tratskas, P.: Stochastic processes evolutionary spectrum estimation via harmonic wavelets. Computer Methods in Applied Mechanics and Engineering 194, 1367–1383 (2005)

Spanos, P.D., Failla, G.: Wavelets: Theoretical concepts and vibrations related applications. The Shock and Vibration Digest 37, 359–375 (2005)

Spanos, P.D., Kougioumtzoglou, I.A.: Harmonic wavelet-based statistical linearization of the Bouc-Wen hysteretic model. In: Faber, et al. (eds.) Proceedings of the 11th International Conference on Applications of Statistics and Probability in Civil Engineering (ICASP 2011), pp. 2649–2656. Taylor and Francis Group (2011) ISBN: 978-0-415-66986-3

Spanos, P.D., Kougioumtzoglou, I.A.: Harmonic wavelets based statistical linearization for response evolutionary power spectrum determination. Probabilistic Engineering Mechanics 27, 57–68 (2012)

Stasjewski, W.J.: Identification of non-linear systems using multi-scale ridges and skeletons of the wavelet transform. Journal of Sound and Vibration 214, 639–658 (1998)

Zeldin, B.A., Spanos, P.D.: Spectral identification of nonlinear structural systems. Journal of Engineering Mechanics 124, 728–733 (1998)

An Upscaling Approach for Uncertainty Quantification of Stochastic Contaminant Transport through Fractured Media

Edoardo Patelli

University of Liverpool, Liverpool L693GQ, UK
edoardo.patelli@liverpool.ac.uk
http://www.liv.ac.uk/risk-and-uncertainty/

Abstract. The assessment of waste repositories are based on predictive models that are able to forecast the migration of contaminants within groundwater. The heterogeneity and stochasticity of the media in which the dispersion phenomenon takes place, renders classical analytical-numerical approaches scarcely adequate in practice. Furthermore, these approaches are computationally intensive and limited to small scale applications.

In this paper, the contaminant transport is described by the linear Boltzmann integral transport equation and solved by means of a Monte Carlo particle tracking based approach. The governing parameters of the stochastic model are calibration on a small-scale analysis based on the Discrete Fracture Network simulation. The proposed approach is very flexible and computationally efficient. It can be adopted as an upscaling procedure and to predict the migration of contaminants through fractured media at scale of practical interest.

Keywords: Monte Carlo Simulation, Upscaling, Uncertainty Quantification, Contaminant Transport, Boltzmann Transport Equation.

1 Introduction

The performance of a deep repository for nuclear waste relies on its capacity to isolate the radionuclides from the biosphere for long time scale [9]. This is achieved through the use of multi-barrier systems as the waste package, the engineered barriers and the natural barriers formed by the rock where the repository is built. Fractures are the principal pathways for the groundwater-driven dispersion of radioactive contaminants which may escape from subsurface waste repositories. Hence, it is of utmost importance to study the phenomena of transport of radionuclides in the fractured natural rock matrix at field scale.

The heterogeneity and stochasticity of the media in which the dispersion phenomenon takes place, renders classical analytical-numerical approaches scarcely adequate in practice. In fractured media, fluid flows mainly through fractures so that considering the host matrix as a homogeneous continuum is an unacceptable modelling simplification. In fact, it has been shown (see e.g. [5,6,16]) that the use

E. Hüllermeier et al. (Eds.): SUM 2012, LNAI 7520, pp. 261–272, 2012.

of effective flow parameters, representing the average properties of a set of independent transport pathways can not capture the complexity of the phenomena involved during the transport, which may lead to unjustified conservatism.

Approaches to explicitly incorporate the geometry and properties of discrete fracture network have been developed since the mid-1990s as a framework for characterization of contaminated sites on fractured rock (see e.g. [1,2]). However, a potential practical difficulty lies in the determination of the values of some model parameters. The task of relating analytically the fractures geometric properties (i.e. distributions of the fracture lengths, orientations, apertures etc.) with the velocity field in the fracture network is impracticable in realistic settings. Furthermore, the characterization of geological media on an engineering scale is not feasible and the estimated properties are associated with a large spatial variability and uncertainty.

Therefore, stochastic models are required in order to capture the effect of the uncertainties associated with the parameters and characteristics of the media. Particle tracking in stochastically generated networks of discrete fractures represents an alternative to the conventional advection-dispersion description of transport phenomena [2,4,15,16]. Numerous discrete fracture network models have been developed and tested (see e.g. [14,19]). However, the particle tracking simulation in discrete fracture networks becomes computationally intensive even for small scales [2,16]. To overcome these difficulties, in the last years a category of approaches, referred to as "hybrid methods" [1], has emerged for modelling both the water flow and the contaminant transport. In a hybrid method, a detailed simulation for a sub-domain of the field-scale domain is used to deduce parameters for a simpler and less computationally demanding model such as e.g. continuum model.

In this paper a continuum stochastic approach, based on an analogy with neutron transport is adopted to treat the transport of (radioactive) contaminants through fractured media. The resulting stochastic model [8] is evaluated by means of the Monte Carlo simulation technique whose flexibility allows considering explicitly the different processes that govern the transport as well as different boundary conditions. The model parameters are derived directly from small-scale geometric configuration of the fractures and from the dynamics of the driving fluid system, as quantitatively described by means of the Discrete Fracture Network simulation (see e.g. [19]). This hybrid approach provides a computationally efficient approach for up-scaling the uncertainty quantification for the diffusion of contaminant in fractured media on the scale of interest needed for practical applications.

The paper is organized as follows. In Section 2 the stochastic model adopted for the simulation of contaminant transport through fractured media is introduced. In Section 3 a strategy to estimate the model parameters of the stochastic model is presented. In Section 4 the proposed approach is applied for upscaling the uncertainty quantification of the radionuclide transport in fractured media. Finally, some final remarks are listed in Section 5.

2   The Stochastic Model

The transport of radionuclides through fractured media can be modelled as a series of straight movements with sudden changes of direction and speed. The straight movements correspond to the transport in a fracture, and between fracture intersections, where the water flow is constant. At fracture intersection the contaminant can enter in a new fracture characterised by different orientation, aperture and water flow. Consequently, the contaminant is changing its direction and speed.

An analogy can be made with the transport of neutrons in a non-multiplying media. Hence, the linear Boltzmann transport equation that is widely used for neutron transport in reactor design and radiation shielding calculations can be applied to describe the transport of contaminant through fractured media [8,17]. By doing so, the linear paths in the fractures can be treated as pseudo-mean free paths and the variations in direction and speed at the intersections can be treated as "pseudo-scattering" events.

Let $C(\mathbf{r}, \boldsymbol{\Omega}, v, t) d\mathbf{r} d\boldsymbol{\Omega} dv$ be the mean density of radionuclides (contaminants) at position $\mathbf{r}$, in volume element $d\mathbf{r}$, moving in the solid angle $d\boldsymbol{\Omega}$ centered about the direction $\boldsymbol{\Omega}$, with velocity between v and $v + dv$, at time t. In the case of a single radioactive nuclide or contaminant, the quantity C satisfies the following transport equation [17]:

$$\left[R_d \frac{\partial}{\partial t} + v\boldsymbol{\Omega}.\nabla + v\Sigma(\mathbf{r}, \boldsymbol{\Omega}) + \lambda R_d \right] C(\mathbf{r}, \boldsymbol{\Omega}, v, t) = \tag{1}$$
$$\int d\boldsymbol{\Omega}' \int dv' v' \Sigma(\mathbf{r}, \boldsymbol{\Omega}') g(\mathbf{r}; v' \to v; \boldsymbol{\Omega}' \to \boldsymbol{\Omega}) C(\mathbf{r}, \boldsymbol{\Omega}', v', t) + S(\mathbf{r}, \boldsymbol{\Omega}, v, t)$$

where R_d is the retardation factor, Σ is the probability density of the branching points (fracture intersections) so that $1/\Sigma$ is the pseudo-mean free path, $g(\cdot)$ is the re-distribution in speed and direction of the particles at the intersection, λ is the decay constant of the radionuclide and S is an independent source of radionuclides. The distributions $\Sigma(\cdot)$, $g(\cdot)$ and $S(\cdot)$ in Eq. (1) encode information about the geometry of the fracture network and the hydrodynamics of the water flow. In particular, the transfer function $g(\cdot)$ describes probabilistically the direction and speed of motion of the contaminant after a fracture intersection. The Boltzmann equation (Eq. 1) is an equation that describes the time evolution of the distribution density function, $C(\cdot)dv$, in the phase space.

Unfortunately, the parameters required by the proposed model are difficult to computed. For example, in Ref. [18] an attempt to derive the parameters appearing in the Boltzmann equation analytically from a purely geometric description of the fractures networks has been proposed. However, such kind of approach have very limited applicability. In fact, parameters appearing in the Boltzmann equation depend not only on the network geometry and configuration but also on the fluid dynamics in the interconnected network. Hence, in practical applications these parameters can be only characterised probabilistically.

2.1   Monte Carlo Solution of the Transport Equation

Eq. (1) can be solved analytically only under strong simplifying assumptions. A powerful alternative commonly used in neutron transport problems, is based on Monte Carlo simulation. In this respect, a Monte Carlo particle tracking simulation method has been proposed to simulate the transport of radionuclides and contaminants along fractured media. The proposed approach has been already verified against analytical solution [8]. It is very flexible and applicable to one-, two- and three-dimensional problems. It allows accounting for the different physical processes occurring during the transport and for the inhomogeneities in time and space by properly varying the parameter values. Inhomogeneities in time is a important aspect in analysis, such as the performance assessments of waste repositories, in which the properties of the medium are likely to change over long period of time required by regulation (millions of years).

The Monte Carlo approach consists in simulating a large number of independent travel "histories". Each Monte Carlo history is obtained by simulating the fate of the contaminant through the fractured medium, i.e. a particle is generated from a suitable source and then followed up through the fractured medium until the end of the time interval of concern. In few words, the simulated particles move within fractures dragged by the water flow. If not adsorbed by the fracture wall or transformed by radioactive decay, they arrive at the next intersection points where they perform "collisions" and entering new fractures with different geometric properties and flow features, i.e. direction and speed. Each Monte Carlo history represents one possible function of the ensemble generated by the stochastic process, made up of what happens to the contaminant during the time interval of interest, called mission time.

In the following for simplicity and without loss of generality, the approach is briefly summarized referring to the case of only one kind of contaminant particle, no radioactive decay and no exchange processes with the fractured host matrix. It is important to mention that the proposed approach allows to simulate realistic cases considering radioactive decays (see e.g. [12]), exchange linear non-linear processes with the host matrix (e.g. [7]).

Let k represent the state of the particle (i.e. its position r, velocity v, flight directions Ω, characteristic of the fracture, chemical state, etc...) and $\psi(k,t)$ the incoming transition probability density function that defines the probability density that the particle makes a transition at time between t and $t + dt$ and as a result of the transition it enters state k. The incoming transition probability density function $\phi(k,t)$ can be expanded as a von Neumann series of products of the transport kernels, K, calculating in correspondence of the points of transition in a random walk in the phase space:

$$\psi(k,t) = \sum_{n=0}^{\inf} \psi^n(k,t) = \psi^0(k,t) + \sum_{k'} \int_{t^*}^{t} \psi(k',t')K(k,t|k',t')dt' \qquad (2)$$

It can be demonstrated that $\phi(k,t)$ is a solution of the linear Boltzmann integral (see e.g. [11]).

Estimations of $\psi(k,t)$ can be obtained by means of Monte Carlo method where a large number of random walk histories are simulated. The average of the random walks allows to estimate the transition density which is the unknown in the Boltzmann equation.

Starting from a generic particle identified by the representative point $P' \equiv (k,t)$, in the phase space state Ω, the next representative point (i.e. the collision point) of the particle, $P \equiv (k,t)$, is identified by sampling from the conditional probabilistic density function (pdf) $K(k,t|k',t')$ called "Transport Kernel".

The "Transport Kernel" can be defined as the product of the "Free Flight Kernel", T, and the "Collision Kernel", C:

$$K(t,k|k',t') = T(t|k',t') \cdot C(k|k',t) \tag{3}$$

Eq. (3) is the conditional probability density function that the particle undergoes a collision, i.e. meets an intersection branching point, at time t given that the previous transition has occurred at time t' and that the particle entered state k' as a result of the transition.

In the model considered, the Free Flight Kernel is written as:

$$T(t|k',t') = \Lambda_{k'} \exp^{-\Lambda_{k'}(t-t')} \tag{4}$$

where $\Lambda_{k'} = v_{k'} \Sigma_{k'}$ and $v_{k'}$ represents the contaminant (and water flow) velocity in the fracture. The reciprocal of $\Sigma_{k'}$ can be seen as the mean distance between fracture intersections while the particle is in state k'. Hence, $1/\Lambda_{k'}$ represents the mean time between particle transitions.

The Collision Kernel, $C(k|k',t)$ defines the conditional probability for the particle in the state k' to enter the new state k by effect of the transition at time t and can be written as:

$$C(k|k',t) = \frac{\Lambda_{kk'}}{\Lambda_{k'}} \tag{5}$$

where $\Lambda_{kk'}$ represents a generic transition rate from the state k' to the state k. The transition rates describe the radioactive decays, the physico-chemical transformations (e.g. the adsorption on the matrix) or a change in speed and direction. The latter kind of transition represents a scattering function called "Redistribution Kernel", R:

$$C(k|k',t) = R(t; \Omega' \to \Omega; v' \to v; b' \to b) \tag{6}$$

which gives the conditional probability that the particle by effect of scattering undergone at time t while travelling with velocity v' along the fracture with orientation Ω' and aperture b', enters in a new fracture in direction Ω, with aperture b and flow velocity v.

In the following Section an approach to calculate the transport Kernels from a (detailed) small-scale simulation is presented.

3   Parameter Identification

The parameters appearing in the Boltzmann equation that govern the contaminant transport through fractured media are in general difficult to computed. They depend not only on the network geometry (e.g. the geometry of the fractures, aperture distribution) but also on the fluid dynamics in the interconnected network.

In realistic situations, the transport kernel can not be determined directly but it can be obtained from statistical about the fractured media. More specifically, the Free Flight Kernel, given by Eq. (5), is proportional to the mean distance between fracture intersections and encodes geometric and hydrodynamic properties of the fracture network. It is completely determined by the total transition rate $\Lambda_k = v(\cdot)\Sigma(\cdot)$ where v represents the velocity of the water in the channel and Σ the channel length between fracture intersections. These parameters can depends on other parameters such as the channel aperture b, orientation θ, position, x, etc.

The Redistribution Kernel, $R(v'_g \to v_g, \theta'_g \to \theta_g, b'_g \to b_g)$, encodes information on how the incoming mass of contaminants is distributed among the outgoing fluxes. Different models can be used to model the contaminant redistribution. For instance, there are two different dual approaches for describing mass mixing at the intersection point: "complete mixing", "static mixing" [3]. In the complete mixing model the contaminant concentration is constant in the outgoing fluxes (i.e. the probability to take a channel is proportional to the flux in that channel). In the static mixing the contaminant has the same probability to be redistributed in the outgoing fluxes (i.e. it is independent of the flux in the outgoing channels). The different hypotheses that describe the redistribution of the solute contaminant at the intersection point has a strong influence on the Redistribution Kernel.

The next Section shows how the transport kernels can be estimated from a small-scale simulation based on discrete fractured network.

3.1   The Discrete Fracture Network Approach

The fundamental motivation of Discrete Fracture Network modelling is the recognition that at every scale, groundwater transport in fractured and carbonate rocks tend to be dominated by a limited number of discrete pathways formed by fractures, karts, and other discrete features. However, despite the importance of the applications it is not feasible to map accurately at engineering scales of interest fractures and fracture networks in rock masses not least because accurate field measurement of a single discontinuity is difficult and measurement of all discontinuities is impossible. In practical applications, the available information comes from surveys of analogues, such as rock outcrops, or from direct or indirect observations of the rock mass such as drill cores, borehole imaging, geophysical surveys or seismic monitoring during fracture stimulation. Such information are then used to construct a stochastic model of the fractured network.

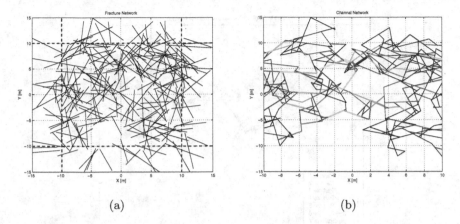

(a)   (b)

Fig. 1. Example of a Discrete Fracture Network simulation over a domain of 20×20 meters. (a) Fracture Network; (b) Channel Network where the line size is proportional to the flux through the fracture.

In the following, we briefly describe the algorithm underlying the Discrete Fracture Network approach. Starting from the statistical distribution of fracture properties, such as length, position, orientation and aperture (or transmissivity), a random realization of fractures is generated by Monte Carlo sampling as shown in Figure 1 (a). This is called "fracture network". The intersections between the sampled fractures are identified as nodes of the network. Then, the fractures connected to both upstream and downstream boundaries are identified forming the so-called "channel network". Hence, isolated fractures and fractures with dead ends are eliminated from the network.

Given the network boundary conditions such as the hydraulic gradient, the hydraulic heads at each connected nodes of the channel network are calculated solving a system of algebraic linear equations obtained from the mass conservation principle applied at the fracture intersections (i.e. Kirchoff's law). The water fluxes in the channels are then computed based on laminar flow law. Hence, the flow rate and fluid velocity in each fracture channel between two nodes of the network are directly computed from the hydraulic heads (Figure 1 (b)).

Finally, a particle tracking algorithm is applied to simulate the contaminant transport through the channel network (Figure 2). Table 1 summarizes the main parameters used in the Discrete Fracture Network simulations. The results of the particle tracking can be used to estimate the probabilistic transition kernels collecting in appropriate counters the quantities of interest.

For instance, it is possible to obtain the velocity distribution of the water flow in the discrete fracture networks or the fracture orientation (see Figure 3). The particle redistribution kernel can be computed recording the number of particles on status k' (i.e. with velocity v' in a channel with aperture b' and orientation θ') entering in a status k (i.e. in a channel with aperture b, orientation θ and velocity v).

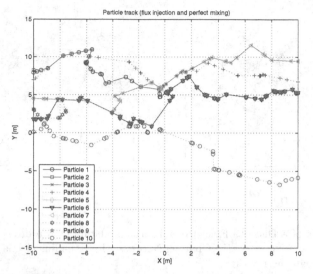

Fig. 2. Particle tracking in the Channel Network of Figure (1) with flux injection boundary condition and complete mixing at each nodes of the network

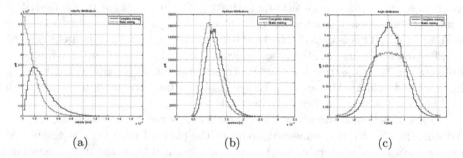

Fig. 3. Probability density functions for (a) the flow velocities, (b), the fracture aperture and (c) the fracture orientation estimate from the Discrete Fracture Network for the complete and static mixing functions

4   Numerical Results

A typical question addressed in long-term geologic repository design is how much of the radionuclides which enter the fractured rock under accidental scenarios is eventually discharged. Particle tracking through stochastically generated network of discrete fractures is largely used to describe the transport in fractured rock. However, discrete fracture network simulations are computationally very intensive and limited to small scales. A hybrid method is here used to upscale the uncertainty quantification on scale of practical interest.

Table 1. Main parameters used in simulation of transport of radionuclides in a fractured media as shown in Figures (1-5)

Discrete Fractured Networks	
Parameter (DFN)	Value
Fracture number	242 [-]
Fracture position	uniform [-10, 10] [m]
Fracture length	lognormal (5, 0.25) [m]
Fracture orientation	uniform [0, 2π) [rad]
Fracture aperture	lognormal (1e-4,0.25) [m]
Domain	2D (parameter estimation): 20x20 [m]
Domain	2D (verification): 70x70 [m]
X Hydraulic gradient	5 [-]
Y Hydraulic gradient	0 [-]
number of realizations	5000 [-]
number of particle tracking	50 [-]
Boundary condition	flux injection
Mixing mode	complete mixing
Monte Carlo - Boltzmann Transport Equation	
Monte Carlo samples	10^5 [-]
Mission time	10^9 [s]

A full Discrete Fracture Network analysis of a subdomain of 20×20 meters is used to estimate the transport kernel for the Monte Carlo random walk simulation. Then, a Monte Carlo simulation is performed to estimated the advection time required by the contaminant to migrate through a 2D fractured media of 70×70 meters. For a verification purpose the particle tracking analysis on Discrete Fracture Network on the same scale has been performed. Table 1 summaries the main parameters used in the Discrete Fracture Network simulations and in the Monte Carlo random walk simulation.

The comparison of the advection time τ required by the radionuclide (contaminant) to travel through a fractured medium of size $L = 70 \times 70$ is reported in Figures 4-5. Figure 4 shows the probability distribution function of τ, while Figure 5 shows the cumulative and the complementary cumulative distribution of τ, respectively.

The results of Monte Carlo random walk simulation have been verified by means of a comparison with the results obtained from particle tracking through stochastically generated networks of discrete fracture in the simplified case of no radioactive decay and no exchange processes with the fractured host matrix. The agreement between the results are very satisfactory.

In is important to notice that the particle tracking simulation through discrete fracture networks is limited to small system sizes due to high computational cost involved in re-solving the governing system of equations (i.e. the identification of the connected fractures and the calculation of the hydraulic heads at each node of the network). On the other hand the computational cost of the proposed approach is independent on the number of fractures present of the media. In fact,

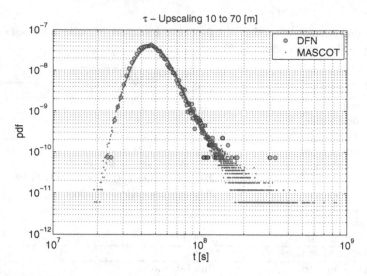

Fig. 4. Probability density functions (PDF) of the advection time required by the contaminant to travel through a fractured medium of 70x70 [m]

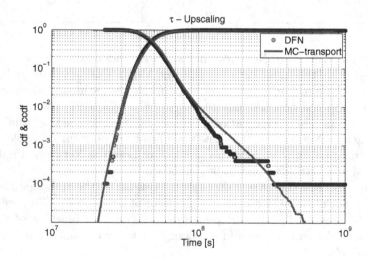

Fig. 5. Cumulative density functions (CDF) and Complementary Cumulative density functions (CCDF) of the advection time required by the contaminant to travel through a fractured medium of 70x70 [m]

the computational cost is proportional to the total number of iterations that the particles undergo during the transport. Furthermore, the computational time of the Monte Carlo random walk can be reduced by resorting to the so-called "forced transport" method that consist in altering the transition probabilities of particles and forcing their movement through the control session during the mission time.

5 Conclusions

The problem of adequately confining the radioactive wastes produced in industrial applications is of paramount importance for the future exploitation of the advantages of ionizing radiation and radioisotopes, in both the energy and non-energy related fields. The goal of a repository for nuclear waste is to isolate the radionuclides from the biosphere for the long time scales required by the radionuclides to decay to levels of negligible radiological significance. Fractures are the principal pathways for the groundwater-driven dispersion of radioactive contaminants which may escape from subsurface waste repositories. It is of utmost importance to study the phenomena of transport of radionuclides in the fractured natural rock matrix at field scale.

In this study, a stochastic approach derived from the transport of neutrons in non-multiplying media and governed by the linear Boltzmann transport equation has been adopted to simulate the transport of radionuclides through fractured media. The corresponding model is evaluated by means of Monte Carlo random walk simulation, where a large number of particles are followed in their travel through the fractures, using appropriate probability distribution functions to characterize their transport processes. Each of this particle history simulate a random walk representative of the life of the contaminant. The main advantage of the proposed model is its flexible structure which allows one to consider multidimensional geometries and to describe a wide range of phenomena, by accounting for the individual interactions which each particle may undergo. Thus, complex retention processes, such as kinetically controlled sorption, radioactive decay and decay chains, can be included in a straightforward manner at the spatial scale of practical interest.

It has been shown that the probability distribution functions that govern the transport of contaminants (i.e. the transport kernels) can be obtained from a discrete fracture network simulation performed on a modest spatial scale. Moreover, the computation requirements for the proposed approach are a small fraction of those required for classical approach based on particle tracking through a statistically generated network of fractures [1].

Furthermore, the proposed framework is very general and can be adopted in different fields as well. As examples, the Boltzmann transport equation, and its evaluation by means of Monte Carlo approach, has been already adopted to study the availability and reliability of engineering systems (see e.g. [10]) and in the design of advanced field effect transistors [13].

References

1. Benke, R., Painter, S.: Modelling conservatve tracer transport in fracture networks with a hybrid approach based on the Boltzmann transport equation. Water Resources Research 39(11) (2003)
2. Berkowitz, B.: Characterizing flow and transport in fractured geological media: A review (2002)

3. Naumann, C., Berkowitz, B., Smith, L.: Mass transfer at fracture intersections: An evaluation of mixing models. Water Resources Research 30(6), 1765–1773 (1994)
4. Bodin, J., Porel, G., Delay, F.: Simulation of solute transport in discrete fracture networks using the time domain random walk method (2003)
5. Cvetkovic, V., Dagan, G.: Transport of Kinetically Sorbing Solute by Steady Random Velocity in Hererogeneus Pouros Formations. Journal of Fluid Mechanics 265, 189–215 (1994)
6. Dverstorp, B., Mendes, B., Pereira, A., Sundstrsm, B.: Data reduction for radionuclide transport codes used in performance assessments: An example of simplification process. Materials Research Society 506, 797–805 (1997)
7. Giacobbo, F., Patelli, E.: Monte carlo simulation of nonlinear reactive contaminant transport in unsaturated porous media. Annals of Nuclear Energy 34(1-2), 51–63 (2007)
8. Giacobbo, F., Patelli, E.: Monte carlo simulation of radionuclides transport through fractured media. Annals of Nuclear Energy 35(9), 1732–1740 (2008)
9. Health and Safety Executive. Safety assessment principles for nuclear facilities. Technical Report 2006 Edition, Revision 1, Redgrave Court Bootle Merseyside L20 7HS (2006)
10. Marseguerra, M., Patelli, E., Zio, E.: Groundwater contaminant transport in presence of colloids i. a stochastic nonlinear model and parameter identification. Annals of Nuclear Energy 28(8), 777–803 (2001)
11. Marseguerra, M., Zio, E.: Basics of the Monte Carlo Method with Application to System Reliability. LiLoLe Publishing, Hagen (2002) ISBN: 3-934447-06-6
12. Marseguerra, M., Zio, E., Patelli, E., Giacobbo, F., Risoluti, P., Ventura, G., Mingrone, G.: Modeling the effects of the engineered barriers of a radioactive waste repository by monte carlo simulation. Annals of Nuclear Energy 30(4), 473–496 (2003)
13. Meinerzhagen, B., Pham, A.T., Hong, S.-M., Jungemann, C.: Solving boltzmann transport equation without monte-carlo algorithms - new methods for industrial tcad applications. In: 2010 International Conference on Simulation of Semiconductor Processes and Devices (SISPAD), pp. 293–296 (September 2010)
14. Mo, H., Bai, M., Lin, D., Roegiers, J.-C.: Study of flow and transport in fracture network using percolation theory (1998)
15. Moreno, L., Neretnieks, I.: Fluid flow and solute transport in a network of channels. Journal of Contaminant Hydrology 14(3-4), 163–192 (1993)
16. Sahimi: Flow and Transport in Porous Media and Fractured Rock: from classical methds to modern approaches. VCH Verlagsgesellschaft mbH (1995)
17. Williams, M.M.R.: A new model for Desribing the Transport of Radionuclide through Fracturated Rock. Ann. Nucl. Energy 19, 791 (1992)
18. Williams, M.M.R.: Radionuclide transport in fractured rock a new model: application and discussion. Annals of Nuclear Energy 20(4), 279–297 (1993)
19. Xu, C., Dowd, P.: A new computer code for discrete fracture network modelling. Computers & Geosciences 36(3), 292–301 (2010)

Development of a Reliability-Based Design Optimization Toolbox for the FERUM Software

Luis Celorrio Barragué

Mechanical Engineering Department, Universidad de La Rioja, Logroño, La Rioja, Spain
luis.celorrio@unirioja.es

Abstract. Uncertainties are inherent in any structural system. Traditional design optimization techniques consider uncertainties implicitly by partial safe factors. However, Reliability-Based Design Optimization (RBDO) methods account for uncertainties explicitly. These methods find an optimum design that verifies several reliability constraints. The objective here is to obtain an economic and safe design. This paper describes several RBDO methods and the functions programmed to implements these methods. These functions form a package added to the well known Structural Reliability Toolbox Finite Element Reliability Using Matlab (FERUM). A structural example has been solved with the functions implemented: a transmission tower truss.

Keywords: Structural Reliability, Reliability Based Design Optimization, Double Loop Method, Transmission Tower.

1   Introduction

Design optimization has undergone a substantial progress. Commercial finite element codes have added optimization methods, but, most of these software packages deal only with deterministic parameters. However, uncertainties are inherent in design variables and parameters such as material properties, loading and geometry parameters. Other type of uncertainty is caused by inaccurate knowledge of the model and is named epistemic uncertainty. It is necessary to consider all these type of uncertainties in the design of any engineering system to assure reliability and quality. Traditionally, these uncertainties have been considered through partial safety factors in structural optimization methods. These partial safety factors are established in structural design codes, such as Eurocodes and the Spanish Building Technical Code. These methods are named "semiprobabilistic" because uncertainties of the design variables are considered implicitly. However, safety factors do not provide a quantitative measure of the safety margin in design and are not quantitatively linked to the influence of different design variables and their uncertainties on the overall system performance.

For a rational design to be made it is crucial to account for uncertainties explicitly. Also, any type of dependence or correlation between these uncertainties must be accounted for. The process of design optimization enhanced by the addition of reliability constraints is referred as Reliability-Based Design Optimization (RBDO) or

E. Hüllermeier et al. (Eds.): SUM 2012, LNAI 7520, pp. 273–286, 2012.

Reliabilty-Based Structural Optimization. A large research effort has been carried out about RBDO methods during the last thirty years. The main objective of this effort has been to reduce the large computational cost needed to obtain an optimum design that verifies reliability constraints.

A selection of RBDO methods have been implemented in Matlab for the Finite Element Reliability Using Matlab (FERUM) program. This work was based in version 3.1 of FERUM that was developed at the Department of Civil & Environmental Engineering in the University of California at Berkeley. This new FERUM_RBDO package can solve both analytical and structural problems. This package is ongoing and new features are constantly added. The organization of this paper is as follow. Section 2 presents the formulation of the RBDO problem more frequently used in literature about this topic. A classification of RBDO methods is provided in Section 3. These methods are classified in three groups: double loop methods, single loop methods and decoupled methods. Section 4 contains a brief review of FERUM and explains the new functions and capabilities added to FERUM to solve RBDO problems. Both analytical and structural problems can be solved. Section 5 includes an example of a transmission tower. Finally, section 6 contains the conclusions.

2   Formulation of the RBDO Problem

A typical RBDO problem is formulated as

$$\min_{\mathbf{d},\boldsymbol{\mu}_X} f(\mathbf{d},\boldsymbol{\mu}_X,\boldsymbol{\mu}_P)$$
$$s.t. \quad P_{fi} = P[g_i(\mathbf{d},\mathbf{X},\mathbf{P}) \le 0] \le P_{fi}^t, \quad i = 1,...,n \tag{1}$$
$$\mathbf{d}^L \le \mathbf{d} \le \mathbf{d}^U, \qquad \boldsymbol{\mu}_X^L \le \boldsymbol{\mu}_X \le \boldsymbol{\mu}_X^U$$

where $\mathbf{d} \in R^k$ is the vector of deterministic design variables. $\mathbf{d}^L$ and $\mathbf{d}^U$ are the upper and lower bounds of vector $\mathbf{d}$, respectively. $\mathbf{X} \in R^m$ is the vector of random design variables. $\boldsymbol{\mu}_X$ is the means of random design variables. $\boldsymbol{\mu}_X^L$ and $\boldsymbol{\mu}_X^U$ are the upper and lower bounds of vector $\boldsymbol{\mu}_X$. $\mathbf{P} \in R^q$ is the vector of random parameters. $\boldsymbol{\mu}_P$ is the mean value of $\mathbf{P}$. $f(\cdot)$ is the objective function, n is the number of constraints, k is the number of deterministic design variables, m is the number or random design variables and q is the number of random parameters, P_{fi} is the probability of violating the i-th probabilistic constraint and P_{fi}^t is the target probability of failure for the i-th probabilistic constraint.

If the First Order Reliability Method (FORM) is used, as usually occurs in practical applications, the failure probability P_f of a probabilistic constraint is given as a function of the reliability index β, written as:

$$P_f \approx \Phi(-\beta) \tag{2}$$

where $\Phi(\cdot)$ is the standard Gaussian cumulated distribution function and β is the reliability index defined by Hasofer and Lind (1974), which is evaluated by solving the constrained optimization problem:

$$\beta = \min\|\mathbf{u}\| = \left(\mathbf{u}^T \cdot \mathbf{u}\right)^{\frac{1}{2}}$$
$$s.t. : G(\mathbf{d}, \mathbf{u}) = 0 \qquad (3)$$

The solution of this optimization problem $\mathbf{u}^*$ is the minimum distance of a point $\mathbf{u}$ on the failure surface $G(\mathbf{d}, \mathbf{u}) = 0$ from the origin of the standard normal space $\mathbf{U}$ and is called the Most Probable Point (MPP) or β-point, as $\beta = \|\mathbf{u}^*\|$. A probabilistic transformation $\mathbf{U} = T(\mathbf{X}, \mathbf{P})$ from the original space of physical random variables $(\mathbf{X}, \mathbf{P})$ to the normalized space $\mathbf{U}$ is needed. Usually Nataf transformation or Rosenblatt transformation are used. The image of a performance function $g_i(\mathbf{d}, \mathbf{X}, \mathbf{P})$ is $G_i(\mathbf{d}, T(\mathbf{X}, \mathbf{P})) = G_i(\mathbf{d}, \mathbf{u})$.

The optimization process of Eq. (1) is carried out in the space of the design parameters $(\mathbf{d}, \mathbf{\mu_x})$. In parallel, the solution of the reliability problem of Eq. (3) is performed in the normalized space $\mathbf{U}$ of the random variables.

Traditional RBDO method requires a double loop iteration procedure, where reliability analysis is carried out in the inner loop for each change in the design variables, in order to evaluate the reliability constraints. The computational time for this procedure is extremely high due to the multiplication of the number of iterations in both outer loop and reliability assessment loop, involving a very high number of mechanical analyses.

3   RBDO Methods

Due to the prohibitive computational effort of the traditional double loop RBDO method, researchers have developed several methods to solve this numerical burden. Below, they are briefly described.

3.1   Double-Loop Methods

These RBDO formulations are based on improvements of the traditional double-loop approach by increasing the efficiency of the reliability analysis. Two approaches have been proposed to deal with probabilistic constraints in the double-loop formulation: Reliability Index Approach (RIA) and Performance Measure Approach (PMA).

RIA based RBDO is the traditional or classic RBDO formulation. The RBDO problem is solved in two spaces: the spaces of design variables, corresponding to a deterministic physical space and the space of Gaussian random variables, obtained by probabilistic transformation of the random physical variables. The RIA based RBDO problem is stated as:

$$\min_{d,\mu_x} f(\mathbf{d},\boldsymbol{\mu}_X,\boldsymbol{\mu}_P)$$

$$s.t. \quad \beta_i(\mathbf{d},\mathbf{X},\mathbf{P}) \geq \beta_i^t, \quad i=1,...,n \tag{4}$$

$$\mathbf{d}^L \leq \mathbf{d} \leq \mathbf{d}^U, \qquad \boldsymbol{\mu}_X^L \leq \boldsymbol{\mu}_X \leq \boldsymbol{\mu}_X^U$$

where β_i is the reliability index of the i-th probabilistic constraint for the structure and β_i^t is the target or allowed reliability index. The assessment of the reliability index β_i involves solving the optimization problem stated as equation (3). Both, standard nonlinear constrained optimization methods and FORM methods like the Hasofer-Lind-Rackwitz-Fiesler (HLRF) method or the improved-HLRF method can be used.

The solution of this RBDO problem consists in solving the two nested optimization problems. For each new set of the design parameters, the reliability analysis is performed in order to get the new MPP, corresponding to a given reliability level. It is well established in the literature that RIA-based RBDO method converges slowly, or provides inaccurate results or even fails to converge due to the highly non linear transformations involved.

Lee and Kwak [1] and Tu et al. [2] proposed the use the Performance Measure Approach (PMA) instead of the widely used RIA. In PMA, inverse reliability analysis is performed to search for a point with the lowest positive performance function value on a hypersurface determined by the target reliability index β_i^t. Since the inverse reliability analysis is also performed iteratively, the reliability analysis and optimization loops are still nested.

The PMA-based RBDO problem is stated as

$$\min_{d,\mu_x} f(\mathbf{d},\boldsymbol{\mu}_X,\boldsymbol{\mu}_P)$$

$$s.t. \quad G_{pi} = F_G^{-1}[\Phi(-\beta_{t_i})] \geq 0, \quad i=1,...,n \tag{5}$$

$$\mathbf{d}^L \leq \mathbf{d} \leq \mathbf{d}^U, \qquad \boldsymbol{\mu}_X^L \leq \boldsymbol{\mu}_X \leq \boldsymbol{\mu}_X^U$$

where the performance measure G_p is obtained from the following reliability minimisation problem:

$$G_p = \min_{u} G(\mathbf{d},\mathbf{u})$$

$$s.t. \quad \|\mathbf{u}\| = \beta_t \tag{6}$$

The solution of the optimization problem stated in equation (6) is named the Most Probable Point of Inverse Reliability (MPPIR).

The computational expense caused by the nesting of design optimization loop and reliability analysis loop makes traditional RBDO impractical for realistic problems with large number of variables and constraints. Various techniques have been proposed to improve the efficiency of RBDO. Some techniques improve the efficiency of reliability analysis in the double loop formulation. Other techniques decouple the design optimization and the reliability analysis problems. Other methods carried out design optimization and reliability analysis in a single loop.

3.2   Improvements of Reliability Analysis in Double-Loop Formulation

PMA-based RBDO is shown to be more efficient and robust than RIA-based RBDO, because performance measure methods do not need to obtain the exact value of the probability of failure for each inner loop and this implies that computational cost decreases. However, several numerical examples using PMA show inefficiency and instability in the assessment of probabilistic constraints during the RBDO process. B.D. Youn *et al* [3], [4], [5] have carried out some advanced PMA-based methods. These methods are Hybrid Mean Value (HMV) method [3], Enhanced Hybrid Mean Value (HMV+) method [4] and Enriched Performance Measure Approach (PMA+) [5] and are briefly described here. HMV method is a powerfull method to obtain the MPPIR in the inner loop. This method combines two methods: the Advance Mean Value (AMV) and the Conjugate Mean Value (CMV) methods and provides accurate results for concave and convex nonlinear reliability constraints. However, although the HMV method performs well for convex or concave performance functions, it could fail to converge for highly nonlinear performance functions. An enhanced HMV method, named HMV+ was proposed by B.D. Youn *et al* [4]. This method substantially improves numerical efficiency and stability in reliability analysis of highly nonlinear performance functions. The difference between HMV and HMV+ is that HMV+ introduces interpolation to improve the MPPIR search.PMA+ is an enriched PMA based method, enhancing numerical efficiency while maintaining stability in the RBDO process. PMA+ integrates three key ideas, in addition to the HMV+ method: launching RBDO at a deterministic optimum design, feasibility checks for probabilistic constraints and fast reliability analysis under the condition of design closeness. Two parameters are used in PMA+ method. *e-active* parameter is used to obtain a set of constraints close to be activated constraints. Only violated, active and *e-active* constraints are evaluated in the reliability analysis. The other parameter, named *e-closeness*, is used to determine when two consecutive designs are close. Then, a fast reliability analysis is developed for the *e-active* constraints.

3.3   Decoupled Methods or Sequential Methods

In this kind of RBDO methods the optimization problem and the reliability assessment are decoupled and are carried out sequentially. The reliability constraints are replaced by equivalent deterministic (or pseudo-deterministic) constraints, involving some additional simplifications.

Du and Chen [6] proposed the Sequential Optimization and Reliability Assessment (SORA) method. This method employs a decoupled strategy where a series of cycles of optimization and reliability assessment is employed. In each cycle, design optimization and reliability assessment are decoupled from each other; no reliability assessment is required within the optimization and the reliability assessment is only conducted after the optimization. The key concept is to use the reliability information obtained in the previous cycle to shift the boundaries of the violated deterministic constraints (with low reliability) to the feasible region. Therefore, the design is improved from cycle to cycle and the computation efficiency is improved significantly.

3.4   Single Loop Approaches

RBDO approaches belonging to this group collapse the optimization and the reliability problems within a single-loop dealing with both design and random variables. Both RIA- and PMA-based single-loop strategies have been developed [7], [8], [9].

The single-loop single-vector (SLSV) approach [7], [10] provides the first attempt of a truly single-loop approach. It improves the RBDO computational efficiency by eliminating the inner reliability loops. However, it requires a probabilistic active set strategy for identifying the active constraints, which may hinder its practicality.

Other single-level RBDO algorithms have also been reported in Argarwal *et al.* [11], Kuschel and Rackwitz [8] and Streicher and Rackwitz [12]. These methods introduce the Karush-Kuhn-Tucker (KKT) optimality conditions at the optima of the inner optimization loops as equality constraints in the outer design optimization loop. This helps to adopt a well-known strategy for effectiveness in optimization, i.e., satisfying the constraints only at the optimum and allowing the solution to be infeasible before convergence. However, these RBDO methods based on KKT conditions have a great drawback: the large number of design variables. Here, all variables in the problem are considered design variables. That is, the original design variables, the components of the MPP in standard normal space for each reliability constraints and the Lagrange multipliers for each optimization sub problem. This can increase the computational cost substantially, especially for practical problems with a large number of design variables and a large number of constraints. Furthermore, the approach in [8], [11] and [12] requires second-order derivatives which are computationally expensive and difficult to calculate accurately.

Liang *et al* [13] have development a single-loop RBDO formulation. This method has a main advantage: it eliminates the repeated reliability loops without increasing the number of design variables or adding equality constraints. It does not require second-order derivatives. The KKT optimality conditions of the inner reliability loops are explicitly used to move from the standard normal **U** space to the original **X** space, where the inequality constraints of the outer design optimization loop are evaluated. It converts the probabilistic optimization formulation into a deterministic optimization formulation. This method estimates the MPP for each probabilistic constraint using gradient information from the previous iteration. It therefore, eliminates the reliability optimization loop of the conventional double-loop RBDO approach.

A recent paper [14] provides a benchmark study of different RBDO methods. Several analytical and structural examples are solved and their results are analyzed there.

4   Review of FERUM

The Finite Element Reliability Using Matlab (FERUM) software is a collection of Matlab functions used to run Reliability Analysis. The first version of this toolbox was released in 1999 at the Structural Engineering Mechanics and Material Division of the Department of Civil & Environmental Engineering in the University of California at Berkeley (UCB). Last release from UCB is the version 3.1 in 2002. This code can be downloaded from www.ce.berkeley.edu/FERUM.

A few years later, a new version of the FERUM code has been released. This version is based on the work carried out at the Institut Français de Mécanique Avancée (IFMA) in Clermont-Ferrand, France and is named FERUM 4.1. This software includes additional functions and improved capabilities. FERUM 4.1 package and the FERUM 4.1 User's Guide can be downloaded from the web page: www.ifma.fr/Recherche/Labos/FERUM.

In the present work some functions have been written to carry out Reliability Based Design Optimization (RBDO). Different RBDO methods have been implemented. This collection of functions added to FERUM can be seen as a new package and is named as FERUM_RBDO. It is necessary to remind that this new package is constantly under development. The following RBDO methods have been implemented in the FERUM_RBDO package:

Double loop methods
> RBDO - DL - RIA
> RBDO - DL - HMV
> RBDO - DL - HMV+
> RBDO – DL- PMA+ (Enriched PMA – HMV+)
> RBDO – DL - PMA+ HMV (without interpolation)

Single – loop methods
> RBDO – Single Loop Method (Liang et al)

Decoupled methods
> RBDO - SORA using HMV
> RBDO - SORA using HMV+

5   Space Truss Example

Figure 1 shows a 25 bar truss that is often used as a tower to support telecommunication lines. The problem here is to minimize the volume of the truss subject to twenty-nine reliability constraints. Four of these constraints are displacement constraints: the displacements of the top nodes are limited. The other twenty five constraints are stress constraints.

The bars are manufactured in steel, with tubular sections with diameters ratio $d_i/d_e = 0.8$. The materials properties of the steel are constant: the Young module is $E = 20700 kN/cm^2$ and the allowed stress is $\sigma_a = 27.5 kN/cm^2$. The truss bars are grouped in eight groups. In Figure 1 bars belonging to the same group have the same index written above them. There exist eight random design variables and they are the cross section areas of the bar groups. The truss is subject to random loads. They are assumed independent and are applied mainly on the highest nodes.

Random variables are summarized in Table 1. All the random variables are normally distributed and independent. It is important to note that FERUM has an extensive collection of distributions and other types of distribution could have been considered for the random variables. Coefficients of variation of random variables are considered constant.

Then the formulation of the RBDO problem is stated as:

$$Min \quad V(\mathbf{d}, \mathbf{\mu_x}, \mathbf{\mu_P})$$

$$s.t. \quad \beta_i \geq \beta_i^t = 3.0 \qquad\qquad i = 1,..,29. \qquad\qquad (7)$$

$$2\,cm^2 \leq \mu_{x_i} \leq 40\,cm^2 \qquad i = 1,...,8.$$

where the reliability constraints are the subsequent state limit functions:

Displacement Constraints: Vertical displacements of the nodes 1 and 2 on coordinate axis x and y are limited:

$$g_i(\mathbf{d}, \mathbf{X}, \mathbf{P}) = 1 - \frac{|q_{ij}(\mathbf{d}, \mathbf{X}, \mathbf{P})|}{q^a} \quad i = 1, 2 \;\; j = x, y \qquad (8)$$

where q_{ij} is the displacement of the node i on coordinate axis j and $q^a = 3.0cm$ is the allowed displacement.

Stress constraints: Maximum stress of the element is limited:

$$g_i(\mathbf{d}, \mathbf{X}, \mathbf{P}) = 1 - \frac{|\sigma_{i-4}(\mathbf{d}, \mathbf{X}, \mathbf{P})|}{\sigma_{i-4}^a(\mathbf{d}, \mathbf{X}, \mathbf{P})} \qquad i = 5,...,29 \qquad (9)$$

where σ^a is the maximum stress allowed and, in the case of bars in compression, it regards the buckling of the bar and takes the value of the Euler's critical stress.

A target reliability index $\beta_t = 3.0$ was stated for all the reliability constraints. However, RBDO_FERUM allows assigning a different target reliability index for each reliability constraints.

Several methods have been applied to solve the RBDO problem. The probabilistic optimum obtained for every methods are showed in Table 2 and 3, where gradients can be evaluated using the Direct Difference Method (DDM) and the Finite Forward Difference method (FFD) respectively. The efficiency of the different methods is measured by the number of iterations of the design optimization loop (ITERS OPT) and the number of limit state function evaluations (LSFE) and these data are showed in Table 4 (for DDM) and Table 5 (for FFD).

Although all random variables are normally distributed, RBDO-RIA method does not converge. This is caused by the large number of random variables and constraints. Double loops RBDO methods provide practically the same probabilistic optimum. Single loop and SORA (HMV) methods provide a different probabilistic optimum design than double loop methods. Figure 2 shows the evolution of design variables $A_1,..., A_8$ for HMV+ double loop method using FFD. Design variables take values close to the final optimum in the 6th iteration. Only 14 iterations were needed to search for the optimum with the tolerance required. The SORA (HMV+) method does not converge. A zigzag design vector was obtained.

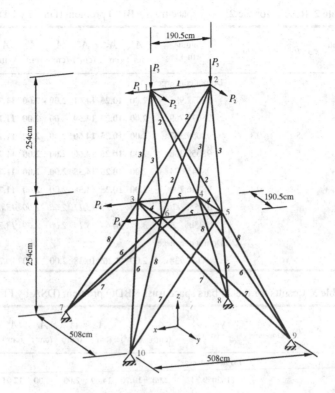

Fig. 1. 25 bars space truss

Table 1. Random Variables in space truss example

Random Variable	Description	Distribution	Initial Mean	CoV or σ	Design Variable
X_1	A_1	N	10.0 cm^2	0.1	μ_{X_1}
X_2	A_2	N	10.0 cm^2	0.1	μ_{X_2}
X_3	A_3	N	10.0 cm^2	0.1	μ_{X_3}
X_4	A_4	N	20.0 cm^2	0.05	μ_{X_4}
X_5	A_5	N	20.0 cm^2	0.05	μ_{X_5}
X_6	A_6	N	25.0 cm^2	0.05	μ_{X_6}
X_7	A_7	N	25.0 cm^2	0.04	μ_{X_7}
X_8	A_8	N	25.0 cm^2	0.04	μ_{X_8}
X_9	P_1	N	30 kN	3 kN	-
X_{10}	P_2	N	50 kN	5 kN	-
X_{11}	P_3	N	100 kN	10 kN	-
X_{12}	P_4	N	30 kN	5 kN	-

Table 2. Results for the 25 bars space truss RBDO problem (DSA by DDM)

RBDO Method (DDM)	e_active	e-closeness	Volume (m³)	A_1 (cm²)	A_2 (cm²)	A_3 (cm²)	A_4 (cm²)	A_5 (cm²)	A_6 (cm²)	A_7 (cm²)	A_8 (cm²)
DL RIA			Do not converge								
DL HMV			98072.985	2.00	10.25	14.61	2.00	2.00	11.77	12.64	16.07
DL HMV+			98079.769	2.00	10.23	14.59	2.00	2.00	11.76	12.71	16.04
DL PMA+ HMV	1.5	0.1	98059.367	2.00	10.25	14.60	2.00	2.00	11.77	12.64	16.07
DL PMA+ HMV	0.5	0.1	98059.366	2.00	10.25	14.60	2.00	2.00	11.77	12.64	16.07
DL PMA+ HMV+	1.5	0.1	98045.694	2.00	10.25	14.58	2.00	2.00	11.77	12.64	16.07
DL PMA+ HMV+	0.5	0.1	98046.790	2.00	10.25	14.58	2.00	2.00	11.77	12.64	16.07
SINGLE LOOP			104556.970	3.03	10.71	13.11	2.00	2.00	13.69	13.91	17.13
SORA (HMV)			103505.720	3.02	10.64	13.14	2.00	2.00	13.79	14.07	16.04
SORA (HMV+)			Do not converge								
DDO			84137.938	2.00	8.26	10.38	2.00	2.00	10.20	11.84	14.35

Table 3. Results for the 25 bars space truss RBDO problem (DSA by FFD)

RBDO Method (FFD)	e_active	e-closeness	Volume (m³)	A_1 (cm²)	A_2 (cm²)	A_3 (cm²)	A_4 (cm²)	A_5 (cm²)	A_6 (cm²)	A_7 (cm²)	A_8 (cm²)
DL RIA			Do not converge								
DL HMV			102059.34	2.00	10.70	14.69	2.00	2.00	12.01	13.87	16.52
DL HMV+			102001.09	2.00	10.71	14.65	2.00	2.00	12.00	13.87	16.51
DL PMA+ HMV	1.5	0.1	102057.77	2.00	10.70	14.69	2.00	2.00	12.01	13.87	16.51
DL PMA+ HMV	0.5	0.1	102057.77	2.00	10.70	14.69	2.00	2.00	12.01	13.87	16.51
DL PMA+ HMV+	1.5	0.1	102063.61	2.00	10.70	14.68	2.00	2.00	12.01	13.88	16.51
DL PMA+ HMV+	0.5	0.1	102064.31	2.00	10.70	14.68	2.00	2.00	12.01	13.88	16.51
SINGLE LOOP			107057.16	3.02	11.11	13.18	2.00	2.00	13.80	14.61	17.44
SORA (HMV)			106176.41	3.01	11.05	13.21	2.00	2.00	13.87	14.75	16.54
SORA (HMV+)			Do not converge								
DDO			84137.86	2.00	8.26	10.38	2.00	2.00	10.20	11.84	14.35

With regard to efficiency, the results represented in Tables 4 and 5 show that PMA+ based RBDO methods are the best. However, a good chose of *e-active* and *e-closeness* parameters must be done. So, using FFD, only 136 evaluations of limit state functions were needed to obtain the probabilistic optimum design using the method PMA+ (HMV+) with *e-active* = 0.5 and *e-closeness* = 0.1.

An empirical rule derived from author experience is proposed: to assign a value into the range 0.7-1.5 to *e-active* and a value one or two orders of magnitude larger than the convergence tolerance of the iterative algorithms to *e-closeness*.

Results include the number of the active probabilistic constraints at the optimum design. Also, some lateral bound constraints are active. That is, optimum design reaches the lower bounds of design variables A_1, A_4 and A_5 in double loop methods.

There are some differences between results obtained using DDM and results obtained using FFD. For instance: the volume at the probabilistic optimum design obtained using FFD is larger than using DDM: 102000 m^3 vs. 98000 m^3 for the double loop methods. Also, design variables at the optimum are different. The large difference takes place on design variable A_7. It changes from 12.64 cm^2 using DDM to 13.87 cm^2 using FFD.

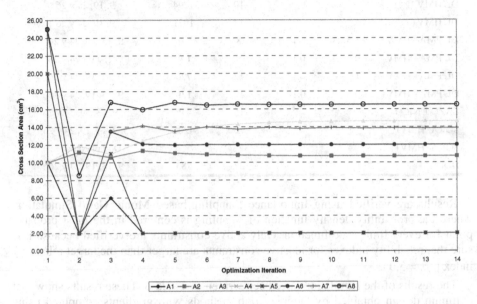

Fig. 2. Evolution of design variables in the DL HMV+ RBDO process

Table 4. Results for the 25 bars space truss RBDO problem (DSA by DDM)

RBDO Method (DDM)	e_active	e_closeness	ITERS OPT	LSFE	Active Constraints (i with $g_i = 0$)
DL RIA					Do not converge
DL HMV			14	1748	9, 13, 19, 24 y 26
DL HMV+			16	2028	9, 13, 18, 19, 24, 25 y 26
DL PMA+ HMV	1.5	0.1	11	1169	9, 13, 19, 24 y 26
DL PMA+ HMV	0.5	0.1	11	353	9, 13, 19, 24 y 26
DL PMA+ HMV+	1.5	0.1	14	958	9, 13, 19, 24 y 26
DL PMA+ HMV+	0.5	0.1	14	214	9, 13, 19, 24 y 26
SINGLE LOOP			13	377	5, 9, 13, 19, 24 y 26
SORA (HMV)			14	9648	5, 9, 13, 19, 24 y 26
SORA (HMV+)					Do not converge
DDO			79	2291	9, 13, 19, 24 y 26

Table 5. Results for the 25 bars space truss RBDO problem (DSA by FFD)

RBDO Method (FFD)	e_active	$e\_closeness$	ITERS OPT	LSFE	Active Constraints $(i$ with $g_i = 0)$
DL RIA					Do not converge
DL HMV			10	1294	9, 19, 24 y 26
DL HMV+			14	1031	9, 19, 24 y 26
DL PMA+ HMV	1.5	0.1	8	828	9, 13, 19, 24 y 26
DL PMA+ HMV	0.5	0.1	8	198	9, 13, 19, 24 y 26
DL PMA+ HMV+	1.5	0.1	12	726	9, 13, 19, 24 y 26
DL PMA+ HMV+	0.5	0.1	12	136	9, 13, 19, 24 y 26
LAZO UNICO			15	435	5, 9, 13, 19, 24 y 26
SORA (HMV)			10	4206	5, 9, 13, 19, 24 y 26
SORA (HMV+)					Do not converge
DDO			27	783	9, 13, 19, 24 y 26

Results are verified using Importance Sampling-based Monte Carlo Simulation, where the probability density function of sampling is centered in the most probable point for every limit state function. Only active constraints are verified because it is sure that reliability indexes for inactive constraints are larger than the target reliability index ($\beta_t = 3.0$).

The results of the simulation have been written in Table 6. These results show that optimum design obtained by Double Loop methods with gradients computed using the FFD method is the most accurate. Errors in reliability index at optimum design are below 0.5%.

Table 6. Reliability index for active constrains computed by IS – MCS

Limit State Function	Actual Reliability Index for active constraints by IS - MCS				β_t
	DDM - DL	DDM - SL	FFD-DL	FFD-SL	
g_5	7.6206	Inf	**7.63**	Inf	3.00
g_9	2.6436	*2.4305*	**2.9859**	2.7245	3.00
g_{13}	2.9460	*2.2883*	**2.9907**	*2.3414*	3.00
g_{19}	2.6200	4.9739	**2.9945**	5.131	3.00
g_{24}	*1.2900*	3.5143	**3.0142**	4.4375	3.00
g_{26}	*2.4071*	3.6578	**2.9846**	4.0222	3.00

6   Conclusions

New functions to solve Reliability Based Design Optimization problems have been added to FERUM software. These functions form a toolbox named FERUM_RBDO.

Several RBDO methods have been implemented. Several double loop methods, a single loop method and the decoupled method called SORA have been included. The author must stress that the implementation of these methods is based only on his own interpretation of algorithms published in research papers.

Both analytical and structural problems can be solved. An example of a transmission tower is included to show the capabilities of the new toolbox. All the RBDO methods implemented have been applied and the results have been analyzed. Also, verification of the results has been carried out using Importance Sampling based Monte Carlo Simulation.

Advanced capabilities of FERUM such as the extensive probability distribution library, the analysis of correlated random variables and different methods of design sensitivity analysis can also be used in FERUM_RBDO. Two types of design variables can be considered: deterministic design variables and random design variables.

Important properties of RBDO algorithms such as stability, convergence and efficiency can be analyzed. User can run several RBDO problem and obtain his o her own conclusions about these properties.

References

1. Lee, T.W., Kwak, B.M.: A Reliability-Based Optimal Design Using Advanced First Order Second Moment Method. Mech. Struct. Mach. 15(4), 523–542 (1987-1988)
2. Tu, J., Choi, K.K.: A New Study on Reliability Based Design Optimization. Journal Mechanical Design ASME 121(4), 557–564 (1999)
3. Youn, B.D., Choi, K.K., Park, Y.H.: Hybrid Analysis Method for Reliability-Based Design Optimization. Journal of Mechanical Design, ASME 125(2), 221–232 (2003)
4. Youn, B.D., Choi, K.K., Du, L.: Adaptive probability analysis using an enhanced hybrid mean value (HMV+) method. Journal of Structural and Multidisciplinary Optimization 29(2), 134–148 (2005)
5. Youn, B.D., Choi, K.K., Du, L.: Enriched Performance Measure Approach (PMA+) for Reliability-Based Design Optimization Approaches. AIAA Journal 43(4), 874–884 (2005)
6. Du, X., Chen, W.: Sequential optimization and reliability assessment method for efficient probabilistic design. ASME J. Mechanical Design 126(2), 225–233 (2004)
7. Chen, X., Hasselman, T.K., Neill, D.J.: Reliability based structural design optimization for practical applications. In: Proceedings, 38th AIAA/ASME/ASCE/AHS/ASC Structures, Structural Dynamics, and Materials Conference and Exhibit, Orlando, FL (1997)
8. Kuschel, N., Rackwitz, R.: A new approach for structural optimization of series systems. In: Melchers, R.E., Stewart, M.G. (eds.) Applications of Statistics and Probability, pp. 987–994. Balkema, Rotterdam (2000)
9. Liang, J., Mourelatos, Z.P., Tu, J.: A single loop method for reliability based design optimization. In: Proceedings of the 30th ASME Design Automation Conference, Salt Lake City, UT, Paper n° DETC 2004/DAC-57255 (September 2004)
10. Wang, L., Kodiyalam, S.: An efficient method for probabilistic and robust design with non-normal distributions. In: Proceedings of 34rd AIAA SDM Conference. Denver, CO (2002)
11. Agarwal, H., Renaud, J., Lee, J., Watson, L.: A unilevel method for reliability based design optimization. In: 45th AAA/ASME/ASCE/AHS/ASC Structures, Structural Dynamics and Material Conference, Palm Springs, CA (2004)

12. Streicher, H., Rackwitz, R.: Time-variant reliability-oriented structural optimization and a renewal model for life-cicle costing. Probabilistic Engineering Mechanics 19(1-2), 171–183 (2004)
13. Liang, J., Mourelatos, Z.P., Tu, J.: A single-loop method for reliability-based design optimization. Int. J. Product Development 5(1-2), 76–92 (2008)
14. Aoues, Y., Chateauneuf, A.: Benchmark study of numerical methods for reliability-based design optimization. Journal of Structural and Multidisciplinary Optimization 41, 277–294 (2010)

Approximating Complex Sensor Quality Using Failure Probability Intervals

Christian Kuka and Daniela Nicklas

OFFIS, Universität Oldenburg
Oldenburg, Germany
christian.kuka@offis.de, dnicklas@acm.org

Abstract. Many pervasive applications depend on data from sensors that are placed in the applications physical environment. In these applications, the quality of the sensor data—e.g., its accuracy or a failed object detection—is of crucial importance for the application knowledge base and processing results. However, through the increasing complexity and the proprietary of sensors, applications cannot directly request information about the quality of the sensor measurements. However, an indirect quality assessment is possible by using additional simple sensors. Our approach uses information from these additional sensors to construct upper and lower bounds of the probability of failed measurements, which in turn can be used by the applications to adapt their decisions. Within this framework it is possible to fuse multiple heterogeneous indirect sensors through the aggregation of multiple quality evidences. This approach is evaluated using sensor data to detect the quality of template matching sensors.

Keywords: Sensor fusion, Measurement uncertainty, Error probability.

1 Introduction

Sensor data is required in almost all pervasive applications. This implies that the applications have to trust the measured sensor data. However, sensors underlay different uncertainties introduced through the physical or chemical principal of measurement or the internal processing through the sensor. Furthermore, due to the complexity of current sensors, a direct access to the measuring process is difficult or even impossible in case of proprietary sensors. The internal function of complex sensors must be seen as a black box. Different approaches try to improve improper sensor data through multisensor fusion techniques [1,2]. However, in this case the application still has no knowledge about the actual quality of the sensor data and the modification through the quality improvement approaches. This can lead to undesired results during the processing inside the application or even lead to failures in connected subsystems. The approach presented in the following tries to approximate the different uncertainties about the sensor data. If an application would know about the quality of a given sensor value, it could use this meta information to adapt its behavior, e.g., by not performing

E. Hüllermeier et al. (Eds.): SUM 2012, LNAI 7520, pp. 287–298, 2012.

an action, by visualizing the uncertainty, or by taking the human in the loop and asking the user what to do.

Our contribution is the mapping of online sensor measurements to failure probability intervals, using additional sensors in the environment that measure related physical or chemical influence factors. In a first step, the sensors are deployed to measure the environmental factors that influence our sensor. In a second step, a failure probability model of the complex is created based on the influence factors. After that, the created model is used to estimate the quality of the complex sensor. In contrast to other approaches [3,4] that estimate the quality of the sensor fusion result through conflicts between aggregated sensors, our approach estimate the quality of the actual sensor reading of one sensor through indirect measurements. Furthermore, our technique allows merging multiple evidences from other heterogeneous sensors during sensor data processing. This allows the annotation of sensor data processing results with the combined failure probability intervals to support an application in the task of decision making.

The rest of the paper is organized as follows: Section 2 explains the problem based on a concrete scenario. In Section 3, we explain our approach, which is based on uncertainties in sensor data and modeling of uncertainties for processing. Furthermore, we show the combination of sensor uncertainties from multiple sources. An evaluation of the approach based on a logistic scenario can be found in Section 4. Finally, we conclude and take an outlook on future work.

2   Scenario

In the following, we explain our approach based on a simple scenario (Fig. 1). A camera vision system monitors the transportation of potential dangerous goods within a depot. The vision system continuously observes the container that is transported by a forklift and forwards the status "in position" or "on ground" to an application that raises an alarm when the container is falling down.

Fig. 1. Observation Scenario

The camera vision system in this scenario is an example for a complex sensor that recognizes the container based on a predefined template. It is assumed that during the monitored transport, the container is always visible to the camera system, so the system always reports the status where it detects the template in the camera image. Thus, the case that the container is not in the camera recorded image is not part of this scenario.

2.1   Uncertainties in Sensor Data

Since a camera vision system heavily depends on environmental influences, results from the vision system could be wrong. To improve wrong measurements, we could measure multiple times and perform the average on the results. However, this is not possible because the application should raise an alarm as fast as possible leaving no time for multiple measurements.

(a) Correct detection of the template   (b) Matching error through the headlights of the forklift

Fig. 2. Template matching

One of the biggest influences of a camera system is light; especially direct light from a forklift that blinds the vision system resulting in an overexposure. Through such an overexposure the vision system could report a wrong status of the container and the application could probably raise a false alarm as shown in Fig. 2. To detect influences through environmental factors we can use indirect sensors in the environment to estimate the quality of the camera vision system.

2.2   Indirect Sensors for Environmental Influence Detection

Indirect sensors in the environment give us additional information about the current context of the complex sensor. In our scenario, we can use cheap sensors like a light dependent resistor (LDR) as an indirect sensor to measure the light influence that results in an overexposure of the camera system. While we could now detect a possible overexposure through the indirect sensors using e.g. a threshold, we cannot do this for sure. Especially at the margins between a correct working state and an overexposure of the vision system an estimation about the correct function of the system is not possible. As the internal processing of the vision system itself is unknown to us, we have to estimate the probability of a failure of our vision system through this influence factor. To estimate the exact probability of a failure of the system at different light settings we have to check the results of the system at every possible light setting. However, testing every setting is time consuming or even impossible in some applications e.g. due to lack of equipment to simulate the settings. Thus, to approximate the failure probability we perform a series of measurements $D = (e^k, x^k)$ with $e^k \in [-1, 1]$ is a binary variable representing the correctness of the complex sensor and $x^k \in x_1 \ldots x_n$ is the value of the indirect sensor in the k-th measurement with reasonable small bins n to build up a failure model that reports the current failure probability based on the evidence of the indirect sensors during runtime as shown in Fig. 3.

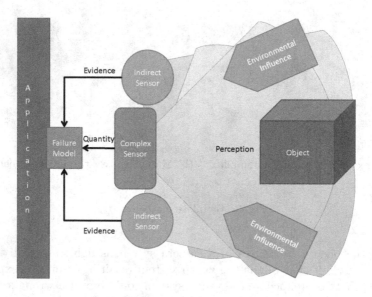

Fig. 3. Architecture

3   Failure Probability Model

Based on the series of measurements for one environmental influence factor e.g. light from a forklift in our scenario, we can estimate the average failure probability at the different light settings $x \in x_i$ from our test series. In this way, it is possible to assign a probability $p(e|x \in x_i)$ of a failure on the result of the vision system based on the information of an indirect sensor. Due to limited data, we have to deal with the uncertainty in the estimation of the failure probability. To this end, we construct a failure probability interval by including the standard error of the mean (σ_{mean}). Thus, our failure probability interval for a given bin i of light influence is given by $p_i = [\langle p_i \rangle - \sigma_i, \langle p_i \rangle + \sigma_i]$ at a given light setting. Under the assumption of a monotone function of the failure probability between two edges of adjacent bins of light influence, the usage of a failure probability interval allows us to bound the failure probability within the bin $x_i \in [\underline{x_i}, \overline{x_{i+1}}]$ from the indirect sensor as follows:

$$p(e|x \in x_i) \in [\min \underline{p}_i, \underline{p}_{i+1}, \max \overline{p}_i, \overline{p}_{i+1}]$$

With $\underline{p}_i$ being the lower bound of the failure probability interval p_i and $\overline{p}_i$ being the upper bound of the failure probability interval p_i respectively. In [5], the authors already showed the approach of indirect measurements through the use of probability intervals. However, instead of using an interval for the measurement of the complex sensor based on the measurement of the indirect sensor, we define an interval on the failure probability based on the measurements of the indirect sensor. The result of defining a lower and upper bound on the failure probability for multiple measurement bins is called a p-box of failure probabilities [6]. A p-box maps a measurement to a probability interval bounded by the probabilities $\underline{p}_i$ and $\overline{p}_i$. This approach allows us to annotate the measurements of our vision system with a lower and upper bound of a failure probability based on the information of the light dependent resistor (LDR). While other approaches assume that all influence factors can be measured directly [7] we make the assumption that other not measured influence factors perform at normal working conditions in the field of application. Thus, the effect of these influence factors are covered by σ_{mean}. Based on this annotation an application can decide through the failure probability bound if the measurement of the complex sensor is sufficient trustworthy for continuing or cancellation of an operation, or if additional information from another source is required, i.e., a human in the loop.

3.1 Variability of Indirect Measurement

Until now, we have described the failure probability of our vision system and collected evidence for the result of the vision system based on the measurements of our LDRs. However, we did not consider that our indirect sensors have also their systematic and random error. Thus, to get a better view on the failure probability of our vision system, we have to take the variability in the measurements of our LDRs into account, too. As the indirect sensors only measure one single unit and the internal function is well known, we can assume that the systematic failure and the bounds of the independent random failure are known a priori. This allows us to model the variability $p(x \in x_i | \hat{x})$ of a measurement $\hat{x}$ of the indirect sensors through a normal distribution [8]. Taking the variability of the indirect sensors into account brings us to the new failure probability of our complex sensor as a combined probability of the affected failure probability intervals of the complex sensor through the variability of the indirect sensors:

$$p(e|\hat{x}) = \sum_{x_i} p(e|x \in x_i)p(x \in x_i|\hat{x})$$

The combination of the variability of the indirect sensors with the failure intervals of the complex sensor delivers us multiple affected failure probability intervals with different weights. The weight can be interpreted as a basic probability assignment (bpa) defined in the Dempster-Shafer Theory of evidence [9] about all possible failure intervals from our failure model.

Thus, we can see every affected failure probability interval $[\underline{p_i}, \overline{p_i}]$ as a bpa, represented as $m([\underline{p_i}, \overline{p_i}]) = p(x \in x_i|\hat{x})$, for this interval expressed by our indirect sensor about the current failure situation. The benefit of using the Dempster-Shafer Theory in our case is the ability to express the believable and the plausible failure probability interval of our complex sensor as the combination of affected failure probability intervals. The belief $Bel(A)$ of a failure probability interval $A = [\underline{p}, \overline{p}]$ is the sum of all other assignments that are contained in this failure probability interval. The plausibility $Pl(A)$ of a failure probability interval $A = [\underline{p}, \overline{p}]$ is the sum of all other bpas that intersects with this failure probability interval.

$$Bel(A) = \sum_{[\underline{p_i}, \overline{p_i}] \subseteq A} m([\underline{p_i}, \overline{p_i}])$$

$$Pl(A) = \sum_{[\underline{p_i}, \overline{p_i}] \cap A \neq \emptyset} m([\underline{p_i}, \overline{p_i}])$$

The belief and the plausibility of a failure probability interval is often interpreted as the minimum and maximum support for a given failure probability interval of the complex sensor through the observation of one indirect sensor. However, there are concerns about this interpretation [10]. Nevertheless, it is widely used in sensor fusion [11,12].

3.2   Multiple Influence Factors

In our scenario we only used indirect sensors for one environmental influence factor; however the data quality of complex sensors can depend on more than one external influence factor that can be observed using other indirect sensors in the environment. Thus, our complex sensor can be described by different failure probability boxes from multiple sensors all delivering an assignment about the current situation. To obtain an accurate approximation of the current quality of our complex sensor we have to combine all information from all sensors using a combination rule $\otimes$ into one associated probability assignment for the complex sensor.

As mention in [13], different combination rules exist for the combination of basic probability assignments. As the size of the failure probability interval is a measure of uncertainty introduced through unmeasurable influence factors for a given failure we want to involve the interval size in our combination. To this end, the combination rule $\otimes$ is realized using the Zhang rule of combination [14]. In the Zhang combination rule, the product of the margins of the bpas are scaled with a measure of intersection $r(A, B)$.

$$(m_1 \otimes m_2)(C) = k \sum_{A \cap B = C} r(A, B) m_1(A) m_2(B)$$

With k being a renormalization factor that provides that the sum of the bpas to add to 1 and a measure of intersection as the ratio of the cardinality of the intersection of two failure probability intervals divided by the product of the cardinality of the individual failure probability intervals.

$$r(A, B) = \frac{\mid A \cap B \mid}{\mid A \parallel B \mid}$$

3.3   Probability Interval Queries

Depending on application, different queries can be performed by the application on the processed failure probability intervals. These queries include:

1. Maximum bpa of a given failure probability: Returning the plausibility of the given failure probability.
2. Minimum bpa of a given failure probability: Returning the belief of the given failure probability.
3. Maximum failure probability of a given bpa: Given a bpa, this query returns the upper bound of a failure probability interval if one exists.
4. Minimum failure probability of a given bpa: Given a bpa, this query returns the lower bound of a failure probability interval if one exists.

Using these queries allows an application now to perform multiple actions instead of just raising an alert. These actions can be e.g. the notification of the forklift driver in our scenario about a bad performance of the vision system or activating another sensor system.

4   Evaluation

To evaluate our approach we build up the scenario described in Section 2. In the scenario we have a surrounding light setting of 75lx. The complex sensor is realized using a Logitech E3500 webcam and a template matching algorithm to detect the container transported by a forklift. For the realization we use the template matching function cvMatchTemplate with the CV_TM_CCORR_NORMED method from the OpenCV library. The environmental influence through the headlights of the forklift is realized through a dimmable 100W lamp under the object that should be detected by the complex sensor. In the evaluation, the complex sensor continuously captures the environment and tries to find the template of the container in the captured image. The template was created through the captured image at an influence of 100lx through the headlight of the forklift. In this evaluation we measure the current lux value using a lux meter at the WebCam and the distance in pixel between the real object in the captured image and the boundaries of the detected object in the captured image. A distance of more than 1px between the real object and the detected object is assumed as a wrong classification of the situation through the complex sensor.

4.1   Calibration

Two light dependent resistors (LDR) near the lens of the camera are acting as indirect sensors that measure the current lux value. To do so, we measured the voltage at the two LDRs using voltage dividers that are connected to an Arduino Mega 2560 mounted at the camera. In this case, the Arduino includes an AC/DC converter returning the voltage value at the LDRs as a digital value between 0 and 1024. First, we measured the DC voltage value from the LDRs at different light settings starting from 85lx up to 205lx in a step size of 5lx to get the variability of the indirect sensors according to the real lux value. As we see in Fig. 4 the measurements of the LDRs are nearly linear allowing us to map the DC value to a lux value using linear regression with $l_1(v) = 4.79v - 3250.32$ for the first LDR and $l_2(v) = 1.21v - 789.89$ for the second LDR. After that, we can model the variability using a Gaussian distribution with $\sigma_1 = 176.90lx$ and $\sigma_2 = 33.31lx$.

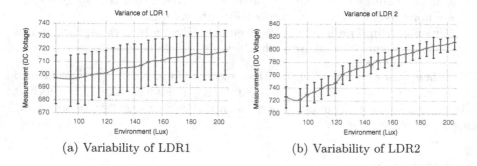

(a) Variability of LDR1   (b) Variability of LDR2

Fig. 4. Variability of the light dependent resistors (LDR) at different light settings

Next, we measure the number of wrong detected positions of the template in the captured image through the complex sensor at different light settings also starting at 85lx up to 205lx in a step size of 5lx. Due to the fact that we cannot simulate each environmental factor in the lab we measured each light setting 600 times. In Fig. 5 the lower line (long dotted green) is the minimal failure probability, the upper line (short dotted red) shows the maximum failure probability of the template matching, and the crosses are the average failure probability at

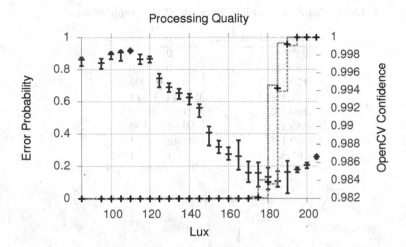

Fig. 5. Failure probability of the complex sensor and confidence of the template matching algorithm according to environmental influence

the given light setting. Furthermore, the figure includes the correlation confidence (blue error bar) of the OpenCV CV_TM_CCORR_NORMED comparison method by values between 0 and 1, with a value of 1.0 for a perfect match, on the right side.

What we see from the resulting data depicted in the figure is that only at a lux value between 85lx and 175lx the algorithm detects the object at the right position in the captured image while the confidence level of the OpenCV algorithm is still high. Furthermore, we have five non-empty failure probability intervals at 170-175lx, 175-180lx, 180-185lx, 185-190lx, and 190-195lx through the measurement of wrong classified situations. With an influence through the headlights above 195lx the algorithm always detects the object at the wrong position. In addition, in the scenario a lux value between 0lx and 85lx could not be measured and so the failure probability in this measurement interval has to be defined as an interval of [0,1], because we have absolute no knowledge about the possible real failure probability.

4.2 Indirect Sensor Measurement

Based on the constructed failure probability intervals we can now express the quality of the complex sensor data through the measurements of the LDRs. For the evaluation we set the value to 185lx resulting in the distribution on the given failure probability intervals depicted in Fig. 6. In addition, the average failure probability (green dotted line) from the light setting is illustrated.

Using the information about the assignments of the different failure probability intervals from the two indirect sensors we can now combine them to one

Table 1. Failure probability intervals of the complex sensor

Measurement (Lux)	$\underline{p}$	$\bar{p}$
[0,85]	0	1
[85,170]	0	0
[170,175]	0	0.0099
[175,180]	0.0033	0.114
[180,185]	0.089	0.70
[185,190]	0.66	0.96
[190,195]	0.948	1
[195,∞]	1	1

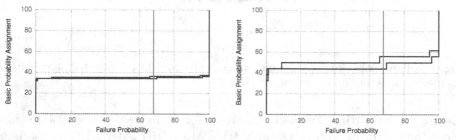

(a) Basic probability assignments of LDR1   (b) Basic probability assignments of LDR2

Fig. 6. Basic probability assignments of the indirect sensors

assignment about the quality of the complex sensor. The result of the combination of both sensors using Zhang combination rule with k=0.0001842 as the renormalization factor is illustrated in Fig. 7.

Based on this information we can now annotate the result of the complex sensor with the knowledge that at least 77.758% of the measurements of the indirect sensor would indicate a total failure of the complex sensor, while 17.446% would indicate no failure at all. Further the annotation would inform that at least 19.367% of the measurements of the indirect sensors would indicate at least a failure probability interval between 0.0 and 0.4 etc. Thus, an adjacent application can use the believe of a failure interval of e.g. [0,0.4] to decide to process the result, [0.4,0.66] to decide to inform the driver of the forklift for feedback, or [0.66,1] to decide to cancel further operation that would use the result of the complex sensor. In this case, if the application is pessimistic it would use the plausibility for the failure interval [0.66,1] which is 80.629% for the decision support. However, if the application is optimistic it would take the believe for the failure probability interval [0.66,1] which is 79.775%. In both cases, the support for the failure probability interval is very high and thus the application should cancel any further operation based on the combined sensor information.

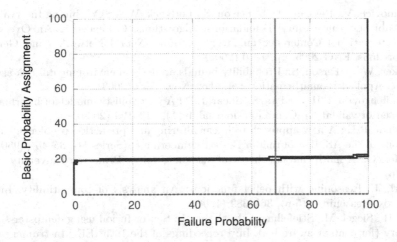

Fig. 7. Combined basic probability assignments of the indirect sensors using Zhang combination rule with k=0.0001842 as the renormalization factor

5   Conclusion and Outlook

The proposed approach to approximate the quality of complex sensors through primitive sensors that measure environmental influences on the complex sensor allows the annotation of measurements through failure probability intervals with a minimum and maximum likelihood. This approach supports pervasive application in decision making through multiple possible queries. The evaluation shows the basic approach based on the described scenario. To do so, the quality of a vision system for object detection is approximated through two LDRs mounted at the vision system that measure the environmental influence through light on the vision system. In an upcoming evaluation, the approach will be evaluated using multiple sensors that measure different influences through the environment on the complex sensor.

Acknowledgment. The authors would like to thank Dr. Sebastian Gerwinn for his superior support for this paper.

References

1. Luo, R., Yih, C.C., Su, K.L.: Multisensor fusion and integration: approaches, applications, and future research directions. IEEE Sensors Journal 2(2), 107–119 (2002)
2. Khaleghi, B., Khamis, A., Karray, F.O., Razavi, S.N.: Multisensor data fusion: A review of the state-of-the-art. Information Fusion (0) (August 2011)
3. Carlson, J., Murphy, R., Christopher, S., Casper, J.: Conflict metric as a measure of sensing quality. In: Proceedings of the 2005 IEEE International Conference on Robotics and Automation, ICRA, pp. 2032–2039 (April 2005)
4. Yu, B., Sycara, K.: Learning the quality of sensor data in distributed decision fusion. In: 9th International Conference on Information Fusion, pp. 1–8 (July 2006)

5. Kreinovich, V., Berleant, D.J., Ferson, S., Lodwick, W.A.: Combining Interval and Probabilistic Uncertainty: Foundations, Algorithms, Challenges – An Overview. In: International Conference on Fuzzy Systems, Neural Networks, and Genetic Algorithms, FNG 2005, pp. 1–10 (2005)
6. Tucker, W.T., Ferson, S.: Probability bounds analysis in environmental risk assessment. Applied Biomathematics, Setauket, New York (2003)
7. Ibargüengoytia, P.H., Vadera, S., Sucar, L.E.: A probabilistic model for information and sensor validation. Computer Journal 49(1), 113–126 (2006)
8. Kalman, R.E.: A new approach to linear filtering and prediction problems. Transactions of the ASME–Journal of Basic Engineering 82(Series D), 35–45 (1960)
9. Shafer, G.: A mathematical theory of evidence, vol. 1. Princeton University Press (1976)
10. Pearl, J.: Reasoning with belief functions: An analysis of compatibility. Int. J. Approx. Reasoning 4(5-6), 363–389 (1990)
11. Wu, H., Siegel, M., Stiefelhagen, R., Yang, J.: Sensor fusion using dempster-shafer theory (for context-aware hci). In: Proceedings of the 19th IEEE Instrumentation and Measurement Technology Conference, IMTC/2002, vol. 1, pp. 7–12 (2002)
12. Pong, P., Challa, S.: Empirical analysis of generalised uncertainty measures with dempster shafer fusion. In: 10th International Conference on Information Fusion, pp. 1–9 (July 2007)
13. Sentz, K., Ferson, S.: Combination of evidence in Dempster-Shafer theory. Technical Report, Sandia National Labs., Albuquerque, NM (April 2002)
14. Zhang, L.: Representation, independence, and combination of evidence in the dempster-shafer theory. In: Yager, R.R., Kacprzyk, J., Fedrizzi, M. (eds.) Advances in the Dempster-Shafer Theory of Evidence, pp. 51–69. John Wiley & Sons, Inc., New York (1994)

Correcting Binary Imprecise Classifiers: Local vs Global Approach

Sébastien Destercke and Benjamin Quost

HEUDIASYC, 6599, Université de Technologie de Compiègne, Centre de Recherches de Royallieu. 60205 COMPIEGNE, France
{desterc,quostben}@hds.utc.fr

Abstract. This paper proposes a simple strategy for combining binary classifiers with imprecise probabilities as outputs. Our combination strategy consists in computing a set of probability distributions by solving an optimization problem whose constraints depend on the classifiers outputs. However, the classifiers may provide assessments that are jointly incoherent, in which case the set of probability distributions satisfying all the constraints is empty. We study different correction strategies for restoring this consistency, by relaxing the constraints of the optimization problem so that it becomes feasible. In particular, we propose and compare a global strategy, where all constraints are relaxed to the same level, to a local strategy, where some constraints may be relaxed more than others. The local discounting strategy proves to give very good results compared both to single classifier approaches and to classifier combination schemes using a global correction scheme.

1 Introduction

In complex multi-class classification problems, a popular approach consists in decomposing the initial problem into several simpler problems, training classifiers on each of these sub-problems, and then combining their results. The advantages are twofold: the sub-problems obtained are generally easier to solve and thus may be addressed with simpler classification algorithms, and their combination may yield better results than using a single classification algorithm.

In this paper, we consider a classical decomposition strategy where each simple problem is binary; then, each classifier is trained to separate two subsets of classes from each other. When the binary classifiers return conditional probabilities estimating whether an instance belongs to a given class subset or not, these conditional probabilities are seldom consistent, due to the fact that they are only approximations of the (admittedly) true but unknown conditional probabilities. Usually, this inconsistency problem is tackled by considering some optimization problem whose solution is a consistent probability whose conditional probabilities are close to each of the estimated ones [5,9]. This consistent probability is then considered as the final predictive model.

Imprecise probabilities are concerned with the cases where the available information is not sufficient (or too conflicting) to identify a single probability,

E. Hüllermeier et al. (Eds.): SUM 2012, LNAI 7520, pp. 299–310, 2012.

and are therefore well adapted to the problem mentioned above. Due to their robustness, imprecise probabilistic models appear particularly interesting in those cases where some classes are difficult to separate, where some classes are poorly represented in the training set or when the data are very noisy. In a previous work [4], we proposed an alternative solution to classifier combination using imprecise probability theory [8]. In this framework, binary classifiers return lower and upper bounds instead of a single evaluation. The case of precise outputs is retrieved when lower and upper bounds coincide.

As even imprecise outputs can turn out to be inconsistent, we initially proposed to apply a global discounting factor (found through a heuristic) to the classifiers. In this paper, we reformulate the problem so that discounting factors can be found by the means of efficient linear programming techniques. This also allows us to easily affect a discounting factor specifically to each classifier, thus adopting a local correction approach. In Section 2, we remind the necessary elements about imprecise probabilities and their use in binary classifiers combination. Section 3 then describes and discusses our discounting strategies, both the global and local one. Finally, we compare in Section 4 the two strategies for the special case of one-vs-one classifiers on several classical real data sets.

2   Imprecise Probability: A Short Introduction

Let $\mathcal{X} = \{x_1, \ldots, x_M\}$ be a finite space of M elements describing the possible values of (ill-known) variables (here, $\mathcal{X}$ represents the set of classes of an instance). In imprecise probability theory, the partial knowledge about the actual value of a variable X is described by a convex set of probabilities $\mathcal{P}$, often called *credal set* [6].

2.1   Expectation and Probability Bounds

A classical way to describe this set consists in providing a set of linear constraints restricting the set of possible probabilities in $\mathcal{P}$ (Walley's lower previsions [8] correspond to bounds of such constraints). Let $\mathcal{L}(\mathcal{X})$ denote the set of all real-valued bounded functions over $\mathcal{X}$, and let $\mathcal{K} \subseteq \mathcal{L}(\mathcal{X})$. Provided $\mathcal{K}$ is not empty, one can compute expectation and probability bounds on a function $f \in \mathcal{K}$.

When one starts from some lower bound $\underline{E} : \mathcal{K} \to \mathbb{R}$, it is possible to associate to it a (convex) set $\mathcal{P}(\underline{E})$ of probabilities such that

$$\mathcal{P}(\underline{E}) = \{p \in \mathbb{P}_{\mathcal{X}} | E(f) \geq \underline{E}(f) \text{ for all } f \in \mathcal{K}\}, \tag{1}$$

where $\mathbb{P}_{\mathcal{X}}$ denotes the set of all probability masses over $\mathcal{X}$.

Alternatively, one can start from a given set $\mathcal{P}$ and compute the lower expectation $\underline{E} : \mathcal{L}(\mathcal{X}) \to \mathbb{R}$ and upper expectation $\overline{E} : \mathcal{L}(\mathcal{X}) \to \mathbb{R}$ such that

$$\overline{E}(f) = \sup_{p \in \mathcal{P}} E(f) \text{ and } \underline{E}(f) = \inf_{p \in \mathcal{P}} E(f),$$

These functions are dual, in the sense that $\overline{E}(f) = -\underline{E}(-f)$.

In Walley's terminology [8], $\underline{E}$ is said to *avoid sure loss* iff $\mathcal{P}(\underline{E}) \neq \emptyset$, and to be *coherent* iff for any $f \in \mathcal{K}$ we have $\underline{E}(f) = \inf_{P \in \mathcal{P}(\underline{E})} E(f)$, i.e. $\underline{E}$ is the lower envelope of $\mathcal{P}(\underline{E})$.

Lower and upper probabilities of an event $A \subseteq \mathcal{X}$ correspond to expectation bounds over the indicator function $\mathbf{1}_{(A)}$ (with $\mathbf{1}_{(A)}(x) = 1$ if $x \in A$, and 0 otherwise). When no confusion is possible, we will denote them $\underline{P}(A)$ and $\overline{P}(A)$ and they are computed as

$$\underline{P}(A) = \inf_{P \in \mathcal{P}} P(A) \quad \text{and} \quad \overline{P}(A) = \sup_{P \in \mathcal{P}} P(A).$$

2.2 Imprecise Probabilities and Binary Classifiers

The basic task of classification is to predict the class or output value x of an object knowing some of its characteristics or input values $y \in \mathcal{Y}$, with $\mathcal{Y}$ the input feature space. Usually, it is assumed that to a given input y correspond a probability mass $p(x|y)$ modeling the class distribution, knowing that the instance y has been observed. Then, classifying the instance amounts to estimating $p(x|y)$ as accurately as possible from a limited set of labeled (training) samples. A binary classifier on a set of classes $\mathcal{X}$ aims at predicting whether an instance class belongs to a subset $A \subseteq \mathcal{X}$ or to a (disjoint) subset $B \subseteq \mathcal{X}$ (i.e., $A \cap B = \emptyset$). For probabilistic classifiers, the prediction takes the form of an estimation of the conditional probability $P(A|A \cup B, y)$ that the instance belongs to A (notice that $P(B|A \cup B, y) = 1 - P(A|A \cup B, y)$ by duality).

In the case of imprecise classifiers, the prediction may be expressed as a set of conditional probabilities, expressed for example as a pair of values bounding $P(A|A \cup B)$ [1]. Let us denote by α_j, β_j the bounds provided by the jth classifier:

$$\alpha_j \leq P(A_j|A_j \cup B_j) \leq \beta_j \tag{2}$$

and, by complementation, we have

$$1 - \beta_j \leq P(B_j|A_j \cup B_j) \leq 1 - \alpha_j. \tag{3}$$

Combining binary classifiers then consists in defining a set $\mathcal{P}$ of probability distributions over $\mathcal{X}$ compatible with the available set of conditional assessments. To get a joint credal set from these constraints, we will turn them into linear constraints over unconditional probabilities. Assuming that $P(A_j \cup B_j) > 0$, we first transform Equations (2) and (3) into

$$\alpha_j \leq \frac{P(A_j)}{P(A_j \cup B_j)} \leq \beta_j \quad \text{and} \quad 1 - \beta_j \leq \frac{P(B_j)}{P(A_j \cup B_j)} \leq 1 - \alpha_j.$$

These two equations can be transformed into two linear constraints over unconditional probabilities:

$$\frac{\alpha_j}{1 - \alpha_j} P(B_j) \leq P(A_j) \quad \text{and} \quad P(A_i) \leq \frac{\beta_j}{1 - \beta_j} P(B_j),$$

---

[1] From now on, we will drop the y in the conditional statements, as the combination always concerns a unique instance which input features remain the same.

or equivalently

$$0 \leq (1 - \alpha_j) \sum_{x_i \in A_j} p_i - \alpha_j \sum_{x_i \in B_j} p_i, \tag{4}$$

$$0 \leq \beta_j \sum_{x_i \in B_j} p_i - (1 - \beta_j) \sum_{x_i \in A_j} p_i, \tag{5}$$

where $p_i := p(x_i)$. Such constraints define the set of probability distributions that are compatible with the classifier outputs. Then, the probability bounds on this set may be retrieved by solving a linear optimization problem under Constraints (4) and (5), for all classifiers. Note that the number of constraints grows linearly with the number N of classifiers, while the number of variables is equal to the number M of classes. As the quantity of classifiers usually remains limited (between M and M^2), the linear optimization problem can be efficiently solved using modern optimisation techniques.

Example 1. Let us assume that $N = 3$ classifiers provided the following outputs:

$$P(\{x_1\}|\{x_1, x_2\}) \in [0.1, 1/3],$$
$$P(\{x_1\}|\{x_1, x_3\}) \in [1/6, 0.4],$$
$$P(\{x_2\}|\{x_2, x_3\}) \in [2/3, 0.8].$$

These constraints on conditional probabilities may be transformed into the following constraints over (unconditional) probabilities p_1, p_2, and p_3:

$$1/9 p_2 \leq p_1 \leq 1/2 p_2, \ 1/5 p_3 \leq p_1 \leq 2/3 p_3, \ 2p_3 \leq p_2 \leq 4p_3,$$

Note that the induced set of probability distributions is not empty, since $p_1 = 0.1, p_2 = 0.6$ and $p_3 = 0.3$ is a feasible solution. Getting the minimal/maximal probabilities for each class then comes down to solve 6 optimization problems (i.e., minimising and maximising each of the unconditional probabilities p_i, under the constraints mentioned above), which yields

$$p_1 \in [0.067, 0.182] \quad p_2 \in [0.545, 0.735] \quad p_3 \in [0.176, 0.31].$$

Here, we can safely classify the instance into x_2. □

Note that, in some cases, the classifiers may provide outputs that are not consistent. This is particularly the case when the classifiers are trained from distinct (non-overlapping) training sets, or when some of them provide erroneous information. Then, $\mathcal{P} = \emptyset$. A solution may still be found provided by (some of) the constraints be relaxed in order to restore the system consistency.

2.3 Vacuous Mixture as Discounting Operator

In some situations, it may be desirable to revise the information provided by a source of information, in particular when the source is known to be unreliable to some extent. Then, the knowledge induced by the source may be weakened according to this degree of unreliability. In most uncertainty theories, this so-called discounting operation consists in combining the original information with a piece of information representing ignorance through a convex combination.

In imprecise probability theory, the piece of information representing ignorance is the *vacuous* lower expectation $\underline{E}_{inf}$, defined such that for any $f \in \mathcal{L}(\mathcal{X})$,

$$\underline{E}_{inf}(f) = \inf_{x \in \mathcal{X}} f(x).$$

Given a state of knowledge represented by a lower expectation $\underline{E}$ on $\mathcal{K}$, the ϵ-discounted lower expectation $\underline{E}^{\epsilon}$ for any $f \in \mathcal{K}$ is

$$\underline{E}^{\epsilon}(f) = (1 - \epsilon)\underline{E}(f) + \epsilon\underline{E}_{inf} \tag{6}$$

with $\epsilon \in [0, 1]$. We may interpret $\underline{E}^{\epsilon}$ as a compromise between the information $\underline{E}$ (which is reliable with a probability $1 - \epsilon$) and ignorance. Note that we retrieve $\underline{E}$ when the source is fully reliable ($\epsilon = 0$), and ignorance when it cannot be trusted ($\epsilon = 1$).

2.4   Decision Rules

Imprecise probability theory offers many ways to make a decision about the possible class of an object [7]. Roughly speaking, classical decision based on maximal expected value can be extended in two ways: the decision rule may result in choosing a single class or in a set of possible (optimal) classes. We will consider the maximin rule, which is of the former type, and the maximality rule, of the latter type.

First, let us remind that for any $x_i \in \mathcal{X}$, the lower and upper probabilities $\underline{P}(\{x_i\}), \overline{P}(\{x_i\})$ are given by the solutions of the constrained optimisation problem

$$\underline{P}(\{x_i\}) = \min p_i \quad \text{and} \quad \overline{P}(\{x_i\}) = \max p_i$$

under the Constraints (4)–(5), and the additional constraints $\sum_{x_i \in \mathcal{X}} p_i = 1$, $p_i > 0$. Then, the maximin decision rule amounts to classify the instance into class $\hat{x}$ such that

$$\hat{x} := \arg \min_{x_i \in \mathcal{X}} \underline{P}(\{x_i\}).$$

Using this rule requires to solve M linear systems with $2N + M + 1$ constraints and to achieve M comparisons.

The maximality rule follows a pairwise comparison approach: a class is considered as possible if it is not dominated by another one. Under the maximality rule, a class x_i is said to dominate x_j, written $x_i \succ_M x_j$, if $\underline{E}(f_{i \to j}) > 0$ with $f_{i \to j}(x_i) = 1$, $f_{i \to j}(x_j) = -1$ and $f_{i \to j}(x) = 0$ for any other element $x \in \mathcal{X}$. The set of optimal classes obtained by this rule is then

$$\hat{X} := \{x_i \in \mathcal{X} | \nexists x_j \text{ s.t. } x_j \succ_M x_i\}.$$

This rule has been justified (and championed) by Walley [8]. Note that finding $\widehat{X}$ requires at most to solve $M^2 - M$ linear programs (one for each pair of classes). Using the maximality rule may seem computationally expensive; however, its computation is easier in a binary framework, as shows the next property.

Proposition 1. $0.5 < P(x_i|x_i \cup x_j) \Rightarrow x_j \notin \widehat{X}$

Proof. If $0.5 < P(x_i|x_i \cup x_j)$, then $p_i > p_j$ according to Equation (4). Assuming constraints (4) and (5) are feasible, i.e. induce a non-empty set $\mathcal{P}$, computing $\underline{E}(f_{i \to j}) > 0$ comes down to solve the following optimisation problem:

$$\min p_i - p_j \tag{7}$$

with $P \in \mathcal{P}$. Since any probability in $\mathcal{P}$ is such that $p_i > p_j$, the value of $p_i - p_j$ is guaranteed to be positive, hence x_j is preferred to x_i ($x_i \succ x_j$).

Note that Proposition 1 only holds if the associated constraint has not been discounted.

3   Discounting Strategies for Inconsistent Outputs

As remarked in Section 2.2, multiple classifiers may provide inconsistent outputs, in which case the constraints induced define an empty credal set $\mathcal{P}$. In this section, we explore various discounting strategies to relax these constraints in order to make the set of probability distributions non-empty.

3.1 ϵ-Discounting of Binary Classifiers

In this paper, we perform an ϵ-discounting for each classifier, in order to relax Constraints (4)–(5) as described by Equation (6). For the jth classifier, we obtain from Constraints (4)–(5) that $\underline{E}_{f_j} = 0$ and $\overline{E}_{\overline{f}_j} = 0$ with

$$\underline{f}_j(x) = \begin{cases} 1 - \alpha_j & \text{if } x \in A_j \\ -\alpha_j & \text{if } x \in B_j \\ 0 & \text{else} \end{cases} \quad \text{and} \quad \overline{f}_j(x) = \begin{cases} 1 - \beta_j & \text{if } x \in A_j \\ -\beta_j & \text{if } x \in B_j \\ 0 & \text{else} \end{cases}.$$

This gives the following discounted equations:

$$\epsilon_j(-\alpha_j) \leq (1 - \alpha_j)P(A_j) - \alpha_j P(B_j), \tag{8}$$
$$\epsilon_j(1 - \beta_j) \geq (1 - \beta_j)P(A_j) - \beta_j P(B_j). \tag{9}$$

Remark that the two discounted equations are here linear in variables p_i and ϵ_j. The constraints become empty when $\epsilon_j = 1$ and are then equivalent to state $P(A_j|A_j \cup B_j) \in [0, 1]$. This means that there always exists a set of coefficients $\{\epsilon_j\}_{j=1,\dots,N}$ that makes the problem feasible.

The question is now how to compute the discounting rates ϵ_j, $j = 1, \ldots, N$ such that the constraints induce a non-empty credal set $\mathcal{P}$ while minimizing the discounting in some sense. We propose the following approach to find the coefficients ϵ_j:

$$\min \sum_{j=1}^{N} \epsilon_j$$

under the constraints

$$\sum_{x_i \in \mathcal{X}} p_i = 1, \quad 0 \leq p_i \leq 1 \text{ for all } i = 1, \ldots, M, \quad 0 \leq \epsilon_j \leq 1 \text{ for all } j = 1, \ldots, N,$$

and Constraints (8)–(9). It is interesting to notice that this new approach is similar to strategies proposed to find minimal sets of infeasible constraints in linear programs [2].

3.2 Credal Discounting vs ϵ-Discounting

In a previous paper [4], we proposed a discounting strategy that was applied to directly to bounds α_j, β_j before transforming Equation (2). The obtained discounted equation for the jth classifier is

$$(1 - \epsilon_j)\alpha_j \leq P(A_j | A_j \cup B_j) \leq \epsilon_j + (1 - \epsilon_j)\beta_j, \quad j = 1, \ldots, N.$$

However, applying such a discounting (or other correction) operation results in quadratic constraints once Equation (2) is "deconditioned" and transformed in Constraints (8)–(9). Discounted constraints on $P(B_j | A_j \cup B_j)$ are obtained by complementation. Remark further that for each constraint, all the coefficients of the square terms p_i^2 and ϵ_j^2 are zero. This implies that the associated quadratic form is indefinite. Therefore, computing the minimum-norm vector of coefficients $\epsilon_1, \ldots, \epsilon_N$ by solving an optimisation problem is very difficult, and searching the space of all solutions is very greedy. To overcome this problem, all the discounting factors were assumed to be equal, and were computed empirically by searching the parameter space (if $\epsilon_1 = \cdots = \epsilon_N$, a dichotomic search can be performed).

In the present approach, we have N discounting rates to compute. However, they may be determined by solving a linear optimization problem under linear constraints, which may be addressed more efficiently than searching the space of discounting coefficients.

3.3 Global vs Local Discounting

In this work, we advocate a local discounting approach, where each classifier is associated with a specific rate ϵ_i. In order to illustrate why this approach seems preferable to a global strategy, where all discounting rates are assumed to be equal, we concentrate on the one-vs-one problem (i.e., each classifier was trained to separate a single class from another). Let us now consider the following simple example:

Example 2. Consider $\mathcal{X} = \{x_1, x_2, x_3, x_4\}$ and the following results:

$$P(x_i|x_i, x_j) \in [0.6, 1]$$

for all pair $1 \le i < j \le 4$, except for $P(x_1|x_1, x_4) \in [0, 0.4]$. Thus, all classifier outputs are consistent with $p_1 > p_2 > p_3 > p_4$, except $P(x_1|x_1, x_4)$ from which one would conclude $p_4 > p_1$. Now, if we were to discount all of them in the same way, we would obtain as a minimal discounting $\epsilon_{ij} = 1/6$ (and $\sum \epsilon_{ij} = 1$, with ϵ_{ij} the discounting value of $P(x_i|x_i, x_j)$), with $p_1 = p_2 = p_3 = p_4 = 1/4$ being the only feasible solution. Thus, in this case, all the information provided by the classifiers is lost, and we are unable to choose between one of the four classes.

Now, assume that each classifier is discounted separately from the others; then, taking $\epsilon_{14} = 1/3$ restores consistency (e.g., $p_1 = 0.5$, $p_2 = 0.31$, $p_3 = 0.2$, $p_4 = 0.09$ is a solution) while still preserving the ordering $p_1 > p_2 > p_3 > p_4$.

4   Experiments

In this section, we present some experiments performed on classical and simulated data sets. We considered both decision rules presented in Section 2.4 to make decisions. Since the maximality rule provides a set of possible classes, we need to define a way to evaluate the accuracy of the decision system in this case. Section 4.1 addresses this topic.

4.1   Evaluating Classifiers Performances

Combined classifiers used with a maximin rule can be directly compared to classical classifiers or to more classical combinations, as both return a single class as output. In this case, accuracy is simply measured as a classical accuracy that will be referred to acc in the following. However, one of the main assets of imprecise probabilistic approaches is the (natural) ability to return sets of classes when information is ambiguous or not precise enough to return a single class. In this case, comparing the imprecise classification output with a classical unique decision is not straightforward.

A first (naive) solution consists in considering the classification as fully accurate whenever the actual class of an evaluated data point belongs to the predicted set of possible classes $\widehat{X}$. It amounts to consider that the final decision is left to the user, who always makes the good choice. The error rate thus computed is an optimistic estimate of the accuracy of the classifier. This estimate will thereafter be referred to as *set accuracy*, or $s - acc$.

Another solution is to use a *discounted* accuracy. Assume we have T observations for which the actual classes $x_i, i = 1, \ldots, T$ are known, and for which T sets of possible classes $\widehat{X}_1, \ldots, \widehat{X}_T$ have been predicted. The discounted accuracy $d - acc$ of the classifier is then

$$d - acc = \frac{1}{T} \sum_{i=1}^{T} \frac{\Delta_i}{g(|\widehat{X}_i|)},$$

with $\Delta_i = 1$ if $x_i \in \widehat{X}_i$, zero otherwise and g an increasing function such that $g(1) = 1$. Although $g(x) = x$ is a usual choice for the discounted accuracy, it has recently been shown [11] that this choice leads to consider imprecise classification as being equivalent to make a random choice inside the set of optimal classes. This comes down to consider that a Decision Maker is risk neutral, i.e., does not consider that having imprecise classification in case of ambiguity is an advantage. This also implies that the robustness of an imprecise classification is rewarded by concave (or risk-averse) functions g.

In our case, we used the function $g(\cdot) = \log_M(\cdot)$ that satisfies $g(1) = 1$ and takes account of the number of classes. Indeed, in the case of two classes, we should have $g(2) = 2$, because predicting two-classes out of two is not informative. However, as M increases, predicting a small number of classes becomes more and more interesting. This is why we pick $\log_M$.

4.2 Datasets and Experimental Setup

We used various UCI data sets that are briefly presented in Table 1. For each of these datasets, we considered the classical one-vs-one decomposition scheme, in which each classifier is trained to separate one class from another. We used as base classifiers CART decision trees (so that comparisons between the precise and the imprecise approaches can be done) and the imprecise Dirichlet model [1] to derive lower and upper conditional probability bounds. This model depends on a hyper-parameter s that settles how quickly the probability converges to a precise value. More precisely, if a_j, b_j are the two classes for the jth binary classifier, and if n_{a_j}, n_{b_j} are the number of training data having respectively a_j and b_j for classes in the leaf of the decision tree reached by the instance, then the bounds are

$$\alpha_j = \frac{n_{a_j}}{n_j + s}, \quad \beta_j = \frac{n_{a_j} + s}{n_j + s},$$

where $n_j = n_{a_j} + n_{b_j}$. Then, s can be interpreted as the number of "unseen" observations, and $\alpha_j = \beta_j$ if $s = 0$.

Table 1. UCI data sets used in experiments

Data set name	#classes M	#input features	#samples
glass	6	9	214
satimage	6	36	6435
segment	7	19	2310
vowel	11	10	990
waveform	3	8	5000
yeast	10	8	1484
zoo	7	18	101
primary tumor	21	17	339
anneal	5	38	898

Table 2 summarises the results obtained for $s = 4$. We compared our method to a single CART decision tree (DT) and to Naive Bayes classifiers (NB). We also displayed the accuracy obtained with maximin rule as well as the set accuracy and the discounted accuracy obtained with the maximality rule, both for the global and local correction methods. We used a 10-fold cross validation. The significance of the differences between the results was evaluated using a Wilcoxon-signed rank test at level 95%. The best results (outside set accuracy) are underlined and results that are not significantly different are printed in bold. Note that we excluded s-acc since it is strongly biased in favor of imprecise decisions.

Table 2. UCI data sets used in experiments

Data set	NB	DT	local			global		
			acc	s-acc	d-acc	acc	s-acc	d-acc
glass	70.55	71.00	**74.77**	79.59	**74.29**	73.73	78.55	**73.79**
satimage	84.34	80.06	87.27	90.43	**88.16**	86.37	89.15	87.78
segment	**96.19**	85.88	**95.58**	97.36	**96.63**	**96.37**	96.88	**96.50**
vowel	**78.88**	72.10	**80.32**	82.84	**81.32**	77.58	77.88	77.27
waveform	71.10	**80.94**	73.24	81.70	75.11	73.44	81.86	75.28
yeast	48.84	45.33	57.13	68.32	**61.94**	55.11	62.39	59.19
zoo	95.17	90.17	**96.17**	96.17	94.59	**96.17**	96.17	94.59
p. tumor	38.05	**48.38**	45.75	45.75	45.75	42.22	42.22	42.22
anneal	**95.99**	93.10	81.74	81.74	81.74	81.86	81.86	81.86

Two main remarks can be made. First, the one-versus-one decomposition strategy provides good results for most data sets, as it gives better results on 6 data sets out of 9. Second, it is clear that the local discounting strategy gives significantly better results than the global discounting strategy. The local strategy dominates the global one on most data sets and gives results very close to the global one otherwise (here, for the "waveform" and "anneal" datasets).

Let us remark that the parameter s, which is directly proportional to the amount of imprecision, has remained the same for all data sets. However, the resulting imprecision also depends on the data. This partly explains the differences between the set accuracy and the discounted accuracy obtained on the datasets: for instance, the resulting imprecision is moderate for "glass" and "yeast", but zero for "anneal" and "primary tumor").

In order to provide an idea of the impact of increasing the overall degree of imprecision, Figure 1 shows the evolution of the discounted accuracy as a function of $\log_2(s)$ (let us remind that since $g(|\widehat{X}_i|) = \log_M(|\widehat{X}_i|)$, the classifier reaches a score of 0.5 when it retains all the classes for all the instances). It shows that moderately increasing the imprecision can give better results (the maximum is reached for $s = 8$) and that the discounted accuracy starts to decrease once the degree of imprecision becomes too large. Similar behaviors could be observed for other data sets.

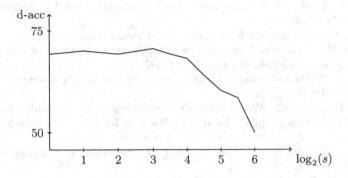

Fig. 1. Evolution of $d - acc$ for data set glass

5   Conclusions

We addressed the problem of pattern classification using binary classifier combination. We adopted imprecise probability theory as a framework for representing the imprecise outputs of the classifiers. More particularly, we consider classifiers that provide sets of conditional probability distributions. It encompasses both cases of precise and imprecise probabilistic outputs (including possibilistic, evidential [3] and credal classifiers [10]). The combination of such classifiers is done by considering classifier outputs as constraints. We presented a local discounting approach for relaxing some of these constraints when the classifiers provide inconsistent outputs. Our strategy computes the discounting rates by solving linear optimization problems, which can be efficiently solved by standard techniques.

Experiments demonstrate that our method give good results compared to the single classifier approach. Moreover, it performs almost always better than the global discounting approach that was presented in a former paper. In future works, we wish to extend our experimentation (by using precise classifiers and genuine imprecise classifiers) and to make a deeper analysis of their results (e.g., checking in which cases inconsistencies happen, verifying that imprecise classification correspond to instances that are hard to classify). We also wish to investigate on the properties of our approach from the point of view of decision making under uncertainty.

References

1. Bernard, J.-M.: An introduction to the imprecise dirichlet model. Int. J. of Approximate Reasoning 39, 123–150 (2008)
2. Chinneck, J.W.: Finding a useful subset of constraints for analysis in an infeasible linear program. INFORMS Journal on Computing 9, 164–174 (1997)
3. Denoeux, T.: A k-nearest neighbor classification rule based on dempster-shafer theory. IEEE Trans. Syst. Man. Cybern. 25, 804–813 (1995)
4. Destercke, S., Quost, B.: Combining Binary Classifiers with Imprecise Probabilities. In: Tang, Y., Huynh, V.-N., Lawry, J. (eds.) IUKM 2011. LNCS, vol. 7027, pp. 219–230. Springer, Heidelberg (2011)

5. Hastie, T., Tibshirani, R.: Classification by pairwise coupling. The Annals of Statistics, 507–513 (1996)
6. Levi, I.: The Enterprise of Knowledge. MIT Press, London (1980)
7. Troffaes, M.: Decision making under uncertainty using imprecise probabilities. Int. J. of Approximate Reasoning 45, 17–29 (2007)
8. Walley, P.: Statistical reasoning with imprecise Probabilities. Chapman and Hall, New York (1991)
9. Wu, T.-F., Lin, C.-J., Weng, R.C.: Probability estimates for multi-class classification by pairwise coupling. Journal of Machine Learning Research 5, 975–1005 (2004)
10. Zaffalon, M.: The naive credal classifier. J. Probabilistic Planning and Inference 105, 105–122 (2002)
11. Zaffalon, M., Corani, G., Maua, D.: Utility-based accuracy measures to empirically evaluate credal classifiers. In: ISIPTA 2011 (2011)

Density-Based Projected Clustering
of Data Streams

Marwan Hassani[1], Pascal Spaus[1], Mohamed Medhat Gaber[2],
and Thomas Seidl[1]

[1] Data Management and Data Exploration Group
RWTH Aachen University, Germany
{hassani,spaus,seidl}@cs.rwth-aachen.de
[2] School of Computing
University of Portsmouth, UK
mohamed.gaber@port.ac.uk

Abstract. In this paper, we have proposed, developed and experimentally validated our novel subspace data stream clustering, termed *Pre-
DeConStream*. The technique is based on the two phase mode of mining
streaming data, in which the first phase represents the process of the
online maintenance of a data structure, that is then passed to an offline
phase of generating the final clustering model. The technique works on
incrementally updating the output of the online phase stored in a microcluster structure, taking into consideration those micro-clusters that are
fading out over time, speeding up the process of assigning new data points
to existing clusters. A density based projected clustering model in developing *PreDeConStream* was used. With many important applications
that can benefit from such technique, we have proved experimentally the
superiority of the proposed methods over state-of-the-art techniques.

1 Introduction

Data streams represent one of the most famous forms of massive uncertain data.
The continuous and endless flow of streaming data results in a huge amount of
data. The uncertainty of these data originates not only from the uncertain characteristics of their sources as in most of the scenarios, but also from their ubiq-
uitous nature. The most famous examples of uncertain data streams are sensor
streaming data which are available in everyday applications. These applications
start from home scenarios like the smart homes to environmental applications
and monitoring tasks in the health sector [12], but do not end with military
and aerospace applications. Due to the nature, the installation and the running
circumstances of these sensors, the data they collect is in most cases uncertain.
Furthermore, the pervasive flow of data and the communication collision additionally leverage the certainty of collected data. The latter fact is the reason why
some other types of data streams can also be uncertain although they are not
produced from sensor data (e.g. streaming network traffic data).

Clustering is a well known data mining technique that aims at grouping similar objects in the dataset together into same clusters, and dissimilar ones into

E. Hüllermeier et al. (Eds.): SUM 2012, LNAI 7520, pp. 311–324, 2012.
© Springer-Verlag Berlin Heidelberg 2012

different clusters, where the similarity is decided based on some distance function. Thus, objects separated by far distances are dissimilar and thus belong to different clusters. In many applications of streaming data, objects are described by using multiple dimensions (e.g. the Network Intrusion Dataset [1] has 42 dimensions). For such kinds of data with higher dimensions, distances grow more and more alike due to an effect termed *curse of dimensionality* [5] (cf. the toy example in Figure 1). Applying traditional clustering algorithms (called in this context: *full-space* clustering algorithms) over such data objects will lead to useless clustering results. In Figure 1, the majority of the black objects will be grouped in a single-object cluster (outliers) when using a full-space clustering algorithm, since they are all dissimilar, but apparently they are not as dissimilar as the gray objects. The latter fact motivated the research in the domain of *subspace* and *projected clustering* in the last decade which resulted in an established research area for static data. For streaming data on the other hand, although a considerable research has tackled the full-space clustering (cf. Section 2.2), very limited work has dealt with subspace clustering (cf. Section 2.3).

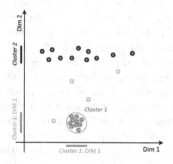

Fig. 1. An example of subspce clustering

Most full-space stream clustering algorithms use a two-phased (online-offline) model (e.g CluStream [2] and DenStream [7], cf. Section 2.2). While the online part summarizes the stream into groups called *microclusters*, the offline part performs some well-known clustering algorithm over these summaries and gives the final output as clustering of the data stream. Usually, the offline part represents the bottleneck of the clustering process, and when considering the projected clustering which is inherently more complicated compared to the full-space clustering, the efficiency of a projected or subspace stream clustering algorithm becomes a critical issue. In this paper we present a density-based projected clustering over streaming data. Our suggested algorithm *PreDeConStream* tries to find clusters over subspaces of the evolving data stream instead of searching over the full space merely. The algorithm uses the famous (online-offline) model, where in the offline phase, it efficiently maintains the final clustering by localizing the part of the clustering result which was affected by the change of the stream input within a certain time, and then sustaining only that part.

Additionally, the algorithm specifies the time intervals within which a guaranteed no-change of the clustering result can be given.

The remainder of this paper is organized as follows: Section 2 gives a short overview of the related work from different neighboring areas. Section 3 introduces some required definitions and formulations to the problem. Our algorithm PreDeConStream is introduced in Section 4, and then thoroughly evaluated in Section 5. Then we conclude the paper with a short outlook in Section 6.

2 Related Work

In this section, we list the related work from three areas: subspace clustering of static data, full-space stream clustering, and finally subspace stream clustering.

2.1 Subspace Clustering Algorithms over Static Data

According to [16], one can differentiate between two main classes of subspace clustering algorithms that deal with static data:

- *Subspace* **clustering algorithms** [14] which aim at detecting all possible clusters in all subspaces. In this algorithm class, each data object can be part of multiple subspace clusters.
- *Projected* **clustering algorithms** [6] which assign each data object to at most one cluster. For each cluster, a subset of projected dimensions is determined which represents the projected subspace.

SubClu [14] is a subspace clustering algorithm that uses the DBSCAN [9] clustering model of density connected sets. SubClu computes for each subspace all clusters which DBSCAN would have found as well if applied on that specific subspace. The subspace clusters are generated in a bottom-up way and for the sake of efficiency, a monotonicity criteria [14] is used. If a subspace T does not contain a cluster, then no higher subspace S with $T \subseteq S$ can contain a cluster.

PreDeCon [6] is a projected clustering algorithm which adapts the concept of density based clustering [9]. It uses a specialized similarity measure based on the subspace preference vector (cf. Definition 6) to detect the subspace of each cluster. Different to DBSCAN, a preference weighted core point is defined in PreDeCon as the point whose number of preference dimensions is at most λ and the preference weighted neighborhood contains at least μ points.

IncPreDeCon [15] is an incremental version of the algorithm PreDeCon [6] designed to handle accumulating data. It is unable to handle evolving stream data since it does not perform any removal or forgetting of aging data. Additionally, the solution performs the maintenance after each insertion, which makes it considerably inefficient, especially for applications with limited memory. The algorithm we present in this paper adopts in some parts of its offline phase the insertion method of IncPreDeCon, but fundamentally differs from IncPreDeCon by maintaining the summaries of drifting streaming data, applying

PreDeCon on the microcluster level, including a novel deletion method, and carefully performing the maintenance of the clustering after some time interval and not after each receiving of an object.

2.2   Full-Space Clustering Algorithms over Streaming Data

There is a rich body of literature on stream clustering. Approaches can be categorized from different perspectives, e.g. whether convex or arbitrary shapes are found, whether data is processed in chunks or one at a time, or whether it is a single algorithm or it uses an online component to maintain data summaries and an offline component for the final clustering. Convex stream clustering approaches are based on a k-center clustering [2,11]. Detecting clusters of arbitrary shapes in streaming data has been proposed using kernels [13], fractal dimensions [17] and density based clustering [7,8]. Another line of research considers the anytime clustering with the existence of outliers [10].

2.3   Subspace Clustering Algorithms over Streaming Data

Similar to the offline clustering algorithms, two types of stream clustering algorithms exist: subspace and projected stream clustering algorithms. However there is, to the best of our knowledge, only one subspace clustering algorithm and two projected clustering ones over streaming data.

Sibling Tree [19] is a grid-based *subspace* clustering algorithm where the streaming distribution statistics is monitored by a list of grid-cells. Once a grid-cell is dense, the tree grows in that cell in order to trace any possible higher dimensional cluster.

HPStream [3] is a k-means-based *projected* clustering algorithm for high dimensional data stream. The relevant dimensions are represented by a d-dimensional bit-vector D, where 1 marks a relevant dimension and 0 otherwise. HPStream uses a projected distance function, called Manhattan Segmental distance MSD [4], to determine the nearest cluster. HPStream cannot detect arbitrary cluster shapes and a parameter for the number of cluster k has to be given by the user, which is in not intuitive in most scenarios. Additionally, as a k-means based approach, HPStream is a bit sensitive to outliers. The model described in this paper is able to detect arbitrarily shaped and numbered clusters in subspaces and, due to its density-based method, is less sensitive to outliers.

HDDStream [18] is a recent density-based projected stream clustering algorithm that was developed simultaneously with PreDeConStream, and published after the first submission of this paper. HDDStream performs an online summarization of both points and dimensions and then, similar to PreDeConStream it performs a modified version of PreDeCon in the offline phase. Different from our algorithm, HDDStream does not optimize the offline part which is usually the bottleneck of subspace-stream clustering algorithm. In the offline phase, our algorithm localizes effects of the stream changes and maintains the old clustering results by keeping non-affected parts. Additionally, our algorithm defines the time intervals where a guaranteed no-change of the clustering result exists, and organizes the online summaries in multiple lists for a faster update.

3   Problem Formulation and Definitions

In this section, we formulate our related problems and give some definitions and data structures that are needed to introduce our PreDeConStream algorithm. In its online phase, our algorithm adopts the microcluster structure used in most other streaming algorithms [2], [7] with an adaptation to fit our problem (cf. Definitions 2-4). Later, we introduce a data structure and some definitions which are related to the offline phase (cf. Definition 5). Since the algorithm uses a density-based clustering over its online and offline phases, similar notations that appear in both phases are differentiated with an F subscript for the offline phase and N for the online phase.

3.1   Basic Definitions

Definition 1. *The Decaying Function* *The fading function [7] used in Pre-DeConStream is defined as* $f(t) = 2^{-\lambda t}$, *where* $0 < \lambda < 1$ *The weight of the data stream points decreases exponentially over time, i.e the older a point gets, the less important it gets. The parameter* λ *is used to control the importance of the historical data of the stream.*

Definition 2. *Core Microcluster* *A core microcluster at time t is defined as a group of close points* $p_1, \ldots, p_n$ *with timestamps* $t_1, \ldots, t_n$. *It is represented by a tuple* $CMC(w, c, r)$ *with:*

1. *Weight,* $w = \sum_{j=1}^{n} f(t - t_j)$, *with* $w \geq \mu_N$
2. *Center,* $c = \frac{\sum_{j=0}^{n} f(t-t_j) p_j}{w}$
3. *Radius,* $r = \frac{\sum_{j=0}^{n} f(t-t_j) dist(p_j, c)}{w}$, *with* $r \leq \varepsilon_N$

The weight and the statistical information about the stream data decay according to the fading function (cf. Definition 1). The maintenance of the microclusters is discussed in Definition 4. Two additional types of microclusters are also given, the potential microcluster and the outlier microcluster, to allow the algorithm to quickly recognize changes in the data stream.

Definition 3. *Potential and Outlier microcluster* *A **potential** microcluster* $PMC = (\overline{CF^1}, \overline{CF^2}, w, c, r)$ *is defined as follows:*

1. *Weight,* $w = \sum_{j=1}^{n} f(t - T_j)$ *with* $w \geq \beta \mu_N$
2. *Linear weighted sum of the points,* $\overline{CF^1} = \sum_{j=1}^{n} f(t - T_j) p_j$
3. *linear weighted squared sum of the points,* $\overline{CF^2} = \sum_{j=1}^{n} f(t - T_j) p_j^2$
4. *Center* $c = \frac{\overline{CF^1}}{w}$
5. *Radius* $r = \sqrt{\frac{|\overline{CF^2}|}{w} - \left(\frac{|\overline{CF^1}|}{w}\right)^2}$

An **outlier** microcluster $OMC = (\overline{CF^1}, \overline{CF^2}, w, c, r, t_0)$ is defined as PMC with the following modifications:

1. Weight $w = \sum_{j=1}^n f(t - T_j)$ with $w < \beta\mu N$
2. An additional entry with the creation time t_0, to decide whether the outlier microcluster is being evolving or is fading out.

The parameter β controls how sensitive the algorithm is to outliers.

Definition 4. **Microclusters Maintenance** With the progress of the evolving stream, any core, potential, or outlier microcluster at time t $MC_t = (\overline{CF^1}, \overline{CF^2}, w)$ is maintained as follows: If a point p hits MC at time $t+1$ then its statistics become: $MC_{t+1} = (2^{-\lambda} \cdot \overline{CF^1} + p, 2^{-\lambda} \cdot \overline{CF^2} + p^2, 2^{-\lambda} \cdot w + 1)$ Otherwise, if no point was added to MC for any time interval δt, the microcluster can be updated after any time interval δt as follows: $MC_{t+\delta t} = (2^{-\lambda\delta t} \cdot \overline{CF^1}, 2^{-\lambda\delta t} \cdot \overline{CF^2}, 2^{-\lambda\delta t} \cdot w)$.

It should be noted that this updating method is different from that in DenStream [7]. The modification considers the decaying of the other old points available in MC, even if MC was updated. This makes the algorithm faster in adapting to the evolving stream data. Additionally, this gives our microcluster structure an upper bound for the weight (w_{max}) of the microcluster which will be useful for the maintenance of the offline part as we will see in Section 3.2.

Lemma 1. The maximum weight w_{max} of any microcluster MC is $\frac{1}{1-2^{-\lambda}}$.

Proof. Assuming that all the points of the stream hit the same microcluster MC. The definition of the weight $w = \sum_{t'=0}^t 2^{-\lambda(t-t')}$ can be transformed with the sum formula for geometric series as following:

$$w = \sum_{t'=0}^t 2^{-\lambda(t-t')} = \frac{1 - 2^{-\lambda(t+1)}}{1 - 2^{-\lambda}} \tag{1}$$

Thus, the maximum weight of a microcluster is:
$w_{max} = \lim_{t\to\infty} w = \lim_{t\to\infty} \frac{1-2^{-\lambda(t+1)}}{1-2^{-\lambda}} = \frac{1}{1-2^{-\lambda}}$.

Any newly created microcluster needs a minimum time T_p to grow into a potential microcluster, during this time the microcluster is considered as an outlier microcluster. Similarly, there is a minimum time T_d needed for a potential microcluster to fade into an outlier microcluster.

Lemma 2. **A)** The minimum timespan for a newly created microcluster to grow into a potential microcluster is: $T_p = \left\lceil \frac{1}{\lambda} \log_2 \left(\frac{1}{1-\beta\mu N(1-2^{-\lambda})} \right) - 1 \right\rceil$.
B) the minimum timespan needed for a potential microcluster to fade into an outlier microcluster is: $T_d = \left\lceil \frac{1}{\lambda} \log_2(\beta\mu N) \right\rceil$.

Proof. **A)**The minimum timespan needed for a newly created microcluster to become potential is $T_p = t_p - t_0$, where t_p is the first timestamp where the

microcluster becomes potential and t_0 the creation time of the *outlier* microcluster. According to Def. 3, a microcluster becomes potential when its weight w becomes $w \geq \beta\mu_N$. Thus, from Equation 1: $w = \sum_{t'=t_0}^{T_p} 2^{-\lambda(t-t')} = \frac{1-2^{-\lambda(T_p+1)}}{1-2^{-\lambda}} \geq \beta\mu_N$.
$\Rightarrow T_p = \left\lceil \frac{1}{\lambda} \log_2\left(\frac{1}{1-\beta\mu_N(1-2^{-\lambda})}\right) - 1 \right\rceil$.

B) Let $T_d = t_d - t_p$ be the minimum timespan needed for a potential microcluster to be deleted, where t_p is the last timestamp where the microcluster was still potential, and t_d is the time when it is deleted. For the deletion, the weight of an outlier microcluster has to be less than $w_{min} = 1$, because the start weight of a newly created microcluster is 1. Let w_p be the last time when the microcluster was potential, according to Def. 4, T_d is the smallest *no-hit* interval that is needed for a potential microcluster to become outlier. Thus: T_d is the smallest value which makes: $w_p \cdot 2^{-\lambda T_d} < 1$. But we know that $w_p = \beta\mu_N \Rightarrow T_d = \left\lceil \frac{1}{\lambda} \log_2(\beta\mu_N) \right\rceil$.

Definition 5. *Minimum Offline Clustering Validity Interval The minimum validity interval of an offline clustering T_v defines the time within which PreDeConStream does not need to update the offline clustering since it is still valid because no change of the status of any microcluster status happened. It is defined as:* $T_v = \min\{T_p, T_d\}$

Definition 6. *Subspace Preference Vector w_c [6] For each dimension i, if the variance of the microclusters c of the Euclidean ε-neighborhood $\mathcal{N}_{\varepsilon_F}(c)$ is below a user defined threshold δ, then the i-th entry of the preference subspace vector w_c is set to a constant $\kappa \gg 1$, otherwise the entry is set to 1.*

3.2   A Data Structure to Manage the Microclusters

A data structure is needed to manage the updated and non-updated microclusters at each timestamp in an efficient and effective way. The main idea is that the algorithm does not need to check for all *potential* microclusters, at each timestamp, whether the potential microcluster remains potential or fades into a *deleted* microcluster. Therefore a data structure is introduced where only a subset of all the potential microclusters needs to be checked. We group the microclusters into multiple lists according to their weight. The borders between these lists are selected as below (cf. also Figure 2) such that all microclusters in a list that are not hit in the previous timestamp will fade to the lower-weighted list. Thus, only the weight of the one which was hit needs to be checked. There are two types of lists: outlier lists l_j^o, and potential lists: l_i^p. The borders of the lists are: $W_d = 1, W_{min} = \beta\mu_N, W_{max} = \frac{1}{1-2^{-\lambda}}$. The internal borders are selected as: $w_i^p = \frac{w_{i-1}^p}{2^{-\lambda}}$ for the potential lists, and $w_i^o = 2^{-\lambda}w_{i-1}^o + 1$ for the outlier lists. It should be noted, that in this case, only the lists around W_{min} from both outlier and potential sides (cf. Figure 2) need to be checked each T_v to see whether the current offline clustering is still valid as we will see in Section 4.

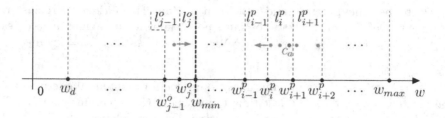

Fig. 2. An example of outlier and potential lists visualized w.r.t. their weights

4   The PreDeConStream Algorithm

Initialisation Phase. In lines 1-4 of Algorithm 1, the minimum timespan T_v based on the user's parameter setting is computed. Furthermore, PreDeConStream needs an initial set of data stream points to generate an initial set of microclusters for the online part. Therefore a certain amount of stream data is buffered and on this initial data, the points are found, whose neighborhood contains at least $\beta\mu_N$ points in its ε_F-neighborhood. If a point p is found, a potential microcluster is created by p and all the points in its neighborhood and they are removed from the initial points. This is repeated until no new potential microcluster is found. Finally the generated initial potential microclusters are inserted into the corresponding lists and the initial clustering is computed with an adapted version of PreDeCon [6].

Algorithm 1. PreDeConStream$(DS, \varepsilon_N, \mu_N, \lambda, \varepsilon_F, \mu_F, \beta, \tau)$

1: $T_p \leftarrow \left\lceil \frac{1}{\lambda} \log_2 \left(\frac{1}{1 - \beta\mu_N(1 - 2^{-\lambda})} \right) - 1 \right\rceil$;
2: $T_d \leftarrow \left\lceil \frac{1}{\lambda} \log_2(\beta\mu_N) \right\rceil$;
3: $T_v \leftarrow \min\{T_p, T_d\}$;
4: initialisation phase
5: **repeat**
6:    get next point $p \in DS$ with the current timestamp t_c
7:    *process(p)*;
8:    maintain microclusters in data structure
9:    **if** $(t_c \mod T_v) == 0$ **then**
10:       $C \leftarrow updateClustering(C)$;
11:    **end if**
12:    **if** user request clustering is received **then**
13:       return clustering C
14:    **end if**
15: **until** data stream terminates

Offline: Maintenance of the Resulting Subspace Clustering. In Algorithm 2, the new arriving data points $p \in DS$ of the stream data within timestamp t are merged with the existing microclusters. In Lines 1-4 of Algorithm 2, the nearest potential microcluster c_p is searched for in all the lists of the potential microclusters l^p. The algorithm clusters the incoming point $p \in DS$ tentatively to c_p to check, if the point actually fits into the potential microcluster c_p. If the radius r_p of the temporary microcluster c_p is still less than ε_F, the point can be clustered into c_p without hesitation. If p does not fit into the nearest potential

---

Algorithm 2. process(data point p)

1: search nearest potential microcluster c_p in all the lists l^p
2: merge p tentatively into c_p
3: **if** $r_p \leq \varepsilon_N$ **then**
4:    insert p into c_p
5: **else**
6:    search nearest outlier microcluster c_o in all the lists l^o
7:    merge p tentatively into c_o
8:    **if** $r_o \leq \varepsilon_N$ **then**
9:      insert p into c_p
10:     **if** $w_o \geq \beta\mu_N$ **then**
11:       insert c_o into potential list $l^p_{\min}$ and remove it from outlier list $l^o_{\max}$
12:     **end if**
13:    **else**
14:     create new outlier microcluster with p
15:    **end if**
16: **end if**

---

microcluster c_p, (cf. Lines 5-12 of Algorithm 2), the algorithm searches for the nearest outlier microcluster c_o in the outlier lists l^o. The algorithm checks again if its radius r_o of c_o is still less than ε_N, when the point is tentatively added to c_o. If the point fits into c_o and it is in the highest list of the outliers l^o, the algorithm checks if the weight w_o is greater than or equal to $\beta\mu_N$. If that is the case, the microcluster c_o is inserted into the lowest list $l^p_{\min}$ of the potential microclusters and has to be considered in the offline part. If the point does not fit into any existing microcluster, a new outlier microcluster is created with this point and is inserted into the outlier microcluster list l^o_0, (cf. Line 14).

Online: Processing of the Data Stream. In Lines 1-10 of Algorithm 3, for each newly created potential microcluster c_p, its subspace preference vector w_{c_p} is computed. Furthermore, for each potential microcluster $c_q \in \mathcal{N}_{\varepsilon_F}(c_p)$, the preference subspace vector of each c_q is updated and checked if its core member property has changed. If that is the case, it is added to the $UPDSEED_i$ set. In Lines 11-19 of Algorithm 3, all the potential microclusters are found which are affected by removing the potential microclusters which faded out in the online part. For each potential microcluster $c_q \in \mathcal{N}_{\varepsilon_F}(c_d)$, the preference subspace vector of each c_q is updated and added to $UPDSEED_d$ if the core member property of c_q has changed because of deleting c_d out of its ε_F-neighborhood. If all the affected potential microclusters were found, $UPDSEED_i$ and $UPDSEED_d$ can be merged to $UPDSEED$. Finally in Lines 20-22, the potential microclusters

---

Algorithm 3. updateClustering(C)

1: **for all** $c_p \in Inserted\_PMC$ **do**
2:    compute the subspace preference vector w_{c_p}
3:    **for all** $c_q \in \mathcal{N}_{\varepsilon_F}(c_p)$ **do**
4:       update the subspace preference vector of c_q
5:       **if** core member property of c_q has changed **then**
6:          add c_q to $AFFECTED\_CORES_i$
7:       **end if**
8:    **end for**
9:    compute $UPDSEED_i$ based on $AFFECTED\_CORES_i$
10: **end for**
11: **for all** $c_d \in Deleted\_PMC$ **do**
12:    **for all** $c_q \in \mathcal{N}_{\varepsilon_F}(c_d)$ **do**
13:       update the subspace preference vector of c_q
14:       **if** core member property of c_q has changed **then**
15:          add c_q to $AFFECTED\_CORES_d$
16:       **end if**
17:    **end for**
18:    compute $UPDSEED_d$ based on $AFFECTED\_CORES_d$
19: **end for**
    $UPDSEED \leftarrow UPDSEED_i \cup UPDSEED_d$
20: **for all** $c_p \in UPDSEED$ **do**
21:    call $expandCluster()$ of PreDeCon [6] by considering the old cluster structure;
22: **end for**

---

of $UPDSEED$ need to be reinserted into the clustering. Starting from a potential microcluster in $UPDSEED$, the function $expandClusters()$ of the algorithm PreDeCon [6] is called under consideration of the existing clustering. This is repeated until all the potential microclusters $c_p \in UPDSEED$ are clustered into a cluster or marked as noise.

5 Experimental Evaluation

In this section, the experimental evaluation of PreDeConStream is presented. PreDeConStream, as well as the two comparative algorithms, HPStream as a k-means based projected algorithm and DenStream as a fullspace density based algorithm, were implemented in Java. All the experiments were done on a Linux operating system with a 2.4 GHz processor and 3GB of memory.

Datasets. For the evaluation of PreDeConStream several datasets were used:
1. Synthetic Dataset: SynStream3D consists of 3-dimensional 4000 objects without noise that form at the beginning two arbitrarily shaped clusters over full space. After some time, the data stream evolves so that for each cluster different dimensions of both clusters become irrelevant.
2. Synthetic Dataset: N100kC3D50R40 generated similar to [7] with 100000 data objects forming 3 clusters with 40 relevant dimensions out of 50.

3. *Real Dataset: Network Intrusion Detection data set KDD CUP'99 (KDDcup)[1]* used to evaluate several stream clustering algorithms [3,7] with 494021 TCP connections, each represents either a normal connection, or any of 22 different types of attacks (cf. Figure 4). Each connection consists of 42 dimensions.

4. *Real Dataset: Physiological Data, ICML'04 (PDMC)* is a collection of activities which was collected by test subjects with wearable sensors over several months. The data set consists of 720792 data objects, each data object has 15 attributes and consists of 55 different labels for the activities and one additional label if no activity was recorded.

Evaluation Measure and Parameter Settings. To evaluate the quality of the clustering results, the cluster purity measure [3,7] is used. For the efficiency, the runtime in seconds was tested. In PreDeConStream the offline parameters ε_F and μ_F specify the density threshold that the clusters must exceed in the offline algorithm part. A lower bound for ε_F is the online parameter ε_N. In the experiments, ε_F was set to at least $2 \times \varepsilon_N$. Unless otherwise mentioned, the parameters for PreDeConStream were set similar to [7] as follows: decay factor $\lambda = 0.25$, initial data object $Init = 2000$, and horizon $H = 5$.

Experiments. Purity and runtime were tested for the three algorithms.

A. Evaluation of Clustering Quality: Using the SynStream3D dataset for both PreDeconStream and Denstream, the online parameters are set to $\varepsilon_N = 4$, $\mu_N = 5$ and $\lambda = 0.25$. For PreDeConStream, the maximal preference dimensionality is set to $\tau = 2$ and $\mu_F = 3$. The stream speed was set to 100 points per time unit. With this speed setting, the data stream evolves at timestamp 26, i.e. one dimension for each cluster becomes irrelevant. It can be seen from Figure 3(a) that the cluster purity of PreDeConStream and DenStream is 100% until the data stream changes, which is not the case for HPStream. This is because both can detect clusters with arbitrarily cluster shapes. Beginning from time unit: 26 the stream evolvs such that in each cluster one dimension is no longer relevant and thus DenStream as a fullspace clustering algorithm, does not detect any cluster. Similarly, Figure 3(b) shows the purity results of both algorithms over the N100C3D50R40 dataset. It can be seen that PreDeConStream outperforms HPStream. The time units are selected in such a way that the changes of the cluster purity can be observed when the stream evolves. It can be observed that HPStream has problems with detecting the changes in the stream. That is because the radius of the projected clusters might be too high and the new points are clustered wrong. PreDeConStream adapts to the changes in the stream fast and keeps a high cluster purity. On the Network Intrusion data set, the stream speed was set to 1000 points per time unit. Since the Network Intrusion data set was already used in [3,7], the same parameter settings are chosen for DenStream and HPStream as in [3,7]. Since PreDeConStream also builds on a microcluster structure, similar parameters settings for the online part of PreDeConStream are chosen, to have a fair comparison. For PreDeConStream: $\beta = 1.23$, $\mu_F = 5$, and $\tau = 32$. For all the three algorithms, the decaying factor λ is set to 0.25.

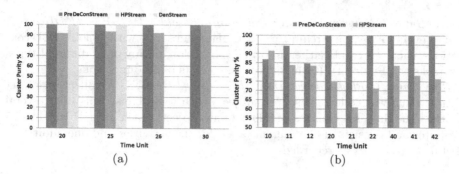

Fig. 3. Clustering purity for: (a) SynStream3D dataset, (b) N100C3D50R40 dataset

Figure 5(a) shows the purity results for the KDDcup Dataset. It can be seen that PreDeConStream produces the best possible clustering quality. For the evaluation, measurements at timestamps where some attacks exist were selected. The data recordings at timestamp 100 and all the recordings within the horizon 5 were only attacks of the type "smurf". At this time unit any algorithm could achieve 100% purity. The attacks that appeared within horizon $H = 5$ in different timestamps are listed in Figure 4. By comparing Figures 4 and 5(a), one can

Normal or attack Type	Objects within horizon $H = 5$ at time unit			
	150	350	373	400
normal	4004	4097	892	406
satan	380	0	0	0
buffer overflow	7	1	2	0
teardrop	99	99	383	0
smurf	143	0	819	2988
ipsweep	52	182	0	0
loadmodule	6	0	0	1
rootkit	1	0	0	1
warezclient	307	0	0	0
multihop	0	0	0	0
neptune	0	618	2688	1603
pod	0	1	99	0
portsweep	0	1	117	1
land	0	1	0	0
sum	5000	5000	5000	5000

Fig. 4. Labels of KDDcup data stream within the horizon $H = 5$, stream speed $= 1000$

observe that PreDeConStream is also resistant against outliers. At the timestamp 350 and 400 there were some outlier attacks within the horizon which affected other algorithms less than PreDeConStream.

Figure 5(b) shows the purity results over the Physiological dataset. The stream speed $= 1000$ and $H = 1$. Again, the timestamps were selected in such a way that there are different activity labels within one time unit. It can be seen from Figure 5(b) that PreDeConStream has the highest purity.

B. Evaluation of Efficiency: The real datasets are used to test the efficiency of PreDeConStream against HPStream. The parameters were set the same way as for the previous experiments on these datasets and the results are shown in Figure 6. Although it is unfair to compare the runtime of a completely density-based approach against a k-means based one, but Figures 6(a) and 6(b) show

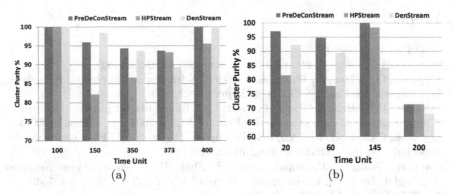

Fig. 5. Clustering purity for: (a) KDDcup dataset, (b) PDMC dataset

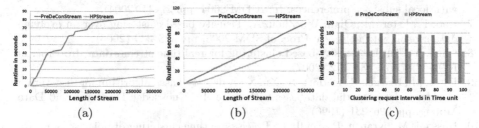

Fig. 6. Runtime results for: (a) KDDcup dataset with a clustering request at each time unit, (b) PDMC dataset with a clustering request at each 20th time unit, and (c) PDMC dataset with different clustering request intervals

a considerable positive effect of our clustering maintenance model when the clustering requests frequency decreases. Usually, clustering requests are not extremely performed at each timestamp or even at each 20th timestamp. This fact motivated a further experiment where we tested the performance of the two algorithms for different clustering frequencies. The result which is depicted in Figure 6(c) confirms our assumption. Due to its clustering maintenance model, PreDeConStream performs better with higher clustering requests intervals. A further experimental evaluation of PreDeConStream was done in [20].

6   Conclusions and Future Work

In this paper we presented a novel algorithm termed *PreDeConStream*. Based on the two phase process of mining data streams, our technique builds a micro-cluster-based structure to store an online summary of the streaming data. The technique is based on subspace clustering targeting applications with high dimensionality of data. For the first time we have utilised projection, density-based clustering, and cluster fading. As a result our technique has proved experimentally its superiority over state-of-the-art techniques. In the future we plan to deploy the technique in a real sensor network testbed, in order to prove its feasibility. Furthermore, a thorough experimental study with different configuration of clusters in the network is planned.

References

1. KDD Cup 1999 Data,
   http://kdd.ics.uci.edu/databases/kddcup99/kddcup99.html
2. Aggarwal, C.C., Han, J., Wang, J., Yu, P.S.: A framework for clustering evolving data streams. In: Proc. of VLDB 2003, pp. 81–92 (2003)
3. Aggarwal, C.C., Han, J., Wang, J., Yu, P.S.: A framework for projected clustering of high dimensional data streams. In: Proc. of VLDB 2004, pp. 852–863 (2004)
4. Aggarwal, C.C., Wolf, J.L., Yu, P.S., Procopiuc, C., Park, J.S.: Fast algorithms for projected clustering. SIGMOD Rec., 61–72 (1999)
5. Beyer, K., Goldstein, J., Ramakrishnan, R., Shaft, U.: When Is Nearest Neighbor Meaningful? In: Beeri, C., Bruneman, P. (eds.) ICDT 1999. LNCS, vol. 1540, pp. 217–235. Springer, Heidelberg (1998)
6. Bohm, C., Kailing, K., Kriegel, H.-P., Kroger, P.: Density connected clustering with local subspace preferences. In: ICDM 2004, pp. 27–34 (2004)
7. Cao, F., Ester, M., Qian, W., Zhou, A.: Density-based clustering over an evolving data stream with noise. In: Proc. of SDM 2006, pp. 328–339 (2006)
8. Chen, Y., Tu, L.: Density-based clustering for real-time stream data. In: Proc. of KDD 2007, pp. 133–142 (2007)
9. Ester, M., Kriegel, H.-P., Jörg, S., Xu, X.: A density-based algorithm for discovering clusters in large spatial databases with noise. In: Knowledge Discovery and Data Mining, pp. 226–231 (1996)
10. Hassani, M., Kranen, P., Seidl, T.: Precise anytime clustering of noisy sensor data with logarithmic complexity. In: Proc. SensorKDD 2011 Workshop in conj. with KDD 2011, pp. 52–60 (2011)
11. Hassani, M., Müller, E., Seidl, T.: EDISKCO: energy efficient distributed in-sensor-network k-center clustering with outliers. In: Proc. SensorKDD 2010 Workshop in conj. with KDD 2009, pp. 39–48 (2009)
12. Hassani, M., Seidl, T.: Towards a mobile health context prediction: Sequential pattern mining in multiple streams. In: Proc. MDM 2011, pp. 55–57 (2011)
13. Jain, A., Zhang, Z., Chang, E.Y.: Adaptive non-linear clustering in data streams. In: Proc. of CIKM 2006, pp. 122–131 (2006)
14. Kailing, K., Kriegel, H.-P., Kröger, P.: Density-connected subspace clustering for high-dimensional data. In: SDM 2004, pp. 246–257 (2004)
15. Kriegel, H.-P., Kröger, P., Ntoutsi, I., Zimek, A.: Towards subspace clustering on dynamic data: an incremental version of predecon. In: Proc. of StreamKDD workshop in conj. with KDD 2010, pp. 31–38 (2010)
16. Kriegel, H.-P., Kröger, P., Zimek, A.: Clustering high-dimensional data: A survey on subspace clustering, pattern-based clustering, and correlation clustering. ACM Trans. on Knowledge Discovery from Data, 1:1–1:58 (2009)
17. Lin, G., Chen, L.: A grid and fractal dimension-based data stream clustering algorithm. In: ISISE 2008, pp. 66–70 (2008)
18. Ntoutsi, I., Zimek, A., Palpanas, T., Kröger, P., Kriegel, H.-P.: Density-based projected clustering over high dimensional data streams. In: Proc. of SDM 2012, pp. 987–998 (2012)
19. Park, N.H., Lee, W.S.: Grid-based subspace clustering over data streams. In: Proc. of CIKM 2007, pp. 801–810 (2007)
20. Spaus, P.: Density based model for subspace clustering on stream data. Bachelor's thesis, Dept. of Computer Science, RWTH Aachen University (May 2011)

Credit-Card Fraud Profiling Using a Hybrid Incremental Clustering Methodology

Marie-Jeanne Lesot and Adrien Revault d'Allonnes

LIP6, Université Pierre et Marie Curie-Paris 6, UMR7606
4 place Jussieu
Paris cedex 05, 75252, France

Abstract. This paper addresses the task of helping investigators identify characteristics in credit-card frauds, so as to establish fraud profiles. To do this, a clustering methodology based on the combination of an incremental variant of the linearised fuzzy c-medoids and a hierarchical clustering is proposed. This algorithm can process very large sets of heterogeneous data, i.e. described by both categorical and numeric features. The relevance of the proposed approach is illustrated on a real dataset containing next to one million fraudulent transactions.

Keywords: Incremental clustering, Hybrid Clustering, Bank Fraud, Credit Card Security.

1 Introduction

With the generalisation of credit and debit cards as modes of payment, credit-card frauds in e-commerce and other mail-order or distant transactions have become a major issue for all banks and card-issuers. For instance, according to the 2011 annual Banque de France report [1], whereas the overall 2010 fraud rate in France is as low as 0.074%, corresponding to an amount of € 368.9 million, the frauds in domestic card-not-present payments (i.e. made online, over the phone or by post) represent 0.26% of this type of transaction, about three and a half times more. These frauds represent 62% of all fraud cases in terms of value.

As a consequence, the analysis and automatic detection of fraudulent transactions has become a largely studied field in the machine learning community [2–4], in particular in the case of e-commerce. This task is both essential from an application point of view, as mentioned above, and scientifically highly challenging, because of its difficulty, part of which is due to the quantity of data that must be processed and the extreme class imbalance.

From a machine-learning standpoint two problems should be separated, namely fraud detection and fraud characterisation. The former aims at predicting whether or not a given transaction should be accepted, so as to decline tentative frauds as they take place. Its objective is, therefore, to differentiate fraudulent and genuine transactions and it is, thus, part of the supervised-learning framework. As such, it should be formulated as a discrimination task in a highly imbalanced

E. Hüllermeier et al. (Eds.): SUM 2012, LNAI 7520, pp. 325–336, 2012.

two-class setting. This particular problem can use the card history to identify frauds as transactions that differ from the card-holder habits.

The second machine-learning problem, fraud characterisation, endeavours to identify distinct fraudster profiles which can then be conceived as operational procedures and used as investigative tools in the apprehension of fraudsters or to assist in fraud detection. The objective is, therefore, to identify, in a set of frauds, distinct subtypes of frauds exhibiting similar properties. It is part of the unsupervised-learning framework and is essentially a clustering task applied to the set of all fraudulent transactions. It should be noted that, in this case, card history is of no use, since fraudster profiles are independent from card-holder habits and, therefore, frauds need only be compared to other frauds and not to legitimate transactions.

In this paper, we consider the latter type of approach and propose a hybrid incremental clustering methodology to address this task. This method exhibits the following characteristics: first, it ensures a rich description of the identified fraudster profiles and allows a multi-level analysis, through the *hierarchical structure* of the extracted clusters it yields. Second, the method allows the processing of large datasets. Indeed, even if frauds represent a small minority of all transactions, they are still very numerous. This is the reason why we propose to combine a hierarchical clustering step, to organise the identified clusters in a dendrogram but with a very high computational cost, to a preliminary data-decomposition step, through an efficient partitioning step. For the partitioning step, we propose an *incremental approach* which processes the dataset in smaller subsets. Third, the method can deal with *heterogeneous data*, i.e. data whose features can be either categorical or numeric. Indeed, apart from the amount which is a number, transactions are, for instance, described by the country where they take place or the general category of transacted product. Finally, on a more general level, the method is *robust*, that is, it does not suffer from its random initialisation.

In the next section, we describe the methodology proposed to address this task. Section 3 then presents the experimental results obtained on real data.

2   Proposed Methodology

Clustering data that are both in vast amounts and of a hybrid nature imposes constraints on candidate algorithms. In this section, we outline the main approaches dealing with these issues and we describe in more detail the linearised Fuzzy c-Medoids [5] on which the proposed methodology relies. We then detail the proposed methodology.

2.1   Related Work

Clustering Large Data Sets. Very large datasets, having become more and more common, have given rise to a large diversity of scalable clustering algorithms.

One way to tackle the problem is to make existing algorithms go faster with specific optimisations. For instance, acceleration of the k-means method and its variants can be achieved using improved initialisation methods to reduce the number of iterations [6, 7]. For the Partitioning Around Medoids (PAM [8]) approach, CLARANS [9] or the linearised fuzzy c-medoids algorithm [5], detailed below, alleviate computational costs by updating medoids in their vicinity.

Another approach, incremental clustering, iteratively applies a clustering algorithm on data subsamples which are processed individually. The samples are extracted from the dataset (e.g. randomly) or they can be imposed by a temporal constraint, when the data is available as time goes by. One variant proposes to build a single sample, guaranteeing its representativeness, so that the results of a single application of a clustering algorithm can be considered as meaningful for the entire dataset [10].

In the general case, however, each sample is clustered and these partial results are then merged into the final partition of the dataset [11]. This fusion can be performed progressively by including in the clustering step of a given sample the results from the previous steps: a sample is summarised by the extracted clusters and this summary is processed together with the next sample [12–14]. The incremental variant of DBSCAN [15] proposes an efficient strategy to determine the region of space where the cluster structure identified in the previous samples should be updated and then locally applies the DBSCAN algorithm. Alternatively, the fusion can be performed at the end, when all samples have been processed, for instance by applying an additional clustering step to the centres obtained from each sample. BIRCH [16], for example, incrementally performs a preclustering step to build a compact representation of the dataset, based on structured summaries that optimise memory usage along user-specified requirements. The centres obtained after scanning the whole dataset then undergo a clustering process. CURE offers a compromise between hierarchical and partitioning clustering, by both using cluster representative points and applying cluster fusion [17]. Both approaches, progressive and closing fusions, can also be combined [18].

Clustering Heterogeneous Data. Hybrid data, data described by both numeric and categorical attributes, define another case where specific clustering algorithms are required. Such data rule out the usage of all mean-centred clustering techniques, in particular the very commonly applied k-means and its variants.

Two main approaches can be distinguished for this problem: so-called relational methods which rely on the pairwise dissimilarity matrix (e.g. based on the pairwise distances) and not on vector descriptions of the data. This type of approach includes, in particular, hierarchical clustering methods, density-based methods [19] as well as relational variants of classic algorithms [20, 21].

On the other hand, medoid-based methods [5, 8] constitute variants of the mean-centered methods that do not define the cluster representative as the average of its members, but as its medoid, that is, the data point that minimises the possibly weighted distance to cluster members.

Linearised Fuzzy c-Medoids. The linearised fuzzy c-medoids algorithm, denoted *l-fcmed* in the following, combines several properties of the previously listed algorithms: it can process data that are both in vast amounts and of a hybrid nature [5]. Indeed, it belongs to both accelerated techniques and medoid-based methods. Moreover, being a fuzzy variant of such algorithms, it offers properties of robustness and independence from random initialisation.

More formally, if x_i, $i = 1, \ldots, n$ are the data points, c the desired number of clusters, v_r, $r = 1, \ldots, c$ the cluster centres and u_{ir} the membership degree of datum x_i to cluster r, the algorithm alternatively updates the membership degrees and the cluster centres using the following equations:

$$
u_{ir} = \left[\sum_{s=1}^{c} \left(\frac{d(x_i, v_r)}{d(x_i, v_s)} \right)^{\frac{2}{m-1}} \right]^{-1} \qquad v_r = \operatorname*{argmin}_{k \in N(v_r)} \sum_{i=1}^{n} u_{ri}^m d(x_k, x_i) \qquad (1)
$$

where m is the so-called fuzzifier, d a suitable metric and $N(v_r)$ the neighbourhood of centre v_r. The latter is defined as the p data maximising membership to cluster r.

The membership degrees are, thus, updated as in the fuzzy c-means and the cluster centres as the data points that minimise the weighted distance to cluster members. To reduce the computational cost, l-$fcmed$ searches for a suitable medoid update close to each current medoid, in $N(v_r)$, instead of computing the minimum over the whole dataset. Both updates are iterated until medoid positions stabilise.

The l-$fcmed$ parameters are c, the number of clusters, m, the fuzzifier, and p, the neighbourhood size. The algorithm also depends on the chosen metric d.

2.2   Global Architecture

To cluster fraudulent transactions, we propose a two-step methodology, illustrated in Figure 1, inspired from the existing approaches described above: before performing a hierarchical clustering, because of its high computational cost, we operate a segmentation using a partitioning algorithm. Because of its advantages, listed in the previous section, we choose to use the linearised fuzzy c-medoids, or rather we propose an incremental extension to further limit computational strains, which we describe in the following.

The second step then uses a hierarchical clustering method to generalise the obtained clusters. Its output dendrogram allows the data analyst to choose the desired level of compromise between homogeneity and generality.

2.3   Incremental Partitioning Step

Following the classic incremental methodology, instead of performing the partitioning clustering on the whole dataset, we operate l-$fcmed$ iteratively on randomly selected samples of size n_l. As detailed below, we propose to introduce two substeps to improve its efficiency in the considered global architecture: medoid selection and unaffected fraud allocation.

Fraud Extraction Step   Partitioning Step   Hierarchical Step

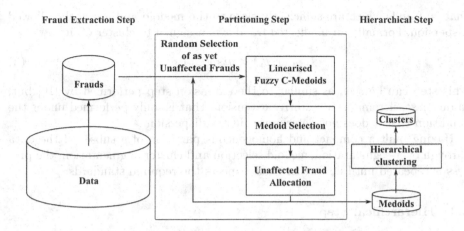

Fig. 1. Global architecture of the proposed methodology

Medoid Selection. The aim of the partitioning step is to purposefully build an over-segmentation to summarise the data while minimising the loss of information, as it is a preliminary step to the hierarchical clustering step. We thus force a compactness constraint on the over-segmented clusters, in order to keep only the most homogeneous, discarding the rest.

To select the clusters C_r, $r = 1, \ldots, c$, whose medoids are highly representative of their assigned data, we propose to keep those of sufficient size and exhibiting a very high homogeneity level. The latter is evaluated by a measure of the dispersion of members of C_r that could, for instance, be relative to the cluster diameter, $diam(C_r) = \max_{x_i, x_j \in C_r} d(x_i, x_j)$. The selection criterion can thus be formalised as:

$$size(C_r) = |C_r| > \tau \quad \text{and} \quad disp(C_r) \leq \xi \tag{2}$$

where τ is the minimal acceptable size and ξ a user-set compactness threshold.

All data in the discarded clusters is then put back into the general pool of transactions to be clustered. They thus become candidates for the random sample selection of following iterations of the partitioning step.

Unaffected Fraud Allocation. Before iterating to the next sample, we scan the data that are yet to be clustered, so as to add unaffected frauds to the identified clusters. This cluster augmentation has a double advantage: first, it avoids the discovery in subsequent iterations of clusters similar to the selected ones, i.e. it avoids cluster duplication or redundancy. It therefore simplifies the posterior fusion step. Moreover, it further alleviates computational costs by reducing the size of the frauds to cluster in following iterations.

This is done by selecting, from the pool of unclustered data, those frauds which can be allocated to the selected clusters without degrading their quality, that is, frauds which are sufficiently close to the medoid and are in the allowed dispersion. Formally an unaffected fraud x is assigned to cluster C_r if:

$$d(x, v_r) < disp(C_r) \tag{3}$$

This step can be seen as similar to the extension step performed by [10] but, in our case, it remains a tentative extension, that is, only performed under the condition that it does not deteriorate cluster dispersion.

Having built a compact and homogeneous partition of a subset of the data through over-segmentation, medoid selection and cluster augmentation, the process is repeated until no further cluster meets the required standards.

2.4   Hierarchical Step

Once all homogeneous clusters satisfying the constraints have been created, a hierarchical clustering with complete linkage is operated on the resulting medoids. Because we have heavily reduced the volume of data with our partitioning, this step is computationally acceptable. The resulting hierarchy offers a progressive agglomeration of clusters and allows for the selection of a suitable compromise between cluster density and number of clusters. The selection of this compromise is made by the visual examination of the hierarchy dendrogram.

3   Experimental Results

3.1   Data and Experimental Setup

We applied the proposed hybrid incremental clustering methodology to a dataset made of 958 828 fraudulent transactions. These correspond to transactions that were rejected by the actual card-holders who, in this way, label the data and identify the transactions to be considered as frauds.

Each fraud in the dataset is described as a combination of numeric and categorical features. The first type of attributes includes the amount of the transaction in euros, a positive real number. The second type includes the country where the transaction took place and the merchant category code of the product.

The distance d between two transactions t_1 and t_2, represented as vectors of their features, is defined as the average of the distances on each attribute, i.e. $d(t_1, t_2) = 1/q \sum_{i=1}^{q} d_i(t_{1i}, t_{2i})$, where d_i is the distance for attribute A_i. This is either $d_i = d_{cat}$, if A_i is categorical, or $d_i = d_{num}$ if it is numeric. Each is defined as follows:

$$d_{cat}(x, y) = \begin{cases} 1 & \text{if } x \neq y \\ 0 & \text{otherwise} \end{cases} \qquad d_{num}(x, y) = \frac{|x - y|}{\max(x, y)}$$

The distance for categorical attributes d_{cat} is binary: it equals 0 if the two values to be compared are identical, and 1 in all other cases. The distance for numeric attributes is defined as a relative gap: the assumption behind this is that a difference of 2€ in amount, for instance, should not have the same impact if the compared amounts are around 5€ or if they are closer to 1 000€.

Regarding the parameters , we use the following setup: each sample contains $n_l = 50\,000$ randomly selected data. *l-fcmed* is initialised randomly and applied with the high value $c = 4\,000$ because we want an over-segmentation. We use the common value $m = 2$ for the fuzzifier. The size of the neighbourhood around the medoids in which the following is selected is $p = \lfloor 50\,000/4\,000 \rfloor$.

Medoid selection, as presented in equation 2, depends on cluster size and dispersion. The minimal size required is set at $\tau = 10$. Since it bears on all attributes, we write the dispersion metric $disp(C)$ as the vector of its attributes' dispersions. For categorical attributes, the number of different values represents dispersion; for numeric attributes, dispersion is best represented by their standard-deviation normalised by their mean value. Formally supposing that for attribute A we write its value in x as $A(x)$, and its average value in cluster C as $\overline{A_C} = \sum_{x \in C} A(x)/|C|$, we may write these dispersions as:

$$disp_{cat}(C) = |\{A(x)|x \in C\}| \qquad disp_{num}(C) = \frac{1}{|C|} \frac{\sqrt{\sum_{x \in C}(A(x) - \overline{A_C})^2}}{\overline{A_C}} \qquad (4)$$

In this way, $disp(C)$ is the vector of all dispersions and ξ is also a vector giving local thresholds, which we set at $\xi_{cat} = 1$, limiting the categorical attributes to a single value in the selected clusters, and $\xi_{num} = 0.01$.

The cut threshold, for the hierarchical clustering, is set to 0.5, based on a visual inspection of the dendrogram (see Figure 4) to achieve an acceptable compromise between the final number of clusters and their homogeneity.

3.2   Incremental Partitioning Step: Results

The evolution of the incremental partitioning step is illustrated on Figures 2 and 3. Figure 2 shows the number of frauds assigned at each iteration, before the allocation of unaffected frauds. It should be observed that this number, starting at 18 373, represents 36.7% of the processed 50 000 points. This small proportion of selected data illustrates how the compactness constraint rejects a large amount of clusters and data. After this initial high value, the amount of selected fraudulent transactions decreases very rapidly. The curve presents an obvious asymptote around 500, i.e. around less than 1% of the considered 50 000 lines of data. This, in itself, is a satisfactory justification for stopping the partitioning step after the illustrated 50 iterations.

The left graph in Figure 3 shows the cumulative number of clusters selected at each iteration. At the first iteration, 394 clusters are selected as being quality clusters. By the end of the process 4 214 have been identified. The inflexion of the curve indicates that the number of newly discovered quality clusters decreases with the number of iterations, another explanation for stopping after fifty iterations of the partitioning step.

The right graph in Figure 3 shows the cumulative number of assigned data after reallocation at each iteration, i.e. the total number of transactions that have

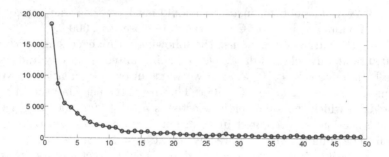

Fig. 2. Evolution of the incremental partitioning step: number of assigned transactions (on the y-axis), at each iteration (on the x-axis)

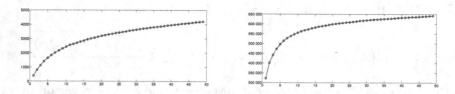

Fig. 3. Evolution of the incremental partitioning step: (left) cumulative number of selected clusters, at each iteration, (right) cumulative number of assigned data after reassignment at each iteration

been assigned to any of the created clusters. 322 115 transactions are assigned at the very first iteration, which indicates a high redundancy in the considered data: many transactions are identical or very close one to another in the considered description space, and thus fulfil the strict homogeneity condition imposed on the assignment step. At each further iteration, the number of assigned transactions drastically decreases, the last iterations bringing very little gain. This curve, thus, also argues in favour of ending iterations of the partitioning.

As a result of these different choices, the incremental partitioning step produces, in the end, 4 214 clusters containing 642 054 frauds.

3.3   Hierarchical Step: Results

Figure 4 shows the dendrogram obtained after applying a hierarchical clustering to the medoids obtained in the partitioning step. The same distance is used for medoids as for transactions.

Visual inspection of the dendrogram prompts a cut above a cost of 0.5. Indeed, cutting the dendrogram at 0.5 yields 156 clusters, which represents a reasonable compromise between cluster homogeneity and number of clusters. The resulting clusters have a small diameter and their number, 156 as compared to the original 958 828 transactions, is a cognitively manageable amount of data to study for the human analyst.

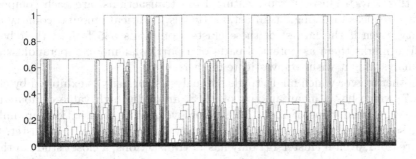

Fig. 4. Dendrogram from the hierarchical clustering step: medoids from the partitioning step on the x-axis, cost of fusion on the y-axis

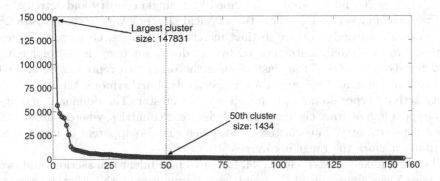

Fig. 5. Sizes of the final clusters on the y-axis, size-ordered cluster-index on the x-axis

3.4   Final Results

To take a closer look at the resulting clusters, we start off by studying the distribution of cluster sizes globally.

Figure 5 shows the cluster sizes in decreasing order and also shows a high disparity. In particular, the greatest cluster represents 23% of affected frauds on its own, whereas the second largest is only one third of the largest. More generally, the fifty largest clusters cover 92% of all affected frauds. In the following we successively comment the hundred and six smallest and the fifty largest clusters.

Analysis of the Smaller Clusters. On average the smaller clusters, the ones after the first fifty, only contain 471 frauds, i.e. 7‰ of the affected frauds. It could, therefore, be argued that they contain too little information and that it not necessary to study them further. Indeed, the fraud profiles they are associated to may seem too anecdotal. However, some of these clusters, or groups of clusters, may still warrant the analyst's attention because of some striking characteristics.

In this way, 9 clusters, representing 1 745 transactions, are each composed of exact replicas of a single transaction, that is, identical amount, country and activity. Even if the largest of these clusters only has 535 frauds, their being identical marks them as potential parts of a particular modus operandi, which any analyst might wish to investigate further.

Another trigger which might tingle an analyst's curiosity is exhibited by cluster 147, which has a mean amount of 914€, an oddly high value compared to the average fraud value of about 112€. Moreover, all its transactions are linked to the same country and activity. This activity only appears in this cluster. For these reasons, even if cluster 147 only has 72 transactions, a closer study of these transactions seems appropriate to find other similarities, such as the identity of the seller or the dates on which they took place, for instance.

Analysis of the Larger Clusters. The fifty largest clusters represent the most notable fraudster profiles. Indeed, they are singularly homogeneous: just two of them, clusters 26 and 44, are not described by a single country and activity.

If we take a closer look at these two atypical clusters, we see that cluster 26 associates 4 546 frauds to three distinct merchant activities. Of these three activities, two are heavily outnumbered by the dominant one, the latter having 4 402 frauds, or 96.8% of the cluster, when the other two represent 80 and 64 transactions, that is, 1.8% and 1.4% respectively. Furthermore, these two minority activity types do not appear in any other cluster. The dominant activity does appear in five other clusters but with different countries, whereas cluster 26 only has one country. This cluster's homogeneity is also apparent in its amount distribution, since the range it covers is $[10, 13.16]$.

The other mixed cluster, cluster 44, has 1 670 fraudulent transactions and two countries. Once again, one of the countries is dominated by the other, representing just 104 individuals or 6.2% of the cluster population, and is only present in this cluster. These compactness anomalies are, therefore, slight and explainable.

If we turn back to the general population of large clusters, the 96% with only one activity and one country, we see that compactness does not constrain size. Indeed, the largest of all clusters, with 147 831 frauds, equivalent to 23% of all affected frauds, belongs to this category. Regarding the amounts, this cluster spreads over the $[1, 480]$ interval, a reasonable size compared to the global span of the data over $[0, 10 276]$. The left part of Figure 6 shows the histogram of its amounts and also shows how the distribution is concentrated on small values.

Another illustration of the first fifty clusters' quality, is given by cluster 12. This cluster, still defined over a single country and activity, has 8 828 frauds with amounts on the interval $[0.99, 83.73]$. The right part of Figure 6 offers a view on their dispersion. Looking at this distribution, we see that it could be subdivided into homogeneous subintervals, probably the ones given by the partitioning step, later joined by the hierarchical clustering. Cluster 12, thus, illustrates the use of this fusion step: instead of studying individually all 93 original clusters – the clusters fused during the hierarchical step to form cluster 12 – the analyst can focus on a generalised view, yet still be able to identify potentially interesting subgroups. The analyst may yet explore these subgroups by choosing to cut

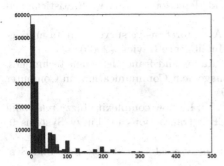 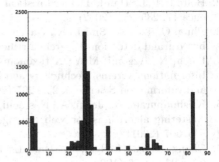

Fig. 6. Histogram of the amounts (left) for the largest cluster (population: 147 831; bin-width: 9.58€), (right) for 12th largest cluster (population: 8 828; bin-width: 1.65€)

the dendrogram at a lower value. This inspection can be made locally by concentrating on the branch which actually contains the interesting clusters. This granularity refocusing ability, local or not, emanating from the cluster hierarchy, is an added benefit and justification for the proposed global architecture of the clustering method.

4   Conclusion

In this paper, we proposed a methodology for the identification of the characteristics of credit-card frauds, through the identification of distinct fraud profiles. It is based on the combination of an incremental variant of the fuzzy c-medoids with hierarchical clustering, and it is thus able to process very large heterogeneous data. We illustrated the relevance of the proposed approach on a real dataset describing next to one million online fraudulent transactions.

Ongoing work aims at enriching the interpretation of the obtained profiles, in particular by the construction of typical transactions representing each fraud profile, so as to ease their characterisation. To that aim, the use of fuzzy prototypes is considered, in order to underline the specificity of each profile as opposed to the others.

Acknowledgments. This work was supported by the project eFraudBox funded by ANR-CSOSG 2009. The authors also thank Nizar Malkiya for his help in implementing and testing the methodology.

References

1. Banque de France: Annual Report of the Observatory for Payment Card Security (2010),
   http://www.banque-france.fr/observatoire/telechar/gb/2010/
   rapport-annuel-OSCP-2010-gb.pdf

2. Bolton, R.J., Hand, D.J.: Statistical fraud detection: a review. Statistical Science 17, 235–255 (2002)
3. Phua, C., Lee, V., Smith, K., Gayler, R.: A comprehensive survey of data mining-based fraud detection research. Artificial Intelligence Review (2005)
4. Laleh, N., Azgomi, M.A.: A taxonomy of frauds and fraud detection techniques. Information Systems, Technology and Management Communications in Computer and Information Science 3, 256–267 (2009)
5. Krishnapuram, R., Joshi, A., Nasraoui, O., Yi, L.: Low complexity fuzzy relational clustering algorithms for web mining. IEEE Transactions on Fuzzy Systems 9, 595–607 (2001)
6. Cheng, T.W., Goldgof, D., Hall, L.: Fast fuzzy clustering. Fuzzy Sets and Systems 93, 49–56 (1998)
7. Altman, D.: Efficient fuzzy clustering of multi-spectral images. In: Proc. of the IEEE Int. Conf. on Fuzzy Systems, FUZZ-IEEE 1999 (1999)
8. Kaufman, L., Rousseeuw, P.: Finding groups in data, an introduction to cluster analysis. John Wiley & Sons, Brussels (1990)
9. Ng, R., Han, J.: Efficient and effective clustering methods for spatial data mining. In: Proc. of the 20th Very Large DataBases Conference, VLDB 1994, pp. 144–155 (1994)
10. Hathaway, R., Bezdek, J.: Extending fuzzy and probabilistic clustering to very large data sets. Computational Statistics & Data Analysis 51, 215–234 (2006)
11. Hore, P., Hall, L., Goldgof, D.: A cluster ensemble framework for large data sets. Pattern Recognition 42, 676–688 (2009)
12. Farnstrom, F., Lewis, J., Elkan, C.: Scalability for clustering algorithms revisited. SIGKDD Explorations 2(1), 51–57 (2000)
13. Hore, P., Hall, L., Goldgof, D.: Single pass fuzzy c means. In: Proc. of the IEEE Int. Conf. on Fuzzy Systems, FUZZ-IEEE 2007, pp. 1–7 (2007)
14. Hore, P., Hall, L., Goldgof, D., Cheng, W.: Online fuzzy c means. In: Proc. of NAFIPS 2008, 1–5 (2008)
15. Ester, M., Kriegel, H.P., Sander, J., Wimmer, M., Xu, X.: Incremental clustering for mining in a data warehousing environment. In: Proc. of the 24th Very Large DataBases Conference, VLDB 1998, pp. 323–333 (1998)
16. Zhang, T., Ramakrishnan, R., Livny, M.: Birch: an efficient data clustering method for very large databases. In: Proc. of the ACM Int. Conf on Management of Data, SIGMOD 1996, pp. 103–114. ACM Press (1996)
17. Guha, S., Rastogi, R., Shim, K.: CURE: an efficient clustering algorithm for large databases. In: Proc. of the ACM Int. Conf on Management of Data, SIGMOD 1998, pp. 73–84 (1998)
18. Bradley, P., Fayyad, U., Reina, C.: Scaling clustering algorithms to large databases. In: Proc. of KDD 1998, pp. 9–15. AAAI Press (1998)
19. Sander, J., Ester, M., Kriegel, H.P., Xu, X.: Density-based clustering in spatial databases: the algorithm DBSCAN and its application. Data Mining and Knowledge Discovery 2(2), 169–194 (1998)
20. Hathaway, R., Bezdek, J.: Nerf c-means: non euclidean relational fuzzy clustering. Pattern Recognition 27, 429–437 (1994)
21. Hathaway, R., Bezdek, J., Davenport, J.: On relational data versions of c-means algorithms. Pattern Recognition Letters 17, 607–612 (1996)

Comparing Partitions by Means of Fuzzy Data Mining Tools

Carlos Molina[1], Belén Prados[2], María-Dolores Ruiz[3], Daniel Sánchez[4],
and José-María Serrano[1]

[1] Dept of Computer Science, University of Jaén, Spain
{carlosmo,jschica}@ujaen.es
[2] Dept. Software Engineering, University of Granada, Spain
belenps@ugr.es
[3] Dept. Computer Science and A.I., University of Granada, Spain
mdruiz@decsai.ugr.es
[4] European Centre for Soft Computing, Mieres, Spain
daniel.sanchezf@softcomputing.es

Abstract. Rand index is one of the most popular measures for comparing two
partitions over a set of objects. Several approaches have extended this measure
for those cases involving fuzzy partitions. In previous works, we developed a
methodology for correspondence analysis between partitions in terms of data
mining tools. In this paper we discuss how, without any additional cost, it can
be applied as an alternate computation of Rand index, allowing us not only to
compare both crisp and fuzzy partitions, but also classes inside these partitions.

Keywords: fuzzy partitions, fuzzy data mining tools, Rand index.

1 Introduction

Fuzzy models have been extensively used in pattern recognition. In particular, clustering techniques have been extended to determine a finite set of groups or categories,
that can be fuzzy (elements being associated to each cluster with a degree of membership), to describe a set of objects with similar features. Developed algorithms have
been successfully applied in a wide range of areas including image recognition, signal
processing, market segmentation, document categorization and bioinformatics.

The main problems arising the comparison of two fuzzy partitions of a given set are
the following: (1) the number of clusters in both partitions are not necessarily the same,
(2) the measures for comparing two equivalent partitions, that can be represented by
matrices A and B, must be invariant under row permutations.

As far as we know, most of current approaches are only suitable for comparing a
fuzzy partition with a crisp one, where the latter represents the "true" partition of data.
But in nearly all real cases, there is no such crisp partition giving a perfect matching.

There are three kinds of approaches for evaluating the partitions quality: internal,
external and relative criteria [27]. *Internal* criterion is used for evaluating a partition
separately, usually for measuring the grade of fit between the partition and the input
data. *External* measures compare the obtained partition with a reference partition that

E. Hüllermeier et al. (Eds.): SUM 2012, LNAI 7520, pp. 337–350, 2012.

pertains to the data but which is independent of it. *Relative* measures, also known as relative indices, assess the similarity between two partitions computed by different methods. Our approach belongs to this last group. Our goal in this paper is to define an alternative to the popular Rand index [29], by means of a family of data mining tools, applied to both crisp and fuzzy correspondence analysis between partitions.

The paper is organized as follows. In the following section, we mention some comparison methods between partitions, specially those related to fuzzy cases. Then, we summarize the models for data mining employed as tools for analyzing some types of correspondences, described in the next section. After this, there is our problem approach in terms of the data mining measures applied to correspondence analysis. Finally, some future trends in this work to come are defined as well as we present our conclusions.

2   Rand Index and Other Comparison Measures

Comparison methods include those measures that compare two partitions. When comparing a resulting partition by a clustering process with a referential one, which is considered to be the "true" partition, we will call it an external method. External indices [27] give the expert an indication of the quality of the resulting partition, while when comparing two different partitions we obtain a grade of how similar they are. If both partitions come from different clustering processes the method is considered as relative.

There are many indices to be reviewed for crisp partitions [22] (see also [2]). For fuzzy partitions we will refer to the most important approaches developed until now. Many of them are generalizations of crisp measures.

The Rand index [29] proposed by Rand in 1971 is given in terms of the number of pairwise comparisons of data objects. It is one of the most popular indices. Given A and B two crisp clusters we set:

- a, pairs belonging to the same cluster in A and to the same cluster in B.
- b, pairs belonging to the same cluster in A but to a different cluster in B.
- c, pairs belonging to a different cluster in A but to the same cluster in B.
- d, pairs belonging to different clusters in both A and B.

Then, the Rand index is given by the proportion between the number of agreements and the total number of pairs:

$$I_R(A,B) = \frac{a+d}{a+b+c+d} \tag{1}$$

Campello [13] extends the Rand index for comparing fuzzy partitions. For that purpose, he rewrites the original formulation in terms of the fuzzy partitions. Let X and Y be two fuzzy partitions defined over the set of objects O, we consider:

- $X_1 = \{(o,o') \in O \times O$ that belong to the same cluster in $X\}$.
- $X_0 = \{(o,o') \in O \times O$ that belong to different clusters in $X\}$.
- $Y_1 = \{(o,o') \in O \times O$ that belong to the same cluster in $Y\}$.
- $Y_0 = \{(o,o') \in O \times O$ that belong to the different clusters in $Y\}$.

The Rand index is rewritten in terms of the previous four quantities: $a = |X_1 \cap Y_1|, b = |X_1 \cap Y_0|, c = |X_0 \cap Y_1|, d = |X_0 \cap Y_0|$. In the fuzzy case the sets X_i, Y_i are defined by means of a t-norm $\otimes$ and a t-conorm $\oplus$. Let $X_i(o) \in [0,1]$ the degree of membership of element $o \in O$ in the i-th cluster of X. Analogously $Y_i(o) \in [0,1]$ is the degree of membership of element $o \in O$ in the i-th cluster of Y

- $X_1(o,o') = \displaystyle\bigoplus_{i=1}^{k} X_i(o) \otimes X_i(o')$ • $X_0(o,o') = \displaystyle\bigoplus_{1 \le i \ne j \le k} X_i(o) \otimes X_j(o')$

- $Y_1(o,o') = \displaystyle\bigoplus_{i=1}^{l} Y_i(o) \otimes Y_i(o')$ • $Y_0(o,o') = \displaystyle\bigoplus_{1 \le i \ne j \le l} Y_i(o) \otimes Y_j(o')$

The four frequencies taking part in equation (1) are then formulated in terms of the intersection of these sets using the sigma-count principle:

$$a = \sum_{(o,o') \in O \times O} X_1(o,o') \otimes Y_1(o,o') \qquad b = \sum_{(o,o') \in O \times O} X_1(o,o') \otimes Y_0(o,o')$$
$$c = \sum_{(o,o') \in O \times O} X_0(o,o') \otimes Y_1(o,o') \qquad d = \sum_{(o,o') \in O \times O} X_0(o,o') \otimes Y_0(o,o') \tag{2}$$

This is not the only generalization of the Rand index. We can find in the literature the approaches of Frigui et al. [21], Brouwer [11], Hüllermeier and Rifqi [23] and Anderson et al. [2]. In [3,2] the reader may find a more extensive comparison of the cited indices. We will resume the main differences between them:

- Campello was interested in comparing a fuzzy partition with a non-fuzzy one, but its proposal is formulated for comparing two fuzzy partitions.
- Frigui et al. present generalizations for several indices including the Rand index. They also restrict the approach when one of the partitions is a crisp one. When using product for the t-norm and sum for the t-conorm for the Campello's approach we obtain this particular case [21].
- Brouwer presents another generalization by defining a relationship called bonding that describes the degree to which two objects are in the same cluster. Then, bonding matrices are built using previous relation and the cosine distance [11].
- Hüllermeier and Rifqi's approach is defined for every two fuzzy partitions by defining a fuzzy equivalence relation on the set of objects O. This fuzzy relation is then used for defining the degree of concordance or discordance between two objects $o, o' \in O$. The distance obtained using the resulting index satisfies the desirable properties for a pseudo-metric and in some special cases it is a metric [23].

A very similar index was proposed by several authors: the so-called Jaccard coefficient [24] where the participation of the quantity d is suppressed in Campello's index.

The Fowlkes-Mallows index proposed in [20] can be defined as in equation (3) obtaining a value of 1 when clusters are good estimates of the groups.

$$I_F(A,B) = \frac{a}{\sqrt{(a+b)(a+c)}} \tag{3}$$

Previous indices, as well as the Adjusted Rand index of Hubert and Arabie [22], the C statistics [25] and the Minkowski measure [26] can be defined in terms of the four frequencies a, b, c, d and they are related to Rand index. All these indices allow uniquely the evaluation of hard (crisp) clustering partitions, but some authors [13,14,2] have extended all of them in a unified formulation. The first attempt of Campello [13] relies solely on the redefinition of the four frequencies using basic fuzzy set concepts, but it has the shortcoming that one of the partitions must to be hard for keeping the important property of reaching their maximum (unit value) when comparing equivalent partitions. A more recent approach [14] settles this shortcoming by defining a *fuzzy transfer distance* between two fuzzy partitions. In addition, Campello also addresses the problem of how to compare two partitions from different subsamples of data.

Anderson et al. [2] developed a method to generalize comparison indices to all possible cases concerning two different partitions: crisp, fuzzy, probabilistic and possibilistic and for every index that can be expressed in terms of the four mentioned frequencies.

A different proposal by Di Nuovo and Catania [27], called DNC index, is based on a defined measure called *degree of accuracy* which is intended to measure the degree of association of a partition with its reference partition representing the real group.

A quite different approach is that developed by Runkler [31] which is based on the similarities between the resultant subsets by the partitions. The *subset similarity index* is computed in terms of the similarities between all the partitions subsets. This index is reflexive and invariant under row permutations which are desirable properties.

3   Crisp and Fuzzy Data Mining Tools

3.1   Association Rules

Given a set I ("set of items") and a database D constituted by set of transactions ("T-set"), each one being a subset of I, association rules [1] are "implications" of the form $A \Rightarrow B$ that relate the presence of itemsets A and B in transactions of D, assuming $A, B \subseteq I, A \cap B = \emptyset$ and $A, B \neq \emptyset$.

The ordinary measures proposed in [1] to assess association rules are *confidence* (the conditional probability $P(B|A)$) and *support* (the joint probability $P(A \cup B)$). An alternative framework [8,16] measures accuracy by means of Shortliffe and Buchanan's certainty factors [33], showing better properties than confidence, and helping to solve some of its drawbacks. Let $\text{supp}(B)$ be the support of the itemset B, and let $\text{Conf}(A \Rightarrow B)$ be the confidence of the rule. The *certainty factor* of the rule is defined as

$$CF(A \Rightarrow B) = \begin{cases} \frac{\text{Conf}(A \Rightarrow B) - \text{supp}(B)}{1 - \text{supp}(B)} & \text{if } \text{Conf}(A \Rightarrow B) > \text{supp}(B \\ \frac{\text{Conf}(A \Rightarrow B) - \text{supp}(B)}{\text{supp}(B)} & \text{if } \text{Conf}(A \Rightarrow B) < \text{supp}(B) \\ 0 & \text{otherwise.} \end{cases} \tag{4}$$

The certainty factor yields a value in the interval [-1, 1] and measures how our belief that B is in a transaction changes when we are told that A is in that transaction.

3.2 Formal Model for Mining Fuzzy Rules

Many definitions for fuzzy rule can be found in the literature, but in this work, we will apply the formal model developed in [17,19], which allows us to mine fuzzy rules in a straightforward way extending the accuracy measures from the crisp case. Its formalization basically underlies in two concepts: the representation by levels associated to a fuzzy property (RL for short) and the four fold table associated to the itemsets A and B in database D, noted by $\mathcal{M} = 4ft(A,B,D)$.

A RL associated to a fuzzy property P in a universe X is defined as a pair (Λ_P, ρ_P) where $\Lambda_P = \{\alpha_1, \ldots, \alpha_m\}$ is a finite set of levels verifying that $1 = \alpha_1 > \cdots > \alpha_m > \alpha_{m+1} = 0$ and $\rho_P : \Lambda_P \to \mathscr{P}(X)$ is a function which applies each level into the crisp realization of P in that level [32]. The set of crisp representatives of P is the set $\Omega_P = \{\rho_P(\alpha) \,|\, \alpha \in \Lambda_P\}$. The values of Λ_P can be interpreted as values of possibility for a possibility measure defined for all $\rho_P(\alpha_i) \in \Omega_P$ as $Pos(\rho_P(\alpha_i)) = \alpha_i$. Following this interpretation we define the associated probability distribution $m : \Omega_P \to [0,1]$ as in equation (5) which give us information about how representative is each crisp set of the property P in Ω_P.

$$m_P(Y) = \sum_{\alpha_i \,|\, Y = \rho(\alpha_i)} \alpha_i - \alpha_{i+1} \tag{5}$$

For each $Y \in \Omega_P$, the value $m_P(Y)$ represents the proportion to which the available evidence supports claim that the property P is represented by Y. From this point of view, a RL can be seen as a basic probability assignment in the sense of the theory of evidence, *plus a structure indicating dependencies between the possible representations of different properties.*

The four fold table associated to the itemsets involved in a rule $A \Rightarrow B$ detaches the number of transactions in D satisfying the four possible combinations between A and B using the logic connectors $\wedge$ (conjunction) and $\neg$ (negation). So, $\mathcal{M} = 4ft(A,B,D) = \{a,b,c,d\}$ where a is the number of rows of D satisfying $A \wedge B$, b the number of rows satisfying $A \wedge \neg B$, c represents those satisfying $\neg A \wedge B$ and d those satisfying the last possibility $\neg A \wedge \neg B$ [30,18]. Note that $|D| = a + b + c + d = n$. The validity of an association rule is assessed by using $\mathcal{M}$ by means an operator $\approx$ (interestingness measure) called 4ft-quantifier. In particular, known measures of support and confidence are 4ft-quantifiers defined as follows:

$$\begin{aligned} \text{Supp}(A \Rightarrow B) = \approx_S (a,b,c,d) = \frac{a}{a+b+c+d} \\ \text{Conf}(A \Rightarrow B) = \approx_C (a,b,c,d) = \frac{a}{a+b} \end{aligned} \tag{6}$$

and we can use them to define the certainty factor $\approx_{CF} (a,b,c,d)$ in terms of the four frequencies of $\mathcal{M}$ (see [18] for its shorter form).

Using these two models we have proposed [19] a framework for fuzzy rules that ables us to extend the interestingness measures for their validation from the crisp to the fuzzy case. Summarizing the model, we can represent the fuzzy sets appearing in the fuzzy rule by the associated RLs $(\Lambda_{\tilde{A}}, \rho_{\tilde{A}})$, $(\Lambda_{\tilde{B}}, \rho_{\tilde{B}})$ and for every level in $\Lambda_{\tilde{A}} \cup \Lambda_{\tilde{B}}$ we define the associated four fold table as $\mathcal{M}_{\alpha_i} = (a_i, b_i, c_i, d_i)$ whose values are computed using the previous RLs (see [19] for more details).

Using $\mathcal{M}_{\alpha_i}$ and the probability distribution of equation (5) we extend the accuracy measures for fuzzy rules from the crisp case [19]:

$$\sum_{\alpha_i \in \Lambda_{\tilde{A}} \cup \Lambda_{\tilde{B}}} (\alpha_i - \alpha_{i+1}) \left(\approx (a_i, b_i, c_i, d_i)\right) \tag{7}$$

The model is a good generalization of the crisp case, allowing the use of equation (7) in the fuzzy definition of measures in equation (6) as, respectively, FSupp$(A \Rightarrow B)$, FConf$(A \Rightarrow B)$, and FCF$(A \Rightarrow B)$ (see [19] for a complete discussion).

3.3 Approximate Dependencies

Let $RE = \{At_1, \ldots, At_m\}$ be a relational scheme and let r be an instance of RE such that $|r| = n$. Also, let $V, W \subset RE$ with $V \cap W = \emptyset$. A functional dependency $V \to W$ holds in RE if and only if

$$\forall t, s \in r \text{ if } t[V] = s[V] \text{ then } t[W] = s[W] \tag{8}$$

Approximate dependencies can be roughly defined as functional dependencies with exceptions. The definition of approximate dependence is then a matter of how to define exceptions, and how to measure the accuracy of the dependence [10]. We shall follow the approach introduced in [15,9], where the same methodology employed in mining for association rules is applied to the discovery of approximate dependencies.

Since a functional dependency '$V \to W$' can be seen as a rule that relates the equality of attribute values in pairs of tuples (see equation (8)), and association rules relate the presence of items in transactions, we can represent approximate dependencies as association rules by using the following interpretations of the concepts of item and transaction:

- An item is an object associated to an attribute of RE. For every attribute $At_k \in RE$ we note it_{At_k} the associated item.
- We introduce the itemset I_V to be $I_V = \{it_{At_k} \mid At_k \in V\}$
- T_r is a T-set that, for each pair of tuples $< t, s > \in r \times r$ contains a transaction $ts \in T_r$ verifying $it_{At_k} \in ts \Leftrightarrow t[At_k] = s[At_k]$ It is obvious that $|T_r| = |r \times r| = n^2$.

Then, an approximate dependence $V \to W$ in the relation r is an association rule $I_V \Rightarrow I_W$ in T_r [15,9]. The support and certainty factor of $I_V \Rightarrow I_W$ measure the interest and accuracy of the dependence $V \to W$.

3.4 Fuzzy Approximate Dependencies

In [7] a definition integrating both approximate and fuzzy dependencies features is introduced. In addition to allowing exceptions, the relaxation of several elements of equation (8) is considered. In particular, we associate membership degrees to pairs $< attribute, value >$ as in the case of fuzzy association rules, as well as the equality of the rule is smoothed as a fuzzy similarity relation.

Extending the crisp case above, fuzzy approximate dependencies in a relation are defined as fuzzy association rules on a special fuzzy T-set obtained from that relation.

Let $I_{RE} = \{it_{At_k} | At_k \in RE\}$ be the set of items associated to the relational schema RE. We define a fuzzy T-set $\widetilde{T}_r$ as follows: for each pair of rows $< t, s >$ in $r \times r$ we have a fuzzy transaction ts in $\widetilde{T}_r$ defined as

$$\forall it_{At_k} \in \widetilde{T}_r, ts(it_{At_k}) = min(At_k(t), At_k(s), S_{At_k}(t(At_k), s(At_k))) \qquad (9)$$

This way, the membership degree of a certain item it_{At_k} in the transaction associated to tuples t and s takes into account the membership degree of the value of At_k in each tuple $(At_k(t))$ and the similarity between them (S_{At_k}). The latter represents the degree to which tuples t and s agree in At_k. According to this, let $X, Y \subseteq RE$ with $X \cap Y = \emptyset$ and $X, Y \neq \emptyset$. The fuzzy approximate dependence[7] $X \rightarrow Y$ in r is defined as the fuzzy association rule $I_X \Rightarrow I_Y$ in $\widetilde{T}_r$.

Analogously to the crisp case, we measure the importance and accuracy of the fuzzy approximate dependence $X \rightarrow Y$ as the support and certainty factor of the fuzzy association rule $I_X \Rightarrow I_Y$ (see section 3.2).

4   Correspondence Analysis in Terms of Data Mining Tools

Correspondence analysis [6] describes existing relations between two nominal variables, by means of a contingency table, obtained as the cross-tabulation of both variables. It can be applied to reduce data dimension, prior to a subsequent statistic processing (classification, regression, discriminant analysis, ...). In particular, it can be helpful in the integration or matching of different partitions over a set of objects.

4.1   Crisp Correspondences

In [5], we introduced an alternate methodology to classic correspondence analysis, centered in the interpretation of a set of rules and/or dependencies. For that, we represent the possible correspondences between objects as a relational table, where the value of a cell for a given object (row) and partition (column) means the class in the partition where the object is.

Let O be a finite set of objects, and $\mathscr{A} = \{A_1, A_2, \ldots, A_p\}$, $\mathscr{B} = \{B_1, B_2, \ldots, B_q\}$ two partitions of O, i.e., $A_i, B_j \subseteq O$ and $A_i, B_j \neq \emptyset$, $A_{i_1} \cap A_{i_2} = \emptyset$ $\forall i_1, i_2 \in \{1, \ldots, p\}$ and $B_{j_1} \cap B_{j_2} = \emptyset$ $\forall j_1, j_2 \in \{1, \ldots, q\}$. Also, $\bigcup_{A_i \in \mathscr{A}} A_i = \bigcup_{B_j \in \mathscr{B}} B_j = O$.

We represent partitions $\mathscr{A}$ and $\mathscr{B}$ by means of a table, $r_{\mathscr{A}\mathscr{B}}$ (see table 1), and we shall use the notation for relational databases. Each row (tuple) and column (attribute) of $r_{\mathscr{A}\mathscr{B}}$ will be associated to an object and a partition, respectively. This way, we assume $|r_{\mathscr{A}\mathscr{B}}| = |O|$.

We shall note t_o the tuple associated to object o, and $X_{\mathscr{P}}$ the attribute associated to partition $\mathscr{P}$. The value for tuple t_o and attribute $X_{\mathscr{P}}$, $t_o[X_{\mathscr{P}}]$, will be the class for o following $\mathscr{P}$, i.e., $t_o[X_{\mathscr{P}}] \in \mathscr{P}$.

Let us remark that we are interested not only in perfect correspondences, but also in those with possible exceptions. Hence, we are concerned with measuring the accuracy of correspondences between partitions.

Definition 1 ([5]). Local correspondence. *Let $A_i \in \mathscr{A}$ and $B_j \in \mathscr{B}$. There exists a local correspondence from A_i to B_j when $A_i \subseteq B_j$.*

Table 1. Table $r_{\mathscr{A}\mathscr{B}}$

Object	tuple	$X_{\mathscr{A}}$	$X_{\mathscr{B}}$
o_1	t_{o_1}	A_1	B_2
o_2	t_{o_2}	A_2	B_2
o_3	t_{o_3}	A_1	B_1
$\cdots$	$\cdots$	$\cdots$	$\cdots$

The analysis of local correspondences can be performed by looking for association rules in the table $r_{\mathscr{A}\mathscr{B}}$. Rules $[X_{\mathscr{A}} = A_i] \Rightarrow [X_{\mathscr{B}} = B_j]$ and $[X_{\mathscr{B}} = B_j] \Rightarrow [X_{\mathscr{A}} = A_i]$ tell us about possible local correspondences between classes A_i and B_j.

Definition 2 ([5]). Partial correspondence. *There exists a partial correspondence from $\mathscr{A}$ to $\mathscr{B}$, noted $\mathscr{A} \Rightarrow \mathscr{B}$, when $\forall A_i \in \mathscr{A} \; \exists B_j \in \mathscr{B}$ such that $A_i \subseteq B_j$.*

Definition 3 ([5]). Global correspondence. *There exists a global correspondence between $\mathscr{A}$ and $\mathscr{B}$, noted $\mathscr{A} \equiv \mathscr{B}$, when $\mathscr{A} \Rightarrow \mathscr{B}$ and $\mathscr{B} \Rightarrow \mathscr{A}$.*

The analysis of partial correspondences can be performed by looking for approximate dependencies in $r_{\mathscr{A}\mathscr{B}}$ [5]. If the dependence $X_{\mathscr{A}} \rightarrow X_{\mathscr{B}}$ holds, there is a partial correspondence from $\mathscr{A}$ to $\mathscr{B}$. The certainty factor of the dependence measures the accuracy of the correspondence. As we are interested in using the same measure to assess global correspondences, this leads to define the certainty factor of $\mathscr{A} \equiv \mathscr{B}$ as the minimum between $CF(\mathscr{A} \Rightarrow \mathscr{B})$ and $CF(\mathscr{B} \Rightarrow \mathscr{A})$, since it is usual to obtain the certainty factor of a conjunction of facts as the minimum of the certainty factors of the facts.

4.2 Fuzzy Correspondences

Consider the case of establishing correspondences between diseases and symptoms. A certain disease can be described by several symptoms, at a given degree, and also a symptom can be related to different diseases. Since the original correspondence analysis [6] is not able to manage such cases in which partitions boundaries are not so clear, we extended the alternate methodology discussed in section 4.1 in order to manage correspondences between fuzzy partitions [12].

Let $O = \{o_1, \ldots, o_n\}$ be again a finite set of objects. Let $\widetilde{\mathscr{A}} = \{\widetilde{A}_1, \ldots, \widetilde{A}_p\}$ and $\widetilde{\mathscr{B}} = \{\widetilde{B}_1, \ldots, \widetilde{B}_q\}$ be two fuzzy partitions over O. Let $\widetilde{T}_{\widetilde{\mathscr{A}}\widetilde{\mathscr{B}}}$(Table 2) be the fuzzy transactional table associated to O, each transaction representing an object, that is, $|\widetilde{T}_{\widetilde{\mathscr{A}}\widetilde{\mathscr{B}}}| = |O|$. Given $o \in O, \widetilde{A}_i \in \widetilde{\mathscr{A}}$ and $\widetilde{B}_j \in \widetilde{\mathscr{B}}$, we noted for $\widetilde{A}_i(o)$ (respectively, $\widetilde{B}_j(o)$) the membership degree of o in $\widetilde{A}_i$ (respectively, $\widetilde{B}_j$). Each object must belong to at least one class of each partition, that is, $\forall o \in O, \exists \widetilde{P}_i \in \widetilde{\mathscr{P}}/\widetilde{P}_i(o) > 0$, and each class must contain at least one object, that is, $\widetilde{A}_i, \widetilde{B}_j \neq \emptyset$.

As we manage fuzzy partitions, we can relax the condition of disjoint classes within a partition. Also, we do not consider the case of partitions being necessarily normalized.

Definition 4 ([12]). *Fuzzy local correspondence.* *Let $\widetilde{A}_i \in \widetilde{\mathscr{A}}$ and $\widetilde{B}_j \in \widetilde{\mathscr{B}}$. There exists a fuzzy local correspondence from $\widetilde{A}_i$ to $\widetilde{B}_j$, noted $\widetilde{A}_i \Rightarrow \widetilde{B}_j$, if $\widetilde{A}_i \subseteq \widetilde{B}_j$, that is, $\forall o \in O,$*

Table 2. Fuzzy transactional table $\widetilde{T}_{\widetilde{\mathscr{A}}\widetilde{\mathscr{B}}}$

Object	$\widetilde{A}_1$ ... $\widetilde{A}_p$	$\widetilde{B}_1$ ... $\widetilde{B}_q$
o_1	$\widetilde{A}_1(o_1)$ ... $\widetilde{A}_p(o_1)$	$\widetilde{B}_1(o_1)$ ... $\widetilde{B}_q(o_1)$
o_2	$\widetilde{A}_1(o_2)$ ... $\widetilde{A}_p(o_2)$	$\widetilde{B}_1(o_2)$ ... $\widetilde{B}_q(o_2)$
o_3	$\widetilde{A}_1(o_3)$ ... $\widetilde{A}_p(o_3)$	$\widetilde{B}_1(o_3)$ ... $\widetilde{B}_q(o_3)$
...	... ...	... ...

$\widetilde{A}_i(o) \leq \widetilde{B}_j(o)$. *This time, we can obtain fuzzy local correspondences in terms of fuzzy association rules.*

When analyzing fuzzy partial and global correspondences, we must manage not classes, but partitions. It would be necessary to define a membership degree of an object in a partition, that is, $\widetilde{\mathscr{A}}(o)$. This defines a multidimensionality problem, already addressed in [12], and it is a pending task currently under researching. For sake of simplicity, we will reduce to the case in which an object is associated to only one class in every partition, for example, that with the highest membership degree.

We shall represent partitions $\widetilde{\mathscr{A}}$ and $\widetilde{\mathscr{B}}$ by means of a fuzzy relational table, $\widetilde{r}_{\widetilde{\mathscr{A}}\widetilde{\mathscr{B}}}$ (Table 3). Each row (object) is related to a column (partition) with a certain membership degree. The value corresponding to tuple t_o and attribute $X_{\widetilde{\mathscr{A}}}$, $t_o[X_{\widetilde{\mathscr{A}}}]$, will be the class for o according to partition $\widetilde{\mathscr{A}}$, that is, $t_o[X_{\widetilde{\mathscr{A}}}] \in \widetilde{\mathscr{A}}$. We shall note as $X_{\widetilde{\mathscr{A}}}(o)$ the membership degree of o in $t_o[X_{\widetilde{\mathscr{A}}}]$. As discussed before, we shall note this as $\widetilde{\mathscr{A}}(o)$.

Table 3. Fuzzy relational table, $\widetilde{r}_{\widetilde{\mathscr{A}}\widetilde{\mathscr{B}}}$

Object	$X_{\widetilde{\mathscr{A}}}$	$X_{\widetilde{\mathscr{B}}}$
t_{o_1}	$\widetilde{A}_{i1}, \widetilde{\mathscr{A}}(o_1)$	$\widetilde{B}_{j1}, \widetilde{\mathscr{B}}(o_1)$
t_{o_2}	$\widetilde{A}_{i2}, \widetilde{\mathscr{A}}(o_2)$	$\widetilde{B}_{j2}, \widetilde{\mathscr{B}}(o_2)$
t_{o_3}	$\widetilde{A}_{i3}, \widetilde{\mathscr{A}}(o_3)$	$\widetilde{B}_{j3}, \widetilde{\mathscr{B}}(o_3)$
...	...	...

Definition 5 ([12]). *Fuzzy partial correspondence. There exists a fuzzy partial correspondence from $\widetilde{\mathscr{A}}$ to $\widetilde{\mathscr{B}}$, noted $\widetilde{\mathscr{A}} \Rightarrow \widetilde{\mathscr{B}}$, when $\forall \widetilde{A}_i \in \widetilde{\mathscr{A}} \ \exists \widetilde{B}_j \in \widetilde{\mathscr{B}}$ such that $\widetilde{A}_i \subseteq \widetilde{B}_j$, that is, $\forall o \in O/t_o[\widetilde{\mathscr{A}}] = \widetilde{A}_i$ implies $t_o[\widetilde{\mathscr{B}}] = \widetilde{B}_j$ and $\widetilde{\mathscr{A}}(o) \lesssim \widetilde{\mathscr{B}}(o)$.*

$\lesssim$ defines a vectorial order relation that, for this particular case, corresponds to a classic order relation.

Definition 6 ([12]). *Fuzzy global correspondence. There exists a fuzzy global correspondence between $\widetilde{\mathscr{A}}$ and $\widetilde{\mathscr{B}}$, noted $\widetilde{\mathscr{A}} \equiv \widetilde{\mathscr{B}}$, when $\widetilde{\mathscr{A}} \Rightarrow \widetilde{\mathscr{B}}$ and $\widetilde{\mathscr{B}} \Rightarrow \widetilde{\mathscr{A}}$.*

Fuzzy partial and global correspondences relate fuzzy partitions, and both can be obtained by means of fuzzy approximate dependencies.

5 Our Proposal. Discussion

In this section, we give an alternate approach to Rand index which is valid for both crisp and fuzzy partitions, in terms of the measures employed in the tools described in section 3, and using the tabular representation described in section 4. Hence, let O be again a finite set of objects, $|O| = n$, with $\mathscr{A} = \{A_1, A_2, \ldots, A_p\}$ and $\mathscr{B} = \{B_1, B_2, \ldots, B_q\}$, two different partitions over O.

Analogously to the approach described in [15,9] for approximate dependencies, let T be a transactional table where each row represents an ordered pair of objects $(o, o') \in O \times O$, $|T| = \frac{n(n-1)}{2}$. Let $I_\mathscr{A}$ (resp., $I_\mathscr{B}$) be an item that indicates that both objects o, o' belong to the same class in partition $\mathscr{A}$ (resp., $\mathscr{B}$). According to this, we can redefine the Rand index parameters, a, b, c, d in terms of the support measure as:

– $a = |T| \cdot supp(I_\mathscr{A} \cap I_\mathscr{B})$,
– $b = |T| \cdot (supp(I_\mathscr{A}) - supp(I_\mathscr{A} \cap I_\mathscr{B}))$,
– $c = |T| \cdot (supp(I_\mathscr{B}) - supp(I_\mathscr{A} \cap I_\mathscr{B}))$, and
– $d = |T| \cdot (1 - supp(I_\mathscr{A} \cup I_\mathscr{B})) = |T| \cdot (1 - supp(I_\mathscr{A}) - supp(I_\mathscr{B}) + supp(I_\mathscr{A} \cap I_\mathscr{B}))$.

Thus, we can rewrite the Rand index as,

$$
I_R(\mathscr{A}, \mathscr{B}) = \frac{a+d}{a+b+c+d} = \frac{|T| \cdot (supp(I_\mathscr{A} \cap I_\mathscr{B}) - (1 - supp(I_\mathscr{A} \cup I_\mathscr{B})))}{|T|} = \tag{10}
$$
$$
= 1 - (supp(I_\mathscr{A}) + supp(I_\mathscr{B}) - 2supp(I_\mathscr{A} \cap I_\mathscr{B}))
$$

Let us notice in first place, how, from this expression, it is trivial that $I_R(\mathscr{A}, \mathscr{B}) = I_R(\mathscr{B}, \mathscr{A})$, since only support is involved in equation 10. Moreover, it is easy to see that these parameters are proportionally equivalent to those of the four fold table used in the model described in section 3.2, allowing us to relate $I_R(\mathscr{A}, \mathscr{B})$ in some way with the measures of support, confidence, and certainty factor (see definitions for 4ft-quantifiers $\approx_S (a,b,c,d)$, $\approx_C (a,b,c,d)$, and $\approx_{CF} (a,b,c,d)$, respectively).

Following this, and taking into account our approach for correspondence analysis (section 4.1), we can establish a direct relation between Rand index and the measurement of partial and global correspondences between two different partitions (by means of approximate dependencies). Even more, we can define a similar measure to analyze not only these types of correspondences, but also local correspondences (by means of association rules).

As for the case of fuzzy partitions, our model for fuzzy correspondences (section 4.2) also allows to obtain measures as informative as the Rand index, again considering both partitions (in terms of fuzzy approximate dependencies) as well as classes (as relations expressed as fuzzy association rules).

Then, according to equation (10), we can redefine the Rand index in terms of the support measure, allowing us to distinguish several interpretations of this measure, based on the data mining tool used as source. Let us consider the following family of indices:

– $I_{AR}(A_i, B_j)$, for comparing classes $A_i \in \mathscr{A}$ and $B_j \in \mathscr{B}$, from association rules,
– $I_{AD}(\mathscr{A}, \mathscr{B})$, for comparing partitions $\mathscr{A}$ and $\mathscr{B}$, from approximate dependencies,

- $I_{FAR}(\widetilde{A}_i, \widetilde{B}_j)$, for comparing fuzzy classes $\widetilde{A}_i \in \widetilde{\mathscr{A}}$ and $\widetilde{B}_j \in \widetilde{\mathscr{B}}$, from fuzzy association rules, and
- $I_{FAD}(\widetilde{\mathscr{A}}, \widetilde{\mathscr{B}})$, for comparing fuzzy partitions $\widetilde{\mathscr{A}}$ and $\widetilde{\mathscr{B}}$, defined in terms of fuzzy approximate dependencies.

Let us remark that our approach allows not only to take into account these proposed measures, but also the already well-defined and popular measures of support, confidence and certainty factor. A very interesting issue could be a deeper study of the combined information obtained by all these values.

From our point of view, these definitions open a new framework in the problem of partition comparison, specially in those cases where the boundaries between classes are unclear, able to be managed by means of fuzzy partitions. In [5], we briefly discussed some interesting properties as, for example, that of how our approach can be applied to the study of relations between more than two partitions. The analysis of the relevance of this and some other properties is a pending task, and future works will be devoted to their study and development.

5.1   A Brief Example

In order to illustrate our proposal, but due to lack of space, we are showing a little example of our methodology, extending the results over the same dataset used in [12]. Here, fuzzy correspondence analysis between different partitions over a set of 211 agricultural zones is addressed and discussed. The first fuzzy partition was obtained as widely discussed in [4] from users (farmers) knowledge, and classified the examples into 19 classes. Let $userclass = \{\widetilde{A}_1, \ldots, \widetilde{A}_{19}\}$ be this classification. A scientific classification was previously presented in [28]. Here, a total of 21 land types, called soil maps units, are found, only 19 being suitable for olive trees cultivation. Let $sciclass = \{\widetilde{B}_1, \ldots, \widetilde{B}_{21}\}$ be this other classification.

Fuzzy local correspondences between classes were computed between $userclass$ and $sciclass$, and those more interesting ($CF > 0.65$) are shown in table 4. Each cell in the table shows the CF for the fuzzy local correspondence (fuzzy association rule) of the type $\widetilde{B}_j \Rightarrow \widetilde{A}_i$ (as discussed in [12], the inverse fuzzy local correspondences were found to be not interesting regarding CF). It must be remarked that these results were validated and properly interpreted by soil experts.

Table 5 shows the I_{FAR} value for the same correspondences in table 4. Let us recall that $I(\mathscr{A}, \mathscr{B}) = I(\mathscr{B}, \mathscr{A})$ for any two partitions (or classes, as it is the case). That is, this index tells us about the relation between $\mathscr{A}$ and $\mathscr{B}$, but gives no information about the direction of this relation. From our point of view, this relation is not necessarily symmetric, since one partition class can be partially included in other partition class, but the opposite might not hold. In this sense, these first results suggest that CF measure seems to be more valuable than I_{FAR}. Hence, a more exhaustive and complete analysis of the relation between these measures, considering additional sets of examples, appears to be necessary, and will be properly addressed in a future extension of this work.

Table 4. Fuzzy local correspondences between *sciclass* (rows) and *userclass* (columns) classes, $\widetilde{B}_j \Rightarrow \widetilde{A}_i$ ($CF > 0.65$)

	A_1	A_2	A_3	A_4	A_5	A_6	A_7	A_8	A_{10}	A_{11}	A_{12}	A_{14}	A_{15}	A_{16}	A_{17}
B_1			0.722	0.783				0.687	0.672						0.690
B_3	0.870	0.859	0.793	0.660		0.689	0.655	0.771		0.783			0.759		
B_5	0.684	0.733	0.795	0.744				0.686		0.736					
B_6	0.895	0.913	0.792	0.675		0.661		0.784		0.803			0.813		
B_8	0.886	0.932													
B_9		0.650	0.719	0.687				0.712		0.675					
B_{10}	0.653		0.718	0.728		0.715		0.705	0.711						
B_{11}	0.691	0.764	0.743	0.721	0.653		0.676	0.670		0.813			0.666		
B_{13}				0.700	0.662				0.687						
B_{15}	0.761	0.760	0.842	0.737				0.687	0.681	0.683			0.674		0.664
B_{16}	0.744	0.853	0.812	0.756		0.729	0.734	0.769	0.688	0.696			0.808		
B_{20}			0.803	0.868	0.871			0.667	0.802	0.742	0.700	0.706		0.811	0.770

Table 5. Rand index $I_{FAR}(\widetilde{A}_i, \widetilde{B}_j)$ for fuzzy local correspondences between *sciclass* (rows) and *userclass* (columns) classes

	A_1	A_2	A_3	A_4	A_5	A_6	A_7	A_8	A_{10}	A_{11}	A_{12}	A_{14}	A_{15}	A_{16}	A_{17}
B_1			0.241	0.292				0.292	0.341						0.416
B_3	0.354	0.331	0.283	0.299		0.402	0.438	0.333		0.336			0.398		
B_5	0.346	0.341	0.326	0.354				0.346		0.363					
B_6	0.297	0.277	0.234	0.271		0.366		0.289		0.291			0.358		
B_8	0.288	0.270													
B_9		0.317	0.305	0.339				0.357		0.346					
B_{10}	0.289		0.241	0.286		0.381		0.294	0.346						
B_{11}	0.293	0.281	0.244	0.285	0.422		0.420	0.290		0.307			0.357		
B_{13}				0.287	0.427				0.347						
B_{15}	0.332	0.311	0.293	0.315				0.316	0.365	0.316			0.380		0.434
B_{16}	0.292	0.280	0.242	0.281		0.376	0.419	0.293	0.337	0.288			0.364		
B_{20}			0.235	0.284	0.428			0.281	0.341	0.287	0.626	0.568		0.560	0.411

6   Further Works and Concluding Remarks

Many measures based on the Rand index have been proposed and developed for the study of partitions comparison. A subset of them can be applied also in those cases involving fuzzy partitions. In this work, we have applied a previously developed methodology for correspondence analysis, in terms of fuzzy data mining tools, to the problem of partition comparison, expressed in the form of a measure such as the Rand index. Our approach offers the advantage of being capable of managing both crisp and fuzzy partitions, and, in addition, it allows to compare not only different partitions, but also classes inside these partitions. We have shown an example combining an accuracy measure as CF with the Rand index. Moreover, we have seen how CF, in comparison to Rand index, allows to determine the direction in which the relation between partitions (or classes) is stronger.

Finally, some interesting properties arise from the proposed measures, and a deeper and more complete study and development will be the main topic in future extensions of this work. Practical works covering the discussion of our methodology applied to real world problems, as fuzzy image segmentation (for the comparison of different methods), and classification in medical cases, are also in progress..

Acknowledgements. The research reported in this paper was partially supported by the Andalusian Government (Junta de Andalucía) under project P07-TIC03175 and from the former Spanish Ministry for Science and Innovation by the project grants TIN2006-15041-C04-01 and TIN2009-08296.

References

1. Agrawal, R., Imielinski, T., Swami, A.: Mining Association Rules between Sets of Items in Large Databases. In: Procs. of ACM SIGMOD Conf., Washington DC, USA, pp. 207–216 (1993)
2. Anderson, D.T., Bezdek, J.C., Popescu, M., Keller, J.M.: Comparing fuzzy, probabilistic, and possibilistic partitions. IEEE Transactions on Fuzzy Systems 18(5), 906–918 (2010)
3. Anderson, D.T., Bezdek, J.C., Keller, J.M., Popescu, M.: A Comparison of Five Fuzzy Rand Indices. In: Hüllermeier, E., Kruse, R., Hoffmann, F. (eds.) IPMU 2010. CCIS, vol. 80, pp. 446–454. Springer, Heidelberg (2010)
4. Aranda, V., Calero, J., Delgado, G., Sánchez, D., Serrano, J., Vila, M.A.: Flexible land classification for olive cultivation using user knowledge. In: Proceedings of 1st. Int. ICSC Conf. On Neuro-Fuzzy Technologies (NF 2002), La HaBana, Cuba, Enero 16–19 (2002)
5. Aranda, V., Calero, J., Delgado, G., Sánchez, D., Serrano, J.M., Vila, M.A.: Using Data Mining Techniques to Analyze Correspondences Between User and Scientific Knowledge in an Agricultural Environment. In: Enterprise Information Systems IV, pp. 75–89. Kluwer Academic Publishers (2003)
6. Benzécri, J.P.: Cours de Linguistique Mathématique. Université de Rennes, Rennes (1963)
7. Berzal, F., Blanco, I., Sánchez, D., Serrano, J.M., Vila, M.A.: A definition for fuzzy approximate dependencies. Fuzzy Sets and Systems 149(1), 105–129 (2005)
8. Berzal, F., Delgado, M., Sánchez, D., Vila, M.A.: Measuring accuracy and interest of association rules: A new framework. Intelligent Data Analysis 6(3), 221–235 (2002)
9. Blanco, I., Martín-Bautista, M.J., Sánchez, D., Serrano, J.M., Vila, M.A.: Using association rules to mine for strong approximate dependencies. Data Mining and Knowledge Discovery 16(3), 313–348 (2008)
10. Bosc, P., Lietard, L., Pivert, O.: Functional Dependencies Revisited Under Graduality and Imprecision. In: Annual Meeting of NAFIPS, pp. 57–62 (1997)
11. Brouwer, R.K.: Extending the rand, adjusted rand and jaccard indices to fuzzy partitions. Journal of Intelligent Information Systems 32, 213–235 (2009)
12. Calero, J., Delgado, G., Sánchez, D., Serrano, J.M., Vila, M.A.: A Proposal of Fuzzy Correspondence Analysis based on Flexible Data Mining Techniques. In: Soft Methodology and Random Information Systems, pp. 447–454. Springer (2004)
13. Campello, R.J.G.B.: A fuzzy extension of the Rand index and other related indexes for clustering and classification assessment. Pattern Recognition Letters 28, 833–841 (2007)
14. Campello, R.J.G.B.: Generalized external indexes for comparing data partitions with overlapping categories. Pattern Recognition Letters 31, 966–975 (2010)
15. Delgado, M., Martín-Bautista, M.J., Sánchez, D., Vila, M.A.: Mining strong approximate dependencies from relational databases. In: Procs. of IPMU 2000 (2000)
16. Delgado, M., Marín, N., Sánchez, D., Vila, M.A.: Fuzzy Association Rules: General Model and Applications. IEEE Transactions on Fuzzy Systems 11(2), 214–225 (2003)
17. Delgado, M., Ruiz, M.D., Sánchez, D.: A restriction level approach for the representation and evaluation of fuzzy association rules. In: Procs. of the IFSA-EUSFLAT, pp. 1583–1588 (2009)

18. Delgado, M., Ruiz, M.D., Sánchez, D.: Studying Interest Measures for Association Rules through a Logical Model. Int. J. of Uncertainty, Fuzziness and Knowledge-Based Systems 18(1), 87–106 (2010)
19. Delgado, M., Ruiz, M.D., Sánchez, D., Serrano, J.M.: A Formal Model for Mining Fuzzy Rules Using the RL Representation Theory. Information Sciences 181, 5194–5213 (2011)
20. Fowlkes, E.B., Mallows, C.L.: A method for comparing two hierarchical clusterings. J. of American Statistical Society 78, 553–569 (1983)
21. Frigui, H., Hwang, C., Rhee, F.C.H.: Clustering and aggregation of relational data with applications to image database categorization. Pattern Recognition 40, 3053–3068 (2007)
22. Hubert, L.J., Arabie, P.: Comparing partition. J. Classification 2, 193–218 (1985)
23. Hüllermeier, E., Rifqi, M., Henzgen, S., Senge, R.: Comparing fuzzy partitions: A generalization of the Rand index and related measures. IEEE Transactions of Fuzzy Systems 20(3), 546–556 (2012)
24. Jaccard, P.: Étude comparative de la distribution florale dans une portion des Alpes et des Jura. Bulletin de la Société Vaudoise des Sciences Naturelles 37, 547–579 (1901)
25. Jain, A., Dubes, R.: Algorithms for Clustering Data. Prentice Hall (1988)
26. Jiang, D., Tang, C., Zhang, A.: Cluster analysis for gene-expression data: A survey. IEEE Trans. Knowledge Data Engineering 16, 1370–1386 (2004)
27. Di Nuovo, A.G., Catania, V.: On External Measures for Validation of Fuzzy Partitions. In: Melin, P., Castillo, O., Aguilar, L.T., Kacprzyk, J., Pedrycz, W. (eds.) IFSA 2007. LNCS (LNAI), vol. 4529, pp. 491–501. Springer, Heidelberg (2007)
28. Pérez-Pujalte, A., Prieto, P.: Mapa de suelos 1:200000 de la provincia de Granada y memoria explicativa. Technical report, CSIC (1980)
29. Rand, W.M.: Objective criteria for the evaluation of clustering methods. J. of the American Statistical Association 66(336), 846–850 (1971)
30. Rauch, J., Simunek, M.: Mining for 4ft Association Rules. In: Morishita, S., Arikawa, S. (eds.) DS 2000. LNCS (LNAI), vol. 1967, pp. 268–272. Springer, Heidelberg (2000)
31. Runkler, T.A.: Comparing Partitions by Subset Similarities. In: Hüllermeier, E., Kruse, R., Hoffmann, F. (eds.) IPMU 2010. LNCS, vol. 6178, pp. 29–38. Springer, Heidelberg (2010)
32. Sánchez, D., Delgado, M., Vila, M.A., Chamorro-Martínez, J.: On a non-nested level-based representation of fuzziness. Fuzzy Sets and Systems 192, 159–175 (2012)
33. Shortliffe, E., Buchanan, B.: A model of inexact reasoning in medicine. Mathematical Biosciences 23, 351–379 (1975)

An Iterative Scaling Algorithm for Maximum Entropy Reasoning in Relational Probabilistic Conditional Logic*

Marc Finthammer

Dept. of Computer Science, FernUniversität in Hagen

Abstract. Recently, different semantics for relational probabilistic conditionals and corresponding maximum entropy (ME) inference operators have been proposed. In this paper, we study the so-called aggregation semantics that covers both notions of a statistical and subjective view. The computation of its inference operator requires the calculation of the ME-distribution satisfying all probabilistic conditionals, inducing an optimization problem under linear constraints. We demonstrate how the well-known Generalized Iterative Scaling (GIS) algorithm technique can be applied to this optimization problem and present a practical algorithm and its implementation.

1 Introduction

There exist many approaches which combine propositional logic with probability theory to express uncertain knowledge and allow uncertain reasoning, e. g. Bayes Nets and Markov Nets [17] or probabilistic conditional logic [18]. Some of these approaches have been extended to first-order logic, e. g. Bayesian logic programs [14], Markov logic networks [8], and relational probabilistic conditional logic [10,13] which introduces relational probabilistic conditionals. Recently, different semantics for relational probabilistic conditionals and corresponding maximum entropy (ME) inference operators have been proposed. One of these approaches is the so-called aggregation semantics presented in [13]. This semantics has some nice properties, as it allows to cover both notions of a statistical and a subjective point of view. It can also handle statements about exceptional individuals without running into imminent inconsistencies. The following example taken from [13] (and inspired by [7]) illustrates some difficulties which can arise from a knowledge base in probabilistic first-order logic.

Example 1. Let X, Y denote variables and let $el(X)$ mean that X is an elephant, $ke(X)$ means that X is a zookeeper, and $likes(X, Y)$ expresses that X likes Y.

$r_1 : (likes(X, Y) \mid el(X) \wedge ke(Y))\ [0.6]$

$r_2 : (likes(X, fred) \mid el(X) \wedge ke(fred))\ [0.4]$

$r_3 : (likes(clyde, fred) \mid el(clyde) \wedge ke(fred))\ [0.7]$

---

* The research reported here was partially supported by the DFG (BE 1700/7-2).

E. Hüllermeier et al. (Eds.): SUM 2012, LNAI 7520, pp. 351–364, 2012.
© Springer-Verlag Berlin Heidelberg 2012

The first probabilistic rule (or probabilistic conditional) r_1 expresses that for an arbitrary chosen elephant and keeper (from some given population), there is a 0.6 probability that the elephant likes the keeper. But r_2 states that there is an (exceptional) keeper *fred*, for whom there is just a 0.4 probability that an arbitrary elephants likes him. Rule r_3 makes a statement about two exceptional individuals, i. e. the probability that the elephant *clyde* likes the keeper *fred* is even 0.7. Rule r_1 expresses statistical knowledge which holds in some given population, whereas r_3 expresses individual belief, and r_2 is a mixture of both. A simple semantical approach would be to ground all first-order rules (according to a given universe) and define semantics on grounded probabilistic rules. But in the above example, this would already cause severe problems, because the grounding $(likes(clyde, fred) \mid el(clyde) \wedge ke(fred))\,[0.6]$ of r_1 is in conflict with r_3 and also with the grounding of $(likes(clyde, fred) \mid el(clyde) \wedge ke(fred))\,[0.4]$ of r_2. However, the aggregation semantics is capable of handling such conflicts if an appropriate universe is provided, so that the probabilities of exceptional and generic individuals can be balanced.

The model-based inference operator $\mathrm{ME}_{\odot}$ for aggregation semantics presented in [13] is based on the principle of maximum entropy. The principle of maximum entropy exhibits excellent properties for commonsense reasoning [16,11,12] and allows to complete uncertain and incomplete information in an information-theoretic optimal way, and also the ME-based inference operator $\mathrm{ME}_{\odot}$ features many desirable properties of a rational inference operator [13]. However, up to now no practical implementation of $\mathrm{ME}_{\odot}$ inference has been developed. The determination of $\mathrm{ME}_{\odot}(\mathcal{R})$ for a set $\mathcal{R}$ of probabilistic conditionals requires the calculation of the ME-distribution satisfying all probabilistic conditionals in $\mathcal{R}$. This induces a convex optimization problem, so general techniques for solving convex optimization problems could be applied to compute a solution. Instead of employing such general techniques, in this paper we present the first practical algorithm for computing $\mathrm{ME}_{\odot}(\mathcal{R})$ which is tailor-made for the problem. We employ and adapt the technique of the well-known Generalized Iterative Scaling (GIS) algorithm [6], which allows to compute the ME-distribution under a set linear constraints. We use the general GIS technique as a template to develop a concrete algorithm which takes a set $\mathcal{R}$ as input and computes $\mathrm{ME}_{\odot}(\mathcal{R})$. We also present a practical implementation of our algorithm.

The rest of the paper is structured as follows. In Sec. 2 we give a brief overview of aggregation semantics and introduce feature functions in Sec. 3. Using these feature functions, in Sec. 4 the ME-inference operator $\mathrm{ME}_{\odot}$ is defined in terms of an optimization problem. In Sec. 5 we demonstrate how this optimization problem can be transformed into a normalized form so that the Generalized Iterative Scaling (GIS) technique can be applied, yielding a concrete algorithm determining $\mathrm{ME}_{\odot}$. In Sec. 6 we present a practical implementation of our algorithm, and in Sec. 7 we conclude and point out further work. Due to lack of space, full technical proofs for all propositions of this paper are given in [9].

2   Background: Aggregation Semantics

We consider a first-order signature $\Sigma := (Pred, Const)$ consisting of a set of first order predicates $Pred$ and a finite set of constants $Const$. So Σ is a restricted signature since it only contains functions with an arity of zero. Let p/k denote the predicate $p \in Pred$ with arity k. The set of atoms $\mathcal{A}$ over $Pred$ with respect to a set of variables Var and $Const$ is defined in the usual way by $p(t_1, \ldots, t_k) \in \mathcal{A}$ iff $p/k \in Pred, t_i \in (Var \cup Const)$ for $1 \leq i \leq k$. For better readability we will usually omit the referring indices.

Let $\mathcal{L}$ be a quantifier-free first-order language defined over Σ in the usual way, that is, $A \in \mathcal{L}$ if $A \in \mathcal{A}$, and if $A, B \in \mathcal{L}$ then $\neg A, A \wedge B \in \mathcal{L}$. Let $A \vee B$ be the shorthand for $\neg(\neg A \wedge \neg B)$. If it is clear from context, we use the short notation AB to abbreviate a conjunction $A \wedge B$.

Let $\mathrm{gnd}(A)$ be a grounding function which maps a formula A to its respective set of ground instances in the usual way.

Definition 1 (Conditional). *Let $A(\boldsymbol{X}), B(\boldsymbol{X}) \in \mathcal{L}$ be first-order formulas with $\boldsymbol{X}$ containing the variables of A and B. $(B(\boldsymbol{X})|A(\boldsymbol{X}))$ is called a conditional. A is the antecedence and B the consequence of the conditional. The set of all conditionals over $\mathcal{L}$ is denoted by $(\mathcal{L}|\mathcal{L})$.*

Definition 2 (Probabilistic Conditional). *Let $(B(\boldsymbol{X})|A(\boldsymbol{X})) \in (\mathcal{L}|\mathcal{L})$ be a conditional and let $d \in [0,1]$ be a real value. $(B(\boldsymbol{X})|A(\boldsymbol{X}))\,[d]$ is called a probabilistic conditional with probability d. If $d \in \{0,1\}$ then the probabilistic conditional is called* hard, *otherwise it is a called* soft. *The set of all probabilistic conditionals over $\mathcal{L}$ is denoted by $(\mathcal{L}|\mathcal{L})^{prob}$.*

A set of probabilistic conditionals is also called a *knowledge base*. (Probabilistic) conditionals are also called *(probabilistic) rules*. If it is clear from context, we will omit the "probabilistic" and just use the term "conditional".

Let $\mathcal{H}$ denote the Herbrand base, i.e. the set containing all ground atoms constructible from $Pred$ and $Const$. A Herbrand interpretation ω is a subset of the ground atoms, that is $\omega \subseteq \mathcal{H}$. Using a closed world assumption, each ground atom $p_{gnd} \in \omega$ is interpreted as true and each $p_{gnd} \notin \omega$ is interpreted as false; in this way a Herbrand interpretation is similar to a complete conjunction in propositional logic. Let Ω denote the set of all possible worlds (i. e. Herbrand interpretations), that is, $\Omega := \mathfrak{P}(\mathcal{H})$ (with $\mathfrak{P}$ denoting the power set).

Definition 3 (Set of Grounding Vectors). *For a conditional $(B(\boldsymbol{X})|A(\boldsymbol{X})) \in (\mathcal{L}|\mathcal{L})$, the set of all constant vectors $\boldsymbol{a}$ which can be used for proper groundings of $(B(\boldsymbol{X})|A(\boldsymbol{X}))$ is defined as:*
$$\mathcal{H}^{\boldsymbol{x}(A,B)} := \{\boldsymbol{a} = (a_1, \ldots, a_s) \mid a_1, \ldots, a_s \in Const$$
$$\text{and } (B(\boldsymbol{a})|A(\boldsymbol{a})) \in \mathrm{gnd}\,((B(\boldsymbol{X})|A(\boldsymbol{X})))\}$$

Let $P : \Omega \to [0,1]$ be a probability distribution over possible worlds and let $\mathcal{P}_\Omega$ be the set of all such distributions. P is extended to ground formulas $A(\boldsymbol{a})$, with $\boldsymbol{a} \in \mathcal{H}^{\boldsymbol{x}(A)}$, by defining $P(A(\boldsymbol{a})) := \sum_{\omega \models A(\boldsymbol{a})} P(\omega)$.

Definition 4 (Aggregation Semantics Entailment Relation [13]). *The entailment relation $\models_\odot$ between a probability distribution $P \in \mathcal{P}_\Omega$ and a probabilistic conditional $(B(\boldsymbol{X})|A(\boldsymbol{X}))\,[d] \in (\mathcal{L}|\mathcal{L})^{prob}$ with $\sum_{\boldsymbol{a} \in \mathcal{H}^{\boldsymbol{x}(A,B)}} P(A(\boldsymbol{a})) > 0$ is defined as:*

$$P \models_\odot (B(\boldsymbol{X})|A(\boldsymbol{X}))\,[d] \quad \text{iff} \quad \frac{\displaystyle\sum_{\boldsymbol{a} \in \mathcal{H}^{\boldsymbol{x}(A,B)}} P(A(\boldsymbol{a})B(\boldsymbol{a}))}{\displaystyle\sum_{\boldsymbol{a} \in \mathcal{H}^{\boldsymbol{x}(A,B)}} P(A(\boldsymbol{a}))} = d \qquad (1)$$

Note that both sums of the fraction run over the same set of grounding vectors and therefore the same number of ground instances, i.e. a particular probability $P(A(\boldsymbol{a}))$ can be contained multiple times in the denominator sum. If $P \models_\odot r$ holds for a conditional r, we say that P *satisfies* r or that P *is a model of* r.

Thus, the aggregation semantics resembles the definition of a conditional probability by summing up the probabilities of all respective ground formulas. The entailment relation $\models_\odot$ is extended to a set $\mathcal{R}$ of probabilistic conditionals by defining

$$P \models_\odot \mathcal{R} \quad \text{iff} \quad \forall r \in \mathcal{R} : P \models_\odot r$$

Let $\mathcal{S}(\mathcal{R}) := \{P \in \mathcal{P}_\Omega : P \models_\odot \mathcal{R}\}$ denote the set of all probability distributions which satisfy $\mathcal{R}$. $\mathcal{R}$ is *consistent* iff $\mathcal{S}(\mathcal{R}) \neq \emptyset$, i.e. there exists a probability distribution which satisfies all conditionals in $\mathcal{R}$. Accordingly, a probabilistic conditional r is called consistent (or *satisfiable*), if there exists a distribution which satisfies r.

3   Feature Functions

For propositional conditionals, the satisfaction relation can be expressed by using feature functions (e.g. [10]). The following definition introduces feature functions for the relational case where the groundings have to be taken into account.

Definition 5 (Feature Function). *For a probabilistic conditional $r_i := (B_i(\boldsymbol{X})|A_i(\boldsymbol{X}))\,[d_i]$ define the functions $v_i^{\#}, f_i^{\#} : \Omega \to \mathbb{N}_0$ with*

$$v_i^{\#}(\omega) := \left| \left\{ \boldsymbol{a} \in \mathcal{H}^{\boldsymbol{x}(A_i,B_i)} \mid \omega \models A_i(\boldsymbol{a})B_i(\boldsymbol{a}) \right\} \right| \quad \text{and}$$

$$f_i^{\#}(\omega) := \left| \left\{ \boldsymbol{a} \in \mathcal{H}^{\boldsymbol{x}(A_i,B_i)} \mid \omega \models A_i(\boldsymbol{a})\overline{B_i(\boldsymbol{a})} \right\} \right|. \qquad (2)$$

$v_i^{\#}(\omega)$ *indicates the number of groundings which* verify r_i *for a certain* $\omega \in \Omega$, *whereas* $f_i^{\#}(\omega)$ *specifies the number of groundings which* falsify r_i.

The linear function function $\sigma_i : \Omega \to \mathbb{R}$ with

$$\sigma_i(\omega) := v_i^{\#}(\omega)(1 - d_i) - f_i^{\#}(\omega)d_i \qquad (3)$$

is called the feature function *of the probabilistic conditional r_i.*

Proposition 1. *Let $(B_i(\boldsymbol{X})|A_i(\boldsymbol{X}))\,[d_i]$ be a probabilistic conditional and let σ_i be its feature function according to Definition 5. Then it holds for $P \in \mathcal{P}_\Omega$ with $\sum_{\boldsymbol{a} \in \mathcal{H}^{\boldsymbol{x}}(A,B)} P\left(A(\boldsymbol{a})\right) > 0$:*

$$P \models_{\odot} (B_i(\boldsymbol{X})|A_i(\boldsymbol{X}))\,[d_i] \quad \text{iff} \quad \sum_{\omega \in \Omega} P(\omega)\sigma_i(\omega) = 0 \tag{4}$$

Proof. For ease of readability, we omit the index i of conditional r_i.

$$P \models_{\odot} (B(\boldsymbol{X})|A(\boldsymbol{X}))\,[d]$$

$$\text{iff} \sum_{\boldsymbol{a} \in \mathcal{H}^{\boldsymbol{x}}(A,B)} P\left(A(\boldsymbol{a})B(\boldsymbol{a})\right) = d \sum_{\boldsymbol{a} \in \mathcal{H}^{\boldsymbol{x}}(A,B)} P\left(A(\boldsymbol{a})\right)$$

$$\text{iff} \sum_{\boldsymbol{a} \in \mathcal{H}^{\boldsymbol{x}}(A,B)} P\left(A(\boldsymbol{a})B(\boldsymbol{a})\right) = d \sum_{\boldsymbol{a} \in \mathcal{H}^{\boldsymbol{x}}(A,B)} \left[P\left(A(\boldsymbol{a})B(\boldsymbol{a})\right) + P\left(A(\boldsymbol{a})\overline{B(\boldsymbol{a})}\right) \right]$$

$$\text{iff} \sum_{\boldsymbol{a} \in \mathcal{H}^{\boldsymbol{x}}(A,B)} \left[(1-d)P\left(A(\boldsymbol{a})B(\boldsymbol{a})\right) - dP\left(A(\boldsymbol{a})\overline{B(\boldsymbol{a})}\right) \right] = 0$$

$$\text{iff} \left((1-d) \sum_{\boldsymbol{a} \in \mathcal{H}^{\boldsymbol{x}}(A,B)} \sum_{\substack{\omega \in \Omega: \\ \omega \models A(\boldsymbol{a})B(\boldsymbol{a})}} P(\omega) \right) - d \sum_{\boldsymbol{a} \in \mathcal{H}^{\boldsymbol{x}}(A,B)} \sum_{\substack{\omega \in \Omega: \\ \omega \models A(\boldsymbol{a})\overline{B(\boldsymbol{a})}}} P(\omega) = 0$$

$$\text{iff} \left((1-d) \sum_{\omega \in \Omega} \sum_{\substack{\boldsymbol{a} \in \mathcal{H}^{\boldsymbol{x}}(A,B): \\ \omega \models A(\boldsymbol{a})B(\boldsymbol{a})}} P(\omega) \right) - d \sum_{\omega \in \Omega} \sum_{\substack{\boldsymbol{a} \in \mathcal{H}^{\boldsymbol{x}}(A,B): \\ \omega \models A(\boldsymbol{a})\overline{B(\boldsymbol{a})}}} P(\omega) = 0$$

$$\text{iff} \sum_{\omega \in \Omega} \left(v^{\#}(\omega)(1-d)P(\omega) - f^{\#}(\omega)dP(\omega) \right) = 0$$

$$\text{iff} \sum_{\omega \in \Omega} \left(v^{\#}(\omega)(1-d) - f^{\#}(\omega)d \right) P(\omega) = 0$$

$$\text{iff} \sum_{\omega \in \Omega} \sigma(\omega)P(\omega) = 0 \tag{5}$$

Proposition 1 shows that under aggregation semantics a conditional induces a linear constraint which has to be met by a satisfying probability distribution. The *expected value* $\mathbb{E}(\sigma_i, P)$ of a function σ_i under a distribution P is defined as $\mathbb{E}(\sigma_i, P) := \sum_{\omega \in \Omega} \sigma_i(\omega)P(\omega)$. Thus (5) states that the expected value of the feature function σ_i must be 0 under every satisfying distribution.

4   ME-Inference for Aggregation Semantics

The *entropy* $H(P) := -\sum_{\omega \in \Omega} P(\omega) \log P(\omega)$ of a probability distribution P measures the indifference within the distribution. From all distributions satisfying $\mathcal{R}$, the principle of *maximum entropy* (*ME*) [16,11] chooses the unique distribution with maximum entropy as a model for $\mathcal{R}$. The ME-model of $\mathcal{R}$

is employed to perform model-based inference and therefore it is also called *ME-inference operator*. The ME-inference operator $ME_\odot$ based on aggregation semantics is introduced in [13] as follows:

Definition 6 ($ME_\odot$ Inference Operator [13]). *Let $\mathcal{R}$ be a consistent set of probabilistic conditionals. The ME-inference operator $ME_\odot$ based on aggregation semantics is defined as*

$$ME_\odot(\mathcal{R}) := \arg \max_{P \in \mathcal{P}_\Omega : P \models_\odot \mathcal{R}} H(P) \tag{6}$$

In [21] it is shown that (6) has a unique solution and induces a *convex optimization problem*, since the solutions to $P \models_\odot \mathcal{R}$ form a convex set and $H(P)$ is a strictly concave function. Thus, $ME_\odot(\mathcal{R})$ is well defined.

To avoid cumbersome distinctions of cases, in the following we consider only soft probabilistic conditionals. Therefore, for the rest of the paper let

$$\mathcal{R} := \{r_1, \ldots, r_m\} \tag{7}$$

be a consistent set of m soft probabilistic conditionals

$$r_i = (B_i(\boldsymbol{X})|A_i(\boldsymbol{X}))[d_i], \text{ with } d_i \in (0,1), 1 \le i \le m \tag{8}$$

and let σ_i denote the feature function of r_i according to Definition 5.

Proposition 2. *For any consistent set of soft probabilistic conditionals $\mathcal{R}$ as given by (7) and (8), there exists a positive probability distribution which satisfies $\mathcal{R}$.*

From Definition 6 and Proposition 1 it follows that the determination of $ME_\odot(R)$ requires to solve the following optimization problem with objective function $H(P)$ and m linear constraints induced by the m conditionals of R:

Definition 7 (Optimization Problem OptAgg($\mathcal{R}$)). *Let $\sigma_i, 1 \le i \le m$, be the feature functions for $\mathcal{R}$. Then the optimization problem $\textsc{OptAgg}(\mathcal{R})$ is defined as:*

$$\text{maximize } H(P)$$
$$\text{subject to } \sum_{\omega \in \Omega} P(\omega)\sigma_i(\omega) = 0, 1 \le i \le m$$
$$\sum_{\omega \in \Omega} P(\omega) \qquad = 1 \tag{9}$$
$$P(\omega) \qquad \ge 0, \forall \omega \in \Omega$$

The two latter constraints ensure that the solution is a proper probability distribution.

Proposition 3. *The optimization problem $\textsc{OptAgg}(\mathcal{R})$ for a given set $\mathcal{R}$ has a unique solution that coincides with $ME_\odot(\mathcal{R})$.*

5   Computing the Maximum Entropy Distribution

There exist several algorithms to calculate the solution of a general convex optimization problem [4], i. e. these algorithms can be applied to a convex optimization problem with an arbitrary (convex) objective function. In this paper, we deal with an algorithm which is tailor-made for a convex optimization problem of the form (9), i. e. for a convex optimization problem with entropy $H(P)$ as objective function. Since this algorithm is specialized to entropy optimization, it can take advantage of certain characteristics of the entropy function, whereas general algorithms for convex optimization problems can just utilize the convexity of the objective function.

5.1   Generalized Iterative Scaling

The so-called *Generalized Iterative Scaling (GIS)* algorithm presented in [6] computes the ME-distribution under linear constraints, i. e. it iteratively calculates a sequence of distributions which converges to the solution. To be precise, the GIS algorithm allows to compute the distribution with minimum relative entropy (also see [5] for an alternative proof of the algorithm). The *relative entropy* (also called *Kullback-Leibler divergence* or *information divergence*) between two distributions P and Q is defined as $K(P,Q) := \sum_{\omega \in \Omega} P(\omega) \log \frac{P(\omega)}{Q(\omega)}$. Let $P_U(\omega) := \frac{1}{|\Omega|}$ for all $\omega \in \Omega$ denote the uniform distribution over Ω. It is easy to see that $K(P, P_U) = \log |\Omega| - H(P)$ holds, i. e. entropy is just a special of relative entropy.

Proposition 4. *Let S be a set of probability distributions over Ω. Then it holds:* $\arg \min_{P \in S} K(P, P_U) = \arg \max_{P \in S} H(P)$

Therefore, instead of maximizing the entropy of a distribution, we will consider minimizing the relative entropy of a distribution with respect to the uniform distribution. The general from of the optimization problem solved by the GIS algorithm is as follows:

Definition 8 (Optimization Problem OptGis(EQ)). *Let $Q \in \mathcal{P}_\Omega$ be a given probability distribution. For $i = 1, \ldots, m$, let $a_i : \Omega \to \mathbb{R}$ be a given function and let $h_i \in \mathbb{R}$ be its given expected value, so that the equation system (denoted by EQ) of linear constraints $\sum_{\omega \in \Omega} P(\omega) a_i(\omega) = h_i$, $i = 1, \ldots, m$, induced by the functions and their expected values can be satisfied by a positive probability distribution. Then the optimization problem* OPTGIS(EQ) *is defined as:*

$$\text{minimize } K(P,Q)$$

$$\begin{aligned}
\text{subject to } &\sum_{\omega \in \Omega} P(\omega) a_i(\omega) = h_i, i = 1, \ldots, m \\
&\sum_{\omega \in \Omega} P(\omega) = 1 \\
&P(\omega) > 0, \ \forall \omega \in \Omega
\end{aligned} \tag{10}$$

If the preconditions of Definition 8 are met, the GIS algorithm can be applied to compute the solution to the optimization problem $\text{OPTGIS}(EQ)$. Since the constraints in (9) have been induced by a consistent set of soft probabilistic conditionals, they can be satisfied by a positive distribution according to Proposition 2. So in principle, the GIS algorithm can be applied to compute the solution to the optimization problem $\text{OPTAGG}(\mathcal{R})$, since this matches the form of an optimization problem $\text{OPTGIS}(EQ)$.

5.2   Transforming OptGis(EQ) into a Normalized Form

The concrete application of the GIS algorithm to $\text{OPTGIS}(EQ)$ requires a transformation into a normalized form meeting some additional requirements.

Definition 9 (Optimization Problem OptGisNorm($\hat{EQ}$)). *Let $Q \in \mathcal{P}_\Omega$ be a given probability distribution. For $i = 1, \ldots, \hat{m}$, let $\hat{a}_i : \Omega \to \mathbb{R}$ be a given function and let $\hat{h}_i \in \mathbb{R}$ be its given expected value, so that the induced equation system (denoted by $\hat{EQ}$) of linear constrains $\sum_{\omega \in \Omega} P(\omega)\hat{a}_i(\omega) = \hat{h}_i, i = 1, \ldots, \hat{m}$, can be satisfied by a positive probability distribution. If*

$$\hat{a}_i(\omega) \geq 0, \forall \omega \in \Omega,\ i = 1, \ldots, \hat{m}, \tag{11}$$

$$\sum_{i=1}^{\hat{m}} \hat{a}_i(\omega) = 1, \qquad \forall \omega \in \Omega \tag{12}$$

$$\hat{h}_i > 0, \qquad i = 1, \ldots, \hat{m} \tag{13}$$

$$\sum_{i=1}^{\hat{m}} \hat{h}_i = 1 \tag{14}$$

hold, then the optimization problem $\text{OPTGISNORM}(\hat{EQ})$ is defined as:

$$\textit{minimize } K(P, Q)$$

$$\textit{subject to } \sum_{\omega \in \Omega} P(\omega)\hat{a}_i(\omega) = \hat{h}_i, i = 1, \ldots, \hat{m}$$
$$\sum_{\omega \in \Omega} P(\omega) \qquad = 1 \tag{15}$$
$$P(\omega) \qquad > 0,\ \forall \omega \in \Omega$$

In [6] it is shown in general that an optimization problem $\text{OPTGIS}(EQ)$ can always be transformed appropriately to meet the requirements (11) – (14) of $\text{OPTGISNORM}(\hat{EQ})$. Note that in [6] for this transformation the additional requirement is made that each function a_i in $\text{OPTGIS}(EQ)$ has at least one non-zero value, thereby assuring that (13) holds in $\text{OPTGISNORM}(\hat{EQ})$ after the transformation. Here, we do not need this additional requirement on the functions a_i since we will show that in our context (13) holds anyway.

In the following, we will demonstrate how the constraints from (9) in $\text{OPTAGG}(\mathcal{R})$ can be transformed into a set of normalized constraints meeting the normalization requirements (11) to (14).

Let
$$G_i^\# := |\mathcal{H}^{\boldsymbol{x}(A_i, B_i)}|$$
be the number of groundings of a conditional $r_i \in \mathcal{R}$, $1 \leq i \leq m$, and let
$$\mathcal{G}^\# := \sum_{i=1}^m G_i^\# \tag{16}$$
denote the total number of groundings of all conditionals in $\mathcal{R}$.

Definition 10 (Non-Negative Feature Function). *For each feature function σ_i of a conditional $r_i \in \mathcal{R}$ of the optimization problem $\text{OptAgg}(\mathcal{R})$, let the non-negative feature function $\sigma_i' : \Omega \to \mathbb{R}_0^+$ be defined as*
$$\sigma_i'(\omega) := \sigma_i(\omega) + d_i G_i^\#, \ \forall \omega \in \Omega \tag{17}$$
and the expected value of σ_i', denoted by ε_i', is set to
$$\varepsilon_i' := d_i G_i^\# \tag{18}$$

Proposition 5. *For a feature function σ_i' and its expected value ε_i' according to Definition 10 the following holds:*
$$0 \leq \sigma_i'(\omega) \leq G_i^\#, \ \forall \omega \in \Omega \tag{19}$$
$$\sum_{\omega \in \Omega} P(\omega)\sigma_i(\omega) = 0 \ \Leftrightarrow \ \sum_{\omega \in \Omega} P(\omega)\sigma_i'(\omega) = \varepsilon_i' \tag{20}$$

Definition 11 (Normalized Feature Function). *For each feature function σ_i of a conditional $r_i \in \mathcal{R}$ of the optimization problem $\text{OptAgg}(\mathcal{R})$, let the normalized feature function $\hat{\sigma}_i : \Omega \to [0, 1]$ be defined as*
$$\hat{\sigma}_i(\omega) := \frac{\sigma_i'(\omega)}{\mathcal{G}^\#} = \frac{\sigma_i(\omega) + d_i G_i^\#}{\mathcal{G}^\#}, \ \forall \omega \in \Omega \tag{21}$$
and the expected value of $\hat{\sigma}_i$, denoted by $\hat{\varepsilon}_i$, is set to
$$\hat{\varepsilon}_i := \frac{\varepsilon_i'}{\mathcal{G}^\#} = \frac{d_i G_i^\#}{\mathcal{G}^\#} \tag{22}$$

Proposition 6. *For a feature function $\hat{\sigma}_i$ and its expected value $\hat{\varepsilon}_i$ according to Definition 11 the following holds:*
$$0 \leq \hat{\sigma}_i(\omega) \leq 1, \ \forall \omega \in \Omega \tag{23}$$
$$0 \leq \sum_{i=1}^m \hat{\sigma}_i(\omega) \leq 1, \ \forall \omega \in \Omega \tag{24}$$
$$\sum_{\omega \in \Omega} P(\omega)\sigma_i(\omega) = 0 \ \Leftrightarrow \ \sum_{\omega \in \Omega} P(\omega)\hat{\sigma}_i(\omega) = \hat{\varepsilon}_i \tag{25}$$
$$\hat{\varepsilon}_i > 0 \tag{26}$$
$$0 < \sum_{i=1}^m \hat{\varepsilon}_i < 1 \tag{27}$$

The definition of $\hat{\varepsilon}_i$ in (22) immediately implies that (26) must hold, because d_i, $G_i^\#$, and $\mathcal{G}^\#$ are positive values.

Definition 12 (Correctional Feature Function). *Let* $\hat{\sigma}_1, \ldots, \hat{\sigma}_m$ *and* $\hat{\varepsilon}_1, \ldots, \hat{\varepsilon}_m$ *be as in Definition 11. Then the additional correctional feature function* $\hat{\sigma}_{\hat{m}}$ *with* $\hat{m} := m + 1$ *is defined as*

$$\hat{\sigma}_{\hat{m}}(\omega) := 1 - \sum_{i=1}^{m} \hat{\sigma}_i(\omega) \tag{28}$$

and the corresponding additional correctional expected value $\hat{\varepsilon}_{\hat{m}}$ *is set to*

$$\hat{\varepsilon}_{\hat{m}} := 1 - \sum_{i=1}^{m} \hat{\varepsilon}_i \tag{29}$$

Proposition 7. *For the additional correctional feature function* $\hat{\sigma}_{\hat{m}}$ *and expected value* $\hat{\varepsilon}_{\hat{m}}$ *from Definition 12 it holds:*

$$0 \le \hat{\sigma}_{\hat{m}}(\omega) \le 1, \ \forall \omega \in \Omega \tag{30}$$

$$\sum_{i=1}^{\hat{m}} \hat{\sigma}_i(\omega) = 1, \ \forall \omega \in \Omega \tag{31}$$

$$\hat{\varepsilon}_{\hat{m}} > 0 \tag{32}$$

$$\sum_{i=1}^{\hat{m}} \hat{\varepsilon}_i = 1 \tag{33}$$

5.3   Normalized Optimization Problem

The above definitions of normalized feature functions $\hat{\sigma}_1, \ldots, \hat{\sigma}_m$ and a correctional feature function $\hat{\sigma}_{\hat{m}}$ (and their expected values $\varepsilon_1, \ldots, \varepsilon_m$ and $\hat{\varepsilon}_{\hat{m}}$) allows us to define the following optimization problem, which represents the optimization problem $\text{OPTAGG}(\mathcal{R})$ in a normalized form, meeting all requirements to apply the GIS algorithm technique:

Definition 13 (Optimization Problem OptAggNorm($\mathcal{R}$)). *Let* $\hat{\sigma}_i(\omega)$, $\hat{\varepsilon}_i$, $\hat{m}$, $\hat{\sigma}_{\hat{m}}(\omega)$, *and* $\hat{\varepsilon}_{\hat{m}}$ *be as in Definition 11 and 12. Then the optimization problem* $\text{OPTAGGNORM}(\mathcal{R})$ *is defined as:*

$$\begin{aligned} &minimize \ K(P, P_U) \\ &subject \ to \ \sum_{\omega \in \Omega} P(\omega)\hat{\sigma}_i(\omega) = \hat{\varepsilon}_i, \ 1 \le i \le \hat{m} \\ &\qquad\qquad \sum_{\omega \in \Omega} P(\omega) \quad = 1 \\ &\qquad\qquad P(\omega) \qquad\qquad > 0, \ \forall \omega \in \Omega \end{aligned} \tag{34}$$

Proposition 8. *The optimization problems* OPTAGG($\mathcal{R}$) *and* OPTAGGNORM($\mathcal{R}$) *have the same solution.*

Proposition 9. *The optimization problem* OPTAGGNORM($\mathcal{R}$) *yields an instance of the optimization problem* OPTGISNORM($\hat{EQ}$), *i. e. in particular, the feature functions and expected values of* OPTAGGNORM($\mathcal{R}$) *satisfy the corresponding requirements (11) – (14) in Definition 9.*

The following proposition is a direct consequence of Propositions 8 and 9:

Proposition 10. *For any consistent set $\mathcal{R}$ of soft conditionals, the GIS algorithm technique can directly be applied to the optimization problem* OPTAGGNORM($\mathcal{R}$) *to compute its solution P^*. Since P^* is also the solution to the optimization problem* OPTAGG($\mathcal{R}$), *the computation delivers the inference operator* $\mathrm{ME}_\odot(\mathcal{R}) = P^*$.

5.4   GIS Algorithm for Aggregation Semantics

Based on the basic template for a GIS algorithm in [6], we present a practical GIS algorithm which computes the solution P^* of the optimization problem OPTAGGNORM($\mathcal{R}$). So according to Proposition 10, the algorithm delivers $\mathrm{ME}_\odot(\mathcal{R})$ as result. The pseudo-code of the GIS algorithm for aggregation semantics is depicted in Fig. 1.

The algorithm starts with the uniform distribution as initial distribution. In the k-th iteration step, for each feature function $\hat{\sigma}_i$ the current ratio $\beta_{(k),i}$ between its given expected value $\hat{\varepsilon}_i$ and its current expected value $\sum_{\omega \in \Omega} P_{(k-1)}(\omega)\hat{\sigma}_i(\omega)$ under the current distribution $P_{(k-1)}$ is determined. So $\beta_{(k),i}$ is the factor required to scale $P_{(k-1)}$ appropriately so that the expected value $\hat{\varepsilon}_i$ of $\hat{\sigma}_i$ would be met exactly. Since the actual scaling of $P_{(k-1)}$ has to be performed with respect to all scaling factors $\beta_{(k),1}, \ldots, \beta_{(k),\hat{m}}$, the scaled distribution $P_{(k)}$ cannot fit all expected values immediately, but it is guaranteed by the GIS approach that a distribution iteratively computed that way converges to the correct solution. Note that the constraint $\sum_{\omega \in \Omega} P(\omega) = 1$, which is contained in each of the above optimization problems, is not explicitly encoded as a constraint in the GIS algorithm in Fig. 1. Instead, the scaled probability values $P'_{(k)}(\omega)$ are normalized in each iteration step, so that $P_{(k)}$ is a proper probability distribution (which is important to determine the correct $\beta_{(k+1),i}$ with respect to $P_{(k)}$). The GIS algorithm iteratively calculates a sequence of distributions which converges to the solution of the optimization problem. So in practice, an abortion condition must be defined which allows to stop the iteration if the solution has been approximated with a sufficient accuracy. A practical abortion condition is, e. g. to stop after iteration step k if $|1 - \beta_{(k),i}| < \delta_\beta$ holds for all $1 \leq i \leq \hat{m}$, with δ_β being an appropriate accuracy threshold, i. e. if there is no more need to scale any values because all scaling factors are almost 1 within accuracy δ_β. Alternatively, the iteration could stop after step k if $|P_{(k)}(r_i) - d_i| < \delta_r, 1 \leq i \leq m$ holds (with $P_{(k)}(r_i)$ denoting the probability of conditional r_i under distribution $P_{(k)}$), i. e. if the probability of each conditional under the current distribution $P_{(k)}$ matches its prescribed probability d_i within accuracy δ_r.

Input: a consistent set $\mathcal{R}$ of m soft probabilistic conditionals

Output: inference operator $\mathrm{ME}_\odot(\mathcal{R}) := P^*$

Algorithm

- Let $P_{(0)} := P_U$
- Let $\hat{\sigma}_1, \ldots, \hat{\sigma}_m, \hat{\sigma}_{\hat{m}}$ be normalized feature functions
  and $\hat{\varepsilon}_1, \ldots, \hat{\varepsilon}_m, \hat{\varepsilon}_{\hat{m}}$ be expected values constructed from $\mathcal{R}$ (according to Def. 13)
- Initialize iteration counter: $k := 0$
- Repeat until an abortion condition holds:
  - Increase iteration counter: $k := k + 1$
  - Calculate the scaling factor $\beta_{(k),i}$ of each feature function $\hat{\sigma}_i$:

  $$\beta_{(k),i} := \frac{\hat{\varepsilon}_i}{\sum\limits_{\omega \in \Omega} P_{(k-1)}(\omega)\hat{\sigma}_i(\omega)}, \quad 1 \le i \le \hat{m}$$

  - Scale all probabilities appropriately:

  $$P'_{(k)}(\omega) := P_{(k-1)}(\omega) \prod_{i=1}^{\hat{m}} \left(\beta_{(k),i}\right)^{\hat{\sigma}_i(\omega)}, \qquad \forall \omega \in \Omega$$

  - Normalize the probability values:

  $$P_{(k)}(\omega) := \frac{P'_{(k)}(\omega)}{\sum\limits_{\omega \in \Omega} P'_{(k)}(\omega)}, \quad \forall \omega \in \Omega$$

- Let $P^* := P_{(k)}$ denote the final distribution of the iteration.

Fig. 1. GIS Algorithm for Aggregation Semantics

6 Implementation

The GIS algorithm for aggregation semantics has successfully been implemented as a plugin for the KREATOR system [2], which is an integrated development environment for representing, reasoning, and learning with relational probabilistic knowledge. Our first implementation of the GIS algorithm is a straight-forward implementation of the pseudo-code from Fig. 1, with further optimizations being referred to further refinements. The following example will be used to illustrate the runtime behavior of the implementation:

Example 2. Suppose we have a zoo with a population of monkeys. The predicate $Fe(X, Y)$ expresses that a monkey X feeds another monkey Y and $Hu(X)$ says that a monkey X is hungry. The knowledge base $\mathcal{R}_{\mathrm{mky}}$ contains conditionals which express generic knowledge as well as one conditional stating exceptional knowledge about a monkey *Charly*:

$r_1 : (Fe(X, Y) \mid \neg Hu(X) \wedge Hu(Y))\ [0.80]$ $r_2 : (Fe(X, Y) \mid Hu(X))\ [0.001]$

$r_3 : (Fe(X, charly) \mid \neg Hu(X))\ [0.95]$ $r_4 : (Fe(X, X) \mid \top)\ [0.001]$

$r_5 : (Fe(X, Y) \mid \neg Hu(X) \wedge \neg Hu(Y))\ [0.10]$

Considering the above example together with set $Const = \{andy, bobby, charly\}$ of constants, the corresponding Herbrand base contains 12 ground atoms. Therefore $|\Omega| = 2^{12} = 4,096$ elementary probabilities have to be computed in every iteration step of the algorithm. Using the values of all scaling factors as abortion condition (as suggested in Sec. 5.4) with an accuracy threshold of $\delta_\beta = 0.001$, the GIS algorithm requires 20,303 steps to compute a solution with sufficient accuracy. On a computer with an Intel Core i5-2500K CPU (4 cores, 3.3 Ghz) the computation of the ME-distribution P^* takes 14 seconds. To additionally check the accuracy of the calculated distribution P^*, the probabilities of the conditionals from $\mathcal{R}_{mky}$ have been recalculated under P^*. Comparing these probabilities with the prescribed probabilities of the conditionals reveals an deviation of $\delta_r = 0.0017$ at most. Performing the same computation with an improved accuracy of $\delta_\beta = 0.0001$ results in 81,726 iteration steps taking 59 seconds and revealing an improved deviation of $\delta_r = 0.00017$ as well. Once the distribution P^* has been computed, it can be (re-)used for probabilistic inference. That way, arbitrary queries like e. g. $q := (Fe(andy, bobby) \mid Hu(charly))$ can be addressed to the knowledge base and the answer (i. e. the queries's probability under P^*) is determined immediately, e. g. $P^*(q) = 0.21$.

7   Conclusion and Future Work

In this paper, we investigated the aggregation semantics for first-order probabilistic logic and its ME-based inference operator $ME_\odot$. We illustrated how the convex optimization problem induced by $ME_\odot$ can be expressed in terms of feature functions. We developed an approach allowing us to use a GIS algorithm technique for solving this optimization problem and presented the pseudo-code of a concrete algorithm which employs GIS to calculate $ME_\odot(\mathcal{R})$, i. e. the ME-optimal probability distribution which satisfies all conditionals in $\mathcal{R}$. A realization of this algorithm has been integrated in a plugin for aggregation semantics in the KREATOR system, being the first practical implementation of $ME_\odot$ inference. In our $ME_\odot$ inference algorithm and its implementation there are many opportunities for extensions and improvements. For instance, in a future version, we will consider also hard conditionals so that deterministic knowledge can be expressed in $\mathcal{R}$. Our current work also includes considering decomposed distributions by employing junction trees and sophisticated propagation techniques, as proposed in [1,20,15,19], or using the *Improved Iterative Scaling* (IIS) algorithm from [3]. Furthermore, we will exploit the equivalence of worlds regarding their ME-probability, thereby allowing to significantly reduce the size of the distribution which has to be computed by the GIS algorithm; a prototypical implementation of this approach already shows very promising results.

References

1. Badsberg, J.H., Malvestuto, F.M.: An implementation of the iterative proportional fitting procedure by propagation trees. Computational Statistics & Data Analysis 37(3), 297–322 (2001)

2. Beierle, C., Finthammer, M., Kern-Isberner, G., Thimm, M.: Automated Reasoning for Relational Probabilistic Knowledge Representation. In: Giesl, J., Hähnle, R. (eds.) IJCAR 2010. LNCS, vol. 6173, pp. 218–224. Springer, Heidelberg (2010)
3. Berger, A.L., Della Pietra, S., Della Pietra, V.J.: A maximum entropy approach to natural language processing. Computational Linguistics 22(1), 39–71 (1996)
4. Boyd, S., Vandenberghe, L.: Convex Optimization. Cambridge University Press, New York (2004)
5. Csiszar, I.: A geometric interpretation of Darroch and Ratcliff's generalized iterative scaling. Annals of Statistics 17(3), 1409–1413 (1989)
6. Darroch, J.N., Ratcliff, D.: Generalized iterative scaling for log-linear models. In: Annals of Mathematical Statistics, vol. 43, pp. 1470–1480. Institute of Mathematical Statistics (1972)
7. Delgrande, J.: On first-order conditional logics. Artificial Intelligence 105, 105–137 (1998)
8. Domingos, P., Lowd, D.: Markov Logic: An Interface Layer for Artificial Intelligence. Synthesis Lectures on Artificial Intelligence and Machine Learning. Morgan & Claypool Publishers (2009)
9. Finthammer, M.: A Generalized Iterative Scaling Algorithm for Maximum Entropy Reasoning in Relational Probabilistic Conditional Logic Under Aggregation Semantics. Informatik-Bericht 363, Fakultät für Mathematik und Informatik, FernUniversität in Hagen (April 2012)
10. Fisseler, J.: Learning and Modeling with Probabilistic Conditional Logic. Dissertations in Artificial Intelligence, vol. 328. IOS Press, Amsterdam (2010)
11. Kern-Isberner, G.: Conditionals in Nonmonotonic Reasoning and Belief Revision. LNCS (LNAI), vol. 2087. Springer, Heidelberg (2001)
12. Kern-Isberner, G., Lukasiewicz, T.: Combining probabilistic logic programming with the power of maximum entropy. Artificial Intelligence, Special Issue on Nonmonotonic Reasoning 157(1-2), 139–202 (2004)
13. Kern-Isberner, G., Thimm, M.: Novel semantical approaches to relational probabilistic conditionals. In: Lin, F., Sattler, U., Truszczyński, M. (eds.) Proceedings of the Twelfth International Conference on the Principles of Knowledge Representation and Reasoning (KR 2010), pp. 382–392. AAAI Press (May 2010)
14. Kersting, K., De Raedt, L.: Bayesian Logic Programming: Theory and Tool. In: Getoor, L., Taskar, B. (eds.) An Introduction to Statistical Relational Learning. MIT Press (2007)
15. Meyer, C.H.: Korrektes Schließen bei unvollständiger Information. Ph.D. thesis, FernUniversität Hagen (1998)
16. Paris, J.: The uncertain reasoner's companion – A mathematical perspective. Cambridge University Press (1994)
17. Pearl, J.: Probabilistic Reasoning in Intelligent Systems: Networks of Plausible Inference. Morgan Kaufmann (1998)
18. Rödder, W.: Conditional logic and the principle of entropy. Artificial Intelligence 117, 83–106 (2000)
19. Rödder, W., Meyer, C.H.: Coherent Knowledge Processing at Maximum Entropy by SPIRIT. In: Proceedings of the Twelfth Conference on Uncertainty in Artificial Intelligence (UAI 1996), pp. 470–476 (1996)
20. Teh, Y., Welling, M.: On improving the efficiency of the iterative proportional fitting procedure. In: Proc. of the 9th Int'l. Workshop on AI and Statistics (AISTATS 2003) (2003)
21. Thimm, M.: Probabilistic Reasoning with Incomplete and Inconsistent Beliefs. Ph.D. thesis, Technische Universität Dortmund (2011)

Probabilistic Conditional Independence under Schema Certainty and Uncertainty[*]

Joachim Biskup[1],[**], Sven Hartmann[2], and Sebastian Link[3]

[1] Fakultät für Informatik, Technische Universität Dortmund, Germany
[2] Department of Informatics, Clausthal University of Technology, Germany
[3] Department of Computer Science, The University of Auckland, New Zealand

Abstract. Conditional independence provides an essential framework to deal with knowledge and uncertainty in Artificial Intelligence, and is fundamental in probability and multivariate statistics. Its associated implication problem is paramount for building Bayesian networks. Saturated conditional independencies form an important subclass of conditional independencies. Under schema certainty, the implication problem of this subclass is finitely axiomatizable and decidable in almost linear time. We study the implication problem of saturated conditional independencies under both schema certainty and uncertainty. Under schema certainty, we establish a finite axiomatization with the following property: every independency whose implication is dependent on the underlying schema can be inferred by a single application of the so-called symmetry rule to some independency whose implication is independent from the underlying schema. Removing the symmetry rule from the axiomatization under schema certainty results in an axiomatization for a notion of implication that leaves the underlying schema undetermined. Hence, the symmetry rule is just a means to infer saturated conditional independencies whose implication is truly dependent on the schema.

1 Introduction

The concept of conditional independence is important for capturing structural aspects of probability distributions, for dealing with knowledge and uncertainty in Artificial Intelligence, and for learning and reasoning in intelligent systems [12,13]. Application areas include natural language processing, speech processing, computer vision, robotics, computational biology, and error-control coding. A conditional independence (CI) statement represents the independence of two sets of attributes relative to a third: given three mutually disjoint subsets X, Y, and Z of a set S of attributes, if we have knowledge about the state of X, then knowledge about the state of Y does not provide additional evidence for

---

[*] This research is supported by the Marsden Fund Council from Government funding, administered by the Royal Society of New Zealand.
[**] This author has been supported by the Deutsche Forschungsgemeinschaft under grant SFB 876/A5 for the Collaborative Research Center 876 "Providing Information by Resource-Constrained Data Analysis".

E. Hüllermeier et al. (Eds.): SUM 2012, LNAI 7520, pp. 365–378, 2012.

the state of Z and vice versa. Traditionally, the notation $I(Y, Z \mid X)$ has been used to specify such CI statement. An important problem is the S-implication problem, which is to decide for an arbitrary set S, and an arbitrary set $\Sigma \cup \{\varphi\}$ of CI statements over S, whether every probability model that satisfies every CI statement in Σ also satisfies φ. The significance of this problem is due to its relevance for building a Bayesian network [11,12,13]. The S-implication problem for CI statements is not finitely axiomatizable by a set of Horn rules [14]. However, it is possible to express CI statements using polynomial likelihood formulae, and reasoning about polynomial inequalities is axiomatizable [8]. Recently, the S-implication problem of stable CI statements has been shown to be finitely axiomatizable, and $coNP$-complete to decide [12]. Here, stable means that the validity of $I(Y, Z \mid X)$ implies the validity of every $I(Y, Z \mid X')$ where $X \subseteq X' \subseteq S - YZ$. An important subclass of stable CI statements are saturated conditional independence (SCI) statements. These are CI statements $I(Y, Z \mid X)$ over $S = XYZ$ where S is the union of X, Y and Z. Geiger and Pearl have established a finite axiomatization for the S-implication problem [7]. For SCI statements we prefer to write $I(Y \mid X)$ instead of $I(Y, Z \mid X)$ since $Z = S - XY$. The notion of saturated conditional independence $I(Y \mid X)$ over S is very closely related to that of a multivalued dependency (MVD) $X \twoheadrightarrow Y$ over S, studied in the framework of relational databases [1,9]. Here, a set X of attributes is used to denote the X-value of a tuple over S, i.e., those tuple components that appear in the columns associated with X. Indeed, $X \twoheadrightarrow Y$ expresses the fact that an X-value uniquely determines the set of associated Y-values independently of joint associations with Z-values where $Z = S - XY$ [4]. Thus, given a specific occurrence of an X-value within a tuple, so far not knowing the specific association with a Y-value and a Z-value within that tuple, and then learning about the specific associated Y-value does not provide any information about the specific associated Z-value. Previous research has established an equivalence between the S-implication problem for SCI statements $I(Y \mid X)$ and the S-implication problem for MVDs $X \twoheadrightarrow Y$ [15]. Utilizing results for MVDs [1,5] the S-implication problem of SCI statements is decidable in almost linear time [6]. In this paper we will refine the known equivalences between the implication of SCI statements and MVDs. The following example provides a motivation for our studies.

Example 1. Suppose S consists of the attributes *Emp, Child, Project*. Declaring $I(Child \mid Emp)$ means that given an employee, we do not learn anything new about her children by learning anything new about a project she is involved in. Due to the symmetry rule [7] the SCI statement $I(Project \mid Emp)$ is S-implied by $I(Child \mid Emp)$. Strictly speaking, however, S-implication here means that $I(Project \mid Emp)$ is implied by both $I(Child \mid Emp)$ and S together, but not by $I(Child \mid Emp)$ alone. Indeed, if S' results from S by adding the attribute *Hobby*, then $I(Project \mid Emp)$ is not S'-implied by $I(Child \mid Emp)$.   □

The last example illustrates the need to distinguish between different notions of semantic implication. The first notion is that of S-implication. The literature on CI statements has dealt with this notion exclusively. An alternative notion addresses uncertainty about the underlying set S of attributes. The associated implication problem asks if for every given set $\Sigma \cup \{\varphi\}$ of SCI statements, every probability model – for any suitable set S – that satisfies Σ also satisfies φ.

Example 2. The SCI statement $I(Child \mid Emp)$ $\{Emp, Child, Project\}$-implies $I(Project \mid Emp)$, but $I(Child \mid Emp)$ does not imply $I(Project \mid Emp)$.   □

Contribution. In this paper we make three main contributions. Firstly, we show that Geiger and Pearl's finite axiomatization $\mathfrak{G}$ for the S-implication of SCI statements cannot distinguish between implied and S-implied SCI statements. That is, there are implied SCI statements for which every inference by $\mathfrak{G}$ applies the symmetry rule; giving incorrectly the impression that the implication of the SCI statements depends on S. Moreover, there are S-implied SCI statements for which every inference by $\mathfrak{G}$ requires applications of the symmetry rule at times different from the last inference step only. Secondly, we establish a finite axiomatization $\mathfrak{C}_S$ such that every implied SCI statement can be inferred without any application of the symmetry rule; and every S-implied SCI statement can be inferred with only a single application of the S-symmetry rule, and this application is done in the last step of the inference. Finally, we establish a finite axiomatization $\mathfrak{C}$ for the implication of SCI statements. As $\mathfrak{C}$ results from $\mathfrak{C}_S$ by removal of the symmetry rule, the results show that the symmetry rule is only necessary to infer those SCI statements that are S-implied but not implied. These results are in analogy to those developed for relational MVDs [2,3,10].

Organization. We review the framework and some results on CI statements in Section 2. We show that Geiger and Pearl's axiomatization does not reflect the role of the symmetry rule to infer S-implied SCI statements that are not implied. In Section 3 we establish an axiomatization for S-implication that does reflect the role of the symmetry rule. The axiomatization for the implication of SCI statements under schema uncertainty is established in Section 4. In Section 5 we outline results for an alternative definition of SCI statements, motivated by the definition of MVDs. We conclude in Section 6.

2   Geiger and Pearl's Axiomatization for SCI Statements

In this section we adopt the framework for conditional independence statements from [7]. We adapt their axiomatization $\mathfrak{G}$ for the S-implication of saturated statements by rewriting it into a format suitable to study the role of the symmetry rule. Finally, we show that the axiomatization does not reflect the role of the symmetry rule to infer S-implied saturated statements that are not implied.

We denote by $\mathfrak{A}$ a countably infinite set of distinct symbols $\{a_1, a_2, \ldots\}$, called *attributes*. A finite subset $S \subseteq \mathfrak{A}$ is called a *schema*. Each attribute a_i can be assigned a finite set $dom(a_i)$. This set is called the *domain* of a_i and each of its elements is a *value* for a_i. For $X = \{s_1, \ldots, s_k\} \subseteq S$ we say that $\mathbf{x}$ is a value of X, if $\mathbf{x} \in dom(s_1) \times \cdots \times dom(s_k)$. For a value $\mathbf{x}$ of X we write $\mathbf{x}(y)$ for the projection of $\mathbf{x}$ onto $Y \subseteq X$. For finite $X, Y \subseteq \mathfrak{A}$ we write XY for the set union of X and Y.

A *probability model* over $S = \{s_1, \ldots, s_n\}$ is a pair (dom, P) where dom is a domain mapping that maps each s_i to a finite domain $dom(s_i)$, and $P : dom(s_1) \times \cdots \times dom(s_n) \to [0, 1]$ is a probability distribution having the Cartesian product of these domains as its sample space.

Definition 1. *The expression* $I(Y, Z \mid X)$ *where* X, Y *and* Z *are disjoint subsets of* S *such that* $S = XYZ$ *is called a* saturated conditional independence *(SCI) statement over* S. *We say that the SCI statement* $I(Y, Z \mid X)$ *holds for* (dom, P) *if for every values* $\mathbf{x}, \mathbf{y},$ *and* $\mathbf{z}$ *of* X, Y *and* Z, *respectively,*

$$P(\mathbf{y}, \mathbf{z}, \mathbf{x}) \cdot P(\mathbf{x}) = P(\mathbf{y}, \mathbf{x}) \cdot P(\mathbf{z}, \mathbf{x}).$$

Equivalently, (dom, P) *is said to* satisfy $I(Y, Z \mid X)$. □

SCI statements interact with one another, and these interactions have been formalized by the following notion of semantic implication.

Definition 2. *Let* $\Sigma \cup \{\varphi\}$ *be a set of SCI statements over* S. *We say that* Σ S-*implies* φ *if every probability model over* S *that satisfies every SCI statement* $\sigma \in \Sigma$ *also satisfies* φ. □

The S-*implication problem* is the following problem.

PROBLEM:	S-implication problem
INPUT:	Schema S, Set $\Sigma \cup \{\varphi\}$ of SCI statements over S
OUTPUT:	Yes, if Σ S-implies φ; No, otherwise

For Σ we let $\Sigma_S^* = \{\varphi \mid \Sigma \ S\text{-implies} \ \varphi\}$ be the *semantic closure* of Σ, i.e., the set of all SCI statements S-implied by Σ. In order to determine the S-implied SCI statements we use a syntactic approach by applying inference rules. These inference rules have the form

$$\frac{\text{premises}}{\text{conclusion}}$$

and inference rules without any premises are called axioms. An inference rule is called S-*sound*, if the premises of the rule S-imply the conclusion of the rule. We let $\Sigma \vdash_{\mathfrak{R}} \varphi$ denote the *inference* of φ from Σ by the set $\mathfrak{R}$ of inference rules. That is, there is some sequence $\gamma = [\sigma_1, \ldots, \sigma_n]$ of SCI statements such that $\sigma_n = \varphi$ and every σ_i is an element of Σ or results from an application of an inference rule in $\mathfrak{R}$ to some elements in $\{\sigma_1, \ldots, \sigma_{i-1}\}$. For Σ, let $\Sigma_{\mathfrak{R}}^+ = \{\varphi \mid \Sigma \vdash_{\mathfrak{R}} \varphi\}$ be its *syntactic closure* under inferences by $\mathfrak{R}$. A set $\mathfrak{R}$ of inference rules is said to be S-*sound* (S-*complete*) for the S-implication of SCI statements, if for every S and for every set Σ of SCI statements over S, we have $\Sigma_{\mathfrak{R}}^+ \subseteq \Sigma_S^*$ ($\Sigma_S^* \subseteq \Sigma_{\mathfrak{R}}^+$). The (finite) set $\mathfrak{R}$ is said to be a (finite) *axiomatization* for the S-implication of SCI statements if $\mathfrak{R}$ is both S-sound and S-complete.

Table 1 contains the set $\mathfrak{G}$ of inference rules that Geiger and Pearl established as a finite axiomatization for the S-implication of SCI statements.

Instead of writing $I(Y, Z \mid X)$ for an SCI statement over S it suffices to write $I(Y \mid X)$ since $Z = S - XY$. The symmetry rule in $\mathfrak{G}$ may then be replaced by

Table 1. Geiger and Pearl's axiomatization $\mathfrak{G}$ of S-implication for SCI statements

$\overline{I(S - X, \emptyset \mid X)}$ (saturated trivial independence, $\mathcal{T}'$)	$\dfrac{I(Y, Z \mid X)}{I(Z, Y \mid X)}$ (symmetry, $\mathcal{S}'$)
$\dfrac{I(ZW, Y \mid X) \qquad I(Z, W \mid XY)}{I(Z, YW \mid X)}$ (weak contraction, $\mathcal{W}'$)	$\dfrac{I(Y, ZW \mid X)}{I(Y, Z \mid XW)}$ (weak union, $\mathcal{U}'$)

Table 2. A different axiomatization $\mathfrak{G}_S$ of S-implication for SCI statements

$\overline{I(\emptyset \mid X)}$ (saturated trivial independence, $\mathcal{T}$)	$\dfrac{I(Y \mid X)}{I(S - XY \mid X)}$ (S-symmetry, $\mathcal{S}_S$)
$\dfrac{I(Y \mid X) \qquad I(Z \mid XY)}{I(Z \mid X)}$ (weak contraction, $\mathcal{W}$)	$\dfrac{I(Y \mid X)}{I(Y - Z \mid XZ)}$ (weak union, $\mathcal{U}$)

a rule that infers $I(S - XY \mid X)$ from $I(Y \mid X)$. Indeed, the set $\mathfrak{C}_S$ of inference rules from Table 2 forms a finite axiomatization for the S-implication of SCI statements.

Proposition 1. $\mathfrak{G}_S$ *is a finite axiomatization for the S-implication of SCI statements.*

Proof (Sketch). Let $S \subseteq \mathfrak{A}$ be a finite set of attributes. Let $\Sigma = \{I(Y_1 \mid X_1), \ldots, I(Y_n \mid X_n)\}$ and $\varphi = I(Y \mid X)$ be a (set of) SCI statement(s) over S. We can show by an induction over the inference length that $\Sigma \vdash_{\mathfrak{G}_S} \varphi$ if and only if $\Sigma' = \{I(Y_1, S - X_1 Y_1 \mid X_1), \ldots, I(Y_n, S - X_n Y_n \mid X_n)\} \vdash_{\mathfrak{G}} I(Y, S - XY \mid X)$. Hence, the S-soundness (S-completeness) of $\mathfrak{G}_S$ follows from the the S-soundness (S-completeness) of $\mathfrak{G}$.   □

Example 3. Consider the SCI statement $I(Child \mid Emp)$ over $S = \{Emp, Project, Child\}$. It models the semantic property that for a given employee we do not learn anything new about the employee's child by learning something new about any other attribute (here: $Project$) of the employee modeled by S. Now, $I(Child \mid Emp)$ S-implies the SCI statement $I(Project \mid Emp)$. It can be inferred by a single application of the S-symmetry rule $\mathcal{S}_S$ to $I(Child \mid Emp)$. However, while $I(Child \mid Emp)$ expresses a semantic property, the SCI statement $I(Project \mid Emp)$ does not *necessarily* correspond to any semantically meaningful constraint. It may merely be a consequence that results from the symmetry between $Project$ and $Child$, given Emp and S.   □

The last example motivates the following definition. It addresses the property of an inference system to first infer all those SCI statements, implied by a set of SCI statements alone, without any application of the symmetry rule, and, subsequently, apply the symmetry rule once to some of these SCI statements to infer any implied SCI statement that also depends on the underlying schema.

Definition 3. *Let $\mathfrak{S}_S$ denote a set of inference rules that is S-sound for the S-implication of SCI statements, and in which the S-symmetry rule $\mathcal{S}_S$ is the only inference rule that is dependent on S. We say that $\mathfrak{S}_S$ is symmetry-preserving, if for every S, and every set $\Sigma \cup \{\varphi\}$ of SCI statements over S such that φ is S-implied by Σ there is some inference of φ from Σ by $\mathfrak{S}_S$ such that the S-symmetry rule $\mathcal{S}_S$ is applied at most once, and, if it is applied, then it is applied in the last step of the inference only.*   □

One may ask whether Geiger and Pearl's system $\mathfrak{G}_S$ is symmetry-preserving.

Theorem 1. *$\mathfrak{G}_S$ is not symmetry-preserving.*

Proof. Let $S = \{A, B, C, D\}$ and $\Sigma = \{I(B \mid A), I(C \mid A)\}$. One can show that $I(BC \mid A) \notin \Sigma^+_{\{\mathcal{T}, \mathcal{W}, \mathcal{U}\}}$. Moreover, for all Y such that $D \in Y$, $I(Y \mid A) \notin \Sigma^+_{\{\mathcal{T}, \mathcal{W}, \mathcal{U}\}}$, see Lemma 1 from Section 4.

However, $I(BC \mid A) \in \Sigma^+_{\mathfrak{G}_S}$. Consequently, in any inference of $I(BC \mid A)$ from Σ by $\mathfrak{G}_S$ the S-symmetry rule $\mathcal{S}_S$ must be applied at least once, but is not just applied in the last step as $\{D\} = S - \{A, B, C\}$.   □

Example 4. Consider the schema $S = \{E(mp), C(hild), P(roject), H(obby)\}$. Let Σ consist of the two SCI statements $I(C \mid E)$ and $I(P \mid E)$. Using $\mathfrak{G}_S$ we can infer $I(CP \mid E)$ from Σ as follows:

$$\cfrac{I(C \mid E) \qquad \cfrac{\cfrac{I(P \mid E)}{\mathcal{U} : I(P \mid CE)} \qquad \mathcal{S}_S : I(H \mid CE)}{I(H \mid E)} \; \mathcal{S}_S :}{\mathcal{S}_S : \qquad I(CP \mid E)} \; \mathcal{W} :$$

This inference leaves open the question whether $I(CP \mid E)$ is implied by Σ alone, i.e., whether the applications of the S-symmetry rule are unnecessary to infer $I(CP \mid E)$.   □

3   A Finite Symmetry-Preserving Axiomatization

Theorem 1 has shown that axiomatizations are, in general, not symmetry-preserving. We will now establish a finite symmetry-preserving axiomatization for the S-implication of SCI statements. For this purpose, we consider the two S-sound inference rules of *additive contraction* $\mathcal{A}$ and *intersection contraction* $\mathcal{I}$, as shown in Table 3.

Table 3. A finite symmetry-preserving axiomatization $\mathfrak{C}_S$

$$
\begin{array}{ll}
\dfrac{}{I(\emptyset \mid X)} & \dfrac{I(Y \mid X)}{I(S - XY \mid X)} \\[2mm]
\text{(saturated trivial independence, } \mathcal{T}) & (S\text{-symmetry, } \mathcal{S}_S) \\[4mm]
\dfrac{I(Y \mid X) \qquad I(Z \mid XY)}{I(Z \mid X)} & \dfrac{I(YZ \mid X)}{I(Y \mid XZ)} \\[2mm]
\text{(weak contraction, } \mathcal{W}) & \text{(weak union, } \mathcal{U}) \\[4mm]
\dfrac{I(Y \mid X) \qquad I(Z \mid XY)}{I(YZ \mid X)} & \dfrac{I(Y \mid X) \qquad I(Z \mid XW)}{I(Y \cap Z \mid X)} \, Y \cap W = \emptyset \\[2mm]
\text{(additive contraction, } \mathcal{A}) & \text{(intersection contraction, } \mathcal{I})
\end{array}
$$

Theorem 2. *Let Σ be a set of SCI statements over S. For every inference γ from Σ by the system*

$$\mathfrak{G}_S = \{\mathcal{T}, \mathcal{W}, \mathcal{U}, \mathcal{S}_S\}$$

there is an inference ξ from Σ by the system

$$\mathfrak{C}_S = \{\mathcal{T}, \mathcal{W}, \mathcal{U}, \mathcal{A}, \mathcal{I}, \mathcal{S}_S\}$$

with the following properties

1. *γ and ξ infer the same SCI statement,*
2. *in ξ the S-symmetry rule $\mathcal{S}_S$ is applied at most once, and*
3. *if $\mathcal{S}_S$ is applied in ξ, then it is applied as the last rule.*

Proof. The proof is done by induction on the length l of γ. For $l = 1$, the statement $\xi := \gamma$ has the desired properties. Suppose for the remainder of the proof that $l > 1$, and let $\gamma = [\sigma_1, \ldots, \sigma_l]$ be an inference of σ_l from Σ by $\mathfrak{G}_S$. We distinguish between four different cases according to how σ_l is obtained from $[\sigma_1, \ldots, \sigma_{l-1}]$.

Case 1. σ_1 is obtained from the saturated trivial independence axiom $\mathcal{T}$, or is an element of Σ. In this case, $\xi := [\sigma_l]$ has the desired properties.

Case 2. We obtain σ_l by an application of the weak union rule $\mathcal{U}$ to a premise σ_i with $i < l$. Let ξ_i be obtained by applying the induction hypothesis to $\gamma_i = [\sigma_1, \ldots, \sigma_i]$. Consider the inference $\xi := [\xi_i, \sigma_l]$. If in ξ_i the S-symmetry rule $\mathcal{S}_S$ is not applied, then ξ has the desired properties. If in ξ_i the S-symmetry rule $\mathcal{S}_S$ is applied as the last rule, then the last two steps in ξ are of the following form:

$$
\dfrac{\dfrac{I(Y \mid X)}{\mathcal{S}_S : \; I(S - XY \mid X)}}{\mathcal{U} : \; I(S - XYZ \mid XZ)} \; .
$$

However, these steps can be replaced as follows:

$$\frac{I(Y \mid X)}{\begin{array}{l}\mathcal{U}: I(Y - Z \mid XZ) \\ \hline \mathcal{S}_S: I(S - XYZ \mid XZ)\end{array}} \quad .$$

The result of this replacement is an inference with the desired properties.

Case 3. We obtain σ_l by an application of the weak contraction rule $\mathcal{W}$ to premises σ_i and σ_j with $i, j < l$. Let ξ_i and ξ_j be obtained by applying the induction hypothesis to $\gamma_i = [\sigma_1, \ldots, \sigma_i]$ and $\gamma_j = [\sigma_1, \ldots, \sigma_j]$, respectively. Consider the inference $\xi := [\xi_i, \xi_j, \sigma_l]$. We distinguish between four cases according to the occurrence of the S-symmetry rule $\mathcal{S}_S$ in ξ_i and ξ_j.

Case 3.1. If the S-symmetry rule $\mathcal{S}_S$ does not occur in ξ_i nor in ξ_j, then ξ has the desired properties.

Case 3.2. If the S-symmetry rule $\mathcal{S}_S$ occurs in ξ_i as the last rule but does not occur in ξ_j, then the last step of ξ_i and the last step of ξ are of the following form:

$$\frac{\dfrac{I(Y \mid X)}{\mathcal{S}_S: I(S - XY \mid X)} \qquad I(Z \mid X(S - XY))}{\mathcal{W}: \qquad\qquad I(Z \mid X)} \quad .$$

Note that $Z \subseteq Y$. However, these steps can be replaced as follows:

$$\frac{I(Y \mid X) \qquad I(Z \mid X(S - XY))}{\mathcal{I}: \qquad\qquad I(Z \mid X)} \quad .$$

The result of this replacement is an inference with the desired properties.

Case 3.3. If the S-symmetry rule $\mathcal{S}_S$ occurs in ξ_j as the last rule but does not occur in ξ_i, then the last step of ξ_j and the last step of ξ are of the following form:

$$\frac{I(Y \mid X) \qquad \dfrac{I(Z \mid XY)}{\mathcal{S}_S: I(S - XYZ \mid XY)}}{\mathcal{W}: \qquad\qquad I(S - XYZ \mid X)} \quad .$$

However, these steps can be replaced as follows:

$$\frac{\dfrac{I(Y \mid X) \qquad I(Z \mid XY)}{\mathcal{A}: \qquad I(ZY \mid X)}}{\mathcal{S}_S: \qquad I(S - XYZ \mid X)} \quad .$$

The result of this replacement is an inference with the desired properties.

Case 3.4. If the S-symmetry rule $\mathcal{S}_S$ occurs in ξ_i as the last rule and occurs in ξ_j as the last rule, then the last steps of ξ_i and ξ_j and the last step of ξ are of the following form:

$$\frac{\dfrac{I(Y \mid X)}{\mathcal{S}_S: I(S - XY \mid X)} \qquad \dfrac{I(Z \mid X(S - XY))}{\mathcal{S}_S: I(S - (XZ(S - XY)) \mid X(S - XY))}}{\mathcal{W}: \qquad\qquad I(\underbrace{S - (XZ(S - XY))}_{=Y - Z} \mid X)} \quad .$$

However, these steps can be replaced as follows:

$$\frac{I(Y \mid X) \qquad I(Z \mid X(S - XY))}{\begin{array}{l} \mathcal{I}: \qquad\qquad I(Y \cap Z \mid X) \\ \mathcal{W}: \qquad\qquad I(\underbrace{Y - (Y \cap Z)}_{=Y-Z} \mid X) \end{array}} \qquad \frac{I(Y \mid X)}{\mathcal{U}: I(Y - (Y \cap Z) \mid X(Y \cap Z))} .$$

The result of this replacement is an inference with the desired properties.

Case 4. We obtain σ_l by an application of the S-symmetry rule $\mathcal{S}_S$ to a premise σ_i with $i < l$. Let ξ_i be obtained by applying the induction hypothesis to $\gamma_i = [\sigma_1, \ldots, \sigma_i]$. Consider the inference $\xi := [\xi_i, \sigma_l]$. If in ξ_i the S-symmetry rule $\mathcal{S}_S$ is not applied, then ξ has the desired properties. If in ξ_i the S-symmetry rule $\mathcal{S}_S$ is applied as the last rule, then the last two steps in ξ are of the following form.

$$\frac{\dfrac{I(Y \mid X)}{\mathcal{S}_S: \ I(S - XY \mid X)}}{\mathcal{S}_S: \ I(\underbrace{S - (X(S - XY))}_{=Y} \mid X)}$$

The inference obtained from deleting these steps has the desired properties.   □

Example 5. Recall Example 4 where $S = \{E(mp), C(hild), P(roject), H(obby)\}$, and Σ consists of the two SCI statements $I(C \mid E)$ and $I(P \mid E)$. The inference of $I(CP \mid E)$ from Σ by $\mathfrak{G}_S$ did not reveal whether applications of the S-symmetry rule are necessary to infer $I(CP \mid E)$ from Σ. Indeed, no inference of $I(CP \mid E)$ from Σ by $\mathfrak{G}_S$ can provide this insight, see the proof of Theorem 1. Using $\mathfrak{C}_S$ we can obtain the following inference of $I(CP \mid E)$ from Σ:

$$\frac{I(C \mid E) \qquad \dfrac{I(P \mid E)}{\mathcal{U}: \ I(P \mid EC)}}{\mathcal{A}: \qquad\quad I(CP \mid E)} .$$

Indeed, the S-symmetry rule unnecessary to infer the SCI statement $I(CP \mid E)$ from Σ.   □

Examples 4 and 5 indicate that the implication of $I(CP \mid E)$ by Σ does not depend on the fixed schema S. Next we will formalize a stronger notion of implication for SCI statements. Theorem 3 shows that the set $\mathfrak{C} := \mathfrak{C}_S - \{\mathcal{S}_S\}$ of inference rules is nearly S-complete for the S-implication of SCI statements.

Theorem 3. *Let $\Sigma \cup \{I(Y \mid X)\}$ be a set of SCI statements over schema S. Then $I(Y \mid X) \in \Sigma^+_{\mathfrak{C}_S}$ if and only if $I(Y \mid X) \in \Sigma^+_{\mathfrak{C}}$ or $I(S - XY \mid X) \in \Sigma^+_{\mathfrak{C}}$.*   □

Theorem 3 indicates that $\mathfrak{C}$ can be used to infer every SCI statement whose implication is independent from the schema S. Another interpretation of Theorem 3 is the following. In using $\mathfrak{C}$ to infer S-implied SCI statements, the fixation of the underlying schema can be deferred until the last step of an inference.

Table 4. An axiomatization $\mathfrak{C}$ of implication for SCI statements

$$\frac{}{I(\emptyset \mid X)}$$
(saturated trivial independence, $\mathcal{T}$)

$$\frac{I(Y \mid X) \quad I(Z \mid XY)}{I(YZ \mid X)}$$
(additive contraction, $\mathcal{A}$)

$$\frac{I(Y \mid X) \quad I(Z \mid XY)}{I(Z \mid X)}$$
(weak contraction, $\mathcal{W}$)

$$\frac{I(YZ \mid X)}{I(Y \mid XZ)}$$
(weak union, $\mathcal{U}$)

$$\frac{I(Y \mid X) \quad I(Z \mid XW)}{I(Y \cap Z \mid X)} \; Y \cap W = \emptyset$$
(intersection contraction, $\mathcal{I}$)

4   Axiomatizing a Stronger Notion of Implication

In this section we will introduce a notion of implication for SCI statements that is independent from the underlying schema. We will then show that the set $\mathfrak{C}$ of inference rules forms a finite axiomatization for this notion of implication. On the one hand, this allows us to distinguish between those SCI statements that are S-implied and those that are implied. On the other hand, the notion of implication can be applied whenever there is some uncertainty about the underlying schema, e.g., whether additional attributes will be required in the future, or some attributes are unknown. This degree of uncertainty may commonly be required in practice.

A *probability model* is a triple (S, dom, P) where $S = \{s_1, \ldots, s_n\} \subseteq \mathfrak{A}$ is a finite set of attributes, dom is a domain mapping that maps each s_i to a finite domain $dom(s_i)$, and $P : dom(s_1) \times \cdots \times dom(s_n) \to [0,1]$ is a probability distribution having the Cartesian product of these domains as its sample space.

The expression $I(Y \mid X)$ where X and Y are finite, disjoint subsets of $\mathfrak{A}$ is called a *saturated conditional independence* (SCI) *statement*. We say that the SCI statement $I(Y \mid X)$ *holds for* (S, dom, P) if $XY \subseteq S$ and for every values $\mathbf{x}, \mathbf{y}$, and $\mathbf{z}$ of X, Y and $Z = S - XY$, respectively,

$$P(\mathbf{y}, \mathbf{z}, \mathbf{x}) \cdot P(\mathbf{x}) = P(\mathbf{y}, \mathbf{x}) \cdot P(\mathbf{z}, \mathbf{x}).$$

Equivalently, (S, dom, P) is said to *satisfy* $I(Y \mid X)$. For an SCI statement $I(Y \mid X)$ let $Attr(\varphi) := XY$, and for a finite set Σ of SCI statements let $Attr(\Sigma) := \bigcup_{\sigma \in \Sigma} Attr(\sigma)$.

Definition 4. *Let $\Sigma \cup \{\varphi\}$ be a finite set of SCI statements. We say that Σ implies φ if every probability model (S, dom, P) with $Attr(\Sigma \cup \{\varphi\}) \subseteq S$ that satisfies every SCI statement $\sigma \in \Sigma$ also satisfies φ.*   □

In this definition the underlying schema is left undetermined. The only requirement is that the SCI statements must apply to the probability model. The implication problem for SCI statements can be stated as follows.

PROBLEM:	Implication problem
INPUT:	Set $\Sigma \cup \{\varphi\}$ of SCI statements
OUTPUT:	Yes, if Σ implies φ; No, otherwise

Implication is a stronger notion than S-implication.

Proposition 2. *Let $\Sigma \cup \{\varphi\}$ be a finite set of SCI statements, such that $Attr(\Sigma \cup \{\varphi\}) \subseteq S$. If Σ implies φ, then Σ S-implies φ, but the other direction does not necessarily hold.*

Proof. The first statement follows directly the definitions of implication and S-implication. For the other direction, let $S = \{Emp, Child, Project\}$, $\Sigma = \{I(Child \mid Emp)\}$ and let φ be $I(Project \mid Emp)$. Clearly, Σ S-implies φ. However, Σ does not imply φ as the following probability model shows. Let $S' = S \cup \{Hobby\}$ and P(Homer, Lisa, Cooling System, TV) $= 0.5 = P$(Homer, Lisa, Safety, Beer). This model satisfies Σ, but violates φ. ☐

The notions of soundness and completeness with respect to the notion of implication from Definition 4 are simply adapted from the corresponding notions in the context of fixed schemata S by dropping the reference to S. While the saturated trivial independence axiom $\mathcal{T}$, the weak contraction rule $\mathcal{W}$, the weak union rule $\mathcal{U}$, the additive contraction rule $\mathcal{A}$, and the intersection contraction rule $\mathcal{I}$ are all sound, the S-symmetry rule $\mathcal{S}_S$ is S-sound, but not sound.

We shall now prove that $\mathfrak{C}$ forms a finite axiomatization for the implication of SCI statements. For this purpose, we prove two lemmata in preparation. The correctness of the first lemma can easily be observed by inspecting the inference rules in $\mathfrak{C}$. For each of the rules, every attribute that occurs on the left-hand side of the bar in the conclusion of the rule, already appears on the left-hand side of the bar in at least one premise of the rule.

Lemma 1. *Let $\Sigma = \{I(Y_1 \mid X_1), \ldots, I(Y_n \mid X_n)\}$ be a finite set of SCI statements. If $I(Y \mid X) \in \Sigma_{\mathfrak{C}}^+$, then $Y \subseteq Y_1 \cup \ldots \cup Y_n$.* ☐

For the next lemma one may notice that the attributes that do not occur in $Attr(\Sigma)$ can always be introduced in the last step of an inference, by applying the weak union rule $\mathcal{U}$.

Lemma 2. *Let Σ be a finite set of SCI statements. If $I(Y \mid X) \in \Sigma_{\mathfrak{C}}^+$, then there is an inference $\gamma = [\sigma_1, \ldots, \sigma_l]$ of $I(Y \mid X)$ from Σ by $\mathfrak{C}$ such that every attribute occurring in $\sigma_1, \ldots, \sigma_{l-1}$ is an element of $Attr(\Sigma)$.*

Proof. Let $W := Attr(\Sigma)$, and $\bar{\xi} = [I(V_1 \mid U_1), \ldots, I(V_{l-1} \mid U_{l-1})]$ be an inference of $I(Y \mid X)$ from Σ by $\mathfrak{C}$. Consider the sequence

$$\xi := [I(V_1 \cap W \mid U_1 \cap W), \ldots, I(V_{l-1} \cap W \mid U_{l-1} \cap W)].$$

We claim that ξ is an inference of $I(Y \cap W \mid X \cap W)$ from Σ by $\mathfrak{C}$. For if $I(V_i \mid U_i)$ is an element of Σ or was obtained by an application of the saturated trivial independence axiom $\mathcal{T}$, then $I(Y \cap W \mid X \cap W) = I(Y \mid X)$. Moreover, one can verify that if $I(V_i \mid U_i)$ is the result of applying one of the rules $\mathcal{U}, \mathcal{W}, \mathcal{A}, \mathcal{I}$, then $I(V_i \cap W \mid U_i \cap W)$ is the result of the same rule applied to the corresponding premises in ξ.

Now by Lemma 1 we know that $Y \subseteq W$, hence $Y \cap W = Y$. However, this means that we can infer $I(Y \mid X)$ from $I(Y \cap W \mid X \cap W)$ by a single application of the weak union rule $\mathcal{U}$:

$$\frac{I(Y \cap W \mid X \cap W)}{I(\underbrace{(Y \cap W) - X}_{=Y} \mid \underbrace{(X \cap W) \cup X}_{=X})}.$$

Hence, the inference $[\xi, I(Y \mid X)]$ has the desired properties. $\square$

We are now prepared to prove that $\mathfrak{C}$ forms a finite axiomatization for the implication of SCI statements.

Theorem 4. *The set $\mathfrak{C} = \{\mathcal{T}, \mathcal{W}, \mathcal{U}, \mathcal{A}, \mathcal{I}\}$ of inference rules forms a finite axiomatization for the implication of SCI statements.*

Proof. Let $\Sigma = \{I(Y_1 \mid X_1), \ldots, I(Y_n \mid X_n)\}$ be a finite set of SCI statements and $I(Y \mid X)$ an SCI statement. We have to show that

$$I(Y \mid X) \in \Sigma^* \quad \text{if and only if} \quad I(Y \mid X) \in \Sigma_{\mathfrak{C}}^+.$$

Let $T := X \cup Y \cup Attr(\Sigma)$. In order to prove the soundness of $\mathfrak{C}$ we assume that $I(Y \mid X) \in \Sigma_{\mathfrak{C}}^+$ holds. Let (S, dom, P) be a probability model that satisfies every element of Σ, and where $T \subseteq S$ holds. We must show that (S, dom, P) also satisfies $I(Y \mid X)$. According to Lemma 2 there is an inference γ of $I(Y \mid X)$ from Σ by $\mathfrak{C}$ such that $U \cup V \subseteq T \subseteq S$ holds for each SCI statement $I(V \mid U)$ that occurs in γ. Since each rule in $\mathfrak{C}$ is sound we can conclude (by induction) that each SCI statement occurring in γ is satisfied by (S, dom, P). In particular, (S, dom, P) satisfies $I(Y \mid X)$.

In order to prove the completeness of $\mathfrak{C}$ we assume that $I(Y \mid X) \notin \Sigma_{\mathfrak{C}}^+$. Let $S \subseteq \mathfrak{A}$ be a finite set of attributes such that T is a proper subset of S, i.e., $T \subset S$. Consequently, $S - XY$ is not a subset of T. Hence, by Lemma 1, $I(S - XY \mid X) \notin \Sigma_{\mathfrak{C}}^+$. Now from $I(Y \mid X) \notin \Sigma_{\mathfrak{C}}^+$ and from $I(S - XY \mid X) \notin \Sigma_{\mathfrak{C}}^+$ we conclude that $I(Y \mid X) \notin \Sigma_{\mathfrak{C}_S}^+$ by Theorem 3. Since $\mathfrak{C}_S$ is S-complete for the S-implication of SCI statements it follows that Σ does not S-imply $I(Y \mid X)$. Hence, Σ does not imply $I(Y \mid X)$ by Proposition 2. $\square$

Example 6. Recall Example 5 where $S = \{E(mp), C(hild), P(roject), H(obby)\}$, and Σ consists of the two SCI statements $I(C \mid E)$ and $I(P \mid E)$. The inference of $I(CP \mid E)$ from Σ by $\mathfrak{C}_S$ in Example 5 is actually an inference by $\mathfrak{C}$. Hence, $I(CP \mid E)$ is implied by Σ, as one would expect intuitively. $\square$

5   Results for Alternative SCI Statements

We will now outline a different strategy to prove desirable properties of axiomatizations. The previous proofs are developed from the point of view of SCI statements and the axiomatization for their S-implication problem, as established by Geiger and Pearl. An alternative proof strategy can be developed from the point of view of multivalued dependencies (MVDs). For this purpose, we define that an *alternative SCI statement* over S is an expression $I(Y, Z \mid X)$ where $X, Y, Z \subseteq S$ such that $S = XYZ$. Here, the attribute sets are not necessarily disjoint. The definition of the satisfaction of an alternative SCI statement by a probability model is the same as for SCI statements. As before we write $I(Y \mid X)$ instead of $I(Y, S - Y \mid X)$. One may show that the set $\overline{\mathfrak{G}}_S = \{\overline{\mathcal{T}}, \overline{\mathcal{W}}, \overline{\mathcal{U}}, \overline{\mathcal{S}}_S\}$ in Table 5 forms a finite axiomatization for S-implication of alternative SCI statements.

Table 5. Inference rules for the S-implication of alternative SCI statements

$\dfrac{}{I(\emptyset \mid \emptyset)}$ (saturated trivial independence, $\overline{\mathcal{T}}$)	$\dfrac{I(Y \mid X)}{I(S - Y \mid X)}$ (S-symmetry, $\overline{\mathcal{S}}_S$)
$\dfrac{I(Y \mid X) \qquad I(Z \mid Y)}{I(Z - Y \mid X)}$ (weak contraction, $\overline{\mathcal{W}}$)	$\dfrac{I(Y \mid X)}{I(YU \mid XUV)}$ (weak union, $\overline{\mathcal{U}}$)
$\dfrac{I(Y \mid X) \qquad I(Z \mid Y)}{I(YZ \mid X)}$ (additive contraction, $\overline{\mathcal{A}}$)	$\dfrac{I(Y \mid X) \qquad I(Z \mid W)}{I(Y \cap Z \mid X)} \, Y \cap W = \emptyset$ (intersection contraction, $\overline{\mathcal{I}}$)

The set $\overline{\mathfrak{G}}_S$, where each alternative SCI statement $I(Y \mid X)$ is interpreted as the MVD $X \twoheadrightarrow Y$ over S, is the exact axiomatization of S-implication for MVDs from [2]. The results in [2], established for MVDs, then establish the following analogue results for alternative SCI statements.

Theorem 5. *Let Σ be a set of alternative SCI statements over S. For every inference γ from Σ by the system $\overline{\mathfrak{G}}_S = \{\overline{\mathcal{T}}, \overline{\mathcal{W}}, \overline{\mathcal{U}}, \overline{\mathcal{S}}_S\}$ there is an inference ξ from Σ by the system $\overline{\mathfrak{C}}_S = \{\overline{\mathcal{T}}, \overline{\mathcal{W}}, \overline{\mathcal{U}}, \overline{\mathcal{A}}, \overline{\mathcal{I}}, \overline{\mathcal{S}}_S\}$ with the following properties*

1. *γ and ξ infer the same alternative SCI statement,*
2. *in ξ the S-symmetry rule $\overline{\mathcal{S}}_S$ is applied at most once, and*
3. *if $\overline{\mathcal{S}}_S$ is applied in ξ, then it is applied as the last rule.* □

One may then also introduce alternative SCI statements and their implication over undetermined schemata.

Theorem 6. *The set $\overline{\mathfrak{C}} = \{\overline{\mathcal{T}}, \overline{\mathcal{W}}, \overline{\mathcal{U}}, \overline{\mathcal{A}}, \overline{\mathcal{I}}\}$ of inference rules forms a finite axiomatization for the implication of alternative SCI statements.* □

6 Conclusion

SCI statements form an important subclass of CI statements, as their implication problem is finitely axiomatizable and decidable in almost linear time. We have observed a difference between SCI statements S-implied by a given set of SCI statements, and those implied by the given set of SCI statements alone. Geiger and Pearl's axiomatization for S-implication cannot distinguish between these two notions. We have established an axiomatization for S-implication in which implied SCI statements can be inferred without applications of the symmetry rule. We have further introduced a notion of implication in which the underlying schema is left undetermined, and established a finite axiomatization for this notion. The results show that the symmetry rule is a mere means to infer SCI statements that are S-implied, but not implied by a given set of SCI statements. These results have analogues in the theory of MVDs in relational databases.

References

1. Beeri, C., Fagin, R., Howard, J.H.: A complete axiomatization for functional and multivalued dependencies in database relations. In: Smith, D. (ed.) SIGMOD Conference, pp. 47–61. ACM (1977)
2. Biskup, J.: Inferences of multivalued dependencies in fixed and undetermined universes. Theor. Comput. Sci. 10(1), 93–106 (1980)
3. Biskup, J., Link, S.: Appropriate inferences of data dependencies in relational databases. Ann. Math. Artif. Intell. 63(3-4), 213–255 (2012)
4. Biskup, J.: Foundations of Information Systems. Vieweg, Wiesbaden (1995) (in German)
5. Fagin, R.: Multivalued dependencies and a new normal form for relational databases. ACM Trans. Database Syst. 2(3), 262–278 (1977)
6. Galil, Z.: An almost linear-time algorithm for computing a dependency basis in a relational database. J. ACM 29(1), 96–102 (1982)
7. Geiger, D., Pearl, J.: Logical and algorithmic properties of conditional independence and graphical models. The Annals of Statistics 21(4), 2001–2021 (1993)
8. Halpern, J.: Reasoning about uncertainty. MIT Press, Cambridge (2005)
9. Hartmann, S., Link, S.: The implication problem of data dependencies over SQL table definitions. ACM Trans. Datab. Syst. 37(2), Article 13 (May 2012)
10. Link, S.: Charting the completeness frontier of inference systems for multivalued dependencies. Acta Inf. 45(7-8), 565–591 (2008)
11. Malvestuto, F.: A unique formal system for binary decompositions of database relations, probability distributions, and graphs. Inf. Sci. 59(1-2), 21–52 (1992)
12. Niepert, M., Van Gucht, D., Gyssens, M.: Logical and algorithmic properties of stable conditional independence. Int. J. Approx. Reasoning 51(5), 531–543 (2010)
13. Pearl, J.: Probabilistic Reasoning in Intelligent Systems: Networks of Plausible Inference. Morgan Kaufmann, San Francisco (1988)
14. Studený, M.: Conditional independence relations have no finite complete characterization. In: Kubik, S., Visek, J. (eds.) Transactions of the 11th Prague Conference on Information Theory, pp. 377–396. Kluwer (1992)
15. Wong, S., Butz, C., Wu, D.: On the implication problem for probabilistic conditional independency. IEEE Transactions on Systems, Man, and Cybernetics, Part A 30(6), 785–805 (2000)

On Dependence in Second-Order Probability

David Sundgren[1] and Alexander Karlsson[2]

[1] Department of Computer and Systems Sciences
Stockholm University, Sweden
dsn@dsv.su.se

[2] Infofusion/Informatics Research Center
University of Skövde, Sweden
alexander.karlsson@his.se

Abstract. We present a notion, relative independence, that models independence in relation to a predicate. The intuition is to capture the notion of a minimum of dependencies among variables with respect to the predicate. We prove that relative independence coincides with conditional independence only in a trivial case. For use in second-order probability, we let the predicate express first-order probability, i.e. that the probability variables must sum to one in order to restrict dependency to the necessary relation between probabilities of exhaustive and mutually exclusive events. We then show examples of Dirichlet distributions that do and do not have the property of relative independence. These distributions are compared with respect to the impact of further dependencies, apart from those imposed by the predicate.

Keywords: Imprecise probability, second-order probability, dependency.

1 Introduction

One commonly used model of *imprecise probability* [1] is to utilize *convex sets of probability distributions*, also known as *credal sets* [2, 3]. The main idea is that the imprecision models *epistemic uncertainty* [4] and that more information usually means less epistemic uncertainty. When using credal sets for *belief updating* [5], i.e., when a *prior credal set* is updated to a *posterior* based on some piece of information, one does not usually consider the existence of a *second-order probability distribution* over these sets, i.e., probability distributions over probability distributions [6–8].

Still, such type of information is often implicitly used. As an example, one proposed method for deciding on a single state[1] based on a posterior credal set is to use the *centroid* distribution of the set [9, 10], i.e., the expected value of a uniform second-order distribution. However, even though one adopts the uniform distribution as ones' second-order belief over the prior credal set, the corresponding belief over the posterior is not likely to be uniform [11]. Hence, there is in principle no reason for using the centroid distribution as a basis for decision making.

---

[1] Imprecision usually means that there exist several optimal decisions.

E. Hüllermeier et al. (Eds.): SUM 2012, LNAI 7520, pp. 379–391, 2012.

With its potential for expressing structure, second-order probabilities can be useful for decision making. It can however be problematic to define a distribution that correctly represents the information at hand; singling out a particular distribution that corresponds to imprecise information can appear as too hard. But some structural knowledge may be available, e.g. that there are no contingent dependencies among the probabilities.

In Section 4, we will see that second-order distributions, e.g. the uniform distribution, that do not factorise into marginals have dependencies other than those caused by probabilities summing to one. A second-order probability distribution can never be the product of its marginal distributions, i.e. the joint distribution and the marginals are not coupled by the product copula [12], since first-order probability values *are* dependent by definition. But we argue that dependence can be restricted and that such *relative* independence is tantamount to the joint distribution being a normalized product of marginals. It is shown in [13] that there exists one unique distribution within the *Dirichlet family* that factors into marginals.

In this paper we motivate factorisation into marginals through the concept of *relative independence*. In particular, we will explore and describe the effects of conditioning with respect to a number of members within the Dirichlet family, including the uniform Bayes-Laplace [14], Perks [15] and Jeffreys [16]. In the cases presented in this paper there are three outcomes and Jeffreys prior coincides with the Dirichlet distribution that factors into marginals. We will see that when relative independence does not hold, dependencies take the form of a bias towards probability vectors with certain relations among the variables.

To select a second-order distribution among many for expressing some imprecise probabilities we need to know something about the dependencies that may or may not hold among the probabilities. We argue that the unique family of second-order distributions that factor into marginals models lack of contingent dependencies, or independence relative to summing to one. For contrast we demonstrate some contingent dependencies that appear with other distributions.

The paper is organized as follows: in Section 2, we make acquaintance with the Dirichlet distribution as second-order probability distribution and take note of the special case of the Dirichlet distribution that factors into marginals. In Section 3 we generalize the idea of independence to independence relative to a predicate and in Section 4 we show concrete examples of how dependencies show themselves when first-order probabilities are not relatively independent and in what sense this type of independence in second-order probability retains important aspects of *absolute independence*, i.e. independence in the classical sense.

2 Preliminaries

In Section 4 we will explore some commonly used Dirichlet distributions, e.g. Bayes-Laplace[14], Jeffreys [16] and Perks [15]. The Dirichlet family of distributions is characterized by the following definition.

Definition 1. *The Dirichlet distribution of order n with parameters $\alpha_1, \ldots, \alpha_n >$ 0 has probability density function with respect to Lebesgue measure on $\mathbb{R}^{n-1}$ given by*

$$f(x_1, \ldots, x_{n-1}; \alpha_1, \ldots, \alpha_n) = \frac{\Gamma\left(\sum_{i=1}^{n} \alpha_i\right) \prod_{i=1}^{n} x_i^{\alpha_i - 1}}{\prod_{i=1}^{n} \Gamma(\alpha_i)},$$

where $x_n = 1 - \sum_{i=1}^{n-1} x_i$. The marginal distributions are Beta distributions with density functions

$$f_i(x_i) = \frac{\Gamma\left(\sum_{i=1}^{n} \alpha_i\right) x_i^{\alpha_i - 1} (1 - x_i)^{\left(\sum_{j \neq i} \alpha_j\right) - 1}}{\Gamma(\alpha_i) \Gamma\left(\sum_{j \neq i} \alpha_j\right)}.$$

The Perks prior has $\sum_{i=1}^{n} \alpha_i = 1$, in Jeffreys $\sum_{i=1}^{n} \alpha_i = n/2$ and Bayes-Laplace has $\alpha_i = 1$. All these are symmetric, i.e. all α_i:s are equal.

The Dirichlet family of distributions are then second-order probability distributions by definition as soon as the variables are interpreted as probability values.

It was shown in [13] that the only continuous second-order probability distributions that factors into marginals are Dirichlet distributions with $\alpha_i = 1/(n-1)$ and the corresponding contracted distributions that comes from restricting the support to $a_i < x_i < 1 - \sum_{j \neq i} a_j$, where a_i are arbitrary lower bounds of x_i with $\sum_{i=1}^{n} a_i \leq 1$. For the purposes of this paper it serves no purpose to have lower bounds $a_i > 0$, so we set $a_i = 0, i = 1, \ldots, n$ and use only a non-shifted Dirichlet distribution that is also a Dirichlet distribution in the proper sense, with parameter values $\alpha_i = 1/(n-1)$. In our examples we have $n = 3$ and the Dirichlet distribution that factors into marginals then have $\alpha_i = 1/2$ which makes it coincide with Jeffreys prior.

3   Relative Independence

When there are three possible exhaustive and mutually exclusive outcomes of an event and the respective probabilities of the outcomes are x_1, x_2 and x_3, we have that $x_1 + x_2 + x_3 = 1$, ruling out independence. If x_1 increases, $x_2 + x_3$ must decrease by the same amount. There might be other dependencies among the x_i, e.g. that $x_1 = x_2/2$ or that x_1 and x_2 are approximately equal. But what if one wishes to model that $x_1 + x_2 + x_3 = 1$ is the *only* dependency? Is it possible to isolate a dependency and claim a form of independence apart from this?

The question is what the characteristic traits of distributions of independent random variables are and whether these traits can be preserved in distributions of dependent random variables. We approach this question by inspiration from the uniform joint distribution of independent variables. Two particular features of the uniform joint distribution is that it (1) *factors into marginals* and (2) *the shape of the distribution is preserved* in the sense that the marginals themselves are uniform. The idea is now to introduce only the constraint that variables sum to one and keep as much as possible of the original structure of a distribution of independent variables, in line with the two mentioned features. In order to achieve this we introduce the concept of relative independence with the more general constraint of a dyadic predicate $Q(A, B)$. In general, since $\Pr(R(A)|A) = R(A)$ for a predicate $R(\cdot)$ and event A, we get

$$
\begin{aligned}
\Pr(A \cap B | Q(A, B)) &= \frac{\Pr(Q(A,B)|A \cap B)\Pr(A \cap B)}{\Pr(Q(A,B))} \\
&= \frac{Q(A,B)\Pr(A \cap B)}{\Pr(Q(A,B))}.
\end{aligned}
\tag{1}
$$

But if the events A and B are independent unconditioned on Q, then

$$
\Pr(A \cap B) = \Pr(A)\Pr(B),
\tag{2}
$$

hence

$$
\Pr(A \cap B | Q(A, B)) = \frac{Q(A,B)\Pr(A)\Pr(B)}{\Pr(Q(A,B))}.
\tag{3}
$$

Definition 2. *Two events A and B are independent relative to a predicate Q, where $\Pr(Q(A,B)|A \cap B) = 0$ or 1, iff the prior product of marginals $\Pr(A)\Pr(B)$ is conjugate when updating on the predicate Q, i.e.*

$$
\Pr(A \cap B | Q(A, B)) = \frac{Q(A,B)\Pr(A)\Pr(B)}{\Pr(Q(A,B))}.
$$

We recall that conjugate means that the posterior $\Pr(A \cap B | Q(A, B))$ and prior $\Pr(A \cap B) = \Pr(A)\Pr(B)$ belong to the same family of distributions.

The concept we want to capture here is that updating on a predicate makes a minimal difference so that properties held by the prior distribution are at least in some degree held by the posterior. The predicate symbol Q is to be interpreted as a 0/1-valued function, the indicator function of the set $\{(A, B)|Q(A, B)\}$, i.e. $Q(A, B) = 1$ if the predicate Q holds for A and B, and 0 otherwise.

Now, if A and B are independent relative to some predicate, A and B are independent only when the predicate is true independently of A and B. Otherwise if the predicate $Q(A, B)$ can be true or false depending on A and B, conditioning on Q causes a dependency between A and B. In other words, $\Pr(Q(A, B)) = 1$ if and only if A and B are absolutely independent, i.e., absolute independence is a special case of relative independence with

$$
\Pr(A \cap B | Q(A, B)) = \Pr(A)\Pr(B).
\tag{4}
$$

If A and B are independent, $\Pr(A|B) = \Pr(A)$, but if A and B are independent relative to a predicate we can only say that the conditional joint probability is *proportional* to the product of marginals so that $\Pr(A|B) \propto \Pr(A)$. That is, conditioning on B makes a difference but the conditional probability is *independent of B:s probability*.

In Figure 1 there are 16 squares of which 8 are marked with black and 4 with grey. Two squares are both black and grey. Hence $\Pr(B) = 1/2, \Pr(G) = 1/4$

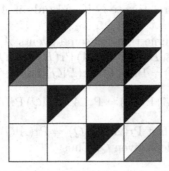

Fig. 1. Unconditional independence and relative independence

and $\Pr(B \cap G) = 1/8$. Further, $\Pr(B)\Pr(G) = \frac{1}{2}\frac{1}{4} = \frac{1}{8} = \Pr(B \cap G)$, so B for black and G for grey are independent.

But if we condition on being in the second row (the predicate Q in Definition 2) independence is lost; $\Pr(B|\text{row } 2) = 3/4, \Pr(G|\text{row } 2) = 1/2$ and $\Pr(B \cap G|\text{row } 2) = \frac{1}{2} \neq \frac{3}{4}\frac{1}{2}$. However,

$$\Pr(B \cap G|\text{row } 2) = \frac{\Pr(B)\Pr(G)}{\Pr(\text{row } 2)} = \frac{\frac{1}{4}\frac{1}{2}}{\frac{1}{4}} = \frac{1}{2}.$$

We say that black and grey are independent relative to being on row 2.

3.1 The Relation of Relative Independence to Conditional Independence

Since relative independence is a type of independence that deals with conditioning it may come natural to consider when relative independence can coincide with conditional independence. Simply put, it can be said that in the case of relative independence, events are independent before conditioning but absolute independence is lost when events are conditioned. But with conditional independence there may or may not be unconditional independence but the events are independent after conditioning.

In fact, if we condition on a predicate Q, independence relative to Q coincides with conditional independence given Q if and only if Q is a tautology, i.e. $\Pr(Q) = 1$. Recall that events A and B are conditionally independent given Q if $\Pr(A \cap B|Q) = \Pr(A|Q)\Pr(B|Q)$.

Theorem 1. *Two events A and B are both conditionally independent on Q and independent relative to Q if and only if* $\Pr(Q) = 1$.

Proof. If $\Pr(Q) = 1$ then $\Pr(C|Q) = \Pr(C)$ for all events C. Therefore conditional and relative independence are equivalent when conditioning is on Q such that $\Pr(Q) = 1$. Since

$$\Pr(A \cap B|Q) = \Pr(A|Q)\Pr(B|Q) = \Pr(A)\Pr(B) = \frac{\Pr(A)\Pr(B)}{\Pr(Q)},$$

A and B are independent relative to Q if and only if A and B are conditionally independent given Q.

On the other hand, assume that for some events A, B and predicate Q, relative independence $\Pr(A \cap B|Q) = \Pr(A) \Pr(B) / \Pr(Q)$, as well as conditional independence $\Pr(A \cap B|Q) = \Pr(A|Q) \Pr(B|Q)$ holds. Then

$$\Pr(A) \Pr(B) = \Pr(A \cap B|Q) \Pr(Q)$$

by relative independence. But $\Pr(A \cap B|Q) = \Pr(A|Q) \Pr(B|Q)$ since A and B are conditionally independent given Q. Thus

$$\Pr(A) \Pr(B) = \Pr(A|Q) \Pr(B|Q) \Pr(Q) = \Pr(A|Q) \Pr(B \cap Q). \tag{5}$$

Likewise we see that $\Pr(A) \Pr(B) = \Pr(B|Q) \Pr(A \cap Q)$.

That is,

$$\Pr(A) \Pr(B) = \Pr(A|Q) \Pr(B \cap Q) = \Pr(A \cap Q) \Pr(B|Q).$$

But $\Pr(A \cap Q) \Pr(B|Q) = \Pr(A \cap Q) \Pr(B \cap Q) / \Pr(Q)$, so

$$\Pr(Q) = \frac{\Pr(A \cap Q) \Pr(B \cap Q)}{\Pr(A) \Pr(B)} = \frac{\Pr(Q|A) \Pr(A) \Pr(Q|B) \Pr(B)}{\Pr(A) \Pr(B)}$$
$$= \Pr(Q|A) \Pr(Q|B). \tag{6}$$

Now, since $\Pr(A) \Pr(B) = \Pr(A|Q) \Pr(B \cap Q)$ by Equation (5) and further $\Pr(B \cap Q) \leq \Pr(B)$ we have that $\Pr(A|Q) \geq \Pr(A)$. But then

$$\Pr(Q|A) = \frac{\Pr(A|Q) \Pr(Q)}{\Pr(A)} \geq \Pr(Q). \tag{7}$$

In the same way we achieve

$$\Pr(Q|B) \geq \Pr(Q). \tag{8}$$

Thus, by Equations (6), (7) and (8), $\Pr(Q) \geq (\Pr(Q))^2$ and either $\Pr(Q) = 0$ or $\Pr(Q) = 1$. But the conditional probability $\Pr(A \cap B|Q)$ is undefined if $\Pr(Q) = 0$, so if both relative and conditional independence holds conditioned on Q, $\Pr(Q) = 1$.   □

4 Examples

Now we turn our attention to the problem of how different second-order probability distributions contain dependencies between first-order probability values. First-order probability variables can never be independent but as we already stated they can display independence relative to summing to one. In this section we try to illustrate both relative independence and cases where first-order probabilities are not relatively independent.

For ease of presentation we limit ourselves to the case of three possible outcomes. Let the probabilities of the three outcomes be x_1, x_2 and x_3, then a second-order probability distribution for these is a probability distribution with support on the set $\{(x_1, x_2, x_3) : x_1, x_2, x_3 > 0, x_1 + x_2 + x_3 = 1\}$. If $x_i = 0$ we could just exclude the corresponding outcome and remove x_i from the set of variables. In the special case where the second-order distribution that factors into marginals is a proper Dirichlet distribution the parameters are equal to $1/(n-1)$, where n is the number of variables s.t. $\sum_{i=1}^{n} x_i$. With $n = 3$ we therefore have $\alpha_1 = \alpha_2 = \alpha_3 = 1/2$ for the distribution that factors into marginals and hence represents relative independence. We will compare this distribution to some other Dirichlet distributions and for simplicity we choose symmetric distributions, that is with $\alpha_1 = \alpha_2 = \alpha_3$, in those cases also and refer to the parameters as simply α. The probability density function of a Dirichlet distribution with parameters $\alpha_1 = \alpha_2 = \alpha_3$ will below be denoted by superscript α, f^{α}.

Let us compare a Dirichlet distribution with $\alpha = 1/2$ with Dirichlet distributions with parameters less than and greater then $1/2$. Recall that with $\alpha = 1/2$, we have relative independence with respect to the predicate $x_1 + x_2 + x_3 = 1$, for other values of α relative independence does not hold. It might be that $\alpha = 1/2$ is a balancing point and that dependence works in one direction for $\alpha < 1/2$ and in an other direction for $\alpha > 1/2$. Parameters $\alpha = 1$ is an interesting special case since it amounts to the uniform joint distribution. So we look at the two special parameter values $\alpha = 1/2$ (i.e. Jeffreys [16], factors into marginals when $n = 3$) and $\alpha = 1$ (Bayes-Laplace [14]), and two outliers, as it were, $\alpha = 1/3 < 1/2$ ($\alpha = 1/3$ being Perks prior [15]) and $\alpha = 2 > 1$. We have the symmetric Dirichlet distributions with $n = 3$ and $\alpha_1 = \alpha_2 = \alpha_3 = \alpha^2$:

$$f^{\alpha}(x_1, x_2, x_3) = \frac{\Gamma(3\alpha)}{\Gamma(\alpha)^3 (x_1 x_2 x_3)^{1-\alpha}} \,. \tag{9}$$

Then with $\alpha = 1/3, 1/2, 1$ and 2 we obtain

$$f^{1/3}(x_1, x_2, x_3) = \frac{1}{\Gamma(1/3)^3 (x_1 x_2 x_3)^{2/3}} \,, \tag{10}$$

$$f^{1/2}(x_1, x_2, x_3) = \frac{\Gamma(3/2)}{\Gamma(1/2)^3 \sqrt{x_1 x_2 x_3}} \,, \tag{11}$$

$$f^{1}(x_1, x_2, x_3) = \frac{\Gamma(3)}{\Gamma(1)^3 (x_1 x_2 x_3)^0} = 2 \,, \tag{12}$$

$$f^{2}(x_1, x_2, x_3) = \frac{\Gamma(6) x_1 x_2 x_3}{\Gamma(2)^3} = 120 x_1 x_2 x_3 \,. \tag{13}$$

The corresponding marginal densities for x are

$$f_i^{\alpha}(x) = \frac{x^{\alpha-1}(1-x)^{2\alpha-1}\Gamma(3\alpha)}{\Gamma(\alpha)\Gamma(2\alpha)} \,, \tag{14}$$

---

[2] Note that $x_3 = 1 - x_1 - x_2$, x_3 is written for convenience.

$$f_i^{1/3}(x) = \frac{1}{x^{2/3}(1-x)^{1/3}\Gamma(1/3)\Gamma(2/3)}, \tag{15}$$

$$f_i^{1/2}(x) = \frac{1}{2\sqrt{x}}, \tag{16}$$

$$f_i^1(x) = 2(1-x), \tag{17}$$

$$f_i^2(x) = 20x(1-x)^3. \tag{18}$$

In these four examples of second-order probability distributions we want to show different dependence behavior. Recall that the only distribution that fulfills relative independence with respect to $x_1 + x_2 + x_3 = 1$ is when $\alpha = 1/2$. One question of interest is how this type of independence is manifested. In the case of independence in the strict sense, or absolute rather than relative, the product of the marginal distributions is enough to build the joint distribution and conditioning on a variable makes no difference. We would have $f(x_1, x_2, x_3) = f_1(x_1)f_2(x_2)f_3(x_3)$ and $f(x_1, x_2|x_3) = f_1(x_1)f_2(x_2)$ if x_1, x_2, x_3 were independent. If on the contrary x_1, x_2 and x_3 are dependent, the ratio $f(x_1, x_2|x_3)/f_1(x_1)f_2(x_2)$ is not equal to one; the ratio tells us in a sense what is missing from the conditional probability $f(x_1, x_2|x_3)$ when the marginal densities f_1 and f_2 have been taken into account.

For ease of presentation we reduce dimensions by considering the conditional probability $f(x_1, x_2|x_3)$ with $x_3 = 0.2$ arbitrarily chosen, i.e. $f(x_1, 0.8 - x_1|x_3 = 0.2)$. Let us then use the second-order derivative to explore if there is any bias towards particular probability vectors after dividing with the product of marginals, i.e.

$$g^\alpha(x_1) = \frac{f^\alpha(x_1, 0.8 - x_1|x_3 = 0.2)}{f_1^\alpha(x_1)f_2^\alpha(0.8 - x_1)} = \frac{f^\alpha(x_1, 0.8 - x_1, x_3 = 0.2)}{f_1^\alpha(x_1)f_2^\alpha(0.8 - x_1)f_3^\alpha(x_3 = 0.2)}. \tag{19}$$

Since

$$g^\alpha(x_1) = \frac{\Gamma(2\alpha)^3}{\Gamma(3\alpha)^2}(0.8(1-x_1)(0.2 + x_1))^{1-2\alpha} \tag{20}$$

and

$$\frac{dg^\alpha}{dx_1} = \frac{\Gamma(2\alpha)^3 0.8^{1-2\alpha}}{\Gamma(3\alpha)^2}(1-2\alpha)(0.8 - 2x_1)\left((1-x_1)(0.2 + x_1)\right)^{-2\alpha} \tag{21}$$

the second-order derivative of g^α with respect to x_1 equals

$$\frac{d^2 g^\alpha}{dx_1^2} = \frac{\Gamma(2\alpha)^3 0.8^{1-2\alpha}}{\Gamma(3\alpha)^2} 2(2\alpha - 1)((1-x_1)(x_1 + 0.2))^{-2\alpha-1}$$
$$\left[4\alpha(0.4 - x_1)^2 + (1-x_1)(x_1 + 0.2)\right]. \tag{22}$$

And since $0 < x_1 < 1$, $1 - x_1 > 0$, $\alpha > 0$ and $0.2 + x_1 > 0.2 > 0$, the second-order derivative in Equation (22) is negative when $\alpha < 1/2$ and positive when $\alpha > 1/2$. In other words, when $\alpha < 1/2$, $g^\alpha(x_1)$ is concave, giving a bias for values of x_1 and x_2 that are close. But when $\alpha > 1/2$ the ratio $g^\alpha(x_1)$ is convex with a resulting bias for values that are far apart.

Let us show some figures that demonstrates this bias. The graphs in Figure 2 below show $g^\alpha(x_1)$ in Equation (19).

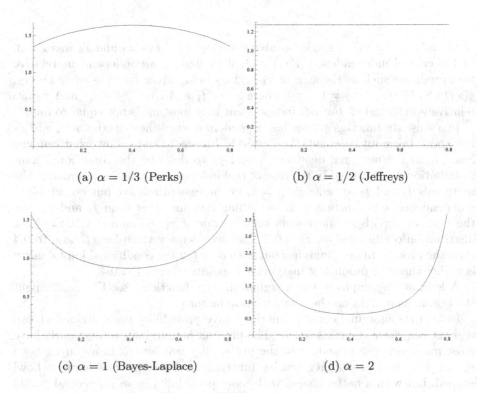

(a) $\alpha = 1/3$ (Perks) (b) $\alpha = 1/2$ (Jeffreys)

(c) $\alpha = 1$ (Bayes-Laplace) (d) $\alpha = 2$

Fig. 2. $g^\alpha(x_1)$ plotted over x_1

In these four examples of second-order probability distributions it is only when $\alpha = 1/2$ in Figure 2(b) that $f^\alpha(x_1, x_2 | x_3 = 0.2) \propto f_1^\alpha(x_1) f_2^\alpha(x_2)$. What does this proportionality mean? That x_3 has a given value restricts the possible values of x_1 and x_2 so that $x_1 + x_2 = 1 - x_3$, so x_1 and x_2 cannot be independent, but there is no dependence on the proportion of x_1 and x_2 in the case where the joint distribution factors into marginals. That the value of x_3 is given restricts the remaining variables in the same way as the sum of variables otherwise is restricted by $x_1 + x_2 + x_3 = 1$. x_1 and x_2 are independent relative to $x_1 + x_2 = 1 - x_3$ in the parlance of Definition 2. The normalising denominator is in this case the probability that x_1 and x_2 have a given sum when drawn from their respective marginal distribution f_1 and f_2.

If x_1, x_2 and x_3 were independent, $f(x_1, x_2|x_3)$ would be equal to $f_1(x_1)f_2(x_2)$. We can always write $f(x_1, x_2|x_3)$ as $f_1(x_1)f_2(x_2)g(x_1)$, where

$$g(x_1) = \frac{f(x_1, x_2|x_3)}{f_1(x_1)f_2(x_2)} \tag{23}$$

needs only x_1 (or x_2) as single variable since $x_2 = 1 - x_3 - x_1$ and x_3 was given. In the case of independence, $g(x_1) = 1$, if x_1 and x_2 are independent relative to a predicate such as the sum of x_1 and x_2 being given by $x_1 + x_2 = 1 - x_3$, $g(x_1) = 1/\Pr(x_1 + x_2 = 1 - x_3)$ where $x_1 \sim f_1$ and $x_2 \sim f_2$. If x_1 and x_2 are relatively independent but not independent this constant is not equal to one.

In a sense the function g describes dependency, when the densities of x_1 and x_2 have been taken into account, the graphs in Figure 2 aim to show what remains apart from the marginal densities f_1 and f_2 to describe the conditional joint distribution $f(x_1, x_2|x_3)$. In the case of real independence, nothing remains, the marginals f_1 and f_2 are enough, $g = 1$, in the case of relative but not absolute independence, g is constant $\neq 1$, something remains other than f_1 and f_2, but the value of $g(x_1)$ is not dependent on the value of x_1. When $\alpha < 1/2$, the joint distribution conditioned on $x_3 = 0.2$ has lower density value for x_1 close to 0.4 than the product of marginals has but with $\alpha > 1/2$ the conditional joint denisty is higher then the product of marginals at points where $x_1 \approx 0.4$.

A look at the graphs of the marginal density functions, see Figure 3, might shed some more light on the compensation factor g.

In 3(a) the marginal density functions have probability mass divided at two regions away from the mass centre. The product of marginals consequently have most mass near the boundary of the probability simplex, as in a convex bowl shape. The joint probability density function for Perks prior is likewise bowl shaped, but with a flatter shape, with more probability mass for central points than the product of marginals would indicate, hence the concavity of $g^{1/3}$.

In contrast, with high parameter values such as $\alpha = 2$ and a concave joint distribution with a peak at the center of the simplex the variables are dependent in the sense that the marginal density functions are so concentrated on the mean $1/n$ that the product of marginals gives more weight to the simplex centroid than the joint distribution does. The convex shape of g^2 compensates by favouring distant variable pairs.

That g^1 is convex just as g^2 means that the Bayes-Laplace prior models the same type of dependency as a Dirichlet distribution with $\alpha = 2$ does even though the joint distribution is flat and uniform without a peak. But since the marginals of Bayes-Laplace, in Figure 3(c) the density for $n = 3$ is shown, disappear at $x_i = 1$, the product of marginals has density zero at the simplex corners and is concave. To bridge the gap between the concave product of marginals and the uniform joint, the compensating factor g^1 must be convex, see Figure 2(c).

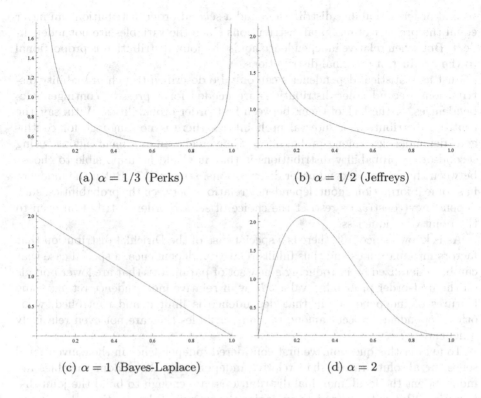

(a) $\alpha = 1/3$ (Perks)

(b) $\alpha = 1/2$ (Jeffreys)

(c) $\alpha = 1$ (Bayes-Laplace)

(d) $\alpha = 2$

Fig. 3. Marginal density functions f_i^α

5   Summary and Conclusions

We have introduced the notion of relative independence, i. e. that random variables are displaying some of the properties of independent random variables without necessarily actually being independent. Variables are then independent relative to a predicate that may cause dependency, but dependency is in a sense restricted precisely to the predicate. More specifically, the joint probability conditioned on a predicate is proportional to the product of marginal probabilities where the normalising factor is equal to the probability of fulfilling the predicate. Since the notion of relative independence involves conditioning it might come natural to believe that relative independence is either a special case or a generalisation of conditional independence. But we proved that relative and conditional independence coincides only in a trivial case. In short, in conditional independence, independence holds under conditioning but in relative independence there is unconditional independence.

Relative independence was then applied to second-order probability. The random variables being first-order probability values of an exhaustive and mutually exclusive set of events and the predicate the consequent constraint that the variables sum to one. A joint probability distribution over such variables is a second-order probability distribution, and a second-order distribution can never equal the product of marginal distributions since the variables are not independent. But when relative independence holds the joint distribution is proportional to the product of marginal distributions.

Just as statistical dependencies can only be described trough probability distributions, second-order distributions are needed for expressing contingent dependencies, or the lack of them, between first-order probabilities. With, say, the centroid distribution or interval probabilities there is no language for contingent dependencies among probabilities. Conversely, an argument against using second-order probability distributions is that it would be impossible to choose between the feasible second-order distributions, see e.g. [17]. But if the modeller has some information about dependency relations between the probabilities, such dependency constraints restrict the choice of second-order distribution even to the point of uniqueness.

As is known since [13], there is a special case of the Dirichlet distribution that factors into marginals and thus fulfills relative independence, a special case that can be generalized by introducing a new set of parameters that are lower bounds on the first-order probability values. Now, if relative independence retains some features of independence in that dependence is limited and controlled, what other dependencies occur among random variables that are not even relatively independent?

To answer this question, we first considered independence in the conventional sense, or absolute rather than relative independence. If random variables are independent the local, marginal distributions, are enough to build the joint distribution. But with dependent variables the marginal distributions do not tell the whole story, something remains to bind together the variables in whatever dependency they share. This binding together can be described by copulas, [12], in the particular case of relative independence, product copulas. But for the purposes of this article we merely wish to point out the gap between joint and marginal distributions. The ratio of the joint distribution and the product of marginals tell us about the dependency, what more that is needed to describe the joint distribution when the marginals have been accounted for.

Similarly, with relatively independent random variables, the marginal distributions alone are not enough to build the joint distribution (unless the relative independence is absolute as in Equation (4), p. 382), but together with the constant probability of the predicate being fulfilled we do have enough information for the joint distribution. The ratio between the joint and the product of marginal distributions is not equal to one, but it is constant. We looked at second-order probability distributions with three first-order variables under the constraint of summing to one. Then independence relative to summing to one means that after taking the product of marginal distributions, you just have to normalize to produce the joint distribution. It does not matter what proportion the variables have to each other, e. g. if they are all nearly equal, or if one of them is nearly zero. But with the second-order distributions that do not factor into marginals there was a bias for or against variables that had values close to each other after having taken the product of marginals, i.e. displaying a form of dependency apart from what is required by summing to one.

Acknowledgements. This work was supported by Formas and the Information Fusion Research Program (University of Skövde, Sweden) in partnership with the Swedish Knowledge Foundation under grant 2010-0320 (URL: http://www.infofusion.se, UMIF project).

References

1. Walley, P.: Towards a unified theory of imprecise probability. International Journal of Approximate Reasoning 24, 125–148 (2000)
2. Levi, I.: The enterprise of knowledge. The MIT press (1983)
3. Cozman, F.G.: Credal networks. Artificial Intelligence 120, 199–233 (2000)
4. Parry, G.W.: The characterization of uncertainty in probabilistic risk assessments of complex systems. Reliability Engineering & System Safety 54, 119–126 (1996)
5. Karlsson, A.: Evaluating Credal Set Theory as a Belief Framework in High-Level Information Fusion for Automated Decision-Making. PhD thesis, Örebro University, School of Science and Technology (2010)
6. Ekenberg, L., Thorbiörnson, J.: Second-order decision analysis. International Journal of Uncertainty, Fuzziness and Knowledge-Based Systems 9(1), 13–38 (2001)
7. Utkin, L.V., Augustin, T.: Decision making with imprecise second-order probabilities. In: ISIPTA 2003 - Proceedings of the Third International Symposium on Imprecise Probabilities and Their Applications, pp. 547–561 (2003)
8. Nau, R.F.: Uncertainty aversion with second-order utilities and probabilities. Management Science 52(1), 136–145 (2006)
9. Arnborg, S.: Robust Bayesianism: Relation to evidence theory. Journal of Advances in Information Fusion 1(1), 63–74 (2006)
10. Cozman, F.: Decision Making Based on Convex Sets of Probability Distributions: Quasi-Bayesian Networks and Outdoor Visual Position Estimation. PhD thesis, The Robotics Institute, Carnegie Mellon University (1997)
11. Karlsson, A., Johansson, R., Andler, S.F.: An Empirical Comparison of Bayesian and Credal Set Theory for Discrete State Estimation. In: Hüllermeier, E., Kruse, R., Hoffmann, F. (eds.) IPMU 2010. CCIS, vol. 80, pp. 80–89. Springer, Heidelberg (2010)
12. Nelsen, R.B.: An introduction to copulas. Lecture Notes in Statistics, vol. 139. Springer (1999)
13. Sundgren, D., Ekenberg, L., Danielson, M.: Shifted dirichlet distributions as second-order probability distributions that factors into marginals. In: Proceedings of the Sixth International Symposium on Imprecise Probability: Theories and Applications, pp. 405–410 (2009)
14. Bernard, J.M.: An introduction to the imprecise dirichlet model for multinomial data. Int. J. Approx. Reasoning 39(2-3), 123–150 (2005)
15. Perks, W.: Some observations on inverse probability including a new indifference rule (with discussion). J. Inst. Actuaries 73, 285–334 (1947)
16. Jeffreys, H.: Theory of Probability. Oxford University Press (1961)
17. Walley, P.: Statistical reasoning with Imprecise Probabilities. Chapman and Hall (1991)

Uncertain Observation Times

Shaunak Chatterjee and Stuart Russell

Computer Science Division,
University of California, Berkeley
Berkeley, CA 94720, USA
{shaunakc,russell}@cs.berkeley.edu

Abstract. Standard temporal models assume that observation times are correct, whereas in many real-world settings (particularly those involving human data entry) noisy time stamps are quite common. Serious problems arise when these time stamps are taken literally. This paper introduces a modeling framework for handling uncertainty in observation times and describes inference algorithms that, under certain reasonable assumptions about the nature of time-stamp errors, have linear time complexity.

1 Introduction

Real-world stochastic processes are often characterized by discrete-time state-space models such as hidden Markov models, Kalman filters, and dynamic Bayesian networks. In all of these models, there is a hidden (latent) underlying Markov chain and a sequence of observable outputs, where (typically) the observation variables depend on the corresponding state variables. Crucially, the *time* of an observation variable is not considered uncertain, so that the observation is always attached to the right state variables.

In practice, however, the situation is not always so simple—particularly when human data entry is involved. For example, a patient in an intensive care unit (ICU) is monitored by several sensors that record physiological variables (e.g., heart rate, breathing rate, blood pressure); for these sensors, the time stamps are reliable. In addition, the ICU nurse records annotated observations of patient state ("agitated," "coughing," etc.) and events ("suctioned," "drew blood," "administered phenylephrine," etc.). Each such annotation includes an accurate *data entry time* (generated by the data recording software) and a manually reported *event time* that purports to measure the *actual* event time. For example, at 11.00 the nurse may include in an hourly report the assertion that phenylephrine was administered at 10.15, whereas in fact the event took place at 10.05.

Such errors matter when their magnitude is non-negligible compared to the time-scale of the underlying process. For example, phenylephrine is a fast-acting vasopressor that increases blood pressure in one or two minutes. In the situation described above, a monitoring system that takes the reported event time of 10.15 literally would need to infer another explanation for the rapid rise in blood pressure at 10.06 (perhaps leading to a false diagnosis) and might also infer that

E. Hüllermeier et al. (Eds.): SUM 2012, LNAI 7520, pp. 392–405, 2012.

the drug injected at 10.15 was not in fact phenylephrine, since it had no observed effect on blood pressure in the ensuing minutes. Such errors in observation times would also cause serious problems for a learning system trying to learn a model for the dynamical system in question; moreover, reversals in the apparent order of events can confuse attempts to learn causal relations or expert policy rules. It would be undesirable, for example, to learn the rule that ICU nurses inject phenylephrine in response to an unexplained rise in blood pressure.

Similar examples of potentially noisy time stamps are found in manual data entry in biological labs, industrial plants, attendance logs, intelligence operations, and active warfare. These examples share a common trait—a sequence of manually entered observations complements continually recorded observations (spectrometer readings, CCTV footage, surveillance tapes, etc) that are temporally accurate. The process of reconstructing historical timelines suffers from "time-stamp" errors in all observation sequences—carbon dating, co-located artifacts, and contemporary sources may give incorrect or inexact ("near the end of the reign of ...") dates for events.

In this work, we present an extension of the hidden Markov model that allows for time-stamp errors in some or all observations. As one might expect, we include random variables for the data entry time, the manually reported event time, and the actual event time, and these connect the observation variable itself to the appropriate state variables via multiplexing. Of particular interest are the assumptions made about the errors—for example, the assumption that *event ordering* among manually reported events in a given reporting stream is not jumbled. We show that, under certain reasonable assumptions, inference in these models is tractable—the complexity of inference is $O(MS^2T)$, where M is the window size of the time stamp uncertainty, S is the state space size of the HMM and T is the length of the observation sequence.

There has been a lot of work on state space models with multiple output sequences. Some authors have modeled observation sequences as non-uniform subsamples of single latent trajectory ([6,4]) and thereby combined information sources. Others, namely [1,2] (asynchronous HMMs (AHMMs)) and [5] (pair HMMs), have proposed alignment strategies for the different sequences using a common latent trajectory. AHMMs ([1]) are closely related to our work. However, the assumptions they make for the generative model of the less frequent observation sequence are different from ours and are not suited to the applications we have described. Also, in our case, the annotations come with noisy time stamps, which help us localize our search for the true time stamp. We also handle missing reports and false reports, which cannot be modeled in AHMMs.

The paper begins (Section 2) with the basic modeling framework for uncertainty in observation times. Section 3 presents a modified forward–backward algorithm for the basic model. Section 4 extends the model to accommodate unreported events and false reports of events, and Section 5 describes an exact inference algorithm for this extended model. The complexity of the exact algorithm is analyzed in Section 6 and some simplifications and approximations are proposed. Section 7 presents some experiments to highlight the performance of the different algorithms.

2   Extending HMMs

A hidden Markov model (HMM) is a special case of the state space model where the latent variable is discrete. As shown in Figure 1(a), $\mathbf{X} = \{X_1, X_2, \ldots, X_T\}$ is a Markov process of order one and X_t is the hidden (latent) variable at time step t. There are two different observation sequences. $\mathbf{Y}$ is the variable observed at every time step (and is assumed to have the correct time stamp). Y_t corresponds to the observation at time t. $Y_{t_1:t_2}$ refers to the sequence of Y_t from $t = t_1$ to $t = t_2$ $(t_1 \leq t_2)$. In the ICU, $\mathbf{Y}$ could be the various sensors hooked up to the patient. The other sequence of observations is less frequent and can be thought of as analogous to *annotations* or manual entries of *events*. In a sequence of T time-steps, there are K annotations $(K < T)$ which mark K events. m_k represents the (potentially erroneous) time stamp of the *report* corresponding to the kth event. a_k represents the actual time of occurrence of the kth event. d_k is the time at which the time stamp data for the kth (i.e. m_k) event was entered. In the ICU example from Section 1, m_k is 10:15, a_k is 10:05 and d_k is 11:00. d_k can be a parameter for the error model of the time stamp (i.e. $p(m_k|m_{k-1}, a_k)$). For instance, the noisy time stamp m_k can be no greater than d_k, if we exclude anticipatory data entry. M_k is the window of uncertainty of the kth event and denotes the possible values of a_k (around m_k). So, if we assume that the nurse can err by at most 15 minutes, M_k is from 10:00 to 10:30.

A key assumption is that *the time stamps of events are chronologically ordered*. This restriction is analogous to the monotonicity of the mapping in time imposed

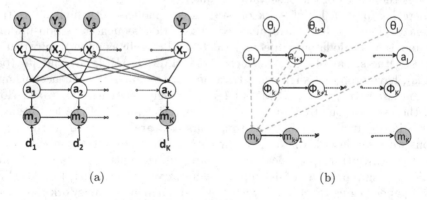

(a)   (b)

Fig. 1.   (a) The extended hidden Markov model with actual and measured times of events. All X's are potential parents of each a_k and the connections depend on the values of X_i. Certain dependencies are denoted by solid lines, while value-dependent ones are dotted. (b) The generalized noisy time stamp hidden Markov model with actual and measured times of events. $\mathbf{X}$ and $\mathbf{Y}$ have been omitted for simplicity (they are identical to Figure 1(a)). Dependencies are only shown completely for ϕ_j and m_j. Color coding indicates definite dependencies (black) and value-dependent ones (gray).

on sequence matching in dynamic time warping [9]. Thus, m_k is strictly greater than m_{k-1}. This assumption holds vacuously if the events are identical and non-distinguishable. It also holds in several real-life scenarios.

The next important point is that there is a deterministic relationship between a_k's and $\mathbf{X}$. For clarity of presentation, let us consider the case where X_t is a binary random variable. $X_t = 1$ is the state corresponding to an event or annotation and $X_t = 0$ is the state representing a non-event. The generalization to the case where the state space is of size S is straight-forward and is presented in the supplementary material. a_k is the smallest i such that $\sum_{j=1}^{i} X_j = k$. The complete likelihood model is as follows:

$$p(X_{1:T}, Y_{1:T}, a_{1:K}, m_{1:K}) =$$

$$p(X_1)p(Y_1|X_1) \prod_{t=2}^{T} p(X_t|X_{t-1})p(Y_t|X_t)p(a_{1:K}|X_{1:T}) \prod_{k=1}^{K} p(m_k|m_{k-1}, a_k)$$

For notational convenience, assume $m_0 = 0$. Since the a_k's are deterministically determined by the sequence $\mathbf{X}$, $p(a_{1:K}|X_{1:T})$ is zero for all $a_{1:K}$ instantiations except the one which corresponds to the given $X_{1:T}$ chain. Also, only those $X_{1:T}$ instantiations which have exactly K events will have non-zero probability support from the evidence $m_{1:K}$. In a later model, we will relax these constraints.

For now, we assume that every annotation corresponds to an event and every event has been recorded/annotated. Thus, it is justified to only consider latent variable trajectories with exactly K events. The inference task is to compute the posterior distributions of X_i and a_k conditioned on all the evidence available (namely $\mathbf{Y}$ and $m_{1:K}$). In the next section we describe an efficient algorithm for this task.

3   The Modified Forward-Backward Algorithm

The notation used in this section will be very similar to the standard notation used in the $\alpha - \beta$ forward backward algorithm as presented in [3]. $\alpha(a_k = t) = p(a_k = t, Y_{1:t}, m_{0:k})$ and will be simply written as $\alpha(a_k)$ when the context is clear. Thus, $\alpha(a_k)$ denotes the joint probability of all given data upto time a_k and the value of a_k itself. $\beta(a_k) = p(Y_{a_k+1:T}, m_{k+1:K}|a_k, m_k)$ represents the conditional probability of all future data given the value of a_k and m_k. Let $\mathcal{L}(a_k, a_{k+1}) = p(a_{k+1}, Y_{a_k+1:a_{k+1}}|a_k)$. This likelihood term can be simplified by

$$\mathcal{L}(a_k, a_{k+1}) = p(a_{k+1}, Y_{a_k+1:a_{k+1}}|a_k)$$

$$= \sum_{X_{a_k+1:a_{k+1}}} p(a_{k+1}, X_{a_k+1:a_{k+1}}, Y_{a_k+1:a_{k+1}}|a_k)$$

$$= \prod_{t=a_k+1:a_{k+1}} p(Y_t|X_t)p(X_t|X_{t-1}),$$

where $X_{a_k+1:a_{k+1}} = \{0, 0, \ldots, 0, 1\}$ since $p(a_{k+1}|a_k, X_{a_k+1:a_{k+1}}) = 0$ for every other $X_{a_k+1:a_{k+1}}$ sequence. The α update step is

$$\alpha(a_k) = p(a_k, Y_{1:a_k}, m_{0:k})$$

$$= \sum_{a_{k-1}} \sum_{X_{1:a_k}} p(a_k, a_{k-1}, X_{1:a_k}, Y_{1:a_k}, m_{0:k})$$

$$= p(m_k | m_{k-1}, a_k) \sum_{a_{k-1}} \alpha(a_{k-1}) \mathcal{L}(a_{k-1}, a_k).$$

The backward (smoothing) step is as follows:

$$\beta(a_k) = p(Y_{a_k+1:T}, m_{k+1:K} | a_k, m_k)$$

$$= \sum_{a_{k+1}} p(a_{k+1}, Y_{a_k+1:T}, m_{k+1:K} | a_k, m_k)$$

$$= \sum_{a_{k+1}} \beta(a_{k+1}) p(m_{k+1} | m_k, a_{k+1}) \mathcal{L}(a_k, a_{k+1}).$$

Given these definitions, the standard rule for computing the posterior still holds.

$$\gamma(a_k) = p(a_k | Y_{1:T}, m_{0:K}) \propto \alpha(a_k) \beta(a_k).$$

3.1 Computing $\gamma(X_i)$ from $\gamma(a_k)$

The final step of the algorithm would be to compute the conditional distributions of the hidden state variables X_i from $\gamma(a_k)$. This computation is straight-forward since X_i can only be 1 if in a chain, there exists a k such that $a_k = i$. It should also be noted that $a_k = i$ denotes that X_i is the kth 1 in the sequence. Therefore, the **X** sequences contributing to $\gamma(a_k = i)$ and $\gamma(a_{k'} = i)$ are disjoint when $k \neq k'$. So the probability of an event at time i is just equal to the probability of any of the K events occurring at time i. The posterior distribution of X_i is given by

$$\gamma(X_i) = p(X_i = 1 | Y_{1:T}, m_{0:K}) = \sum_{k=1}^{K} \gamma(a_k = i)$$

3.2 Tractable Error Models for m_k

In our analysis, we have conditioned the error model of m_k on the time stamp of the previous report m_{k-1} and the actual time of the kth event a_k. The time of data entry d_k can also be a parameter in this conditional distribution and we could additionally condition on a_{k-1}. We cannot include any previous events or reports since that would destroy the first-order Markovian dynamics that we need for our analysis. However, with the allowed parameters, very flexible error models can be created. m_{k-1} as a parent can be used to model an expected gap between two reports. a_k, a_{k-1} and m_{k-1} together could be used to specify a (stochastic) relationship between the relative timings of events and their reports. Two sample error models are shown in Figure 2.

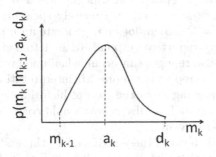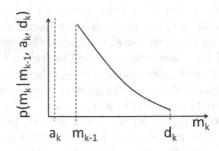

Fig. 2. Sample probability distributions for $p(m_k|m_{k-1}, a_k)$. The possible values of m_k are bounded by m_{k-1} (to satisfy the monotonicity constraint) and d_k (to exclude anticipatory entries), with some bias for a_k.

3.3 Complexity of the Algorithm

Our analysis has assumed that there are K events in T time steps. Let us also assume (for simplicity) that all uncertainty windows are of the same size, i.e. $\forall k, |M_k| = M$. Let the maximum possible interval between $t_1 \in M_k$ and $t_2 \in M_{k+1}$ be I_k. Then the computation of $(\mathcal{L}(t_1, t_2)$ for all values of $\{t_1, t_2 : t_1 \in M_k, t_2 \in M_{k+1}, t_1 < t_2\}$ is an $O(M^2 + I_k)$ operation.

Once the relevant $\mathcal{L}(t_1, t_2)$ and $\alpha(a_k)$ values have been computed, the computation of $\alpha(a_{k+1})$ is an $O(M^2)$ operation. Thus the total complexity of the modified forward step is $O(KM^2 + \sum_k I_k)$. If we assume that only a constant number of uncertainty windows can overlap, then $\sum_k I_k = O(T)$ and $MK \leq O(T)$. Thus, the total complexity expression simplifies to $O(MT)$. The modified backward (or β) step has a similar complexity. Computing $\gamma(a_k)$ and $\gamma(X_i)$ are both $O(MK)$ operations. Thus, the overall complexity is $O(MT)$.

If we consider an HMM with $S + 1$ states, where state S corresponds to the annotation state, then the computation of $\mathcal{L}(t_1, t_2)$ becomes an $(M^2 S^2 + M S^2 I_k)$ operation. The other steps have the same complexity as in the previous analysis, so the overall complexity becomes $O(MS^2 T)$. Thus, we see an M-fold increase in the inference complexity over a regular HMM.

The space complexity is $O(KM^2)$ for storing the relevant $\mathcal{L}(t_1, t_2)$ values and $O(KM)$ for storing the α, β and γ values. Thus, it is independent of the state space size. The algorithm can be trivially extended to handle cases with more than one type of event.

4 Unreported Events and False Reports

The model in section 2 assumes that every event is reported (with a possibly erroneous time stamp). However, in real life, events often go unreported. An example of this in the ICU setting would be a nurse forgetting to make an entry of a drug administration because the recording was done in a batch fashion. Many events in history might go unreported by a historian if she does not come

across sufficient evidence which warrants a report. Thus, negligence and ignorance would be primary causes for unreported events. Precisely speaking, an
unreported event is an event (some $X_t = 1$) which does not generate an m_k.

Previously, we also assumed that every report corresponds to an actual event.
This is also often violated in reality. False reports can occur when one event is
entered twice (making one of them a false report) or more. Misinterpretation of
observations could also lead to false reporting as in the case of historians often
drawing contentious conclusions. In the model, a false report would correspond
to an m_k which was not generated by any event.

We wish to extend our model to handle both of these artifacts. To this end, we
introduce some new variables in the original model. Let us still assume that there
are K reports of events. In addition, let us hypothesize I actual events. I can be
chosen using prior knowledge about the problem (the rate of false reports and
missed reports). For each hypothesized event a_i, we introduce a binary variable
θ_i. $\theta_i = 0$ indicates that the event a_i is unreported, while $\theta_i = 1$ indicates that a_i
has been reported and thus generates some m_k. $\Theta = \{\theta_1, \ldots, \theta_I\}$ denotes the set
of all θ_i. Now, for each report m_k, we introduce a new variable ϕ_k whose range
is $\{0, 1, \ldots, I\}$. If the report m_k is generated by the event a_i then $\phi_k = i$. In
other words, ϕ_k is the index of the (reported) event corresponding to the report
m_k. As is obvious, $p(\phi_k = i|\theta_i = 0) = 0$. $\phi_k = 0$ means m_k is a false report. Φ
is the set of all ϕ_j. The generalized model is shown in Figure 1(b).

The deterministic relationship between $\mathbf{X}$ and $\mathbf{a}$ remains unaffected. The prior
on θ_i can be problem-specific. For our analysis, we assume it is a constant. Let
$p(\theta_i = 0) = \delta_i$. The conditional probability table for ϕ_j is as follows:

$$p(\phi_k|\phi_{k-1}, \theta_{1:I}) = \begin{cases} \epsilon_k, & \text{if } \phi_k = 0 \\ 1 - \epsilon_k, & \text{if } \phi_k = i, \theta_i = 1, \theta_{\phi_{k-1}:i-1} = 0 \end{cases}$$

The prior probability of a false report (modeled currently with a constant ϵ_k)
can also be modeled in more detail to suit a specific problem. However, if m_k
is not a false report (currently an event with probability $1 - \epsilon_k$), then ϕ_k is
deterministically determined by ϕ_{k-1} and Θ. When $\phi_k = 0$, m_k is no longer
parameterized by a_{ϕ_k}. The new distribution is represented as $\tilde{p}(m_k|m_{k-1})$.

5   Exact Inference Algorithm for the Generalized Model

We shall briefly explore the effect of a particular choice of I in Section 7. For inference in this generalized model, there is an added layer of complexity. We now
have to enumerate all possible instances of Θ and Φ. A meaningful decomposition
of the posterior distribution (in the lines of the the standard forward-backward
algorithm) and using dynamic programming could be a potential solution. All elements of Θ are independent and hence enumerating all possibilities is infeasible.
Φ is a better proposition because there are dependencies that can be exploited -
either the report is false (i.e. ϕ_k is 0) or it corresponds to an event after the previous reported actual event (i.e. $\phi_k > \phi_{k-1}$). We will use this key fact to divide all
possible instantiations of Φ into some meaningful sets. Our main objective is to
compute the posterior distribution $p(a_i|Y_{1:T}, m_{1:K})$, from which we can compute

the posterior distribution of each X_i as described in Section 3.1. The posterior distribution for a_i is

$$
\begin{aligned}
\gamma(a_i) &= p(a_i|Y_{1:T}, m_{0:K}) \\
&\propto p(a_i, Y_{1:T}, m_{0:K}) \\
&= \sum_{\Phi} p(a_i, \Phi, Y_{1:T}, m_{0:K}).
\end{aligned}
$$

Now we will describe a way to partition the possible instantiations of Φ which will then be used to formulate the forward and backward steps.

5.1 Partitioning the Φ Sequences

Theorem 1. *For any i, such that $0 < i \leq I$, consider the following sets of ϕ sequences:* $\mathcal{S}_0 = \{\phi_1 > i\};$ $\mathcal{S}_1 = \{\phi_1 \leq i \text{ and } \phi_2 > i\};$ $\mathcal{S}_2 = \{\phi_2 \leq i \text{ and } \phi_3 > i\}; \ldots$ $\mathcal{S}_K = \{\phi_K \leq i\}$

The sets $\mathcal{S}_0, \mathcal{S}_1, \ldots, \mathcal{S}_K$ are disjoint and exhaustively cover all valid instantiations of Φ.

Proof. Intuitively, the set $\mathcal{S}_k$ corresponds to all the cases where the first i events generate the first k reports and the $k+1^{th}$ report is a true report.

Clearly, any sequence in $\mathcal{S}_0$ cannot belong to any other set. Any sequence ϕ belonging to $\mathcal{S}_1$ will have $\phi_2 > i$ and hence cannot belong to $\mathcal{S}_2$. Also, any sequence belonging to $\mathcal{S}_k$ will have $\phi_k \leq i$, which would imply $\phi_2 \leq i$. Thus ϕ cannot be in any $\mathcal{S}_k$ for $k \geq 2$. Similar arguments can be presented to show that $\forall k_1, k_2, \mathcal{S}_{k_1} \cap \mathcal{S}_{k_2} = \emptyset$. One important point to note is that all sequences in $\mathcal{S}_1$ have $\phi_2 \neq 0$, which means that ϕ_2 is not a false report in those cases.

Let ϕ be a valid instantiation of Φ. Now we have to show that every ϕ lies in some $\mathcal{S}_k$. The sequence $\phi = \{0, 0, \ldots, 0\}$ lies in $\mathcal{S}_K$. In every other sequence, there is at least some $\phi_j > 0$. If $\phi_j > i$, then that sequence belongs to $\mathcal{S}_{j-1}$. Thus, we have proved that the proposed partition of all valid instances of Φ is both disjoint and exhaustive. Note that this partition is not unique, and the pivot (currently set to i) can be any value between 1 and I. $\qquad\square$

5.2 Defining Forward-Backward Steps

Now we can use the partitions $\mathcal{S}_k$ to define an efficient dynamic program to compute the posterior distribution of a_i. As we saw earlier,

$$
\begin{aligned}
\gamma(a_i) &\propto \sum_{\Phi} p(a_i, \Phi, Y_{1:T}, m_{1:K}) \\
&= \sum_{k=0}^{K} \sum_{\phi \in \mathcal{S}_k} p(a_i, \phi, Y_{1:T}, m_{1:K}) \\
&= \sum_{\phi \in \mathcal{S}_0} p(a_i, \phi, Y_{1:T}, m_{1:K}) + \sum_{k=1}^{K} \sum_{\phi \in \mathcal{S}_k} p(a_i, \phi, Y_{1:T}, m_{1:K})
\end{aligned}
$$

Let us denote $\sum_{\phi \in S_k} p(a_i, \phi, Y_{1:T}, m_{1:K})$ by P_k. We can compute P_k by further decomposing it.

$$P_k = \sum_{\phi_k \leq i} \sum_{\phi_{k+1} > i} p(a_i, \phi_k, \phi_{k+1}, Y_{1:T}, m_{1:K})$$

$$= \sum_{\phi_k \leq i} p(a_i, \phi_k, Y_{1:a_i}, m_{1:k}) \sum_{\phi_{k+1} > i} p(Y_{a_i+1:T}, m_{k+1:K}, \phi_{k+1} | a_i, m_k)$$

$$= \alpha(a_i, m_k)\beta(a_i, m_k)$$

Due to lack of space, we skip the detailed derivation of the update equations for the α and β expressions. Intuitively $\alpha(a_i, m_k)$ is the probability of the trajectories where the first k reports are associated with the first i events, whereas $\beta(a_i, m_k)$ is the probability of the trajectories where the last $K - k$ reports are associated with the last $I - i$ events. The initialization steps for α and β are straightforward. The order in which the α and β variables are computed is identical to other well-known dynamic programs of a similar structure ([7,10]).

5.3   Computing $\gamma(a_i)$ and $\gamma(X_t)$

Once $\alpha(a_i, m_k)$ and $\beta(a_i, m_k)$ are computed for $\forall i, k$ s.t. $i \in \{1, 2, \ldots, I\}$ and $k \in \{0, 1, \ldots, K\}$, we can compute $\gamma(a_i)$ and $\gamma(X_t)$ by the following

$$\gamma(a_i) = \sum_{k=0}^{K} \alpha(a_i, m_k)\beta(a_i, m_k); \qquad \gamma(X_t) = \sum_{i=1}^{I} \gamma(a_i = t)$$

5.4   Multiple Report Sequences

Consider a scenario where there are R historians and each of them have their own set of annotations of historical events replete with time stamp conflicts. Since all historians do not concur on which events took place, there are instances of missed reports as well as false reports (assuming there is a set of actual events that took place). A simplifying assumption we make is that the historians reach their conclusions independently based solely upon the latent state sequence (**X**) and do not consult one another.

In this case, the addition to the generalized model from Section 4 is that the single report sequence $m_{1:K}$ is now replaced by R report sequences $m^{(r)} = m_{1:K_r}^{(r)}$, where $r \in \{1, 2, \ldots, R\}$. The key feature of the model which makes inference tractable (and very similar to the single report sequence case) is that *given the hidden state sequence* **X**, *$m^{(r_1)}$ is independent of $m^{(r_2)}$*.

The posterior distribution of a_i is computed as follows:

$$\gamma(a_i) = \sum_{\Phi^{(1):(\mathbf{R})}} p(a_i, \Phi^{(1):(\mathbf{R})}, m^{(1):(R)}, Y_{1:T})$$

$$= p(a_i, Y_{1:T}) \prod_{r=1}^{R} \sum_{\Phi^{(\mathbf{r})}} p(\Phi^{(\mathbf{r})}, m^{(r)} | a_i, Y_{1:T})$$

$$\propto p(a_i, Y_{1:T})^{-R+1} \prod_{r=1}^{R} \sum_{\Phi^{(\mathbf{r})}} p(\Phi^{(\mathbf{r})}, m^{(r)}, a_i, Y_{1:T})$$

$$= p(a_i, Y_{1:T})^{-R+1} \prod_{r=1}^{R} \gamma(a_i^{(r)})$$

The $\gamma(a_i^{(r)})$ will be computed as before. $p(a_i, Y_{1:T})$ is proportional to $p(a_i | Y_{1:T})$ which can be computed using a standard forward-backward algorithm. $\gamma(X_t)$ is computed as before.

6   Complexity and Simplifications

The algorithm presented in the previous section, while exact, is computationally very expensive. We now analyze the computational complexity of the exact algorithm and present some simplifications and possible approximation schemes.

6.1   Complexity Analysis

In the model where a_i corresponded to m_i, an uncertainty window resulted from the error model $p(m_i | m_{i-1}, a_i)$. If the error model suggested that m_i could only be within $M/2$ time units of a_i on either side, then this resulted in an uncertainty window of size M for a_i centered at m_i. However, when events can go unreported and reports can be false, the uncertainty window of a_i becomes much larger since we no longer know which (if any) m_k it corresponds to. The safe bet is to assume that $0 < a_i < T$ as long as it satisfies the monotonicity constraint (i.e. $a_{i-1} < a_i < a_{i+1}$). Thus, the uncertainty window in the worst case is $O(T)$.

If there are I hypothesized events and K reports (in the single report sequence case), then the complexity of the α computation step is $O(IKT^2)$. This is of course prohibitively expensive. However, there is a simplifying case.

6.2   Shifts in Data Entry

In the ICU setting, the nurse often enters data once an hour. A safe assumption is that all report(s) generated during the period between consecutive data entries correspond to the events in that same period. Let there be $\bar{I}$ hypothesized events and $\bar{K}$ reports in the time span $\bar{T}$ between two data entries. Then we can run the exact inference algorithm locally within the time span. The computational complexity will be $O(\bar{I}\bar{K}\bar{T}^2)$ for one time span. Over the entire time period T there will be $T/\bar{T}$ such time spans. Thus, the total computational complexity reduces to $O(\bar{I}\bar{K}\bar{T}T)$ which is much more tractable. *If the time span $\bar{T}$ is a constant, then the inference complexity is linear in T.*

6.3   Approximate Inference

Another possible way to reduce the inference complexity is to not consider all possible trajectories of $\boldsymbol{\Phi}$. If the probability of a missed report (δ_i) and false report (ϵ_k) are small, then $\alpha(a_i, m_k)$ and $\beta(a_i, m_k)$ have significant non-zero values only when i and k are close to one another. We could potentially zero out all α and β values corresponding to $|i - k| > c$ where c is some threshold. This would naturally reduce the uncertainty window of event a_i which can now only be associated with some m_k where $i - c \leq k \leq i + c$. This means it suffices if $I = K + c$. If the reduced window size is $O(M_c)$ then the computational complexity of the algorithm becomes $O((K + c)KM_c^2 + T)$.

7   Experiments

For our experiments, we have primarily focused on simulations. The main reason for this choice is that we do not know the ground truth (correct time stamp of events) in the ICU data that we have been working on. Conversely we know the ground truth for our simulations and hence can evaluate our posterior inference results.

7.1   Simple Model Simulations

We set up an HMM with two states whose emission distributions were Gaussian with means $\mu_0 = -1$ and $\mu_1 = -3$ and standard deviation 0.5. The error model was $p(m_k | m_{k-1}, a_k) \sim \mathcal{N}(a_k, \sigma)$ with the window-size $M = 15$. The Gaussian was truncated at m_{k-1} and d_k. A standard HMM treats the time stamps as accurate (equivalent to a noise model with $\sigma = 0$). With this model, we generated data for $T = 1000$ for different values of σ (increasing steps of 0.5 from 0 to 5. We repeated this exercise 20 times to generate more simulations and remove random effects. All results are averaged over the 20 simulations and the bars indicate one standard deviation.

Increase in Likelihood. One objective of using the HMM with the noisy time stamp error model extension is to provide a better explanation for the data. This can be measured in terms of the likelihood. The average log likelihood of the data computed by a standard HMM and the noisy time stamp model are shown in Figure 3(a). The inference algorithms were run with the same transition and observation parameters used to generate the data.

The difference in the two likelihoods increases as the variance of the time stamp noise increases, since this makes noisy time stamps more likely. The trend

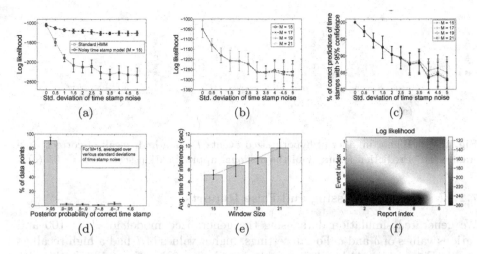

Fig. 3. (a) Average log likelihood of the simulated data using the two models. (b) Average log likelihood for models with various window sizes M. (c) % of high confidence correct time stamp inferences for varying window sizes. (d) Almost all time stamps are predicted with at least 60% accuracy. (e) Average time taken by the inference algorithms for different M. (f) Heat map of Q.

was similar for other values of $\{\mu_0, \mu_1\}$ and the two plots came closer as the two means became similar. Also, noteworthy is the fact that the likelihood of the data under the noisy model changes very little even as the noise increases - thus indicating robustness.

Next, we ran the inference algorithm with different window sizes ($M = 17$, 19 and 21). The likelihood of the data did not change significantly as shown in Figure 3(b). The time taken by the inference algorithm is also linear in the window size M as shown in Figure 3(e).

Accuracy of Posterior Inference. Another objective of the model is to accurately infer the correct time stamps of events. This will lead to better learning of the event characteristics. After computing the posterior distribution $\gamma(\mathbf{X})$, we looked at $\gamma(X_t)$ corresponding to all t which were correct time stamps of events (i.e. a_i). Figure 3(c) shows the percentage of events where the $\gamma(X_t)$ value exceeds .95. The percentage varies between 85% and 100% with the performance degrading as the noise in the time stamp increases. *We can also see that the accuracy is not sensitive to the window size used in the inference.*

Figure 3(d) shows that there are almost no correct time stamps t where the $\gamma(X_t)$ value goes below .6. Thus, we do not miss any event completely. However, there are also some rare false positives. These result because the observation at the event's correct time stamp is not peaked enough to warrant a time stamp movement hypothesis in terms of likelihood.

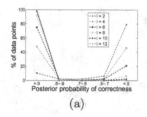 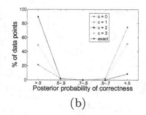 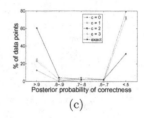

(a)   (b)   (c)

Fig. 4. (a) Higher number of hypothesized events I has a high recall of correct time stamps. (b) Prediction accuracy of the diagonal approximation scheme. $\delta = .05$ (c) $\delta = .3$.

7.2   Model with Missing and False Reports

We generated simulation data using the generalized model for $T = 100$ and various values of δ and ϵ. For all settings, higher values of I had a high recall as seen in Figure 4(a). Although it would seem like a safe bet to set I high, this also leads to a lot of false positives, since a lot of events have to be hypothesized and accounted for. One possible approach to find a good I could be the likelihood measure. We consistently found that the data likelihood peaked at the value of I which corresponded to the correct number of events.

Another observation we made for small values of δ and ϵ was that most of the α and β values were concentrated along the diagonal. e take a look at the following matrix Q where $Q(i,k) = \sum_{t=1:T} \alpha(a_i = t, m_k)\beta(a_i = t, m_k)$ in Figure 3(f).

α and β entries were only considered along the (skewed) diagonal and c diagonals around that - the scheme described in Section 6.3. As we increase c from 0 (only the skewed diagonal) to larger values (more diagonals), our time stamp prediction accuracy increases as shown in Figure 4(b) and (c). However, the accuracy in the presence of these approximations is more when δ is smaller.

8   Conclusion

In this paper, we have proposed two model extensions of the HMM to deal with noisy time stamps of events. These models have inference algorithms quite similar in structure to the forward-backward algorithm used for inference in HMMs. It is easy to see how this model can be used in an EM setup to learn the error model $p(m_k|m_{k-1}, a_k)$ or the transition model for **X** or the emission model for **Y**. The algorithm is linear in T with one-to-one correspondence between events and reports. In other cases, certain reasonable assumptions can get it back to linear time. Noisy time stamps are pervasive in data - especially data recorded by humans (machines can also occasionally have logging errors). Algorithms which try to learn about human expertise will always have to deal with such data.

Looking ahead, it will be interesting to consult with doctors and run experiments on real data from the ICU. Another interesting direction is to model events which have a finite duration (and hence potential overlap). Such events could also be modeled with continuous time Bayesian networks (CTBNs) [8].

Acknowledgements. We would like to acknowledge NSF (IIS-0904672 RI: Hierarchical Decision Making for Physical Agents) and DARPA (DSO contract FA8650-11-1-7153: Open-Universe Theory for Bayesian Information and Decision Systems) for their support and the anonymous reviewers for their comments and suggestions.

References

1. Bengio, S.: An asynchronous hidden markov model for audio-visual speech recognition. In: Advances in Neural Information Processing Systems, NIPS 15. MIT Press (2003)
2. Bengio, S., Bengio, Y.: An EM algorithm for asynchronous input/output hidden markov models (1996)
3. Bishop, C.M.: Pattern Recognition and Machine Learning. Springer (2006)
4. Coates, A., Abbeel, P., Ng, A.Y.: Learning for control from multiple demonstrations. In: Proceedings of 25th International Conference on Machine Learning (2008)
5. Durbin, R., Eddy, S., Krogh, A., Mitchison, G.: Biological sequence analysis: probabilistic models of proteins and nucleic acids. Cambridge Univ. (1998)
6. Listgarten, J., Neal, R.M., Roweis, S.T., Emili, A.: Multiple alignment of continuous time series. In: Advances in Neural Information Processing Systems, pp. 817–824. MIT Press (2005)
7. Needleman, S.B., Wunsch, C.D.: A general method applicable to the search for similarities in the amino acid sequence of two proteins. Journal of Molecular Biology (1970)
8. Nodelman, U., Shelton, C.R., Koller, D.: Continuous time bayesian networks. In: Proceedings of the Eighteenth Conference on Uncertainty in Artificial Intelligence (UAI), pp. 378–387 (2002)
9. Sakoe, H., Chiba, S.: Dynamic programming algorithm optimization for spoken word recognition. IEEE Transactions on Acoustics, Speech and Signal Processing 26(1), 43–49 (1978)
10. Smith, T.F., Waterman, M.S.: Identification of common molecular subsequences. Journal of Molecular Biology (1981)

Center-Wise Intra-Inter Silhouettes

Mohammad Rawashdeh and Anca Ralescu

Machine Learning and Computational Intelligence Laboratory
School of Computing Sciences and Informatics
University of Cincinnati, ML 0008
Cincinnati, OH 45221, USA
rawashmy@mail.uc.edu, Anca.Ralescu@uc.edu

Abstract. Silhouettes were defined as measures of clustering quality in the context of crisp partitions. This study extends the work that generalized silhouettes to fuzzy partitions in a natural profound manner. As opposed to constructing silhouettes for each data point, described here is the construction of silhouettes for each cluster center in terms of center-to-point distances rather than point-to-point distances.

Keywords: intra-distances, inter-distances, intra-scores, inter-scores, validity index, clustering, cluster validity, fuzzy, crisp, compactness, separation, intra-inter, hcm, fcm, point-wise, ground truth clustering, silhouette.

1 Introduction

Cluster analysis is concerned with the structure of a given dataset or more precisely: the grouping structure of the objects. Clustering algorithms, equipped with some notion of *similarity*, group the data points (objects) in clusters such that *points within one cluster are similar to each other and dissimilar to points in other clusters.* Such goal has become a common description of the clustering problem [10, 3, 11, 20]. However, beyond this description and in most clustering applications, the end-goal is to identify meaningful groups of points that have something in common, for example, proteins of similar *functionality*, documents on the same *topic* or images which share common *content*. Manual (expert) grouping in such applications is substantially labor and time intensive especially for large datasets. Rather, a complete grouping of the points might be unknown; as in the case of protein functionality that predicts the functional behavior of a protein based on its similarity to other proteins of known functionality. In this respect it is of interest to note the following. Pairwise similarities-dissimilarities are usually modeled by a *distance measure*. A clustering algorithm either takes, as input, the set of pairwise distances, or a combination of data representation and a measure. The granularity of the clustering process depends on whether the algorithm operates on point pairwise distances, point cluster-prototype distances or just cluster-wise distances. Cluster analysis output is thus determined by the distance measure.

E. Hüllermeier et al. (Eds.): SUM 2012, LNAI 7520, pp. 406–419, 2012.
© Springer-Verlag Berlin Heidelberg 2012

The unknown correct clustering consistent with the end-goal is the only ground truth assumed about the problem. Hence it is referred as the *ground truth clustering*[1]. Obtaining satisfactory results with respect to the end-goal, clearly, depends on the choice of the distance measure. For instance, in the application of clustering images by facial expression, a measure that binds images because they show the same expression should be employed rather than a measure based on individuals appearing in the photo collection. This has encouraged investigators to treat the clustering problem in a semi-supervised setting to learn the measure useful for achieving the ground truth clustering [8, 2]. The properties of the distance measure that can be used by a clustering algorithm to cluster well are given as part of a theoretical framework in [1]. Consider for example, the trivial property, namely *single cutoff*, with all distances being below a given cutoff value having their endpoints in the same ground cluster or in different clusters if they are larger. A distance measure with such property is the most efficient in recovering the ground truth clustering by a simple greedy algorithm. Other properties beside single cutoff are briefly described in [5]. Relevant to the task of learning a distance measure is the selection of certain features to define the feature space. Different measures impose different weights on the features, in turn, determining feature contribution in the pairwise similarities. Thus, one can focus on selecting the features relevant to the task instead of learning a distance measure. Changing the features definitely changes the grouping structure of a given dataset, an observation well investigated in [15].

With a distance measure, it is convenient to talk about *compactness* and *separation* instead of similarity and dissimilarity. The goal of clustering can be restated as the search for optimally *compact* and *separated* clusters. In its essence is the desire to have member data points of each cluster within small proximity from, equivalently similar to, each other and only points that are farther apart separated in different clusters. A clustering algorithm normally realizes a measure of compactness, separation or both as a clustering criterion and implements a mechanism to optimize this measure, mainly in cluster assignment of data points. For instance in the c-means approach to clustering, Lloyd's algorithm [12] is popularly used to minimize a within-cluster variance criterion, a measure of compactness. The term, c-means (k-means), was first coined with theoretical analysis of the asymptotic behavior of the model in [13]. Minimizing c-means criterion is identical to maximizing a measure of separation for the *same number of clusters* [21]. Since the algorithm produces a clustering into c disjoint subsets characterized by hard (crisp) membership values, it is also referred as hard c-means (HCM). Modeling structures with *overlapping clusters* is possible by performing fuzzy c-means (FCM) clustering that generates fuzzy membership values [7, 4]. The fuzzy values, ranging between zero and one, indicate grades of membership. A partition, either crisp or fuzzy, can be succinctly represented as a membership matrix of rows and columns that correspond to clusters and data points respectively.

---

[1] According to Google Scholar, the first occurrence of the term is in 1993 with more frequent use in the recent years.

2   Cluster Validity

Clustering is complicated by several issues related to data representation, similarity measure, clustering scheme (criterion and mechanism) and mainly cluster number. Most clustering algorithms require the user to specify *a priori* the number of clusters [9]. The problem should benefit from domain knowledge to guide the selection of each design element in the context of the end-goal so the results do not fall short of expectations. However, what is known about the ground-truth clustering might be vague; for instance, it is uncertain which topic taxonomy to adopt in a topic-driven clustering application. Even if the representation or the similarity measure, with respect to the end-goal, is universally known the cluster number remains an open challenge of an answer that varies from a given dataset to another. Consider again clustering images by facial expression, a collection might contain images only of smiling faces. Choosing the cluster number too large causes the separation of similar points. On the other hand, specifying a too small cluster number causes the grouping of dissimilar points in the same cluster. Moreover, different types of clustering, based on different models, can be obtained on the same dataset, crisp versus fuzzy for example. Despite these challenging issues, the partitions of interest are those composed of compact and well separated clusters that best meet our expectations i.e. the ground-truth clustering, if any.

When performing cluster analysis, there are different grouping structures to consider. On one hand, there is the *ground-truth* structure, the desired clustering. On the other, the *underlying* structure is established once the pairwise distances or a combination of data representation and a measure are selected. In the absence of an end-goal, identifying the underlying structure becomes the main goal behind clustering the dataset. A clustering algorithm takes as input the pairwise distances or the data representation in some feature space with a measure and fits the underlying structure by the best concrete model possible, a partition. When performing HCM clustering, the produced partition is the Voroni tessellation that minimizes HCM criterion for the specified number of clusters c (centroidal Voroni tessellation). The produced partition is one *candidate* structure that might align with the underlying structure of the dataset. Different candidate partitions, can be obtained in the same setting (representation, measure, distances, algorithm, parameters, etc.) or in a different setting. HCM convergence is determined by the initialization step since it is based on local search optimization; thus it might generate different partitions in the same setting. The clustering function, the criterion, to guide the generation of plausible clustering results should account for compactness and separation. Thus, evaluating a measure of compactness and separation on a given partition is a typical assessment of its clustering quality; such measure becomes the validation function. The partition that is best supported by the underlying structure can be selected from a pool of candidate partitions by means of a *validity index* (validation function). Of interest to this study is the validation of fuzzy partitions, using silhouette-based indices in particular.

3   Silhouette-Based Cluster Validity

The notion of silhouette as a measure of clustering quality was first introduced by Rousseeuw in [18]. The measure is evaluated at each data point and therefore it is the finest in granularity in its assessment compared to other indices. The clustering quality of one cluster is measured by taking the average silhouette over the member points. When it comes to the validation of fuzzy partitions, the construction of silhouettes is not directly applicable since it requires partitions into disjoint clusters. Still, silhouettes can be evaluated on fuzzy partitions by carrying out a defuzzification of the membership matrix. However, this discards the modeling of cluster overlapping, defeating the purpose of performing fuzzy clustering rather than crisp clustering. Moreover, it is possible for different fuzzy partitions on the same dataset to evaluate to the same silhouette values, provided they are converted to the same crisp partition by means of defuzzification. Such fuzzy partitions can be obtained on one dataset by slightly changing the exponent parameter, m, when performing FCM clustering. The following sections present the basic average silhouette index, an extension of the average index to fuzzy partitions, a generalization of the individual silhouettes to fuzzy partitions and then a less rigorous but computationally appealing version of the generalized index which is the main contribution of this study.

3.1   The Average Silhouette Index

To illustrate the construction of silhouettes [18], consider the data point x_j with an HCM clustering of the dataset shown in Fig. 1. Based on the arrangement of the data points in the figure, it can be said that x_j is well-clustered since it is grouped in cluster A with points within *relatively* small proximity. Also x_j is separated from the points that are relatively distant, which are assigned to clusters different than A, namely B and C. Such clustering is satisfactory since it conforms literally to the goal as stated in the very beginning of this article. Again, and from the perspective of x_j, the goal is two-fold: to group x_j with its similar points in the same cluster, and to assign its dissimilar points to the other clusters. Two measures are defined to make an assessment of each sub-goal: a measure of compactness a_j and a measure of separation b_j. The average distance between x_j and the points in cluster A, is assigned to a_j and the average distance between x_j and the points in B is assigned to b_j, provided it is smaller than the average distance obtained over the points in C. Note that a_j is an average over *intra-distances* (within-cluster, within A) and b_j is an average over a subset of *inter-distances* (between-cluster, between A and B).

Since both measures are defined in terms of distances, a good clustering of x_j is assumed only if the compactness distance a_j is much smaller than the separation distance b_j. Hence, a clustering algorithm should minimize a_j and maximize b_j.

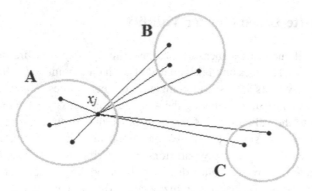

Fig. 1. HCM clustering of a small dataset of nine points

Remark 1. If A denotes the cluster to which x_j has been assigned, several separation distances (averages) can be found with respect to each cluster that is different from A but only the smallest is assigned to b_j. This is critical for b_j to detect the case of separating x_j from its similar points when the cluster number starts to become relatively large.

Remark 2. Only a subset of inter-distances is involved in the computation of b_j, as opposed to the computation of a_j, this is necessary in order to avoid a monotone behavior with the cluster number since more distances become inter-distances.

The silhouette s_j of x_j combines both a_j and b_j in one measure

$$s_j = \frac{b_j - a_j}{\max\{a_j, b_j\}}. \tag{1}$$

Equation (1) evaluates to values in $[-1, +1]$. The difference $(b_j - a_j)$, if positive, indicates good clustering of x_j and poor clustering if negative. Values about zero either emerge from misclustering or the point being in overlapping regions. Dividing the difference by the maximum among the terms facilitates ease of interpretation (strong, weak or no structure) and reflects the fact that similarity and dissimilarity are perceived from distance values *relative* to each other rather than *absolute* magnitudes. For instance, a distance of 100 in one dataset might indicate similarity, or dissimilarity in another dataset. Since partitions produced by clustering algorithms cannot be any better than the underlying structure, silhouettes near +1 are attained only in the presence of a strong underlying structure. Let $U = [u_{ij}]$ be a crisp membership matrix; u_{ij} indicates the membership of x_j in cluster u_i. The number of points and clusters are denoted by n and c respectively. The average silhouette over the whole dataset X is given by

$$Sil(X) = \frac{\sum_{j=1}^{n} s_j}{n}. \tag{2}$$

While the average silhouette over a cluster u_i is given by

$$Sil(u_i) = \frac{\sum_{x_j \in u_i} s_j}{|u_i|} = \frac{\sum_{j=1}^{n} u_{ij} \cdot s_j}{\sum_{j=1}^{n} u_{ij}} ; \quad u_{ij} \in \{0, 1\}. \tag{3}$$

Average silhouettes can be used to judge the behavior of clustering algorithms. Algorithms that tend to produce partitions of zero average silhouettes just carry random clustering of the points. 'Mischievous' algorithms are the ones with output of negative averages, clustering inconsistent with the pairwise distances.

3.2   The Extended Average Silhouette Index

Equation (3) is a reasonable start when considering the extension of the silhouette approach to the fuzzy case i.e. fuzzy partitions. But first, one must consider how to calculate s_j in the case of fuzzy clusters, in particular, how to calculate the quantities a_j and b_j . Crisp clustering, exclusive *hard* cluster assignment of the data points, is necessary for the computation of a_j and b_j. A fuzzy partition $U = [u_{ij}]$; $u_{ij} \in [0,1]$, can be converted into a crisp partition $U' = [u'_{ij}]$; $u'_{ij} \in \{0,1\}$, by means of defuzzification. For example, this can be done by exclusively assigning each point to the cluster that has the highest fuzzy membership value. That is, with respect to the data point x_j

$$u'_{ij} = \begin{cases} 1; & u_{ij} \geq u_{rj} , \ \forall r = 1, \dots, c. \\ 0; & \text{otherwise.} \end{cases} \tag{4}$$

Effectively then, this discards information of cluster overlapping. An attempt to incorporate the fuzzy membership values is given by [6]. The average silhouette given in (2) is redefined as the center of gravity of silhouettes evaluated using U'. Each silhouette is weighted by the difference of the two highest fuzzy membership values of the associated point. More precisely, let $p(j)$ and $q(j)$ denote cluster indices with the two highest fuzzy membership values associated with x_j. Then, the extended average index is defined as

$$eSil(X) = \frac{\sum_{j=1}^{n}(u_{p(j)j} - u_{q(j)j}) \cdot s_j}{\sum_{j=1}^{n}(u_{p(j)j} - u_{q(j)j})}. \tag{5}$$

In [19], it was made clear that the partition coefficient [14], a measure based merely on membership values, is irrelevant to the problem of cluster validity. This result also applies to the weighting terms, $(u_{p(j)j} - u_{q(j)j})$. More on the performance of the extended index is given in the experimental section.

3.3   Generalized Intra-Inter Silhouettes

A natural generalization of the construction of silhouettes to fuzzy partitions is given in [16]. The rationale behind this generalization is based on a *distance view of the clustering problem* and *the problem of cluster validity*. A crisp clustering of the data points is essentially a clustering of the associated pairwise distances into *intra-distances* and *inter-distances*. This clustering can be modeled by associating two scores to each distance, namely intra-score and inter-score that assume only the values 0 and 1. It is straightforward to draw these values from a crisp clustering of the points; a distance becomes intra-distance, hence intra-score of 1 and inter-score of 0, if the endpoints are assigned to the same cluster and inter-distance otherwise. Since compactness is determined by the set of intra-distances while separation is determined by inter-distances, it is desirable to have small intra-distances and large inter-distances, as much as possible. The intuition suggests that a fuzzy clustering of the data points should impose a fuzzy clustering of the distances i.e. each distance is both intra-distance and inter-distance with some grades of membership (scores). Obtaining fuzzy scores seems plausible considering that fuzzy logic and sets generalize their classical counterparts. Indeed, evaluating fuzzy logical connectives on point membership values, treating them as truth values, computes the intra- and inter-scores either crisp or fuzzy as shown in the definition below.

Definition 1. Let $d_{jk} = d(x_j, x_k)$ denote the distance between the data points x_j and x_k ; $1 \leq j \neq k \leq n$. Let u_i ; $1 \leq i \leq c$, denote a cluster and u_{ij} be the membership of x_j to cluster u_i . The intra-score for d_{jk} with respect to cluster u_i is defined as

$$intra_i(d_{jk}) = (u_{ij} \wedge u_{ik}).\tag{6}$$

The inter-score for d_{jk} with respect to clusters u_r and u_s ; $1 \leq r < s \leq c$, is defined as

$$inter_{rs}(d_{jk}) = (u_{rj} \wedge u_{sk}) \vee (u_{sj} \wedge u_{rk}).\tag{7}$$

Remark 3. It is assumed that each d_{jj} has zero intra- and inter-scores. Accordingly, the scores can be represented by ($n \times n$) zero diagonal matrices, namely, $\text{IntraDist}_i = [intra_i(d_{jk})]$ and $\text{InterDist}_{rs} = [inter_{rs}(d_{jk})]$; $1 \leq i \leq c, 1 \leq r < s \leq c$ and $1 \leq j, k \leq n$.

The compactness distance a_j and the separation distance b_j of each data point x_j are now defined as weighted means of the associated pairwise distances using

$$a_j = \min \left\{ \frac{\sum_{k=1}^n intra_i(j,k) \cdot d_{jk}}{\sum_{k=1}^n intra_i(j,k)} \mid \sum_{k=1}^n intra_i(j,k) > 0, \quad 1 \leq i \leq c \right\},\tag{8}$$

$$b_j = \min \left\{ \frac{\sum_{k=1}^n inter_{rs}(j,k) \cdot d_{jk}}{\sum_{k=1}^n inter_{rs}(j,k)} \mid \sum_{k=1}^n inter_{rs}(j,k) > 0, \quad 1 \leq r < s \leq c \right\}.\tag{9}$$

At this point the silhouette s_j of x_j can be computed using (1).

Remark 4. The standard min/max operators are used in (6) and (7) for the fuzzy conjunction and disjunction.

The average generalized silhouette, $gSil$, computed using (2) over all of the data points is a measure of the dataset clustering quality. Those interested in the validation of a particular cluster might consider computing a weighted mean over the whole dataset or just an average over some alpha-cut of the associated fuzzy subset. The performance of the generalized index is tested and compared with the performance of other cluster validity indices in [17].

4   Center-Wise Intra-Inter Silhouettes

An index that involves fewer distances and hence fewer scores, necessary for silhouette construction, is proposed here. Silhouettes are constructed only for each cluster center over center-to-point distances. The scores given by (6) and (7) are directly fetched from the membership matrix. Each center v_i is treated as a core point of the associated fuzzy subset u_i i.e. $u_i(v_i) = 1$. Accordingly, each center-to-point distance has at most one nonzero intra-score; since using (6)

$$intra_r\left(d(v_i, x_j)\right) = u_r(v_i) \wedge u_r(x_j) = 0 \;\wedge u_r(x_j) = 0; \quad r \neq i.$$

For the same reason, v_i is a core point, each center-to-point distance has at most $(c - 1)$ nonzero intra-scores; using (7)

$$\begin{aligned}
inter_{rs}\left(d(v_i, x_j)\right) &= \left(u_r(v_i) \wedge u_s(x_j)\right) \vee \left(u_s(v_i) \wedge u_r(x_j)\right) \\
&= \left(0 \;\wedge u_s(x_j)\right) \vee \left(0 \wedge u_r(x_j)\right) \\
&= 0; \quad r, s \neq i.
\end{aligned}$$

Given the fuzzy $(c \times n)$ membership matrix, the distances associated with v_i can be directly assigned values that appear in the ith row of the matrix as intra-scores while the remaining rows give the inter-scores. Thus, (8) and (9) can be reformulated to take advantage of the fact that (6) and (7) are not needed anymore. With respect to the center of cluster u_i, v_i, a_i and b_i are directly computed by

$$a_i = \frac{\sum_{j=1}^{n} u_{ij} \cdot d(v_i, x_j)}{\sum_{j=1}^{n} u_{ij}}, \tag{10}$$

$$b_i = \min\left\{\frac{\sum_{j=1}^{n} u_{rj} \cdot d(v_i, x_j)}{\sum_{j=1}^{n} u_{rj}} \;|\; 1 \leq r \neq i \leq c\right\}. \tag{11}$$

The silhouette s_i of v_i is then computed by (1).

Example 1. Consider the dataset and its FCM clustering shown in Fig. 2. The membership matrix and center-to-point distance matrix are given in Table 1 and Table 2 respectively. Consider the distance (v_2, x_5), the associated inter-scores are computed using (6) as the following

$$\begin{aligned}
intra_1\left(d(v_2, x_5)\right) &= u_1(v_2) \wedge u_1(x_5) = 0 \wedge 0.50 = 0, \\
intra_2\left(d(v_2, x_5)\right) &= u_2(v_2) \wedge u_2(x_5) = 1 \wedge 0.32 = 0.32, \\
intra_3\left(d(v_2, x_5)\right) &= u_3(v_2) \wedge u_3(x_5) = 0 \wedge 0.18 = 0.
\end{aligned}$$

The inter-scores using (7) are

$$\begin{aligned}
inter_{12}\left(d(v_2, x_5)\right) &= \left(u_1(v_2) \wedge u_2(x_5)\right) \vee \left(u_2(v_2) \wedge u_1(x_5)\right) \\
&= (0 \wedge 0.32) \vee (1 \wedge 0.50) = 0.50, \\
inter_{13}\left(d(v_2, x_5)\right) &= \left(u_1(v_2) \wedge u_3(x_5)\right) \vee \left(u_3(v_2) \wedge u_1(x_5)\right) \\
&= (0 \wedge 0.18) \vee (0 \wedge 0.50) = 0,
\end{aligned}$$

$$inter_{23}\big(d(v_2, x_5)\big) = \big(u_2(v_2) \wedge u_3(x_5)\big) \vee \big(u_3(v_2) \wedge u_2(x_5)\big)$$
$$= (1 \wedge 0.18) \vee (0 \wedge 0.32) = 0.18.$$

Thus $d(v_2, x_5)$ has only one nonzero intra-score that is the membership of x_5 in u_2 (u_{25}) and two nonzero inter-scores (u_{15} and u_{35}). This is the case for all distances associated with v_2. Thus a_2 is the weighted mean with weights obtained directly from u_2 and b_2 is the minimum among weighted means obtained using u_1 and u_3. Similarly, with respect to distances associated with v_3, the values of u_3 become intra-scores while u_1 and u_2 give the inter-scores. It is straightforward to compute the center-wise intra-inter silhouettes. The silhouettes and the involved terms are given in Table 3. An individual center silhouette is a rough assessment of clustering quality of the associated cluster. The average, denoted by $gSilV$, measures the clustering quality of the whole dataset.

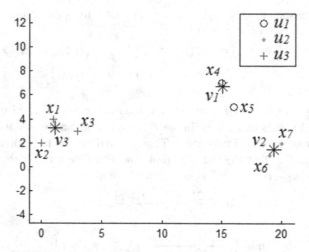

Fig. 2. FCM clustering of 7 points with cluster centers shown as stars

Table 1. Membership matrix of the FCM clustering shown in Fig. 2

U	x_1	x_2	x_3	x_4	x_5	x_6	x_7	v_1	v_2	v_3
u_1	0.17	0.20	0.22	0.74	0.50	0.20	0.23	1	0	0
u_2	0.14	0.18	0.20	0.15	0.32	0.68	0.64	0	1	0
u_3	0.69	0.62	0.58	0.11	0.18	0.12	0.13	0	0	1

Table 2. Center-to-Point distance matrix of FCM clustering shown in Fig. 2

D	x_1	x_2	x_3	x_4	x_5	x_6	x_7
v_1	14.35	15.81	12.65	0.30	1.94	6.92	6.81
v_2	18.52	19.35	16.41	7.03	4.86	0.58	0.84
v_3	0.79	1.68	1.86	14.35	14.96	17.99	18.89

Table 3. Silhouettes of FCM clsuter centers shown in Fig. 2, the values are rounded

v_i	a_i	b_i	s_i
v_1	5.54	min {7.41, 11.99}	0.25
v_2	5.58	min {8.22, 14.86}	0.32
v_3	4.75	min {11.91, 13.87}	0.60

5 Experimental Results and Discussion

This section investigates the performance of the proposed index, compared to other indices in conjunction with FCM clustering. A final remark on the use of silhouette-based indices is to point out that they measure the quality (compactness and separation) of a given clustering with respect to the distance (similarity-dissimilarity) measure used in clustering, see Fig. 3. It is the job of the practitioner to choose a distance measure appropriate for achieving the desired clustering while it is the job of the clustering algorithm to find the optimal clustering with respect to the chosen distance measure. Clustering structures of varying qualities are found considering different settings or even in the same setting (parameter values and constraints) where silhouettes are used to find the best clustering among them with respect to the clustering distance.

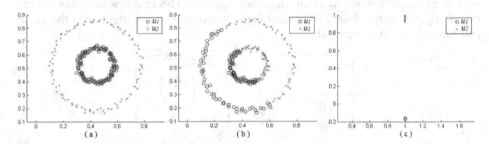

Fig. 3. (a) The desired clustering of points arranged in two circular rings, with average silhouette value of 0.22. (b) HCM clustering of the same dataset in the same space that has a better average silhouette value of 0.32. (c) Same dataset represented by the 1st two eigenvectors of the graph Laplacian obtained using a Gaussian similarity function. Points of the inner ring and the outer ring from (a) appear at the bottom and the top of (c) respectively. The structure in (c) has the best average silhouette value, of 0.99, and hence it is straightforward for HCM to detect the two rings in (a) using the Euclidean distance on the dataset representation given by (c).

Example 2. Different FCM partitions using $c = 2, ... , 9$ and a fuzzifier (FCM exponent) $m = 1.5$ were obtained on a dataset sampled from a mixture of four bivariate Gaussians. Only the clustering into 3 and 4 clusters are shown in Fig. 4.

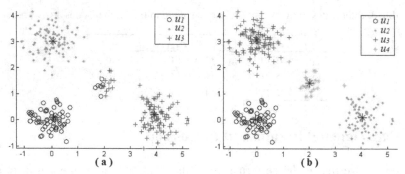

Fig. 4. FCM clustering ($m = 1.5$) of the dataset described in Example 2 into (a) 3 and (b) 4 clusters

The performance of the basic average silhouette Sil, the extended average silhouette $eSil$, the average generalized silhouette $gSil$ and the average generalized center-wise silhouette $gSilV$ is shown in Fig. 5. Despite the fact that the basic index and the extended index are based on the same silhouettes, computed from the defuzzified partition (U'), they disagree in how they score the clustering into $c = 3$ and $c = 4$ clusters. The extended index scores $c = 3$ higher than $c = 4$ due to the incorporated weights. The points in the middle region (Fig. 4a) do not have significant membership to any of the three clusters, hence the low weights and the insignificant contribution to the final average silhouette. Clearly, the extended index does not provide a meaningful assessment of clustering quality. The center-wise index scores $c = 4$ slightly higher than $c = 3$, showing a performance consistent with the basic and generalized point-wise indices.

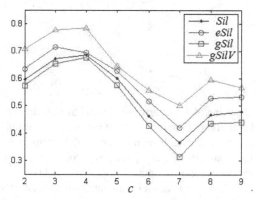

Fig. 5. The performance of silhouette-based indices, number of clusters versus index score, on FCM output described in Example 2

Example 3. FCM partitions into $c = 2, ..., 9$ clusters on the same dataset from the previous example were obtained but using a fuzzifier $m = 3$. Again, only the clustering into 3 and 4 clusters are shown in Fig. 6.

FCM, by specifying higher values for m, is expected to produce 'fuzzier' partitions, justified by the limiting behavior of the FCM model as m approaches infinity [14], mainly,

$$\lim_{m \to \infty} u_{ij} = \frac{1}{c} \; ; \quad 1 \leq i \leq c, \quad 1 \leq j \leq n.$$

Moreover, this limit is valid regardless the underlying structure latent in the dataset. In other words, partitions near the fuzziest partition are obtained on any dataset for any number of clusters by specifying too high values for m. The generalized silhouette-based indices detect well such increase in the amount of overlapping, caused by specifying a higher value for m, and evaluate the corresponding clustering to lower scores. This explains why the curves of both $gSil$ and $gSilV$ in Fig. 7 lie entirely below their counterparts in Fig. 5 while Sil and $eSil$ barely show any change. Such change in performance is consistent with the fact that increasing m allows FCM to carry out some sort of loose computation of the membership values, not fully accounting for the *geometric similarities-dissimilarities* among the points. In turn, FCM produces partitions of increased amount of overlapping (fuzziness). According to $gSil$ (Fig. 7) the clustering into $c = 3$ is slightly better than $c = 4$ justified by noting that the setting tolerates more overlapping in the dataset. Similar evaluation is shown by the rough, less accurate, index $gSilV$. It is apparent that Sil just ignores the effect of increasing m due to defuzzifcation, while, $eSil$ is not reliable in its assessment since it treats points unequally by ignoring those of bad clustering quality.

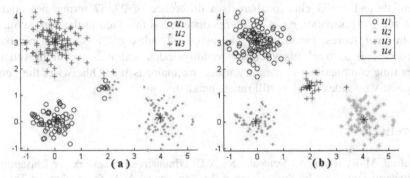

Fig. 6. FCM clustering ($m = 3$) of the same dataset from Example 2 into (a) 3 and (b) 4 clusters

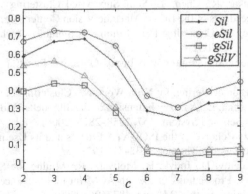

Fig. 7. The performance of silhouette-based indices, number of clusters versus index score, on FCM output described in Example 3

Table 4. Comparing number of distances, scores and silhouette terms involved in $gSil$ and $gSilV$, $|X|$ denotes the size of set X

Index	|distances|	|intra-scores|	|inter-scores|	|s|
$gSil$	$\binom{n}{2}$	$\binom{n}{2} \cdot c$	$\binom{n}{2} \cdot \binom{c}{2}$	n
$gSilV$	$c \cdot n$	$c \cdot n$	$c \cdot n \cdot (c-1)$	c

6 Conclusions

The silhouette approach provides a powerful and effective device to address the issue of cluster validity. Its performance comes from the fact that it is computed for each data point. At the same time, for the same reason its complexity tends to be rather large, as it considers all the pairwise distances in a dataset. It is then quite natural to seek to improve its complexity by defining silhouettes over cluster centers. This paper considers the generalized center-wise silhouettes. This approach requires fewer terms compared to the point-wise silhouettes as shown in Table 4. For instance, for $n = 100$ points and $c = 3$ clusters there is a difference of 23547 terms, not counting compactness (a), separation (b) or intermediate terms involved in the computation of the intra-inter scores. Nevertheless, the proposed index $gSilV$ shows a reasonable performance as a 'true' fuzzy cluster validity index and it is useful in situations wheres time complexity and memory usage are major issues. Otherwise, the generalized point-wise index $gSil$ is still recommended.

References

1. Balcan, M.-F., Blum, A., Vempala, S.: A Discriminative Framework for Clustering via Similarity Functions. In: Proceedings of the 40th Annual ACM Symposium on Theory of Computing, STOC (2008)
2. Batra, D., Sukthankar, R., Chen, T.: Semi-Supervised Clustering via Learnt Codeword Distances. In: Proceedings of the British Machine Vision Conference, BMVC (2008)
3. Berkhin, P.: Survey of Clustering Data Mining Techniques. Technical report, Accrue Software, San Jose, CA (2002)
4. Bezdek, J.C.: Pattern Recognition with Fuzzy Objective Function Algorithms. Plenum Press (1981)
5. Blum, A.: Thoughts on clustering. In: NIPS Workshop Clustering: Science or Art (2009)
6. Campello, R., Hruschka, E.: A Fuzzy Extension of the Silhouette Width Criterion for Cluster Analysis. Fuzzy Sets and Systems 157, 2858–2875 (2006)
7. Dunn, J.C.: A Fuzzy Relative of the ISODATA Process and its Use in Detecting Compact Well-Separated Clusters. J. Cybernet. 3, 32–57 (1973)
8. Gavrilov, M., Anguelov, D., Indyk, P., Motwani, R.: Mining the Stock Market: Which Measure is Best? In: Proceedings of the 6th ACM Int'l. Conference on Knowledge Discovery and Data Mining, Boston, MA, pp. 487–496 (2000)
9. Halkidi, M., Batistakis, Y., Vazirgiannis, M.: On Clustering Validation Techniques. Journal of Intelligent Information Systems 17, 107–145 (2001)

10. Jain, A., Murty, M., Flynn, P.: Data Clustering: A Review. ACM Computing Surveys 31(3), 264–323 (1999)
11. Kleinberg, J.: An Impossibility Theorem for Clustering. Proceedings of Advances in Neural Information Processing Systems 15, 463–470 (2002)
12. Lloyd, S.: Least Squares Quantization in PCM. IEEE Transactions on Information Theory 28(2), 129–137 (1982)
13. MacQueen, J.: Some Methods for Classification and Analysis of Multivariate Observations. In: Proceedings of the 5th Berkeley Symposium on Mathematics. Statistics and Probability, Berkeley, CA, vol. 2, pp. 281–297 (1967)
14. Pal, N., Bezdek, J.: On Cluster Validity for the Fuzzy c-Means Model. IEEE Transactions on Fuzzy Systems 3(3), 370–379 (1995)
15. Pampalk, E., Goebl, W., Widmer, G.: Visualizing Changes in the Inherent Structure of Data for Exploratory Feature Selection. In: Proceedings of the Ninth ACM SIGKDD International Conference on Knowledge Discovery and Data Mining, pp. 157–166. ACM, Washington, DC (2003)
16. Rawashdeh, M., Ralescu, A.: Crisp and Fuzzy Cluster Validity: Generalized Intra-Inter Silhouette Index. To appear in the Proceedings of the 2012 NAFIPS Annual Meeting (2010)
17. Rawashdeh, M., Ralescu, A.: Fuzzy Cluster Validity with Generalized Silhouettes. In: Proceedings of the 23rd Annual Midwest Artificial Intelligence and Cognitive Science Conference, Cincinnati, Ohio, April 11-18. CEUR-WS.org (2012),
    http://CEUR-WS.org/Vol-841/submission_33.pdf
18. Rousseeuw, P.J.: Silhouettes: A Graphical Aid to the Interpretation and Validation of Cluster Analysis. Computational and Applied Mathematics 20, 53–65 (1987)
19. Trauwaert, E.: On the Meaning of Dunn's Partition Coefficient for Fuzzy Clusters. Fuzzy Sets and Systems 25, 217–242 (1988)
20. Xu, R., Wunsch, D.I.I.: Survey of Clustering Algorithms. IEEE Transactions on Neural Networks 16(3), 645–678 (2005)
21. Zhao, Y., Karypis, G.: Criterion Functions for Document Clustering: Experiments and Analysis. Technical Report, CS Dept., Univ. of Minnesota (2001)

Clustering Sets of Objects Using Concepts-Objects Bipartite Graphs

Emmanuel Navarro[1], Henri Prade[1], and Bruno Gaume[2]

[1] IRIT, Université de Toulouse III,
118 Route de Narbonne; 31062 Toulouse Cedex 9, France
{navarro,prade}@irit.fr
[2] CLLE-ERSS, Université de Toulouse II,
5, allées Antonio Machado; 31058 Toulouse Cedex 9, France
gaume@univ-tlse2.fr

Abstract. In this paper we deal with data stated under the form of a binary relation between objects and properties. We propose an approach for clustering the objects and labeling them with characteristic subsets of properties. The approach is based on a parallel between formal concept analysis and graph clustering. The problem is made tricky due to the fact that generally there is no partitioning of the objects that can be associated with a partitioning of properties. Indeed a relevant partition of objects may exist, whereas it is not the case for properties. In order to obtain a conceptual clustering of the objects, we work with a bipartite graph relating objects with formal concepts. Experiments on artificial benchmarks and real examples show the effectiveness of the method, more particularly the fact that the results remain stable when an increasing number of properties are shared between objects of different clusters.

Keywords: formal concept analysis, bipartite graph, graph clustering.

1 Introduction

For making sense of complex data, one may need to cluster them, and if possible, to provide labels for the clusters. In this paper we are interested in data that take the form of a binary relation between a set of objects and a set of properties. Several families of approaches exist for such a task: one may use bi-clustering (or two-mode clustering) approaches [3], formal concept analysis (FCA for short) methods, and hybridization of them.

In previous work, the authors have emphasized the parallelism between FCA operators and two views of graph clustering, referring respectively to the search for maximal bi-cliques and to the search of maximal connected components [12]. Moreover, since the number of formal concepts is usually very large, we have proposed a preliminary approach for providing an approximate conceptual view of data by taking inspiration from the recent literature on graph clustering (often called community detection problem). More precisely, we have proposed a two-step procedure: i) random walks are used for providing an approximate and

E. Hüllermeier et al. (Eds.): SUM 2012, LNAI 7520, pp. 420–432, 2012.

more robust view of the formal context leading to a smaller number of formal concepts, ii) these concepts are then fused when they have a sufficient overlap [13]. However, this two-step method requires the tuning of threshold parameters.

In this paper we propose a new approach based on bipartite graphs between objects and concepts, rather than on bipartite graphs between objects and properties, as it was the case in the step i) of the previous method. Moreover no threshold are any longer needed. Our goal is now to look for a partition of the set of objects, while properties may remain shared between different clusters of objects. The paper is organized as follows. After a background on FCA and its bipartite graph counterpart (Section 2), we present the new approach in Section 3, and suggest a way of labeling the clusters of objects in Section 3.3. Experiments are reported in Section 4 that show the effectiveness of the method on artificial benchmarks and on a real dataset. Comparison with related works (Section 5) and concluding remarks (Section 6) end the paper.

2   Background: From Formal Concept Analysis to Clustering

In this section we first recall the standard notion of FCA, as well as the notion of independent sub-contexts, and then give their counterpart in the setting of bipartite graphs where we interpret them in clustering terms.

2.1   Formal Concepts and Independent Subcontexts

Let R be a *binary relation* between a set $\mathbf{O}$ of objects and a set $\mathbf{P}$ of Boolean properties. We note $\mathcal{R} = (\mathbf{O}, \mathbf{P}, R)$ the tuple formed by these objects and properties sets and the binary relation. It is called a *formal context* [11]. The notation $(x, y) \in R$ means that object x has property y. Let $R(x) = \{y \in \mathbf{P} | (x, y) \in R\}$ be the set of properties of object x. Similarly, $R^{-1}(y) = \{x \in \mathbf{O} | (x, y) \in R\}$ is the set of objects having property y.

Formal concept analysis [11] defines two set operators, here denoted $(.)^{\Delta}$ and $(.)^{-1\Delta}$, called *intent* and *extent* operators respectively, s.t. $\forall Y \subseteq \mathbf{P}$ and $\forall X \subseteq \mathbf{O}$:

$$X^{\Delta} = \{y \in \mathbf{P} | \forall x \in X, (x, y) \in R\} \tag{1}$$
$$Y^{-1\Delta} = \{x \in \mathbf{O} | \forall y \in Y, (x, y) \in R\} \tag{2}$$

X^{Δ} is the set of properties possessed by all objects in X. $Y^{-1\Delta}$ is the set of objects having all properties in Y. These two operators induce an antitone Galois connection between $2^{\mathbf{O}}$ and $2^{\mathbf{P}}$. This means that the following property holds

$$X \subseteq Y^{-1\Delta} \Leftrightarrow Y \subseteq X^{\Delta}.$$

A pair such that $X^{\Delta} = Y$ and $Y^{-1\Delta} = X$ is called a *formal concept*[11]. X is its extent and Y its intent. In other words, a formal concept is a pair (X, Y)

such that X is the set of objects having all properties in Y and Y is the set of properties shared by all objects in X. It can be shown that formal concepts correspond to *maximal pairs* (X, Y) such that

$$X \times Y \subseteq R.$$

A recent parallel between formal concept analysis and possibility theory[8] has led to emphasize the interest of an other remarkable set operator $(.)^{\Pi}$, and their two respective duals. The new operator and the already defined intent operator can be written as follows, $\forall X \subset \mathbf{O}$:

$$X^{\Pi} = \{y \in \mathbf{P} | R^{-1}(y) \cap X \neq \emptyset\} \tag{3}$$
$$X^{\Delta} = \{y \in \mathbf{P} | R^{-1}(y) \supseteq X\} \tag{4}$$

Note that (4) is equivalent to the definition of operator $(.)^{\Delta}$ in (1). X^{Π} is the set of properties that are possessed by at least *one* object in X. X^{Δ} is the set of properties shared by all objects in X.

Operators $(.)^{-1\Pi}$, $(.)^{-1\Delta}$ are defined similarly on a set Y of properties by substituting R^{-1} to R and by inverting $\mathbf{O}$ and $\mathbf{P}$. $(Y)^{-1\Pi}$, $(Y)^{-1\Delta}$ are respectively, the set of objects having at least one property in Y and the set of objects that have all the properties in Y.

This new operator lead to consider a new connection[9] that corresponds to pairs (X, Y) such that $X^{\Pi} = Y$ and $Y^{-1\Pi} = X$ (while $(.)^{\Delta}$ leads to formal concepts, as already said). Pairs (X, Y) such that $X^{\Pi} = Y$ and $Y^{-1\Pi} = X$ do not define formal concept, but *independent sub-contexts*. Indeed, it has been recently shown[9] that pairs (X, Y) of sets exchanged through the new connection operator, are subsets such that

$$(X \times Y) \cup (\overline{X} \times \overline{Y}) \supseteq R,$$

just as formal concepts correspond to maximal pairs (X, Y) such that

$$X \times Y \subseteq R.$$

In Figure 1, two examples of formal concepts are the pairs $(\{a1, a2, a3, a4, b1\}, \{2, 7\})$ and $(\{c1, c2\}, \{4, 5, 6, 8\})$. On the other hand, if we forget the fact that the object $a2$ verify the property 10, the pairs $(\{a1, a2, a3, a4, b1, b2, b3, b4, c1, c2\}, \{1, 2, 3, 4, 5, 6, 7, 8\})$ and $(\{d1, d2\}, \{9, 10, 11\})$ are two independent sub-contexts.

Thus, in the setting of formal concept analysis, by means of two companion connections, two key aspects of the idea of clustering are at work. On the one hand, independent sub-contexts are characterized, and on the other hand inside each sub-context, formal concepts (X, Y) are identified where *each* pair (x, y) such that $x \in X, y \in Y$ are in relation (while *no* pair (x, y) such that $x \in \overline{X'}, y \in Y'$ or $x \in X', y \in \overline{Y'}$ are in relation if (X', Y') and $(\overline{X'}, \overline{Y'})$ are two independent subcontexts). In particular, two formal concepts belonging to two different sub-contexts are clearly well-separated. The relation with clustering is made still clearer in the next sub-section by providing a bipartite graph reading of FCA.

	1	2	3	4	5	6	7	8	9	10	11
a1	×	×	×				×	×			
a2		×	×				×	×		×	
a3	×	×					×	×			
a4	×	×	×				×				
b1			×	×	×	×		×			
b2		×	×		×			×			
b3			×	×	×						
b4		×	×	×	×						
c1					×	×	×		×		
c2				×	×	×	×	×			
d1								×	×	×	
d2								×	×		

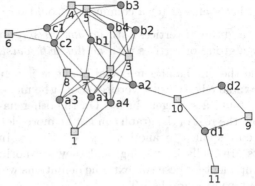

Fig. 1. A formal context R and the corresponding bipartite graph

2.2   Formal Concept Analysis, Bipartite Graphs and Clustering

For every formal context $\mathcal{R} = (\mathbf{O}, \mathbf{P}, R)$, one can build an undirected bi-graph $\mathcal{G} = (V_o, V_p, E)$ s.t. there is a direct correspondence between: the set of objects $\mathbf{O}$ and a set V_o of "o-vertices", the set of properties $\mathbf{P}$ and a set V_p of "p-vertices", and between the binary relation R and a set of edges E. In other words, there is one o-vertex for each object, one p-vertex for each property, and one edge between an o-vertex and a p-vertex if and only if the corresponding object possesses the corresponding property (according to R).

The operators $(.)^{\Pi}$ and $(.)^{\Delta}$ can then be rewritten in the following way:

$$X^{\Pi} = \cup_{x \in X} \Gamma(x) \qquad (5)$$
$$X^{\Delta} = \cap_{x \in X} \Gamma(x) \qquad (6)$$

where $\Gamma(x)$ denotes the set of neighbors of the vertex x. These notations are interesting since only the neighborhood of vertices of X is involved. It permits to immediately understand operators $(.)^{\Pi}$ and $(.)^{\Delta}$ in terms of neighborhood in the bi-graph: X^{Π} is the union of neighbors of vertices of X whereas X^{Δ} is the intersection of these neighbors. The same expressions apply to $(.)^{-1\Pi}$ and $(.)^{-1\Delta}$, changing X by Y (and x by y).

The connections induced by $(.)^{\Delta}$ and $(.)^{\Pi}$ can also be understood in the graph setting framework: the first connection corresponds to maximal bi-cliques whereas the second one two maximal connected components [12]. Indeed on the bi-graph $\mathcal{G} = (V_o, V_p, E)$, with $X \subseteq V_o$ and $Y \subseteq V_p$, we have:

Proposition 1. $X = Y^{-1\Delta}$ and $Y = X^{\Delta}$, iff $X \cup Y$ is a maximal bi-clique.

Proposition 2. For a pair (X, Y) the two following propositions are equivalent:

1. $X = Y^{-1\Pi}$ and $Y = X^{\Pi}$ and there is no strict subset $X' \subset X$ and $Y' \subset Y$ such that $X' = Y'^{-1\Pi}, Y' = X'^{\Pi}$.
2. $X \cup Y$ is a maximal connected component (which counts at least 2 vertices).

It is worth noticing that the two connections correspond to extreme definitions of what a cluster (or a community) could be:

1. a group of vertices with *no link missing inside*.
2. a group of vertices with *no link with outside*.

One the one hand a maximal bi-clique is a maximal subset of vertices with a maximal edge density. Vertices cannot be moved closer, and in that sense one can not build a stronger cluster. On the other hand, a set of vertices disconnected from the rest of the graph can not be more clearly separated from other vertices. It corresponds to another type of cluster. In fact, only the smallest of such sets are really interesting, and they are nothing else than maximal connected components. These two extreme definitions were already pointed out for clusters in unipartite graphs [19].

3   Looking for Meaningful Clusters of Objects

In this section we motivate the need for a new clustering procedure which enables us to obtain meaningful clusters of objects, even if the objects in different clusters share many properties.

3.1   Preliminary Discussion

As said in the introduction, our primary purpose is to cluster the set of considered objects into distinct subsets on the basis of their properties. However the application of a graph clustering method on the bipartite graph (associated to the formal context) generally fails. It is due to the fact that the method when tentatively gathering objects in separate clusters, often fails to do it since objects in different potential clusters usually share many common properties. In other words, bipartite graph clustering looks for a partition of the graph vertices. When applied to the object-properties graph it puts into correspondence subsets of objects with subsets of properties, i.e. they look for a partition of objects and a partition of properties such that each set of objects is in correspondence with a set of properties. This is illustrated on the Figure 2(a) for the formal context example of Figure 1. As can be seen, the method isolates the cluster $\{d1, d2\}$, but fails to discriminate more, leaving the rest of the objects in the same cluster. Indeed, it will have been desirable to separate these remaining objects in 3 clusters, namely $\{a1, a2, a3, a4\}$, $\{b1, b2, b3, b4\}$ and $\{c1, c2\}$, as revealed by a careful examination of the formal context of Figure 1.

Besides, it can be checked that there are 30 formal concepts in the formal context of Figure 1. Note that it is usually observed that FCA returns a rather large number of formal concepts, in particular with noisy data or when exceptions are present. Moreover there is no immediate way of using the lattice of concepts for building a partition of the objects. However, as can been seen in Figure 1 the 3 subsets of objects that the method have failed to separate (Figure 2(a)) form the "approximate" concepts ($\{a1, a2, a3, a4\}, \{1, 2, 3, 7, 8\}$),

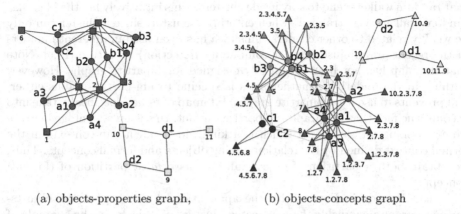

(a) objects-properties graph,   (b) objects-concepts graph

Fig. 2. For the relation given in Figure 1, results of Infomap [18] graph clustering method either on the basic objects-properties graph (a) or on the objects-concepts graph (b). On the two graphs circles are objects, for (a) squares are properties and for (b) triangles are concepts.

$(\{b1, b2, b3, b4\}, \{2, 3, 4, 5\})$ and $(\{c1, c2\}, \{4, 5, 6, 7, 8\})$. By approximate concepts [9], we mean that, up to a few missing crosses, we have large formal concepts (X, Y) (i.e., they correspond in Figure 1 to "approximate" $X \times Y$ rectangles). This suggests to investigate a "conceptual" clustering of the objects by dealing with the objects-formal concepts bipartite graph.

3.2   Clustering Objects-Concepts Bipartite Graphs

We now describe the method we propose more precisely. First, a preliminary step consists in building all the formal concepts associated to the objects-properties graph, using a formal concept extraction method, e.g. [10].

Second, a bipartite graph between objects and concepts is built such that each object $o \in \mathbf{O}$ is connected to a concept (X, Y) iff $o \in X$, then the corresponding edge is weighted by $w = |Y|$ the number of properties of the corresponding concept. This weighting is introduced in order to favor "large" concepts, which are expected to be more "meaningful". Indeed concepts with a small number of properties are likely to connect "too many" objects. Note that the top and bottom concepts are ignored, if they contain zero objects or zero properties.

The vertices of this bipartite graph are then partitioned by using the graph clustering Infomap method [18]. Infomap is recognized as one of the best methods of graph clustering [16]. It consists in searching for the clusters that best compress the description length of the trajectory of a random walk through the whole graph. This trajectory is described in a two-level way in function of the clusters: when the walker enters a cluster, the name of the cluster is used, but then only the name of the current vertex inside the cluster is retained. In this way, short length names may be used for naming different vertices that are in different clusters leading to shorter trajectory descriptions, at the condition that clusters are such that random walkers tend to stay inside clusters. This intuitively fits the idea that random walkers are "trapped" when entering a cluster, since a cluster can only be weakly related to other clusters. This idea has been used in different manners in the recent graph clustering (or community detection) literature [19,6]. Note that Infomap has not been specifically designed for bipartite graphs. However, nothing in the underlying mathematics is specific to uni-partite graphs either, and prevents to use it for bipartite graphs. Infomap does not specifically take into account the fact that the graph is bipartite. In fact, this is an advantage because we are looking for something which is a kind independent sub-contexts in the formal context defined by the relation linking objects and formal concepts. Thus, we obtain both a partition of objects and an *associated* partition of (formal) concepts.

As can be seen in Figure 2(b), the application of Infomap on the objects-concepts graph now yields the 4 expected clusters of objects, in the example of Figure 1.

3.3   Labeling Clusters

In order to label each cluster of objects with a subset of relevant properties, we use the following simple method.

For each cluster of objects we look for two particular concepts: namely the concept (X^*, Y^*) which is associated with the largest subset of objects (of the corresponding objects cluster) and the concept (X_*, Y_*) which is associated with the smallest superset of objects. In formal terms, let $\mathcal{C} = (X, S)$ be a cluster of objects X with the associated set S of concepts, i.e. $S = \{(X_1', Y_1'), (X_2', Y_2'), ...\}$. Let be T the set of all formal concepts. Then we compute the two noticeable formal concepts that are defined as follows:

$$(X^*, Y^*) \in T \quad \text{s.t.} \quad \begin{cases} X^* \supseteq X \\ \nexists (X_j, Y_j) \in T \quad \text{s.t.} \quad X^* \supset X_j \supseteq X \end{cases} \tag{7}$$

$$(X_*, Y_*) \in T \quad \text{s.t.} \quad \begin{cases} X_* \subseteq X \\ \nexists (X_j, Y_j) \in T \quad \text{s.t.} \quad X_* \subset X_j \subseteq X \end{cases} \tag{8}$$

One can check that $X_* \subseteq X^*$, and $Y_* \supseteq Y^*$. Therefore the two sets of properties Y^* and Y_* can be used for labeling the cluster. Note that we are sure that all the properties of Y^* are shared by *all* the objects of the cluster.

4   Experiments and Discussions

For evaluating (and illustrating) the proposed procedure we consider two kinds of benchmark, one generated artificially and a real example available in the literature.

4.1   Evaluation on Artificial Benchmarks

In order to build a benchmark for object clustering procedure, we built formal contexts in the following way. We take n groups of k objects, each group is associated with m_{own} properties that only objects of this group may satisfy, and with m_{shared} properties that may be verified by objects of s other groups. For each group of objects, an object of the group satisfies each property in the group with a probability μ. An example of such a context is given in Table 1.

Table 1. An example of formal context artificially generated by the procedure described in Section 4.1, with $n = 3$, $k = 3$, $m_{own} = 2$, $m_{shared} = 4$, $s = 1$, $\mu = 0.8$

	A0	A1	B0	B1	C0	C1	AB0	AB1	BC0	BC1	CA0	CA1
a0	×						×	×			×	×
a1		×					×	×				×
a2	×	×					×	×			×	
b0				×				×	×	×		
b1			×	×			×	×	×	×		
b2				×					×	×		
c0						×			×	×	×	×
c1					×	×			×	×	×	×
c2					×	×			×	×		

The Figures 3(a) and 3(b) present the results of the clustering on the objects-properties graph (the curve O ↔ P, in blue) and on the objects-concepts graph (the curve O ↔ C, in red). To evaluate the accuracy of our algorithm against the correct partition of objects we use the normalised mutual information (NMI). A value of 0 indicate that the two partitions are totally dissimilar, whereas a value of 1 indicate that the two partitions are identical. This is a commonly use measure in graph clustering literature [5]. Each point indicated the average value obtained on 50 realizations, the standard deviation is indicated by the vertical error bar on each point. As shown in Figure 3(a), the results remain stable with our approach when an increasing number of properties are shared between objects in different clusters, while it is not the case if we work with the objects-properties graph only.

4.2   The UCI Zoo Dataset

The UCI Zoo dataset describes 101 animals on 16 Boolean-valued attributes and one numerical attribute (the number of legs). We transformed this numerical attribute in 7 Boolean attributes (no legs, one leg, two legs, ...). For each animal the type is indicated, there are 7 types of animals: mammal, bird, reptile, fishes, amphibians, insects, invertebrates. This data set can be downloaded from the UCI Machine Learning Repository[1].

---

[1] http://archive.ics.uci.edu/ml/datasets/Zoo

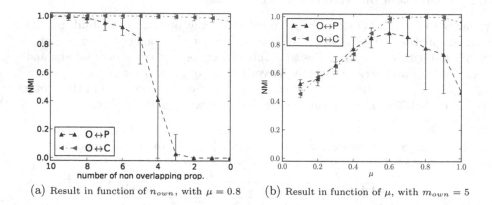

(a) Result in function of n_{own}, with $\mu = 0.8$ (b) Result in function of μ, with $m_{own} = 5$

Fig. 3. Normalised Mutual Information (NMI) [5] value of Infomap clustering method on objects-properties graph (curve $O \leftrightarrow P$ in blue) and on objects-concepts graph (curve $O \leftrightarrow C$ in red). Benchmark contexts are built with the following parameters: $n = 5, k = 10, m_{shared} = 6, s = 2$

Table 5 shows the result of the clustering over the objects-properties graph, while Table 5 shows the results on the objects-concepts graph. One can see that the clustering method fails on the objects-properties graph, whereas the partition given on the objects-concepts graph retrieves the types of animals almost exactly. Moreover, the labels Y^* and Y_* coincide in several cases.

5   Related Works

We focus our discussion on the literature either related to bipartite graph clustering or to clustering of objects according to a binary relation between objects and properties. The first group makes explicit reference to the graph representation whereas the second one doesn't.

Let us start with representatives of the vast amount of literature in the second group. In [14], the authors use a measure of quality for clustering objects based on Kullback-Leibler entropy which is optimized by means of a genetic algorithm. However, such a black box method does not provide a means for labeling the clusters. In [1] A FCA-based method is proposed, where potentially interesting concepts are selected and then the underlying formal context is revised. It enables the extraction of new descriptors which allows for the reuse of concepts in an incremental way. This leads to a method taking inspiration from inverse resolution in inductive logic programming which enables the extraction of clusters with associated properties, in the Zoo dataset example. Note that there exist a lot of methods that look for bi-clustering (also named co-clustering, or two-mode clustering) which consist in finding a partition of objects that is in direct correspondence with a partition of properties, see [3] for a state of the art.

Table 2. Results of the clustering of the objects-concepts graph ($NMI = 0.81$)

$Y^* = \{backbone, breathes, hair, milk, toothed\}$ $Y_* = \{backbone, breathes, hair, milk, tail, toothed\}$
Mammals: aardvark, lynx, leopard, bear, boar, puma, lion, cheetah, raccoon, mink, pussycat, mongoose, wolf, polecat, antelope, calf, elephant, oryx, goat, deer, reindeer, buffalo, pony, giraffe, vole, mole, hare, cavy, hamster, opossum, sealion, girl, wallaby, gorilla, fruitbat, squirrel, vampire
$Y^* = \{0legs\}$ $Y_* = \{0legs, aquatic, eggs\}$
Fishes: stingray, pike, piranha, catfish, herring, dogfish, tuna, chub, bass, sole, seahorse, carp, haddock   *Invertebrates:* clam, seawasp   *Reptiles:* seasnake
$Y^* = \{2legs, backbone, breathes, eggs, feathers, tail\}$ $Y_* = \{2legs, backbone, breathes, eggs, feathers, predator, tail\}$
Birds: flamingo, gull, skimmer, sparrow, wren, skua, hawk, crow, duck, vulture, lark, swan, pheasant, kiwi, rhea, ostrich, penguin
$Y^* = Y_* = \{4legs, eggs\}$
Amphibians: newt, frog2, frog1, toad   *Reptiles:* tortoise, tuatara   *Mammals:* platypus   *Invertebrates:* crab
$Y^* = Y_* = \{0legs, aquatic, backbone, breathes, catsize, fins, milk,$ $predator, toothed\}$
Mammals: porpoise, dolphin, seal
$Y^* = Y_* = \{6legs, breathes, eggs\}$
Insects: flea, ladybird, moth, gnat, wasp, honeybee, housefly, termite
$Y^* = Y_* = \{0legs, breathes, eggs\}$
Reptiles: slowworm, pitviper   *Invertebrates:* worm, slug
$Y^* = Y_* = \{2legs, airborne, backbone, breathes, domestic, eggs, feathers, tail\}$
Birds: chicken, parakeet, dove
$Y^* = \{aquatic, eggs, predator\}$ $Y_* = \{6legs, aquatic, eggs, predator\}$
Invertebrates: crayfish, starfish, lobster
$Y^* = Y_* = \{8legs, breathes, predator, tail, venomous\}$
Invertebrates: scorpion
$Y^* = Y_* = \{8legs, aquatic, catsize, eggs, predator\}$
Invertebrates: octopus

Table 3. Results of the clustering of the objects-properties graph ($NMI = 0.02$)

Mammals:	aardvark, antelope, bear, boar, buffalo, calf, cavy, cheetah, deer, dolphin, elephant, fruitbat, giraffe, girl, goat, gorilla, hamster, hare, leopard, lion, lynx, mink, mole, mongoose, opossum, oryx, platypus, polecat, pony, porpoise, puma, pussycat, raccoon, reindeer, seal, sealion, squirrel, vampire, vole, wallaby, wolf
Birds:	chicken, crow, dove, duck, flamingo, gull, hawk, kiwi, lark, ostrich, parakeet, penguin, pheasant, rhea, skimmer, skua, sparrow, swan, vulture, wren
Fishes:	bass, carp, catfish, chub, dogfish, haddock, herring, pike, piranha, seahorse, sole, stingray, tuna
Invertebrates:	clam, crab, crayfish, lobster, octopus, scorpion, seawasp, slug, worm
Insects:	flea, gnat, honeybee, housefly, ladybird, moth, termite, wasp
Reptiles:	pitviper, seasnake, slowworm, tortoise, tuatara
Amphibians:	frog1, frog2, newt, toad
Invertebrates:	starfish

Another family of methods came from the literature concerning bipartite graph clustering. In [7] a spectral method is used for finding a partition of a bipartite graph that minimized the cut size, i.e. the number of edges running between clusters. The main drawback of such approaches is that the number of clusters has to be known in advance, and the methods tend to create clusters having almost the same size, which rarely makes sense with real data sets.

In [2], the authors proposed an adaptation of Newman modularity [4] to the case of bipartite graph. The Newman modularity is a measure of quality of a partition of a graph vertices, a relevant partitioning is usually found by optimizing this quality measure using various heuristics.

Most of these methods lead to a partition of objects and properties, and therefore do not manage to partition objects when properties are shared between many clusters. Note that this issue has been partially addressed in [17], where the authors proposed a measure of quality (inspired from the Newman modularity) of a bipartite graph clustering that allows the fact that there is no direct correspondence between properties cluster and object clusters.

Finally, note that in [15] the authors propose an approach that consists in partitioning a bipartite graph between objects and hypercliques (which can be understood as a set of properties that are satisfied by almost the same objects). This method is in a spirit similar to the method we proposed. However they use a partitioning method that amounts to minimizing a *cut* measure, which suffers from the main drawbacks as the one used in [7].

6   Conclusion

Starting with a binary relation linking objects and properties, formal concept analysis enables us to obtain formal concepts on the one hand, but also independent sub-contexts on the other hand, as recalled at the beginning of this paper.

Then, the independent sub-contexts may be viewed as separating clusters of objects and properties, inside which formal concepts identify homogeneous families of objects. But due to noisy data, due to the existence of exceptions, and more generally due to the fact that the same property may be shared by a variety of objects, it is difficult to cluster a set of objects in a meaningful way directly on a formal context. In the paper, we have proposed to handle the problem in a new formal context where the properties are replaced the formal concept obtained from the initial formal context. Then we have shown on artificial benchmarks and on a real data set that looking for clusters in this higher level formal context makes possible to obtain clusters that can then be interpreted in terms of two nested sets of properties where the smallest one contains only properties that are shared by all the objects in the cluster. As can be seen on the real data set, the two nested sets of properties may be equal, and then a perfect characterization of the cluster is obtained. More experiments would be necessary to evaluate possible variants of this general approach.

References

1. Bain, M.: Structured Features from Concept Lattices for Unsupervised Learning and Classification. In: McKay, B., Slaney, J.K. (eds.) Canadian AI 2002. LNCS (LNAI), vol. 2557, pp. 557–568. Springer, Heidelberg (2002)
2. Barber, M.J.: Modularity and community detection in bipartite networks. Physical Review E (Statistical, Nonlinear, and Soft Matter Physics) 76(6) (December 2007)
3. Busygin, S., Prokopyev, O., Pardalos, P.M.: Biclustering in data mining. Computers and Operations Research 35(9), 2964–2987 (2008)
4. Clauset, A., Newman, M.E.J., Moore, C.: Finding community structure in very large networks. Phys. Rev. E 70(6) (December 2004)
5. Danon, L., Díaz-Guilera, A., Duch, J., Arenas, A.: Comparing community structure identification. Journal of Statistical Mechanics: Theory and Experiment 2005(09), P09008 (2005)
6. Delvenne, J.-C., Yaliraki, S.N., Barahona, M.: Stability of graph communities across time scales. Proc. of the National Academy of Sciences of the USA 107(29), 12755–12760 (2010)
7. Dhillon, I.S.: Co-clustering documents and words using bipartite spectral graph partitioning. In: Proceedings of the Seventh ACM SIGKDD International Conference on Knowledge Discovery and Data Mining, pp. 269–274. ACM, San Francisco (2001)
8. Dubois, D., Dupin de Saint-Cyr, F., Prade, H.: A possibility theoretic view of formal concept analysis. Fundamenta Informaticae 75(1), 195–213 (2007)
9. Dubois, D., Prade, H.: Possibility theory and formal concept analysis: Characterizing independent sub-contexts. Fuzzy Sets and Systems 196, 4–16 (2012)
10. Fu, H., Nguifo, E.M.: A Parallel Algorithm to Generate Formal Concepts for Large Data. In: Eklund, P. (ed.) ICFCA 2004. LNCS (LNAI), vol. 2961, pp. 394–401. Springer, Heidelberg (2004)
11. Ganter, B., Wille, R.: Formal Concept Analysis. Springer (1999)
12. Gaume, B., Navarro, E., Prade, H.: A Parallel between Extended Formal Concept Analysis and Bipartite Graphs Analysis. In: Hüllermeier, E., Kruse, R., Hoffmann, F. (eds.) IPMU 2010. LNCS (LNAI), vol. 6178, pp. 270–280. Springer, Heidelberg (2010)

13. Gaume, B., Navarro, E., Prade, H.: Clustering bipartite graphs in terms of approximate formal concepts and sub-contexts. IJCIS (to be published, 2012)
14. Gonçalves, T., Moura-Pires, F.: An Attribute Redundancy Measure for Clustering. In: Mercer, R.E. (ed.) Canadian AI 1998. LNCS, vol. 1418, pp. 273–284. Springer, Heidelberg (1998)
15. Hu, T., Qu, C., Lim, C., Yuan Sung, T.S., Zhou, W.: Preserving patterns in bipartite graph partitioning. In: 18th IEEE International Conference on Tools with Artificial Intelligence (ICTAI 2006), pp. 489–496. IEEE Computer Society, Washington, USA (2006)
16. Lancichinetti, A., Fortunato, S.: Community detection algorithms: A comparative analysis. Phys. Rev. E 80(5), 056117 (2009)
17. Liu, X., Murata, T.: Evaluating community structure in bipartite networks. In: Elmagarmid, A.K., Agrawal, D. (eds.) Proceedings of the 2010 IEEE Second International Conference on Social Computing, SocialCom / IEEE International Conference on Privacy, Security, Risk and Trust, PASSAT 2010, Minneapolis, USA, pp. 576–581. IEEE Computer Society (2010)
18. Rosvall, M., Bergstrom, C.T.: Maps of random walks on complex networks reveal community structure. Proceedings of the National Academy of Sciences 105(4), 1118–1123 (2008)
19. Schaeffer, S.E.: Graph clustering. Computer Science Review 1(1), 27–64 (2007)

Evaluating Indeterministic Duplicate Detection Results

Fabian Panse and Norbert Ritter

University of Hamburg, Vogt-Kölln Straße 30, 22527 Hamburg, Germany
{panse,ritter}@informatik.uni-hamburg.de
http://vsis-www.informatik.uni-hamburg.de/

Abstract. Duplicate detection is an important process for cleaning or integrating data. Since real-life data is often polluted, detecting duplicates usually comes along with uncertainty. To handle duplicate uncertainty in an appropriate way, indeterministic duplicate detection approaches, i.e. approaches in which ambiguous duplicate decisions are probabilistically modeled in the resultant data, have been developed. To rate the goodness of a duplicate detection approach, its detection results need to be evaluated in their quality. In this paper, we propose several semantics to apply traditional quality evaluation measures to indeterministic duplicate detection results and exemplarily present an efficient evaluation for one of these semantics. Finally, we present some experimental results.

Keywords: indeterministic duplicate detection, probabilistic duplicate detection, quality evaluation, probabilistic clustering, entity resolution.

1 Introduction

Duplicate detection [4,8] is an important task in cleaning a single data source or in meaningfully combining data from different sources. Due to deficiencies like missing data, typos or data obsolescence, it often cannot be determined with absolute certainty from the data itself that two or more representations belong to the same real-world entity. This principally hinders duplicate detection and is a crucial source of uncertainty. Most current duplicate detection approaches [4] acknowledge many kinds of uncertainty and often apply fuzzy matching techniques, but in the end they still are deterministic: finally an absolute decision needs to be taken either by (1) deferring the situation to domain experts which is expensive and time consuming, or (2) choose the most likely configuration thereby risking a wrong choice with all consequences this may have.

To better deal with uncertainty in duplicate detection, several approaches [2,6,9] have been proposed that avoid ambiguous decisions, but instead try to model all significantly likely configurations in the resultant data. Hence any query answer or other derived data will reflect the inherent uncertainty. Since in such approaches duplicate decisions are handled in an indeterministic way, we refer to them as an indeterministic duplicate detection. This concept may protect against negative impact resulting from false duplicate decisions made under ambiguous circumstances.

For effectively comparing deterministic- and indeterministic duplicate detection approaches new methods for quality evaluation are required, because existing evaluation

E. Hüllermeier et al. (Eds.): SUM 2012, LNAI 7520, pp. 433–446, 2012.

measures are not designed to deal with indeterministic results. As we think, the quality of an indeterministic duplicate detection result generally depends on the intended handling of the datas' inherent uncertainty. For that reason, in this paper we define different semantics for evaluating the quality of indeterministic duplicate detection results and propose strategies to compute these evaluations in an efficient way.

The paper is structured as follows. In Section 2, we formally introduce the concepts of deterministic duplicate detection and indeterministic duplicate detection. Moreover, we present measures for evaluating the quality of deterministic duplicate detection results. In Section 3, we introduce different semantics on how the quality of an indeterministic duplicate detection result can be scored and exemplarily discuss one of them in detail. In Section 4, we present an efficient quality computation for this semantics. Section 5 shows some experimental results. In Section 6 we present related work. Finally, Section 7 concludes the work.

2   Duplicate Detection

Duplicate detection [8,4] is the process of identifying multiple representations in a database relation referring to the same real-world entity.

Definition 1 (Real World): *We postulate a real world, denoted by $\mathfrak{W}$, as the set of all existing real-world entities. The mapping $\omega : \mathcal{R} \to \mathfrak{W}$ maps tuples of a database relation $\mathcal{R}$ on entities of $\mathfrak{W}$.*

In our linguistic use, two tuples $t_1, t_2 \in \mathcal{R}$ are called duplicates, iff $\omega(t_1) = \omega(t_2)$.

2.1   Deterministic Duplicate Detection

Deterministic duplicate detection is a partitioning of the input relation into clusters (equivalence classes or partition classes) such that all tuples of one cluster refer to the same real-world entity and hence are duplicates.

Definition 2 (Deterministic Duplicate Detection): *Deterministic duplicate detection is a function δ_{det} that maps a relation $\mathcal{R}$ to a clustering $\mathcal{C} = \{C_1, \ldots, C_l\}$ such that $\bigcup \mathcal{C} = \mathcal{R}$ (each tuple is assigned to a cluster) and $(\forall C_1, C_2 \in \mathcal{C}) : C_1 \cap C_2 = \emptyset$ (the clusters are disjoint). The duplicate detection is considered to be* perfect, *iff:*

- $(\forall C \in \mathcal{C} \ \forall t_1, t_2 \in C) : \omega(t_1) = \omega(t_2)$, *i.e., all tuples of one cluster represent the same real-world entity (the duplicate detection is correct $\Rightarrow$ precision=1)*

- $(\forall C_1, C_2 \in \mathcal{C} \ \forall t_1 \in C_1 \ \forall t_2 \in C_2) : C_1 \neq C_2 \Rightarrow \omega(t_1) \neq \omega(t_2)$, *i.e., all tuples of different clusters represent different real-world entities (the duplicate detection is complete $\Rightarrow$ recall=1)*

To evaluate the quality of a duplicate detection process performed on $\mathcal{R}$, its resultant clustering $\mathcal{C}$ is compared with the clustering $\mathcal{C}_{gold}$ which would result from a perfect duplicate detection process (called the gold standard) on $\mathcal{R}$.

As a running example throughout this paper, we consider a duplicate detection on a relation $\mathcal{R}_{Ex}$ with the ten tuples $t_1, \ldots, t_{10}$. Figure 1 presents the gold standard and a certain clustering resultant from a non-perfect deterministic duplicate detection process.

Fig. 1. The gold standard and a non-perfect deterministic clustering result on $\mathcal{R}_{\text{Ex}}$

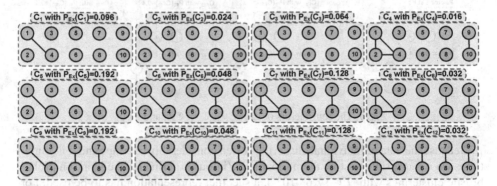

Fig. 2. The possible clusterings $\mathcal{C}_i \in \Gamma_{\text{Ex}}$ of our sample probabilistic clustering $\mathfrak{C}_{\text{Ex}} = (\Gamma_{\text{Ex}}, P_{\text{Ex}})$

2.2   Indeterministic Duplicate Detection

In contrast to a deterministic duplicate detection approach where two tuples have to be declared as duplicates or not, in an indeterministic approach duplicate decisions can be made in a probabilistic way, i.e. tuples can be declared as duplicates with a given probability. For example, the two tuples t_1 and t_2 can be declared to be duplicates with a probability of 60% (and hence to be non-duplicates with a probability of 40%).

The result of an indeterministic duplicate detection is a probability distribution on a set of possible clusterings where each clustering corresponds to a deterministic duplicate detection result.

Definition 3 (**Indeterministic Duplicate Detection**): *Indeterministic duplicate detection is a function δ_{idet} that maps a relation $\mathcal{R}$ to a probabilistic clustering $\mathfrak{C}=(\Gamma, P)$ where:*

- *Γ is a set of possible clusterings so that $(\forall \mathcal{C} \in \Gamma) : (\exists \delta_{det}) : \mathcal{C} = \delta_{det}(\mathcal{R})$,*

- *$P : \Gamma \to (0, 1], \sum_{\mathcal{C} \in \Gamma} P(\mathcal{C}) = 1$ is a probability distribution on Γ*

A sample probabilistic clustering $\mathfrak{C}_{\text{Ex}} = (\Gamma_{\text{Ex}} = \{\mathcal{C}_1, \ldots, \mathcal{C}_{12}\}, P_{\text{Ex}})$ of our sample input relation $\mathcal{R}_{\text{Ex}} = \{t_1, \ldots, t_{10}\}$ is graphically presented in Figure 2.

Definition 4 (**Cross Product of Probabilistic Clusterings**): *The cross product of two probabilistic clusterings $\mathfrak{C}_i = (\Gamma_i, P_i)$ and $\mathfrak{C}_j = (\Gamma_j, P_j)$ is the probabilistic clustering $\mathfrak{C}_{ij} = \mathfrak{C}_i \times \mathfrak{C}_j = (\Gamma_{ij}, P_{ij})$ where $\Gamma_{ij} = \{\mathcal{C}_i \cup \mathcal{C}_j \mid \mathcal{C}_i \in \Gamma_i, \mathcal{C}_j \in \Gamma_j\}$ and the probability of each resultant possible clustering $\mathcal{C} = \mathcal{C}_i \cup \mathcal{C}_j$ is $P_{ij}(\mathcal{C}) = P_i(\mathcal{C}_i) \cdot P_j(\mathcal{C}_j)$.*

The n-ary cross product is defined accordingly.

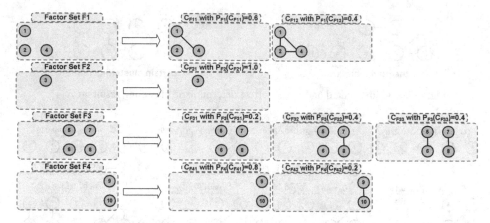

Fig. 3. The four factors of $\mathfrak{C}_{Ex}$: $\mathfrak{C}_{F1}$, $\mathfrak{C}_{F2}$, $\mathfrak{C}_{F3}$ and $\mathfrak{C}_{F4}$

Because the data of individual clusterings often considerably overlaps and it is sometimes even impossible to store them separately (in our experiments we work on probabilistic clusterings with $|\Gamma| \to 7.6 \cdot 10^{44}$), a succinct representation has to be used. For that reason, a probabilistic clustering is usually represented in a factorized way.

Definition 5 (Factorization of a Probabilistic Clustering): A factorization of a probabilistic clustering $\mathfrak{C} = (\Gamma, P)$ defined on a relation $\mathcal{R}$ is a set of probabilistic clusterings (called factors) $\mathcal{F}(\mathfrak{C}) = \{\mathfrak{C}_{F1}, \ldots, \mathfrak{C}_{Fn}\}$ where each factor is defined on a tuple set $Fi \subset \mathcal{R}$ (called factor set) so that the following three requirements are satisfied:

- Each tuple $t \in \mathcal{R}$ is covered by a factor (the factorization is lossless): $\bigcup_{\mathfrak{C}_F \in \mathcal{F}(\mathfrak{C})} F$

- The overall probabilities of the individual clusterings are preserved (the factorization is probability correct): $(\forall \mathfrak{C}_F \in \mathcal{F}(\mathfrak{C}) \; \forall C_F \in \mathfrak{C}_F) : P_F(C_F) = \sum_{C \in \Gamma, C_F \subseteq C} P(C)$

- Each two factors $\mathfrak{C}_{Fi}$ and $\mathfrak{C}_{Fj}$ are independent to each other (the factorization is correct): $(\forall C_1 \in \mathfrak{C}_{Fi} \; \forall C_2 \in \mathfrak{C}_{Fj}) : P_{Fi}(C_1) \cdot P_{Fj}(C_2) = \sum_{C \in \Gamma, C_1 \cup C_2 \subseteq C} P(C)$. This implies that each two factors $\mathfrak{C}_{Fi}$ and $\mathfrak{C}_{Fj}$ are defined on disjoint factor sets, i.e. $F_i \cap F_j = \emptyset$.

The factorization is complete, iff none of its factors can be further factorized. Due to a factorization is correct and lossless, Theorem 1 is valid:

Theorem 1. *A probabilistic clustering $\mathfrak{C} = (\Gamma, P)$ can be rebuilt from the cross product of its factors:* $\mathfrak{C} = \times_{\mathfrak{C}_F \in \mathcal{F}(\mathfrak{C})} \mathfrak{C}_F$

Proof. The proof directly results from the definition of the cross product and the definition of a correct and lossless factorization.

Due to the number of possible clusterings is usually overwhelming, existent approaches of indeterministic duplicate detection [2,6,9] are designed in a way that they already produce a factorized representation as output.

Figure 3 shows the four factors of our sample probabilistic clustering $\mathfrak{C}_{Ex} = (\Gamma_{Ex}, P_{Ex})$ along with their factor sets and their sets of possible clusterings.

2.3 Quality Evaluation Measures

Existing quality measures for deterministic duplicate detection [5,7,11] can be classified into decision-based evaluation measures and cluster-based evaluation measures.

Decision-Based Evaluation. Traditional approaches for duplicate detection [8] are based on pairwise tuple comparisons. For that reason, the quality of a duplicate detection approach is often measured based on the pairwise duplicate decisions made by this approach. The two most known decision-based evaluation measures are recall and precision [11] which originate from the area of information retrieval.

Before a decision-based evaluation measures can be applied to a clustering, the clustering needs to be transformed into a set of pairwise decisions. A transformation from a clustering $\mathcal{C}$ to the corresponding set of duplicate decisions[1] (the set of proposed matches M and the set of proposed unmatches U) can be defined as:

$$M(\mathcal{C}) = \{(t_i, t_j) \mid t_i, t_j \in \mathcal{R} \wedge (\exists C \in \mathcal{C}) : \{t_i, t_j\} \subseteq C\} \tag{1}$$

$$U(\mathcal{C}) = \{(t_i, t_j) \mid t_i, t_j \in \mathcal{R} \wedge (\nexists C \in \mathcal{C}) : \{t_i, t_j\} \subseteq C\} \tag{2}$$

From these two sets three decision classes, the true positives (TP), the false positives (FP), and the false negatives (FN) can be derived as:

$$\text{TP}(\mathcal{C}, \mathcal{C}_{\text{gold}}) = M(\mathcal{C}) \cap M(\mathcal{C}_{\text{gold}}) \tag{3}$$

$$\text{FP}(\mathcal{C}, \mathcal{C}_{\text{gold}}) = M(\mathcal{C}) \cap U(\mathcal{C}_{\text{gold}}) = M(\mathcal{C}) - M(\mathcal{C}_{\text{gold}}) \tag{4}$$

$$\text{FN}(\mathcal{C}, \mathcal{C}_{\text{gold}}) = U(\mathcal{C}) \cap M(\mathcal{C}_{\text{gold}}) = M(\mathcal{C}_{\text{gold}}) - M(\mathcal{C}) \tag{5}$$

Using these three classes, recall (Rec) and precision (Prec) can be defined as:

$$\text{Rec}(\mathcal{C}, \mathcal{C}_{\text{gold}}) = \frac{|\text{TP}(\mathcal{C}, \mathcal{C}_{\text{gold}})|}{|M(\mathcal{C}_{\text{gold}})|} = \frac{|\text{TP}(\mathcal{C}, \mathcal{C}_{\text{gold}})|}{|\text{TP}(\mathcal{C}, \mathcal{C}_{\text{gold}})| + |\text{FN}(\mathcal{C}, \mathcal{C}_{\text{gold}})|} \tag{6}$$

$$\text{Prec}(\mathcal{C}, \mathcal{C}_{\text{gold}}) = \frac{|\text{TP}(\mathcal{C}, \mathcal{C}_{\text{gold}})|}{|M(\mathcal{C})|} = \frac{|\text{TP}(\mathcal{C}, \mathcal{C}_{\text{gold}})|}{|\text{TP}(\mathcal{C}, \mathcal{C}_{\text{gold}})| + |\text{FP}(\mathcal{C}, \mathcal{C}_{\text{gold}})|} \tag{7}$$

A third measure that combines precision and recall into a single quality score by computing their harmonic mean is the F_1-score:

$$F_1\text{-score}(\mathcal{C}, \mathcal{C}_{\text{gold}}) = 2 \cdot \frac{\text{Rec}(\mathcal{C}, \mathcal{C}_{\text{gold}}) \cdot \text{Prec}(\mathcal{C}, \mathcal{C}_{\text{gold}})}{\text{Rec}(\mathcal{C}, \mathcal{C}_{\text{gold}}) + \text{Prec}(\mathcal{C}, \mathcal{C}_{\text{gold}})} \tag{8}$$

$$= \frac{2 \cdot |\text{TP}(\mathcal{C}, \mathcal{C}_{\text{gold}})|}{2 \cdot |\text{TP}(\mathcal{C}, \mathcal{C}_{\text{gold}})| + |\text{FP}(\mathcal{C}, \mathcal{C}_{\text{gold}})| + |\text{FN}(\mathcal{C}, \mathcal{C}_{\text{gold}})|} \tag{9}$$

Further decision-based evaluation measures are proposed in [5,7]. When clear from context, we often simply use TP, FP, FN, Rec, Prec and F_1-score instead of $\text{TP}(\mathcal{C}, \mathcal{C}_{\text{gold}})$, $\text{FP}(\mathcal{C}, \mathcal{C}_{\text{gold}})$, $\text{FN}(\mathcal{C}, \mathcal{C}_{\text{gold}})$, $\text{Rec}(\mathcal{C}, \mathcal{C}_{\text{gold}})$, $\text{Prec}(\mathcal{C}, \mathcal{C}_{\text{gold}})$ and F_1-score$(\mathcal{C}, \mathcal{C}_{\text{gold}})$.

---

[1] Note, most often only the positive duplicate decisions need to be computed.

Cluster-Based Evaluation. In measures for cluster-based evaluation the quality of a duplicate detection process is scored by the similarity of its final clustering to the perfect clustering. The more similar both clusterings (partitions) are, the better is the process's quality. The most of these approaches [11], e.g. the Rand Index, the Adjusted Rand Index, and the Talburt-Wang Index, are based on the *partition overlap* of the two clusterings to be compared. According to [11], the partition overlap V of two partitions $\mathcal{C}_A$ and $\mathcal{C}_B$ is the set of all nonempty intersections between the clusters of $\mathcal{C}_A$ and the clusters of $\mathcal{C}_B$ and is defined as:

$$V(\mathcal{C}_A, \mathcal{C}_B) = \{A_i \cap B_j \mid A_i \in \mathcal{C}_A, B_j \in \mathcal{C}_B \wedge A_i \cap B_j \neq \emptyset\} \qquad (10)$$

Whereas the Rand Index and the Adjusted Rand Index are computationally intensive, the Talburt-Wang Index (short TWI) is simply to calculate, because it does not use the size of the overlaps, but only the number of overlaps:

$$\mathrm{TWI}(\mathcal{C}_A, \mathcal{C}_B) = \frac{\sqrt{|\mathcal{C}_A| \cdot |\mathcal{C}_B|}}{|V(\mathcal{C}_A, \mathcal{C}_B)|} \qquad (11)$$

In this paper, we will use the TWI as a representative for cluster-based evaluation measures. Table 1 depicts the quality scores of the twelve possible clusterings of our sample probabilistic clustering $\mathfrak{C}_{\mathrm{Ex}}$ w.r.t. the four presented evaluation measures.

Table 1. Quality scores of the possible clusterings of $\mathfrak{C}_{\mathrm{Ex}}$

	C_1	C_2	C_3	C_4	C_5	C_6	C_7	C_8	C_9	C_{10}	C_{11}	C_{12}	min	max	exp
Rec	.333	.333	.333	.333	.667	.667	.667	.667	.667	.667	.667	.667	.333	.670	.600
Prec	1.00	.500	.333	.250	1.00	.667	.500	.400	.667	.500	.400	.333	.250	1.00	.648
F_1-score	.500	.400	.333	.286	.800	.667	.571	.500	.667	.571	.500	.444	.286	.800	.592
TWI	.882	.881	.831	.778	.935	.875	.875	.810	.875	.810	.810	.740	.740	.935	.864

3   Quality of Indeterministic Duplicate Detection Results

In this section, we analyze the different types of on-top applications which can process indeterministic duplicate detection results and define specific quality semantics for each of them (Section 3.1). For exemplary reasons, we go then into detail with one of these semantics in Section 3.2. A closer consideration of the other semantics is intended for future publications.

3.1   Quality Semantics

The quality of data generally depends on its intended use, i.e. a database can be of good quality w.r.t. a given application and can be of bad quality w.r.t. another one. The same holds for the quality of a duplicate detection process, because its goodness is automatically a quality yardstick of the data resulting from deduplication, i.e. the more error the duplicate detection process produces, the worse is the quality of the resultant data.

We identify four different ways of handling uncertainty in data processing and hence classify four different types of on-top applications (see Figure 4):

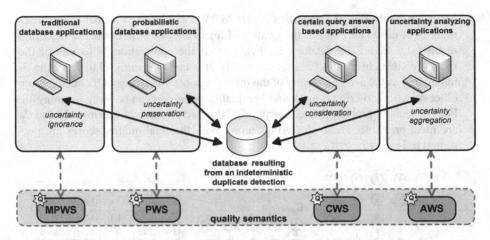

Fig. 4. The four types of database applications along with their corresponding quality semantics

1. **Traditional Database Applications (Uncertainty Ignorance):** Most of traditional database applications cannot process probabilistic data and need certain data as input. In this case, uncertainty must be ignored by evaluating queries only on one of the possible worlds (most meaningful: one of the most probable worlds).
2. **Probabilistic Database Applications (Uncertainty Preservation):** Another way an application can handle data uncertainty is to consider any kind of uncertainty during query evaluation and to produce an uncertain query result. Evaluating a query on a probabilistic database follows the principles of the possible world semantics [10]. This means that the query is evaluated in each world individually and each result represents a possible world of the probabilistic query answer.
3. **Certain Query Answer based Applications (Uncertainty Consideration):** A lot of applications require query answers as input, which are (nearly) dead certain. For that reason *consistent query answering* [1] (also known as *certain query answering* or *sure information answering*) need to be applied to the uncertain data. In this case, uncertainty is resolved by processing a query only on the certain facts, or at least on the facts which are certain with a given level of tolerance, of the probabilistic database. It is important to note that indeterministic duplicate detection allows a more correct evaluation of certain query answering than it is possible by querying a deterministic duplicate detection result, because the query answering algorithm can distinct between ambiguous duplicate decisions and certain duplicate decisions.
4. **Uncertainty Analyzing Applications (Uncertainty Aggregation):** The last type includes applications which are designed to directly analyze the uncertainty of the data, as for example to compute the minimal/maximal/expected number of database tuples which satisfy a specific selection criterion (e.g.: *What is the minimal/maximal/expected number of persons living in Germany*). In this case, aggregation functions are used to resolve data uncertainty.

Since the quality of an indeterministic duplicate detection result essentially depends on the way the resultant datas' inherent uncertainty is processed by an on-top application, we define the following four corresponding quality semantics:

1. **Most Probable World Semantics (short MPWS):** The most probable world semantics is designed to score the quality of an indeterministic duplicate detection result w.r.t. traditional database applications. If the application picks one of the most probable clusterings (worlds) randomly, it is most meaningful to score the final quality as the average quality of the most probable clusterings. Of course, if any other selection criterion is used, another quality definition can be more meaningful. For our sample probabilistic clustering $\mathfrak{C}_{Ex}$ the two possible clusterings C_5 and C_9 are most probable. These clusterings along with the final quality scores are presented in Figure 5.

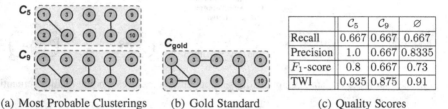

	C_5	C_9	$\varnothing$
Recall	0.667	0.667	0.667
Precision	1.0	0.667	0.8335
F_1-score	0.8	0.667	0.73
TWI	0.935	0.875	0.91

(a) Most Probable Clusterings   (b) Gold Standard   (c) Quality Scores

Fig. 5. $\mathfrak{C}_{Ex}$: its most probable clusterings (a), its gold standard (b) and the quality scores (c)

2. **Possible World Semantics (short PWS):** If the data is processed according to the possible world semantics, it seems most meaningful to define the quality of an indeterministic duplicate detection result as a probability distribution on all possible scores (the possible worlds of the datas' quality). Moreover, this approach allows a subtle analysis of the probabilistic clustering's quality. For example, it allows to determine to what probability is the quality score greater than a user specific threshold. Nevertheless, this semantics is computationally intensive.
For our sample probabilistic clustering $\mathfrak{C}_{Ex}$, the resultant probability distribution on possible F_1-scores is presented in Figure 6.

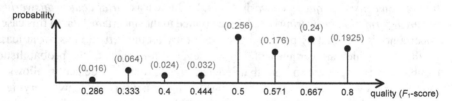

Fig. 6. Probability distribution on possible F_1-scores of $\mathfrak{C}_{Ex}$

3. **Certain World Semantics (short CWS):** The certain world semantics is tailor-made for applications based on certain query answers. Thus, in the CWS the quality of the duplicate detection result is only scored on clustering information which is postulated to be certain. The CWS is extensively discussed in Section 3.2.

4. **Aggregated World Semantics (short AWS):** For uncertainty analyzing applications, it is most meaningful to score the final quality of an indeterministic duplicate detection result by the quality of the analysis result and hence by aggregating (e.g. by min, max or exp) the quality scores of all its possible clusterings. This semantics

allows a rough analysis of the datas' quality. For example it allows to query: *What is the worst case scenario (minimal quality score), the expected scenario (expected quality score), and the best case scenario (maximal quality score)*. The aggregated quality scores of $\mathfrak{C}_{Ex}$ are listed in Table 1.

It is important to note that these semantics are no competitors in general, but each of them fits best for a specific application scenario.

3.2 Certain World Semantics

In the certain world semantics we evaluate the quality only on the facts postulated to be dead certain (or more probable than $1 - \epsilon$ respectively). We have to differentiate between a cluster-based interpretation and a decision-based interpretation. Whereas the cluster-based interpretation considers only clusters with a certain existence, the decision-based interpretation considers certain duplicate decisions.

Cluster-Based Interpretation. Intuitively, a cluster C with $|C| > 1$ can be considered to be certain, if in each possible clustering there exists a cluster C' with $C \subseteq C'$. However, under a closer consideration, we will see that this intuitive definition is not appropriate, because a certain cluster $\{t_1, t_2, t_3\}$ not only means that t_1, t_2 and t_3 are certainly duplicates, but it also implicitly means that these three tuples are certainly no duplicates with any other tuple what in reality muss not be a certain fact at all.

As a consequence, we consider the certain clustering component $C_{cert.\epsilon}(\mathfrak{C})$ of a probabilistic clustering $\mathfrak{C} = (\Gamma, P)$ to be the set of clusters which belong to every possible clustering of $\mathfrak{C}$ with a probability equal to $1 - \epsilon$ or greater.

Definition 6 (Certain Clustering Component): *Let $\mathfrak{C} = (\Gamma, P)$ be a probabilistic clustering. The certain clustering component of $\mathfrak{C}$ with the tolerance setting ϵ is a traditional clustering defined as:*

$$C_{cert.-\epsilon}(\mathfrak{C}) = \{C \mid \sum_{C \in \Gamma, C \in C} P(C) \geq 1 - \epsilon\} \tag{12}$$

By definition, the certain clustering component of a probabilistic clustering $\mathfrak{C} = (\Gamma, P)$ can be contain less tuples than $\mathcal{R}$ and hence less tuples than the gold standard. To enable a meaningful execution of a cluster-based evaluation measure as the Talburt-Wang index, we have to modify C_{gold} so that it shares the same tuples as $C_{cert.-\epsilon}(\mathfrak{C})$. For that purpose, we discard all clusters from C_{gold} which do not have a tuple belonging to $C_{cert.-\epsilon}(\mathfrak{C})$ and then drop from the remaining clusters all tuples which do not belong to any cluster of $C_{cert.-\epsilon}(\mathfrak{C})$. Formally, the modifications are defined as:

$$T_{cert.-\epsilon}(\mathfrak{C}) = \bigcup C_{cert.-\epsilon}(\mathfrak{C}) = \{t \mid (\exists C \in C_{cert.-\epsilon}(\mathfrak{C})) : t \in C\} \tag{13}$$

$$C_{gold}^* = \{C \cap T_{cert.-\epsilon}(\mathfrak{C}) \mid C \in C_{gold}\} - \emptyset \tag{14}$$

Let q be the cluster-based quality measure to score, the quality of the probabilistic clustering $\mathfrak{C}$ to C_{gold} using q under the CWS with the tolerance setting ϵ is scored as:

$$q(\mathfrak{C}, C_{gold})_{CWS,\epsilon} = q(C_{cert.-\epsilon}(\mathfrak{C}), C_{gold}^*) \tag{15}$$

As an example consider Figure 7. The certain clustering component of $\mathfrak{C}_{\text{Ex}}$ with the tolerance setting $\epsilon = 0.3$ is $C_{\text{cert.-0.3}}(\mathfrak{C}) = \{\{t_3\}, \{t_7, t_8\}, \{t_9\}, \{t_{10}\}\}$. The modified gold standard is $C^*_{\text{gold}} = \{\{t_3\}, \{t_7, t_8\}, \{t_9\}, \{t_{10}\}\}$. Thus, the F_1-score as well as the TWI of $\mathfrak{C}_{\text{Ex}}$ are 1.0. In contrast, by using the tolerance setting $\epsilon = 0.4$ the certain clustering component is equivalent to C_9 and the gold standard remained unchanged. Thus, the resultant scores of $\mathfrak{C}_{\text{Ex}}$ are 0.667 (F_1-score) and 0.875 (TWI).

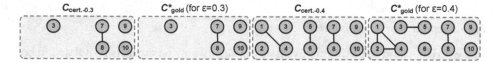

Fig. 7. $C_{\text{cert.-}\epsilon}(\mathfrak{C})$ and C^*_{gold} for the tolerance settings $\epsilon = 0.3$ and $\epsilon = 0.4$

Decision-Based Interpretation. The set of certain decisions of a probabilistic clustering $\mathfrak{C}$ is the set of decisions which are postulated with a probability greater than $1 - \epsilon$.

Definition 7 (Certain Duplicate Decisions): *Let $\mathfrak{C} = (\Gamma, P)$ be a probabilistic clustering. The set of certain positive (negative) duplicate decisions of $\mathfrak{C}$ with the tolerance setting ϵ are defined as:*

$$M_{\text{cert.-}\epsilon}(\mathfrak{C}) = \{(t_i, t_j) \mid t_i, t_j \in \mathcal{R} \wedge \sum_{C \in \Gamma, (t_i, t_j) \in M(C)} P(C) \geq 1 - \epsilon\} \quad (16)$$

$$U_{\text{cert.-}\epsilon}(\mathfrak{C}) = \{(t_i, t_j) \mid t_i, t_j \in \mathcal{R} \wedge \sum_{C \in \Gamma, (t_i, t_j) \in U(C)} P(C) \geq 1 - \epsilon\} \quad (17)$$

$$= \{(t_i, t_j) \mid t_i, t_j \in \mathcal{R} \wedge \sum_{C \in \Gamma, (t_i, t_j) \in M(C)} P(C) \leq \epsilon\} \quad (18)$$

The three decision classes TP, FP, and FN can be then computed as follows:

$$\text{TP}_{\text{cert.-}\epsilon}(\mathfrak{C}, C_{\text{gold}}) = M_{\text{gold}} \cap M_{\text{cert.-}\epsilon}(\mathfrak{C}) \quad (19)$$

$$\text{FP}_{\text{cert.-}\epsilon}(\mathfrak{C}, C_{\text{gold}}) = M_{\text{cert.-}\epsilon}(\mathfrak{C}) - M_{\text{gold}} \quad (20)$$

$$\text{FN}_{\text{cert.-}\epsilon}(\mathfrak{C}, C_{\text{gold}}) = M_{\text{gold}} \cap U_{\text{cert.-}\epsilon}(\mathfrak{C}) \quad (21)$$

Recall, precision and F_1-score are then computed according to Equations 6-9 by using $\text{TP}_{\text{cert.-}\epsilon}$, $\text{FP}_{\text{cert.-}\epsilon}$, and $\text{FN}_{\text{cert.-}\epsilon}$ instead of TP, FP, and FN.

Let q be the decision-based quality measure to score, let m be the evaluation method performed in q having the three decision classes TP, FN, and FP as input: $q(C, C_{\text{gold}}) = m(\text{TP}(C, C_{\text{gold}}), \text{FP}(C, C_{\text{gold}}), \text{FN}(C, C_{\text{gold}}))$. The quality of $\mathfrak{C}$ to C_{gold} using q under the CWS with the tolerance setting ϵ is scored as:

$$q(\mathfrak{C}, C_{\text{gold}})_{\text{CWS},\epsilon} = m(\text{TP}_{\text{cert.-}\epsilon}(\mathfrak{C}, C_{\text{gold}}), \text{FP}_{\text{cert.-}\epsilon}(\mathfrak{C}, C_{\text{gold}}), \text{FN}_{\text{cert.-}\epsilon}(\mathfrak{C}, C_{\text{gold}})) \quad (22)$$

The sets of certain decisions of $\mathfrak{C}_{\text{Ex}}$ with $\epsilon = 0.2$ are (for illustration see Figure 8):

$$M_{\text{cert.-0.2}}(\mathfrak{C}_{\text{Ex}}) = \{(t_1, t_4), (t_7, t_8)\}, \text{ and}$$

$$U_{\text{cert.-0.2}}(\mathfrak{C}_{\text{Ex}}) = \{(a, b) \mid a, b \in \mathcal{R}_{\text{Ex}}\} - \{(t_1, t_2), (t_1, t_4), (t_2, t_4), (t_5, t_6), (t_7, t_8)\}$$

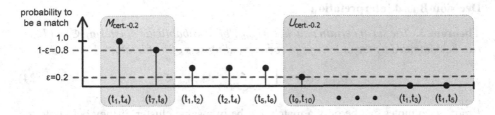

Fig. 8. $M_{\text{cert.-}0.2}(\mathfrak{C}_{\text{Ex}})$ and $U_{\text{cert.-}0.2}(\mathfrak{C}_{\text{Ex}})$ for the tolerance setting $\epsilon = 0.2$

Hence: $\text{TP}_{\text{cert.-}0.2} = \{(t_1, t_4), (t_7, t_8)\}$, $\text{FP}_{\text{cert.-}0.2} = \emptyset$, and $\text{FN}_{\text{cert.-}0.2} = \{(t_3, t_5)\}$. The F_1-score of $\mathfrak{C}_{\text{Ex}}$ using the CWS with the tolerance setting $\epsilon = 0.2$ is thus 0.8.

An important fact of the CWS is that the equations $M = \text{FN} + \text{TP}$ and $U = \text{FP} + \text{TN}$ are not valid anymore and other quality measures, e.g. the number of false decisions $(\text{FN} + \text{FP})$, can capture quality aspects which are not captured by precision, recall or F_1-score anymore.

Tolerance Setting. It is to note that a tolerance setting $\epsilon \geq 0.5$ has to be used carefully, because it can lead to inconsistent duplicate clusterings, i.e. a tuple can belong to multiple clusters (cluster-based interpretation) or two tuples can be declared as a match and as an unmatch at the same time (decision-based interpretation).

4   Efficient Quality Computation for the Certain World Semantics

The certain world semantics proposed in the previous section is defined on a complete probabilistic clustering $\mathfrak{C}$. However, as discussed in Section 3.1, instead of $\mathfrak{C}$ usually its factors are available. To rebuild $\mathfrak{C}$ from its factors is most often not practical. For that reason, in this section, we figure out how the quality of a probabilistic clustering $\mathfrak{C}$ can be scored based on the quality scores of its factors.

Cluster-Based Interpretation

Theorem 2. *The certain clustering component of a probabilistic clustering $\mathfrak{C} = (\Gamma, P)$ can be computed by the union of the certain clustering components of its factors:*

$$C_{\text{cert.-}\epsilon}(\mathfrak{C}) = \bigcup_{\mathfrak{C}_F \in \mathcal{F}(\mathfrak{C})} C_{\text{cert.-}\epsilon}(\mathfrak{C}_F) \tag{23}$$

Proof. All factor sets are disjoint. Hence any possible cluster belongs to a single factor. By Definition 5, for every subset (cluster) C of the factor set F of a factor $\mathfrak{C}_F = (\Gamma_F, P_F)$ holds: $\sum_{C \in \Gamma_F, C \in \mathcal{C}} P_F(\mathcal{C}) = \sum_{C \in \Gamma, C \in \mathcal{C}} P(\mathcal{C})$. Hence a cluster is certain for $\mathfrak{C}$, if it is certain for its corresponding factor.

Due to Theorem 2, the costs for computing the certain clustering component of a probabilistic clustering can be reduced to the costs required for computing the certain clustering component of its factor with the greatest number of possible clusterings.

Decision-Based Interpretation

Theorem 3. *The set of certain matches $M_{cert.\text{-}\epsilon}$ of a probabilistic clustering $\mathfrak{C} = (\Gamma, P)$ can be computed by the union of the corresponding sets of its factors:*

$$M_{cert.\text{-}\epsilon}(\mathfrak{C}) = \bigcup_{\mathfrak{C}_F \in \mathcal{F}(\mathfrak{C})} M_{cert.\text{-}\epsilon}(\mathfrak{C}_F) \tag{24}$$

Proof. Two tuples can be only a match, i.e. be in a same cluster, if they belong to the same factor set. By Definition 5, for every two tuples $\{t_i, t_j\}$ of the factor set F of a factor $\mathfrak{C}_F = (\Gamma_F, P_F)$ holds: $\sum_{C \in \Gamma_F, \{t_i, t_j\} \in C} P_F(C) = \sum_{C \in \Gamma, \{t_i, t_j\} \in C} P(C)$. Hence a tuple pair is certainly a match in $\mathfrak{C}$, if it is certainly a match in its corresponding factor.

Theorem 4. *The decision classes $TP_{cert.\text{-}\epsilon}$ and $FP_{cert.\text{-}\epsilon}$ of a probabilistic clustering $\mathfrak{C} = (\Gamma, P)$ can be computed by the union of the corresponding classes of its factors:*

$$TP_{cert.\text{-}\epsilon}(\mathfrak{C}, C_{gold}) = \bigcup_{\mathfrak{C}_F \in \mathcal{F}(\mathfrak{C})} TP_{cert.\text{-}\epsilon}(\mathfrak{C}_F, C_{gold}) \tag{25}$$

$$FP_{cert.\text{-}\epsilon}(\mathfrak{C}, C_{gold}) = \bigcup_{\mathfrak{C}_F \in \mathcal{F}(\mathfrak{C})} FP_{cert.\text{-}\epsilon}(\mathfrak{C}_F, C_{gold}) \tag{26}$$

Proof. We prove the theorem only for $TP_{cert.\text{-}\epsilon}$ ($FP_{cert.\text{-}\epsilon}$ can be proved accordingly).

$$TP_{cert.\text{-}\epsilon}(\mathfrak{C}, C_{gold}) = M_{gold} \cap M_{cert.\text{-}\epsilon}(\mathfrak{C}) \overset{(\text{Theorem 3})}{=} M_{gold} \cap \bigcup_{\mathfrak{C}_F \in \mathcal{F}(\mathfrak{C})} M_{cert.\text{-}\epsilon}(\mathfrak{C}_F)$$

$$= \bigcup_{\mathfrak{C}_F \in \mathcal{F}(\mathfrak{C})} (M_{gold} \cap M_{cert.\text{-}\epsilon}(\mathfrak{C}_F)) = \bigcup_{\mathfrak{C}_F \in \mathcal{F}(\mathfrak{C})} TP_{cert.\text{-}\epsilon}(\mathfrak{C}_F, C_{gold})$$

In contrast to $TP_{cert.\text{-}\epsilon}$ and $FP_{cert.\text{-}\epsilon}$, the decision class $FN_{cert.\text{-}\epsilon}$ cannot be restricted to the individual factors, because it could happen that some true duplicates do not belong to the same factor set. However, we can distinct between inter-factor false negatives (FN^{inter}), i.e. a not detected duplicate pair which tuples belong to different factors (e.g. the tuple pair (t_3, t_5) in our running example) and intra-factor false negatives (FN^{intra}), i.e. a not detected duplicate pair which tuples belong to the same factor.

All inter-factor false negatives are dead certain decisions, because they do not belong to a same cluster in any possible clustering $C \in \Gamma$. Thus, the set (and hence number) of inter-factor false negatives is the same for all possible clusterings of $\mathfrak{C}$ and can be simply computed from the factor sets: $FN^{inter}(\mathfrak{C}, C_{gold}) = FN(\{F \mid \mathfrak{C}_F \in \mathcal{F}(\mathfrak{C})\}, C_{gold})$. In contrast, the set of intra-factor false negatives results per definition from the union of the false negative decisions of each factor where each factor $\mathfrak{C}_F$ is only compared with the tuple-equivalent part of the gold standard: $C_{gold}^F = \{C \cap F \mid C \in C_{gold}\}$.

Using FN^{inter} and $FN^{intra}_{cert.\text{-}\epsilon}$, the class of certain false negatives can be computed by:

$$FN_{cert.\text{-}\epsilon}(\mathfrak{C}, C_{gold}) = FN^{inter}(\mathfrak{C}, C_{gold}) \cup FN^{intra}_{cert.\text{-}\epsilon}(\mathfrak{C}, C_{gold}) \tag{27}$$

$$= FN(\{F \mid \mathfrak{C}_F \in \mathcal{F}(\mathfrak{C})\}, C_{gold}) \cup \bigcup_{\mathfrak{C}_F \in \mathcal{F}(\mathfrak{C})} FN_{cert.\text{-}\epsilon}(F, C_{gold}^F) \tag{28}$$

Thus, the number of certain TP, certain FP and certain FN can be simply computed as:

$$|\text{TP}_{\text{cert.-}\epsilon}(\mathfrak{C}, C_{\text{gold}})| = \sum_{\mathfrak{C}_F \in \mathcal{F}(\mathfrak{C})} |\text{TP}_{\text{cert.-}\epsilon}(\mathfrak{C}_F, C_{\text{gold}})| \tag{29}$$

$$|\text{FP}_{\text{cert.-}\epsilon}(\mathfrak{C}, C_{\text{gold}})| = \sum_{\mathfrak{C}_F \in \mathcal{F}(\mathfrak{C})} |\text{FP}_{\text{cert.-}\epsilon}(\mathfrak{C}_F, C_{\text{gold}})| \tag{30}$$

$$|\text{FN}_{\text{cert.-}\epsilon}(\mathfrak{C}, C_{\text{gold}})| = |\text{FN}^{\text{inter}}(\mathfrak{C}, C_{\text{gold}})| + \sum_{\mathfrak{C}_F \in \mathcal{F}(\mathfrak{C})} |\text{FN}_{\text{cert.-}\epsilon}(\mathfrak{C}_F, C_{\text{gold}}^F)| \tag{31}$$

In summary, the costs for computing the number of certain TP, certain FP and certain FN and hence the costs for computing the recall, the precision and the F_1-score of the certain facts of a probabilistic clustering can be reduced to the costs required for computing the certain TP, certain FP and certain FN of its factor with the greatest number of possible clusterings.

5   Experimental Evaluation

To experimental evaluate our quality semantics, we use a duplicate detection scenario extensively discussed in [9]. In this scenario, on a real-life CD data set several indeterministic duplicate detection processes have been performed. Each process is characterized by its number of indeterministically handled decisions (#inDec). In Figure 9, we present the F_1-score and the TWI of these processes each scored with our different quality semantics (PWS is not considered, because it does not supply a single value).

The most probable world was the same for all processes and hence the MPWS returned a constant result. Moreover, the quality of the MPWS was equivalent to the quality of a deterministic approach (#inDec =0). The minimal (maximal) possible score was always lower (greater) than by using MPWS and decreased (increased) with growing uncertainty. The CWS without tolerance ($\epsilon = 0$) performed the better than the MPWS, the more uncertainty was modeled in the indeterministic result. Only for less uncertainty, the CWS was worse than MPWS using the F_1-score. In general, the TWI was not that restrictive than the F_1-score (all scores were between 0.984 and 0.998), but show similar results than the F_1-score. Solely the CWS was always better than the MPWS (sometimes even better than the maximal possible quality score).

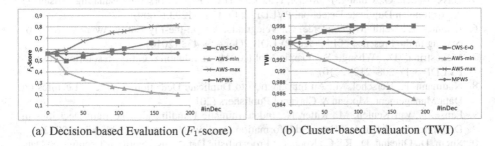

(a) Decision-based Evaluation (F_1-score)   (b) Cluster-based Evaluation (TWI)

Fig. 9. F_1-score and TWI of several indeterministic duplicate detection results

6 Related Work

Quality evaluation of deterministic duplicate detection results has been considered in several works [5,7,11], but none of them take an uncertain clustering into account. De Keijzer et al. propose measures for scoring the quality of uncertain data [3]. The expected precision and the expected recall for scoring the quality of uncertain query results are similar to a variant of our aggregated world semantics. Further semantics, especially the certain world semantics are not covered by this work. Moreover, they restrict themselves to probabilistic results with independent events (in our case, decisions), which is not useful for duplicate detection scenarios.

7 Conclusion

Duplicate detection usually comes along with a high degree of uncertainty and often it cannot be determined with absolute certainty whether two tuples are duplicates or not. Indeterministic duplicate detection approaches have been proposed to handle uncertainty on duplicate decisions by storing multiple possible duplicate clusterings in the resultant data. In this paper, we introduced a framework for scoring the quality of indeterministic duplicate detection results. For that purpose, we presented four different quality semantics, each defined for a special class of data processing tasks.

In this paper, we only went into computation details for a single semantics. In future work, we aim to focus on an efficient computation of the remaining three semantics.

References

1. Arenas, M., Bertossi, L.E., Chomicki, J.: Consistent Query Answers in Inconsistent Databases. In: PODS, pp. 68–79 (1999)
2. Beskales, G., Soliman, M.A., Ilyas, I.F., Ben-David, S.: Modeling and Querying Possible Repairs in Duplicate Detection. PVLDB 2(1), 598–609 (2009)
3. de Keijzer, A., van Keulen, M.: Quality Measures in Uncertain Data Management. In: Prade, H., Subrahmanian, V.S. (eds.) SUM 2007. LNCS (LNAI), vol. 4772, pp. 104–115. Springer, Heidelberg (2007)
4. Elmagarmid, A.K., Ipeirotis, P.G., Verykios, V.S.: Duplicate Record Detection: A Survey. IEEE Trans. Knowl. Data Eng. 19(1), 1–16 (2007)
5. Hassanzadeh, O., Chiang, F., Miller, R.J., Lee, H.C.: Framework for Evaluating Clustering Algorithms in Duplicate Detection. PVLDB 2(1), 1282–1293 (2009)
6. Ioannou, E., Nejdl, W., Niederée, C., Velegrakis, Y.: On-the-Fly Entity-Aware Query Processing in the Presence of Linkage. PVLDB 3(1), 429–438 (2010)
7. Menestrina, D., Whang, S., Garcia-Molina, H.: Evaluating entity resolution results. PVLDB 3(1), 208–219 (2010)
8. Naumann, F., Herschel, M.: An Introduction to Duplicate Detection. Synthesis Lectures on Data Management. Morgan & Claypool Publishers (2010)
9. Panse, F., van Keulen, M., Ritter, N.: Indeterministic Handling of Uncertain Decisions in Deduplication. Journal of Data and Information Quality (accepted for publication, 2012)
10. Suciu, D., Olteanu, D., Ré, C., Koch, C.: Probabilistic Databases. Synthesis Lectures on Data Management. Morgan & Claypool Publishers (2011)
11. Talburt, J.R.: Entity Resolution and Information Quality. Morgan Kaufmann (2011)

Merging Interval-Based Possibilistic Belief Bases

Salem Benferhat[1], Julien Hué[2], Sylvain Lagrue[1], and Julien Rossit[3]

[1] Université Lille - Nord de France
CRIL - CNRS UMR 8188
Artois, F-62307 Lens
{benferhat,lagrue}@cril.fr
[2] Institut für Informatik
Albert-Ludwigs-Universität Freiburg
Georges-Köhler-Allee 52
79110 Freiburg, Germany
hue@informatik.uni-freiburg.de
[3] Université Paris Descartes
LIPADE - France
julien.rossit@parisdescartes.fr

Abstract. In the last decade, several approaches were introduced in literature to merge multiple and potentially conflicting pieces of information. Within the growing field of application favourable to distributed information, *data fusion strategies* aim at providing a global and consistent point of view over a set of sources which can contradict each other. Moreover, in many situations, the pieces of information provided by these sources are uncertain.

Possibilistic logic is a well-known powerful framework to handle such kind of uncertainty where formulas are associated with real degrees of certainty belonging to $[0, 1]$. Recently, a more flexible representation of uncertain information was proposed, where the weights associated with formulas are in the form of intervals. This interval-based possibilistic logic extends classical possibilistic logic when all intervals are singletons, and this flexibility in representing uncertain information is handled without extra computational costs. In this paper, we propose to extend a well known approach of possibilistic merging to the notion of interval-based possibilistic knowledge bases. We provide a general semantic approach and study its syntactical counterpart. In particular, we show that convenient and intuitive properties of the interval-based possibilistic framework hold when considering the belief merging issue.

1 Introduction

The problem of *belief merging* [Lin96] arises when a situation requires to take into account several pieces of information obtained from distinct and often conflicting sources (or agents). This kind of situations frequently appears in many usual frameworks, such as distributed databases, multi-agent systems, or distributed information in general (e.g., semantic web), and leads to perform some combination operations on available information to extract a global and coherent point of view. Roughly speaking, merging operators introduced in literature strongly rely on the representation of available

E. Hüllermeier et al. (Eds.): SUM 2012, LNAI 7520, pp. 447–458, 2012.
© Springer-Verlag Berlin Heidelberg 2012

information. In the last decade, several approaches were proposed to merge pieces of information provided without explicit priority [KPP02, EKM10, EKM12] or to the contrary to merge prioritized information [BK03, GLB06].

From a semantic point of view, these approaches are generally divided in two steps: first locally rank interpretations using some scales (depending on the considered framework, possibilistic distributions or κ-functions for instance), then aggregating these local rankings among all the bases to obtain a global total pre-order over considered interpretations (see [KPP02] for more details). The result of merging is finally obtained by considering preferred interpretations according to this global total pre-order.

In our framework, the pieces of information provided by each source may be uncertain. In this paper, these pieces of information, encoded by the means of propositional formulas, are called *beliefs*. Possibilistic logic [DLP94] is a well-known framework which allows to conveniently represent and reason with such uncertain pieces of information: uncertainty is represented by real numbers, belonging to [0,1], associated with each piece of information. Moreover, uncertainty is also represented at the semantic level by associating a possibility degree with each possible world (or interpretation). An inference mechanism was proposed in [Lan00] to derive plausible conclusions from a possibilistic knowledge base K, which needs $log_2(m)$ calls to the satisfiability test of a set of propositional clauses (SAT), where m is the number of different degrees used in K.

However, in many situations, providing a precise weight to evaluate the certainty associated with a belief can be a difficult problem (e.g., when scales are provided by an expert). In [BHLR11], a flexible representation was introduced to allow the expression of an imprecision on possibilistic degrees associated with beliefs, where weights associated with formulas are in the form of intervals of $[0, 1]$. An interesting result is that handling this flexibility is done without extra computational costs with respect to the classical framework. A natural question concerns now the ability of this framework to keep such properties while considering more sophisticated issues, like the belief merging problem.

Several approaches to merge classical possibilistic belief bases were introduced in [BDPW99, BK03, QLB10]. Resulting possibilistic merging operators were analyzed in [BDKP00], where they are sorted into different classes depending on the configuration of the bases to merge. We can distinguish:

- conjunctive operators, exploiting symbolic complementarities between sources;
- disjunctives operators, which deal with conflicting but equally reliable sources;
- idempotent operators, suitable when sources to merge are not independent;
- reinforcement operators, which consider the repetition of pieces of information among sources to merge as a confirmation;
- adaptive and average operators, which adopt a disjunctive attitude in case of conflicts and a reinforcement behaviour in the other cases.

In this paper, we extend this approach to the framework of interval-based possibilistic logic. More precisely, we extend the strategies introduced in [BDKP02] by adapting possibilistic aggregation operators to deal with intervals. In particular, we show that intuitive intervals characterization principles and computational properties introduced in [BHLR11] still stand when considering the problem of belief merging. Section 3 introduces a general semantic approach, which relies on aggregation operations over intervals at the level of possibility distributions. In particular, we focus on the well-known *minimum*-based, *maximum*-based and *product*-based operations. Finally, Section 4 provides a syntactic counterpart to our semantic approach.

2 Background and Notations

In this paper, we consider a finite propositional language $\mathcal{L}$. We denote by Ω the finite set of interpretations of $\mathcal{L}$ and by ω an element of Ω.

2.1 Possibilistic Logic

Possibility Distributions. A possibility distribution, denoted by π, is a function from Ω to $[0, 1]$. $\pi(\omega)$ represents the degree of compatibility (or consistency) of the interpretation ω with the available knowledge. $\pi(\omega) = 1$ means that ω is fully consistent with the available knowledge, while $\pi(\omega) = 0$ means that ω is impossible. $\pi(\omega) > \pi(\omega')$ simply means that ω is more compatible than ω'. A possibility distribution π is said to be normalized if there exists an interpretation ω such that $\pi(\omega) = 1$. Otherwise, the distribution is inconsistent and is called subnormalized.

A possibility distribution allows to define two functions from $\mathcal{L}$ to $[0, 1]$ called possibility and necessity measures, denoted by Π and N, and defined by:

$$\Pi(\varphi) = \max\{\pi(\omega) : \omega \in \Omega, \omega \models \varphi\} \qquad \text{and} \qquad N(\varphi) = 1 - \Pi(\neg\varphi)$$

$\Pi(\varphi)$ measures to what extent the formula φ is compatible with the available knowledge while $N(\varphi)$ measures to what extent it is entailed.

Given a possibility distribution π encoding some available knowledge, a formula φ is said to be a consequence of π, denoted by $\pi \models_\pi \varphi$, iff $\Pi(\varphi) > \Pi(\neg\varphi)$.

Possibilistic Knowledge Bases. A possibilistic formula is a tuple $\langle \varphi, \alpha \rangle$ where φ is an element of $\mathcal{L}$ and $\alpha \in (0, 1]$ is a valuation of φ representing $N(\varphi)$. Note that no formula can be of type $\langle \varphi, 0 \rangle$ as it brings no information. Then, a possibilistic base $K = \{\langle \varphi_i, \alpha_i \rangle, 1 \leq i \leq n\}$ is a set of possibilistic formulas.

An important notion that plays a central role in the inference process in the one of strict α-cut. A strict α-cut, denoted by K_α, is a set of propositional formulas defined by $K_\alpha = \{\varphi : \langle \varphi, \beta \rangle \in K \text{ and } \beta > \alpha\}$. The strict α-cut is useful to measure the inconsistency degree of K defined by $Inc(K) = \max\{\alpha : K_\alpha \text{ is inconsistent }\}$.

If $Inc(K) = 0$ then K is said to be completely consistent. If a possibilistic base is partially inconsistent, then $Inc(K)$ can be seen as a threshold below which every formula is considered as not enough entrenched to be taken into account in the inference process. More precisely, we define the notion of core of a knowledge base as the set of formulas with a necessity value greater than $Inc(K)$, i.e.,

$$\mathcal{C}ore(K) = K_{Inc(K)} = \{\varphi : \langle \varphi, \alpha \rangle \in K \text{ and } \alpha > Inc(K)\}$$

A formula φ is a consequence of a possibilistic base K, denoted by $K \vdash_\pi \varphi$, iff $Core(K) \vdash \varphi$.

Given a possibilistic base K, we can generate a unique possibility distribution where interpretations ω satisfying all propositional formulas in K have the highest possible degree $\pi(\omega) = 1$ (since they are fully consistent), whereas the others are pre-ordered w.r.t. highest formulas they falsify. More formally:

$$\forall \omega \in \Omega, \pi_K(\omega) = \begin{cases} 1 \text{ if } \forall \langle \varphi, \alpha \rangle \in K, \ \omega \models \varphi \\ 1 - max\{\alpha_i : \langle \varphi_i, \alpha_i \rangle \in K, \omega \nvDash \varphi_i\} \text{ otherwise.} \end{cases}$$

The following completeness and soundness result holds:

$$K \vdash_\pi \varphi \text{ iff } \pi_K \models_\pi \varphi.$$

2.2 Merging Possibilistic Belief Bases

Let us consider a multi-set of possibilistic belief bases $E = \{K_1, \ldots, K_n\}$ and their associated possibilistic distributions $\pi_1, \ldots, \pi_n$, each of these bases representing the local point of view associated with a single source. The aim of belief merging is to compute an unique possibilistic distribution, denoted $\pi_\oplus$, representing a global and consistent point of view among pieces of information provided by sources, even if some of these sources contradict each others. The most common approach to merge possibilistic knowledge bases is the one presented in [BDKP00, BDKP02]. These strategies are close to the ones introduces in [KPP02] in the case of merging classical non prioritized propositional belief bases.

Generally speaking, most common belief merging operators, denoted Δ, are divided in two steps. First, all interpretations are rank ordered with respect to individual sources. In the framework of possibilistic logic, this step is performed quite straightforwardly since each possibilistic belief base induces an unique possibilistic distributions over interpretations. Then, ranks individually computed are aggregated among all belief bases to merge, using an aggregation operator denoted $\oplus$, to associate a global rank to each considered interpretations: these ranks allow to induce a global order, denoted in this paper $<_{\pi_\oplus}$, where preferred interpretations are usually considered as models of the result of merging, denoted $\Delta_\oplus(E)$. This distribution finally induces a possibilistic belief base, denoted $K_\oplus$, representing the result of merging. Obviously, several aggregation operators are possible, depending on expected properties for the result of merging.

In the context of possibilistic logic, several aggregation function were discussed in [BDKP02] to compute the value of $\pi_\oplus(w)$ from the $\nu_E(w) = \langle \pi_1(w), \ldots, \pi_n(w) \rangle$ vector. These operators were divided into several categories: conjunctive (adequate when the sources are consistent), disjunctive (adequate when the sources are conflicting), idempotent (ignoring the redundancies) and reinforcement (seeing redundancies as confirmation).

Generally speaking, any function $\oplus$ which respects the following conditions can be considered a possibilistic aggregating function:

1. $\oplus(1, \ldots, 1) = 1$
2. If $\forall 1 \leq i \leq n, a_i \geq b_i$ then $\oplus(a_1, \ldots, a_n) \geq \oplus(b_1, \ldots, b_n)$

Note that clearly, many aggregation operators are possible to combine initial distributions, offering different behaviours in computing the result of merging. The most common operators used in the context of the possibilistic merging are the following ones:

- the minimum: $\oplus_{min}(\pi_1, \ldots, \pi_n) = min(\pi_1, \ldots, \pi_n)$;
- the product: $\oplus_{prd}(\pi_1, \ldots, \pi_n) = \sqrt[n]{\Pi \pi_i}$;
- the maximum: $\oplus_{max}(\pi_1, \ldots, \pi_n) = max(\pi_1, \ldots, \pi_n)$;
- the dual product: $\oplus_{dpr}(\pi_1, \ldots, \pi_n) = 1 - \sqrt[n]{\Pi(1 - \pi_i)}$;
- the probabilistic sum: $\oplus_{prs}(\pi_1, \ldots, \pi_n) = 1 - \Pi(1 - \pi_i)$;
- the averaging: $\oplus_{ave}(\pi_1, \ldots, \pi_n) = \Sigma \pi_i / n$.

In particular, this paper focuses on the *minimum*-based, the *maximum*-based and the *product*-based merging operators.

Moreover, a syntactic counterpart of possibilistic merging operators was introduced in [BDPW99, BDKP02]. Namely, authors show that the result of merging can be characterized by the belief base $\mathcal{B}_{\oplus}$ defined as follows:

$$\mathcal{B}_{\oplus} = \{(D_j, 1 - \oplus(x_1, \ldots, x_n)) \; : \; j = 1, \ldots, n\}$$

where D_j are disjunctions of size j between formulas ϕ_i obtained from each K_i and $x_i = 1 - \alpha_i$ if $\phi_i \in D_j$, $x_i = 1$ otherwise.

Desterecke et al. [DDC09] proposed another way of merging possibilistic bases on the ground of maximal coherent subsets which is closer to what is usually done in propositional logic.

2.3 Interval-Based Possibilistic Logic

Interval-based possibilistic logic was introduced in [BHLR11]. This framework can be described as a generalization of possibilistic logic, where uncertainty associated to beliefs is represented by the means of an interval of $I = [\alpha, \beta] \subseteq [0, 1]$ instead of a number. The intuitive meaning behind this interval is that the real value of uncertainty is unknown and belong to the interval. A more detailed discussion on imprecise possibilities can be found in [CRP12].

The set of intervals of $[0, 1]$ is denoted by $\mathcal{I}$. An interval based possibility distribution, denoted by $\pi_{\mathcal{I}}$, is then also described by the means of intervals of $\mathcal{I}$. This induces a partial pre-ordering among the set of interpretations of Ω. More precisely, an interpretation ω is said to be preferred to ω', denoted by $\omega \lhd \omega'$, iff $\beta < \alpha'$ where $\pi_{\mathcal{I}}(\omega) = [\alpha, \beta]$ and $\pi_{\mathcal{I}}(\omega') = [\alpha', \beta']$.

A first approach to compute possibility and necessity measures is to use the notion of compatible possibility distribution. Formally, a classical possibility distribution π is said to be compatible with $\pi_{\mathcal{I}}$ iff $\forall \omega \in \Omega, \pi(\omega) \in \pi_{\mathcal{I}}(\omega)$. The non finite set of all compatible possibility distributions obtained from $\pi_{\mathcal{I}}$ is denoted by $\mathcal{C}mp(\pi_{\mathcal{I}})$. Possibility and necessity measures are then defined as follows:

$$\Pi_{\mathcal{I}}(\varphi) = [min_{\pi \in \mathcal{C}mp(\pi_{\mathcal{I}})} \Pi(\varphi), max_{\pi \in \mathcal{C}mp(\pi_{\mathcal{I}})} \Pi(\varphi)]$$
$$N_{\mathcal{I}}(\varphi) = [min_{\pi \in \mathcal{C}mp(\pi_{\mathcal{I}})} N(\varphi), max_{\pi \in \mathcal{C}mp(\pi_{\mathcal{I}})} N(\varphi)]$$

As it is shown in [BHLR11], these measures can be characterized by the means of operations on intervals:

$$\Pi_{\mathcal{I}}(\varphi) = \mathcal{M}\{\pi_{\mathcal{I}}(\omega) : \omega \in \Omega, \omega \models \varphi\}$$
$$N_{\mathcal{I}}(\varphi) = 1 \ominus \Pi_{\mathcal{I}}(\neg\varphi)$$

where $\mathcal{M}\{I_1, \ldots, I_n\} = [\max\{\alpha_1, \ldots, \alpha_n\}, \max\{\beta_1, \ldots, \beta_n\}]$ with $I_i = [\alpha_i, \beta_i]$ and $1 \ominus [\alpha, \beta] = [1 - \beta, 1 - \alpha]$.

Interval-Based Possibilistic Bases. A syntactic representation of interval-based possibilistic logic is obtained by associating necessity-values, in the form of intervals, to formulas. An interval-based possibilistic base, denoted IK, is thus defined as:

$$IK = \{\langle \varphi, I \rangle, \varphi \in \mathcal{L} \text{ and } I \in \mathcal{I}\}$$

Likewise possibility distributions, a compatible possibilistic base can be obtained from an interval-based possibilistic base by replacing each interval-based possibilistic formula $\langle \varphi, I \rangle$ by a standard possibilistic formula $\langle \varphi, \delta \rangle$ where $\delta \in I$. The non-finite set of all possible standard possibilistic bases compatible with an interval-based possibilistic base IK is denoted by $Cmp(IK)$. Two particular compatible bases are $IK_{lb} = \{\langle \varphi, \alpha \rangle : \langle \varphi, [\alpha, \beta] \rangle \in IK\}$, $IK_{ub} = \{\langle \varphi, \beta \rangle : \langle \varphi, [\alpha, \beta] \rangle \in IK\}$, which are respectively obtained by selecting either lower or upper bounds of intervals.

Like in the standard possibilistic logic, interval-based possibilistic bases can be partially inconsistent. As it is shown in [BHLR11], the interval-based inconsistency degree can be equivalently computed in the two following ways:

$$Inc(IK) = \{Inc(K) : K \in Cmp(IK)\}$$
$$= [Inc(IK_{lb}), Inc(IK_{ub})]$$

This central notion allows to characterize the interval-based syntactic inference which can intuitively be defined by considering all compatible bases:

$$IK \models_c \phi \text{ iff } Core(IK) \vdash \phi \text{ iff } \forall K \in Cmp(IK), Core(K) \vdash \phi$$

where $Core(IK) = \{\varphi : \langle \varphi, I \rangle \in IK \text{ and } Inc(IK) \lhd I\}$.

3   A Semantic Approach to Interval-Based Possibilistic Merging

In this section, we propose a general semantic approach to merge interval-based possibility distributions. Let $E = \{IK_1, \ldots, IK_n\}$ be a multi-set of n interval-based possibilistic bases. From E, we can derive a family of interval-based possibility distributions $\pi_1^{\mathcal{I}} \ldots \pi_n^{\mathcal{I}}$, each IK_i inducing an unique interval-based possibility distribution $\pi_i^{\mathcal{I}}$. This step allows to locally rank order each interpretation ω of Ω with respect to each IK_i. To compute the result of merging E, we now need to aggregate all intervals associated to each interpretation ω to obtain a global ordering over Ω. We introduce the notion of interval-based aggregation operator, denoted by $\oplus^{\mathcal{I}}$, and then denote by $\pi_\oplus^{\mathcal{I}}$ the interval-based possibility distribution obtained by aggregating all distributions obtained from E with $\oplus$.

In this framework, the real uncertainty value associated to a formula is unknown, and may be any value of the interval. Therefore, a first approach to define an aggregation operator on intervals is to take into account each possible combination of all existing standard distributions compatible with interval-based distributions to consider. Namely:

Definition 1. *Let $\pi_1^{\mathcal{I}}, \ldots, \pi_n^{\mathcal{I}}$ be interval-based possibility distributions and let $\oplus$ be a possibilistic aggregation operator, then an interval-based possibilistic aggregation operator $\oplus^{\mathcal{I}}$ based on $\oplus$ can be defined as follows:*

$$\oplus^{\mathcal{I}}(\pi_1^{\mathcal{I}}(\omega), \ldots, \pi_n^{\mathcal{I}}(\omega)) = \bigcup_{\pi_i \in Cmp(\pi_i^{\mathcal{I}})} \{\oplus(\pi_1(\omega), \ldots, \pi_n(\omega))\}.$$

However, the previous definition is not helpful computationally speaking. Indeed, the number of compatible possibility distributions being infinite, computing the result of merging is a difficult problem. We introduce a first characterisation of aggregating intervals, when considering the *minimum*-based, the *maximum*-based and the *product*-based operations, which only relies on considering lower and upper bounds of intervals. Moreover, the following proposition shows that in these cases, the value respectively associated to each interpretation ω of Ω by $\pi_\oplus^{\mathcal{I}}$ constitutes an interval. Namely:

Proposition 1. *Let $\pi_1^{\mathcal{I}}, \ldots, \pi_n^{\mathcal{I}}$ be n interval-based possibility distributions, let $\oplus^{\mathcal{I}}$ be an interval-based aggregation operator relying on the minimum-based, the maximum-based or the product-based classical possibilistic operator and let $\pi_\oplus^{\mathcal{I}}$ be the interval-based possibility distribution obtained by considering $\oplus^{\mathcal{I}}$, then:*

$$\pi_\oplus^{\mathcal{I}}(\omega) = \oplus^{\mathcal{I}}(\pi_1^{\mathcal{I}}(\omega), \ldots, \pi_n^{\mathcal{I}}(\omega)) = [\oplus(\alpha_1, \ldots, \alpha_n), \oplus(\beta_1, \ldots, \beta_n)] .$$

where for each $i = 1 \ldots n$, $\pi_i^{\mathcal{I}}(\omega) = [\alpha_i, \beta_i]$.

Obviously, many other definitions of $\oplus^{\mathcal{I}}$ are possible. We then propose three intuitive requirements for an interval-based aggregation operator. Formally:

1. $\oplus^{\mathcal{I}}(I_1, \ldots, I_n)$ is an interval;
2. $\oplus^{\mathcal{I}}([1,1], \ldots, [1,1]) = [1,1]$;
3. If $\forall 1 \leq i \leq n, I_i \lhd I_i'$ then $\oplus^{\mathcal{I}}(I_1, \ldots, I_n) \lhd \oplus^{\mathcal{I}}(I_1', \ldots, I_n')$.

The first requirement means that the result of aggregating some intervals should also be an interval. The second requirement says that if each source agrees that ω is fully possible, then the result of aggregation should confirm it. The last says that if each source prefers ω to ω', then the result of aggregation should prefer so. The following result shows that any interval-based aggregation operator based on a classical aggregation operator straightforwardly ensures the two last requirements. Namely:

Proposition 2. *Let $\oplus$ be a n-ary function from $[0, 1]^n$ to $[0, 1]$. If $\oplus$ is a possibilistic aggregation operator, then $\oplus^{\mathcal{I}}$ (given by Definition 1) satisfies conditions 2 and 3.*

In a general case, Condition 1 is not guaranteed. However, when considering the *minimum*-based, the *maximum*-based or the *product*-based aggregation operations, the three requirements are satisfied. Namely:

Proposition 3. *Let $\oplus_{min}$, $\oplus_{max}$ and $\oplus_{prd}$ be respectively the n-ary minimum-based, the maximum-based and the product-based classical aggregation operations. Then $\oplus_{min}^{\mathcal{I}}$, $\oplus_{max}^{\mathcal{I}}$ and $\oplus_{prd}^{\mathcal{I}}$ satisfy conditions 1-3, and are interval-based possibilistic aggregation operators.*

One can remark that this approach of aggregating interval-based possibility distributions generalizes aggregation operations defined in the classical case, since in the extreme case, where all intervals provided by sources are only singletons, the possibility distribution $\oplus^{\mathcal{I}}$ recovers the results provided by $\oplus$. More formally:

Proposition 4. *In the case where intervals within each $\pi_i^{\mathcal{I}}$ obtained from E only consist in singletons (namely for all $\pi_i^{\mathcal{I}}$ obtained from E, for all ω, $\pi_i^{\mathcal{I}}(\omega) = [\alpha, \alpha]$) then: i) each $\pi_i^{\mathcal{I}}$ obtained from E has a unique compatible possibility distribution π_i and ii) $\pi_{\oplus}^{\mathcal{I}}(\omega) = \oplus^{\mathcal{I}}(\pi_1^{\mathcal{I}}(\omega), \ldots, \pi_n^{\mathcal{I}}(\omega)) = \oplus(\pi_1(\omega), \ldots, \pi_n(\omega))$, where is $\oplus^{\mathcal{I}}$ an interval-based aggregation operator based on the classical aggregation operator $\oplus$, and each π_i is the unique classical distribution compatible with the respective interval-based distribution $\pi_i^{\mathcal{I}}$.*

Let us illustrate these definitions with the following example:

Example 1. *Let π_1, π_2 and π_3 be three possibility distributions such that:*

ω	π_1	π_2	π_3
ω_1	$[.1, .3]$	$[.4, .6]$	$[.7, .9]$
ω_2	$[.4, .5]$	$[.4, .5]$	$[.4, .5]$
ω_3	$[.4, .5]$	$[.7, .8]$	$[1, 1]$
ω_4	$[.1, .2]$	$[.1, .5]$	$[.1, .8]$

Considering merging operators relying respectively on the minimum-based, the product-based and the maximum-based operations, we obtain the following results:

ω	$\oplus_{min}$	$\oplus_{prd}$	$\oplus_{max}$
ω_1	$[.1, .3]$	$[.303, .545]$	$[.7, .9]$
ω_2	$[.4, .5]$	$[.4, .5]$	$[.4, .5]$
ω_3	$[.4, .5]$	$[.654, .736]$	$[1, 1]$
ω_4	$[.1, .2]$	$[.1, .430]$	$[.1, .8]$

Note that considering the interval-based comparative relation $\lhd$, $\pi_{\oplus}^{\mathcal{I}}$ only induces a partial pre-order over interpretations ω of Ω (different comparative relations are possible but are out of the scope of this paper, see [BHLR11] for more details). This result allows to provide a result for the merging of E without focusing on each combination of all possible classical distributions compatible with considered interval-based distributions. As a corollary, this result also shows that aggregating interval-based possibility distribution can be achieved within only two calls to classical aggregating operations, respectively on lower and upper bounds of intervals.

In the classical possibilistic case, merging operators, relying on standard aggregation operators, are divided into several non-exclusive families. Since these definitions do not hold anymore when considering interval-based possibilistic degrees, we thus provide their counterparts in the interval based possibilistic framework. Formally:

Definition 2. *Let $I = [\alpha, \beta]$ and $I' = [\alpha', \beta']$ be intervals and let $\oplus^{\mathcal{I}}$ be a possibilistic merging operator, then $\oplus^{\mathcal{I}}$ is said to be:*

1. *conjunctive iff $\forall I \in \mathcal{I}, I \oplus^{\mathcal{I}} [1, 1] = [1, 1] \oplus^{\mathcal{I}} I = I$*
2. *disjunctive iff $\forall I \in \mathcal{I}, I \oplus^{\mathcal{I}} [1, 1] = [1, 1] \oplus^{\mathcal{I}} I = [1, 1]$*
3. *idempotent iff $\forall I \in \mathcal{I}, I \oplus^{\mathcal{I}} I = I$*
4. *reinforcement iff $\forall I, I' \in \mathcal{I}$ s.t. $I, I' \neq [1, 1]$ and $I, I' \neq [0, 0]$ then*
   $(I \oplus^{\mathcal{I}} I')_\alpha \leq min(I_\alpha, I'_\alpha)$ *and* $(I \oplus^{\mathcal{I}} I')_\beta \leq min(I_\beta, I'_\beta)$
5. *averaging iff $\forall I \in \mathcal{I}, min(I_\alpha, I'_\alpha) \leq (I \oplus^{\mathcal{I}} I')_\alpha \leq max(I_\alpha, I'_\alpha)$ and*
   $min(I_\beta, I'_\beta) \leq (I \oplus^{\mathcal{I}} I')_\beta \leq max(I_\beta, I'_\beta)$

where I_α and I_β are respectively the lower bound and the upper bound of an interval I.

From these definitions, and from previous results introduced in this paper, one can remark that several families of interval-based aggregation operators extend properties associated with the classical families on which they are based. Namely:

Proposition 5. *Let $\pi_1, \ldots, \pi_n$ be n interval-based possibility distributions and let $\oplus^{\mathcal{I}}$ be a possibilistic merging operator. If $\oplus$ is conjuctive (resp. disjunctive, or idempotent) then $\oplus^{\mathcal{I}}$ is conjuctive (resp. disjunctive, or idempotent).*

Let us illustrate this fact with the following example:

Example 2. *Let us consider again Example 1. On this example, we have:*

- *The interpretation ω_3 is associated with merged value of $[1, 1]$ for the disjunctive operator $\oplus^{\mathcal{I}}_{max}$;*
- *The interpretation ω_2 has a merged value of $[.4, .5]$ for the idempotent operators $\oplus^{\mathcal{I}}_{min}, \oplus^{\mathcal{I}}_{prd}, \oplus^{\mathcal{I}}_{max}$.*

4   A Syntactic Counterpart

In this section, we provide some syntactic counterparts to the general semantic approach introduced previously.

The definitions given in the previous section allow to define, from the sources, a possibility value for every interpretations. The result of the merging operation is thus define as the interpretations maximal according to their possibility values.

Definition 3. *Let $IK_1, \ldots, IK_n$ be interval-based possibilistic knowledge bases where $\pi_1, \ldots, \pi_n$ being their respective interval-based possibility distributions and let $\oplus^{\mathcal{I}}_o$ be a possibilistic merging operator, then:*

$$\Delta^{\mathcal{I}}_o(IK_1, \ldots, IK_n) = \{\pi(\omega) = \oplus^{\mathcal{I}}_o(\pi_1, \ldots, \pi_n)\}$$

There is also a syntactic counterpart to this definition. One can build an interval-based possibilistic base out of the sources the following way:

Definition 4. *Let $IK_i = \{\langle \varphi, I_i^j \rangle\}$ be interval-based possibilistic knowledge bases and let $\oplus_o^{\mathcal{I}}$ be a possibilistic merging operator, then:*

$$\blacktriangle_o^{\mathcal{I}}(IK_1, \ldots, IK_n) = \{(D_j, 1 \ominus \oplus_o^{\mathcal{I}}(x_1, \ldots, x_n)) \ : \ j = 1, \ldots, n\}$$

and D_j are disjunctions of size j between formulas φ_i taken from different IK_i and $x_i = 1 \ominus [\alpha_i, \beta_i]$ if $\varphi_i \in D_j$ and $x_i = [1,1]$ otherwise.

The consequences from the syntactic merging operation are equivalent to the results of the semantic merging operation.

Proposition 6. *Let $IK_i = \{\langle \varphi, I_i^j \rangle\}$ be interval-based possibilistic knowledge bases, let ϕ be a formula and let $\oplus_o^{\mathcal{I}}$ be a possibilistic merging operator, then:*

$$\Delta_o^{\mathcal{I}}(IK_1, \ldots, IK_n) \models \phi \text{ iff } \blacktriangle_o^{\mathcal{I}}(IK_1, \ldots, IK_n) \vdash \phi$$

Now, let us instantiate three particular cases of Definition 4, namely with $\oplus_{min}^{\mathcal{I}}$ (interval-based idempotent conjunctive merging), $\oplus_{max}^{\mathcal{I}}$ (interval-based idempotent disjunctive merging), $\oplus_{prd}^{\mathcal{I}}$ (interval-based product-based conjunctive merging).

We restrict ourselves to the case of two knowledge IK_1 and IK_2 (since all operations are associative and commutative).

For the interval-based idempotent conjunctive operation, we have:

$$
\begin{aligned}
\Delta_{min}^{\mathcal{I}}(IK_1, IK_2) = {} & \{\langle \varphi_i, [1,1] \ominus min([1,1] \ominus I_i, [1,1]) \rangle : (\varphi_i, I_i) \in IK_1\} \\
& \cup \{\langle \psi_j, [1,1] \ominus min([1,1] \ominus I_j, [1,1]) \rangle : (\psi_j, I_j) \in IK_2\} \\
& \cup \{\langle \varphi_i \vee \psi_j, [1,1] \ominus min([1,1] - I_i, [1,1] - I_j) \rangle : \\
& \quad (\varphi_i, I_i) \in IK_1 \text{ and } (\psi_j, I_j) \in IK_2\} \\
= {} & \{\langle \varphi_i, I_1 \rangle : \langle \varphi_i, I_1 \rangle \in IK_1\} \\
& \cup \{\langle \psi_j, I_j \rangle : \langle \psi_j, I_j \rangle \in IK_2\} \\
& \cup \{\langle \varphi_i \vee \psi_j, max(I_i, I_j) \rangle : \\
& \quad \langle \varphi_i, I_1 \rangle \in IK_1 \text{ and } \langle \psi_j, I_j \rangle \in IK_2\}
\end{aligned}
$$

we can check that is equivalent to

$$\Delta_{min}^{\mathcal{I}}(IK_1, IK_2) = IK_1 \cup IK_2$$

For the interval-based idempotent disjunctive operation, we have:

$$
\begin{aligned}
\Delta_{max}^{\mathcal{I}}(IK_1, IK_2) = {} & \{\langle \varphi_i, [1,1] \ominus max([1,1] \ominus I_i, [1,1]) \rangle : (\varphi_i, I_i) \in IK_1\} \\
& \cup \{\langle \psi_j, [1,1] \ominus max([1,1] \ominus I_j, [1,1]) \rangle : (\psi_j, I_j) \in IK_2\} \\
& \cup \{\langle \varphi_i \vee \psi_j, [1,1] \ominus max([1,1] - I_i, [1,1] - I_j) \rangle : \\
& \quad (\varphi_i, I_i) \in IK_1 \text{ and } (\psi_j, I_j) \in IK_2\} \\
= {} & \{\langle \varphi_i, [0,0] \rangle : \langle \varphi_i, I_1 \rangle \in IK_1\} \\
& \cup \{\langle \psi_j, [0,0] \rangle : \langle \psi_j, I_j \rangle \in IK_2\} \\
& \cup \{\langle \varphi_i \vee \psi_j, min(I_i, I_j) \rangle : \\
& \quad \langle \varphi_i, I_1 \rangle \in IK_1 \text{ and } \langle \psi_j, I_j \rangle \in IK_2\}
\end{aligned}
$$

we can check that is equivalent to

$$\Delta_{max}^{\mathcal{I}}(IK_1, IK_2) = \{\langle \varphi_i \vee \psi_j, min(I_i, I_j) \rangle : \langle \varphi_i, I_1 \rangle \in IK_1 \text{ and } \langle \psi_j, I_j \rangle \in IK_2\}$$

For the interval-based product-based conjunctive merging operation, we have:

$$\Delta_{prd}^{\mathcal{I}}(IK_1, IK_2) = \{\langle \varphi_i, [1,1] \ominus (([1,1] \ominus I_i) \times [1,1]) \rangle : (\varphi_i, I_i) \in IK_1\}$$
$$\cup \{\langle \psi_j, [1,1] \ominus ([1,1] \times ([1,1] \ominus I_j)) \rangle : (\psi_j, I_j) \in IK_2\}$$
$$\cup \{\langle \varphi_i \vee \psi_j, [1,1] \ominus (([1,1] \ominus I_i) \times ([1,1] \ominus I_j)) \rangle :$$
$$(\varphi_i, I_i) \in IK_1 \text{ and } (\psi_j, I_j) \in IK_2\}$$
$$= IK_1 \cup IK_2$$
$$\cup \{\langle \varphi_i \vee \psi_j, [\alpha_i + \alpha_j - \alpha_i \times \alpha_j, \beta_i + \beta_j - \beta_i \times \beta_j] \rangle :$$
$$\langle \varphi_i, I_1 \rangle \in IK_1 \text{ and } \langle \psi_j, I_j \rangle \in IK_2\}$$

Example 3. *Let* $E = \{IK_1, IK_2\}$ *be a belief profile with* $IK_1 = \{\langle a, [.5, .7] \rangle,$ $\langle \neg a \vee b, [.4, .8] \rangle\}$ *and* $IK_2 = \{\langle \neg b \vee a, [.2, .3] \rangle, \langle \neg b, [.6, .7] \rangle\}$. *The interval-based possibility distribution is given in the following table.*

ω_i	$\pi_{\mathcal{I}}(IK_1)$	$\pi_{\mathcal{I}}(IK_2)$	$\oplus_{min}^{\mathcal{I}}$	$\oplus_{max}^{\mathcal{I}}$	$\oplus_{prd}^{\mathcal{I}}$
$a \wedge b$	[1,1]	[.3, .4]	[.3, .4]	[1,1]	[.547, .632]
$a \wedge \neg b$	[.2, .6]	[1,1]	[.2, .6]	[1,1]	[.447, .774]
$\neg a \wedge b$	[.3, .5]	[.3, .4]	[.3, .4]	[.3, .5]	[.3, .447]
$\neg a \wedge \neg b$	[.3, .5]	[1,1]	[.3, .5]	[1,1]	[.547, .707]

In the following, we give the resulting base for our three main operators, the result of the disjunction of $a \vee \neg b$ *and* $\neg b \vee a$ *and* $\neg a \vee b$ *and* $\neg b$ *are not given as they produce* $\top$.

- $\blacktriangle_{min}^{\mathcal{I}}(IK_1, IK_2) = IK_1 \cup IK_2 \cup \{\langle a \vee \neg b \vee a, [.5, .7] \rangle, \langle a \vee \neg b, [.6, .7] \rangle\}$
- $\blacktriangle_{max}^{\mathcal{I}}(IK_1, IK_2) = \{\langle a \vee \neg b \vee a, [..2, .3] \rangle, \langle a \vee \neg b, [.5, .7] \rangle\}$
- $\blacktriangle_{prd}^{\mathcal{I}}(IK_1, IK_2) = IK_1 \cup IK_2 \cup \{\langle a \vee \neg b \vee a, [.368, .542] \rangle, \langle a \vee \neg b, [.553, .7] \rangle\}$

One can easily verify that the syntactic and semantic operators have the same consequences.

5   Conclusion

This paper addressed a first approach for merging interval-based possibilistic belief bases. More precisely, we have extended the possibilistic merging operators introduced in the classical case to handle the concept of interval-based possibilistic degrees. This way, our study shown that convenient and intuitive properties associated to this framework still hold when dealing with more tricky issues, in particular the problem of belief merging.

As a work on progress, we study the links between our framework and results introduced in [DP11]. A future work is to consider the belief revision problem in the context of interval-based possibilistic logic. This problem consists in integrating a higher priority information in a belief base, such that this information must be deduced from the base after the process. Despite this problem is a particular case of merging, namely a belief base is merged with a higher priority piece of information, it still raises some difficult issues.

References

[BDKP00] Benferhat, S., Dubois, D., Kaci, S., Prade, H.: A principled analysis of merging operations in possibilistic logic. In: Proceedings of Uncertainty in Artificial Intelligence, pp. 24–31 (2000)

[BDKP02] Benferhat, S., Dubois, D., Kaci, S., Prade, H.: Possibilistic merging and distance-based fusion of propositional information. Annal of Mathematical Artificial Intelligence 34(1-3), 217–252 (2002)

[BDPW99] Benferhat, S., Dubois, D., Prade, H., Williams, M.-A.: A practical approach to fusing prioritized knowledge bases. In: Proceedings of Portuguese Conference on Artificial Intelligence, pp. 223–236. Springer (1999)

[BHLR11] Benferhat, S., Hué, J., Lagrue, S., Rossit, J.: Interval-based possibilistic logic. In: Proceedings of International Joint Conferences on Artificial Intelligence, pp. 750–755 (2011)

[BK03] Benferhat, S., Kaci, S.: Fusion of possibilistic knowledge bases from a postulate point of view. International Journal of Approximate Reasoning 33(3), 255–285 (2003)

[CRP12] Chemin, M., Rico, A., Prade, H.: Decomposition of Possibilistic Belief Functions into Simple Support Functions. In: Melo-Pinto, P., Couto, P., Serôdio, C., Fodor, J., De Baets, B. (eds.) Eurofuse 2011. AISC, vol. 107, pp. 31–42. Springer, Heidelberg (2011)

[DDC09] Destercke, S., Dubois, D., Chojnacki, E.: Possibilistic information fusion using maximal coherent subsets. Transactions on Fuzzy Systems 17(1), 79–92 (2009)

[DLP94] Dubois, D., Lang, J., Prade, H.: Possibilistic logic. In: Handbook of Logic in Artificial Intelligence and Logic Programming, vol. 3, pp. 439–513 (1994)

[DP11] Dubois, D., Prade, H.: Generalized Possibilistic Logic. In: Benferhat, S., Grant, J. (eds.) SUM 2011. LNCS, vol. 6929, pp. 428–432. Springer, Heidelberg (2011)

[EKM10] Everaere, P., Konieczny, S., Marquis, P.: Disjunctive merging: Quota and gmin merging operators. Artificial Intelligence Journal 174(12-1), 824–849 (2010)

[EKM12] Everaere, P., Konieczny, S., Marquis, P.: Compositional belief merging. In: Proceedings of International Conference on Principles of Knowledge Representation and Reasoning (to appear, 2012)

[GLB06] Guilin, Q., Liu, W., Bell, D.A.: Merging stratified knowledge bases under constraints. In: Proceedings of AAAI Conference on Artificial Intelligence, pp. 348–356 (2006)

[KPP02] Konieczny, S., Pino-Pérez, R.: Merging information under constraints: a logical framework. Journal of Logic and Computation 12(5), 773–808 (2002)

[Lan00] Lang, J.: Possibilistic logic: complexity and algorithms. In: Handbook of Defeasible Reasoning and Uncertainty Management Systems, vol. 5, pp. 179–220. Kluwer Academic (2000)

[Lin96] Lin, J.: Integration of weighted knowledge bases. Artificial Intelligence 83(2), 363–378 (1996)

[QLB10] Qi, G., Liu, W., Bell, D.: A Comparison of Merging Operators in Possibilistic Logic. In: Bi, Y., Williams, M.-A. (eds.) KSEM 2010. LNCS, vol. 6291, pp. 39–50. Springer, Heidelberg (2010)

Comparing and Fusing Terrain Network Information

Emmanuel Navarro[1], Bruno Gaume[2], and Henri Prade[2]

[1] IRIT, Université de Toulouse III,
118 Route de Narbonne, 31062 Toulouse Cedex 9, France
{navarro,prade}@irit.fr
[2] CLLE-ERSS, Université de Toulouse II,
5, allées Antonio Machado, 31058 Toulouse Cedex 9, France
gaume@univ-tlse2.fr

Abstract. Terrain networks (or complex networks) is a type of relational information that is encountered in many fields. In order to properly answer questions pertaining to the comparison or to the merging of such networks, a method that takes into account the underlying structure of graphs is proposed. The effectiveness of the method is illustrated using real linguistic data networks and artificial networks, in particular.

1 Introduction

Complex networks [1,19] are graphs with non-trivial topological features. In the following we prefer to call them "terrain networks" to emphasize the fact that they represent practical data, supposed to have some underlying structure. Moreover, it is a counterpart of the French "graphe de terrain". Such networks can be observed in many areas ranging from computer sciences to biology, linguistics, and social sciences. Examples of such graphs are synonymy networks between words, social relation networks between people, or protein interaction networks. One of their main features is to be globally sparse and locally dense. In other words, while their number of edges is relatively small, they exhibit a rather high transitivity (or clustering) coefficient (defined by the ratio of the number of 3-cliques over the number of paths of length 2). Moreover their diameter, i.e. the average minimal path length between pairs of vertices, is very small [19] and the degree distribution follows approximately a power law [1].

Since terrain networks are more and more common pieces of information, general information processing issues, such as comparison or fusion of two networks, make sense for them and become increasingly important. In this paper, we consider the particular case of special interest where the two graphs have the same vertices. This means that the two graphs represent data pertaining to the same items, objects, or agents. Generally speaking, the comparison of graphs may be envisaged in different ways. One may compare two graphs either at the edges and vertices level [8,11,16,18,20], or in terms of global structural property measures [10,12]. None of these two classes of methods appear to be fully satisfactory for comparing terrain networks sharing the same vertices. Indeed, the former do not take into account the latent similarity information since they work in a too local way, while the latter only deals with global properties without any reference to the fact that the graphs share common vertices.

E. Hüllermeier et al. (Eds.): SUM 2012, LNAI 7520, pp. 459–472, 2012.

Terrain networks depart from other graphs often encountered in AI. Indeed, graph representations are associated with taxonomies or ontologies, or with Bayesian nets. They encode various forms of generic knowledge, possibly pervaded with uncertainty, which can be applied to factual pieces information describing the particular situations to reason about. This contrasts with terrain networks which gather what may be called data information. They are made of collections of pieces of factual information, but we are no longer primarily interested in just answering requests pertaining to particular individuals. The emphasis is rather on the way the pieces of information are related together and are organized in cluster-like structures. Thus, for instance, the proximity between two graphs is not only a matter of identity of edges, but should also take into account the neighborhood structures of vertices. For example, a non-edge may "virtually" exist as an edge if there are short paths linking its vertices.

In this paper, we propose a general procedure that labels each pair of vertices in a graph, i.e., each edge, as well each non-edge, in terms of two categories: the edge (or the non-edge) is "confirmed", or is "not-confirmed" (in Section 2). Thus, the existence, or the non-existence of an edge between two vertices is confirmed, or not according to their neighborhood situation that in some sense support or not this existence, or non-existence. Then, we show the interest of such labeled graphs for comparing (in Section 3) or merging (in Section 4) terrain network information. Related work is discussed in Section 5.

2   What a Data Information Graph May Mean

In this section, data information graphs, issued from terrain networks, are considered as knowledge representation entities, which can be manipulated in order to lay bare some hidden part of the information. In such graphs, the information conveyed is not just made of a collection of links existing between certain pairs of vertices, but should also take into account the graph topology in the neighborhood of pairs of vertices. Before presenting a labeling procedure whose purpose is to confirm (or not) each edge and each non-edge in a graph in order to bring back the graph topology information, we first restate general knowledge representation concerns by examining in what respect a graph may be correct or complete.

2.1   Correctness and Completeness of a Graph

If the information given by a graph is correct and complete, any edge expresses the certainty of the existence of a relation between the two associated vertices, and the absence of edge between two vertices asserts that there is no relation between them. However, if a graph is only correct, each edge is there for sure, but the absence of an edge may be as much the result of missing information as acknowledging the certainty of the absence of link. Conversely, if a graph is only complete, no edge are missing, but some may be questionable. Then the absence of an edge reflects the certainty that there is no relation.

Also note that in case some prior knowledge exists about the graph, it may be used for revising it. Thus knowing, for instance, that the graph should represent a transitive relation, two situations would be of interest. If the graph is correct but incomplete, then it can be replaced by its transitive closure. If the graph is incorrect but complete, we may try to remove a minimal number of edges to make the relation transitive (but in general the solution is not unique!). However, in the following we do not assume the availability of such strong prior knowledge.

When comparing or merging two graphs, assuming that the information conveyed by each of them is correct, and/or complete is a crucial issue. Indeed in such operations, knowing of which edge, or non-edge one may be certain is clearly important. When a graph is correct and complete, any edge (resp. non-edge) is certain and has a status denoted 1! (resp. 0!). When a graph is incorrect and incomplete, any edge (resp. non-edge) is uncertain and has a status denoted 1? (resp. 0?). Table 1 sums up the four possible cases. More generally, the status of edges or non-edges in a graph may differ from one pair of vertices to another. Indeed it may be interesting to have such a binary "uncertainty" information for each edge and non-edge. Thus, for instance, a graph may be complete and correct, except for some pairs of vertices.

Table 1. Four possible cases of graph correctness and completeness, and there counterpart in terms of edges and non-edges certainty

	edge	non-edge
correct *and* complete	1!	0!
incorrect *but* complete	1?	0!
correct *but* incomplete	1!	0?
incorrect *and* incomplete	1?	0?

In a similar spirit, in the next section, we propose a method for providing a similar type of status to each edge, or non-edge in a graph, and thus laying bare information that is not explicitly given with the graph. According to the neighborhood (possibly taken in a broad sense) of each pair of vertices, the corresponding edge (resp. non-edge) will be labeled 1? (resp. 0?) and regarded as "uncertain", or will be labeled 1! (resp. 0!) and regarded as "confirmed". Mind however that this is not genuine uncertainty information, but rather a way to bring back some "global information" to a local level. Indeed, roughly speaking the idea is to label with 0? the non-edge that are inside clusters, and to label with 1? the edges outside clusters, thus acknowledging the "imperfect transitivity" that may exist in the graph (and which is at work in the clusters).

2.2 Labeling Edges and Non-edges for Reflecting the Graph Topology

In a graph, two vertices may be regarded as being "close" according to the graph topology between them, independently of the existence or not of a direct edge between them. For example, in the Figure 1 the pair a is not an edge, but the two vertices are close in the graph in the sense that there are 3 paths of length 2 between them. This contrasts with the situation of the non-edge b. Conversely the pair d is an edge, but the two vertices are relatively distant since there is no path between them other than this edge itself. Lastly, the edge c is "strengthened" by the existence of 3 paths of length 2 between its two vertices.

The above observation is important when comparing and fusing two graphs (the problems considered in the next sections). Indeed, if a pair of vertices is an edge in one

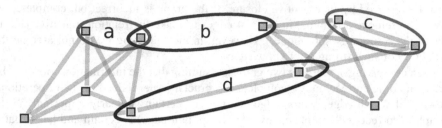

Fig. 1. A toy example of non-edge labeled "0?" (a) or "0!" (b), and edges labeled "1?" (d) or "1!" (c)

graph but not in the other one, the situation is not the same if this edge is like d or like c in Figure 1 (and similarly for the non-edge, if it is like a or like b). So, we propose to label each pair of vertices according to their closeness to be judged from the topology of the graph in the neighborhood of the two vertices, using the conventions summarized in Table 2. Several ways of evaluating closeness may be considered.

Table 2. Labeling procedure of edges and non-edges, according to a closeness evaluation of pairs of vertices

edge	closeness	label	
0	0	0!	not an edge, and not close in the graph.
0	1	0?	not an edge, *but* close in the graph.
1	0	1?	an edge, *but* not close in the graph.
1	1	1!	an edge, and close in the graph.

Evaluating Closeness. We now describe two methods that one may think of for evaluating closeness of a pair of vertices on an undirected graph $G = (V, E)$ (with V the vertex set and E the edge set).

Triangle. A very simple method could be to consider as "close" every pair of vertices that are connected by a path of length 2. An edge will be confirmed (i.e. 1!) if it is supported by at least one path of length 2 ; or "unconfirmed" if there is no path of length 2 between the two corresponding vertices (i.e. 1?). Similarly a pair of non adjacent vertices, will be labeled as unconfirmed (i.e. 0?) if they are connected by at least one path of length 2, or as "confirmed" (i.e. 0!) if they are not connected by a path of length 2.

Confluence. Short length random walks may provide a more accurate method for measuring the closeness of two vertices in a graph [6,7,13]. Let $G = (V, E)$ be an undirected and reflexive[1] graph. Let us imagine a walker wandering on G:

- At a time $t \in \mathbb{N}$, the walker is on one vertex $u \in V$;
- At time $t + 1$, the walker can reach any neighboring vertex of u, with a uniformly distributed probability.

---

[1] i.e. each vertex is connected to itself. If such self-loops do not exist in the data, they may be generally added without loss of information.

Table 3. Number and proportion of each label for edges and non-edges on 3 different terrain networks (*V.rob*, *V.pwn*, *V.wikt*) and one random network (rob^R, Erdös Rényi random graph of the same size than *V.rob*, average on 20 realizations)

		V.rob		V.pwn		V.wikt		rob^R	
triangle	1!	20442	76.9%	37473	91.6%	2886	34.8%	187.0	0.7%
	1?	6125	23.1%	3446	8.4%	5407	65.2%	26380.0	99.3%
	0?	395555	1.5%	527336	0.8%	33685	0.1%	190910.8	0.7%
	0!	26636924	98.5%	65884901	99.2%	26884813	99.9%	26841568.2	99.3%
5-confl.	1!	22726	85.5%	36760	89.8%	2864	34.5%	4032.2	15.2%
	1?	3841	14.5%	4159	10.2%	5429	65.5%	22534.8	84.8%
	0?	2844964	10.5%	3744489	5.6%	250177	0.9%	4795066.4	17.7%
	0!	24187515	89.5%	62667748	94.4%	26668321	99.1%	22237412.6	82.3%
10-confl.	1!	23143	87.1%	36887	90.1%	2980	35.9%	4657.2	17.5%
	1?	3424	12.9%	4032	9.9%	5313	64.1%	21909.8	82.5%
	0?	5282868	19.5%	7350176	11.1%	513214	1.9%	8292114.4	30.7%
	0!	21749611	80.5%	59062061	88.9%	26405284	98.1%	18740364.6	69.3%
20-confl.	1!	22405	84.3%	36741	89.8%	3056	36.9%	39.8	0.1%
	1?	4162	15.7%	4178	10.2%	5237	63.1%	26527.2	99.9%
	0?	8055282	29.8%	12241791	18.4%	948772	3.5%	10567375.4	39.1%
	0!	18977197	70.2%	54170446	81.6%	25969726	96.5%	16465103.6	60.9%

This process is called a simple random walk [3]. It can be defined by a Markov chain on V with a $|V| \times |V|$ transition matrix $[G]$:

$$[G] = (g_{u,v})_{u,v \in V}, \quad \text{with} \quad g_{u,v} = \begin{cases} \dfrac{1}{d_G(u)} & \text{if } (u,v) \in E, \\ 0 & \text{else.} \end{cases}$$

where $d_G(u) = |\{v \in V / (u,v) \in E\}|$ is the degree of vertex u in the graph G. Since G is reflexive, each vertex has at least one neighbor (itself) thus $[G]$ is well-defined. Furthermore, by construction, $[G]$ is a stochastic matrix: $\forall u \in V, \sum_{v \in V} g_{u,v} = 1$. The probability $P_G^t(u \rightsquigarrow v)$ of a walker starting on vertex u to reach a vertex v after t steps is:

$$P_G^t(u \rightsquigarrow v) = ([G]^t)_{u,v} \tag{1}$$

One can then prove [7], with the Perron-Frobenius theorem [17], that if G is connected (i.e., there is always at least one path between any two vertices), reflexive and undirected, then $\forall u, v \in V$:

$$\lim_{t \to \infty} P_G^t(u \rightsquigarrow v) = \lim_{t \to \infty} ([G]^t)_{u,v} = \frac{d_G(v)}{\sum_{x \in V} d_G(x)} \tag{2}$$

It means that when t tends to infinity, the probability of being on a vertex v at time t does not depend on the starting vertex but only on the degree of v. In the following we will refer to this limit as $\pi_G(v)$. If G is composed of several connected components then for any pair (u, v) of vertices, we have two possible cases:

- u and v are in the same connected component $G' = (V', E')$, with $V' \subseteq V$ and $E' \subseteq E$, then equation 2 applies to this subgraph:

$$\lim_{t \to \infty} P_G^t(u \leadsto v) = \lim_{t \to \infty} ([G]^t)_{u,v} = \frac{d_G(v)}{\sum_{x \in V'} d_G(x)} \tag{3}$$

- u and v are in distinct components, then for all t, $P_G^t(u \leadsto v) = 0$, therefore $\lim_{t \to \infty} P_G^t(u \leadsto v) = 0$.

So the probability $P_G^t(u \leadsto v)$ converges to a limit that only depends of vertex v degree. However the way this probability converges to the limit heavily depends on the topology of the graph between the two vertices. If u and v are connected by many short paths the probability will converge to the limit by above, whereas if there is no short path between the two vertices it will converge to the limit by below. Indeed when t is small the more interconnections there are between u and v, the higher the probability of reaching v from u. Therefore we define the t-confluence $\Gamma(G, u, v, t)$ between two vertices u, v on a graph G as follows:

$$\Gamma(G, u, v, t) = \begin{cases} \dfrac{P_G^t(u \leadsto v)}{\pi_G(v)} & \text{if } u \text{ and } v \text{ are in the same} \\ & \qquad \text{connected component,} \\ 0 & \textbf{else.} \end{cases} \tag{4}$$

We propose to consider as "close" each pair of non adjacent vertices (u, v) having a t-confluence greater than 1. In other words, we consider u and v as close if the probability of reaching v from u in a t step random walk is greater than the probability to be on v after an infinite walk. (u, v) is then labeled 0?. Conversely non-adjacent vertices (u, v) having a t-confluence smaller than 1 are labeled 0!.

In order to measure the closeness of an edge (u, v), the t-confluence is computed on the graph G where the considered edge has been removed. This removal is important, otherwise almost all edges would have a strong confluence, as the edge may be used by the random walker to go from u to v in few steps. The idea is to measure the closeness of the two vertices according to the graph structure and, this independently of the existence of an edge between them. Therefore, an edge (u, v) is labeled 1! if it has a t-confluence on the graph $G' = (V, E \setminus \{(u, v)\})$ greater than 1. In other words, without going through this edge, a random walker is more likely to be in v after t steps starting from u, than to be on v after un infinite walk. Conversely an edge (u, v) is labeled 1? if the t-confluence of (u, v) on the graph $G' = (V, E \setminus \{(u, v)\})$ is smaller than 1.

There are other possible ways of evaluating the closeness. Any measure of similarity between two vertices in a graph may be use, and in particular the ones developed to address the problem of link prediction [9]. However we are interested in binary evaluation of similarity between to vertices, and there is rarely a natural threshold of gradual similarity measure. Also any robust graph clustering method [15] may be used: two vertices can be considered as close if and only if they are in a common cluster. However, note that the idea of short random walks proposed by [7] has been used in graph clustering method [13].

Illustration. On the toy example of Figure 1, with the t-confluence labeling procedure, as one may expect, all edges are confirmed (1!), except edge d, and all non-edges are confirmed (0!), except the pair a. It has been verified for t between 2 and 20. With the triangle method, the results are the same except that many other non-edges are labeled 0?. Indeed edge d creates many paths of length 2 between pairs of vertices that are not adjacent. Note that, for the same pairs of vertices, these paths of length 2 do not lead to a value of the t-confluence larger than 1.

Table 3 gives the number of pairs for each label on 3 different terrain networks[3,4] (the graph characteristics are given in table 10) and on one random network. Labels are computed according to the "triangle" method, abd with the 5, 10 and 20-confluence methods. As can be seen, the orders of magnitude of the four different labeled categories of pairs of edges are similar with the different methods. We note that most of the edges are confirmed in the two first terrain networks. This is not the case on the 3rd one where only about one third are confirmed. This is due to the fact that the synonyms network extracted from Wiktionary is very incomplete [14], which is not the case for the two previous networks that are based on linguistic resources that have been established for a long time. In the case of random network, the reported results are the average of the results obtained for 20 random networks of the same size, and we can notice that almost none of the edges are confirmed.

Note that if a vertex is connected to a large part of all the vertices, the triangle method would abusively consider as close all the pairs of neighbors of this vertex. This would not be generally the case with the random walk method.

Note also that these labeling methods may be restricted if one know, for instance, that the graph is fully correct. Indeed it will mean that every edge exists even if it is not confirmed by the topology. Therefore the labeling procedure could then be only applied to non-edges, and all edges are labeled 1!. Conversely, if one knows that the graph is complete, and thus all non-edges are certain and labeled 0!, while edges are labeled according to the graph topology.

3 Comparing Graphs Having the Same Vertex Set

Comparing graphs is important in order to determine to what extent they contain the same information. In the following, we assume that the two graphs have *the same set of vertices*. In practice, this assumption mean that we compare two pieces of network information pertaining to the same set of objects or agents. For example, if a first graph represents friendship relation among a set of people, and a second one represents co-working relation inside the same set of people, one may be interested to know to what extent these two relations are similar, or if one relation is included (or "almost" included) in the other.

In the following subsection, we propose a naive method for comparing two graphs by counting the number of matches "at the edge level". We shall see the limitation of this method. We then use the labeling method described in the previous section to compare graphs in a more robust way.

3.1   Classical Agreement Measure between Edges

A simple method for comparing two graphs (having the same set of vertices) is to count on how many edges and no-edges they agree. Table 4 summarizes the 4 different cases: ok^+ is the number of edges present in both graphs, ok^- the number of non-edges present in both graphs, whereas ko_1 is the number of pairs that are linked by an edge in the first graph, but not in the second one, and ko_2 the number of pairs that are linked by an edge in the second graph, but not in the first one.

Table 4. Fusion of two graphs $G_1 = (V, E_1)$ and $G_2 = (V, E_2)$

	E_1	$\overline{E_1}$
E_2	ok^+	ko_1
$\overline{E_2}$	ko_2	ok^-

We use Cohen's kappa coefficient [4] as a simple measure of agreement between two graphs. It is a inter-judge agreement measure. Here we consider each graph as a judge that annotates each pair of vertices either as "edge" or "non-edge". It is defined as follows:

$$Kappa(G_1, G_2) = \frac{p_0 - p_e}{1 - p_e} \qquad (5)$$

with:

$$p_0 = \frac{1}{\omega}.(ok^+ + ok^-) = \frac{1}{\omega}.(|E_1 \cap E_2| + |\overline{E_1} \cap \overline{E_2}|) \qquad (6)$$

$$p_e = \frac{1}{\omega^2}.(|E_1|.|E_2| + |\overline{E_1}|.|\overline{E_2}|) \qquad (7)$$

It has the advantage to take into account the agreement on edges (ok^+) and on non-edges (ok^-), without being influenced by the strong difference that exists in a terrain network between the size of these two sets (graphs are usually sparse, and thus there are many more non-edges than edges). Another alternative could be to measure the agreement only on edges, by using Jaccard coefficient

$$Jaccard(G_1, G2) = \frac{|E_1 \cap E_2|}{|E_1 \cup E_2|} = \frac{ok^+}{ko_1 + ko_2 + ok^+} \qquad (8)$$

between the two sets of edges. However we observe that these two measures behave in similar ways in the experimentations.

The column "edges" in Table 7 gives the values of the kappa and Jaccard coefficients on two pairs of synonymy networks. One can already note that this value are low, which seems to attest a low agreement between synonymy networks. We comment these results more in detail in the section 3.3.

To demonstrate that the two above coefficients alone are insufficient for accounting for a global topological similarity of the graphs beyond the exact comparison pair of vertices by pair of vertices, we consider the following experiment. We build a graph

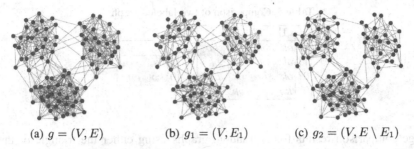

(a) $g = (V, E)$ (b) $g_1 = (V, E_1)$ (c) $g_2 = (V, E \setminus E_1)$

Fig. 2. Artificial graphs with 3 clusters. g_1 and g_2 are subgraph of g and share the same 3 clusters but they have no edge in common.

$g = (V, E)$ with 3 groups of 30 vertices, where edges are built randomly with a probability 0.4 between two vertices of the same group, and 0.01 between vertices of two different groups. We then build a new graph $g_1 = (V, E_1)$ by randomly choosing half of the edges of g, and a new graph $g_2 = (V, E_2)$ such that $E_2 = E \setminus E_1$. These 3 graphs are plotted in Figure 2. The kappa measure between the two graphs g_1 and g_2 is negative (≈ -7.5 on 20 realizations) and the Jaccard measure equals 0. This would mean that these two graphs are completely dissimilar, which is true in the sense that they have no edges in common, however it is clearly wrong with respect to the topological "organization" they share. Indeed two vertices that are in the same group in the first graph will also be in the same group in the two other graphs. The above comparison methods have the drawback of only comparing graphs as "bag of edges", thus ignoring the topological structure created by these edges. We propose in the next section to use the labeling method presented in section 2.2 in order to propose a similar comparison method which does not suffer of this drawback.

3.2   Using the Graph Topology Information

The labeling procedure described in Section 2.2 brings back topology information on each pair of vertices. We use this labeling procedure for comparing the two graphs, "pairs of vertices by pairs of vertices", without now missing the graph topology information. More precisely, if a pair is an edge confirmed by the structure in a graph (label 1!), but is a non-edge not confirmed by the structure in the other graph (label 0?) we consider that the two graphs do not disagree on this pair. Indeed the two vertices are topologically "close" in both graphs, even if they are adjacent in one, but not in the other. Similarly, if a pair is an unconfirmed edge in one graph (label 1?) and a confirmed non-edge in the other (label 0!), we consider that the two graphs agree on this pair as the two vertices are not "close" in any of the two graphs. The table 5 summarized the 16 different possible cases for a pair of vertices. We can now use the same kappa or Jaccard coefficients as in the previous section, but now by counting as agreeing pairs those labeled 0? in a graph and 1! in the other, or 0! in one and 1? in the other.

When we compare the two random graphs described in the previous subsection (see Figure 2) with this method that takes into account topology information, they appear to have a kappa (and a Jaccard) coefficient much higher than initially. Table 6 gives

Table 5. Comparison of two labeled graph

	1!	1?	0?	0!
1!	ok^+	ok^+	ok^+	ko_1
1?	ok^+	ok^+	ko_1	ok^-
0?	ok^+	ko_2	ok^-	ok^-
0!	ko_2	ok^-	ok^-	ok^-

$\Rightarrow$

	1	0
1	ok^+	ko_1
0	ko_2	ok^-

average comparison results for 20 random graphs using either the triangle or the 5-confluence method. We can see that they have no edges in common (by construction) but many of the edges present and confirmed (1!) in one graph are pairs of "close" vertices in the other (0?). We can also note that, as expected, the confluence method gives better results.

Table 6. Robust comparison of the graphs g_1 and g_2 of the Figure 2. Average value on 20 realizations.

(a) triangle, $kappa = 0.651$

	1!	1?	0?	0!
1!	0.0	0.0	127.3	58.9
1?	0.0	0.0	58.1	35.2
0?	128.7	59.8	390.2	302.3
0!	55.2	35.5	313.5	2440.2

(b) 5-confluence, $kappa = 0.881$

	1!	1?	0?	0!
1!	0.0	0.0	229.2	5.0
1?	0.0	0.0	31.9	13.3
0?	207.2	54.2	775.1	67.5
0!	3.1	14.4	58.1	2545.7

3.3 Comparison of Synonymy Networks

We illustrate the method proposed here on the comparison of pairs of synonymy networks. In such networks, one may expect that almost all edges are correct, even if few ones are "questionable", and that a large part of the non-edges are not related at all, even if some pairs of words are very close (but not really synonymous). We consider the networks $V.rob$ and $V.lar$, two synonymy networks between French verbs[2] and the networks $V.wikt$ and $V.pwn$, two synonymy networks between English verbs[3]. Table 7 gives the comparison results for the 2 French synonymy networks ($V.rob$, $V.lar$) and the 2 English synonymy networks ($V.wikt$ and $V.pwn$). Since these different networks do not have exactly the same lexical coverage, the comparison is based on the common sets of vertices. As can be seen, there is only a weak agreement between pairs of graphs, when they are compared by the classical agreement measure. This may not be

---

[2] $V.rob$ and $V.lar$ are two synonymy networks between French verbs. There where digitalized from paper dictionaries (Robert and Larousse dictionaries) by an IBM/ATILF research unit partnership http://www.atilf.fr/spip.php?article208

[3] $V.wikt$ and $V.pwn$ two synonymy networks between English verbs. $V.wikt$ has been extracted from the English wiktionary by [14] whereas $V.pwn$ is built from Princeton Wordnet [5] synsets. A synset is a set of interchangeable words that denotes a meaning or a particular usage. The vertices of the network $V.pwn$ are the lemmas of the verbs present in Wordnet, and there is an edge $(x, y) \in E$ if and only if x and y belong to at least one common synset

expected, especially for the French graphs since they are obtained from authoritative general purpose dictionaries. Once the topology information is taken into account, we observe a strong agreement between the French graphs (up to 95 %) in the sense that almost all the edges of one graph are retrieved in the other as 0? labeled non-edges. In other words, most of the initial disagreements pertain to pairs of vertices that are close in both graphs (even if they are not adjacent in one). For the English graph, the agreement remains relatively weak. This is due to the fact that the wiktionary is very sparse, and not built at all in the same way as Wordnet. In Wordnet each edge reflects a common belonging to a synset, while the wiktionary graph edges are built by non expert contributors (without special care about synsets).

Table 8 provides similar comparisons on fictitious random graphs having the same overlaps as the French and English pairs of graphs previously considered. The obtained results strongly contrast with the previous ones, as expected. The random graphs still finally reach another form of strong agreement (now on "non-edges"), *but* only because the initial disagreement pertain to pairs that are *not* close in both graphs even if they are adjacent in one graph.

Table 7. Synonymy network comparison. Column "edges" gives the measures without the labeling procedure. $(0?, 1!)$ (resp. $(0!, 1?)$) indicates the number of pairs of vertices labeled 0? (resp. 0!) in one graph and 1! (resp. 1?) in the other.

		edges	triangle	5-confl.	10-confl.	20-confl.
V.rob vs. *V.lar*	*Kappa*	0.518	0.876	0.937	0.953	0.946
	Jaccard	0.350	0.781	0.882	0.910	0.898
	$(0?, 1!)$	-	11769	16310	17401	16878
	$(0!, 1?)$	-	2860	1129	881	1050
V.wikt vs. *V.pwn*	*Kappa*	0.202	0.498	0.600	0.636	0.673
	Jaccard	0.113	0.332	0.429	0.467	0.507
	$(0?, 1!)$	-	2511	3878	4485	5027
	$(0!, 1?)$	-	2246	1919	1667	1641

Table 8. Graphs comparison results on Erdös Rényi random network having the same initial overlaps as the real networks. Average on 20 realizations.

		edges	triangle	5-confl.	10-confl.	20-confl.
rob^R vs. lar^R	*kappa*	0.518	0.850	0.803	0.763	0.719
	Jaccard	0.351	0.739	0.671	0.617	0.562
	$(0?, 1!)$	-	801.4	1193.0	1097.8	0.2
	$(0!, 1?)$	-	16001.4	13734.2	12235.0	12209.9
$wikt^R$ vs. pwn^R	*Kappa*	0.203	0.716	0.592	0.541	0.595
	Jaccard	0.113	0.558	0.420	0.371	0.424
	$(0?, 1!)$	-	13.1	68.5	165.1	106.5
	$(0!, 1?)$	-	11865.0	10734.4	9919.4	10678.6

4   Fusing Graphs

The same idea can also be used when merging two graphs. Let $G_1 = (V, E_1)$ and $G_1 = (V, E_2)$ be two graphs. A first type of merging could be to add, to the intersection of the two edge sets, the pairs of vertices that are labeled 1! in one graph and 0? in the other, i.e.,

$$E'_\cap = (E_1 \cap E_2) \cup \{\text{pairs labeled } (1!, 0?)\}$$

Another type of merging (more tolerant) consists in removing the pairs of vertices labeled 1? in one graph and 0! in the other from the union of the edge sets of the two graphs, i.e.,

$$E'_\cup = (E_1 \cup E_2) \setminus \{\text{pairs labeled } (1?, 0!)\}$$

These two fusion procedures are such that the resulting edge sets $E'_\cap$ and $E'_\cup$ satisfy the following inclusions:

$$E_1 \cap E_2 \subset E'_\cap \subset E'_\cup \subset E_1 \cup E_2$$

This is illustrated with two graphs about e-mail relations between people. More precisely, we build an ego-centric social network from someone mailbox: each e-mail address u (which means more or less a person) is connected to another e-mail address v, iff u is the author of -at least- one mail having v as recipient ("To" or "CC"). It may be worth of interest to fuse such a graph built from all e-mails during a given year with the same graph built from e-mails of the previous year: we can then see which parts of the graph have been stable during these two years. The results corresponding to two social networks built from the e-mails of one person are shown in Table 9. Note that

Table 9. Example of fusion of two social networks built from the e-mails of one of the paper's author for two different years. Pedigrees of these graphs are in Table 10: $mail10$ and $mail11$.

	1!	0?	0!
1!	65	16	32
0?	13	80	99
0!	38	110	1377

$|E_1 \cap E_2| = 65$
$|E'_\cap| = 94$
$|E'_\cup| = |E_1 \cup E_2| = 164$

Table 10. Pedigrees of 6 different terrain networks, n and m are respectively the number of vertices and edges, $\langle k \rangle$ is the average degree of vertices, n_{lcc} and m_{lcc} are the number of vertices and edges in the largest connected component, C is the transitivity coefficient of the graph, L_{lcc} is the average shortest path between any two nodes of the largest connected component, λ is the coefficient of the best fitting power law of the degree distribution and r^2 is the correlation coefficient of the fit.

	n	m	$\langle k \rangle$	n_{lcc}	m_{lcc}	C	L_{lcc}	λ	r^2
$V.rob$	7357	26567	7.48	7056	26401	0.12	4.59	-2.01	0.93
$V.lar$	5377	22042	8.44	5193	21926	0.17	4.61	-1.94	0.88
$V.wikt$	7339	8353	2.84	4285	6093	0.11	8.98	-2.40	0.94
$V.pWN$	11529	40919	8.16	9674	39459	0.24	4.66	-2.10	0.92
$mail10$	385	603	3.14	383	602	0.10	3.71	-1.11	0.73
$mail11$	391	671	3.45	389	671	0.06	3.32	-0.93	0.55

here the second fusing method $E'_\cup$, will give the same result as $E_1 \cup E_2$ since there is no edges labeled 1? as all the edges may be considered as "sure" since they rely on at least one existing e-mail. As can be seen 29 edges $(16 + 13)$ are restored on top of the edge sets intersection. Thus, they should not be count among the real change that took place between the two years.

5   Related Work

In the literature, the idea of graph comparison may refer to various problems and approaches. A first group of works deals with approaches that evaluate to what extent two graphs are isomorphic, or looks for approximate isomorphisms between two graphs. Measuring how two graphs are similar is a common problem for querying graph databases. Some methods [8,18] use an edit distance between graphs. Other approaches measure the size of the maximal common subgraph[16,20]. A related problem is to find a matching, or approximate matching between two graphs [11]. It consists in looking for a correspondence between vertices of one graph and vertices of the other such that the two graphs appears as similar as possible. The kappa and Jaccard measures (between not-labeled graphs) proposed in section 3.1 are comparable to such approaches in the "very" particular case where graphs have exactly the same vertices, and where each vertex cannot be put in correspondence with another one but itself. Besides, [2] proposes a different way of measuring graph similarity. This method gives a similarity score between any vertex of one graph and any vertex of a second graph. It applies between any pair of graphs, and does not consider any correspondence between vertices of the two graphs. So it may be applied when the two graphs are on the same set of vertices, however this knowledge is not taken into account by the method. A second group of works proposes to compare graphs by global statistical features [10], or compare graphs by measuring the number of occurrences of small particular sub-graphs [12]. To the best of our knowledge there was no work interested in comparing two graphs having the same set of vertices and taking into account the graph structure, if we except maybe [6].

6   Concluding Remarks

This paper has presented a method that provides an augmented view of a undirected graph which acknowledges its underlying structure. This augmented view turns to be useful when comparing or fusing graphs, as illustrated in this paper, when we need to go beyond a purely "edge" by "edge" pairing. An obvious line for further research is the extension of the approach to weighted and/or directed graphs.

References

1. Albert, R., Barabási, A.: Statistical mechanics of complex networks (2001)
2. Blondel, V.D., Gajardo, A., Heymans, M., Senellart, P., Dooren, P.V.: A measure of similarity between graph vertices: Applications to synonym extraction and web searching. SIAM Rev. 46, 647–666 (2004)

3. Bollobas, B.: Modern Graph Theory. Springer (October 2002)
4. Cohen, J.: A coefficient of agreement for nominal scales. Educ. Psychol. Meas. 20(1), 37–46 (1960)
5. Fellbaum, C. (ed.): WordNet: An Electronic Lexical Database. MIT Press (1998)
6. Gaillard, B., Gaume, B., Navarro, E.: Invariants and variability of synonymy networks: Self mediated agreement by confluence. In: TextGraphs-6, ACL, pp. 15–23 (2011)
7. Gaume, B.: Balades Aléatoires dans les Petits Mondes Lexicaux. I3: Information Interaction Intelligence 4(2) (2004)
8. He, H., Singh, A.K.: Closure-tree: An index structure for graph queries. In: Proc. of the 22th IEEE Int. Conf. on Data Engineering (ICDE), p. 38 (April 2006)
9. Liben-Nowell, D., Kleinberg, J.: The link-prediction problem for social networks. Journal of the American Society for Information Science and Technology 58(7), 1019–1031 (2007)
10. Macindoe, O., Richards, W.: Graph comparison using fine structure analysis. In: Second IEEE Int. Conf. on Social Computing, pp. 193–200 (August 2010)
11. Melnik, S., Garcia-Molina, H., Rahm, E.: Similarity flooding: a versatile graph matching algorithm and its application to schema matching. In: Proceedings of the 18th International Conference on Data Engineering 2002, pp. 117–128 (2002)
12. Milo, R., Itzkovitz, S., Kashtan, N., Levitt, R., Shen-Orr, S., Ayzenshtat, I., Sheffer, M., Alon, U.: Superfamilies of evolved and designed networks. Science 303(5663), 1538–1542 (2004)
13. Pons, P., Latapy, M.: Computing communities in large networks using random walks (long version). Journal of Graph Algorithms and Applications (JGAA) 10(2), 191–218 (2006)
14. Sajous, F., Navarro, E., Gaume, B., Prévot, L., Chudy, Y.: Semi-automatic enrichment of crowdsourced synonymy networks: the wisigoth system applied to wiktionary. Language Resources and Evaluation, 1–34 (to appear)
15. Schaeffer, S.E.: Graph clustering. Computer Science Review 1(1), 27–64 (2007)
16. Shang, H., Zhu, K., Lin, X., Zhang, Y., Ichise, R.: Similarity search on supergraph containment. In: Proc. of the 26th IEEE Int. Conf. on Data Engineering (ICDE), pp. 637–648 (March 2010)
17. Stewart, G.W.: Perron-frobenius theory: a new proof of the basics. Technical report, College Park, MD, USA (1994)
18. Tian, Y., Patel, J.M.: Tale: A tool for approximate large graph matching. In: Proc. of the 24th IEEE Int. Conf. on Data Engineering (ICDE), pp. 963–972 (2008)
19. Watts, D., Strogatz, S.: Collective dynamics of 'small-world' networks. Nature 393, 440–442 (1998)
20. Yan, X., Yu, P.S., Han, J.: Substructure similarity search in graph databases. In: Proc. of the 2005 ACM Int. Conf. on Management of Data (SIGMOD), pp. 766–777 (2005)

A Characteristic Function Approach to Inconsistency Measures for Knowledge Bases

Jianbing Ma, Weiru Liu, and Paul Miller

School of Electronics, Electrical Engineering and Computer Science,
Queen's University Belfast, Belfast BT7 1NN, UK
{jma03,w.liu,p.miller}@qub.ac.uk

Abstract. Knowledge is an important component in many intelligent systems. Since items of knowledge in a knowledge base can be conflicting, especially if there are multiple sources contributing to the knowledge in this base, significant research efforts have been made on developing inconsistency measures for knowledge bases and on developing merging approaches. Most of these efforts start with flat knowledge bases. However, in many real-world applications, items of knowledge are not perceived with equal importance, rather, weights (which can be used to indicate the importance or priority) are associated with items of knowledge. Therefore, measuring the inconsistency of a knowledge base with weighted formulae as well as their merging is an important but difficult task. In this paper, we derive a numerical characteristic function from each knowledge base with weighted formulae, based on the Dempster-Shafer theory of evidence. Using these functions, we are able to measure the inconsistency of the knowledge base in a convenient and rational way, and are able to merge multiple knowledge bases with weighted formulae, even if knowledge in these bases may be inconsistent. Furthermore, by examining whether multiple knowledge bases are dependent or independent, they can be combined in different ways using their characteristic functions, which cannot be handled (or at least have never been considered) in classic knowledge based merging approaches in the literature.

Keywords: Knowledge Bases, Characteristic Function, Inconsistency Measure, Merging, Evidence Theory.

1 Introduction

Logic based knowledge representation is used in many cases, such as software requirements [20], expert systems [22], belief merging [9]. In most of the applications, logic based knowledge bases (KB) are flat, that is all formulae in the base are equally important. However, in some applications, such as requirement engineering [21], some formulae can be more important than others. So ranked or stratified knowledge bases are commonly deployed. The importance of a formula can also be modelled by attaching a numerical value to the formula, which in some case is explained as a *weight*.

When a numerical value is attached to a logical formula, this value can be explained in many different ways according to the semantics of this value. Some typical explanations are belief degrees, preference degrees, truth degrees, trust degrees. In this paper, we consider a numerical value as a weight indicating the importance (or priority) of

E. Hüllermeier et al. (Eds.): SUM 2012, LNAI 7520, pp. 473–485, 2012.

this formulae w.r.t other formulae in the same knowledge base and we study both flat KBs in which all the logical formulae are viewed as equally important and weighted knowledge bases in which each formula is associated with a weight.

It should be noted that the meaning of a knowledge base having weights attached to them is different from the *weighted knowledge bases* discussed in [10], in which a weighted knowledge base means that a knowledge base as a whole is attached with a weight representing the relative degree of importance (or reliability) of the source from which the knowledge base is derived.

There are two frequently studied topics for knowledge bases. One is on measuring the inconsistency of a knowledge base (e.g., [20,8]) and the other is the merging of multiple knowledge bases e.g., [9,12,13]). Most of these approaches deploy logic-based formalisms, even when priorities or degrees of certainty are involved. An interesting phenomenon is that although these two issues are closely related (if we view the aim of merging is to obtain a consistent knowledge base), their solutions usually follow different paths. Therefore, a natural question is: can we develop an underlying formalism which can handle both of the issues simultaneously?

Another potential problem on merging multiple knowledge bases is that, so far in the literature, merging of knowledge bases does not really consider any possible dependency relationship among knowledge bases, which subsequently causes the difficulty of justifying a merging result. For example, if two experts have provided their knowledge in terms of weighted formulae, and one expert has been heavily influenced by another, then their knowledge bases are not totally independent. That is if K_1, K_2 are the two knowledge bases from these two experts, and they are represented as $K_1 = K_2 = \{(\alpha, 0.8), (\beta, 0.2)\}$, then the merging result K should be identical to either of them, if K_1 and K_2 are dependent. On the other hand, if K_1 and K_2 are independent, then we should expect that the weight on α being increased and the weight on β being decreased. Formally, K should be $\{(\alpha, x), (\beta, y)\}$ such that $x > 0.8$ and $y < 0.2$. That is, whether some knowledge bases have a dependent relationship should influence how merging should be carried out. However, this issue has not been explicitly discussed in classical logic based approaches for merging knowledge bases in the literature. Thus, another question could be asked here is: can we reflect the information on dependency relationship among knowledge bases when performing merging?

In this paper, we provide positive answers for both questions mentioned above. We propose a *characteristic function* for a knowledge base with formulae having weights, and a flat knowledge base is treated as a special case where all formulae having the same weight. The characteristic function entails all the information of a knowledge base provides and hence can be used to measure the inconsistency of the knowledge base and to handle the merging of multiple knowledge bases. Characteristic functions are defined in the form of basic probability assignments in the Dempster-Shafer (DS) theory [3,4,23,15,16].

Example 1. *In [18,19,17], an intelligent surveillance system was designed and developed. In this project, interested events are recognized from analyzing data coming from different sources (e.g., cameras) and these event descriptions usually contain uncertain information, such as, an gender-profile event describes a passenger as a Male with 70% probability and the rest is unknown. Hence DS theory is introduced to model the* *uncertainty implied in the events and combination approaches are applied to combine gender recognition events descriptions from multiple sources. Furthermore, a knowledge base is developed which contain inference rules for inferring high-level events from a set of elementary recognized events.*

Rules are elicited from domain experts. Each expert provides its knowledge containing multiple rules, each of which is attached with a weight indicating its importance. Examples of simple rules are as follows:

Let a denote a passenger A is a male, b denote A is shouting, c denote A is dangerous, and d denote A is an old lady. Then an expert's knowledge could contain $K = \{(a \wedge b \to c, 0.8), (d \wedge b \to \neg c, 0.2)\}$ which is semantically explained as if a passenger A is a Male and A is shouting, then passenger A is dangerous, if a passenger A is an old lady and A is shouting, then passenger A is not dangerous.

Here, numerical values 0.8 and 0.2 are the weights of these two rules, which says that a rule indicating a dangerous event is more important than otherwise[1].

When different experts provide their knowledge in terms of such knowledge base with weighted formulae, we need to merge them. This can be achieved by using their corresponding characteristic functions.

This characteristic function approach also absorbs some nutrients from papers such as [1,13,14], etc., in which methods of inducing probability measures from knowledge bases are studied that demonstrates the demand and usefulness of quantitative methods on managing knowledge bases.

The main contribution of this paper is as follows:

- From a knowledge base with weighted formulae, a unique basic belief assignment (bba) can be recovered. To the best of our knowledge, there is no paper having emphasized this point.
- We show that the associated bba could be used to measure the inconsistency of the knowledge base and to merge multiple knowledge bases in a quantitative way that is beyond the usual approaches in the knowledge base inconsistency measure / merging fields.
- We show that merging of knowledge bases can take into account dependencies between knowledge bases using this approach.

The rest of the paper is organized as follows. In Section 2, we recall some basic concepts and notations of propositional language, knowledge bases and the evidence theory. In Section 3, we define the characteristic function of a knowledge base. In Section 4, we provide an inconsistency measure of a knowledge base and show some rational properties of this inconsistency measure. In Section 5, we discuss various merging methods of knowledge bases using their characteristic functions. Finally, in Section 6, we conclude the paper.

---

[1] It should be noted that this weight about the importance of the rule should not be confused with a statistical value (such as 0.8) showing the likelihood of how dangerous a shouting male passenger could be. The later is explained as that when a male passenger is shouting, there is 80% chance this will lead to a dangerous consequence.

2   Preliminaries

2.1   Knowledge Bases

Here we consider a propositional language $\mathcal{L}_\mathcal{P}$ defined from a finite set $\mathcal{P}$ of propositional atoms, which are denoted by p, q, r etc (possibly with sub or superscripts). A proposition ϕ is constructed by atoms with logical connectives $\neg, \wedge, \vee$ in the usual way. An interpretation w (or possible world) is a function that maps $\mathcal{P}$ onto the set $\{0, 1\}$. The set of all possible interpretations on $\mathcal{P}$ is denoted as W. Function $w \in W$ can be extended to any propositional sentence in $\mathcal{L}_\mathcal{P}$ in the usual way, $w : \mathcal{L}_\mathcal{P} \to \{0, 1\}$. An interpretation w is a model of (or satisfies) ϕ iff $w(\phi) = 1$, denoted as $w \models \phi$. We use $Mod(\phi)$ to denote the set of models for ϕ.

For convenience, let $form(\{w_1, \cdots, w_n\})$ be the formula whose models are exactly $w_1, \cdots, w_n$, and also $form(A)$ denote a formula μ such that $Mod(\mu) = A$.

A *flat knowledge base* K is a finite set of propositional formulas. K is consistent iff there is at least one interpretation that satisfies all the propositional formulas in K.

A *weighted knowledge base* K is a finite set of propositional formulas, each of which has a numerical value, called weight, attached to it i.e., $\{(\mu_1, x_1), \cdots, (\mu_n, x_n)\}$ where $\forall i, 0 < x_i \leq 1$. In fact, the requirement $x_i \in [0, 1]$ can be relaxed to allow x_i taking any positive numerical value. In that case, a normalization step will reduce each x_i to a value within $[0, 1]$. Obviously, if all x_is are equivalent, then a weighted knowledge base is reduced to a flat knowledge base. Conversely, a flat knowledge base can be seen as a weighted knowledge base, e.g., $K = \{(\mu_1, \frac{1}{n}), \cdots, (\mu_n, \frac{1}{n})\}$. Therefore, for convenience, in the rest of the paper, we refer to all knowledge bases as weighted knowledge bases.

For a weighted knowledge base $K = \{(\mu_1, x_1), \cdots, (\mu_n, x_n)\}$, we let $\hat{K} = \{\mu_1, \cdots, \mu_n\}$ be its corresponding flat knowledge base in which the weights of all the formulae of K are removed. For a formula μ, we write $\mu \in K$ if and only if $\mu \in \hat{K}$. In addition, K is consistent if and if $\hat{K}$ is consistent.

If a classical knowledge base $\hat{K}$ is inconsistent, then we can define its minimal inconsistent subsets as follows [2,7]:

$$MI(\hat{K}) = \{\hat{K}' \subseteq \hat{K} | \hat{K}' \vdash \bot \text{ and } \forall \hat{K}'' \subset \hat{K}', \hat{K}'' \not\vdash \bot\}.$$

A *free formula* of a knowledge base $\hat{K}$ is a formula of $\hat{K}$ that does not belong to any minimal inconsistent subset of the knowledge base $\hat{K}$ [2,7]. A free formula in a weighted knowledge base K is defined as a free formula of $\hat{K}$.

2.2   Evidence Theory

We also recall some basic concepts of Dempster-Shafer's theory of evidence.

Let Ω be a finite set called the frame of discernment (or simply frame). In this paper, we denote $\Omega = \{w_1, \ldots, w_n\}$.

Definition 1. *A basic belief assignment (bba for short) is a mapping* $m : 2^\Omega \to [0, 1]$ *such that* $\sum_{A \subseteq \Omega} m(A) = 1$.

A bba m is also called a mass function when $m(\emptyset) = 0$ is required. A vacuous bba m is such that $m(\Omega) = 1$.

If $m(A) > 0$, then A is called a focal element of m. Let $\mathscr{F}(m)$ denote the set of the focal elements of m. That is, if A is a focal elements of m, then $A \in \mathscr{F}(m)$.

Let $\oplus$ be the conjunctive combination operator (or Smets' operator [24]) for any two bbas m, m' over Ω such that

$$(m \oplus m')(C) = \sum_{A \subseteq \Omega, B \subseteq \Omega, A \cap B = C} m(A)m'(B), \forall\, C \subseteq \Omega.$$

Particularly, we have:

$$(m \oplus m')(\emptyset) = \sum_{A \subseteq \Omega, B \subseteq \Omega, A \cap B = \emptyset} m(A)m'(B). \tag{1}$$

A simple bba m such that $m(A) = x, m(\Omega) = 1 - x$ for some $A \neq \Omega$ is denoted as A^x. The vacuous bba can thus be denoted as A^0 for any $A \subset \Omega$. By abuse of notations, we also use μ^x to denote the simple bba A^x where $A = Mod(\mu)$. Note that this notation, i.e., A^x, is a different from the one defined in [5] such that A^x in our paper should be denoted as A^{1-x} based on explanations in [5].

3   Characteristic Functions

In this section, we define characteristic functions for knowledge bases.

A first and direct thought for characteristic function of a knowledge base $K = \{(\mu_1, x_1), \cdots, (\mu_n, x_i)\}$ is to define it as follows.

$$m'_K(Mod(\mu_i)) = x_i, 1 \leq i \leq n.$$

But this characteristic function definition brings many problems. For example, it is difficult to use this characteristic function to measure inconsistency of a single knowledge base. Since the usual way of defining inconsistency of a mass function is its *empty mass*. Hence, a simple definition of the inconsistency of K could be:

$$Inc'(K) = (m'_K \oplus m'_K)(\emptyset).$$

That is, the internal inconsistency of K is the empty mass when K interacts with itself, i.e., $m'_K \oplus m'_K$.

But it does not give reasonable results. For instance, if m'_K is such that $m'_K(\{w_1, w_2\}) = m'_K(\{w_1, w_3\}) = m'_K(\{w_2, w_3\}) = \frac{1}{3}$, then we have $(m'_K \oplus m'_K)(\emptyset) = 0$. But the corresponding knowledge base $\hat{K} = \{(form(\{w_1, w_2\}), \frac{1}{3}), (form(\{w_1, w_3\}), \frac{1}{3}), (form(\{w_2, w_3\}), \frac{1}{3})\}$ is not consistent since $form(\{w_1, w_2\}) \wedge form(\{w_1, w_3\}) \wedge form(\{w_2, w_3\}) \vdash \bot$.

Therefore, we define our characteristic function for a weighted knowledge base as follows.

Definition 2. *For any weighted knowledge base $K = \{(\mu_1, x_1), \cdots, (\mu_n, x_i)\}$, its corresponding characteristic function is m_K such that $m_K = \oplus_{i=1}^{n} \mu_i^{x_i}$.*

This characteristic function makes use of both the formulae (μ_i) in K and their weights (x_i), which produces a bba that is a kind of characterization of K. More precisely, the defined characteristic function m_K is unique for K. That is, if K_1 and K_2 are logically different[2], then m_{K_1} should also be different from m_{K_2}. This is ensured by the following result [5]:

A bba m such that $m(\Omega) > 0$ can be uniquely decomposed into the following form:

$$m = \oplus_{\phi:Mod(\phi) \subset \Omega} \phi^{x_\phi}, x_\phi \in [0,1]. \tag{2}$$

That is, if $m_{K_1} = m_{K_2}$, then based on Equation 2, they have the same decomposition and hence the same knowledge base (in a sense that vacuous information is ignored). Equation 2 also demonstrates that the bba m_K encodes all the information contained in K. In fact, we can easily recover K from m_K with a few steps. Since recovering K is not the main focus of this paper, and due to the space limitation, the details of how to recover μ_i and x_i from m_K is omitted here. Interested readers could be refer to [5].

A simple result about this characteristic function is shown as follows.

Proposition 1. *For any weighted knowledge base K and any x such that $0 < x \le 1$, we have $m_K = m_{K \cup \{\top, x\}}$.*

That is, vacuous information does not change the characteristic function. This is an intuitive result and it does not contradict the former statement that m_K induces a unique K, since vacuous information in K could somehow be ignored. For example, if someone tells you: tomorrow will be either sunny or not sunny. Obviously this piece of vacuous information could be ignored.

4   Inconsistency Measure

In this section, we use the characteristic function of a knowledge base to measure its inconsistency. In addition, we prove that this inconsistency measure satisfies a set of rational properties proposed in [7].

Definition 3. *For any weighted knowledge base K, the inconsistency measure of K is defined as:*
$$Inc(K) = m_K(\emptyset).$$

Taking $m_K(\emptyset)$ as the inconsistency measure for K is very natural since in DS theory, $m_K(\emptyset)$ is a largely used to measure the degree of conflict between beliefs of agents[3].

Now we show that this definition of inconsistency satisfies some intuitive properties. In [7], a set of properties that an inconsistency measure I for a knowledge base shall have is proposed as follows.

Definition 4. *([7]) An inconsistency measure I is called a basic inconsistent measure if it satisfies the following properties:*
for any flat knowledge bases K, K' and any two formulae α, β:

---

[2] That is, two different but logically equivalent formulas are considered equivalent here.

[3] While in several papers, most notably in [11], it is argued that $m_K(\emptyset)$ is not enough for an inconsistency measure for bbas. However, in most applications, $m_K(\emptyset)$ is still being used to measure the inconsistency between bbas.

Consistency. $I(K) = 0$ *iff K is consistent*
Normalization. $0 \leq I(K) \leq 1$
Monotony. $I(K \bigcup K') \geq I(K)$
Free Formula Independence. *If α is a free formula of $K \bigcup \{\alpha\}$, then $I(K \bigcup \{\alpha\}) = I(K)$*
Dominance. *If $\alpha \vdash \beta$ and $\alpha \not\vdash \bot$, then $I(K \bigcup \{\alpha\}) \geq I(K \bigcup \{\beta\})$*

For weighted knowledge bases, the above definition should be adapted to:

Definition 5. *An inconsistency measure I is called* a basic inconsistent measure for weighted knowledge bases *if it satisfies the following properties:*
for any weighted knowledge bases K, K', any two formulae α, β and any real x, $0 < x \leq 1$:

Consistency. $Inc(K) = 0$ *iff K is consistent.*
Normalization. $0 \leq Inc(K) \leq 1$.
Monotony. $Inc(K \bigcup K') \geq Inc(K)$.
Free Formula Independence. *If α is a free formula of $K \bigcup \{(\alpha, x)\}$, then $Inc(K \bigcup \{(\alpha, x)\}) = Inc(K)$.*
Dominance. *If $\alpha \vdash \beta$ and $\alpha \not\vdash \bot$, then $Inc(K \bigcup \{(\alpha, x)\}) \geq Inc(K \bigcup \{(\beta, x)\})$.*

We prove that our inconsistency measure satisfies all the above properties. In addition, we show that our inconsistency measure satisfies a *Strong Free Formula Independence* property as follows.

Strong Free Formula Independence. $Inc(K \bigcup \{(\alpha, x)\}) = Inc(K)$ if and only if α is a free formula of $K \bigcup \{(\alpha, x)\}$.

Proposition 2. *For any weighted knowledge base K, $Inc(K)$ is a basic inconsistency measure. In addition, $Inc(K)$ satisfies the* Free Formula Independence *property.*

Proof of Proposition 2

Consistency. $Inc(K) = 0$ iff K is consistent.
$\quad Inc(K) = 0$ iff $\forall \mu_i \in K, \bigcap_{i=1}^{n} \mu_i \not\vdash \bot$ iff K is consistent.
Normalization. $0 \leq Inc(K) \leq 1$.
$\quad$ Obvious.
Monotony. $Inc(K \bigcup K') \geq Inc(K)$.
$\quad$ We show that for any formula α and $0 < x \leq 1$, $Inc(K \bigcup \{(\alpha, x)\}) \geq Inc(K)$. In fact, we have

$$Inc(K \bigcup \{(\alpha, x)\}) = m_{(K \bigcup \{(\alpha, x)\})}(\emptyset)$$

$$= m_K(\emptyset) + \sum_{A \in \mathscr{F}(m_K), A \neq \emptyset, A \cap Mod(\alpha) = \emptyset} m_K(A) \times x$$

$$\geq m_K(\emptyset)$$

$$= Inc(K),$$

hence without loss of generality, assume $K' = \{(\alpha_1, x_1), \cdots, (\alpha_n, x_n)\}$, we then have

$$Inc(K) \leq Inc(K \bigcup \{(\alpha_1, x_1)\})$$
$$\leq \cdots$$
$$\leq Inc(K \bigcup K')$$

Free Formula Independence. If α is a free formula of $K \bigcup \{(\alpha, x)\}$, then $Inc(K \bigcup \{(\alpha, x)\}) = Inc(K)$.

If α is a free formula of $K \bigcup \{(\alpha, x)\}$, then there does not exist $A \in \mathscr{F}(m_K)$ and $A \neq \emptyset$, s.t., $A \cap Mod(\alpha) = \emptyset$, hence we have

$$Inc(K \bigcup \{(\alpha, x)\}) = m_K(\emptyset) + \sum_{A \in \mathscr{F}(m_K), A \neq \emptyset, A \cap Mod(\alpha) = \emptyset} m_K(A) \times x$$
$$= m_K(\emptyset)$$
$$= Inc(K).$$

Dominance. If $\alpha \vdash \beta$ and $\alpha \nvdash \bot$, then $Inc(K \bigcup \{(\alpha, x)\}) \geq Inc(K \bigcup \{(\beta, x)\})$.

It is straightforward from a simple fact that for any $A \in \mathscr{F}(m_K)$ and $A \neq \emptyset$, if $A \cap Mod(\beta) = \emptyset$, then $A \cap Mod(\alpha) = \emptyset$, hence

$$Inc(K \bigcup \{(\beta, x)\}) = m_K(\emptyset) + \sum_{A \in \mathscr{F}(m_K), A \neq \emptyset, A \cap Mod(\beta) = \emptyset} m_K(A) \times x$$
$$\leq m_K(\emptyset) + \sum_{A \in \mathscr{F}(m_K), A \neq \emptyset, A \cap Mod(\alpha) = \emptyset} m_K(A) \times x$$
$$= Inc(K \bigcup \{(\alpha, x)\})$$

Strong Free Formula Independence. $Inc(K \bigcup \{(\alpha, x)\}) = Inc(K)$ if and only if α is a free formula of $K \bigcup \{(\alpha, x)\}$.

It is obvious from the prove of **Free Formula Independence** property. $\square$

It is worth pointing out that our inconsistency measure can be naturally used to deal with weighted knowledge bases in addition to flat knowledge bases, while the existing inconsistency measures based on minimal inconsistency subsets or from a classic logic-based approach are incapable of tackling with weighted knowledge bases.

Example 2. *Let* $K_1 = \{(\alpha, 0.8), (\alpha \vee \beta, 0.2)\}$ *and* $K_2 = \{(\alpha, 0.6), (\beta, 0.3), (\alpha \vee \beta, 0.1)\}$. *Then the characteristic function of* K_1, *i.e.,* m_{K_1}, *is such that*

$$m_{K_1}(\alpha) = 0.8, m_{K_1}(\alpha \vee \beta) = 0.2.$$

Hence it is easy to see that $Inc(K_1) = 0$.

Then the characteristic function of K_2, *i.e.,* m_{K_2}, *is such that*

$$m_{K_2}(\emptyset) = 0.18, m_{K_2}(\alpha) = 0.42, m_{K_2}(\beta) = 0.12, m_{K_2}(\alpha \vee \beta) = 0.28.$$

Hence we easily get $Inc(K_2) = 0.18$.

5 Merging

In this section, we discuss the merging of weighted knowledge bases using their characteristic functions. Since characteristic functions are in the form of bbas, the merging methods are hence based on combination rules of bbas in DS theory. There are many different combination rules for bbas, which are similar to the many different merging strategies for knowledge bases. For simplicity, here we only mention Dempster's rule and Didier & Prade's hybrid rule.

For convenience and convention, if there is no confusion, in the following we may use both sets and formulae, e.g., $m(\emptyset) = 0.4, m(\phi) = 0.6$, etc., just be noted that any propositional formulae used in these situations, like ϕ, are in fact standing for $Mod(\phi)$. Conversely, when we write $K = \{(A, 0.5), (B, 0.3)\}$, it simply means $K = \{(form(A), 0.5), (form(B), 0.3)\}$. This short-hand notation is simply for making the mathematical formulas shorter and does not suggest any technical changes.

The subsequent definitions use combination rules of bbas, but as our aim is to merge knowledge bases, we will call them merging methods. Therefore, in the following, each definition defines a merging method for merging weighted knowledge bases using their characteristic functions. Also note that the merging methods only give the characteristic function for the merged knowledge base since from this function, its corresponding knowledge base can be easily induced (Equation (2), cf. [5] for details).

Definition 6. *(Dempster's Merging, [3,4,23]) Let K_1, K_2 be two knowledge bases and m_{K_1}, m_{K_2} be their characteristic functions, respectively, then the characteristic function $m_{K_{12}}$ of the merged knowledge base by Dempster's combination rule is such that:*

$$m_{K_{12}}^{Dem}(A) = \frac{\sum_{B,C \subseteq \Omega, B \cap C = A} m_1(B)m_2(C)}{1 - \sum_{B,C \subseteq \Omega, B \cap C = \emptyset} m_1(B)m_2(C)}, \forall A \subseteq \Omega, A \neq \emptyset,$$

$$m_{K_{12}}^{Dem}(\emptyset) = 0.$$

Definition 7. *(Dubois and Prade's Merging, [6]) Let K_1, K_2 be two knowledge bases and m_{K_1}, m_{K_2} be their characteristic functions, respectively, then the characteristic function $m_{K_{12}}$ of the merged knowledge base by DP's combination rule is such that:*

$$m_{DP}(\emptyset) = 0$$

$$m_{DP}(A) = \sum_{B,C \subseteq \Omega, B \cap C = A} m_1(B)m_2(C)$$

$$+ \sum_{B,C \subseteq \Omega, B \cup C = A, B \cap C = \emptyset} m_1(B)m_2(C), \forall A \subseteq \Omega, A \neq \emptyset$$

In Dempster's merging, weights of conflicting formulae are proportionally distributed to formulae resulted from intersection of non-conflicting formulae. Instead, in Dubois & Prade's Merging, weights of conflicting formulae are added to the disjunction of the conflicting formulae.

Example 3. *(Example 2 Continued) In Example 2, we have that the characteristic function of K_1, i.e., m_{K_1}, is such that $m_{K_1}(\alpha) = 0.8, m_{K_1}(\alpha \vee \beta) = 0.2$, and the* *characteristic function of K_2, i.e., m_{K_2}, is such that $m_{K_2}(\emptyset) = 0.18, m_{K_2}(\alpha) = 0.42, m_{K_2}(\beta) = 0.12, m_{K_2}(\alpha \vee \beta) = 0.28$, then we can get the following merging results using the above merging methods.*

Dempster's Merging. $m_{K_{12}}^{Dem}$ *is such that*

$$m_{K_{12}}^{Dem}(\alpha) = 0.89, m_{K_{12}}^{Dem}(\beta) = 0.03, m_{K_{12}}^{Dem}(\alpha \vee \beta) = 0.08$$

DP's Merging. $m_{K_{12}}^{DP}$ *is such that*

$$m_{K_{12}}^{DP}(\alpha) = 0.788, m_{K_{12}}^{DP}(\beta) = 0.024, m_{K_{12}}^{DP}(\alpha \vee \beta) = 0.188$$

Note that the above merging methods are not idempotent but have a reinforcement effect. That is, in general, we do not have $\Delta(K, K) = K$ when Δ is a merging operator defined by one of the above merging methods. Reinforcement merging is rational when the knowledge bases to be merged are from distinct sources. However, for knowledge bases from nondistinct sources (i.e., sources providing possibly overlapping knowledge [5]), we intuitively require the merging to be idempotent. To the best of our knowledge, we do not see any idempotent merging methods for knowledge bases in the literature, here we provide an idempotent merging method based on the cautious rule of combination introduced in [5].

Definition 8. *(Denœux's Cautious Merging) Let K_1, K_2 be two knowledge bases and m_{K_1}, m_{K_2} be their characteristic functions, s.t., $m_{K_1} = \oplus_{A \subset \Omega} A^{x_A^1}$ and $m_{K_2} = \oplus_{A \subset \Omega} A^{x_A^2}$, respectively, then the characteristic function $m_{K_{12}}$ of the merged knowledge base by Denœux's cautious combination rule is such that (again notice that our A^{x_A} setting is different from Denœux's):*

$$m_{K_{12}}^{Den} = \oplus_{A \subset \Omega} A^{max(x_A^1, x_A^2)}, \forall A \subset \Omega.$$

Example 4. *(Example 2 Continued) In Example 2, the characteristic function of K_1, i.e., m_{K_1}, is such that $m_{K_1}(\alpha) = 0.8, m_{K_1}(\alpha \vee \beta) = 0.2$, and the characteristic function of K_2, i.e., m_{K_2}, is such that $m_{K_2}(\emptyset) = 0.18, m_{K_2}(\alpha) = 0.42, m_{K_2}(\beta) = 0.12, m_{K_2}(\alpha \vee \beta) = 0.28$, then the characteristic function of the merging result using Denœux's merging method, i.e., $m_{K_{12}}^{Den}$, is as follows.*

$$m_{K_{12}}^{Den}(\emptyset) = 0.24, m_{K_{12}}^{Den}(\alpha) = 0.56, m_{K_{12}}^{Den}(\beta) = 0.06, m_{K_{12}}^{Den}(\alpha \vee \beta) = 0.14.$$

From the characteristic functions, the corresponding knowledge base for $m_{K_{12}}^{Den}$ is:

$$K^{Den} = \{(\alpha, 0.8), (\beta, 0.3)\}.$$

For Denœux's merging method, we have the following result.

Proposition 3. *Let two knowledge bases K_1, K_2 be $K_1 = \{(\mu_1, x_1), \cdots, (\mu_n, x_n)\}$ and $K_2 = \{(\phi_1, y_1), \cdots, (\phi_m, y_m)\}$, then the merging result of K_1 and K_2 using Denœux's merging method is $K_{12} = \{(\psi_1, z_1), \cdots, (\psi_t, z_t)\}$ such that K_{12} is a subset of $K_1 \bigcup K_2$ satisfying the following conditions:*

- if $\mu_i \equiv \top$, $1 \leq i \leq n$, then $(\mu_i, x_i) \notin K_{12}$; if $\phi_j \equiv \top$, $1 \leq j \leq m$, then $(\phi_j, y_j) \notin K_{12}$,
- if $\mu_i \equiv \phi_j$, $1 \leq i \leq n$, $1 \leq j \leq m$, and $x_i > y_j$ (resp. $y_j > x_i$), then $(\phi_j, y_j) \notin K_{12}$ (resp. $(\mu_i, x_j) \notin K_{12}$),
- all other elements of K_1 or K_2 are in K_{12}.

Proof of Proposition 3: From $K_1 = \{(\mu_1, x_1), \cdots, (\mu_n, x_n)\}$ and $K_2 = \{(\phi_1, y_1), \cdots, (\phi_m, y_m)\}$, we get $m_{K_1}^{Den} = \oplus_{i=1}^{n} \mu_i^{x_i}$ and $m_{K_2}^{Den} = \oplus_{j=1}^{m} \phi_j^{y_j}$.

Let $\{\phi_{j_1}, \cdots, \phi_{j_a}\}$ be the set of all formulae each of which is in K_2 but not in K_1, then we have $m_{K_1}^{Den} = \oplus_{i=1}^{n} \mu_i^{x_i} \oplus \phi_{j_1}^{0} \oplus \cdots \oplus \phi_{j_a}^{0}$.

Similarly, let $\{\mu_{i_1}, \cdots, \mu_{i_b}\}$ be the set of all formulae each of which is in K_1 but not in K_2, we have $m_{K_2}^{Den} = \oplus_{j=1}^{m} \phi_j^{y_j} \oplus \mu_{i_1}^{0} \oplus \cdots \oplus \mu_{i_b}^{0}$.

Hence from Definition 8, it is straightforward to see that the merged characteristic function corresponds to the knowledge base K_{12} which is exactly the same as stated in Proposition 3. □

Proposition 3 makes it convenient to solve the merging of nondistinct knowledge bases. For instance, from Proposition 3, it is easy to obtain $K^{Den} = \{(\alpha, 0.8), (\beta, 0.3)\}$ from $K_1 = \{(\alpha, 0.8), (\alpha \vee \beta, 0.2)\}$ and $K_2 = \{(\alpha, 0.6), (\beta, 0.3), (\alpha \vee \beta, 0.1)\}$.

In the literature, a basic assumption for knowledge base merging is that the knowledge bases to be merged should be consistent. This assumption is often not applicable in practice, as argued in many research work discussing the inconsistency of a knowledge base (e.g., [20,7,8]). An obvious advantage of our merging methods is that they do not require this assumption. That is, even for inconsistent knowledge bases, it is still possible to merge them and obtain rational fusion results. The second advantage is that we can deal with knowledge bases from nondistinct sources. Usually logic-based merging methods do not consider whether the knowledge bases to be combined are from distinct sources. In this paper, however, if the information sources are known to be distinct, then Dempster's merging method, Smets's, Yager's, or Dubois and Prade's merging method can be chosen, whilst if the sources are known to be nondistinct, then Denœux's merging method can be selected. This differentiation of merging methods based on dependency relationship among knowledge bases is obviously more suitable. Of course, proper methods should be developed to judge whether two knowledge bases are dependent or not, but this topic is beyond the scope of this paper.

6   Conclusion

In this paper, we introduced a bba based characteristic function for any weighted knowledge base which take flat knowledge bases as a special case. We then used the characteristic function to measure the inconsistency of the knowledge base and proved that this inconsistency measure follows a set of rational properties. We also deployed the characteristic functions to merge multiple knowledge bases. Different merging methods were provided corresponding to different combination rules of bbas. These merging methods could provide some advantages than the existing merging methods, e.g., the ability to merge inconsistent knowledge bases, the use of distinctness information of knowledge sources, etc.

An obvious future work is to apply these methods in intelligent surveillance applications. In addition, extending this approach to stratified/prioritized/ranked knowledge bases is also an interesting topic with the help of the non-Archimedean infinitesimals [13]. Furthermore, providing comparisons with related works, e.g., our inconsistency measure vs. inconsistency measures in [7]; our merging methods vs. the existing merging methods and merging postulates [9], etc., is a promising issue.

References

1. Bacchus, F., Grove, A., Halpern, J., Koller, D.: From statistical knowledge bases to degrees of belief. Artificial Intelligence 87(1-2), 75–143 (1996)
2. Benferhat, S., Dubois, D., Prade, H.: Representing default rules in possibilistic logic. In: Procs. of KR, pp. 673–684 (1992)
3. Dempster, A.P.: Upper and lower probabilities induced by a multivalued mapping. The Annals of Statistics 28, 325–339 (1967)
4. Dempster, A.P.: A generalization of bayesian inference. J. Roy. Statist. Soc. 30, Series B, 205–247 (1968)
5. Denœux, T.: Conjunctive and disjunctive combination of belief functions induced by nondistinct bodies of evidence. Artifical Intelligence 172(2-3), 234–264 (2008)
6. Dubois, D., Prade, H.: Representation and combination of uncertainty with belief functions and possibility measures. Computational Intelligence 4, 244–264 (1988)
7. Hunter, A., Konieczny, S.: Shapley inconsistency values. In: Procs. of KR 2006, pp. 249–259 (2006)
8. Hunter, A., Konieczny, S.: Measuring inconsistency through minimal inconsistent sets. In: Procs. of KR, pp. 358–366 (2008)
9. Konieczny, S., Pino-Pérez, R.: On the logic of merging. In: Cohn, A.G., Schubert, L., Shapiro, S.C. (eds.) Principles of Knowledge Representation and Reasoning, KR 1998, pp. 488–498. Morgan Kaufmann, San Francisco (1998)
10. Lin, J.: Integration of weighted knowledge bases. Artif. Intel. 83(2), 363–378 (1996)
11. Liu, W.: Analyzing the degree of conflict among belief functions. Artificial Intelligence 170, 909–924 (2006)
12. Ma, J., Liu, W., Hunter, A.: Incomplete Statistical Information Fusion and Its Application to Clinical Trials Data. In: Prade, H., Subrahmanian, V.S. (eds.) SUM 2007. LNCS (LNAI), vol. 4772, pp. 89–103. Springer, Heidelberg (2007)
13. Ma, J., Liu, W., Hunter, A.: The Non-archimedean Polynomials and Merging of Stratified Knowledge Bases. In: Sossai, C., Chemello, G. (eds.) ECSQARU 2009. LNCS, vol. 5590, pp. 408–420. Springer, Heidelberg (2009)
14. Ma, J., Liu, W., Hunter, A.: Inducing probability distributions from knowledge bases with (in)dependence relations. In: Procs. of AAAI, pp. 339–344 (2010)
15. Ma, J., Liu, W., Hunter, A.: Inducing probability distributions from knowledge bases with (in)dependence relations. In: Proceedings of the 24th American National Conference on Artificial Intelligence, AAAI 2010 (2010)
16. Ma, J., Liu, W.: A framework for managing uncertain inputs: an axiomization of rewarding. International Journal of Approximate Reasoning 52(7), 917–934 (2011)
17. Ma, J., Liu, W., Miller, P.: Event Modelling and Reasoning with Uncertain Information for Distributed Sensor Networks. In: Deshpande, A., Hunter, A. (eds.) SUM 2010. LNCS, vol. 6379, pp. 236–249. Springer, Heidelberg (2010)
18. Ma, J., Liu, W., Miller, P., Yan, W.: Event composition with imperfect information for bus surveillance. In: Procs. of AVSS, pp. 382–387. IEEE Press (2009)

19. Miller, P., Liu, W., Fowler, F., Zhou, H., Shen, J., Ma, J., Zhang, J., Yan, W., McLaughlin, K., Sezer, S.: Intelligent sensor information system for public transport: To safely go... In: Procs. of AVSS, pp. 533–538 (2010)
20. Mu, K., Jin, Z., Lu, R., Liu, W.: Measuring Inconsistency in Requirements Specifications. In: Godo, L. (ed.) ECSQARU 2005. LNCS (LNAI), vol. 3571, pp. 440–451. Springer, Heidelberg (2005)
21. Mu, K., Liu, W., Jin, Z.: A blame-based approach to generating proposals for handling inconsistency in software requirements. International Journal of Knowledge and Systems Science 3(1), 1–16 (2012)
22. Rossini, P.: Using expert systems and artificial intelligence for real estate forecasting. In: Procs. of Sixth Annual Pacific-Rim Real Estate Society Conference, Sydney, Australia, January 24-27, pp. 1–10 (2000)
23. Shafer, G.: A Mathematical Theory of Evidence. Princeton University Press (1976)
24. Smets, P.: Belief functions. In: Smets, P., Mamdani, A., Dubois, D., Prade, H. (eds.) Non-Standard Logics for Automated Reasoning, pp. 253–286 (1988)

Comfort as a Multidimensional Preference Model for Energy Efficiency Control Issues[*]

Afef Denguir[1,2], François Trousset[1], and Jacky Montmain[1]

[1] LGI2P, Laboratoire de Génie informatique et d'ingénierie de la production,
EMA Site EERIE –parc scientifique Georges Besse, 30035 – Nîmes, France
[2] Université Montpellier 2, Place Eugène Bataillon, 34095 Montpellier, France
{afef.denguir,francois.trousset,jacky.montmain}@mines-ales.fr

Abstract. The incessant need for energy has raised its cost to unexpected heights. In response to this situation, many projects have been started in order to save energy. In this context, RIDER project tries to develop a weak system dependency of energy management framework which could be applied for different systems. Particularly, our RIDER Decision Support System (DSS) focuses on proposing generic control rules and optimization techniques for energy management systems. Therefore, the DSS aims to compute the most relevant target values (i.e., setpoints) to be provided to the energy control system and then, improving thermal comfort sensation or reducing energy costs. Literature proposes reusable system independent statistical models for thermal comfort. However, they are not easily interpretable in terms of a preference model which makes control not intuitive and tractable. Since thermal comfort is a subjective multi-dimensional concept, an interpretable and reusable preference model is introduced in this paper. Multi Attribute Utility Theory (MAUT) is used for this.

Keywords: Thermal comfort, preference model, energy control, MAUT, Choquet integral.

1 Problematic Introduction

Total building energy consumption accounts for about 40% of total energy demand and more than one half is used for space conditioning: heating, cooling, and ventilation [1] [2] [3]. In the EU, about 57% of total energy consumption is used for space heating, 25% for domestic hot water, and 11% for electricity [4]. In response to this situation, many projects have been started in order to save energy. Recent studies have investigated efficient building control in order to find strategies that provide a comfortable environment from thermal, and indoor-air quality points of views, and minimize energy consumption at the same time [5]. Nevertheless, these optimization systems are strongly dependent on the energy management framework and cannot be applied for other systems. Indeed, they are conceived by the energy manager

---

[*] This research is part of the FUI RIDER project, "Research for IT Driven EneRgy efficiency" (rider-project.com).

E. Hüllermeier et al. (Eds.): SUM 2012, LNAI 7520, pp. 486–499, 2012.
© Springer-Verlag Berlin Heidelberg 2012

depending on one building characteristics. So, its associated optimization routines are directly implemented on its control system and cannot be reused for further energy management. Additionally, these optimization routines are not supposed to be interpreted by human operators since they are integrated in regulation loops which made them necessarily dependent on the SCADA system (supervisory control and data acquisition). In order to solve this problem and satisfy the weak energy system dependency which is required by the RIDER project, control rules should neither be too specific nor integrated in control loops. They must rather be a high level supervision rules which can be suggested to the energy manager. That's why; we propose that the RIDER DSS core functionalities should rather provide qualitative recommendations such as suggesting the most relevant target values to the energy control system. This approach ensures, the control rules interpretability, as well as, the weak dependency of the DSS w.r.t. the energy system and its control.

This research is part of the RIDER project and deals only with its optimization aspects. In this paper, we focus on a specific optimization aspect based on human's thermal sensation. In fact, the notion of comfort is subjective and multidimensional. Subjectivity entails that comfort cannot be modeled in a deterministic way and its multidimensionality comes from the fact that many variables can be considered in its definition: temperature but also hygrometry, radiant temperature and air velocity. These remarks explain why providing efficient energy management for optimal comfort may be considered as a multicriteria decision-making process in uncertain environment, and must be modeled as such [6].

The next sections discuss about the modeling and the implementation of an original thermal comfort function and formalize, as well, some RIDER optimization problem based on the aforementioned comfort function.

This paper is organized as follows. Section 2 discusses about most common thermal comfort models and their relevance when they are used in optimization process. It explains our choice to have a model which interprets the comfort statistical model on the MAUT framework. Section 3 summarizes Labreuche's method to identify our thermal comfort model, the way that this method was applied and extended to build a comfort overall utility function in our complex context, and finally shows the usefulness of this new formalization to infer comfort control rules. Finally, section 4 formulates some control problems based on the new thermal comfort preference model.

2   Optimization and Comfort

Even when no malfunctioning is detected in a heating system, *i.e.*, temperature values in a building match their setpoints, two users may be more or less tolerant with regard to the setpoint variations and thus not equally satisfied. It can be explained by the more or less tolerant user's requirements are but also by other parameters than temperature which may differ from one situation to another and then contribute to different thermal sensation. This illustrates that thermal comfort (and not only temperature) should be the variable to be controlled by the RIDER DSS in order to ensure building occupants' satisfaction. However, comfort is a complex and subjective concept that cannot be modeled as a deterministic variable. That's why, in literature, the most well-known thermal comfort is based upon a statistical approach [7] [8].

2.1   Thermal Comfort Model Overview

2.1.1   Comfort as a Statistical Model

The Predicted Mean Vote *PMV* [7] is the most used statistical thermal comfort index. It defines the mean thermal sensation vote on a standard 7 level scale from a group of approximately1300 persons. It is written as a function of 4 thermal environmental variables: air temperature *Ta*, air humidity *Hy*, air velocity *Va*, and mean radiant temperature *Tr*. It also includes 2 human parameters: metabolic rate *Me* and cloth index *Ci*. The *PPD* (Predicted Percentage Dissatisfied) index is based on the *PMV* one and indicates the percentage of thermal dissatisfied persons. Both *PMV* and *PPD* indexes have been used since 1995 by the NF EN ISO 7730 standard to describe ergonomics of thermal environments [8].

Such a thermal comfort representation verifies the RIDER DSS weak dependency constraint from one hand, and captures the inherent subjectivity and uncertainty related to thermal sensation from the other hand. The statistical based thermal comfort modeling is the result of a sample-ballot which makes it reusable for various application contexts. Whereas comfort is intuitively related to a preference model, the formalism in [7] and [8] is far away from any classical preference modeling framework. *PMV* and *PPD* indexes are considered as if they were outputs of any behavioral model associated to a physical process. In particular, interactions among comfort attributes are considered as if they were physical ones which is not the case. The monotony of *PMV* and *PPD* with regard to attributes variations, is not obvious and can only be numerically computed. As a consequence, interpreting such a model to support control rules design for a human operator is not so intuitive.

2.1.2   Comfort as a Preferential Model

The representation of preferences is a central topic in decision-making and measurement theory [9]. Usually, it amounts to find a real-valued overall utility function U such that for any pair of alternatives $x, x' \in X$ where X is a set of alternatives, $x \succeq x'$ (x is preferred to x') *iff* $U(x) \geq U(x')$. When alternatives are N-dimensional (attribute $i \in N$ takes its values in X_i), *i.e.*, $X = \prod_{i=1}^{n} X_i$, a widely studied model is the decomposable model of Krantz et al. [10], where U has the form $U(x_1,..,x_n) = g(u_1(x_1),..,u_n(x_n))$ where u_i are real-valued functions. Assuming that $\succeq$ is a weak order on X, it is known that a representation with g being strictly increasing can be found *iff* $\succeq$ satisfies independence and X is separable [9]. The MAUT [11] [12] is based upon the utility theory which is a systematic approach to quantify individual preferences. Utility theory consists in interpreting any measurement as a satisfaction degree in $[0,1]$ where 0 is related to the worst alternative and 1 to the best one. Measurements are thus made commensurate and interpretable. In this way, a utility $u_i(x_i)$ is attached to each measurement x_i.

Indirect interviewing methods such as MACBETH (Measuring Attractiveness by a Categorical Based Evaluation TecHnique) are generally applied to identify attribute elementary functions $u_i(x_i)$ in a weighted average aggregation model. However, when aggregation operators do not fulfill the weak difference independence property then constructing elementary utilities functions is more complicated [13]. Indeed, this property allows building the value function on attribute i by asking questions directly regarding the preference of the decision maker on the attribute value range X_i (independently of other attributes values) rather than from questions regarding options in X. An extension of MACBETH for a Choquet integral aggregation function that respect weak difference dependence has been proposed in [14] [15].

When comfort can be written under the decomposable form $U(Ta,...,Me) = g(u_{Ta}(Ta),...,u_{Me}(Me))$ it makes thermal sensation more interpretable w.r.t attributes variations and avoids the coexistence of antagonist behavioral rules. For instance, comfort may be improved when humidity increases for one given ambient temperature whereas it can be disturbed by an increasing humidity for another ambient temperature. The coexistence of such behavioral rules makes difficult for the energy manager to directly imagine attribute variations in order to control the energy system. Whereas co-monotony of comfort U and u_{Hy} holds everywhere in X_{Hy}. Then, identifying the elementary utility functions u_i would greatly facilitate the design of control rules. Moreover, in the real thermal comfort perception, there is no physical correlation between attributes. Interactions between attributes should rather be considered as preferential interactions related to criteria associated to attributes [13] [16]. Fuzzy integrals provide adequate models to capture such interactions. It is then obvious that a preferential model of thermal comfort would be more appropriate for semantic reasons.

2.1.3 Discussion

Let us now introduce these models in optimization issues. Optimization problem (1) and its dual (2) —where $Cost(\delta Ta, \delta Hy, \delta Tr, \delta Va, \delta Ci, \delta Me)$ function evaluates the cost of the attributes variations $(\delta Ta, \delta Hy, \delta Tr, \delta Va, \delta Ci, \delta Me)$ and PPD^* (resp. C^*) is a comfort setpoint (resp. a budget threshold)— formalize efficient comfort improvement issues.

$$\begin{cases} \min Cost(\delta Ta, \delta Hy, \delta Tr, \delta Va, \delta Ci, \delta Me) \\ PPD(Ta + \delta Ta, Hy + \delta Hy, Tr + \delta Tr, \\ Va + \delta Va, Ci + \delta Ci, Me + \delta Me) \leq PPD^* \end{cases} \quad (1)$$

$$\begin{cases} \max 100 - PPD(Ta + \delta Ta, Hy + \delta Hy, \\ Tr + \delta Tr, Va + \delta Va, Ci + \delta Ci, Me + \delta Me) \\ Cost(\delta Ta, \delta Hy, \delta Tr, \delta Va, \delta Ci, \delta Me) \leq C^* \end{cases} \quad (2)$$

Let $\vec{\nabla}PPD$ the gradient when $PPD(Ta,...,Me) = PPD^*$. It provides attributes that their local variations impact the most significantly the comfort variation (maximal component of $\vec{\nabla}PPD$). However, there are some practical and computational drawbacks to this formulation. First, the gradient is generally not of common sense for the energy manager to be use in optimization process. Then, there is no information regarding the neighborhood in which this result is valid: maximal component of $\vec{\nabla}PPD$ may change rapidly *i.e.* it depends on non linearity of *PPD* and this notion is meaningless for the energy manager. Finally, we cannot a priori know whether we have to increase or decrease an attribute value to improve *PPD*. It necessitates computing the derivative. It depends on Ta, Hy, Tr, Va, Ci, Me attribute values and the monotony of *PPD* relatively to these attributes, which is not easily understandable for the energy manager. However, a preferential based thermal comfort modeling solves the aforementioned drawbacks thanks to the co-monotony between utility functions u_i and thermal comfort overall evaluation U, and offers as well a more relevant control system for thermal comfort attributes.

In order to ensure the RIDER DSS weak dependency, the thermal comfort model has also to fulfill this condition. The statistical thermal comfort modeling satisfies already the weak dependency condition and can be applied for different systems whereas the preferential thermal comfort modeling, it depends on the way with which utility functions u_i have been identified *i.e.* utility functions should result from statistical techniques like in [7] and [8] which would roughly make the interviewing method more complex.

So, to grant to the comfort preferential based model the ability to be system independent without having to proceed by the statistical way, we propose to identify utility functions from the existent statistical model *PPD*. Labreuche [17] has proposed an original approach to compute both the utilities and the aggregated overall utility function $U(x_1,...,x_n)$ when U is a Choquet integral without any commensurateness assumption. It is important to highlight that using a Choquet integral facilitates optimization problem solving ((1) and (2)) thanks to its linearity by simplex. Next section describes the Choquet integral and gives a short description of Labreuche's method [17] in order to identify utility functions and the Choquet integral parameters.

2.2   Measurements Overall Utility without Commensurateness Hypothesis

2.2.1   The Choquet Integral

The Choquet integral family provides adequate models to capture decisional behaviors when there are preferential interactions between criteria. They enable accommodating both the relative importance of each criterion and the interactions among them [18] [19]. In our preference model, an interaction occurs between any two criteria once they need to be satisfied simultaneously (i.e., positive synergy) or when their simultaneous satisfaction is seen as a redundancy (negative synergy).

$$U = C_\mu(u_1, u_2, ..., u_n) = \sum_{i=1}^{n}(u_{(i)} - u_{(i-1)}).\mu(A_{(i)}) = \sum_{i=1}^{n}\Delta\mu_{(i)}.u_{(i)} \qquad (3)$$

U, in (3), is the aggregate utility of the elementary utility profile $\vec{u} = (u_1, ..., u_n)$ (to simplify u_i abusively denotes $u_i(x_i)$ when no misinterpretation is possible) where

$\mu : 2^C \rightarrow [0,1]$ is a fuzzy measure on the subsets of criteria in C; (.) indicates a permutation, such that the elementary utilities $u_{()}$ are ranked: $0 \le u_{(1)} \le ... \le u_{(n)} \le 1$ and $A_{(i)}^k = \{c_{(i)}, ..., c_{(n)}\}$. This expression can also be rewritten as in the last part of (3) where $\Delta\mu_{(i)} = \mu_{(i)} - \mu_{(i+1)}$ and $\mu_{(i)} = \mu(A_{(i)})$, $\mu_{(n+1)} = 0$.

Note that a simplex $H_{()} = \{\vec{u} \in [0,1]^n / 0 \le u_{(1)} \le ... \le u_{(n)} \le 1\}$ corresponds to the ranking (.), where the Choquet integral assumes a linear expression. Such a remark proves that optimization problems that involve a Choquet integral can be solved with linear programming techniques within simplexes.

2.3.2   Construction of Choquet Integral and Elementary Utilities without Any Commensurateness Assumption

Since we want to represent PPD with the decomposable model of Krantz, weak separability property has to be first verified. A preference relation $\succeq$ is said weak separable *iff*, it verifies (4) for every attribute $i \in N$ where N denotes the attribute set, $x_i, x'_i \in X_i$ two possible values of i, and $y_{N\backslash i}, y'_{N\backslash i} \in \prod_{j \ne i}^n X_j$ two possible alternatives described for $\forall k \in N$ and $k \ne i$.

$$\forall x_i, x'_i, y_{N\backslash i}, y'_{N\backslash i}, \ (x_i, y_{N\backslash i}) \succeq (x'_i, y_{N\backslash i}) \Leftrightarrow (x_i, y'_{N\backslash i}) \succeq (x'_i, y'_{N\backslash i}) \tag{4}$$

Labreuche [17] supposes that the weak separability property is verified for the overall utility function U (PPD in our case) and suggests a method to check commensurateness among attributes i and k. For this, he proposes to analyze the gradient function related to x_i w.r.t x_k variations. It returns on studying the function $f_i : x_k \mapsto U(x_i + \varepsilon, x_{N\backslash i}) - U(x)$ where $\varepsilon > 0$. If f_i is a constant function, then there is no interaction between attributes i and k (it means that even when there is a ranking change between utilities related to i and k their "weights" in (3) do not change in the new simplex). And, thus, attributes i and k do not interact. Otherwise, if f_i is not a constant function, then attributes i and k interact with each other (the "weight" in (3) depends on their ranking). In this case, i and k are considered as commensurate and it is possible, then to compute the value $x_k^* \in X_k$ for the attribute k where $u_k(x_k^*) = u_i(x_i)$ [17]. At the end of this step, subsets of commensurate attributes $S_j \subseteq N$ are constructed, where $\bigcup_j S_j = N$ and $\forall i,k \in S_j$, i and k are commensurate.

Once S_j are identified, the utility functions u_i and capacities μ_j can be computed. According to [20], u_i cannot be built from one attribute regardless to the other ones. u_i's construction in [17] is thus based on the overall utility U. [17] supposes that U is continuous and all u_i functions are strictly increasing over $\mathbb{R}(\forall i, X_i = \mathbb{R}$ in [17]).

In order to build u_i and μ two reference vectors $\mathbf{O}_{S_j}, \mathbf{G}_{S_j} \in \prod_{k \in S_j} X_k$ should be computed for each attribute subset S_j. They refer respectively to an unacceptable (Null) situation level and a Good situation level. For vectors $\mathbf{O}_{S_j}$ (resp. $\mathbf{G}_{S_j}$), the first attribute value x_l^O (resp. x_l^G) is chosen by the decision maker and the others $x_{k \neq l}^O$ (resp. $x_{k \neq l}^G$) are computed such as $u_{k \neq l}\left(x_{k \neq l}^O\right) = u_l\left(x_l^O\right)$ (resp. $u_{k \neq l}\left(x_{k \neq l}^G\right) = u_l\left(x_l^G\right)$) for $\forall k \in S_j$ and $k \neq l$. To make sure that $\mathbf{G}_{S_j}$ corresponds to a better situation than $\mathbf{O}_{S_j}$, $x_{k \neq l}^G$ must be preferred to $x_{k \neq l}^O$.

Based on the identified reference vectors, a utility function v_i is defined by (5), where $v_i\left(x_i^O\right) = 0$, $v_i\left(x_i^G\right) = 1$, $\mathbf{O}_A$ is the restriction of $\mathbf{O}_N = \left(x_1^O, ..., x_n^O\right)$ to $A \subset N$ (resp. $\mathbf{G}_A$ is the restriction of $\mathbf{G}_N = \left(x_1^G, ..., x_n^G\right)$ to $A \subset N$).

$$\forall x_i \in \left[x_i^O, x_i^G\right], v_i(x_i) = \frac{U(x_i, \mathbf{O}_{N \setminus i}) - U(\mathbf{O}_N)}{U(x_i^G, \mathbf{O}_{N \setminus i}) - U(\mathbf{O}_N)} \tag{5}$$

Since several solutions for $\mathbf{O}_{S_j}$ and $\mathbf{G}_{S_j}$ may be envisaged, a normalization condition is required: when $\sum_{S_j}\left(U(\mathbf{G}_{S_j}, \mathbf{O}_{N \setminus S_j}) - U(\mathbf{O}_N)\right) = 1$ is checked, then normalized utilities u_i and normalized capacities μ_{S_j} are respectively (6) and (7):

$$u_i(x_i) = v_i(x_i) + U(\mathbf{O}_N) \quad (6) \quad \forall A \subset S_j, \mu_{S_j}(A) = U(\mathbf{G}_A, \mathbf{O}_{N \setminus A}) - U(\mathbf{O}_N) \quad (7)$$

Finally, the Choquet integral (see (3)) that represents the overall utility U of the normalized utilities u_i based upon the generalized capacity μ given by (8) is achieved:

$$\forall A \subset N, \mu(A) = \sum_{S_j}\left(\mu_{S_j}(A \cap S_j)\right) \tag{8}$$

3   Decomposable Form of the Aggregation Model of Comfort

Interactions between $Ta, Hy, Tr, Va, Ci,$ and Me are preference interactions rather than physical correlations [16]. In fact, preferences are perfect to model human perception or opinion about comfort which is a subjective concept and cannot be treated like a physical process output as in [8]. Choquet integral is a relevant solution to model preference interactions among thermal comfort attributes and confers to the comfort aggregated concept its semantic interpretability. Also, the simplex piecewise linearity of Choquet integral facilitates optimization processes. So, to solve problems described by (1) and (2), approximating the PPD with a Choquet integral, is then of interest. It first reduces the PPD complexity by giving the possibility to have a linear formulation. Second, it allows easy prediction of PPD variation with regards to one attribute fluctuation since we have elementary utility functions.

In order to simplify the Choquet comfort modeling, we suppose that people hosted by a same building have almost the same activity level and are dressed pretty much the same depending on seasons. Considering those assumptions, Ci and Me attributes can be removed from the model variables (they are seen as constant parameters instead of variables). Thus, depending on seasons and the activity nature of a building, both Ci and Me are evaluated by average values i.e., $Me = 1.2\,met$ for average administrative employees and $Ci = 0.7\,clo$ for a shirt/pant dressing sample [8].

3.1   Weak Separability Assumption and Choquet Integral-Based Local Model

In order to write the PPD function as an overall utility function, it's necessary to check, first, the weak separability property (4) among its attributes $Ta, Hy, Tr,$ and Va which, intuitively, seems to be not the case. Here is a counterexample of the weak separability non-satisfaction computed for the vector (Ta, Hy, Tr, Va), $(23, 50, 23, 0.2)$ $\prec (25, 50, 23, 0.2)$ however $(23, 100, 23, 0.2) \nprec (25, 100, 23, 0.2)$. Therefore, we can say that $PPD(Ta, Hy, Tr, Va)$ defined for $Ta \in [10, 30°]$, $Tr \in [10, 40°]$, $Hy \in [0, 100\%]$, and $Va \in [0, 1m/s]$ [8] is not a weak separable function. Second, the monotony assumption of Labreuche's construction must be checked. Again, intuitively, this assumption cannot be proved for the considered areas of $Ta, Hy, Tr,$ and Va. It is obvious that an increasing temperature is appreciated until an upper threshold. Above this threshold, people get hot and their thermal sensation progressively decreases. This fact implies that the elementary utility function of the ambient temperature $u_{Ta} : [10, 30°] \rightarrow [0, 1]$ has at least one monotony change.

Fig., 1 and 2, illustrate respectively the PPD curve for ($Tr = 23°$, $Hy = 25\%$, $Me = 1.2\,met$, $Ci = 0.7\,clo$, $Ta \in [10, 30°]$ and $Va \in [0, 1m/s]$) and ($Tr = 23°$, $Hy \in [0, 100\%]$, $Me = 1.2\,met$, $Ci = 0.7\,clo$, $Ta \in [10, 30°]$ and $Va = 0.2m/s$). Iso-temperature curves of both figures have the same shape for respectively all Va and Hy values. So, we can realize that the minimal PPD is reached for slightly different Ta values, which means that the weak separability property is not verified in the considered PPD domain. Fig., 1 and 2, show, also, that PPD function has two different monotonies w.r.t. Ta values which means that u_{Ta} cannot be considered as strictly increasing for $Ta \in [10, 30°]$. Since none of the two required assumptions is verified, we cannot build an overall Choquet integral for all PPD attributes domains. However, these assumptions can be checked for different local domains and, then, a Choquet integral can be computed for each of these domains. Based on this, we have to identify domains in which the shape of the PPD function has the same monotony and verifies, as well, the weak separability property i.e., according to figures 1 and 2, for $Ta \in [25, 30°]$, we have both assumptions verified. So it is possible to compute a Choquet integral defined locally for $Ta \in [25, 30°]$. Hence, the PPD function can locally be approximated by a Choquet integral. This technique allows the computation of local preference models for the thermal comfort. It means that depending on situations, attribute utility functions change. In fact, one attribute influence on the thermal comfort becomes more or less important depending on Ta value range. Studying the PPD monotony w.r.t Ta, Va, Hy and Tr attributes and especially its extrema has lead to identify valid domains for the Labreuche's construction.

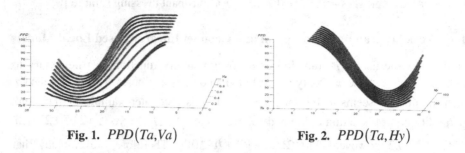

Fig. 1. $PPD(Ta,Va)$ **Fig. 2.** $PPD(Ta,Hy)$

3.2   A Fuzzy Inference System to Estimate Comfort

In practice, the PPD index can only be controlled through Ta, Hy, and Va attributes where Va is equivalent to a room airflow of the heating exchanger. So, more specifically, we need to associate elementary utility functions to these attributes to simplify control issues. Besides, Tr is beyond control except if we close the shutters! Furthermore, it can be checked that interactions with Tr are not preferential ones. Tr interactions are related to physical relationships with Ta which are not semantically considered by the Choquet integral model and do not correspond to the Labreuche's construction. That's why, in order to simplify our model and, also, reduce the complexity of the identification of local validity domains, we decide to remove Tr from our PPD approximation. Therefore, a Choquet integral is computed for a fixed Tr in tridimensional local domains of validity of Ta, Hy, and Va.

In this case, a fuzzy interpolation for Tr is proposed to consider all Tr range. Fig. 3 shows the way the 5 different tridimensional model $U_{comfort}^{Tr=x_{Tr}}$ cover all Tr range. A local Choquet integral model approximates the PPD function for these fixed values of Tr: 15, 20, 23, 25 and 30°. Then, $U_{comfort}^{Tr=x_{Tr}}$ approximates the PPD function only in its associated valid local domain. Comfort can finally be computed for any Tr value thanks to an interpolation between two local models as proposed by the triangular membership functions in figure 3.

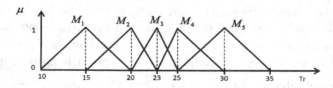

Fig. 3. Five tridimensional $U_{comfort}^{Tr=x_{Tr}}$ based interpolation

Performances of these models and their relative local domain of validity are presented by table 1. Average approximation errors between the local Choquet approximation and the *PPD* function are computed based on 9261 different simulations and it never exceeds 8.6% on $Ta \times Hy \times Va$ valid domains. It is important to bring back that these models are computed for an average activity of employees $Me = 1.2\,met$ and a shirt/pant clothing model $Ci = 0.7\,clo$. $U_{comfort}^{Tr=x_{Tr}}$ described by table 1 approximates the *PPD* function only in its associated valid local domain.

Table 1. Valid local domains and average error for the Choquet approximation

Tridimensional model of comfort		Valid local domain $\mathcal{M}_i$	Choquet Approximation error
$\mathcal{M}_1: U_{comfort}^{Tr=15°C}$	1	$Ta \in [15,25°], Hy \in [50,100\%], Va \in [0.125,1m/s]$ $u_{Ta}(Ta) \nearrow, u_{Hy}(Hy) \nearrow, u_{Va}(Va) \searrow$	7.8%
	2	$Ta \in [25,30°], Hy \in [50,100\%], Va \in [0.25,1m/s]$ $u_{Ta}(Ta) \nearrow, u_{Hy}(Hy) \searrow, u_{Va}(Va) \nearrow$	4.2%
$\mathcal{M}_2: U_{comfort}^{Tr=20°C}$	1	$Ta \in [15,27°], Hy \in [25,100\%], Va \in [0.125,1m/s]$ $u_{Ta}(Ta) \nearrow, u_{Hy}(Hy) \searrow, u_{Va}(Va) \nearrow$	8.6%
$\mathcal{M}_3: U_{comfort}^{Tr=23°C}$	1	$Ta \in [15,22°], Hy \in [25,100\%], Va \in [0.1857,1m/s]$ $u_{Ta}(Ta) \nearrow, u_{Hy}(Hy) \nearrow, u_{Va}(Va) \searrow$	7.2%
	2	$Ta \in [22,28°], Hy \in [50,100\%], Va \in [0.25,1m/s]$ $u_{Ta}(Ta) \searrow, u_{Hy}(Hy) \nearrow, u_{Va}(Va) \searrow$	5.03%
$\mathcal{M}_4: U_{comfort}^{Tr=25°C}$	1	$Ta \in [14,22°], Hy \in [25,100\%], Va \in [0.1875,1m/s]$ $u_{Ta}(Ta) \nearrow, u_{Hy}(Hy) \nearrow, u_{Va}(Va) \searrow$	8.3%
	2	$Ta \in [22,27°], Hy \in [50,100\%], Va \in [0.25,1m/s]$ $u_{Ta}(Ta) \searrow, u_{Hy}(Hy) \nearrow, u_{Va}(Va) \searrow$	3.9%
$\mathcal{M}_5: U_{comfort}^{Tr=30°C}$	1	$Ta \in [10,19°], Hy \in [25,100\%], Va \in [0.1875,1m/s]$ $u_{Ta}(Ta) \nearrow, u_{Hy}(Hy) \nearrow, u_{Va}(Va) \searrow$	4.6%
	2	$Ta \in [19,29°], Hy \in [25,100\%], Va \in [0.1875,1m/s]$ $u_{Ta}(Ta) \searrow, u_{Hy}(Hy) \nearrow, u_{Va}(Va) \searrow$	4.5%

In all computed local approximations, $Ta, Hy,$ and Va have been checked as commensurate which means that utilities and capacity approximations are all based on two reference vectors $\mathbf{O}_{S_1}$ and $\mathbf{G}_{S_1}$ associated to the unique commensurate subset $S_1 = \{Ta, Hy, Va\}$. According to [17], the Choquet integral is unique when commensurate subsets are composed with the coalition of all attributes. In this case, it can be checked that all our local constructions are unique [17]. Utilities and capacity functions require to be normalized. Because there is only one commensurate subset, checking the normalization condition $\alpha.\left(PPD\left(\mathbf{G}_{S_1}\right) - PPD\left(\mathbf{O}_{S_1}\right)\right) = 1$ is easy but must be verified in each local domain.

The next section describes how these 5 tridimensional comfort models can be useful to control one building thermal comfort.

3.3   Interpretable Control Rules

The computation of these local Choquet integrals with [17] implies that we have simultaneously built utility functions for each attribute u_{Ta}, u_{Hy}, and u_{Va} in each domain of validity. These last can then be used in order to build control rules. Thanks to the utility functions, from each local Choquet integral model, the influence of each attribute variation $\delta Ta, \delta Hy$, and δVa on the $U_{comfort}^{Tr=x_{Tr}}$ monotony can be computed. Of course, these influences models are not as precise as we hoped for because they result from the interpolation of two local Choquet integral models; but they still useful to give helpful control recommendations. In fact, the non-existence of a unique overall Choquet integral defined for the whole *PPD* domain inhibits all comparison between utilities in two different local domains. However, it is not such a problematic thing because, for each local domain, we are yet able to identify its valid rules i.e., (9) is an identified thermal comfort control rule for the local model M3 (fig. 3). It models the attribute *Hy* influence upon $U_{comfort}^{Tr=23°}$ for the local domain M3. Then this rule can be formulated as a recommendation when environmental conditions satisfy the local domain M3. The "gains" related to these relationships are the $\Delta\mu_{(i)}$ Choquet integral parameters (3) for M3. Hence, the energy manager knows the negative or positive influence of any attribute upon comfort function in any domain, the polyhedrons in which this influence is valid, and the expected impact from an attribute variation. It allows enunciating control rules such as (9).

$$M_3 : Ta \in [22,28], Hy \in [50,100], \text{and} Va \in [0.25,1]$$

$$\text{if } Hy \nearrow \text{ then } U_{comfort} \searrow \text{ because } u_{Hy}(Hy) \searrow \text{ and}$$

$$\delta U_{comfort} = \Delta\mu_{Hy}^{M_{3.2}} .g^{M_{3.2}} .\delta Hy \tag{9}$$

where $g^{M_{3.2}}$ is the approximation gradient of $u_{Hy}(Hy)$ w.r.t x_{Hy} in $M_{3.2}$.

Hence, thanks to the *PPD* approximation by local Choquet integrals, we obtain a set of rules for the thermal comfort control. These rules can directly be applied by the energy manager as suggested just above because they are interpretable rules in term of satisfaction degrees (like comfort itself) which is part of our work objectives. The Choquet integral based models can also be included in optimization problems to efficiently improve comfort or reduce energy costs automatically as it is explained in the next section. This can be achieved thanks to the Choquet integral linearity by simplex.

4 Some Control Problems Based on the Piecewise Choquet Integral

The model of comfort is now built in the control of the energy system of a building floor. Let us suppose that the control variables are ambient temperature and airflow of all the offices at this floor. There is a General Air Treatment (GAT) —a central heating exchanger—for the whole building and additional individual heating exchangers in the offices. Basic control functions are already implemented in the GAT.

In RIDER DSS, comfort appears as an overall performance of the control problem. It must help the energy manager to satisfy each individual comfort expectation with a minimal cost. Indeed, persons do not have the same expectations w.r.t thermal comfort, on one hand, and one office heat loss depends on its exposure to sunlight, and its neighboring offices isolation characteristics, on the other hand.

RIDER DSS supports the energy manager to manage significantly different temperature setpoints in each office at the floor in order to warranty the comfort levels and minimize as well the energy cost. Then, in order to satisfy both requirements: cost and comfort constraints, RIDER DSS aims to compute adequate setpoints to be provided to the GAT control system. In this paper, we consider that RIDER DSS manages only the energy system performances (utilities related to measurements) without worrying about the way these performances are achieved (GAT control). RIDER DSS aims to prove that reasoning using an aggregated comfort objective function already provides substantial savings. Let us consider some tractable issues by RIDER DSS:

— **Control.** The control issue may be used to adjust the thermal sensation of an unsatisfied officer and whenever any disturbance distracts from the comfort expectation;
— **Adaption.** Thermal sensation is not the same in the north sided offices of the building and the south sided ones. Furthermore the sunlight exposure varies every day and during the day;
— **Anticipation.** Season changes and occupation rates, are proceeding to phenomena that directly impact energy management.

For instance, (10) is a formalization of a simple control problem based on the comfort preference model F that has been identified from the *PPD* model in this paper. This formalization aims to control variables Ta and Va (the offices airflow) in order to improve the thermal sensation of an unsatisfied office occupant $comfort(k)$ without decreasing the comfort of its neighboring offices k' when $Ta(k)$ and $Va(k)$ change.

$$\begin{cases} \min \delta Ta(k) \\ Comfort(k) = F(Ta + \delta Ta(k), Hy(k), Tr(k), Va + \delta Va(k), Ci, Me) = Comfort* \\ \delta Ta(k) \geq 0 \\ \forall k' \neq k, Comfort(k') \geq setpoint(k') \end{cases} \tag{10}$$

Because F has been approximated with Choquet integrals, this optimization problem can be locally linearized and, so, becomes an easily tractable problem [21]. Furthermore, the gain between comfort degree and δTa or δVa variations is locally a constant computed with Ta and Va related utility functions and also the Choquet integral parameters in the simplex search space. This gain value makes the improvement interpretable for control purposes. Finally, domains of validity of the Choquet integral based approximations provide the necessary bounds to reason with a constant gain.

Similarly, the adaptation and anticipation problems can be easily formalized as the control one and their resolution are also simplified thanks to the local linearity of the Choquet integral expression.

5   Conclusion

This work focuses on proposing generic optimization techniques for energy management systems based on a thermal comfort preference model. It explains why and how associating comfort to a MAUT preference model for energy management issues. The introduced thermal comfort model can be easily generalized for different building occupants and simplifies the energy control issues. In fact, thanks to the MAUT, the interpretation of attributes influences on the thermal sensation in term of utility functions makes the multidimensional comfort control process more tractable. The introduction of MAUT techniques in energy control completely shifts the energy control paradigm. For example, the aggregated model for comfort allows designing new lower temperature setpoints that could not be envisaged even in advanced multivariable control techniques. Indeed, relationships between attributes are preferential interactions and not physical influences: each attribute can be controlled independently but any change of an attribute entails a variation of its local utility that may have consequences on the comfort overall utility. RIDER aims to prove that reasoning using an aggregated comfort objective function provides substantial savings. Within the MAUT, it can reasonably be imagined that temperature setpoints of a building could be decreased from one to two degrees. It represents a substantial economic gain that is probably much more significant than any optimization of the energy manager control system. Furthermore, the control recommendations resulting from this model are obviously transferable to any energy facilities.

References

1. Yang, I.H., Yeo, M.S., Kim, K.W.: Application of artificial neural network to predict the optimal start time for heating system in building. Energy Conversion and Management 44, 2791–2809 (2003)
2. Pérez-Lombard, L., Ortiz, J., Pout, C.: A review on buildings energy consumption information. Energy and Buildings 40, 394–398 (2008)
3. Morosan, P.D., Bourdais, R., Dumur, D., Buisson, J.: Building temperature regulation using a distributed model predictive control. Energy and Buildings (2010)
4. Chwieduk, D.: Towards sustainable-energy buildings. Applied Energy 76, 211–217 (2003)

5. Jiangjiang, W., Zhiqiang, J.Z., Youyin, J., Chunfa, Z.: Particle swarm optimization for redundant building cooling heating and power system. Applied Energy 87(12), 3668–3679 (2010)

6. Pohekar, S.D., Ramachandran, M.: Application of multi-criteria decision making to sustainable energy planning-A review. Renewable and Sustainable Energy Reviews 8(4), 365–381 (2004)

7. Fanger, P.O.: Thermal comfort: analysis and applications in environmental engineering. McGraw-Hill, New York (1972)

8. Norme, NF EN ISO 7730. Ergonomie des ambiances thermiques : Détermination analytique et interprétation du confort thermique à l'aide de calculs des indices PMV et PPD et du confort thermique local. AFNOR (2006)

9. Modave, F., Grabisch, M.: Preference representation by a Choquet integral: Commensurability hypothesis. In: IPMU 1998, Paris, France, pp. 164–171 (1998)

10. Krantz, D.H., Luce, R.D., Suppes, P., Tversky, A.: Foundations of measurement. In: Additive and Polynomial Representations, vol. 1. Academic Press (1971)

11. Fishburn, P.C.: Utility Theory for Decision-Making. John Wiley & Sons, New York (1970)

12. Fishburn, P.C.: The foundations of expected utility. Reidel, Dordrecht (1982); Keeney, R.L., Raiffa, H.: Decisions with Multiple Objectives – Preferences and Value Tradeoffs. Cambridge University Press (1976)

13. Montmain, J., Trousset, F.: The translation of will into act: achieving a consensus between managerial decisions and operational abilities. In: Information Control Problems in Manufacturing (INCOM), Moscow, Russia (2009)

14. Labreuche, C., Grabisch, M.: The Choquet integral for the aggregation of interval scales in multicriteria decision making. Fuzzy Sets & Systems 137, 11–26 (2003)

15. Grabisch, M., Labreuche, C.: A decade of application of the Choquet and Sugeno integrals in multi-criteria decision aid. Quarterly Journal of Operations Research 6, 1–44 (2008)

16. Roy, B.: À propos de la signification des dépendances entre critères: quelle place et quels modes de prise en compte pour l'aide à la décision? RAIRO-Oper. Res. 43, 255–275 (2009)

17. Labreuche, C.: Construction of a Choquet integral and the value functions without any commensurateness assumption in multi-criteria decision making. In: European Society of Fuzzy Logic and Technology (EUSFLAT-LFA), Aix-les-Bains, France (2011)

18. Grabisch, M., Roubens, M.: The application of fuzzy integrals in multicriteria decision-making. European Journal of Operational Research 89, 445–456 (1996)

19. Grabisch, M.: k-Ordered Discrete Fuzzy Measures and Their Representation. Fuzzy Sets and Systems 92, 167–189 (1997)

20. Labreuche, C., Grabisch, M.: The Choquet integral for the aggregation of interval scales in multicriteria decision making. Fuzzy Sets & Systems 37, 11–26 (2003)

21. Sahraoui, S., Montmain, J., Berrah, L., Mauris, G.: User-friendly optimal improvement of an overall industrial performance based on a fuzzy Choquet integral aggregation. In: IEEE International Conference on Fuzzy Systems, London, UK (2007)

A Probabilistic Hybrid Logic
for Sanitized Information Systems

Tsan-sheng Hsu, Churn-Jung Liau, and Da-Wei Wang

Institute of Information Science
Academia Sinica, Taipei 115, Taiwan
{tshsu,liaucj,wdw}@iis.sinica.edu.tw

Abstract. As privacy-preserving data publication has received much
attention in recent years, a common technique for protecting privacy is
to release the data in a sanitized form. To assess the effect of sanitization methods, several data privacy criteria have been proposed. Different privacy criteria can be employed by a data manager to prevent
different attacks, since it is unlikely that a single criterion can meet the
challenges posed by all possible attacks. Thus, a natural requirement
of data management is to have a flexible language for expressing different privacy constraints. Furthermore, the purpose of data analysis is to
discover general knowledge from the data. Hence, we also need a formalism to represent the discovered knowledge. The purpose of the paper
is to provide such a formal language based on probabilistic hybrid logic,
which is a combination of quantitative uncertainty logic and basic hybrid
logic with a satisfaction operator. The main contribution of the work is
twofold. On one hand, the logic provides a common ground to express
and compare existing privacy criteria. On the other hand, the uniform
framework can meet the specification needs of combining new criteria as
well as existing ones.

Keywords: Data privacy, information systems, probabilistic logic, hybrid logic, k-anonymity, logical safety.

1 Introduction

Privacy-preserving data publication has received much attention in recent years.
When data is released to the public for analysis, re-identification is considered a
major privacy threat to microdata records, which contain information about specific individuals. Although identifiers, such as names and social security numbers
are typically removed from the released microdata sets, it has been long recognized that several quasi-identifiers, e.g., ZIP codes, age, and sex, can be used
to re-identify individual records. The main reason is that the quasi-identifiers
may appear together with an individual's identifiers in another public database.
Therefore, the problem is how to prevent adversaries inferring private information about an individual by linking the released microdata set to some public or
easy-to-access database.

E. Hüllermeier et al. (Eds.): SUM 2012, LNAI 7520, pp. 500–513, 2012.

To address the privacy concerns about the release of microdata, data is often sanitized before it is released to the public. For example, generalization and suppression of the values of quasi-identifiers are widely used sanitization methods. To assess the effect of sanitization methods, several data privacy criteria have been proposed. One of the earliest criteria was the notion of k-anonymity[12,13]. Although k-anonymity is an effective way to prevent *identity disclosure*, it was soon realized that it was insufficient to ensure protection of sensitive attributes. To address the *attribute disclosure* problem, a *logical safety* criterion was proposed in [6]. The criterion was later expanded to the epistemic model in [14] and the well-known *l-diversity* criterion in [8,9]. Different privacy criteria can be employed by a data manager to prevent different attacks, since it is unlikely that a single criterion can meet the challenges posed by all possible attacks. Thus, a natural requirement of data management is to have a flexible language for expressing different privacy constraints. Furthermore, the purpose of data analysis is to discover general knowledge from the data. Hence, we also need a formalism to represent the discovered knowledge. The purpose of the paper is to provide such a formal language based on probabilistic hybrid logic that can represent discovered knowledge as well as data security constraints. Probabilistic hybrid logic is a fusion of a hybrid logic with a satisfaction operator[1] and a logic for reasoning about quantitative uncertainty[4]. The syntax of the proposed logic is comprised of well-formed formulas of both logics, and its semantics is based on epistemic probability structures with the additional interpretation of nominals. To express privacy requirements, we have to represent knowledge, uncertainty, and individuals. While existing probabilistic logic[5,7,4] can only represent knowledge and uncertainty, the incorporation of hybrid logic into the framework facilitates the representation of individuals.

The remainder of this paper is organized as follows. In Section 2, we review the definitions of microdata and sanitized information systems for privacy-preserving data publication. In Section 3, we introduce the syntax and semantics of probabilistic hybrid logic. An axiomatization of the logic is also presented. While the axioms are valid for general probabilistic hybrid logic models, we need additional specific axioms for released data. In Section 4, and formulate the additional axioms based on sanitized information systems. We also use examples to explain how privacy constraints and discovered knowledge can be expressed with the proposed logic. Finally, Section 5 contains some concluding remarks.

2   Information Systems

In database applications, microdata, such as medical records, financial transaction records, and employee data, are typically stored in information systems. Am *information system* or *data table* [10][1] is formally defined as a tuple $T = (U, A, \{V_f \mid f \in A\})$, where U is a nonempty finite set, called the universe, and A is a nonempty finite set of attributes such that each $f \in A$ is a total function $f : U \to V_f$, where V_f is the domain of values for f. In an information system,

---

[1] Also called knowledge representation systems or attribute-value systems in [10].

the information about an object consists of the values of its attributes. Thus, for a subset of attributes $B \subseteq A$, the *indiscernibility relation* with respect to B is defined as follows:

$$ind_T(B) = \{(x,y) \in U^2 \mid \forall f \in B, f(x) = f(y)\}.$$

Usually, we omit the symbol T in the indiscernibility relation when the underlying information system is clear from the context. We also abbreviate an equivalence class of the indiscernibility relation $[x]_{ind(B)}$ as $[x]_B$. Note that an information system is equivalent to a data matrix defined in [3].

The attributes of an information system can be partitioned into three subsets [11]. First, we have a subset of *quasi-identifiers*, the values of which are known to the public. For example, in [12,13], it is noted that certain attributes like birth-date, gender, and ethnicity are included in some public databases, such as census data or voter registration lists. These attributes, if not appropriately sanitized, may be used to re-identify an individual's record in a medical data table, thereby causing a violation of privacy. The second kind is the subset of *confidential attributes*, the values of which we have to protect. Note that confidential attributes can also serve as quasi-identifiers in some cases. However, since the values of confidential attributes are not easily accessible by the public, in this paper, we simply assume that the set of quasi-identifiers is disjoint with the set of confidential attributes. The remaining attributes are *neutral attributes* (also called Non-confidential outcome attributes in [3]) that are neither quasi-identifying, nor confidential. Hereafter, we assume that the set of attributes $A = Q \cup C \cup N$, where Q, C, N are pairwise disjoint, Q is the set of quasi-identifiers, C is the set of confidential attributes, and N is the set of neutral attributes. Sometimes, the set of attributes is defined such that it contains identifiers that can be used to identify a person's data record. However, for simplicity, we equate each individual with his/her identifier, so the universe U can be considered as the set of identifiers. Furthermore, since identifiers are always removed in a released data table, U simply denotes a set of serial numbers for a de-identified information system.

Example 1. Table 1 is a simple example of an information system. The quasi-identifiers of the information systems are "Date of Birth" and "ZIP". The confidential attributes are "Income" and "Health Status". The values of "Health Status" indicate "normal" (0), "slightly ill" (1), and "seriously ill" (2). "Height" is a neutral attribute.   ∎

A common technique for protecting privacy is to release the information system in a sanitized form. Formally, we define *sanitization* as an operation on information systems.

Definition 1. *Let $T = (U, A, \{V_f \mid f \in A\})$ be an information system. Then, a sanitization operation $\sigma = (\iota, (s_f)_{f \in A})$ is a tuple of mappings such that*

- $\iota : U \to U'$ *is a 1-1 de-identifying mapping, where $|U'| = |U|$, and*
- *for each $f \in A$, $s_f : V_f \to V'_f$ is a sanitizing mapping, where $V'_f \supseteq V_f$ is the domain of generalized values for f.*

Table 1. An information system in a data center

U	Date of Birth	ZIP	Height	Income	Health Status
i_1	24/09/56	24126	160	100K	0
i_2	06/09/56	24129	160	70K	1
i_3	23/03/56	10427	160	100K	0
i_4	18/03/56	10431	165	50K	2
i_5	20/04/55	26015	170	30K	2
i_6	18/04/55	26032	170	70K	0
i_7	12/10/52	26617	175	30K	1
i_8	25/10/52	26628	175	50K	0

The application of σ on T results in a sanitized information system $\sigma T = (U', A', \{V'_f \mid f \in A\})$ such that $A' = \{f' \mid f \in A\}$; and for each $f \in A$, $f' = s_f \circ f \circ \iota^{-1}$, where $\circ$ denotes the functional composition. Note that the de-identifying mapping ι is invertible because it is a bijection.

The universe U' in a sanitized information system is regarded as the set of *pseudonyms* of the individuals. We assume that V'_f is a superset of the original domain V_f, so a sanitizing mapping may be an identity function. A sanitization operation $\sigma = (\iota, (s_f)_{f \in A})$ is *truthful* if for each $f \notin Q$, $s_f = id$ is the identity function; and it is *proper* if $\iota(ind_T(Q)) = \{(\iota(x), \iota(y)) \mid (x, y) \in ind_T(Q)\}$ is a proper subset of $ind_{\sigma T}(Q)$. That is, $ind_T(Q)$ corresponds to a strictly finer partition[2] than $ind_{\sigma T}(Q)$ does. In this paper, we only consider truthful sanitization operations. Moreover, in most cases, proper sanitization is necessary for the protection of privacy. A special sanitization, called *trivial* sanitization, is commonly used as the baseline of privacy assessment[2]. Formally, a sanitization operation is trivial if, for all $f \in Q$, $|s_f(V_f)| = 1$. The suppression of all quasi-identifiers can achieve the effect of trivial sanitization.

Example 2. In privacy research, generalization is a widely-used sanitization operation. For example, the date of birth may only be given as the year and month, or only the first two digits of the ZIP code may be given. A concrete generalization of the information system in Table 1 is presented in Table 2. The first column of the table shows the pseudonyms of the individuals. Note that the sanitization is truthful and proper. ∎

When a sanitized information system is released, the sanitizing mappings are usually known to the public, but the de-identifying mapping must be kept secret. In fact, when a sanitization is truthful and the adversary knows the values of the quasi-identifiers, the adversary can easily infer the sanitizing mappings. For example, in the previous sanitized information system, it is easy to see how "ZIP" and "Date of Birth" are generalized.

---

[2] Recall that each equivalence relation corresponds to a partition.

Table 2. A sanitized information system

d_1	09/56	24***	160	100K	0
d_2	09/56	24***	160	70K	1
d_3	03/56	10***	160	100K	0
d_4	03/56	10***	165	50K	2
d_5	04/55	26***	170	30K	2
d_6	04/55	26***	170	70K	0
d_7	10/52	26***	175	30K	1
d_8	10/52	26***	175	50K	0

3 Probabilistic Hybrid Logic

3.1 Syntax

Hybrid logics are extensions of standard modal logics with *nominals* that name individual states in possible world models[1]. The simplest hybrid language is the extension of the basic modal language with nominals only. More expressive variants can include the existential modality E, the satisfaction operator @, and the binder $\downarrow$. The simplest hybrid language is denoted by $\mathcal{H}$ and its extensions are named by listing the additional operators. For example, $\mathcal{H}(@)$ is the simplest hybrid language extended with the satisfaction operator @. On the other hand, the probabilistic logic $\mathcal{L}_n^{QU}$ proposed in [4] consists of *(linear) likelihood formulas* of the form

$$r_1 l_{a_1}(\varphi_1) + \cdots + r_k l_{a_k}(\varphi_k) > s,$$

where $r_1, \ldots, r_k, s$ are real numbers, $a_1, \ldots, a_k$ are (not necessarily distinct) agents, and $\varphi_1, \ldots, \varphi_k$ are well-formed formulas of the probabilistic language. The proposed probabilistic hybrid logic is a straightforward fusion of $\mathcal{H}(@)$ and $\mathcal{L}_n^{QU}$. The following definition gives the syntax of the language.

Definition 2. *Let* PROP $= \{p_1, p_2, \ldots\}$ *(the propositional symbols),* AGT $= \{a_1, a_2, \ldots\}$ *(the agent symbols), and* NOM $= \{i_1, i_2, \ldots\}$ *(the nominals) be pairwise disjoint, countably infinite sets of symbols. The well-formed formulas of the probabilistic hybrid logic* $\mathcal{PH}(@)$ *in the signature* $\langle$ PROP, AGT, NOM $\rangle$ *are given by the following recursive definition:*

$$\text{WFF} ::= \top \mid p \mid i \mid \neg\varphi \mid \varphi \wedge \psi \mid \langle a \rangle\varphi \mid @_i\varphi \mid r_1 l_{a_1}(\varphi_1) + \cdots + r_k l_{a_k}(\varphi_k) > s,$$

where $p \in$ PROP; $i \in$ NOM; $a, a_1, \ldots, a_k \in$ AGT; $\varphi, \varphi_1, \ldots, \varphi_k \in$ WFF; *and* $r_1, \ldots, r_k, s$ *are real numbers.*

As usual, we abbreviate $\neg(\neg\varphi \wedge \neg\psi)$, $\neg(\varphi \wedge \neg\psi)$, and $\neg\langle a \rangle\varphi$ as $\varphi \vee \psi$, $\varphi \supset \psi$, and $[a]\varphi$ respectively. In addition, $(\varphi \supset \psi) \wedge (\psi \supset \varphi)$ is abbreviated as $(\varphi \equiv \psi)$; and several obvious abbreviations can be applied to likelihood formulas, e.g., $r_1 l_{a_1}(\varphi_1) + \cdots + r_k l_{a_k}(\varphi_k) < s$ denotes $(-r_1) l_{a_1}(\varphi_1) + \cdots + (-r_k) l_{a_k}(\varphi_k) > -s$.

A formula is *pure* if it does not contain any propositional symbols, and *nominal-free* if it does not contain any nominals. In a likelihood formula, $l_a(\varphi)$ is called a *likelihood term*. A formula whose outermost likelihood terms only involve agent a is called an *a-likelihood formula*; that is, an a-likelihood formula is of the form $r_1 l_a(\varphi_1) + \cdots + r_k l_a(\varphi_k) > s$. A set of wffs Σ is called a *theory* in $\mathcal{PH}(@)$.

3.2   Semantics

The semantics of $\mathcal{PH}(@)$ is based on the *epistemic probability frame* introduced in [4]. Before stating the definition, we review the notion of *probability space*.

Definition 3. *A probability space is a triple* $(W, \mathcal{X}, \mu)$, *where* $\mathcal{X}$ *is an algebra[3] over* W *and* $\mu : \mathcal{X} \to [0,1]$ *satisfies the following two properties:*

- *P1.* $\mu(W) = 1$,
- *P2.* $\mu(U \cup V) = \mu(U) + \mu(V)$ *if* U *and* V *are disjoint elements of* $\mathcal{X}$.

The subsets in the algebra $\mathcal{X}$ are called the *measurable* subsets of W. In general, not all subsets of W are measurable; however, for our application, it suffices to consider the case of $\mathcal{X} = 2^W$. Thus, hereafter, we assume all probability spaces are measurable, i.e., $\mathcal{X} = 2^W$; consequently, a probability space is simply written as a pair (W, μ).

Definition 4. *An epistemic probability frame is a tuple* $\mathfrak{F} = (W, (R_a)_{a \in AGT}, (\mathcal{PR}_a)_{a \in AGT})$, *where* W *is a set of possible worlds (states) and for each* $a \in AGT$

- $R_a \subseteq W \times W$ *is a binary relation (the accessibility relation) on* W, *and*
- $\mathcal{PR}_a$ *is probability assignment, i.e., a function that associates a probability space* $(W_{w,a}, \mu_{w,a})$ *with each world* w.

Definition 5. *Let* $\mathfrak{F} = (W, (R_a)_{a \in AGT}, (\mathcal{PR}_a)_{a \in AGT})$ *be an epistemic probability frame. Then, an epistemic probability structure (or* $\mathcal{PH}(@)$ *model) based on* $\mathfrak{F}$ *is a pair* $\mathfrak{M} = (\mathfrak{F}, \pi)$, *where* $\pi : PROP \cup NOM \to 2^W$ *is an interpretation such that for all nominals* $i \in NOM$, $\pi(i)$ *is a singleton. In this case, we also say that* $\mathfrak{F}$ *is the underlying frame of* $\mathfrak{M}$.

By slightly abusing the notation, we can identify a singleton and its element. Thus, when $\pi(i) = \{w\}$, we use $\pi(i)$ to denote both $\{w\}$ and w.

Definition 6. *Let* $\mathfrak{M} = (W, (R_a)_{a \in AGT}, (\mathcal{PR}_a)_{a \in AGT}, \pi)$ *be a* $\mathcal{PH}(@)$ *model and* $w \in W$ *be a possible world. Then, the satisfaction relation is defined as follows:*

1. $\mathfrak{M}, w \models \top$
2. $\mathfrak{M}, w \models p$ *iff* $w \in \pi(p)$ *for* $p \in PROP \cup NOM$
3. $\mathfrak{M}, w \models \neg\varphi$ *iff* $\mathfrak{M}, w \not\models \varphi$

---

[3] That is, $\mathcal{X}$ satisfies the following conditions (i) $\mathcal{X} \subseteq 2^W$, (ii) $W \in \mathcal{X}$, and (iii) $\mathcal{X}$ is closed under union and complementation.

4. $\mathfrak{M}, w \models \varphi \wedge \psi$ iff $\mathfrak{M}, w \models \varphi$ and $\mathfrak{M}, w \models \psi$
5. $\mathfrak{M}, w \models \langle a \rangle \varphi$ iff there is a w' such that $(w, w') \in R_a$ and $\mathfrak{M}, w' \models \varphi$
6. $\mathfrak{M}, w \models @_i \varphi$ iff $\mathfrak{M}, \pi(i) \models \varphi$
7. $\mathfrak{M}, w \models r_1 l_{a_1}(\varphi_1) + \cdots + r_k l_{a_k}(\varphi_k) > s$ iff $r_1 \mu_{w,a_1}(|\varphi_1| \cap W_{w,a_1}) + \cdots + r_k \mu_{w,a_k}(|\varphi_k| \cap W_{w,a_k}) > s$, where $|\varphi_i| = \{u \mid \mathfrak{M}, u \models \varphi_i\}$ is the truth set of φ in the model $\mathfrak{M}$.

Let Σ be a theory. Then, we write $\mathfrak{M}, w \models \Sigma$ if $\mathfrak{M}, w \models \varphi$ for all $\varphi \in \Sigma$. A wff φ is said to be *true* in a model, denoted by $\mathfrak{M} \models \varphi$, if $\mathfrak{M}, w \models \varphi$ for all $w \in W$; and a wff φ is *valid*, denoted by $\models \varphi$, if $\mathfrak{M} \models \varphi$ for all $\mathcal{PH}(@)$ models $\mathfrak{M}$. Moreover, a wff φ is *satisfiable* if there exists a model $\mathfrak{M}$ and a world $w \in W$ such that $\mathfrak{M}, w \models \varphi$; and it is a *local logical consequence* of a theory Σ, denoted by $\Sigma \models^{loc} \varphi$, if for all models $\mathfrak{M}$ and worlds $w \in W$, $\mathfrak{M}, w \models \Sigma$ implies $\mathfrak{M}, w \models \varphi$. Finally, a wff φ is a *global logical consequence* of a theory Σ, denoted by $\Sigma \models^{glo} \varphi$, if for all models $\mathfrak{M}$, $\mathfrak{M} \models \Sigma$ implies $\mathfrak{M} \models \varphi$.

3.3 Axiomatization

The axiomatization of $\mathcal{PH}(@)$ in Figure 1 consists of axioms and rules of both hybrid logic[1] and probabilistic logic[4]. It is essentially a modular combination of the two logics, since no additional interaction axioms are introduced. As usual, a *derivation* in the axiomatic system is a sequence of wffs $\varphi_1, \cdots, \varphi_m$ such that each φ_i is either an instance of an axiom schema, or follows from previous formulas by the application of an inference rule. The derivation $\varphi_1, \cdots, \varphi_m$ is also a *proof* of the last wff φ_m. A wff φ is *provable* in an axiomatization, denoted by $\vdash \varphi$, if there is a proof of φ in the system. A provable wff is also called a *theorem* of the system. Let Σ be a theory and φ be a wff. Then, we write $\Sigma \vdash \varphi$ and say that φ is derivable from Σ if there exists $\varphi_1, \cdots, \varphi_k \in \Sigma$ such that $(\varphi_1 \wedge \cdots \wedge \varphi_k) \supset \varphi$ is a theorem of the system.

4 Information System Theory

4.1 Models of Sanitized Information Systems

To specify an information system and its sanitization, we have to use a fixed language. Let us consider an information system $T = (U, A, \{V_f \mid f \in A\})$, where $A = Q \cup N \cup C$ and a truthful sanitization operation $\sigma = (\iota, (s_f)_{f \in A})$. In addition, let $\sigma T = (U', A', \{V'_f \mid f \in A\})$ be defined as above. We assume that $U = \{i_1, \cdots, i_n\}$ and $U' = \{d_1, \cdots, d_n\}$. Then, the signature of our language comprises

- PROP $= \{(f, v) \mid f \in A, v \in V'_f\}$,
- AGT $= \{a_0, a_1\}$, and
- NOM $= U \cup U'$.

In Pawlak's decision logic[10], a propositional symbol (f, v) is called a *descriptor*, which means that the value of attribute f of an individual is v. We consider

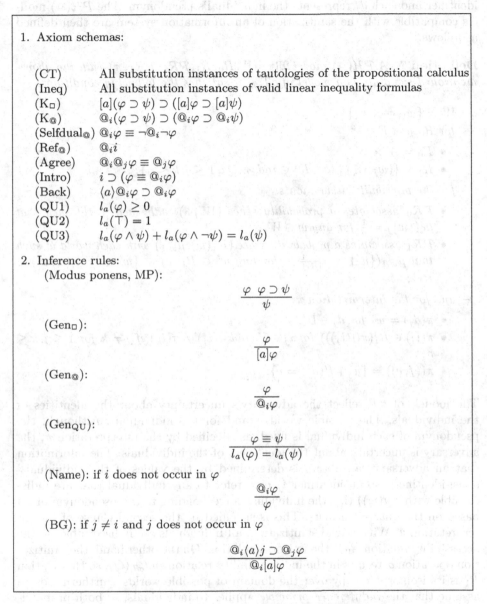

1. Axiom schemas:

(CT)   All substitution instances of tautologies of the propositional calculus
(Ineq)   All substitution instances of valid linear inequality formulas
($K_\square$) $[a](\varphi \supset \psi) \supset ([a]\varphi \supset [a]\psi)$
($K_@$) $@_i(\varphi \supset \psi) \supset (@_i\varphi \supset @_i\psi)$
(Selfdual$_@$) $@_i\varphi \equiv \neg@_i\neg\varphi$
(Ref$_@$) $@_i i$
(Agree) $@_i@_j\varphi \equiv @_j\varphi$
(Intro) $i \supset (\varphi \equiv @_i\varphi)$
(Back) $\langle a\rangle@_i\varphi \supset @_i\varphi$
(QU1) $l_a(\varphi) \geq 0$
(QU2) $l_a(\top) = 1$
(QU3) $l_a(\varphi \wedge \psi) + l_a(\varphi \wedge \neg\psi) = l_a(\psi)$

2. Inference rules:
   (Modus ponens, MP):

$$\frac{\varphi \quad \varphi \supset \psi}{\psi}$$

   (Gen$_\square$):

$$\frac{\varphi}{[a]\varphi}$$

   (Gen$_@$):

$$\frac{\varphi}{@_i\varphi}$$

   (Gen$_{QU}$):

$$\frac{\varphi \equiv \psi}{l_a(\varphi) = l_a(\psi)}$$

   (Name): if i does not occur in φ

$$\frac{@_i\varphi}{\varphi}$$

   (BG): if $j \neq i$ and j does not occur in φ

$$\frac{@_i\langle a\rangle j \supset @_j\varphi}{@_i[a]\varphi}$$

Fig. 1. The axiomatization of $\mathcal{PH}(@)$

two agents a_0 and a_1; and we assume that agent a_0 only receives the trivially sanitized information system, and a_1 receives the system σT. The set of nominals is partitioned into two subsets such that each i_j denotes an individual's identifier and each d_j represents the individual's pseudonym. The $\mathcal{PH}(@)$ models compatible with the sanitization of an information system are then defined as follows.

Definition 7. *A $\mathcal{PH}(@)$ model $\mathfrak{M} = (W, R_0, R_1, \mathcal{PR}_0, \mathcal{PR}_1, \pi)$ with the above-mentioned signature is a model of σT if it satisfies the following conditions:*

- $W = \{w_1, \cdots, w_n\}$;
- *for R_0 and R_1:*
  - $R_0 = W \times W$,
  - $R_1 = \{(w_j, w_k) \mid (d_j, d_k) \in ind_{\sigma T}(Q), 1 \le j, k \le n\}$;
- *for the probability assignments:*
  - $\mathcal{PR}_0$ *associates a probability space (W, μ_0) with each world such that $\mu_0(\{w\}) = \frac{1}{n}$ for any $w \in W$,*
  - $\mathcal{PR}_1$ *associates a probability space $(R_1(w), \mu_{w,1})$ with each world w such that $\mu_{w,1}(\{w'\}) = \frac{1}{|R_1(w)|}$ for any $w' \in R_1(w) = \{w' \in W \mid (w, w') \in R_1\}$;*
- *and for the interpretation π:*
  - $\pi(d_j) = w_j$ for $d_j \in U'$,
  - $\pi(i_j) \in R_1(\pi(\iota(i_j)))$ for $i_j \in U$ and $\pi(i_j) \ne \pi(i_k)$ if $j \ne k$ for $1 \le j, k \le n$,
  - $\pi((f, v)) = \{w_j \mid f'(d_j) = v\}$.

The models of σT reflect the adversary's uncertainty about the identities of the individuals. The possible worlds stand for the individuals. Although, the pseudonym of each individual is fixed, as specified by the interpretation π, the adversary is uncertain about the identifiers of the individuals. The information that an adversary can obtain is determined by the values of the individuals' quasi-identifiers, so an identifier i_j may refer to any individual that are indiscernible with $\pi(\iota(i_j))$ (i.e. the individual corresponding to the pseudonym of i_j) based on the quasi-identifiers. This is specified by the second clause of the interpretation π. With trivial sanitization, all individuals are indiscernible, so the accessibility relation R_0 is the universal relation. On the other hand, the sanitization operation σ results in the indiscernibility relation $ind_{\sigma T}(Q)$, so the relation R_1 is its isomorphic copy over the domain of possible worlds. Furthermore, we assume that the *indifference principle* applies to individuals, so both probability assignments associate a uniform distribution with each possible world. Since the two probability assignments are characterized completely by the accessibility relations and R_0 is simply the universal relation, we can omit these three components from a model of σT and write it as a simple hybrid model (W, R_1, π). A wff φ is *valid* in σT, denoted by $\Vdash_{\sigma T} \varphi$, if it is true in all models of σT.

---

1. General axiom schemas of information systems:

$$(\text{Ref}_\diamond) \quad @_i \langle a \rangle i$$
$$(\text{Sym}_\diamond) \quad @_i \langle a \rangle j \supset @_j \langle a \rangle i$$
$$(\text{Tran}_\diamond) \quad (@_i \langle a \rangle j \wedge @_j \langle a \rangle k) \supset @_i \langle a \rangle k$$
$$(\text{Uni}_\diamond) \quad @_i \langle a_0 \rangle j$$
$$(\text{Cons}) \quad @_i \langle a \rangle j \equiv @_i (l_a(j) > 0)$$
$$(\text{Unif}) \quad @_i \langle a \rangle j \supset @_i (l_a(i) = l_a(j))$$
$$(\text{SVA}) \quad (f, v_1) \supset \neg(f, v_2) \text{ if } v_1 \neq v_2$$

2. Specific axioms for σT:

$$(\text{DC}) \quad (\bigvee_{j=1}^n d_j) \wedge (\bigvee_{j=1}^n i_j)$$
$$(\text{UI}) \quad \neg@_{i_j} i_k \text{ for } 1 \leq j \neq k \leq n$$
$$(\text{UP}) \quad \neg@_{d_j} d_k \text{ for } 1 \leq j \neq k \leq n$$
$$(\text{Rec}) \quad @_{d_j} \bigwedge_{f \in A}(f, f'(d_j)) \text{ for } 1 \leq j \leq n$$
$$(\text{RelP}) \quad @_d(([a_1] \bigvee_{d_j \in [d]_Q} d_j) \wedge \bigwedge_{d_j \in [d]_Q} \langle a_1 \rangle d_j) \text{ for } d \in U'/Q$$
$$(\text{Connect}) \quad @_i[(\bigwedge_{d_j \in [\iota(i)]_Q} \langle a_1 \rangle d_j) \wedge [a_1](\bigvee_{d_j \in [\iota(i)]_Q} d_j)] \text{ for } i \in U$$

Fig. 2. The $\mathcal{PH}(@)$ theory of a sanitized information system

4.2 Theories of Sanitized Information Systems

While the axiomatization for general $\mathcal{PH}(@)$ models is presented in Figure 1, the models for information systems are special kinds of $\mathcal{PH}(@)$ models. Thus, additional axioms are needed to characterize the special constraints imposed on general $\mathcal{PH}(@)$ models. The axioms can be separated into two groups. The first group consists of axiom schemas valid for all information systems, while the second group contains specific axioms for the given information system and its sanitization. The additional axioms are shown in Figure 2, where we use U'/Q to denote the *quotient set* of U' with respect to $ind_{\sigma T}(Q)$. That is, we select a unique representative from each equivalence class of $ind_{\sigma T}(Q)$ arbitrarily and let U'/Q denote the set of all such representatives. In addition, we use $[d]_Q$ to denote the equivalence class of $ind_{\sigma T}(Q)$ that contains d.

Axioms $(\text{Ref}_\diamond)$, $(\text{Sym}_\diamond)$, and $(\text{Tran}_\diamond)$ reflect the fact that each R_a in the models of σT is an equivalence relation. Thus, $[a]$ is an epistemic modality or S5 modality in terms of conventional modal logic systems. Indeed, the intended meaning of $[a]\varphi$ is that agent a knows φ. In traditional modal logic, the S5 modalities are typically characterized by the following three axioms:

T: $[a]\varphi \supset \varphi$
4: $[a]\varphi \supset [a][a]\varphi$
5: $\neg[a]\varphi \supset [a]\neg[a]\varphi$.

However, due to the extra expressivity of nominals in hybrid logic, we can represent these constraints by pure formulas. Furthermore, because the agent a_0 only has minimal information about the trivial sanitization, the accessibility relation R_0 corresponding to a_0's knowledge is the universal relation. The axiom (Uni$_\diamond$) simply indicates this fact.

Axioms (Cons) and (Unif) correspond to constraints on the relationship between R_a and $\mathcal{PR}_a$. The axiom (Cons) represents a consistency constraint. It requires that, if $\mathcal{PR}_a = (W_{w,a}, \mu_{w,a})$, then $W_{w,a} = R_a(w) = \{u \mid (w, u) \in R_a\}$. The axiom (Unif) stipulates that the probability assignments associate a uniform distribution with each possible world. Formally, it means that if $(w, u) \in R_a$, then $\mu_{w,a}(\{w\}) = \mu_{w,a}(\{u\})$. This, together with the domain closure axiom in the second group, captures the fact that $\mu_{w,a}$ is a uniform distribution over $R_a(w)$. These axioms imply the following natural constraints proposed in [4]:

1. uniformity: if $u \in W_{w,a}$, then $\mathcal{PR}_a(w) = \mathcal{PR}_a(u)$;
2. state determined probability (SDP): if $(w, u) \in R_a$, then $\mathcal{PR}_a(w) = \mathcal{PR}_a(u)$; and
3. consistency: $W_{w,a} \subseteq R_a(w)$.

Obviously, (Cons) implies a stronger consistency requirement than that in [4]. Under our consistency requirement, the uniformity and state determined probability are equivalent. Since (Unif) implies that, for each w, $\mu_{w,a}$ is a uniform distribution over $R_a(w)$, it is clear that the state determined probability constraint holds. Consistency, SDP, and uniformity are characterized in [4] by the axioms KP1, KP2, and KP3 respectively:

KP1: $[a]\varphi \supset (l_a(\varphi) = 1)$
KP2: $\varphi \supset [a]\varphi$ if φ is an a-likelihood formula
KP3: $\varphi \supset (l_a(\varphi) = 1)$ if φ is an a-likelihood formula or the negation of an a-likelihood formula.

However, as in the case of S5 modalities, we can also characterize these properties by using pure formulas.

The last axiom (SVA) in the first group is the *single-valued attribute* axiom, which means that each attribute is a single-valued function from the universe of individuals to the domain of values. Sometimes, this axiom can be relaxed if the domain of values has some kind of structure. For example, if the attribute is disease and the domain of values include "AIDS", "Flu", "Cancer", "Infection with virus" and so on, then the value "Flu" implies the value "Infection with virus". In such cases, the values of an attribute may be not completely mutually exclusive and we can represent the relationship between the values by a formula of the form $(f, v_1) \supset (f, v_2)$.

In the second group of axioms, we delimit the domain of individuals and represent their properties according to the sanitized information system. The domain closure axiom (DC) means that each individual is named by an identifier and a pseudonym. The unique identifier axiom (UI) (resp. unique pseudonym axiom (UP)) indicates that two different identifiers (resp. pseudonyms) will not refer to the same individual. The axiom (Rec) represents the information contained in each record of the table σT, and connects each pseudonym with its corresponding attribute values. The axiom (RelP) encodes the indiscernibility relation $ind_{\sigma T}(Q)$. Because of the reflexivity, symmetry, and transitivity axioms in the first group, we only need to specify an axiom for each equivalence class of $ind_{\sigma T}(Q)$ (via each element in the quotient set U'/Q). The last axiom connects identifiers to pseudonyms. Due to the uncertainty caused by the sanitization, an identifier is exactly connected to all pseudonyms that are indiscernible with its own pseudonym.

Let $\Sigma(\sigma T)$ denote the set of all instances of the axioms in Figure 2. Then, it is straightforward to verify the following theorem.

Theorem 1. *A* $\mathcal{PH}(@)$ *model* $\mathfrak{M}$ *with the above-mentioned signature is a model of* σT *iff* $\mathfrak{M} \models \Sigma(\sigma T)$.

The following is a direct corollary of the theorem:

Corollary 1. *Let* φ *be a wff. Then,* $\Vdash_{\sigma T} \varphi$ *iff* $\Sigma(\sigma T) \models^{glo} \varphi$.

4.3 Applications

We have shown that the language of $\mathcal{PH}(@)$ is expressive enough to describe an information system and its sanitization. In the following, we use examples to explain how it can be used to express privacy constraints and knowledge discovered from information systems.

Example 3. According to [12,13], σT satisfies the k-anonymity criterion if $|[d]_Q| \geq k$ for any $d \in U'$. This is easily expressed in $\mathcal{PH}(@)$ language. Formally, a sanitized information system σT satisfies the k-anonymity criterion iff $\Vdash_{\sigma T} (l_{a_1}(i) \leq \frac{1}{k})$ for $i \in U$. The formula means that an individual can be identified with probability at most $\frac{1}{k}$. In particular, it can be derived that $@_d(l_{a_1}(i) \leq \frac{1}{k})$ is valid in σT for any $d \in [\iota(i)]_Q$, which means that, given any record whose quasi-identifiers are indiscernible from i's quasi-identifiers, the adversary will be able to recognize i with probability at most $\frac{1}{k}$.   ■

Example 4. The logical safety criterion was proposed in [6] to prevent homogeneity attacks. Subsequently, it was articulated into an epistemic model for privacy protection in the database linking context [14]. Recall that, in modal logic, the modality-free formulas are called *objective* formulas. Let Γ denote the set of all nominal-free objective formulas, i.e., the set of descriptors closed under Boolean combinations. The logical safety criterion allows a flexible personalized privacy policy, so each individual can specify the information that he/she wants to keep confidential. More precisely, $Sec : U \to 2^{\Gamma}$ is such a specification function. According to the semantics of decision logic[10], a pseudonym d satisfies a descriptor (f, v) with respect to σT, denoted by $d \models_{\sigma T} (f, v)$, if $f'(d) = v$, and the satisfaction relation is extended to all formulas in Γ as usual. We normally omit the subscript σT. It is said that the adversary knows the individual i has property φ, denoted by $i \models K\varphi$ if, for $d \in [\iota(i)]_Q$, $d \models_{\sigma T} \varphi$. Then, σT satisfies the logical safety criterion if $Sec(i) \cap \{\varphi \mid i \models K\varphi\} = \emptyset$ for $i \in U$. Thus, a sanitized information system σT satisfies the logical safety criterion iff $\Vdash_{\sigma T} @_i \neg [a_1]\varphi$ (or equivalently $\Vdash_{\sigma T} @_i l_{a_1}(\varphi) < 1$) for $i \in U$ and $\varphi \in Sec(i)$.   ■

Example 5. To discover general rules from an information system is the main purpose of data analysis. The rules usually exhibit some kind of dependency between different attributes in the following form:

$$\bigwedge_{f \in B} (f, v_f) \longrightarrow (g, v_g),$$

where $B \subset A$ and $g \in A \backslash B$. The rule means that if for each attribute $f \in B$, the value of f is v_f, then the value of attribute g is v_g. The formulas $\bigwedge_{f \in B}(f, v_f)$ and (g, v_g) are respectively called the *antecedent* and *consequent* of the rule. Two important measures, *support* and *confidence*, are used to assess the significance of the discovered rules. The support of a formula is defined as the proportion of objects in the information system that satisfy the formula and the support of a rule is simply the support of the conjunction of its antecedent and consequent. On the other hand, the confidence of a rule is the result of dividing its support by the support of its antecedent. Hence, a rule $\varphi \longrightarrow \psi$ with support at least r_1 and confidence at least r_2 can be represented in our logic as

$$l_{a_1}(\varphi \wedge \psi) \geq r_1 \wedge l_{a_1}(\varphi \wedge \psi) \geq r_2 l_{a_1}(\varphi). \qquad \blacksquare$$

5 Concluding Remarks

In this paper, we propose a probabilistic hybrid logic that can specify data privacy constraints and represent discovered knowledge at the same time. The main contribution of the logic is twofold. On one hand, the *uniformity* of the framework can explicate the common principle behind a variety of privacy requirements and highlights their differences. On the other hand, the *generality* of the framework extends the scope of privacy specifications. In particular, we can specify heterogeneous requirements between different individuals, so it is possible to achieve personalized privacy specification. For example, someone may feel that "slight illness" is sensitive, while others may not care. Consequently, it is desirable that the specification language can express each individual's requirement. Through the use of the satisfaction operator $@_i$, our logic provides such a personalized privacy specification. For example, we can use $@_i \neg [a_1] \varphi \wedge @_j \neg [a_1] \psi$ to express different privacy requirements of individuals i and j.

Although we have shown that the logic is expressive enough to specify many existing privacy policies, we have not utilized its full power yet. For example, nested modalities rarely play a role in our examples, since we only consider a single agent with sanitized information and a dummy baseline agent. However, if we consider the setting of multi-party computation, where several agents share information about the individuals, an agent's knowledge about other agents' knowledge may be crucial for inferring private information. In such a setting, nested modal formulas can represent more sophisticated privacy constraints.

References

1. Areces, C., ten Cate, B.: Hybrid logics. In: Blackburn, P., van Benthem, J., Wolter, F. (eds.) Handbook of Modal Logic, pp. 821–868. Elsevier (2007)
2. Brickell, J., Shmatikov, V.: The cost of privacy: destruction of data-mining utility in anonymized data publishing. In: Proceedings of the 14th ACM SIGKDD International Conference on Knowledge Discovery and Data Mining (KDD), pp. 70–78 (2008)
3. Domingo-Ferrer, J.: Microdata. In: Liu, L., Tamer Özsu, M. (eds.) Encyclopedia of Database Systems, pp. 1735–1736. Springer, US (2009)
4. Halpern, J.: Reasoning about Uncertainty. The MIT Press (2003)
5. Heifetz, A., Mongin, P.: Probability logic for type spaces. Games and Economic Behavior 35(1-2), 31–53 (2001)
6. Hsu, T.-s., Liau, C.-J., Wang, D.-W.: A Logical Model for Privacy Protection. In: Davida, G.I., Frankel, Y. (eds.) ISC 2001. LNCS, vol. 2200, pp. 110–124. Springer, Heidelberg (2001)
7. Larsen, K.G., Skou, A.: Bisimulation through probabilistic testing. Information and Computation 94(1), 1–28 (1991)
8. Machanavajjhala, A., Gehrke, J., Kifer, D., Venkitasubramaniam, M.: l-diversity: Privacy beyond k-anonymity. In: Proc. of the 22nd IEEE International Conference on Data Engineering (ICDE), p. 24 (2006)
9. Machanavajjhala, A., Gehrke, J., Kifer, D., Venkitasubramaniam, M.: l-diversity: Privacy beyond k-anonymity. ACM Transactions on Knowledge Discovery from Data 1(1) (2007)
10. Pawlak, Z.: Rough Sets–Theoretical Aspects of Reasoning about Data. Kluwer Academic Publishers (1991)
11. Samarati, P.: Protecting respondents' identities in microdata release. IEEE Transactions on Knowledge and Data Engineering 13(6), 1010–1027 (2001)
12. Sweeney, L.: Achieving k-anonymity privacy protection using generalization and suppression. International Journal on Uncertainty, Fuzziness and Knowledge-based Systems 10(5), 571–588 (2002)
13. Sweeney, L.: k-anonymity: a model for protecting privacy. International Journal on Uncertainty, Fuzziness and Knowledge-based Systems 10(5), 557–570 (2002)
14. Wang, D.-W., Liau, C.-J., Hsu, T.-s.: An epistemic framework for privacy protection in database linking. Data and Knowledge Engineering 61(1), 176–205 (2007)

Evidential Fusion for Gender Profiling

Jianbing Ma, Weiru Liu, and Paul Miller

School of Electronics, Electrical Engineering and Computer Science,
Queen's University Belfast, Belfast BT7 1NN, UK
{jma03,w.liu}@qub.ac.uk, p.miller@ecit.qub.ac.uk

Abstract. Gender profiling is a fundamental task that helps CCTV systems to provide better service for intelligent surveillance. Since subjects being detected by CCTVs are not always cooperative, a few profiling algorithms are proposed to deal with situations when faces of subjects are not available, among which the most common approach is to analyze subjects' body shape information. In addition, there are some drawbacks for normal profiling algorithms considered in real applications. First, the profiling result is always uncertain. Second, for a time-lasting gender profiling algorithm, the result is not stable. The degree of certainty usually varies, sometimes even to the extent that a male is classified as a female, and vice versa. These facets are studied in a recent paper [16] using Dempster-Shafer theory. In particular, Denoeux's cautious rule is applied for fusion mass functions through time lines. However, this paper points out that if severe mis-classification is happened at the beginning of the time line, the result of applying Denoeux's rule could be disastrous. To remedy this weakness, in this paper, we propose two generalizations to the DS approach proposed in [16] that incorporates time-window and time-attenuation, respectively, in applying Denoeux's rule along with time lines, for which the DS approach is a special case. Experiments show that these two generalizations do provide better results than their predecessor when mis-classifications happen.

Keywords: Gender Profiling, Gender Recognition, Evidence Theory, Cautious Rule.

1 Introduction

Nowadays, CCTV systems are broadly deployed in the present world, e.g., Florida School Bus Surveillance project [1], the First Glasgow Bus Surveillance [21], Federal Intelligent Transportation System Program in the US [20], Airport Corridor Surveillance in the UK [19,17,18,13], etc. However, despite the wide-range use of CCTVs, the impact on anti-social and criminal behaviour has been minimal. For example, assaults on bus and train passengers are still a major problem for transport operators. That is, surveillance systems are not capable of reacting events of interest instantly.

A key requirement for active CCTV systems is to automatically determine the threat posed by each individual to others in the scene. Most of the focus of the computer vision community has been on behaviour/action recognition. However, experienced security analysts profile individuals in the scene to determine their threat. Often they can identify individuals who look as though they may cause trouble before any anti-social

E. Hüllermeier et al. (Eds.): SUM 2012, LNAI 7520, pp. 514–524, 2012.

behaviour has occurred. From criminology studies, the vast majority of offenders are young adolescent males. Therefore, key to automatic threat assessment is to be able to automatically profile people in the scene based on their gender and age. In this paper, we focus on the former.

Although it is a fundamental task for surveillance applications to determine the gender of people of interest, however, normal video algorithms for gender profiling (usually face profiling) have three drawbacks. First, the profiling result is always uncertain. Second, for a time-lasting gender profiling algorithm, the result is not stable. The degree of certainty usually varies, sometimes even to the extent that a male is classified as a female, and vice versa. Third, for a robust profiling result in cases were a person's face is not visible, other features, such as body shape, are required. These algorithms may provide different recognition results - at the very least, they will provide different degrees of certainties. To overcome these problems, in [16], an evidential (Dempster-Shafer's (DS) theory of evidence) approach is proposed that makes use of profiling results from multiple profiling algorithms using different human features (e.g., face, full body) over a period of time, in order to provide robust gender profiling of subjects in video. Experiments show that this approach provides better results than a probabilistic approach.

DS theory [2,22,8,9] is a popular framework to deal with uncertain or incomplete information from multiple sources. This theory is capable of modelling incomplete information through ignorance. For combining difference pieces of information, DS theory distinguishes two cases, i.e., whether pieces of information are from distinct, or non-distinct, sources. Many combination rules are proposed for information from distinct sources, among which are the well-known Dempster's rule [22], Smets' rule [23], Yager's rule [24], and Dubois & Prade's hybrid rule [4], etc. In [3], two combination rules, i.e., the cautious rule and the bold disjunctive rule, for information from non-distinct sources are proposed. Therefore, gender profiling results from the same classifier, e.g. face-based, at different times are considered as from non-distinct sources while profiling results from different classifiers are naturally considered as from distinct sources.

In [16], for gender profiling results from the same classifier at different time points, Denoeux's cautious rule [3] is used to combine them. For profiling results from different classifiers (i.e., face profiling and full body profiling), Dempster's rule [2,22] is introduced to combine them. And finally, the pignistic transformation is applied to get the probabilities of the subject being male or female.

However, if severe mis-classification happens at the beginning of the time line, the result of applying Denoeux's rule could be disastrous. For instance, if a subject is classified as a female with a certainty degree 0.98, and later on it is classified as a male with certainty degrees from 0.85 to 0.95, then by Denoeux's cautious rule, it will be always classified as a female. In order to remedy this weakness, in this paper, we propose two generalizations on applying Denoeux's rule through time lines, in which one uses time-window and the other uses time-attenuation, respectively. In the time-window generalization, Denoeux's rule is applied only for the most recent n frames where n is a pre-given threshold depending on the time length. In the time-attenuation generalization, the certainty degree is reduced gradually by time at a pre-defined attenuation factor. Experiments show that these two generalizations do provide better results when mis-classifications happen, but they have to pay the price of performing less accurate in other situations than the fusion method proposed in [16]. In summary, we can say these two generalizations are more robust than their predecessor.

The rest of the paper is organized as follows. Section 2 provides the preliminaries on Dempster-Shafer theory. Subsequently, Section 3 introduces the two generalizations of the DS approach. In Section 4, we discuss the difficulties in gender profiling in terms of scenarios. Section 5 provides experimental results which shows our generalizations perform better than its predecessor and a classic fusion approach as well as single profiling approaches. Finally, we conclude the paper in Section 6.

2 Dempster-Shafer Theory

For convenience, we recall some basic concepts of Dempster-Shafer's theory of evidence. Let Ω be a finite, non-empty set called the frame of discernment, denoted as, $\Omega = \{w_1, \cdots, w_n\}$.

Definition 1. *A basic belief assignment(bba) is a mapping* $m : 2^\Omega \to [0,1]$ *such that* $\sum_{A \subseteq \Omega} m(A) = 1$.

If $m(\emptyset) = 0$, then m is called a mass function. If $m(A) > 0$, then A is called a focal element of m. Let $\mathscr{F}_m$ denote the set of focal elements of m. A mass function with only a focal element Ω is called a *vacuous* mass function.

From a bba m, *belief* function (Bel) and *plausibility* function (Pl) can be defined to represent the lower and upper bounds of the beliefs implied by m as follows.

$$Bel(A) = \sum_{B \subseteq A} m(B) \text{ and } Pl(A) = \sum_{C \cap A \neq \emptyset} m(C). \quad (1)$$

One advantage of DS theory is that it has the ability to accumulate and combine evidence from multiple sources by using *Dempster's rule of combination*. Let m_1 and m_2 be two mass functions from two distinct sources over Ω. Combining m_1 and m_2 gives a new mass function m as follows:

$$m(C) = (m_1 \oplus m_2)(C) = \frac{\sum_{A \cap B = C} m_1(A)m_2(B)}{1 - \sum_{A \cap B = \emptyset} m_1(A)m_2(B)} \quad (2)$$

In practice, sources may not be completely reliable, to reflect this, in [22], a *discount rate* was introduced by which the mass function may be discounted in order to reflect the reliability of a source. Let r ($0 \leq r \leq 1$) be a discount rate, a discounted mass function using r is represented as:

$$m^r(A) = \begin{cases} (1-r)m(A) & A \subset \Omega \\ r + (1-r)m(\Omega) & A = \Omega \end{cases} \quad (3)$$

When $r = 0$ the source is absolutely reliable and when $r = 1$ the source is completely unreliable. After discounting, the source is treated as totally reliable.

Definition 2. *Let m be a bba on Ω. A pignistic transformation of m is a probability distribution P_m over Ω such that $\forall w \in \Omega, P_m(w) = \sum_{w \in A} \frac{1}{|A|} \frac{m(A)}{1 - m(\emptyset)}$ where $|A|$ is the cardinality of A.*

Let $\oplus$ be the conjunctive combination operator (or Smets' operator [23]) for any two bbas m, m' over Ω such that

$$(m \oplus m')(C) = \sum_{A \subseteq \Omega, B \subseteq \Omega, A \cap B = C} m(A)m'(B), \forall\, C \subseteq \Omega. \tag{4}$$

A simple bba m such that $m(A) = x, m(\Omega) = 1 - x$ for some $A \neq \Omega$ will be denoted as A^x. The vacuous bba can thus be noted as A^0 for any $A \subset \Omega$. Note that this notation, i.e., A^x, is a bit different from the one defined in [3] in which A^x in our paper should be denoted as A^{1-x} in [3].

Similarly, for two sets $A, B \subset \Omega$, $A \neq B$, let $A^x B^y$ denote a bba m such that $m = A^x \oplus B^y$ where $\oplus$ is the conjunctive combination operator defined in Equation (4). For these kinds of bbas, we call them *bipolar* bbas. A simple bba A^x could be seen as a special bipolar bba $A^x B^0$ for any set $B \subseteq \Omega, B \neq A$.

It is easy to prove that any $m = A^x B^y$ is:

$$m(\emptyset) = xy, m(A) = x(1 - y), m(B) = y(1 - x), m(\Omega) = (1 - x)(1 - y) \tag{5}$$

In addition, when normalized, m in Equation 5 is changed to m' as follows.

$$m'(A) = \frac{x(1 - y)}{1 - xy}, m'(B) = \frac{y(1 - x)}{1 - xy}, m'(\Omega) = \frac{(1 - x)(1 - y)}{1 - xy} \tag{6}$$

For two bipolar bbas $A^{x_1} B^{y_1}$ and $A^{x_2} B^{y_2}$, the cautious combination rule proposed in [3] is as follows.

Lemma 1. *(Denœux's Cautious Combination Rule) Let $A^{x_1} B^{y_1}$ and $A^{x_2} B^{y_2}$ be two bipolar bbas, then the combined bba by Denœux's cautious combination rule is also a bipolar bba $A^x B^y$ such that: $x = max(x_1, x_2), y = max(y_1, y_2)$.*

Also, according to [3], for $m_1 = A^{x_1} B^{y_1}$ and $m_2 = A^{x_2} B^{y_2}$, the combined result by Equation (2) is[1]

$$m_{12} = A^{x_1 + x_2 - x_1 x_2} B^{y_1 + y_2 - y_1 y_2} \tag{7}$$

3 Two Generalizations

In this section, we discuss two generalizations for the Cautious rule, i.e., the time-window approach and the time-attenuation approach. Let $\oplus_C$ be the operator defined by the Cautious rule.

Definition 3. *(Time-Window Cautious Combination Rule) Let $A^{x_1} B^{y_1}, \cdots, A^{x_n} B^{y_n}$ be n successive bipolar bbas, then the combined bba by Time-Window cautious combination rule of window size t is $m_t = A^{x_{n-t+1}} B^{y_{n-t+1}} \oplus_C \cdots \oplus_C A^{x_n} B^{y_n}$.*

That is, a time-window cautious rule of window size t only combines the recent t bbas.

---

[1] In [3], the combined result is $m_{12} = A^{x_1 x_2} B^{y_1 y_2}$, but recall that we use a slightly different notation from [3].

Definition 4. *(Time-Attenuation Cautious Combination Rule) Let $A^{x_1}B^{y_1}, \cdots, A^{x_n}B^{y_n}$ be n successive bipolar bbas, then the combined bba by Time-Attenuation cautious combination rule of attenuation factor t, $0 < t < 1$, is $m_t = A^{x_1 t^{n-1}}B^{y_1 t^{n-1}} \oplus_C \cdots \oplus_C A^{x_n}B^{y_n}$.*

That is, in a time-attenuation cautious rule of attenuation factor t, the coefficient is reduced by t each time. Hence if a male is mis-classified as a female with a certainty degree 0.98, and hence is represented as $M^0 F^{0.98}$, will be attenuated gradually that it will not affect the cautious combination result for long since $0.98t^n$ will grow smaller when $0 < t < 1$ and n increases.

4   Gender Recognition Scenario

In this section, we provide a detailed description of a gender profiling scenario, which lends itself naturally to a DS approach.

Figure 1 shows three images taken from a video sequence that has been passed through a video analytic algorithm for gender profiling. In this sequence, a female wearing an overcoat with a hood enters the scene with her back to the camera. She walks around the chair, turning, so that her face becomes visible, and then sits down.

Fig. 1. Three images taken from a video sequence

Fig. 1(a) shows that the subject is recognised by the full body shape profiling as a male. Note that her face is not visible. In Fig. 1(b), the subject is classified as female by the full body shape profiling algorithm. In Fig. 1(c), as she sits down, with her face visible, the face profiling algorithm classifies her as female, whilst the full body profiling classifies her as male. Note that the full body profiling algorithm is not as reliable as the face profiling algorithm. Conversely, full body profiling is always possible whilst the face information can be missing. That is why these two profiling algorithms should be considered together. In addition, as full body profiling is not as robust, discount operations should be performed on the algorithm output (cf. Equation (3)). The discount rate is dependent on the video samples and the training efficiency. For every video frame in which a body (face) is detected, gender recognition results are provided. The full body profiling algorithm and the face profiling algorithm, provided a person's face is detected, report their recognition results for every frame of the video, e.g., male with 95% certainty.

For a frame with only a body profiling result, for instance Fig. 1(a), the corresponding mass function m for body profiling will be M^x where M denotes that *the person is classified as a male* and x is the mass value of $m(\{M\})$. The corresponding mass function for face profiling is $M^0 F^0$ where F denotes that *the person is classified as a female*, or the vacuous mass function. Alternatively, we can refer to this as the vacuous mass function.

Similarly, for a frame with both body profiling and face profiling, for instance Fig. 1(c), the corresponding mass function for body profiling will be M^x (or in a bipolar form $M^x F^0$) and the mass function for face profiling is F^y (or in a bipolar form $M^0 F^y$) where x, y are the corresponding mass values. As time elapses, fusion of bipolar bbas by the cautious rule or its two generalizations are introduced, as shown by Lemma 1 and Definition 3 and Definition 4. And when it comes to present the final profiling result, we use Dempster's rule to combine the two fused bipolar mass functions from the two recognition algorithms, respectively. Namely, for the two bipolar bbas $m_1 = M^{x_1} F^{y_1}$ and $m_2 = M^{x_2} F^{y_2}$, it is easy to get that the combined result m_{12} by Dempster's rule is (normalized from the result of Equation 7):

$$m_{12}(\{M\}) = \frac{(x_1 + x_2 - x_1 x_2)(1 - y_1)(1 - y_2)}{1 - (x_1 + x_2 - x_1 x_2)(y_1 + y_2 - y_1 y_2)},$$

$$m_{12}(\{F\}) = \frac{(1 - x_1)(1 - x_2)(y_1 + y_2 - y_1 y_2)}{1 - (x_1 + x_2 - x_1 x_2)(y_1 + y_2 - y_1 y_2)},$$

$$m_{12}(\Omega) = \frac{(1 - x_1)(1 - x_2)(1 - y_1)(1 - y_2)}{1 - (x_1 + x_2 - x_1 x_2)(y_1 + y_2 - y_1 y_2)}.$$

Finally, we use the pignistic transformation (Def. 2) for the final probabilities. That is, $p(\{M\}) = m_{12}(\{M\}) + m_{12}(\Omega)/2$ and $p(\{F\}) = m_{12}(\{F\}) + m_{12}(\Omega)/2$. Obviously, we will say the subject is a male if $p(\{M\}) > p(\{F\})$, and a female if $p(\{M\}) < p(\{F\})$. In very rare cases that $p(\{M\}) = p(\{F\})$, we cannot know whether it is male or female.

The following example illustrates the computation steps.

Example 1. *Let us illustrate the approach by a simple scenario with four frames, and there is a mis-classification in the first frame. In the first frame, the corresponding both body profiling (m_b^1) and face profiling (m_f^1) results as $m_b^1 = M^{0.6}$ and $m_f^1 = F^{0.9}$ (mis-classification). In the second frame, there is only a body profiling (m_b^2) result which is $m_b^2 = M^{0.7}$. Frame three is associated with body profiling (m_b^3) and face profiling (m_f^3) results as $m_b^3 = F^{0.4}$ and $m_f^3 = M^{0.6}$, and frame four is associated with body profiling (m_b^4) and face profiling (m_f^4) results as $m_b^4 = M^{0.6}$ and $m_f^4 = M^{0.6}$.*

By Lemma 1, the fusion results by the cautious rule are $m_b = M^{0.7} F^{0.4}$ and $m_f = M^{0.6} F^{0.9}$.

By Definition 3 with window size 2, the fusion results by the time-window cautious rule are $m_b^W = M^{0.6} F^{0.4}$ and $m_f^W = M^{0.6}$.

By Definition 4 with attenuation factor 0.95, the fusion results by the time-attenuation cautious rule are $m_b^A = M^{0.6} F^{0.38}$ and $m_f^A = M^{0.6} F^{0.77}$.

Then by Equation 7, we get $m_{bf} = M^{0.88} F^{0.94}$, which, when normalized, is equivalent to $m_{bf}(\{M\}) = \frac{0.88(1 - 0.94)}{1 - 0.88 * 0.94} = 0.31$, $m_{bf}(\{F\}) = \frac{0.94(1 - 0.88)}{1 - 0.88 * 0.94} = 0.65$,

$m_{bf}(\Omega) = \frac{(1-0.88)(1-0.94)}{1-0.88*0.94} = 0.04$. *And finally we get* $p(\{M\}) = 0.33$ *and* $p(\{F\}) = 0.67$ *which indicates that the subject is a female.*

Similarly, we have $m_{bf}^W = M^{0.84}F^{0.4}$, *and hence* $m_{bf}^W(\{M\}) = \frac{0.84(1-0.4)}{1-0.84*0.4} = 0.76$, $m_{bf}^W(\{F\}) = \frac{0.4(1-0.84)}{1-0.84*0.4} = 0.10$, $m_{bf}^W(\Omega) = \frac{(1-0.84)(1-0.4)}{1-0.84*0.4} = 0.14$ *and* $p^W(\{M\}) = 0.83$ *and* $p^W(\{F\}) = 0.17$, *which indicates that the subject is a male.*

Also, we have $m_{bf}^A = M^{0.88}F^{0.857}$, *and hence* $m_{bf}^A(\{M\}) = \frac{0.88(1-0.857)}{1-0.88*0.857} = 0.51$, $m_{bf}^A(\{F\}) = \frac{0.857(1-0.88)}{1-0.88*0.857} = 0.42$, $m_{bf}^A(\Omega) = \frac{(1-0.88)(1-0.857)}{1-0.88*0.857} = 0.07$ *and* $p^A(\{M\}) = 0.55$ *and* $p^A(\{F\}) = 0.45$ *which also supports that the subject is a male.*

5   Experimental Results

In this section we compare fusion results obtained by a classic approach, a Dempster-Shafer theory approach proposed in [16] and two of its generalization approaches. As there are no benchmark datasets for both body and face profiling, we simulate the output of both body and face classifiers on a sequence containing a male subject. For the body classifier, the probability of any frame being correctly classified as male/female is roughly 60-90%. For the face classifier, only 75% of the available frames are randomly allocated as containing a face. For each of these frames the probability of the frame being correctly classified as being male/female is 85-100%. In both cases the values for $m(\{M\})$ and $m(\{F\})$ are uniformly sampled from the ranges 0.6-0.9 and 0.85-1.0 for the body and face classifiers outputs respectively.

As mentioned before, for gender profiling results from the same classifier at different time points, we use the cautious rule to combine them. For profiling results from different classifiers (i.e., face profiling and full body profiling), we use Dempster's rule to combine them. And finally, we apply the pignistic transformation (Def. 2) to get the probabilities of the subject being male or female.

Classic fusion in the computer vision community [25] takes the degrees of certainty as probabilities, i.e., they consider the face profiling and the full body profiling output p_f^t and p_b^t indicating the probabilities of faces and full bodies being recognized as males at time t. Then it uses $p_{b,f}^t = c_f^t p_f^t + c_b^t p_b^t$ to calculate the final probability $p_{b,f}^t$ at time t, where c_f^t and c_b^t are the weights of the face and full body profiling at time t, proportional to the feasibility of the two algorithms in the last twenty frames. As full body profiling is always feasible, suppose face profiling can be applied n times in the last twenty frames, then we have:

$$c_b = \frac{20}{20+n}, c_f = \frac{n}{20+n}.$$

For this experiment, the performance of the DS and classic fusion schemes were characterised by the true positive rate:

$$T_{PR} = \frac{N_{PR}}{N}$$

where N_{PR} is the number of frames in which the gender has been correctly classified and N is the total number of frames in which the body/face is present. According to the training on the sample videos, the discount rate r for the full body profiling is set to

Table 1. Comparison of T_{PR} for body classification, face classification, classic fusion, DS fusion and its two adaptions - Mis-Classification Cases

Methods	N	N_{PR}	T_{PR} (%)
Full Body	2900	1606	55.4
Face	2159	2002	92.7
Classic Method	2900	2078	71.7
DS Approach	2900	2380	82.1
Time-Attenuation (0.95)	2900	2194	75.7
Time-Attenuation (0.99)	2900	2431	83.8
Time-Window (5)	2900	2586	89.2

0.3. For comparison, we calculate the T_{PR} value for the body classifier alone, the face classifier, the DS fusion scheme and the classic fusion scheme.

Here, we first apply the approaches to 58 simulations each with 50 frames (so there are 2900 total frames), where a mis-classification happens at the beginning. The comparison results are presented as follows.

From Table 1, we can see that the two generalizations provide better results than the DS fusion scheme, except when the attenuation factor is 0.95. This may be because setting the attenuation factor to 0.95 reduces the certainty degrees too quickly.

An example simulation result comparing the classic, DS, Time-Attenuation (0.99) and Time-Window (5) approaches is shown in Fig. 2.

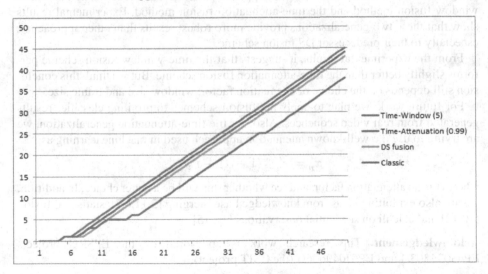

Fig. 2. An Example Simulation

Now we apply the approaches to 20 simulations each with 150 frames (so there are 3000 total frames), where we do not assume mis-classification happened at the beginning. The comparison results are presented as follows.

From Table 2, we can see that the two generalizations perform worse than the DS fusion scheme. This is not surprising since the former do not always hold the highest

Table 2. Comparison of T_{PR} for body classification, face classification, classic fusion, DS fusion and its two adaptions - General Cases

Methods	N	N_{PR}	T_{PR} (%)
Full Body	3000	1792	59.7
Face	2229	2125	95.3
Classic Method	3000	2490	83.0
DS Approach	3000	2899	96.6
Time-Attenuation (0.95)	3000	2126	70.9
Time-Attenuation (0.99)	3000	2401	80.0
Time-Window (5)	3000	2395	79.8
Time-Window (20)	3000	2552	85.1

certainty degree as in the DS fusion scheme. Table 2 also shows that when the attenuation factor or the window size increases, the results improve. Actually, if the attenuation factor is one or the window size equals to the number of frames, then these two generalizations will provide the same results as the DS fusion one, or we can see the DS fusion scheme is a special case of these two generalizations.

6   Conclusion

In this paper, we have proposed two generalized fusion methods to combine gender profiling classifier results by modifying the application of the Cautious rule, i.e., the time-window fusion method and the time-attenuation fusion method. Experimental results show that these two generalizations provide more robust results than other approaches, especially to their predecessor DS fusion scheme.

From the experimental results, it suggests that the time-window fusion scheme performs slightly better than the time-attenuation fusion scheme. But we think this conclusion still depends on the choice of attenuation factor, window size and frame size.

For future work, we plan to apply the fusion schemes to profiling classifier results generated from real video sequences. Also, for the time-attenuation generalization, we are trying to use the well-known attenuation approach used in machine learning as:

$$x'_n = x'_{n-1}(1 - \alpha) + x_n\alpha,$$

where α is an attenuation factor, and see whether this will be a better choice. In addition, we are also exploiting ideas from knowledge base merging [5,11,6,7], statistical fusion [10,12] and calculi on sequential observations [14,15].

Acknowledgement. This research work is sponsored by the EPSRC projects EP/G034303/1 and EP/H049606/1 (the CSIT project).

References

1. Bsia. Florida school bus surveillance,
   http://www.bsia.co.uk/LY8VIM18989_action;displaystudy
   _sectorid;LYCQYL79312_caseid;NFLEN064798
2. Dempster, A.P.: Upper and lower probabilities induced by a multivalued mapping. The Annals of Statistics 28, 325–339 (1967)

3. Denœux, T.: Conjunctive and disjunctive combination of belief functions induced by nondistinct bodies of evidence. Artifical Intelligence 172(2-3), 234–264 (2008)
4. Dubois, D., Prade, H., Yager, R.: Fuzzy set connectives as combinations of belief structures. Information Sciences 66, 245–275 (1992)
5. Konieczny, S., Pino-Pérez, R.: On the logic of merging. In: Cohn, A.G., Schubert, L., Shapiro, S.C. (eds.) Principles of Knowledge Representation and Reasoning, KR 1998, pp. 488–498. Morgan Kaufmann, San Francisco (1998)
6. Ma, J., Liu, W.: Modeling belief change on epistemic states. In: Proc. of 22th Flairs, pp. 553–558. AAAI Press (2009)
7. Ma, J., Liu, W.: A framework for managing uncertain inputs: An axiomization of rewarding. Inter. Journ. of Approx. Reasoning 52(7), 917–934 (2011)
8. Ma, J., Liu, W., Dubois, D., Prade, H.: Revision rules in the theory of evidence. In: Procs. of ICTAI, pp. 295–302 (2010)
9. Ma, J., Liu, W., Dubois, D., Prade, H.: Bridging jeffrey's rule, agm revision and dempster conditioning in the theory of evidence. International Journal on Artificial Intelligence Tools 20(4), 691–720 (2011)
10. Ma, J., Liu, W., Hunter, A.: Incomplete Statistical Information Fusion and Its Application to Clinical Trials Data. In: Prade, H., Subrahmanian, V.S. (eds.) SUM 2007. LNCS (LNAI), vol. 4772, pp. 89–103. Springer, Heidelberg (2007)
11. Ma, J., Liu, W., Hunter, A.: The Non-archimedean Polynomials and Merging of Stratified Knowledge Bases. In: Sossai, C., Chemello, G. (eds.) ECSQARU 2009. LNCS, vol. 5590, pp. 408–420. Springer, Heidelberg (2009)
12. Ma, J., Liu, W., Hunter, A., Zhang, W.: Performing meta-analysis with incomplete statistical information in clinical trials. BMC Medical Research Methodology 8(1), 56 (2008)
13. Ma, J., Liu, W., Miller, P.: Event Modelling and Reasoning with Uncertain Information for Distributed Sensor Networks. In: Deshpande, A., Hunter, A. (eds.) SUM 2010. LNCS, vol. 6379, pp. 236–249. Springer, Heidelberg (2010)
14. Ma, J., Liu, W., Miller, P.: Belief change with noisy sensing in the situation calculus. In: Proceedings of the 27th Conference on Uncertainty in Artificial Intelligence (UAI 2011), pp. 471–478 (2011)
15. Ma, J., Liu, W., Miller, P.: Handling Sequential Observations in Intelligent Surveillance. In: Benferhat, S., Grant, J. (eds.) SUM 2011. LNCS, vol. 6929, pp. 547–560. Springer, Heidelberg (2011)
16. Ma, J., Liu, W., Miller, P.: An evidential improvement for gender profiling. In: Denoeux, T., Masson, M. (eds.) Procs. of the Belief Functions: Theory and Applications, BELIEF 2012 (2012)
17. Ma, J., Liu, W., Miller, P., Yan, W.: Event composition with imperfect information for bus surveillance. In: Procs. of AVSS, pp. 382–387. IEEE Press (2009)
18. Miller, P., Liu, W., Fowler, F., Zhou, H., Shen, J., Ma, J., Zhang, J., Yan, W., McLaughlin, K., Sezer, S.: Intelligent sensor information system for public transport: To safely go.... In: Procs. of AVSS (2010)
19. ECIT Queen's University of Belfast. Airport corridor surveillance (2010), http://www.csit.qub.ac.uk/Research/ResearchGroups/IntelligentSurveillanceSystems
20. US Department of Transportation. Rita - its research program (2010), http://www.its.dot.gov/ITS_ROOT2010/its_program/ITSfederal_program.htm

21. Gardiner Security. Glasgow transforms bus security with ip video surveillance,
    http://www.ipusergroup.com/doc-upload/
    Gardiner-Glasgowbuses.pdf
22. Shafer, G.: A Mathematical Theory of Evidence. Princeton University Press (1976)
23. Smets, P.: Non-standard logics for automated reasoning. In: Smets, P., Mamdani, A., Dubois, D., Prade, H. (eds.) Belief Functions, pp. 253–286 (1988)
24. Yager, R.: On the dempster-shafer framework and new combination rules. Information Sciences 41, 93–138 (1987)
25. Zhou, H., Miller, P., Zhang, J., Collins, M., Wang, H.: Gender classification using facial and full body features. Technical Report, CSIT, Queen's University Belfast, UK (2011)

From Imprecise Probability Laws
to Fault Tree Analysis

Christelle Jacob[1,2] *, Didier Dubois[2], and Janette Cardoso[1]

[1] Institut Supérieur de l'Aéronautique et de l'Espace (ISAE), DMIA Department,
Campus Supaéro, 10 Avenue Édouard Belin - Toulouse
[2] Institut de Recherche en Informatique de Toulouse (IRIT), ADRIA Department,
118 Route de Narbonne 31062 Toulouse Cedex 9, France
{jacob,cardoso}@isae.fr, dubois@irit.fr

Abstract. Reliability studies and system health predictions are mostly based on the use of probability laws to model the failure of components. Behavior of the components of the system under study is represented by probability distributions, derived from failure statistics. The parameters of these laws are assumed to be precise and well known, which is not always true in practice. Impact of such imprecision on the end result can be crucial, and requires adequate sensitivity analysis. One way to tackle this imprecision is to bound such parameters within an interval. This paper investigates the impact of the uncertainty pervading the values of law parameters, specifically in fault tree based Safety analysis.

Keywords: fault trees, Imprecise probabilities, Interval analysis.

1 Introduction

This work takes place in the context of an Airbus project called $@MOST$. The main aim of the $@MOST$ project is to improve the schedule of operational and maintenance activities of the aircrafts. This is achieved by using some extended safety models and by predicting the expected failures. These predictions are based upon the safety analysis of underlying system models.

One of the objectives of safety analysis is to evaluate the probability of undesired events. In our previous work [1], we studied how to evaluate the imprecision of this probability when the undesired event is described by a fault tree, and the probabilities of elementary events are imprecise numbers. In the usual approach, the fault tree is a graphical representation of a Boolean formula F, representing all the conditions of occurrence of the undesired event under study, as a function of some atomic events. Those atomic events represent the failures of the components of the system, or possibly some of its configuration states. All of them are supposed to be stochastically independent. Then the probability of the undesired event can be computed from the probabilities of the atomic events, by

---

* C. Jacob has a grant supported by the @MOST Prototype, a joint project of Airbus, IRIT, LAAS, ONERA and ISAE.

E. Hüllermeier et al. (Eds.): SUM 2012, LNAI 7520, pp. 525–538, 2012.

means of Binary Decision Diagrams (BDDs) [3]: that allows an easy probability computation for very large Boolean functions.

In safety analysis, as well in reliability studies, the probabilities of the atomic events are time-dependent, and generally described by means of some standard probability distributions [4], e.g. exponential or Weibull laws. Their parameters are supposed to be precisely known numbers, but actually, they generally come from statistical observations of failure times. They are derived by means of data fitting methods and regression analysis: for example, the paper [5] explains how to use different methods, like least squares or the actuary method, in order to find the best parameters of a Weibull law that fit some samples.

In this paper, we investigate the impact of imprecision in parameters of probability distributions commonly used in safety analysis, by using intervals values for the parameters. First of all, we study the impact on the probability distributions themselves: p-boxes [6] are obtained, i.e. minimum and maximum probability distributions bounding the real one. In a second step, an extension of the algorithm described in paper [1] is used to evaluate the imprecise probability of a Boolean formula depending on several p-boxes. In this work, we compute the output p-box attached to undesired events.

The paper is organized as follow: section 2 introduces the basic concepts of reliability. Sections 3 and 4 present the resulting ranges of the cumulative distribution of an atomic event for, respectively, exponential law and Weibull law. At last, section 5 explains the computation of the range of undesired event probability across time (cumulative distribution), in function of the distributions of the atomic events leading to this undesired event. A case study illustrates this section. Finally, the last section presents some conclusions and future work.

2   Basics of Reliability Study

The *reliability* $R(t)$ of a system, also called the *survival function*, is the probability that the system does not fail before time t. It can be expressed as:

$$R(t) = P(T > t) \tag{1}$$

where T is a random variable representing the *failure date*.

The *probability of failure* of a system before time t, called *failure distribution* $F_T(t) = P(T \leq t)$, is the complement of its reliability:

$$F_T(t) = 1 - R(t) \tag{2}$$

The *failure density function* $f_T(t)$ expresses the probability that the system fails between t and $t + dt$:

$$f_T(t)dt = P(t \leq T < t + dt) \tag{3}$$

The *failure rate* λ of a system is the frequency of its failure. It is a function of the system health state, and in general it is time dependent. λ is often considered as proportional to the probability that a failure occurs at a specified time point t, given that no failure occurred before this time:

$$\lambda(t)dt = P(t \leq T \leq t + dt \mid T > t) \tag{4}$$

This conditional probability can be written as:

$$\lambda(t)dt = \frac{f_T(t)dt}{R(t)} = \frac{-R'(t)}{R(t)}, \tag{5}$$

where $R'(t)$ is the derivative of $R(t)$ with respect to the time.

The solution of this differential equation is:

$$\ln(R(t)) = \int_0^t \lambda(u)du + c, \text{where } c \text{ is a constant.} \tag{6}$$

Hence, the reliability expressed in terms of the failure rate has the expression:

$$R(t) = e^{-\int_0^t \lambda(u)du} \tag{7}$$

In the following text, we will present two particular cases of failure rates in equation (7), leading to the following distributions:

– exponential distribution,
– Weibull distribution.

Furthermore, we will study the impact of the lack of knowledge about failure rates on those distributions.

In reliability studies, the probabilities of all events are assumed to be well known, which is not always verified in practice. Making this assumption has shown some limitation, therefore, some researchers started to work on other methods, using intervals instead of precise values. Utkin and Coolen, for example, worked on imprecise reliability using imprecise probability theory, with upper and lower expectations instead of a single probability value [2]. They studied imprecise monotonic fault trees, and also the impact of some components failure over the system under study by means of imprecise *importance measures*[].

In this paper, the goal is to compute the probability distribution of an undesired event described by any binary fault tree, monotonic or not. Those kinds of fault trees are often obtained from automatic fault tree generation software, or systems with reconfiguration states. The impact of imprecision about the distribution of the undesired event depends on the architecture of the system, and on the imprecision about the parameters of the probability distributions of its elementary components. Hence, the first step is to study the impact of imprecise parameters on the commonly used probability distributions.

3   The Exponential Distribution

Recall that the reliability analysis of an aircraft takes into account, among others, the electronic components. Their probabilities of failure are modeled with

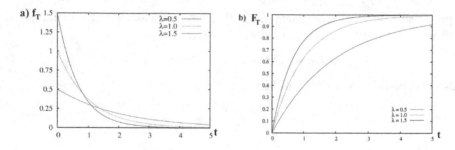

Fig. 1. a) Exponential density function b) Cumulative distribution

constant failure rates $\lambda(t) = \lambda$, because they do not have any burn-in nor any wear-out periods, respectively at their beginning and their end of life. Moreover, when the failure rate is constant, equation (7) becomes $R(t) = e^{-\lambda t}$, i.e., an exponential distribution.

The probability density function is given by:

$$f_T(t) = \lambda e^{-\lambda t} \tag{8}$$

And is represented in Fig. 1.a). Its cumulative distribution, depicted in Fig. 1.b) is given by:

$$F_T(t) = 1 - e^{-\lambda t} \tag{9}$$

3.1   Exponential Law with Imprecise Failure Rate

If the only information available about the failure rate λ is an interval containing it, then there are different probability distributions representing the failure of the component, as will be presented in the sequel.

The goal is to find the range of the cumulative distribution, when the failure rate is imprecise: $\lambda \in [\underline{\lambda}, \overline{\lambda}]$. In interval analysis, knowing the monotonicity of a function makes the determination of its range straightforward.

The function $1 - e^{-\lambda t}$ is strictly increasing with λ, hence the range of the cumulative distribution, when λ is varying, for every $t > 0$ and $\lambda > 0$, is given by the expression:

$$F_T(t) = \{1 - e^{-\lambda t}, s.t. \ \lambda \in [\underline{\lambda}, \overline{\lambda}]\} = [1 - e^{-\overline{\lambda} t}, 1 - e^{-\underline{\lambda} t}] \tag{10}$$

The range of the cumulative distribution with respect to some values of λ and in time interval $t = [0, 10]$ is represented in Fig. 2.a).

For the probability density function, it is a little bit more complex. The derivative with respect to λ of the function $f_T(t)$ is:

$$\frac{\partial}{\partial \lambda} f_T(t) = (1 - \lambda t)e^{-\lambda t} \tag{11}$$

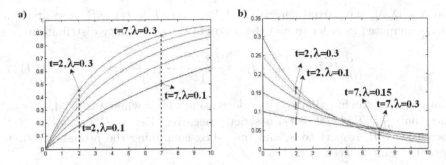

Fig. 2. a) Range of the cumulative distribution b) Range of the probability distribution $(0.1 < \lambda < 0.3, \, 0 < t < 10)$

This means that the function will be increasing with respect to λ when $\lambda t < 1$, and decreasing otherwise. The range of the function will depend on λ and t, as illustrated on Fig. 2.b).

In the following, we give different interpretations of the probability of failure of a component, as used in fault tree analysis.

3.2 Occurrence of an Atomic Failure before Time t

In the quantitative analysis of a safety model, each component (or type of component) of this model will have its own failure rate, and its own failure probability. The main goal of this analysis is to ensure that, at each time t, the probability that the system has failed remains below a certain value. We are interested in the probability of failure of a component or system before time t, hence the cumulative distribution will be used for our computations.

When the parameter λ is imprecise, and its possible values are known to lie within the interval $[\underline{\lambda}, \overline{\lambda}]$, the probability distribution will be contained in the p-box [6]:

$$\{P, P(T < t) \in [1 - e^{-\overline{\lambda}t}, 1 - e^{-\underline{\lambda}t}]\}, \tag{12}$$

where a p-box is an ordered pair of cumulative distributions, representing the probability family. This family (12) contains more probability distributions than those with an exponential distribution. However it is enough to use the p-box when computing probability bounds of events of the form $T < t$. But it will not be the case for computing other indicators, e.g. the variance.

3.3 Occurrence of an Atomic Failure between t_1 and t_2

In some cases, it can also be useful to compute the probability that the event will occur between two dates t_1 and t_2. This can be expressed as the conditional probability $t_1 < T < t_2$ given that T does not occur before t_1:

$$P(T < t_2 | T \geq t_1) = \frac{e^{-\lambda t_1} - e^{-\lambda t_2}}{1 - e^{-\lambda t_1}} \tag{13}$$

When $\lambda \in [\underline{\lambda}, \overline{\lambda}]$, the partial derivative of $P(T < t_2 | T \geq t_1)$ with respect to λ must be computed in order to find the p-box of the probability distribution.

$$\frac{\partial}{\partial \lambda} P(T < t_2 | T \geq t_1) = \frac{t_2 e^{-\lambda t_2} - t_1 e^{-\lambda t_1}}{(1 - e^{-\lambda t_1})^2} \tag{14}$$

By noticing that the function $xe^{-\lambda x}$ is decreasing with x when λ is fixed, we can deduce that $\frac{\partial}{\partial \lambda} P(T < t_2 | T \geq t_1)$ is strictly negative. Hence, $P(T < t_2 | T \geq t_1)$ is decreasing with respect to λ, and the p-box containing the probability that the event occurs between t_1 and t_2 is:

$$P(T < t_2 | T \geq t_1) \in \left[\frac{e^{-\overline{\lambda} t_1} - e^{-\overline{\lambda} t_2}}{1 - e^{-\overline{\lambda} t_1}}, \frac{e^{-\underline{\lambda} t_1} - e^{-\underline{\lambda} t_2}}{1 - e^{-\underline{\lambda} t_1}} \right]. \tag{15}$$

3.4 Case of Periodic Preventive Maintenance

It is also possible to represent schedules of preventive maintenance by means of probability distributions. Indeed, some components are preventively replaced or repaired with a period of length θ: this maintenance task will reset the probability of failure to 0 after θ flight hours (FH). The cumulative distribution representing this probability of failure in this case is a periodic function that can be written as:

$$\text{for } k \in \mathbb{N}, P(T < t) = 1 - e^{-\lambda(t - k\theta)}, \text{ if } t \in [k\theta, (k+1)\theta] \tag{16}$$

If the failure rate λ is imprecise, then the probability of failure is the same as in the section 3.2 on the interval $[0, \theta]$:

$$\text{for } k \in \mathbb{N}, P(T < t) \in [1 - e^{-\overline{\lambda}(t - k\theta)}, 1 - e^{-\underline{\lambda}(t - k\theta)}], \text{ if } t \in [k\theta, (k+1)\theta] \tag{17}$$

If both the failure rate and the period are imprecise, then it is still possible to compute the range of the resulting cumulative distribution: we can consider that the period θ can be any value in the interval of time $[\theta_1, \theta_2]$. In this case, the size of the interval probability will grow very quickly with the size of the interval $[\theta_1, \theta_2]$. The minimum and maximum cumulative distributions, denoted by $\overline{P}(F < t)$ and $\overline{P}(F < t)$, are given for $k \in \mathbb{N}$ by the following expressions:

$$\underline{P}(T < t) = \begin{cases} 0 \text{ for } k\theta_1 < t < k\theta_2 \\ 1 - e^{-\underline{\lambda}(t - k\theta_1)}, \text{ for } t \in [k\theta_2, (k+1)\theta_1] \end{cases}$$

$$\overline{P}(T < t) = 1 - e^{-\overline{\lambda}(t - k\theta_1)}, \text{ for } t \in [k\theta_2, (k+1)\theta_2]$$

An example of those p-boxes for $\lambda \in [0.5, 0.6]$ and $T \in [\theta_1, \theta_2]$ is shown on Fig. 3.

4   The Weibull Distribution and Imprecise Parameters

In the case of a hardware component, it can be useful to model its *burn-in* period (i.e. the fact that the failure rate is high at the beginning but will decrease after

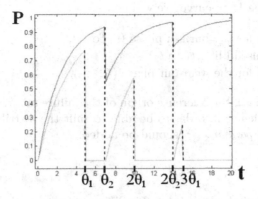

Fig. 3. An example of periodic maintenance with $\theta \in [\theta_1, \theta_2]$ and $\lambda \in [0.5, 0.6]$

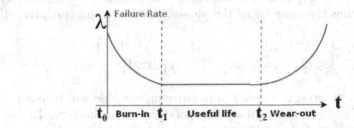

Fig. 4. Bathtub Curve

some time) and its *wear-out* phase (i.e. the fact that after some time, the failure rate of the component increases). Therefore, the failure rate has the shape of a *bathtub curve*, as shown on Fig.4.

In order to model the reliability in this case, the *Weibull* law is used. It is a two parameters law, described by the formula:

$$R(t) = e^{-(\frac{t}{\eta})^{\beta}} \tag{18}$$

where η is the *scale parameter* and β the *shape parameter*.

The probability density function of a Weibull law is given by the expression:

$$f_T(t) = \frac{\beta}{\eta}(\frac{t}{\eta})^{\beta-1}e^{-(\frac{t}{\eta})^{\beta}} \tag{19}$$

And its cumulative distribution is:

$$F_T(t) = 1 - e^{-(\frac{t}{\eta})^{\beta}} \tag{20}$$

From equation (5), the expression of the failure rate as a function of t is:

$$\lambda(t) = \beta.\frac{1}{\eta^{\beta}}.t^{\beta-1} \tag{21}$$

In order to get a bathtub curve, we will chose:

- a value $\beta_1 < 1$ for the burn-in phase (t_0 to t_1),
- $\beta = 1$ for the useful life (t_1 to t_2),
- a value $\beta_2 > 1$ for the wear-out phase ($t > t_2$).

In the wear-out phase, the reference origin of the failure rate and the cumulative function is not 0, hence in order to be able to shift the distribution to starting time t_2, a *location parameter* γ should be added:

$$F_T(t) = 1 - e^{-(\frac{t-\gamma}{\eta})^\beta} \tag{22}$$

Despite the fact that the parameter β is different for each phase of the bathtub curve, the failure rate is a continuous function. Therefore, there will be a constraint for each change of phase, that will ensure the continuity. When the scale parameter η remains the same for all the phases, this constraint is expressed as below:

$$\begin{cases} \beta_1 . \frac{1}{\eta^{\beta_1}} . t_1^{\beta_1-1} = \frac{1}{\eta} \\ \beta_2 . \frac{1}{\eta^{\beta_2}} . (t_2 - \gamma)^{\beta_2-1} = \frac{1}{\eta} \end{cases} \Leftrightarrow \begin{cases} \beta_1 . (\frac{t_1}{\eta})^{\beta_1-1} = 1 \\ \beta_2 . (\frac{(t_2-\gamma)}{\eta})^{\beta_2-1} = 1 \end{cases} \tag{23}$$

Like the failure rate curve, the global cumulative distribution will be composed of three pieces of cumulative distributions with different parameters. To ensure the continuity of the global one, the cumulative distribution of each new phase should start from the last value of the previous phase.

When the parameters of a Weibull law are imprecise, they should still verify the constraints of β for each phase, and the ones expressing the continuity of $\lambda(t)$ (equation 23). Fig. 5 shows the variation of the failure rate with the variation of η for the three different phases of the bathtub curve.

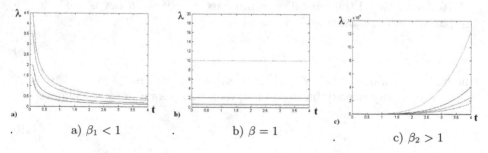

a) $\beta_1 < 1$ b) $\beta = 1$ c) $\beta_2 > 1$

Fig. 5. Variation of the Weibull distribution with η

The imprecision pervading the parameters β and η of the Weibull law affects the value of the time points where the phases change in the bathtub curve (t_1 and t_2), due to equation (23). These time points become themselves intervals.

In order to find the range of the cumulative distribution with the different parameters, the monotonicity study of the function will also be required, as in section 3. In this case, we have a two parameter function, hence we compute its gradient.

$$\vec{\nabla}P(T < t) = \begin{vmatrix} \frac{\partial}{\partial \eta}P(T < t) \\ \frac{\partial}{\partial \beta}P(T < t) \end{vmatrix} = \begin{vmatrix} \beta.\frac{t^\beta}{\eta^{\beta+1}}e^{-(\frac{t}{\eta})^\beta} \\ ln(\frac{t}{\eta}).(\frac{t}{\eta})^\beta e^{-(\frac{t}{\eta})^\beta} \end{vmatrix} \tag{24}$$

By noticing that t, η, β and $e^{-(\frac{t}{\eta})^\beta}$ are always positive, we can conclude that the partial derivative $\frac{\partial}{\partial \eta}P(T < t)$ is positive. But the partial derivative $\frac{\partial}{\partial \beta}P(T < t)$ is positive when $\eta > t$ and negative otherwise, because of the term $ln(\frac{t}{\eta})$. Equation (23) implies that for $\eta > t_1$, hence $P(T < t)$ is decreasing with respect to β for $t < t_1$. This means that the p-box of a Weibull distribution will be:

$$[1 - e^{-(\frac{t}{\eta})^{\overline{\beta}}}, 1 - e^{-(\frac{t}{\eta})^{\underline{\beta}}}], \text{ for } t \in [0, t_1] \tag{25}$$

Between t_1 and t_2, β is fixed to 1, hence the bounds for the cumulative distribution are:

$$[1 - e^{-\frac{t-t_1}{\eta}} + P(T < t_1), 1 - e^{-\frac{t-t_1}{\eta}} + P(T < t_1)], \text{ for } t \in [t_1, t_2] \tag{26}$$

When $t > t_2$, the quantity $(t_2 - \gamma)$ is computed through equation (23). Now the partial derivatives are similar to the ones in equation (24), replacing t by $t - \gamma$. Hence the condition for $\frac{\partial}{\partial \beta}P(T < t)$ being positive is that $t < \gamma + \eta$, so we get the range of the cumulative distribution:

$$[1 - e^{-(\frac{t-\gamma}{\eta})^{\underline{\beta}}} + P(T < t_2), 1 - e^{-(\frac{t-t_2}{\eta})^{\overline{\beta}}} + P(T < t_2)], \text{ for } t < \gamma + \eta \tag{27}$$

$$[1 - e^{-(\frac{t-\gamma}{\eta})^{\overline{\beta}}} + P(T < t_2), 1 - e^{-(\frac{t-t_2}{\eta})^{\underline{\beta}}} + P(T < t_2)], \text{ for } t > \gamma + \eta \tag{28}$$

Once we know the impact of imprecise parameters on one probability distribution, we can use it in the computation of the probability distribution of an undesired event described by a fault tree.

5   Range of a Undesired Event Probability across Time

In the case of fault tree analysis, the probability of a undesired event is described with a Boolean formula F, function of N Boolean variables $V_i, i = 1 \ldots N$ representing the failure (or states) of its components. When the probability of V_i is represented by a probability distribution with an imprecise parameter, we have a p-box for the probability of the undesired event. The variables V_i are supposed to be stochastically independent, and they can follow different probability distributions. Also, their parameters can be of different types: some can be precise, when the information is available and well known, some can be imprecise.

The goal will be to find the p-box describing the undesired event probability across time from the p-boxes of the variables V_i. The best way to carry out this computation is to discretize the time, and to find for each t and for each V_i, the associated interval $I(t, V_i)$. Of course, if all input probability distributions are precise, the probability of the undesired event will be precise. When T_i is a random variable representing the failure time of the component V_i, we have that:

$$I_i(t, V_i) = [\underline{P(T_i < t)}, \overline{P(T_i < t)}]$$

Let us consider two variables V_1 and V_2 with exponential laws, and respective imprecise parameters $\lambda_1 \in [0.1, 0.14]$ and $\lambda_2 \in [0.1, 0.2]$. The two first graphs of Fig. 6 depicts the intervals $I_i(t = 6, V_i)$ associated to these variables.

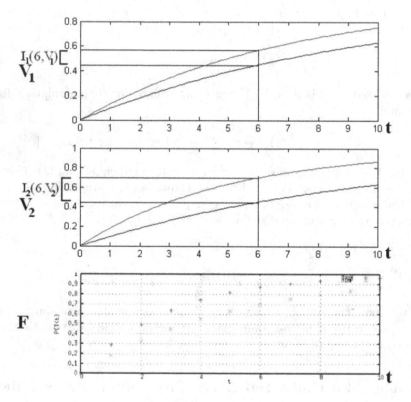

Fig. 6. Example of aggregation of two exponential p-boxes with $F = V_1 \vee V_2$

For the same time point, the range of the probability of variable V_i is given by the interval $I_i(t, V_i)$. So, *for this time t*, the algorithm presented in [1] can be used to compute the probability of the undesired event.

In order to compute the range of the cumulative distribution of the undesired event for *all time instants*, we apply the algorithm for all k time instants, $k = \frac{T_o}{T_s}$, where T_o is the observation interval and T_s is the time step. An example of the result given by the algorithm for an undesired event described by the Boolean formula $F = V_1 \vee V_2$, and for a time step of 1, is depicted on the last graph of Fig. 6.

5.1   Case Study: Safety Model of a Primary/Backup Switch

In this case study, the analysis process of Safety models used in the @MOST project will be described. We will take the example of a small system allowing a reconfiguration: a Primary/Backup Switch. It is constituted of three components:

- A primary supplier
- A back-up supplier
- A switch that selects the active supplier between the primary or the backup one

When a fault occurs in the Primary supplier, then it switches to the Back-up supplier. But it may also happen that the Switch gets stuck: in this case, it will be impossible to switch to Backup supplier.

The software Cécilia OCAS is used to model the architecture and the behavior of the system thanks to the AltaRica language, which is mode-automata based. From this description, some algorithms [7] will extract fault trees or Minimal Cut Sets for any undesired event selected by the user by means of observers. Fig.7 shows the OCAS model of the Primary/Backup Switch.

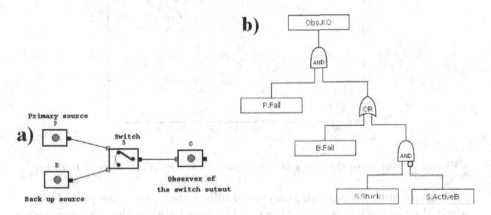

Fig. 7. a) OCAS model of the Primary/Backup Switch b)Fault tree associated to event *Obs.KO*

In the following, we will study the undesired event corresponding to the fact that the whole system is down, written as *Obs.KO*. The OCAS tool extracts the fault tree associated to this event, as displayed on Fig.7.

Therefore, this fault tree is equivalent to the Boolean formula:

$$Obs.KO = P.Fail \wedge (B.Fail \vee (S.stuck \wedge \neg S.activeB)),$$

where *P.Fail* stands for a failure of the Primary supplier and *B.Fail* for a failure of the Backup supplier. *S.stuck* represents the fact that the Switch is stuck and is not able to activate the Backup, and *S.activeB* is the activation order of the Switch.

The sensitivity analysis algorithm is applied to the fault tree, with the following imprecise parameters for the distributions:

- *P.Fail* possesses an exponential distribution with an imprecise failure rate $\lambda = 10^{-4} +/- 50\%$ and a precise periodic maintenance of period $\theta = 30$ FH
- *B.Fail* possesses an exponential distribution with an imprecise failure rate $\lambda = 10^{-4} +/- 10\%$ and a precise periodic maintenance of period $\theta = 35$ FH
- *S.stuck* possesses an exponential distribution with a precise $\lambda = 10^{-5}$
- An activation order of the Switch occurs after $t = 80$ FH

On Fig. 8, we can observe the minimum and maximum cumulative distributions of the event *Obs.KO*, for a duration of 100 flight hours (around three or four months for a commercial aircraft). The picture lays bare the effect of periodic maintenance on those distributions.

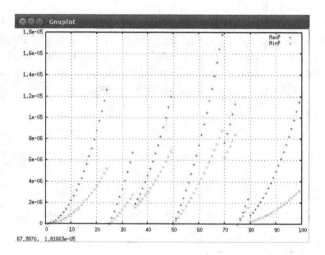

Fig. 8. Evolution of the p-box of the event *Obs.KO* for a duration of 100 FHs

The study of this p-box can give crucial information about the probability of undesired events, such as *Obs.KO*: when the area between the minimum curve and the maximum curve is tight, computations are reliable. The larger it is, the more uncertainty we will get. But even under uncertainty, it can still be possible to ensure safety, if the upper probability of the undesired event is below a legal threshold. For instance, in our case study, the maximum probability is always less than 1.8×10^{-5} for 100 flight hours, with this maintenance schedule.

In safety analysis, the requirements to meet for each failure are described in the Failure Mode and Effects Analysis document (FMEA, [10]). They are classified with respect to their probability of occurrence, their severity and some other features. This classification defines the legal probability threshold to be met. In the example, the event *Obs.KO* meets a requirement of an event occurring less than 10^{-4} over the 100 first flight hours, but not the threshold of 10^{-5}. In case a threshold of 10^{-5} is required by the FMEA for this event, then we must change the maintenance schedule in order to meet this requirement.

The algorithm allows to test easily several scenarios of maintenance, in order to find one that ensures the threshold of 10^{-5}. With a periodic maintenance of the primary supplier every 19 flight hours instead of 25, and of the backup supplier every 22 flight hours instead of 35, this requirement can be met despite the uncertainty about the inputs, as shown on Fig. 9.

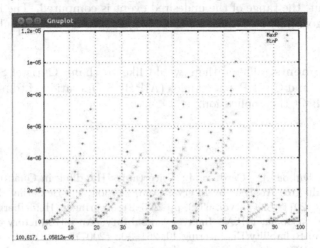

Fig. 9. A different scenario of maintenance schedule

The obtained p-box is also compatible with the fuzzy extension of FMEA [11]. In this case, a linguistic description of the occurrence scale is assumed, and the threshold to meet for the occurrence parameter is a linguistic term described by a membership function over an ad hoc scale. After casting the probability interval on such a scale, we can compute the necessity and the plausibility of a fuzzy level of occurrence at each time t. This pair of evaluation expresses the compatibility of the probability interval and the linguistic occurrence threshold.

6 Conclusion

Being able to model the impact of incomplete information on probabilistic safety analysis is very useful for maintenance management. It allows the user to select the best representation for available data, in order to get a faithful advice. Precise data can be used when they are available, but they do not need to be assumed so when they are not. Consequently, more facets of uncertainty can be taken into account, and especially the difference between the variability of failure times and the lack of knowledge on distribution parameters. This difference can be very crucial in a decision process, where confidence about the results of computations plays a decisive role.

The computation time of this algorithm is exponential with respect to logical variables that appears both in positive and negative forms in the fault tree (in practice there are very few of them [1]). Hence our methodology for risk analysis in maintenance management looks scalable. Further experiments should be run to demonstrate this point.

Future work will take into account the uncertainty of parameters represented by means of fuzzy intervals, using the concept of α-cuts ([9]). In fact a fuzzy set can be considered as a collection of nested (classical) intervals, called α-cuts. For each α-cut, the range of the undesired event is computed. The issue is then to find an algorithm to compose all these ranges into a fuzzy cumulative distribution.

Acknowledgments. The authors would like to thank Christel Seguin (ON-ERA, France) and Chris Papadopoulos (AIRBUS Operations Ltd., UK) for the discussions about the application.

References

1. Jacob, C., Dubois, D., Cardoso, J.: Uncertainty Handling in Quantitative BDD-Based Fault-Tree Analysis by Interval Computation. In: Benferhat, S., Grant, J. (eds.) SUM 2011. LNCS, vol. 6929, pp. 205–218. Springer, Heidelberg (2011)
2. Utkin, L.V., Coolen, F.P.A.: Imprecise reliability: An introductory overview. Intelligence in Reliability Engineering 40, 261–306 (2007)
3. Siegle, M.: BDD extensions for stochastic transition systems. In: Kouvatsos, D. (ed.) Proc. of 13th UK Performance Evaluation Workshop, Ilkley/West Yorkshire, pp. 9/1–9/7 (July 1997)
4. Morice, E.: Quelques modèles mathématiques de durée de vie. Revue de Statistique Appliquée 14(1), 45–126 (1966)
5. Palisson, F.: Détermination des paramètres du modèle de Weibull à partir de la méthode de l'actuariat. Revue de Statistique Appliquée 37(4), 5–39 (1989)
6. Ferson, S., Ginzburg, L., Kreinovich, V., Myers, D., Sentz, K.: Constructing probability boxes and Dempster-Shafer structures. Tech. rep., Sandia National Laboratories (2003)
7. Rauzy, A.: Mode automata and their compilation into into fault trees. Reliability Engineering and System Safety (78), 1–12 (2002)
8. Gauthier, J., Leduc, X., Rauzy, A.: Assessment of Large Automatically Generated Fault Trees by means of Binary Decision Diagrams. Journal of Risk and Reliability. Professional Engineering 221(2), 95–105 (2007)
9. Zadeh, L.: Similarity relations and fuzzy orderings. Information Sciences 3, 177–200 (1971)
10. Stamatis, D.H.: Failure Mode Effect Analysis: FMEA from theory to execution, 2nd edn. (1947)
11. Tay, K.M., Lim, C.P.: A Guided Rule Reduction System for Prioritization of Failures in Fuzzy FMEA. International Journal of Quality and Reliability Management 23(8), 1047–1066 (2006)

CUDA Accelerated Fault Tree Analysis with C-XSC

Gabor Rebner[1] and Michael Beer[2]

[1] University of Duisburg-Essen, Computer and Cognitive Science,
Duisburg, Germany
`rebner@inf.uni-due.de`
[2] Institute for Risk & Uncertainty, University of Liverpool, Liverpool, UK
`mbeer@liverpool.ac.uk`

Abstract. Fault tree analysis is a widespread mathematical method for determining the failure probability of observed real-life systems. In addition to failure probability defined by wear, the system model has to take into account intrinsic and extrinsic system influences. To make allowance for such factors, we draw on an implementation by Rebner et al. to compute the lower and upper bounds of the failure probability of the top event based on interval analysis implemented in MATLAB using INTLAB. We present a new verified implementation in C++ to reduce the trade-off between accuracy and computation time, describe the new implementation by giving an illustrative example based on work by Luther et al. and show the advantages of our new implementation.

Keywords: fault tree analysis, DSI, C-XSC, CUDA.

1 Introduction

Fault tree analysis is a widespread mathematical method for describing and computing the failure probabilities of observed systems, such as nuclear power plants or complex dynamic processes [1, 2]. Because of the deterministic behavior of the underlying failure probabilities, it is not possible to describe intrinsic and extrinsic influences on the observed systems. Real-life systems combine sure and unsure failure probabilistics as shown by various calamities in the last decade. To introduce unsure data into fault tree analysis, we draw on interval arithmetic [3, 4], which serves as a basis for our implementation. It is well known that traditional implementations of fault tree analysis on modern computer systems do not consider failures reasoned in floating-point arithmetic, for example, round-off and approximation errors, which occur because of the finite nature of machine numbers [5, 6]. In [7] we introduced a verified fault tree algorithm written in MATLAB using the interval library INTLAB [8] to serve as a basis for interval arithmetic and directed rounding when computing the lower and upper bounds of failure probability based on the work of Carreras et al. [9]. One disadvantage of this implementation is the computation time needed to arrive at the verified solution. In this paper, we present a new implementation of the

E. Hüllermeier et al. (Eds.): SUM 2012, LNAI 7520, pp. 539–549, 2012.
© Springer-Verlag Berlin Heidelberg 2012

algorithms described in [7, 10] using NVIDIA's Compute Unified Device Architecture (CUDA) [11], which can carry out high-performance computing using multithreading functionality and, in its latest implementation, conduct floating-point computation based on the IEEE754 floating-point standard [12] on the NVIDIA graphics card. This paper is structured as follows. First, we introduce the basic theory of interval arithmetic, fault tree analysis and CUDA to build the basis for Sect. 3, which describes the new implementation. Then, we compare our solutions to those obtained with the implementation in [7]. We close by recapitulating the main results and discussing future research.

2   Basic Theory

In this section, we describe the fundamentals of interval arithmetic and high-performance computing using NVIDIA's Compute Unified Device Architecture (CUDA) to verify computation on the graphical processor unit (GPU). We close this section with a short introduction to fault tree analysis.

2.1   Interval Arithmetic

In this paper, we use the term verification in the sense of using arithmetic that provides mathematical proofs to compute or to include the right solution. One well-established way to obtain verified results is interval arithmetic.

Let $\mathbf{x} = [\underline{\mathbf{x}}, \overline{\mathbf{x}}] \mid \underline{\mathbf{x}} \leq x \leq \overline{\mathbf{x}}$, $\underline{\mathbf{x}}$, $\overline{\mathbf{x}}$ and $x \in \mathbb{R}$ be a real interval from the set $\mathbb{IR}$. The set of arithmetic operations $\circ = \{+, -, \div, \cdot\}$ on the intervals $\mathbf{x} \in \mathbb{IR}$ and $\mathbf{y} \in \mathbb{IR}$ is defined as

$$\mathbf{x} \circ \mathbf{y} = \left[\min \left\{ \underline{\mathbf{x}} \circ \underline{\mathbf{y}}, \underline{\mathbf{x}} \circ \overline{\mathbf{y}}, \overline{\mathbf{x}} \circ \underline{\mathbf{y}}, \overline{\mathbf{x}} \circ \overline{\mathbf{y}} \right\}, \max \left\{ \underline{\mathbf{x}} \circ \underline{\mathbf{y}}, \underline{\mathbf{x}} \circ \overline{\mathbf{y}}, \overline{\mathbf{x}} \circ \underline{\mathbf{y}}, \overline{\mathbf{x}} \circ \overline{\mathbf{y}} \right\} \right].$$

The result of such an operation is an interval that includes all possible combinations of the operation $x \circ y \mid x, y \in \mathbb{R}$ with $x \in \mathbf{x}$ and $y \in \mathbf{y}$. In the case of division, $\mathbf{y}$ must not enclose zero.

To obtain verified enclosures on a computer system based on the IEEE754 floating-point standard, we draw on the set of machine intervals ($\mathbb{IM}$). A machine interval can be obtained from a real interval by using directed rounding, that is, by rounding the infimum of a real interval downwards ($\triangledown$) and the supremum upwards ($\triangle$) to the next floating-point number in the set of the utilized machine numbers ($\mathbb{M}$). Furthermore, we utilize the Boolean intersection function ($inter(\mathbf{x}, y)$) to define an interval mass assignment in Sect. 2.3:

$$inter(\mathbf{x}, y) = \begin{cases} \text{true} & , \text{ if } \mathbf{x} \cap y = y \\ \text{false} & , \text{ if } \mathbf{x} \cap y = \emptyset \end{cases} \text{ for } \mathbf{x} \in \mathbb{IR} \text{ or } \mathbb{IM}, \ y \in \mathbb{R} \text{ or } \mathbb{M}.$$

To obtain such behavior in C++, we utilize the C-XSC library [13, 14, 15], which provides interval arithmetic, rigorous interval standard, vector and matrix functions, and a broad range of verified algorithms based on the C++ standard of 1998.

One disadvantage of using interval arithmetic is the dependency problem [16]. This obstacle leads to an overestimation of the functions range because of the independent occurrence of the same interval several times in the same formula. To avoid such behavior, we draw on floating-point arithmetic with directed rounding. Besides tighter intervals, this approach leads to a faster computation time.

2.2 Compute Unified Device Architecture

CUDA is a GPU computing framework for NVIDIA graphic cards that provides floating-point operations conforming to the IEEE754 standard in double precision for addition, multiplication, division, reciprocity, fast multiply-add and square roots calculations. These operations are available for graphic cards with a compute capability greater than or equal to 2.0. For these operations, the user can define the rounding mode to infinity, minus infinity, zero and the next floating-point number.

The advantage of the CUDA framework is the huge amount of processor cores available for computing arithmetic functions in parallel, in contrast to regular central processor units (CPU). To use this advantage, the algorithm has to be split into sub-problems, which can be distributed to each available thread. Each thread belongs to a block. A block is a logical quantity defining a set of threads, for example, 256 threads belong to one block. Each block has a memory, called *shared memory*, that can be accessed more rapidly than the main memory of the GPU. To obtain a fast computation, the programmer has to make use of these memory structures. Moreover, each block has a unique identification number, which is defined by its Cartesian coordinates. The identification of each thread in a block is also implemented using Cartesian coordinates.

Besides logical graduation of threads and blocks, the basic hardware implementation is based on warps [11]. The Single Instruction, Multiple Threads (SIMT) architecture of CUDA-based GPUs manages and executes threads in groups of 32, which are called *warps*. Because a warp executes one instruction at a time, the greatest efficiency is obtained by using 32 threads for each block. One central problem of such an architecture is the thread-safe execution of operations, which leads to a slower runtime, but gives the user the advantage of writing on the same memory structure, for example, accumulation into one register. CUDA does not include thread-safe arithmetic functions based on the IEEE754 standard for double precision floating-point number with directed rounding. This leads to an implementation of such functions on the CPU. In the case of iterative and non-parallel operations, the CPU computes the solution much faster than an individual thread of the GPU.

In this paper, we use an NVIDIA Geforce GTX 560 with one gigabyte working memory with a compute capability of 2.1 on an Intel Core2 Quad Q9450 @ 2.66GHz based architecture with four gigabytes of main memory. We also use Windows 7 32-bit, Visual Studio 2010, MATLAB 7.12.0 (R2011b), INTLAB V6 and C-XSC v. 2.5.2 to obtain the results in this paper. A higher performance using C-XSC can be obtained by using Linux, such as Ubuntu 11.10. This distinction occurs because MS Visual Studio is not able to use inline assembler code written for the GNU compiler collection (gcc), which leads to increased computation time using software-emulated rounding operations.

2.3 Fault Tree Analysis

Fault tree analysis (FTA) is a mathematical tool for describing and computing system failure probability by dividing the observed system into its subsystem down to the required level of detail. This is done by using a top-down algorithm starting at the top event that describes the observed system failure and following it down through the logically related system parts to the leaves of the tree, which constitute the smallest entities of the subsystems. In our case, we observe logical AND and OR gates. Joining two system parts with an AND gate describes a system failure in which both subsystems have to fail. In the case of an OR gate, only one subsystem has to fail. To describe the logical junctures between subsystems, each subsystem is characterized by an uncertain epistemic failure probability [17], which is expressed as an interval and not as a distribution function of the failure probability. If the user prefers to assume a probability distribution over the interval, we utilize beta- and normal distributions with uncertainty [18] to provide the necessary flexibility. This common approach, however, leads to a kind of aleatory uncertainty since the distribution is a subjective assumption but the lack of knowledge leads to an inexact or uncertain parameter set for the distribution. In this paper, we draw on failure probabilities like those used by Carreras et al. [9] to model failure probabilities for each observed subsystem.

In this paper, we make use of the example defined by Carreras et al. [9, 10, 19], which describes a robot with three joints. Each joint consists of a motor and two independent working sensors describing the actual angle of the joint. The robot is in "failure" state if two or more joints are broken. A joint is broken if the associated sensors are broken or the motor is broken.

Before defining the algorithms to obtain the solutions of the logical gates, we have to introduce the approach to obtain the lower and upper bounds of the failure probability of the top event. We split the interval $p = [0, 1]$ of failure probability into n intervals and the first f intervals into l subintervals. The finer intervals near zero are a compromise between computation time and the accuracy of the solution [19]. The more intervals are defined, the more time will be spent on computation. As described in [7], each interval has a mass assignment. Each logical gate has two inputs representing the scale structure mentioned and the mass assigned to each interval. Let $\mathbf{A} = \{\mathbf{A}_1 \ldots \mathbf{A}_x | x = n - l + (l \cdot f)\}$ be the set of intervals defining the scale; the mass assignment (m) for each interval is defined as:

$$m : \mathbf{A}_i \rightarrow [0, 1], \sum_{i=1}^{n} m(\mathbf{A}_i) = 1, \ m(\emptyset) = 0 \text{ for all } i = 1 \ldots n.$$

In the case of our new verified implementation, we utilize intervals to bound the mass assignment for each interval. In this case, we have to extend the definition of the formula to an interval mass assignment (IMA) $\mathbf{m}$ for all $i = 1 \ldots n$:

$$\mathbf{m} : \mathbf{A}_i \to [0, 1], \; inter \left(\sum_{i=1}^{n} \mathbf{m}(\mathbf{A}_i), \; 1 \right) = \text{true}, \; \mathbf{m}(\emptyset) = [+0, +0].$$

To obtain the solution for each gate, we have to associate each interval of input scale A with input scale B and each mass assignment of A and B as described in [7]. For reasons of clarity and comprehensibility, we use the notation "·" to denote an AND gate and "+" to denote an OR gate. The usual probabilistic rules of computing are used to compute the logical gates. Let $\mathbf{x}$ and $\mathbf{y}$ be scale elements of two independent scales with IMAs $\mathbf{m}_x$ and $\mathbf{m}_y$, which have to be propagated through an AND gate. In this case, the result is the interval $\mathbf{z} = [\underline{\mathbf{x}} \cdot \underline{\mathbf{y}}, \overline{\mathbf{x}} \cdot \overline{\mathbf{y}}]$ with IMA $\mathbf{m}_z = \mathbf{m}_x \cdot \mathbf{m}_y$. This new probability has to be fitted into the scale. An illustrative example is shown in [10]. To obtain a lower bound, the bounds of the interval have to rounded downwards, while the IMA has to be rounded upwards, and vice versa for the upper bound [7]. In the case of the OR gate, only one bound has to be computed, while the IMA is computed in the same way as computing the AND gate. To obtain the lower bound of the above-defined scale elements, we compute $\underline{\mathbf{x}} + \underline{\mathbf{y}} - (\underline{\mathbf{x}} \cdot \underline{\mathbf{y}})$, while computing $\overline{\mathbf{x}} + \overline{\mathbf{y}} - (\overline{\mathbf{x}} \cdot \overline{\mathbf{y}})$ to obtain a lower bound.

Having provided these definitions, we will next illustrate our new implementation.

3   Implementation

In this section, we describe the new implementation of the verified FTA algorithm using high-performance computing supplied by CUDA and verified interval arithmetic provided by C-XSC.

As described in Sect. 2, the main problem with this algorithm is the trade-off between accuracy and computation time. To reduce the computation time while keeping the accuracy, we utilize the ability of CUDA to obtain correctly rounded solutions with error bounds defined by the IEEE754 floating-point standard. To obtain a fast and verified implementation, we assign the computation of the mass and the solution interval for each gate to one CUDA thread, using CUDA compiler intrinsic IEEE754-compliant algorithms, such as $\_dx\_r(\ldots)$. Here, $x \in \{\text{add, mul, fma, rcp, div, sqrt}\}$ represents the arithmetic functions addition, multiplication, fused multiply-add $(x + (y \cdot z))$, reciprocal $(\frac{1.0}{x})$, division and square root, and $r \in \{\text{rn, rz, ru, rd}\}$ representing the rounding modes rounding to the next floating-point number, zero, infinity and minus infinity.

One central disadvantage of GPU computing is the allocation of memory on the GPU and the cloning of the data stored in the main memory into the memory of the GPU. In cases where there is a large amount of data and a small computation effort, computation on the CPU is favored because of the computation time. Moreover, the restriction of memory size on the GPU has to be considered. In our case, we can allocate one gigabyte of data on the GPU, which leads to an implementation optimizing the memory. Where there are a large number of scale intervals, we have to subdivide the algorithm into smaller parts to clear the GPU memory between the computations of each part. To do so, we decided to split the algorithm into two steps: The first is to compute the bounds obtained by the AND and the OR gates; the second is to compute the mass assignment to each interval. This separation of tasks reduces the amount of GPU memory used.

Let us begin by discussing the OR-gate implementation. As shown in [7, 10, 19], we have to distinguish between lower and upper bound computation. We only need to discuss the lower bound computation, because the computation of the upper bound is relevant only in the opposite rounding mode. In contrast to the AND gate, only one bound has to be computed using directed rounding. Let $\mathbf{x}$ and $\mathbf{y}$ be two scale elements and $\mathbf{m}_x$ and $\mathbf{m}_y$ be the corresponding IMAs; then, the lower bound is computed as

$$lb = \triangledown \left[\triangledown \left(\underline{\mathbf{x}} + \underline{\mathbf{y}} \right) - \triangle \left(\underline{\mathbf{x}} \cdot \underline{\mathbf{y}} \right) \right] \text{ with } \underline{\mathbf{x}}, \underline{\mathbf{y}} \in [0, 1],$$

and the mass assignment as

$$m_{lb} = \triangle \left(\overline{\mathbf{m}_x} \cdot \overline{\mathbf{m}_y} \right) \text{ with } \overline{\mathbf{m}_x}, \ \overline{\mathbf{m}_y} \in [0, 1].$$

The lower bound (best-case scenario) is obtained by rounding the corresponding mass assignment towards infinity, which is justified by the fact that a small number of robots (rounding down the bound) with a high failure probability is preferable to a large number of robots with a high failure probability.

The CPU computes the assignment of the mass to scale elements because no thread-safe addition with IEEE754 support is provided by CUDA. Next, we will demonstrate the implementation of the OR kernel in CUDA, which computes the mass and the bound in parallel for each combination of $\mathbf{x}$ and $\mathbf{y}$. A *kernel* is a GPU-related function that is executed n times on n different threads.

```
__global__ void or_Kernel(...){
    // Get coordinates of the actual block
    int blockRow = blockIdx.y;
    int blockCol = blockIdx.x;

    // Get coordinates of the actual thread
    int i = threadIdx.y;
    int j = threadIdx.x;
    ...
    // Create shared variables
    __shared__ double sol_mass[BLOCK_SIZE][BLOCK_SIZE];
    __shared__ double sol_interval[BLOCK_SIZE][BLOCK_SIZE];

    // Compute mass assignment and corresponding bound
    sol_mass[i][j]  = __dmul_ru(m_x[i],m_y[j]);
    sol_bound[i][j] = __dadd_rd(__dadd_rd(x[i],y[j]),
        __dmul_ru(__dmul_ru(-1.0,x[i]),y[j]));
    ...
```

```
    // Write mass assignment and bounds into the solution arrays
    SetElement(sol_mass_output,i,j,sol_mass[i][j]);
    SetElement(sol_interval_output,i,j,index[i][j]);
}
```

Because of the available space, we use "..." to indicate an uninterrupted text line or a section of code of no interest. In this example, we use shared memory to obtain a fast computation. Each thread of the OR kernel computes one element of mass assignment (sol_mass) and the bound (sol_bound) addressed by i and j, which are the Cartesian coordinates of the threads inside the actual block. The solutions of this computation are two arrays filled with $((n - f) + (f \cdot l))^2$ floating-point numbers in double format. The function $setElement(...)$ writes the solution from each thread to the output parameter. Because each thread writes only one element, we do not need thread-safe functions. Another kernel computes the correct position depending on the computed bound to assign the mass to the correct scale element.

To fasten the computation of the AND gate, we draw on three kernels to compute the right mass, bounds and indices. Let $\mathbf{x}$ and $\mathbf{y}$ be two scale elements and $\mathbf{m}_x$ and $\mathbf{m}_y$ be the corresponding IMAs; then, the two bounds are computed as

$$\text{bound}_1 = \nabla(\underline{\mathbf{x}} \cdot \underline{\mathbf{y}}) \text{ with } \underline{\mathbf{x}}, \underline{\mathbf{y}} \in [0,1],$$
$$\text{bound}_2 = \nabla(\overline{\mathbf{x}} \cdot \overline{\mathbf{y}}) \text{ with } \overline{\mathbf{x}}, \overline{\mathbf{y}} \in [0,1],$$

and the mass assignment as

$$m_{lb} = \triangle(\overline{\mathbf{m}_x} \cdot \overline{\mathbf{m}_y}) \text{ with } \overline{\mathbf{m}_x}, \overline{\mathbf{m}_y} \in [0,1].$$

The direction of the rounding modes to obtain bound$_1$ and bound$_2$ are equal to the direction of the computation of the bound for the OR-gate. The following program code illustrates the computation of the mass with CUDA.

```
__global__ void mass_Kernel(...){
    // Get coordinates of the actual block
    int blockRow = blockIdx.y;
    int blockCol = blockIdx.x;

    // Get coordinates of the actual thread
    int i = threadIdx.y;
    int j = threadIdx.x;

    // Get subvector for each input
    double* m_x =   subVector(mass1, blockCol);
    double* m_y =   subVector(mass2, blockRow);

    // Get corresponding part of the solution vector
    double* solution = subVector(sol_mass,blockRow,blockCol);
```

```
// Create shared variable
__shared__ double sol_mass[BLOCK_SIZE][BLOCK_SIZE];

// Compute mass assignment
sol_mass_sh[i][j]=__dmul_ru(m_x[j],m_y[i]);

// Write mass assignment into the solution array
and_SetElement(solution,i,j,sol_mass[i][j]);
}
```

The computation of the lower and upper bounds is straightforward in the example of the OR gate. To obtain verified results, assigning the computed mass to the scale, we draw on interval arithmetic by C-XSC that provides verified bounds on mass assignments for each scale element.

4 Performance

In this section, we report on the accuracy and computation time of the fault tree analysis described above. In addition to failures in computation, which occur when using floating-point arithmetic [5, 6], rounding failures by INTLAB, CUDA and C-XSC were observed. Solutions provided by C-XSC and INTLAB on a x86 computer architecture are always identical, while some solutions differ when a 64-bit version of the libraries is used; this results from different processor registers being used by C-XSC and INTLAB for directed rounding. In this paper, we draw on an x86-based architecture. To compare the solutions of directed rounding by C-XSC and CUDA, we define a small test using one million random floating-point numbers in $[0, 1]$ accumulated by using addition, multiplication and division. If using the GPU provides correct rounding, we expect the same solutions as those computed by INTLAB and C-XSC. Our tests have shown that CUDA provides correct solutions by using directed rounding defined by the IEEE754 floating-point standard.

To test our new implementation, we use two different scales. First, we utilize a scale with $n = 200$, $f = 20$ and $l = 100$ and, to obtain a more significant benchmark, the second scale is more finer partitioned with parameters $n = 5000$, $f = 100$, $l = 60$. In Table 1, we compare the solutions of the first benchmark obtained by our new implementation (Columns 2 and 3) with those obtained by the MATLAB implementation introduced in [7] in the present version 3.5.2 (Columns 4 and 5) for lower bound (LB) and upper bound (UB) computation. The results have to be interpreted as follows. Ninety-five percent of the robots produced have an error probability of $\mathbf{p} = [2.725 \cdot 10^{-2}, 2.805 \cdot 10^{-2}]$. Thus, $\mathbf{p}$ percentage of 95 percentages of all robots produced will fail in the first thousand operation hours. Conversely, $0.95 - 0.95 \cdot \mathbf{p} = [0.9233, 0.9241]$ percent of all robots will not fail in this timeframe.

In this example, we compute the same results as MATLAB with two differences: for $x = 0.45$ and $x = 0.95$. While the new implementation encloses the MATLAB solutions for $x = 0.45$, the new approach leads to a tighter interval for $x = 0.95$.

Table 1. Solutions obtained with $n = 200$, $f = 20$ and $l = 100$

x	LB	UB	MATLAB LB	MATLAB UB
0.16	$1 \cdot 10^{-4}$	$4 \cdot 10^{-4}$	$1 \cdot 10^{-4}$	$4 \cdot 10^{-4}$
0.25	$3 \cdot 10^{-4}$	$6 \cdot 10^{-4}$	$3 \cdot 10^{-4}$	$6 \cdot 10^{-4}$
0.38	$6.5 \cdot 10^{-4}$	$1 \cdot 10^{-3}$	$6.5 \cdot 10^{-4}$	$1 \cdot 10^{-3}$
0.45	$9 \cdot 10^{-4}$	$1.3 \cdot 10^{-3}$	$9 \cdot 10^{-4}$	$1.25 \cdot 10^{-3}$
0.84	$7.05 \cdot 10^{-3}$	$7.7 \cdot 10^{-3}$	$7.05 \cdot 10^{-3}$	$7.7 \cdot 10^{-3}$
0.95	$2.725 \cdot 10^{-2}$	$2.805 \cdot 10^{-2}$	$2.725 \cdot 10^{-2}$	$2.825 \cdot 10^{-2}$

To compare the computation times, we draw on the wall-clock time used by MATLAB and C++ and the accurate timing used by CPU and CUDA, using CUDA events to obtain a reproducible benchmark solution. The solutions obtained are provided in Table 2. We recognize a performance improvement of up to 99 percent for lower and upper bound computation. To compare the computation times, we refer to the wall-clock time spent on computing the results.

Table 2. Comparison of the computation times for lower and upper bound in MATLAB and C++ ($n = 200$, $f = 20$, $l = 100$)

	CPU time [s]	CUDA time [s]	Wall-clock time [s]
C++ implementation (LB)	14	14	7
MATLAB (LB)	5102	–	1685
C++ implementation (UB)	14	14	7
MATLAB (UB)	5154	–	1712

The computation times for the the second benchmark are shown in Table 3.

Table 3. Comparison of the computation time of lower and upper bound in MATLAB and C++ with finer scale elements ($n = 5000$, $f = 100$, $l = 60$)

	CPU time [s]	CUDA time [s]	Wall-clock time [s]
C++ implementation (LB)	721	591	721
MATLAB (LB)	147661	–	48070
C++ implementation (UB)	654	526	654
MATLAB (UB)	142248	–	46160

The solutions for **p** in benchmark two are identical to the solutions obtained by the MATLAB implementation. In this example, we obtain an efficiency enhancement of up to 98 percent in wall-clock time.

To compute the results in this section, we utilized the DSI toolbox in the present version 3.5.2 [18].

5   Conclusion

In this paper, we have presented a new implementation approach to fault tree analysis using the CUDA framework to obtain high-performance computing of verified probabilistics. First, we gave a short introduction to interval arithmetic, CUDA and fault tree analysis to provide a basis for the subsequent sections. We next gave examples of the new implementation followed by a comparison of the computation accuracy of CUDA, C-XSC and INTLAB for random probabilistic values in $[0, 1]$. We ended our paper by benchmarking our new approach against the MATLAB- and INTLAB-based implementation, which showed that, by using the new approach, we obtained a gain in algorithm efficiency.

As future research, we plan to optimize the Dempster Shafer with Intervals toolbox (DSI) for MATLAB [18] using CUDA and C-XSC routines.

References

[1] Ramesh, V., Saravannan, R.: Reliability Assessment of Cogeneration Power Plant in Textile Mill Using Fault Tree Analysis. Journal of Failure Analysis and Prevention 11, 56–70 (2011)

[2] Limbourg, P., Savić, R., Petersen, J., Kochs, H.: Fault tree analysis in an early design stage using the Dempster-Shafer theory of evidence. Risk, Reliability and Societal Safety (2007)

[3] Moore, R., Kearfott, B., Cloud, M.: Introduction to Interval Computation. Society for Industrial and Applied Mathematics, Philadelphia (2009)

[4] Alefeld, G., Herzberger, J.: Introduction to interval computation. Academic Press (1984)

[5] Arnold, D.: Computer Arithmetic Tragedies, http://www.ima.umn.edu/~arnold/455.f96/disasters.html

[6] Goldberg, D.: What Every Computer Scientist Should Know About Floating-Point Arithmetic. ACM Computing Surveys (CSUR), 5–48 (1991)

[7] Rebner, G., Auer, E., Luther, W.: A verified realization of a Dempster–Shafer based fault tree analysis. Computing 94, 313–324 (2012)

[8] Rump, S.: INTLAB - INTerval LABoratory. Developments in Reliable Computing 1, 77–104 (1999)

[9] Carreras, C., Walker, I.: Interval Methods for Fault-Tree Analyses in Robotics. IEEE Transactions on Reliability 50, 3–11 (2001)

[10] Traczinski, H.: Integration von Algorithmen und Datentypen zur validierten Mehrkörpersimulation in MOBILE. Dissertation, Universität Duisburg-Essen. Logos-Verlag, Berlin (2006) ISBN 978-3-8325-1457-0

[11] NVIDIA: Plattform für Parallel-Programmierung und parallele Berechnungen, http://www.nvidia.de/object/cuda_home_new_de.html

[12] IEEE Computer Society: IEEE Standard for Floating-Point Arithmetic. IEEE Std 754-2008 (29 2008) 1–58 (2008)

[13] Hofschuster, W., Krämer, W.: C-XSC 2.0 – A C++ Library for Extended Scientific Computing. In: Alt, R., Frommer, A., Kearfott, R.B., Luther, W. (eds.) Num. Software with Result Verification. LNCS, vol. 2991, pp. 15–35. Springer, Heidelberg (2004)

[14] Zimmer, M.: Using C-XSC in a Multi-Threaded Environment. Preprint BUW-WRSWT 2011/2, Universität Wuppertal (2011),
http://www2.math.uni-wuppertal.de/wrswt/preprints/prep_11_2.pdf

[15] Krämer, W., Zimmer, M., Hofschuster, W.: Using C-XSC for High Performance Verified Computing. In: Jónasson, K. (ed.) PARA 2010, Part II. LNCS, vol. 7134, pp. 168–178. Springer, Heidelberg (2012)

[16] Mayer, G.: Grundbegriffe der Intervallrechnung. Wissenschaftliches Rechnen mit Ergebnisverifikation, 101–117 (1989)

[17] Der Kiureghian, A., Ditlevsen, O.: Aleatory or epistemic? Does it matter? Special Workshop on Risk Acceptance and Risk Communication (March 2007)

[18] Auer, E., Luther, W., Rebner, G., Limbourg, P.: A Verified MATLAB Toolbox for the Dempster-Shafer Theory. In: Proceedings of the Workshop on the Theory of Belief Functions (April 2010), http://www.udue.de/DSIPaperone, http://www.udue.de/DSI

[19] Luther, W., Dyllong, E., Fausten, D., Otten, W., Traczinski, H.: Numerical Verification and Validation of Kinematics and Dynamical Models for Flexible Robots in Complex Environments. In: Perspectives on Enclosure Methods, pp. 181–200. Springer, Wien (2001)

Deterministic Seismic Hazard Analysis Considering Non-controlling Seismic Sources and Time Factors

Duruo Huang[1,*], Jui-Pin Wang[1], Logan Brant[2], and Su-Chin Chang[3]

[1] The Hong Kong University of Science and Technology, Hong Kong
{drhuang,jpwang}@ust.hk
[2] Columbia University, New York City
lcb2002@columbia.edu
[3] The University of Hong Kong, Hong Kong
suchin@hku.hk

Abstract. Deterministic seismic hazard analysis (DSHA) is an approach for evaluating site-specific seismic hazard that is influenced by the maximum hazard from the controlling sources affecting the specific study site. In its conventional form, DSHA does not consider sources other than the largest "controlling" source and it does not account for the time factors owing to the uncertainty of earthquakes occurrences in time. Under certain condition, ignoring these factors can lower the conservatism of the hazard estimate, especially when other non-controlling sources generate hazards nearly equivalent to that of the controlling source or when the structure's design life is longer than the controlling source earthquake's return period. This study discusses several limitations of conventional DSHA and provides a modified approach for DSHA which we believe should supplement the conventional method. An example is presented to demonstrate how conventional DSHA can be un-conservative for certain problem types.

Keywords: DSHA, non-controlling sources, return period, design life.

1 Introduction

Seismic hazard analysis (SHA) is considered an engineering solution to the uncertain earthquake process [1]. Understanding the site-specific seismic hazard is needed for earthquake resistant design. Seismic hazard can be determined from either deterministic or probabilistic frameworks, which include the two most commonly used methods: deterministic seismic hazard analysis (DSHA) and probabilistic seismic hazard analysis (PSHA). These methods have been prescribed for developing site-specific earthquake resistant design by numerous codes and guidelines [2].

The limitations of the two approaches are also well established [3-8]. Debates concerning the pros and cons of each method are common and all too often only weakened the case for one or another method, without offering concrete improvements to either. In reality, as Mualchin [9] previously stated, no seismic hazard analysis is perfect and the key to performing this work successfully is understanding the fundamental differences and limitation of each, before then selecting one or both approaches for your study.

E. Hüllermeier et al. (Eds.): SUM 2012, LNAI 7520, pp. 550–557, 2012.
© Springer-Verlag Berlin Heidelberg 2012

DSHA looks at multiple seismic sources and determines the maximum seismic hazard from the single source which creates the largest hazard for the site. A benchmark DSHA example (Fig. 1) from a well-regarded geotechnical earthquake engineering textbook [10] included three seismic sources surrounding the site. Those sources include a line, an area and a point sources, each resulting in deterministic seismic hazards expressed in peak ground accelerations (PGAs) of 0.74 g, 0.99 g and 0.04 g, respectively, at one standard deviation above the mean (84th percentile). Therefore, the maximum value of 0.99 g is considered the PGA for the DSHA, making the area source the controlling source for this example.

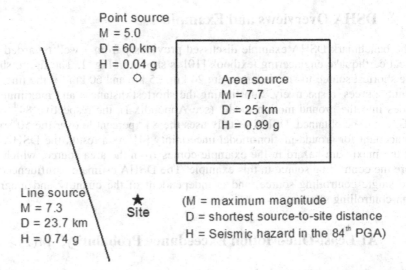

Fig. 1. Benchmark DSHA example (after Kramer, 1996)

However, what would happen if each of the three sources generates seismic hazards with similar PGA, such as 0.99 g, 0.98 g, and 0.97 g? Compared to 0.99 g, 0.74 g and 0.04 g in the benchmark example, obviously this hypothetical situation would be under higher risk in terms of seismic hazard. However, since the conventional DSHA does not consider non-controlling sources, both situations would result in the same PGA using the conventional DSHA. Let's use an example to further demonstrate the potential deficiency in DSHA. What if a site is affected by 30 sources in the surrounding instead of three? It is risky to conclude that two situations have the same seismic hazard, simply because the maximum hazard from the controlling source is the same.

The uncertainty of earthquake occurrence with time is well understood, with an example being the Parkfield section of the San Andreas Fault [9]. The reasons why conventional DSHA chooses not to account for such uncertainty are reasonable considering that this information is not very well understood, but at the same time this approach makes conventional DSHA insufficient for certain situation. For example, consider two sources, each having the same maximum magnitude but with different periods supported by geological evidence. Logically, structures constructed near the source with a shorter return period will demand a more robust design. Apart from earthquake time factors, conventional DSHA also does not account for engineering time factors. Clearly, a "permanent" structure should be designed more conservatively than a "temporary" structure, since the former is expected to experience more earthquakes throughout its service life.

This paper discusses the influence of non-controlling sources and time factors on deterministic seismic hazard analysis, and the characteristics which are not otherwise accounted for using conventional DSHA. A modified DSHA approach is presented considering these important characteristics and an example is presented to illustrate the need and use of the modified approach.

2   DSHA Overviews and Example

The benchmark DSHA example discussed previously from a well-regarded geotechnical earthquake engineering textbook [10] is shown in Fig. 1. The figure shows that the shortest source-to-site distances are 24 km, 25 km and 60 km for the line, area and point sources, respectively. Substituting the shortest distances and maximum magnitudes into the ground motion model (see Appendix I), the respective 84th percentile PGAs can be obtained. DSHA usually uses the 84th percentile over the 50th percentile to account for ground-motion-model uncertainty [4]. As a result, the DSHA estimate of the maximum hazard in the example comes from the area source, which is therefore the controlling source in this example. The DSHA estimate is influenced by only the single controlling source, and is independent of the quantity and magnitude of non-controlling sources.

3   At-Least-One-Motion Exceedance Probability, $Pr(\tilde{Y} > y*)$

Prior to demonstrating the influence of the aforementioned factors on DSHA, we first introduce the criterion to assess a motion (e.g. PGA) by examining its exceedance probability. The exceedance probability is the probability that at least one motion will exceed a given motion. It is calculated using the following expression.

$$Pr(\tilde{Y} > y^*) = 1 - Pr(Y_1 \leq y^*) \times Pr(Y_2 \leq y^*) \times \cdots \times Pr(Y_n \leq y^*) \tag{1}$$

Where Y_i denotes any given possible motion, either from controlling sources or non-controlling sources. Y_i estimated through a ground motion model is known to follow a lognormal distribution, so the probability distribution $Pr(Y_i \leq y^*)$ in Eq. (1) can be easily established (Appendix I).

As a result, the original and modified DSHA estimates can be compared with the same criterion. The criterion is logically sound for engineering designs, since failure of engineered structures is caused when one motion exceeding the design motion.

4   Non-controlling Seismic Sources

For the example shown in Fig. 1, Eq. (1) can be reorganized to account for the three sources described in the example, such that $Pr(\tilde{Y} > y^*)$ is as follows:

$$\Pr(\tilde{Y} > y^*) = 1 - [\Pr(Y_A \le y^*)]^{n_A} \times [\Pr(Y_L \le y^*)]^{n_L} \times [\Pr(Y_P \le y^*)]^{n_P} \qquad (2)$$

Where Y_A, Y_L and Y_P denote the motions from the area, line and point source, respectively; n_A, n_L and n_P (integer) are the respective number of earthquakes, equal to one with conventional DSHA which does not account for the earthquake rate.

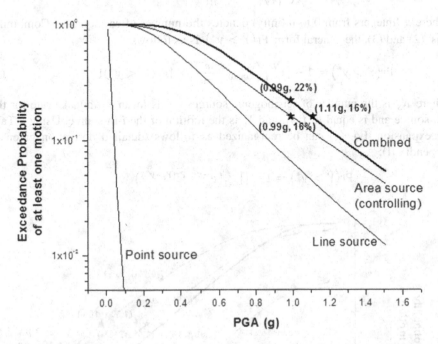

Fig. 2. Comparison in hazard curves with and without accounting for non-controlling sources

Fig. 2 shows the relationship between the exceedance probability and a given PGA, for each source from the example and for the three sources combined. The conventional DSHA estimate of 0.99 g using the 84th percentile corresponds to 16 percent exceedance probability. At that exceedance probability, the PGA estimate from the three combined sources (which includes two non-controlling sources) increases to 1.11 g. Alternatively, the PGA estimate (0.99 g) becomes less conservative, associated with a 22 percent exceedance probability compared to the original 16 percent. Using either method of comparison, the example demonstrates the decreased conservatism in DSHA when non-controlling sources are considered.

5   Time Factors

Conventional DSHA does not consider time factors, such as return period and design life. Given a return period for the maximum earthquake equal to r with design life equal to t, the mean earthquake rate during the structure life becomes t/r. It is common to use a Poisson distribution to model a rare event, such as earthquake, in time or space [10-12]. Given a mean earthquake rate of t/r, the Poisson probability density is as follows:

$$Pr\left(n\left|\frac{t}{r}\right.\right) = \frac{(\frac{t}{r})^n \times e^{-(\frac{t}{r})}}{n!} \tag{3}$$

Where n (integers from 0 to infinity) denotes the number of earthquakes. Combining Eqs. (2) and (3), the general form $Pr(\tilde{Y} > y^*)$ is as follows:

$$Pr(\tilde{Y} > y^*) = 1 - \prod_{i=1}^{N_s}\left(\sum_{n=0}^{\infty}\frac{v_i \times e^{-v_i}}{n!} \times [Pr(Y_i < y^*)]^n\right) \tag{4}$$

Where N_S is the number of seismogenic sources, v_i is mean earthquake rate for the i-th source and is equal to t_i/r_i, and Y_i is the motion of the i-th source. Using a Taylor expansion, Eq. (4) can be reorganized as follows (detailed derivation given in Appendix II):

$$Pr(\tilde{Y} > y^*) = 1 - \prod_{i=1}^{N_s}\left(e^{-v_i \times Pr(Y_i > y^*)}\right) \tag{5}$$

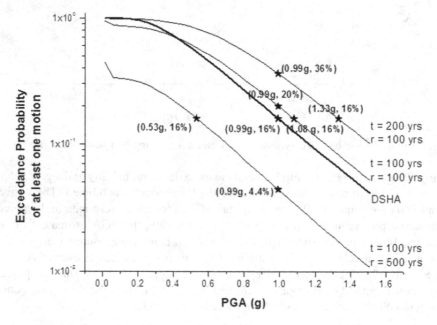

Fig. 3. Comparison in hazard curves of the original and modified DSHA

Since the DSHA example in Fig. 1 does not provide design life (t) or return period (r), we conducted the following calculations using three hypothetical conditions: 1) return period five-time as large as design life, i.e., $r = 500$ yrs; $t = 100$ yrs, 2) return period equal to design life, i.e., $r = 100$ yrs; $t = 100$ yrs, and 3) return period half of design life, i.e., $r = 100$ yrs; $t = 200$ yrs. Using Eq. (5), the hazard curves are shown on

Fig. 3. The exceedance probabilities at 0.99 g correspond to 4.4 percent, 20 percent, and 36 percent, for Conditions 1 through 3, respectively. For reaching the same exceedance probability equal to 16 percent the new PGA estimates are 0.53 g, 1.08 g and 1.33 g, for Conditions 1 through 3, respectively. This calculation also demonstrates and verifies the negative effect on DSHA when certain time factors, such as large t/r, are encountered.

Notice that at low PGA (say 0.06 g) discontinuous points are generated in the hazard curves for Conditions 1 and 2, as shown on Fig. 3. This is believed to be due to the nature of computation, possibly owing to the relatively small maximum magnitude (5.0) from the point source. To verify the postulation, a sensitivity study using a larger magnitude for the point source equal to 7.0 was conducted.

6   Discussions and Conclusions

The benchmark example illustrated that without accounting for these non-controlling sources and time factors, under certain situations the conventional DSHA is not as conservative as the modified approach, especially for Condition 3 with a large mean earthquake rate (t/r) during the service life of structures.

When the return period is five-time as large as the design life, the DSHA estimate (0.99 g) is indeed becoming more conservative (4.4 percent exceedance probability) than originally suggested (16 percent exceedance probability). As a result, the conventional DSHA should provide adequate conservatism when the return period of the maximum-magnitude earthquakes is long relative to the service life of structures being designed, and when the number of seismic sources affecting the site is limited. Under such circumstances, the conventional DSHA can be used either with or without modification.

On the other hand, special attention is needed when the return period is less than or equal to the design life and when many sources are involved in analysis. For this condition, with the conventional DSHA being demonstrated and proved not as conservative as expected, we suggest using the modified DSHA approach presented in this paper or another seismic hazard approach, such as PSHA. Decision makers should not confine themselves to just one approach, since for all seismic hazard analysis methods each present their own challenges and limitations. As is shown in the paper, there are times when the conventional DSHA should be supplement with other methods of analysis.

Appendix I

The ground motion model used in the example is as follows:

$$\ln Y = 6.74 + 0.859M - 1.8 \times \ln(R + 25) \; ; \; \sigma_{\ln Y} = 0.57 \qquad (6)$$

Where M and R are earthquake magnitude and source-to-site distance (km), respectively; Y denotes peak ground acceleration in gal. Since Y follows a lognormal distribution (or $\ln Y$ follows a normal distribution), the cumulative probability for a given motion y^* is as follows [11]:

$$\Pr(Y \le y^*) = \Phi\left(\frac{\ln y^* - \mu_{\ln Y}}{\sigma_{\ln Y}}\right) \tag{7}$$

Where μ_{lnY} and σ_{lnY} are the mean and standard deviation of $\ln Y$, both computed from Eq. (6).

Appendix II

Derivations of Eq. (5) are as follows:

$$\Pr(\tilde{Y} > y^*) = 1 - \prod_{i=1}^{N_S}\left(\sum_{n=0}^{\infty} \frac{v_i^{\,n} \times e^{-v_i}}{n!} \times [\Pr(Y_i \le y^*)]^n\right)$$

$$= 1 - \prod_{i=1}^{N_S}\left(e^{-v_i} \times \sum_{n=0}^{\infty} \frac{[v_i \times \Pr(Y_i \le y^*)]^n}{n!}\right) \tag{8}$$

Owing to the Taylor expansion:

$$\sum_{n=0}^{\infty} \frac{v_i \times [\Pr(Y_i \le y^*)]^n}{n!} = e^{v_i \times \Pr(Y_i \le y^*)} \tag{9}$$

As a result,

$$\Pr(\tilde{Y} > y^*) = 1 - \prod_{i=1}^{N_S}\left(e^{-v_i} \times e^{v_i \times \Pr(Y_i \le y^*)}\right)$$

$$= 1 - \prod_{i=1}^{N_S}\left(e^{-v_i \times (1 - \Pr(Y_i \le y^*))}\right) \tag{10}$$

$$= 1 - \prod_{i=1}^{N_S}\left(e^{-v_i \times \Pr(Y_i > y^*)}\right)$$

References

1. Geller, R.J., Jackson, D.D., Kagan, Y.Y., Mulargia, F.: Earthquake cannot be predicted. Science 275, 1616 (1997)
2. U.S. Nuclear Regulatory Commission, A performance-based approach to define the site-specific earthquake ground motion, Regulatory Guide 1.208, Washington D.C. (2007)

3. Bommer, J.J.: Deterministic vs. probabilistic seismic hazard assessment: An exaggerated and obstructive dichotomy. J. Earthq. Eng. 6, 43–73 (2002)
4. Bommer, J.J.: Uncertainty about the uncertainty in seismic hazard analysis. Eng. Geol. 70, 165–168 (2003)
5. Castanos, H., Lomnitz, C.: PSHA: Is it science? Eng. Geol. 66, 315–317 (2002)
6. Krinitzsky, E.L.: Earthquake probability in engineering: Part 1. The use and misuse of expert opinion. Eng. Geol. 33, 219–231 (1993a)
7. Krinitzsky, E.L.: Earthquake probability in engineering: Part 2. Earthquake recurrence and limitations of Gutenberg-Richter b-values for the engineering of critical structures. Eng. Geol. 36, 1–52 (1993b)
8. Krinitzsky, E.L.: Epistematic and aleatory uncertainty: a new shtick for probabilistic seismic hazard analysis. Eng. Geol. 66, 157–159 (2002)
9. Mualchin, L.: Seismic hazard analysis for critical infrastructures in California. Eng. Geol. 79, 177–184 (2005)
10. Kramer, S.L.: Geotechnical Earthquake Engineering. Prentice Hall Inc., New Jersey (1996)
11. Ang, A.H.S., Tang, W.H.: Probability Concepts in Engineering: Emphasis on Applications to Civil and Environmental Engineering. John Wiley & Sons, New York (2007)
12. Devore, J.L.: Probability and Statistics for Engineering and the Sciences. Duxbury Press, UK (2008)

Towards a Logic of Argumentation

Leila Amgoud and Henri Prade

IRIT, Université Paul Sabatier, 31062 Toulouse Cedex 9, France
{amgoud,prade}@irit.fr

Abstract. Starting from a typology of argumentative forms proposed in linguistics by Apothéloz, and observing that the four basic forms can be organized in a square of oppositions, we present a logical language, somewhat inspired from generalized possibilistic logic, where these basic forms can be expressed. We further analyze the interplay between the formulas of this language by means of two hexagons of oppositions. We then outline the inference machinery underlying this logic, and discuss its interest for argumentation.

1 Introduction

In a work still largely ignored in Artificial Intelligence, the linguist Apothéloz [1] established a catalogue of argumentative forms more than two decades ago; see also [11]. In particular, he advocated the difference between statements such as "y is not a reason for concluding x" and "y is a reason against concluding x", which may be viewed as two different negative forms that are in opposition with the more simple statement "y is a reason for concluding x". It has been noticed in philosophical logic for a long time that the existence of two negation systems gives birth to a square of oppositions [10]. This has led Salavastru [12] to present a reading of Apothéloz' typology of argumentative forms in terms of square of oppositions a few years ago, and to propose a propositional logic translation of the basic argumentative statements. However, propositional logic is not rich enough for offering a representation setting for such a variety of statements.

In the following, after a brief reminder on the square of oppositions, we first reexamine Salavastru's proposal and identify several weaknesses. We first restate the argumentative square of opposition properly, and then introduce the basic elements of a kind of conditional logic language, somewhat inspired by generalized possibilistic logic [5], in which we can get a more suitable translation of the argumentative square. This square can in fact be extended into a more complete hexagon of opposition [3], which makes clear that its underlying structure is based on the trichotomy "y is a reason for concluding x", "y is a reason for concluding $\neg x$", "y is neither a reason for concluding x, nor for concluding $\neg x$". We then outline the inference machinery of the proposed logic, emphasize the difference between "y is a reason for concluding x" and "x follows logically from y", which leads to build another hexagon of opposition. We discuss the potential interest of the proposed logic for argumentation, and finally mention some possible extensions for handling nonmonotonic reasoning and graded argumentative statements in the spirit of possibilistic logic.

E. Hüllermeier et al. (Eds.): SUM 2012, LNAI 7520, pp. 558–565, 2012.
© Springer-Verlag Berlin Heidelberg 2012

2 Argumentative Square

Apothéloz [1] points out the existence of four basic argumentative forms:

- i) "y is a reason for concluding x" (denoted $\mathcal{C}(x) : \mathcal{R}(y)$),
- ii) "y is not a reason for concluding x" ($\mathcal{C}(x) : -\mathcal{R}(y)$),
- iii) "y is a reason against concluding x" ($-\mathcal{C}(x) : \mathcal{R}(y)$), and
- iv) "y is not a reason against concluding x" ($-\mathcal{C}(x) : -\mathcal{R}(y)$).

Interestingly enough, several of these forms have not been considered in Artificial Intelligence research. As can be seen several forms of opposition are present in these statements, where two negations are at work. A key point in this categorization is indeed the presence of two kinds of negation, one pertaining to the contents x or y, and the other to the functions $\mathcal{R}$ or $\mathcal{C}$. It has been observed that such a double system of negations gives birth to a formal logical structure called *square of opposition*, which dates back to Aristotle's time (see, e.g., [10] for a historical and philosophical account). We first briefly recall what this object is, since it has been somewhat neglected in modern logic.

It has been noticed for a long time that a statement (A) of the form "every a is p" is negated by the statement (O) "some a is not p", while a statement like (E) "no a is p" is clearly in even stronger opposition to the first statement (A). These three statements, together with the negation of the last statement, namely (I) "some a is p", give birth to the square of opposition in terms of quantifiers $A : \forall a\ p(a)$, $E : \forall a\ \neg p(a)$, $I : \exists a\ p(a)$, $O : \exists a\ \neg p(a)$, pictured in Figure 1. Such a square is usually denoted by the letters A, I (affirmative half) and E, O (negative half). The names of the vertices come from a traditional Latin reading: **AffIrmo, nEgO**). Another standard example of the square of opposition is in terms of modalities: $A : \Box r$, $E : \Box \neg r$, $I : \Diamond r$, $O : \Diamond \neg r$ (where $\Diamond r \equiv \neg \Box \neg r$). As can be seen from these two examples, different relations hold between the vertices. It gives birth to the following definition:

Definition 1 (Square of opposition). *Four statements A, E, O, I make a square of opposition if and only if the following relations hold:*

- *(a) A and O are the negation of each other, as well as E and I;*
- *(b) A entails I, and E entails O;*
- *(c) A and E cannot be true together, but may be false together, while*
- *(d) I and O cannot be false together, but may be true together.*

Note the square in Fig. 1 pressupposes the existence of some s (non empty domain). $r \not\equiv \bot$, $\top$ is assumed in the modal logic case.

The observation that two negations are at work in the argumentative statements classified by Apothéloz [1] has recently led Salavastru [12] to propose to organize the four basic statements into a square of opposition; see also [9]. However, his proposal is debatable on one point, as we are going to see. Indeed, taking $\mathcal{C}(x) : \mathcal{R}(y)$ for vertex A, leads to take its negation $\mathcal{C}(x) : -\mathcal{R}(y)$ for O. Can we take $-\mathcal{C}(x) : \mathcal{R}(y)$ for E? This first supposes that A and E are mutually exclusive, which is clearly the case. Then, we have to take the negation of E for I, i.e. $-\mathcal{C}(x) : -\mathcal{R}(y)$. We have still to check that A entails I and E entails O, as

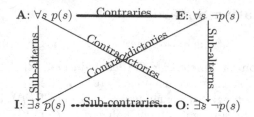

Fig. 1. Square of opposition

well as condition (d) above. If y is a reason for not concluding x, then certainly y is not a reason for concluding x, so E entails O; similarly y is a reason for concluding x entails that y is not a reason for not concluding x, i.e. A entails I. Finally, y may be a reason neither for concluding x nor for not concluding x. This gives birth to the argumentative square of opposition of Figure 2. It can be checked that the contradiction relation (a) holds, as well as the relations (b), (c), and (d) of Definition 1.

Proposition 1. *The four argumentative forms $A = \mathcal{C}(x) : \mathcal{R}(y)$, $E = -\mathcal{C}(x) : \mathcal{R}(y)$, $O = \mathcal{C}(x) : -\mathcal{R}(y)$, $I = -\mathcal{C}(x) : -\mathcal{R}(y)$ make a square of opposition.*

Note that we should assume that $\mathcal{C}(x) : \mathcal{R}(y)$ is not self-contradictory (or self-attacking) in order that the square of opposition really makes sense. In propositional logic, this would mean that $x \wedge y \neq \bot$.

This square departs from the one obtained by Salavastru [12] where vertices A and I as well as E and O are exchanged: In other words the entailments (b) are put in the wrong way. This may come from a misunderstanding of the remark made in [1] that the rejection $\mathcal{C}(x) : -\mathcal{R}(y)$ is itself a reason for not concluding x, which can be written $-\mathcal{C}(x) : \mathcal{R}(\mathcal{C}(x) : -\mathcal{R}(y))$. But this does not mean that $\mathcal{C}(x) : -\mathcal{R}(y)$ entails $-\mathcal{C}(x) : \mathcal{R}(y)$ since it may be the case, for instance, that $\mathcal{C}(-x) : \mathcal{R}(y)$. Salavastru made another similar mistake regarding the link between A and I. He assumed that I entails A. It can be seen on a simple example that this implication is false, and is rather in the other way round: "The fact that Paul is a French citizen (fr) is not a reason for not concluding that he is smart (st). This is clearly a statement of the form $-\mathcal{C}(sm) : -\mathcal{R}(fr)$. The question now is: does this statement entail the argument $\mathcal{C}(sm) : \mathcal{R}(fr)$ (i.e. the fact that Paul is French is a reason to conclude that he is smart)? The answer is certainly no. However, the converse is true. That is $\mathcal{C}(sm) : \mathcal{R}(fr)$ implies $-\mathcal{C}(sm) : -\mathcal{R}(fr)$.

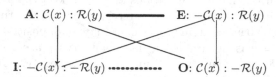

Fig. 2. An informal, argumentative square of opposition

Salavastru [12] also proposed a propositional logic reading of the informal square of opposition of Figure 2. This square is in terms of logical binary connectives and is pictured in Figure 3 (where $\uparrow$ here denotes Sheffer's incompatibility operator, which corresponds to the negation of the conjunction, i.e. $y \uparrow x = \neg y \vee \neg x$). In Salavastru's view "y is a reason for concluding x" is modeled by $y \rightarrow x$, which corresponds to a strong reading of the consequence relation. Then "y is a reason for not concluding x" is understood as the incompatibility of y and x, while "y is not a reason for not concluding x" is just their conjunction, i.e. two symmetrical connectives w.r.t. x and y, which may seem troublesome. Still from a formal point of view, this makes a perfect square of opposition. Indeed $y \wedge x$ entails $y \rightarrow x$ (and $y \not\rightarrow x$ entails $y \uparrow x$), but as already said, "y is not a reason for not concluding x" does not entail "y is a reason for concluding x" (and modeling "y is a reason for concluding x" by the symmetrical formula $y \wedge x$ would look strange). In fact, propositional logic is not enough expressive for providing a logical language for reasoning about arguments.

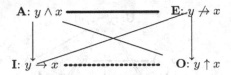

Fig. 3. Salavastru's logical square of opposition for argumentation

3   Towards a Logical Language for Argumentative Reasoning

Let x, y, z, x', y', ... denote any propositional logic formula. The basic building block of the proposed logical language is made of pairs of the form (x, y) to be read "y is a reason for x", or "x is supported by y". It will stand for $\mathcal{C}(x)$: $\mathcal{R}(y)$. In fact, this may be viewed as a formula of the logic of supporters [8], a counterpart of possibilistic logic [4], where the certainty level (usually belonging to an ordered chain) of a proposition x is replaced by its support (belonging to a Boolean lattice of propositions). The logic of supporters is a lattice-based generalization of possibilistic logic. See [8] for a detailed account of its semantics. In particular, if (x, y) and (x, y') hold, $(x, y \vee y')$ holds as well: if y and y' are reasons for concluding x, $y \vee y'$ is also a reason for concluding x.

As in standard possibilistic logic, the logic of supporters only allows for conjunctions of such pairs, and we have $(x \wedge x', y) = (x, y) \wedge (x', y)$. This means that if y is a reason for concluding x and for concluding x', then y is a reason for concluding $x \wedge x'$, and conversely. But, as immediately revealed by the kind of statements we have to handle, we need a *two layer* propositional-like language. Indeed, we need to express negations of such pairs, namely $\neg(x, y)$ to express $\mathcal{C}(x)$: $-\mathcal{R}(y)$. We also need disjunctions between such pairs as we are going to see, by completing the square of oppositions into an hexagon.

Indeed, it is always possible to complete a square of opposition into a hexagon by adding the vertices $Y =_{def} I \wedge O$, and $U =_{def} A \vee E$. This completion of a square of opposition was proposed and advocated by a philosopher and logician, Robert Blanché (see, e.g., [3]). It fully exhibits the logical relations inside a structure of oppositions generated by the three mutually exclusive situations A, E, and Y, where two vertices linked by a diagonal are contradictories, A and E entail U, while Y entails both I and O. Moreover $I = A \vee Y$ and $O = E \vee Y$. The interest of this hexagonal construct has been especially advocated by Béziau [3] in the recent years for solving delicate questions in paraconsistent logic modeling. Conversely, three mutually exclusive situations playing the roles of A, E, and Y always give birth to a hexagon [6], which is made of three squares of opposition: $AEOI$, $AYOU$, and $EYIU$.[1]

Definition 2 (Hexagon of opposition). *Six statements A, U, E, O, Y, I make a hexagon of opposition if and only if A, E, and Y are mutually exclusive two by two, and $AEOI$, $AYOU$, and $EYIU$ are squares of opposition.*

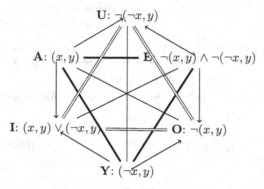

Fig. 4. Possible argumentative relations linking a reason y to a conclusion x

Figure 4 exhibits the six possible epistemic situations (apart complete ignorance) regarding argumentative statements relating y and x. Indeed it provides an organized view of the six argumentative statements, namely: A: "y is a reason for concluding x" represented by (x, y); Y: "y is a reason for concluding $\neg x$" represented by $(\neg x, y)$, O: "y is not a reason for concluding x" represented by $\neg(x, y)$; U: "y is not a reason for concluding $\neg x$" represented by $\neg(\neg x, y)$; completed by I: "y is conclusive about $x/\neg x$" represented by $(x, y) \vee (\neg x, y)$; and E: "y is not conclusive about $x/\neg x$" represented by $\neg(x, y) \wedge \neg(\neg x, y)$. Thus, matching the square of Fig. 2 with the square $AEOI$ in the hexagon of Figure 4 reveals that $-\mathcal{C}(x) : \mathcal{R}(y)$ is represented by $\neg(x, y) \wedge \neg(\neg x, y)$, i.e. "$y$ is a reason for not concluding about x" is also understood here as a reason "y is a reason for not concluding about $\neg x$".

---

[1] Note that, if we complete Salavastru's square of Figure 2 into a hexagon, we obtain $U = y$ and $Y = \neg y$, which corresponds to the simple affirmation and negation of y, where x is no longer involved, which is not very satisfactory.

This leads to consider a logic, we call LA, which is a propositional-like logic where *all* literals are replaced by pairs, e.g. $\neg(x, y)$, $\neg(x, y) \vee (x', y')$, $(x, y) \wedge (x', y')$ are wffs in LA (but not $\neg z \wedge (x, y)$). At the higher level, the pairs (x, y) are manipulated as literals in propositional logic, e.g., $(x, y) \wedge \neg(x, y)$ is a contradiction, and

$$(x, y), \neg(x, y) \vee (x', y') \vdash (x', y')$$

is a valid rule of inference. LA is a two-layer logic, just as the generalized possibilistic logic [5] is w.r.t. the standard possibilistic logic [4].

At the internal level, x, y are themselves propositional variables, and LA behaves as possibilistic logic: $(x, y) \wedge (x', y) \equiv (x \wedge x', y)$ as already said, and

$$(\neg x \vee x', y), (x \vee z, y') \vdash (x' \vee z, y \wedge y')$$

is a valid inference rule.

Moreover, we have: if $x \vdash x'$ then $(x, y) \vdash (x', y)$. So when $x \vdash x'$, if "y is a reason for x" then "y is a reason for x'". Writing $x \vdash x'$ as $(\neg x \vee x', \top)$, the above rule follows from the previous one: $(\neg x \vee x', \top), (x, y) \vdash (x', \top \wedge y)$ and $(x', \top \wedge y) \equiv (x', y)$. Thus $(x, y) \vdash (x \vee x', y)$. Besides, $\vdash ((x, y) \vee (x', y)) \rightarrow (x \vee x', y)$ (where $\rightarrow$ is the material implication). Note that the converse implication does not hold in general. Indeed y may be a reason for $x \vee x'$, without being a reason for x or being a reason for x'.

Note also that $(\neg x, y), (x, y') \vdash (\bot, y \wedge y')$ is a contradiction only if the reasons y and y' are *not* mutually exclusive. Generally speaking, one has to to distinguish, between

- $(\bot, y)$ (with $y \neq \bot$) which is a contradiction;
- $(x, \bot)$ (which can be obtained, e.g. from $(x \vee y, z)$ and $(\neg y, \neg z)$), which expresses that x has no support.

This should not be confused with the case where x would be equally supported by opposite reasons: (x, z) and $(x, \neg z)$ which entails that $(x, \top)$. Besides, $(\top, x)$ holds for any propositional formula $x \neq \bot$. Finally, it can be checked that:

Proposition 2. *The 6 LA formulas in Fig. 4 make a hexagon of opposition.*

Then, it can be seen in Fig. 4 that the argumentative form $\mathcal{C}(x) : -\mathcal{R}(y)$ is equal to the disjunction of the forms $\mathcal{C}(\neg x) : \mathcal{R}(y)$ and $-\mathcal{C}(x) : \mathcal{R}(y)$, which is satisfactory. Indeed $(\neg x, y)$ entails $\neg(x, y)$ (since $(x, y), (\neg x, y) \vdash (\bot, y)$).

We have only outlined how LA behaves. It is worth noticing that LA is expressive enough for allowing the use of negation in three places:

- $(x, \neg y)$ i.e., "$\neg y$ is a reason for x";
- $(\neg x, y)$, i.e., "y is a reason for $\neg x$", and
- $\neg(x, y)$ "y is not a reason for x", i.e., in particular it is possible that $\neg x$ holds while y holds.

Besides, we can also build another hexagon by considering two distinct argumentative relations linking a reason y to a conclusion x positively; see Fig. 5. Indeed,

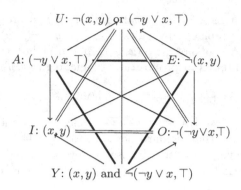

Fig. 5. Hexagon showing the interplay between a strong and a weak argumentative link between y and x

note the difference between (x, y) and $(\neg y \vee x, \top)$ which uncontroversially expresses that x entails y. Due to their structural differences they will not play the same role in LA: $(\neg y \vee x, \top)$ is stronger than (x, y), since $(\neg y \vee x, \top), (y, y) \vdash (x, y)$ (note that (y, y), i.e. "y is a reason for concluding y" always holds).

Before concluding this short paper, let us suggest what kinds of attacks may exist between arguments in this setting. Let us take an example (adapted from [1]). Let us consider an argument such has "Mary little worked ($= y$), she will fail her exam ($= x$)" will be represented by $(y, \top) \wedge (x, y)$. It may be attacked in different ways:

- by adding $(\neg y, \top)$ (No "Mary worked a lot");
- by adding $\neg(x, y)$ ("working little is not a reason for failing one's exam");
- by adding $(\neg x, y)$ (not very realistic here, although one might say "working little is a reason for not failing the exam (since one is not tired)");
- by adding $(\neg x, y \wedge z)$ ("Mary is gifted").

Note that the handling of this latter attack would require a nonmontonic inference mechanism, which may be encoded in a way taking lesson from the possibilistic logic approach [2], here based on the (partial) ordering defined by the propositional entailment: (x, y) should be drown if it exists a reason y', more specific than y, for concluding x' in a way opposite to x, i.e. we have both (x, y) and (x', y'), with x and x' inconsistent, $y' \vdash y$ (and $y \nvdash y'$).

4   Concluding Remarks

Starting from linguistics-based evidence about argumentative statements, we have outlined the description of a two-layer logic, LA, for handling arguments, in conformity with a rich structure of oppositions which has been laid bare in terms of squares and hexagons. LA parallels generalized possibilistic logic. A more formal study of LA is the next step, as well as a comparison with other formal approach to argumentation. More lessons have to be taken from generalized possibilistic logic for handling the strength of arguments in a weighted extension, or for taking advantage of its relation to logic programming [7].

References

1. Apothéloz, D.: Esquisse d'un catalogue des formes de la contre-argumentation. Travaux du Centre de Recherches Sémiologiques 57, 69–86 (1989)
2. Benferhat, S., Dubois, D., Prade, H.: Practical handling of exception-tainted rules and independence information in possibilistic logic. Applied Intelligence 9, 101–127 (1998)
3. Béziau, J.-Y.: The power of the hexagon. Logica Universalis (2012)
4. Dubois, D., Prade, H.: Possibilistic logic: a retrospective and prospective view. Fuzzy Sets and Systems 144, 3–23 (2004)
5. Dubois, D., Prade, H.: Generalized Possibilistic Logic. In: Benferhat, S., Grant, J. (eds.) SUM 2011. LNCS, vol. 6929, pp. 428–432. Springer, Heidelberg (2011)
6. Dubois, D., Prade, H.: From Blanchés hexagonal organization of concepts to formal concept analysis and possibility theory. Logica Universalis 6(1), 149–169 (2012)
7. Dubois, D., Prade, H., Schockaert, S.: Règles et métarègles en théorie des possibilités. De la logique possibiliste à la programmation par ensembles-réponses. Revue d'Intelligence Artificielle 26(1-2), 63–83 (2012)
8. Lafage, C., Lang, J., Sabbadin, R.: A logic of supporters. In: Information, Uncertainty and Fusion. In: Bouchon, B., Yager, R.R., Zadeh, L.A. (eds.) pp. 381–392. Kluwer (1999)
9. Moretti, A.: Argumentation theory and the geometry of opposition (abstract). In: 7th Conf. of the Inter. Soc. for the Study of Argumentation, ISSA 2010 (2010)
10. Parsons, T.: The traditional square of opposition. The Stanford Encyclopedia of Philosophy (Fall 2008 Edition), Zalta, E.N. (ed.) (2008), http://plato.stanford.edu/archives/fall2008/entries/square/
11. Quiroz, G., Apothéloz, D., Brandt, P.: How counter-argumentation works. In: van Eemeren, F.H., Grootendorst, R., Blair, J.A., Willard, C.A. (eds.) Argumentation Illuminated, pp. 172–177. International Centre for the Study of Argumentation (SICSAT), Amsterdam (1992)
12. Salavastru, C.: Logique, Argumentation, Interprétation. L'Harmattan, Paris (2007)

A Structured View on Sources of Uncertainty in Supervised Learning

Andreas Buschermöhle, Jens Hülsmann, and Werner Brockmann

University of Osnabrück, Smart Embedded Systems Group,
Albrechtstr. 28, 49069 Osnabrück, Germany
{Andreas.Buschermoehle,Jens.Huelsmann,Werner.Brockmann}@uos.de
http://www.informatik.uni-osnabrueck.de/techinf/

Abstract. In supervised learning different sources of uncertainty influence the resulting functional behavior of the learning system which increases the risk of misbehavior. But still a learning system is often the only way to handle complex systems and large data sets. Hence it is important to consider the sources of uncertainty and to tackle them as far as possible. In this paper we categorize the sources of uncertainty and give a brief overview of uncertainty handling in supervised learning.

Keywords: Supervised Learning, Uncertainty, Regression, Classification.

1 Introduction

Learning systems are more and more applied to reproduce properties of a set of given data. Two main tasks in this field are regression and classification. The type of the output is the key difference between these two: Whereas in regression tasks the output is continuous and ordered, the output is discrete and has not necessarily an ordinal relation in classification. In both categories often a supervised learning scenario with a set of training data consisting of tuples of input I_T and output O_T is given as a representation of a desired function f. The learning system is hence operated in two modes. First, at training the parameter vector α is determined. Here a set of training data can be incorporated altogether defining an approximation $\widetilde{f} \in \widetilde{F}$ from a set of possible approximations $\widetilde{F}$. Second, at evaluation time the resulting approximator $\widetilde{f}$ is used to assign an output O to a given input I through $O = \widetilde{f}(I, \alpha)$. This is done on the basis of a single input I every time a result is required. Such a learning system is hence subject to different sources of uncertainty, both within the training and evaluation phase which may result in a possibly wrong output at evaluation time and thus increasing the risk of an unsafe system behavior.

2 Sources of Uncertainty

A supervised learning system is subject to uncertainties at different levels, be it aleatoric uncertainties stemming from variability, e.g. from noise or unobserved

E. Hüllermeier et al. (Eds.): SUM 2012, LNAI 7520, pp. 566–573, 2012.

influences, or epistemic uncertainties if no information is available, e.g. due to sensor faults or missing knowledge. A more abstract view regarding uncertainties in clustering and classification is presented in [12]. In this paper we categorize the sources of uncertainty regarding their possible consequences for supervised learning. On the one hand the available data at training and evaluation are subject to uncertainties and on the other hand the learning algorithm and its parameter vector α introduce uncertainties which all accumulate at a given output O. From this perspective the sources can be structured into the following five categories:

Uncertainty of training input I_T: The training input I_T defines where in input space the approximation f is trained to give a certain output value O_T. As these data are often given by sensor readings or as output signals of prior processing modules, the training inputs contain different uncertainties. For sensors it is possible that e.g. sensor-noise, drift or a fault lead to a gradually or totally wrong input I_T. Or, prior processing modules, like simple filter algorithms or even other learning systems in a cascaded structure, that again depend on uncertain information which propagates to their outputs generate the next uncertain input I_T. Thus it gets ambiguous, **where** the approximation should be trained within the input space to a certain value O_T.

Uncertainty of training output O_T: The training output O_T defines an output value, the approximation should give at a certain point I_T in the input space. In regression tasks, these training outputs O_T are subject to similar uncertainties as the training input data I_T. Additionally, in classification often the reference to give correct class labels can be uncertain as well, especially in complex classification tasks. This is related to the confidence in the correctness of a given label. Again, this results in an uncertainty, but about **what** the approximation should be trained to for a certain input.

Uncertainty of approximation $\widetilde{f}$: The approximation $\widetilde{f}$ is done by some structure out of $\widetilde{F}$ which is only capable of representing a limited set of functions, e.g. polynomials of a certain degree or a limited number of Gaussians. Either this structure is not expressive enough, or the structure is too expressive causing the well known over-fitting problem. Additionally the learning algorithm, which chooses $\widetilde{f} \in \widetilde{F}$, and its optimization criterion can be uncertain. The settings of this learning algorithm, e.g. the learning rate or a regularization term, can be chosen suboptimal and thus lead to systematic remaining errors. Consequently, it is uncertain, if a suitable **approximation** $\widetilde{f}$ of the desired function f can be found with the chosen setup.

Uncertainty of parameter vector α: Corresponding to the approximation structure, the parameter vector α determined by training is influenced in many ways. For some algorithms it is necessary to have a random initialization which has impact on the result. And as many training algorithms present the training data in some order and not as a complete batch, this order might influence the resulting parameter vector α as well. Additionally, the training input data I_T need not cover the complete input space, so that some parameters might not be set up properly as not enough information is present. Similarly, the training data may contain conflicts, e.g. overlapping class areas or different output values O_T at close input points. Furthermore, often the learning algorithm is confronted with local minima and its convergence cannot be guaranteed. Hence, the finally determined **parameter vector** α of the approximation $\tilde{f}$ is uncertain.

Uncertainties of evaluation input I: In contrast to training, at evaluation time, the input I defines where in input space the approximation $\tilde{f}$ should be evaluated. These input data again have the same sources of uncertainty as the training input I_T. But now for a fixed parameter vector α, there is uncertainty, **where** the approximation $\tilde{f}$ should be evaluated.

3   Resulting Problems

All sources of uncertainty mentioned in the previous chapter influence the uncertainty of the output value at evaluation time and hence the risk of choosing a wrong output. This risk can be divided in two categories. Either the uncertainties result in (some degree of) ignorance so that it is not known which output O should be given as not enough training data supplied information for the respective point I in input space. Or the uncertainties result in (some degree of) conflict so that multiple outputs O are possible as the available data are contradictory. Both categories are gradual in nature as ignorance increases with distance of evaluation point I to training points I_T in input space and conflict increases with the amount of plausible outputs in output space. This notion of conflict and ignorance was introduced in [11] for classification problems, but can also be easily applied for other supervised learning tasks.

If the output O is uncertain and its uncertainty cannot be compensated within the learning system, an information about its degree of uncertainty is valuable to increase safety by successive processing modules. This way it is possible to explicitly react to uncertainties and maintain a safe operation of the whole system. Hence it is of interest not only to minimize the uncertainty of the approximator, but also to express remaining degrees of uncertainty of the output O. So the important questions are:

1. How do the sources of uncertainty influence the uncertainty of an output O?
2. How can the influence of any uncertainty be measured or modeled?
3. How is it possible to deal with uncertainties within the approximator?
4. How can the uncertainty of the output be measured or modeled?
5. How can information about output uncertainty be used in further processing?

As it is at least necessary to represent the uncertainty of internal parameters to estimate the uncertainty of the output, this information can as well be useful to rate the model. Together with the information about conflict and ignorance hereby it is possible to determine where new data is required and where further refinement of the model is needed. This is closely related to the problem of active learning [24].

4   Explicit Uncertainty Handling in Learning Systems[1]

For **regression**, uncertain input values have been dealt with in feed forward neural networks in a statistical way [25,26]. Missing but also uncertain dimensions of inputs I_T and I were considered at learning as well as evaluation time by integrating the probability weighted output across the uncertain dimension. This way an input with high uncertainty can be blended out, if its influence on the output is low. For the assumption of Gaussian input distribution, an analytical solution of the integral is possible [8] making the approach more feasible. This leads to the field of Gaussian processes, where the uncertainty is treated by assuming the output O to be normally distributed [23] and reflecting its uncertainty by its variance. Recent work on Gaussian processes included a separate treatment of input- and output-uncertainties, thus additionally representing the variance of the training input data I_T [4].

Two uncertainty measures were introduced for radial basis function networks [16]. They reflect the local density of training data I_T and the variability of the output values O_T, depending on the used network, thus representing uncertainty of the approximation $\widetilde{f}$ and the parameter vector α. In the approach of [17], the uncertainty of $\widetilde{f}$ is addressed indirectly by the introduction of a risk averting error criterion as a gradual blending between the minimax and the least-squares criteria.

Another approach of explicitly representing uncertainties in regression uses evidence theory [5]. Here the uncertainty of a neural network output O is expressed by lower and upper expectations. The width of this interval reflects the uncertainty resulting from the relative scarcity of training data, i.e. one kind of approximation uncertainty. A successive work additionally incorporates explicit uncertainties of the training output O_T [20]. Another approach to on-line uncertainty estimation is applied for Takagi-Sugeno fuzzy systems in [19]. It estimates the local training input density and the global training output variance, i.e. the uncertainty of $\widetilde{f}$ and α, to add error bars to the output O.

In previous work [3], we addressed the influence of all uncertainty categories on the local uncertainty of the output O in an incremental learning setup and investigated different measures to reflect those uncertainties in a uniform way. These uncertainty measures are used by a new extended on-line learning method, called *trusted learner*, to enhance the incremental learning process [2]. Thus the approximation is adjusted more quickly when it is uncertain and the on-line learning is not influenced by initial parameters.

For **classification**, the uncertainty of an output O, i.e. the class label, can be estimated by all major classification algorithms [15,18], e.g. by discriminant measures. For neural networks and also for rule based classifiers, a class distribution is the result of the actual approximator and a downstream evaluation algorithm determines the final class label afterwards, e.g. simply taking the class

---

[1] As this is a short paper only some representative methods are discussed here. The general techniques for dealing with uncertainties are neither described in this overview, but a structured overview can be found in [14].

with the highest probability or plausibility value. All this discriminant measures implicitly address the uncertainty in I_T and α but do not distinguish between them in detail. For distance based algorithms like k-nearest neighbor, the information about the uncertainty is inherent in the distances to the involved points. The same holds for the distances to the class borders, e.g. a hyperplane, in all algorithms that fit these borders to the training data (support vector machines, all linear methods, etc.). Nevertheless this uncertainty measure is not as expressive as a complete class distribution. This distribution can be used to combine different or differently trained classifiers. Their combination with respect to their class distribution leads to a more robust classification [9].

Statistical information about uncertainty of the training input data I_T, like the probability density function, are used to adjust the rules of a rule-based classifier in [22]. Certain training data are weighted higher in the rule generation process and during the rule optimization. The uncertainty in the classification structure $\tilde{f}$, i.e. the rule base, of a fuzzy classifier due to conflict and ignorance is explicitly assessed by [10] to build more reliable classifiers. The classification output O is augmented with two gradual measures for conflict and ignorance, which are derived from the firing strength of the fuzzy classification rules. In the context of active learning, the method of uncertainty sampling requests data in parts of the input space with a high degree of uncertainty [21] to deal with aspects of uncertainty in α. The degree of uncertainty is computed here from the entropy of the class membership probabilities.

In the framework of evidence theory, the k-nearest neighbor-algorithm is formulated for an uncertain output O in [6] similar to the regression variant of [5]. For support vector machines, the effects of input data noise are studied in [27]. A measure for this uncertainty is computed via probability modeling. With a reformulation of the support vector machine this information is used to determine the optimal classifier. For noisy input data in neuronal networks the uncertainty (modeled as a Gaussian distribution) can be used as an additional parameter of the perceptrons activation function to improve performance [7]. These approaches deal with training input uncertainty I_T, but only due to noise.

In previous work [1] we extended the support vector machine-algorithm to cope with arbitrary input uncertainty for input I during the evaluation phase by fading out uncertain dimensions gradually. A more advanced approach to deal with epistemic uncertainties in fuzzy classifiers during the evaluation is proposed in [13]. Using the concept of conflict and ignorance and gradual dimension reduction the classifier provides a classification result, as long as its structure can obtain it without conflict. We show there that this approach improves the classification performance for a more general class of uncertainties, namely a gradual ignorance about the input I.

5   Conclusion

In summary, several approaches can be found dealing with different sources of uncertainty. Especially input uncertainties, at training as well as evaluation time, have been subject to research and the uncertainty of the resulting output is represented in many approaches. But no approach deals with all sources of uncertainty at once in one combined framework. Accordingly, it is currently not possible to regard all sources of uncertainty influencing a learning system adequately. Yet, with a combined framework, the uncertainty could be minimized further and an explicit representation of the output and model uncertainty can be determined. This way the user of a learning system (be it another processing module or a human) is able to incorporate additional knowledge in the decision process to make the total system behavior more robust and safe. So more research needs to be done to address all sources of uncertainty in supervised learning as a whole.

Moreover, current approaches are mostly based on statistical analysis of data. This statistical approach often is difficult to follow in practice. With an increasing amount of data the computational complexity grows and the need for on-line methods arises to deal with this complexity. On the other hand, if only little data are available the statistical relevance is low. If, in addition, the process that is modeled by the supervised learning system is time-variant it is hard to decide which data can be used for statistical analysis. Furthermore, tasks such as approximation based adaptive control result in training data that are drawn according to a continuously changing state and thus do not hold the (often made) assumption of independently and identically distributed training data. Thus for several domains with growing interest, e.g. with big or expansive data, non independently drawn data, unbalanced datasets, and dynamic environments, it is essential to develop methods which estimate uncertainty of on-line learning systems themselves on-line along the learning process. We are convinced that the field of on-line learning under uncertainties will be a key concept in upcoming applications.

References

1. Buschermöhle, A., Rosemann, N., Brockmann, W.: Stable Classification in Environments with Varying Degrees of Uncertainty. In: Proc. Int. Conf. on Computational Intelligence for Modelling, Control & Automation (CIMCA), pp. 441–446. IEEE Press (2008)
2. Buschermöhle, A., Schoenke, J., Brockmann, W.: Trusted Learner: An Improved Algorithm for Trusted Incremental Function Approximation. In: Proc. Symp. on Computational Intelligence in Dynamic and Uncertain Environments (CIDUE), pp. 16–24. IEEE Press (2011)
3. Buschermöhle, A., Schoenke, J., Brockmann, W.: Uncertainty and Trust Estimation in Incrementally Learning Function Approximation. In: Greco, S., Bouchon-Meunier, B., Coletti, G., Fedrizzi, M., Matarazzo, B., Yager, R.R. (eds.) IPMU 2012, Part I. CCIS, vol. 297, pp. 32–41. Springer, Heidelberg (2012)
4. Dallaire, P., Besse, C., Chaib-Draa, B.: Learning Gaussian Process Models from Uncertain Data. In: Proc. Int. Conf. on Neural Information Processing (NIPS), pp. 433–440. Springer (2009)
5. Denoeux, T.: Function Approximation in the Framework of Evidence Theory: A Connectionist Approach. In: Proc. Int. Conf. on Neural Networks (ICNN), pp. 199–203. IEEE Press (1997)

6. Denoeux, T.: A k-nearest neighbor classification rule based on Dempster-Shafer theory. IEEE Trans. on Systems, Man and Cybernetics (SMC) 25(5), 804–813 (1995)
7. Ge, J., Xia, Y., Nadungodage, C.: UNN: A Neural Network for Uncertain Data Classification. In: Zaki, M.J., Yu, J.X., Ravindran, B., Pudi, V. (eds.) PAKDD 2010, Part I. LNCS, vol. 6118, pp. 449–460. Springer, Heidelberg (2010)
8. Girard, A., Murray-Smith, R.: Learning a Gaussian Process Model with Uncertain Inputs. Technical Report TR-2003-144, University of Glasgow (2003)
9. Gonçalves, L.M.S., Fonte, C.C., Caetano, M.: Using Uncertainty Information to Combine Soft Classifications. In: Hüllermeier, E., Kruse, R., Hoffmann, F. (eds.) IPMU 2010. LNCS, vol. 6178, pp. 455–463. Springer, Heidelberg (2010)
10. Hühn, J., Hüllermeier, E.: FR3: A Fuzzy Rule Learner for Inducing Reliable Classifiers. Trans. on Fuzzy Systems 17(1), 138–149 (2009)
11. Hüllermeier, E., Brinker, K.: Learning Valued Preference Structures for Solving Classification Problems. Fuzzy Sets and Systems 159(18), 2337–2352 (2008)
12. Hüllermeier, E.: Uncertainty in Clustering and Classification. In: Deshpande, A., Hunter, A. (eds.) SUM 2010. LNCS, vol. 6379, pp. 16–19. Springer, Heidelberg (2010)
13. Hülsmann, J., Buschermöhle, A., Brockmann, W.: Incorporating Dynamic Uncertainties into a Fuzzy Classifier. In: Proc. Conf. of the European Society for Fuzzy Logic and Technology (EUSFLAT), pp. 388–395. Atlantis Press (2011)
14. Klir, G.J.: Uncertainty and Information: Foundations of Generalized Information Theory. Wiley (2006)
15. Kuncheva, L.I.: Combining Pattern Classifiers: Methods and Algorithms, pp. 111–113. Wiley (2004)
16. Leonard, J.A., Kramer, M.A., Ungar, L.H.: Using Radial Basis Functions to Approximate a Function and its Error Bounds. IEEE Trans. on Neural Networks 3(4), 624–627 (1992)
17. Lo, J.T., Bassu, D.: Robust approximation of uncertain functions where adaptation is impossible. In: Proc. Int. Joint Conf. on Neural Networks (IJCNN), pp. 1956–1961. IEEE Press (2002)
18. Lu, D., Weng, Q.: A Survey of Image Classification Methods and Techniques for Improving Classification Performance. Int. J. of Remote Sensing 28(5), 823–870 (2007)
19. Lughofer, E., Guardiola, C.: Applying Evolving Fuzzy Models with Adaptive Local Error Bars to On-Line Fault Detection. In: Proc. of Genetic and Evolving Fuzzy Systems (GEFS), pp. 35–40. IEEE Press (2008)
20. Petit-Renaud, S., Denoeux, T.: Nonparametric Regression Analysis of Uncertain and Imprecise Data Using Belief Functions. Int. J. of Approximate Reasoning 35, 1–28 (2004)
21. Qi, G.J., Hua, X.S., Rui, Y., Tang, J., Zhang, H.J.: Two-dimensional Active Learning for Image Classification. In: Proc. Conf. on Computer Vision and Pattern Recognition (CVPR), pp. 1–8. IEEE Press (2008)
22. Qin, B., Xia, Y., Prabhakar, S., Tu, Y.: A Rule-Based Classification Algorithm for Uncertain Data. In: Proc. Int. Conf. on Data Engineerin (ICDE), pp. 1633–1640. IEEE Press (2009)
23. Rasmussen, C.E., Williams, C.K.I.: Gaussian Processes for Machine Learning. MIT Press (2006)

24. Settles, B.: Active Learning Literature Survey. Computer Sciences Technical Report 1648. University of Wisconsin (2010)
25. Tresp, V., Ahmad, S., Neuneier, R.: Training Neural Networks with Deficient Data. In: Advances in Neural Information Processing Systems (NIPS), pp. 128–135. MIT Press (1993)
26. Tresp, V., Neuneier, R., Ahmad, S.: Efficient Methods for Dealing With Missing Data in Supervised Learning. In: Advances in Neural Information Processing Systems (NIPS), pp. 689–696. MIT Press (1995)
27. Zhang, J.B.T.: Support vector classification with input data uncertainty. In: Advances in Neural Information Processing Systems, vol. 17, pp. 161–169. MIT Press (2004)

On Development of a New Seismic Base Isolation System

Sanjukta Chakraborty, Koushik Roy, Chetan Chinta Arun, and Samit Ray Chaudhuri

Department of Civil Engineering, Indian Institute of Technology Kanpur, Kanpur,
UP-208016, India
{sanjukta,koushik,samitrc}@iitk.ac.in

Abstract. Various base isolation schemes have been implemented to isolate a structure from intense base excitations. In this paper, performance of a friction sliding bearing with nonlinear restoring mechanism is studied on a three-storey steel moment-resisting frame building under varying seismic hazard conditions. The performance of the proposed system is compared with that of the fixed-base and with only the friction sliding bearing. For this purpose, the effectiveness of the proposed isolation system is evaluated in terms of the residual displacement, peak inter-storey drift and the maximum base displacement. It is envisioned that a system as proposed here, if optimized for a target hazard situation, will result in a cost-effective solution.

Keywords: Non-linear spring, Friction sliding bearing.

1   Introduction

Seismic base isolation is becoming a cost effective way to mitigate the seismic vulnerability of various structures and bridges. Over the years, many types of base isolation devices have been proposed. Although these devices have pros and cons, the selection of these devices is mainly decided based on their expected performance under earthquake motions of various hazard levels and more importantly, their cost of installations. Sliding base isolation system has widely been used as a cost effective choice to reduce seismic vulnerability of the structures and bridges. While this type of isolation system is insensitive to dominant frequency of ground motion, it does not possess a restoring mechanism. As a result, a structure isolated with this type of device requires a large base plate to accommodate excessive base displacement in addition to end barriers to prevent the structure from the falling of the plate. Further, high frequency shock waves are generated when the isolator hits the barrier during strong earthquakes resulting in damage to non structural component s and systems. In this study, the concept of nonlinear restoring mechanism is employed to improve the performance of a conventional sliding isolation system. The nonlinear restoring mechanism has been achieved by designing nonlinear springs for which the stiffness increases with an increase in displacement. An extensive parametric study involving time history analysis of structure and subjected to a suite of ground motions with different hazard levels is conducted to evaluate the effectiveness of the proposed isolation system. It is observed from the results of the parametric study that the proposed isolation system

E. Hüllermeier et al. (Eds.): SUM 2012, LNAI 7520, pp. 574–581, 2012.

(Sliding bearing with nonlinear spring) reduces the earthquake response of the structure in terms of residual and peak base displacement demands while keeping the peak inters story drift within the safe limit. It is envisioned that the proposed isolation system when tested experimentally, can be used as a better choice over the conventional sliding bearing.

2   Spring Model

In this section a brief description of the spring model as well as the mathematical formulation is presented. The desired behavior of the spring as described earlier can be achieved by a conical spring with uniform pitch. The conical spring with increasing diameter towards the bottom provides a varying flexibility each loop. Therefore the bottom loop grounds first as the force on the conical spring increases followed by other loops with reducing diameter and also decrease in the active number of loops. Thus the stiffness of the spring increases gradually along with the increase in displacement of the coil. Working characteristics can be divided into two regions- working region with linear characteristics where no coil is grounded and working region with progressive characteristics after the contact of the first active coil.

The variation of the loop diameter of the conical spring along with the length can be considered to be linear or some other types of variation also can be assumed. Here two types of variations are studied. One spring is considered with a linear reduction in the diameter along the length and other one having logarithmic spiral type of variation in diameter along length.

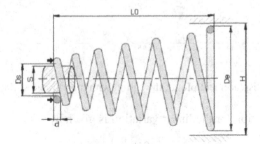

Fig. 1. A typical spring conical spring model showing different component

$$U = \int_0^l \frac{1}{2} T d\emptyset = \int_0^l \frac{T^2}{2GJ} dx \tag{1}$$

From Castigliano's Theorem

$$\delta = \frac{\partial U}{\partial F} = \int_0^l \frac{T\frac{\partial T}{\partial F}}{GJ} dx = \frac{F}{GJ} \int_0^{2\pi N} r^3 d\theta \tag{2}$$

The equation of logarithmic spiral is given by from equation

$$r = \frac{D_s}{2} e^{\ln \left(\frac{D_e}{D_s}\right)\frac{\theta}{2\pi N}} \tag{3}$$

For deflection in between two different angle θ_1 and θ_2 the deflection can be estimated from equation 2 as below.

$$\delta = \frac{F}{GJ} \int_{\theta_1}^{\theta_2} \left[\frac{D_s}{2} e^{\ln \left(\frac{D_e}{D_s}\right)\frac{\theta}{2\pi N}} \right]^3 d\theta \tag{4}$$

$$= \frac{8FD_s^3 N}{Gd^4} \frac{1}{3 \ln \left(\frac{D_e}{D_s}\right)} \left[\left(\frac{D_e}{D_s}\right)^{\frac{3\theta_2}{2\pi N}} - \left(\frac{D_e}{D_s}\right)^{\frac{3\theta_1}{2\pi N}} \right] \tag{5}$$

$$= \frac{8FD_s^3 N}{Gd^4} \frac{1}{3 \ln \left(\frac{D_e}{D_s}\right)} \left[\left(\frac{D_e}{D_s}\right)^{\frac{3n_2}{N}} - \left(\frac{D_e}{D_s}\right)^{\frac{3n_1}{N}} \right] \tag{6}$$

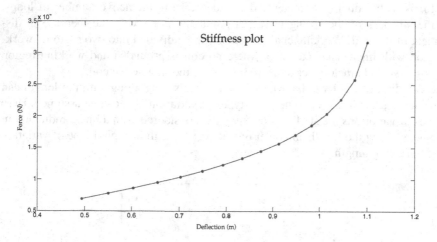

Fig. 2. The force displacement plot of the spring with logarithmic variation of diameter

Similarly the equation for the linear variation is given by

$$r = \left\{ \frac{D_f - D_s}{4\pi N} \theta + \frac{D_s}{2} \right\} \tag{7}$$

$$\delta = \frac{8FD_s^3 N}{Gd^4} \int_{\theta_1}^{\theta_2} \frac{1}{2\pi N} \left\{ 1 + \frac{\left(\frac{D_f}{D_s} - 1\right)}{2\pi N} \theta \right\}^3 d\theta \tag{8}$$

where d=diameter of the wire the spring made of steel, G= Shear modulus, N=Total number of loop, n_1 and n_2 = the loop number the deflection to be calculated, D_s= Diameter of the tapered section, D_e= Diameter of the larger side, and F=Force to be applied.

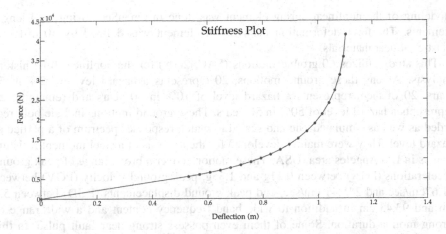

Fig. 3. The force displacement plot of the spring with linear variation of diameter

The two types of springs exhibited similar types of stiffness behavior with a very flat range at the beginning and a very high stiffness at larger deformation. Both the plot of the springs as shown in Figure 2 and Figure 3 considered same parameters as stated above for the spring design. However the spring with linear diameter variation is considered further because of its feasibility and higher stiffness at larger deformation as compared to the spring with logarithmic variation.

3   Numerical Model

Steel moment resisting frame building with four storeys's and originally with fixed base is considered. The buildings consist of American standard steel sections with uniform mass distribution over their height and a non uniform distribution of lateral stiffness. The beams and columns were assigned with various W sections and the materials used were uniaxial material, `Steel01' with (kinematic) hardening ratio of 3%. All these elements were modeled as Beam with hinges with length of plastic hinges taken as 10% of the member length and each of the nodes were lumped with a mass of 16000 kg. The damping was given as 2% Rayleigh damping for the first two modes. The base isolator was modeled using Flat slider bearing element. The nonlinear force deformation behavior is considered by using an elastic bilinear material with the properties as per the spring design as mentioned earlier. The numerical modeling of the FSB isolation was done in OpenSees using flat slider bearing element, defined by two nodes. A zero length bearing element which is defined by two nodes is used to model the same. These two nodes represent the flat sliding surface and the slider. The bearing has unidirectional (2D) friction properties for the shear deformations, and force-deformation behaviors defined by uniaxial materials in the remaining two (2D) directions. Coulomb's friction model is used with a friction coefficient equal to 0.05 yield displacement as 0.002. To capture the uplift behavior of the bearings, user specified no-tension behavior uniaxial material in the axial direction is used. P-Delta moments are entirely transferred to the flat sliding surface. The numerical modeling of the nonlinear spring element was done in OpenSees using zero length elements. The force-deformation behavior of element was defined by 40 different Elastic bilinear materials.

This study utilizes 60 ground motions (SAC, 2008) for the nonlinear time history analysis. Among these ground motions, 20 represents a hazard level of 2% in 50 years, 20 of them represents a hazard level of 10% in 50 years and remaining 20 represents a hazard level of 50% in 50 years. These ground motions include both recorded as well as simulated one and scaled to match response spectrum of a particular hazard level. They were mainly developed for the analysis of a steel moment resisting frames in Los Angeles area, USA. These motions cover a broad range of peak ground accelerations (PGA) between 0.11g and 1.33g, peak ground velocity (PGV) between 21.67cm/sec and 245.41 cm/sec, and peak ground displacements (PGD) between 5.4 cm and 93.43 cm in addition to wide band frequency content and a wide range of strong motion duration. Some of them even possess strong near fault pulse. In this study the steel moment resisting frame building considered was designed in such a way that the fundamental period of the building is similar to that of existing buildings in California. Therefore, the ground motions are reasonably considered here for the structure to be analyzed.

4   Result and Discussion

Nonlinear time history analysis of the structure with flat sliding bearing plus nonlinear spring is carried out for a particular value of coefficient of friction (μ=0.05). The results obtained from the analysis are compared with the nonlinear analysis results of the structure with fixed base and sliding base with no restoring mechanism. The parameters used for the comparisons are -peak inter-storey drift ratio of the structure, residual displacement of the structure and the maximum base displacement. Statistical comparison is performed to observe the pattern of the above parameters. For this statistical comparison mean, maximum and the standard deviation values are calculated and are summarized in a tabular form. The results are further subcategorized in three parts depending on the hazard level. These parameters are the measures of damage that a structure undergoes after an earthquake. FEMA-356 has specified three performance levels, these beings the immediate occupancy (IO), the life safety (LS) and the collapse prevention (CP). These are associated with the inter-storey drift limits of 0.7%, 2.5% and 5%, respectively. These limits are said to be appropriate for the performance evaluation of pre-Northridge steel moment frames. FEMA-356 prescribes a basic safety objective (BSO), which comprises a dual-level performance objective. It requires LS performance for a 10% in 50-year event and CP performance level for a 2% in 50-year earthquake. According to these guidelines, the drifts should be such that for LA01- LA20 ground motions, the structures should be safe in Life Safety; LA21-LA40, the buildings should not be crossing the performance level of Collapse Prevention and Immediate Occupancy in case of LA41-LA60. LA01-LA20, LA21-LA40 and LA41-LA60 are corresponding to ground motion representing hazard level of 2%, 10% and 50% respectively.

Peak Inter-storey Drift Ratio- It is observed (Table) that the inter storey drift ratio is exceeding the allowable range as prescribed by FEMA-356 for the fixed base building. At a particular coefficient of friction, the peak inter-storey drift ratio of the structure having flat sliding bearing with nonlinear spring are almost similar compared to the flat sliding bearing without nonlinear spring and this range is well inside the limit for a specific hazard level. . A rather decreasing pattern is observed for the case of sliding isolator with spring.

Table 1. Peak interstorey drift ratio at different hazard level for different base condition

Coefficient of friction(μ)=0.05	Comparison of Peak inter-storey drift ratio in percentage					
	10% in 50 years (La01-La20)		2% in 50 years (La21-La40)		50% in 50 years (La41-La60)	
	Mean peak inter-storey drift (%)	Mean+SD of peak inter-storey drift (%)	Mean peak inter-storey drift (%)	Mean+SD of peak inter-storey drift(%)	Mean peak inter-storey drift (%)	Mean+SD of peak inter-storey drift (%)
FB	3.22	4.62	6.97	9.68	1.47	2.15
SI_no spring	0.411	0.493	0.478	0.571	0.349	0.411
SI_with spring	0.323	0.373	0.399	0.468	0.291	0.333

Residual displacement- This is an important parameter for this study to unfold the importance of the necessity of nonlinear spring in a sliding bearing. The residual displacement acts as an important measure of post-earthquake functionality in determining whether a structure is safe and usable to the occupants. The large residual displacement alters the new rest position of the structure which results in high cost of repair or replacement of non-structural elements. Restricting the large residual displacements also helps in avoiding the pounding effect. From the results discussed above, it has been observed that the effect of sliding bearing with and without nonlinear spring does not differ much. However there is a large reduction in the residual displacement in all the hazards levels considered for the study. The range of reduction in residual displacement when compared with sliding bearing without spring are 74% to 76% for hazard level of 2% in 50 years, 40% to 60% for hazard level of 10% in 50 years and 54% to 60 % for hazard level of 50% in 50 years. Therefore the non-linear spring is found to be very effective for this factor. Thus the large reduction in the residual displacement of the structure due to the incorporation of nonlinear springs in the sliding bearing systems depicts its importance.

Base Displacement- The comparison of the base displacement at different hazard levels shows reduction in displacement for the non-linear spring as compared with the case without nonlinear spring with sliding bearing. The trend is same as that obtained for residual displacement. However the % of reduction is much less here for all the three hazard level.

Table 2. Peak maximum residual displacement at different hazard level for different base condition

Coefficient of friction(μ)=0.05	Comparison of maximum residual displacement					
	10% in 50 years (La01-La20)		2% in 50 years (La21-La40)		50% in 50 years (La41-La60)	
	Mean of peak base displacement	Mean+SD of peak base displacement	Mean of peak base displacement	Mean+SD of peak base displacement	Mean peak inter-storey drift (%)	Mean+SD of peak base displacement
SI_no spring	0.149	0.269	0.478	0.887	0.082	0.205
SI_with spring	0.054	0.106	0.123	0.215	0.041	0.093

Table 3. Peak maximum base displacement at different hazard level for different base condition

Coefficient of friction (μ)=0.05	Comparison of maximum base displacement					
	10% in 50 years (La01-La20)		2% in 50 years (La21-La40)		50% in 50 years (La41-La60)	
	Mean of peak base displacement	Means + SD of peak base displacement	Mean of peak base displacement	Mean + SD of peak base displacement	Mean peak inter-storey drift (%)	Mean + SD of peak base displacement
SI_no spring	0.327	0.459	0.774	1.158	0.136	0.249
SI_with spring	0.194	0.329	0.725	1.121	0.118	0.200

Table 4. %reduction in the response by the provision of non-linear spring

% decrement in the residual displacement and maximum base displacement considering non-linear spring along with sliding bearing with respect to the base with only sliding bearing			
	10% in 50 years (La01-La20)	2% in 50 years (La21-La40)	50% in 50 years (La41-La60)
Residual displacement	62%	75%	52%
Base displacement	35%	5%	16%

5   Conclusion

In this paper, the seismic performance of a steel moment-resisting frame structure resting on sliding type of bearing with restoring force device as a conical non-linear spring is studied. The results showed that the structure with fixed base is subjected to huge peak inter-storey drift outside the allowable range that needed to be checked. The provision of sliding isolator is effective in reducing inter storey drift. However it results into a large amount of residual and base displacement. The provision of a properly designed non-linear spring with a very small stiffness at the beginning is found to be very effective in reducing residual displacement to a great extent. The reduction in base displacement is also obtained at a lesser extent. The peak storey drift also reduced to a very small extent by the provision of this kind of spring. The proposed isolation system, if optimized for various target performance levels, will result in a cost-effective solution.

References

1. Bhaskar Rao, P., Jangid, R.S.: Experimental Study of Base Isolated Structures. Journal of Earthquake Technology 38(1), 1–15 (2001)
2. Constantinou, M.C., Mokha, A.S., Reinhorn, A.M.: Study of Sliding Bearing and Helicalsteel-Spring Isolation System. Journal of Structural Engineering 117(4), 1257–1275 (1991)
3. Dolce, M., Cardone, D., Croatto, F.: Frictional Behavior of Steel- PTFE Interfaces for Seismic Isolation. Bulletin of Earthquake Engineering 3, 75–99 (2005)
4. Dolce, M., Cardone, D., Palermo, G.: Seismic Isolation of Bridges using Isolation Systems Based on Flat Sliding Bearings. Bulletin of Earthquake Engineering 5, 491–509 (2007)
5. FEMA-356. Prestandard and Commentary for the Seismic Rehabilitation of Buildings. American Society of Civil Engineers (2000)
6. Feng, M.Q., Shinozuka, M., Fujii, S.: Fricrion-Controllable Sliding Isolation System. Journal of Engineering Mechanics 119(9), 1845–1864 (1993)
7. Glenn, J.M., Michael, D.S., Wongprasert, N.: Experimental Verification of Seismic Response of Building Frame with Adaptive Sliding Base-Isolation System. Journal of Structural Engineering 128(8), 1037–1045 (2002)
8. Glenn, J.M., Wongprasert, N., Michael, D.S.: Analytical and Numerical Study of a Smart Sliding Base Isolation System for Seismic Protection of Buildings. Computer-Aided Civil and Infrastructure Engineering 18, 19–30
9. Ray Chaudhuri, S., Villaverde, R.: Effect of Building Nonlinearity on Seismic Response of Nonstructural Components: A Parametric Study. Journal of Structural Engineering 134(4), 661–670 (2008)
10. SAC. Phage lambda: description & restriction map (2008)
11. Juvinall, R.C.: Fundamentals of Machine Component Design, pp. 351–384. John Wiley & Sons, Inc., Delhi (1983)
12. JSSI. Introduction of Base Isolated Structures. Japan Society of Seismic Isolation, Ohmsa, Tokyo (1995)
13. Krishnamoorthy, A.: Response of Sliding Structure with Restoring Force Device to Earthquake Support Motion. Journal of Social Science Education 10(1), 25–39 (2008)
14. Su, L., Goodarz, A., Iradj, G.: Comparative Study of Base Isolation Systems. Journal of Engineering Mechanics 115(9), 1976–1992 (1989)

Trying to Understand How Analogical Classifiers Work

William Fernando Correa[1], Henri Prade[1], and Gilles Richard[1,2]

[1] IRIT, UPS-CNRS, 118 route de Narbonne, 31062 Toulouse Cedex-France
[2] British Institute of Technology and E-commerce - 252 Avicenna House, London UK
{william.corea,prade}@irit.fr, grichard@bite.ac.uk

Abstract. Based on a formal modeling of analogical proportions, a new type of classifier has started to be investigated in the last past years. With such classifiers, there is no standard statistical counting or distance evaluation. Despite their differences with classical approaches, such as naive Bayesian, k-NN, or even SVM classifiers, the analogy-based classifiers appear to be quite successful. Even if this success may not come as a complete surprise, since one may imagine that a general regularity or conformity principle is still at work (as in the other classifiers), no formal explanation had been provided until now. In this research note, we lay bare the way analogy-based classifiers implement this core principle, highlighting the fact that they mainly relate changes in feature values to changes in classes.

Keywords: analogical proportion, classification, regularity.

1 Introduction

An analogical proportion is a statement of the form "a is to b as c is to d" (denoted $a : b :: c : d$) and may be thought as a symbolic counterpart to a numerical proportion of the form $\frac{a}{b} = \frac{c}{d}$. Different proposals have been made for modeling an analogical proportion in a lattice setting, or in terms of a measure of analogical dissimilarity, or in a logical manner [8,3,4]. Based on the second viewpoint and despite a complexity which is cubic in the size of the training set, analogical proportion-based binary classifiers have been implemented providing very good results, with accuracy rate in excess of 95% (with significance at a level of 0.05), on 8 binary attributes benchmarks coming from the UCI repository [3]. A multi-valued logic extension [5] of this binary view has been implemented in a more general classifier [6,7], able to work with discrete and numerical attributes and still providing accurate results (on UCI repository benchmarks). The success of the analogy-based classifiers is quite intriguing since they do not rely on any statistical counting, nor take into account any distance or metrics on the set of items they are dealing with. This paper is a step toward a formal explanation of this state of fact, highlighting that analogical inference mainly relies on the observation of links between changes in descriptive features and changes in classes. Features are assumed to be binary in this study. So, the purpose is not to provide a new algorithm, but rather to get a deeper understanding of the reasons why analogy may be successful for classification. This paper is structured as follows: Section 2 provides the background on the Boolean view of analogical proportions. In section 3, we provide a brief discussion of the classification problem, and then show that the analogy-based inference is equivalent to the application of simple rules acknowledging the fact that a given change in some feature

E. Hüllermeier et al. (Eds.): SUM 2012, LNAI 7520, pp. 582–589, 2012.
© Springer-Verlag Berlin Heidelberg 2012

values always leads to the same change in classification, or on the contrary that a given change in some feature values never affects the classification. In section 4, we illustrate and further discuss how such rules behave for completing the partial specification of some of Boolean functions.

2 Background on Analogical Proportions

Let us briefly recall how an analogical proportion "a is to b as c is to d", can be logically defined by equating the dissimilarities between a and b with the ones between c and d. This equality tells us that "a differs from b as c differs from d" and conversely. Then the analogical proportion is defined as in [4]:

$$a : b :: c : d = (a \wedge \overline{b} \equiv c \wedge \overline{d}) \wedge (\overline{a} \wedge b \equiv \overline{c} \wedge d).$$

With this definition, we recover the fundamental properties of analogical proportions, namely symmetry ($a : b :: c : d$ implies $c : d :: a : b$), and central permutation ($a : b : c : d$ implies $a : c :: b : d$). As a Boolean formula, an analogical proportion can be viewed through its truth table displayed in Figure 1 (the lines leading to false are discarded). It appears that each line of the table encapsulates an identity or absence of change between the 2 pairs of data (a, b) and (c, d): for instance, with 1100 there is no change on both sides, or with 1010 the change is the same on both sides. These patterns correspond to forms of regularity between pairs of data: from this "regularity viewpoint", 1001 or 1000 are not satisfactory for instance. In a numerical proportion,

$a\ b\ c\ d$
0 0 0 0
1 1 1 1
0 1 0 1
1 0 1 0
0 0 1 1
1 1 0 0

$\vec{a}$	0 1 1 0	$\vec{a}$	0 1 1 0 1
$\vec{b}$	1 0 1 0	$\vec{b}$	1 0 1 0 1
$\vec{c}$	0 1 1 0	$\vec{c}$	0 1 1 0 0
$\vec{x}$	1 0 1 0	$\vec{x}$	1 0 1 0 0

Fig. 1. Analogical proportion truth table   **Fig. 2.** Two Boolean vector examples

the 4th number can be computed from the 3 others via the "rule of three". The situation is somewhat similar in the Boolean setting, and it can be checked on the truth table that the analogical equation $a : b :: c : x$ is solvable iff $((a \equiv b) \vee (a \equiv c)) = 1$. In that case, the unique solution x is $a \equiv (b \equiv c)$ [4]. The solution of the equation $a : b :: c : x$, when solvable, is a copy of the value of b or of the value of c. This is due to the fact that we deal with only 2 distinct values, 0 and 1.

Analogical proportions can be extended to Boolean vectors that represent descriptions of situations in terms of n binary features, by

$$\vec{a} : \vec{b} :: \vec{c} : \vec{d} \text{ iff } \forall i \in [1, n], a_i : b_i :: c_i : d_i.$$

In terms of "regularities", we encapsulate in a single formula n regularities. Clearly, $\vec{a} : \vec{b} :: \vec{c} : \vec{x}$ is solvable iff $\forall i \in [1, n], a_i : b_i :: c_i : x_i$ is solvable. But, when solvable, the solution $\vec{x}$ of this equation is not in general $\vec{b}$ or $\vec{c}$ (for $n \geq 2$). This can be seen on the examples in Figure 2, with $n = 4$ and $n = 5$ respectively. Indeed, while in the first case $\vec{x} = \vec{b}$, in the second one, the solution $\vec{x}$ is neither $\vec{a}$, nor $\vec{b}$, nor $\vec{c}$. Note that for solving the equation in this latter example, we use the last line of the analogy truth table. The 2 last lines of this truth table play a special role, since they allow to obtain a solution for an analogical proportion equation which is not necessarily a complete copy of one of the 3 first elements of the proportion: in that case, all the elements of the proportion may be distinct. With the multiple-valued extension of analogical proportions [5] (for handling real-valued attributes), we can observe a similar behavior of the proportions in all respects.

The inference principle in use in any of the analogical proportion-based classifiers states that when the attributes of 4 items $\vec{a}, \vec{b}, \vec{c}, \vec{d}$ are all in analogical proportion, then it should be also the case for the classes of these items, which is formalized as:

$$\frac{\vec{a} : \vec{b} :: \vec{c} : \vec{d}}{cl(\vec{a}) : cl(\vec{b}) :: cl(\vec{c}) : cl(\vec{d})}$$

If $\vec{a}, \vec{b}, \vec{c}$ are just classified examples and $\vec{d}$ a new data to be classified, the predicted label for $\vec{d}$ is the solution of the analogical equation $cl(\vec{a}) : cl(\vec{b}) :: cl(\vec{c}) : x$. In some sense, we assume that the regularities observed on the data attributes and encapsulated via the proportions are preserved for the labels. Implementing this principle for classification purpose is done via the algorithmic scheme of Figure 3. On real datasets,

```
input: a set TS of classified examples,
d new data with cl(d) unknown;
search for all triples (a,b,c) in TS such that:
      cl(a):cl(b)::cl(c):x is solvable with solution x0
      a:b::c:d holds
if all solutions agree with x0 then cl(d)=x0;
output: cl(d)
```

Fig. 3. Algorithmic scheme

this principle has to be relaxed to take into account two facts:

- analogical proportion may hold only for a large part of the attributes, but not for all of them (some attributes could be irrelevant to the problem), i.e. we maximize the number of true proportions, i.e. the number of regularities.
- in the case of real-valued attributes, it is quite rare to get an exact proportion and we have to deal with approximate proportions. We then follow the lines of [7] where real-valued attributes are interpreted as fuzzy logic truth values and the truth value of $a : b :: c : d$ becomes $1 - |(a - b) - (c - d)|$ for the simplest case where $a \leq b$ and $c \leq d$, or $b \leq a$ and $d \leq c$. All the previous explanations remain valid.

Ultimately, the analogy-based classifiers maximize the number of approximate proportions (i.e. regularities) for predicting the class of a new instance. As explained now, the regularities that are considered here depart from the ones underlying other classifiers.

3 How Analogical Proportion-Based Classifiers Work

Let us first restate the classification problem. Let X be a set of data: each piece of data $\overrightarrow{x} = (x_1, \ldots, x_i, \ldots, x_n) \in X$ is a vector of n feature values representing some object wrt the set of considered features (here assumed *binary*). Each piece of data $\overrightarrow{x}$ is to be associated with a *unique* label or class $cl(\overrightarrow{x}) \in C$. The space of classes C is finite with p (≥ 2) elements. cl is known on a subset $E \subset X$. The classification problem amounts to propose a plausible value of $cl(\overrightarrow{y})$ for any $\overrightarrow{y} \in X \setminus E$, on the basis of the available information, namely the training set $\mathcal{TS} = \{(\overrightarrow{x}, cl(\overrightarrow{x})) \mid \overrightarrow{x} \in E\}$ whose elements are called examples. We also suppose that each label of C is present in $\mathcal{TS}$, i.e. $\forall c \in C, \exists(\overrightarrow{x}, cl(\overrightarrow{x})) \in \mathcal{TS} \wedge cl(\overrightarrow{x}) = c$. The extrapolation of the value of the function cl for $\overrightarrow{y} \in X \setminus E$ on the basis of $\mathcal{TS}$, requires 2 minimal assumptions:

- i) the description of the objects is rich enough in terms of features to insure that the knowledge of the n feature values uniquely determines $cl(\overrightarrow{y})$ for any instance $\overrightarrow{y} \in X$, and
- ii) the knowledge of $\mathcal{TS}$ is enough for this task, i.e. cl is not more "complicated" than what $\mathcal{TS}$ may suggest.

In other words, whatever the technique in use, for $\overrightarrow{y} \in X \setminus E$, assigning the label c to $\overrightarrow{y}$ (i.e. considering $cl(\overrightarrow{y})$ as c) should not make $\mathcal{TS} \bigcup \{(\overrightarrow{y}, c)\}$ less 'regular' than $\mathcal{TS}$. The new classified data $(\overrightarrow{y}, c)$ should be "conform" to $\mathcal{TS}$.

3.1 General Analysis of the Binary Features Case

A simple idea for guessing $cl(\overrightarrow{y})$ for $\overrightarrow{y} = (y_1, \ldots, y_i, \ldots, y_n) \in X \setminus E$ is to consider how examples having similar feature values are classified. This is valid for both binary and real-valued features. Let us focus on the binary case (where feature similarity reduces to identity), and denote

$$R_i(y) = \{c \mid \exists(\overrightarrow{x}, cl(\overrightarrow{x})) \in \mathcal{TS} \wedge x_i = y \wedge cl(\overrightarrow{x}) = c\}$$

the set of observed classes associated to the value y for feature i (x_i is the i-th component of $\overrightarrow{x}$). Similarly, let $R_i^{-1}(c) = \{y \mid \exists(\overrightarrow{x}, cl(\overrightarrow{x})) \in \mathcal{TS} \wedge x_i = y \wedge cl(\overrightarrow{x}) = c\}$ the set of observed values for feature i associated with class c. Given an instance $\overrightarrow{y} \in X \setminus E$, the principle that items having feature values identical to the values of classified examples should be classified in the same way, leads us to assume that only the class(es) in $R_i(y_i)$ is / are possible for $\overrightarrow{y}$ if it has value y_i for feature i. Taking into account all the features, we get $cl(\overrightarrow{y}) \in \bigcap_i R_i(y_i)$ (or equivalently $cl(\overrightarrow{y}) \notin \bigcup_j \{c \mid y_j \notin R_j^{-1}(c)\}$), provided that $\bigcap_i R_i(y_i) \neq \emptyset$, which means in particular that for any i, $R_i(y_i) \neq \emptyset$, i.e., the value y_i has been observed for attribute i in $\mathcal{TS}$. Naive Bayesian classifiers refine this view by choosing the class that maximizes a product of likelihood functions, which may be understood as an intersection of fuzzy sets [2].

When features are not independent (non-interactive in the sense of possibility theory [10]), instead of dealing feature by feature, one should rather group features and consider the set of possible classes associated to the value of tuples of features, e.g. in the case of 2 features, $R_{i,j}(y, z) = \{c \mid \exists(\overrightarrow{x}, cl(\overrightarrow{x})) \in \mathcal{TS} \wedge x_i = y \wedge x_j = z \wedge cl(\overrightarrow{x}) = c\}$

where $R_{i,j}(y,z) \subseteq R_i(y) \cap R_j(z)$. At the extreme, $cl(\overrightarrow{y}) = \{c | \exists(\overrightarrow{x}, cl(\overrightarrow{x})) \in TS \wedge x_1 = y_1 \wedge \ldots \wedge x_n = y_n \wedge cl(\overrightarrow{x}) = c\}$, which means, if this set is not empty, that a vector fully identical to $\overrightarrow{y}$ is already in TS. For going beyond this trivial situation, one is led to relax the requirement of the identity of the feature values to the largest possible subset of features: this is the idea of k-nearest neighbors methods (see [9] for an analogy-oriented discussion of such methods).

3.2   Change Analysis within Pairs of Examples

In the analogical proportion-based approach to classification, we no longer consider the function $cl : X \to C$ in a static way, from a pointwise viewpoint, but rather how this function varies over TS. The main idea is to establish a link between some change in feature values for *pairs* of examples together with the change (or the absence of change) for their class. Dealing with such pairs $(x, cl(x))$ and $(x', cl(x'))$ (from now on for alleviating the notation, we no longer make the vectors explicit), we relate differences between $cl(x)$ and $cl(x')$ to differences between x and x', thus dealing with some counterpart of the idea of derivative![1]

Given two examples $(x, cl(x))$ and $(x', cl(x'))$, let $Dis(x, x')$ denote the set $\{i \in [1,n] | x_i \neq x'_i\}$ of indexes where x and x' differ. $Dis(x, x')$ is the disagreement set between x and x', while its complement $Ag(x, x') = [1, n] \setminus Dis(x, x')$ is the set of indexes where x and x' coincide (agreement set). For instance when $n = 5, x = 10101, x' = 11000, Dis(x, x') = \{2, 3, 5\}$ and $Ag(x, x') = \{1, 4\}$. $Dis(x, x')$ enjoys obvious properties: i) $Dis(x, x') = Dis(x', x)$, ii) $Dis(x, x') = \emptyset$ iff $x = x'$, iii) knowing $Dis(x, x')$ and x allows us to compute x'. $Dis(x, x')$ is the counterpart of the indicators $x \wedge \overline{x'}$ and $\overline{x} \wedge x'$ that appear in the definition of the analogical proportion. Associated to $Dis(x, x')$ is a unique disagreement pattern $p_d(x, x')$ which is just the sequence of pairs (feature value, feature index) in x which differ in x': in our previous example this is $0_2 1_3 1_5$ where 0, 1, 1 are the respective values of features 2, 3, 5 in x.

Let us now suppose that $cl(x) \neq cl(x')$: one may think that the set $Dis(x, x')$ is in some sense the cause for the label change between x and x'. But, it would be too adventurous since we have exactly $2^{|Dis(x,x')|}$ potential disagreement patterns, and the observation of x and x' does not give any hint about patterns other than the current one $p_d(x, x')$. It is more cautious to consider that the particular change in the pattern, i.e. going from $p_d(x, x') = 0_2 1_3 1_5$ in x to $1_2 0_3 0_5$ in x' is the real cause for the class change between x and x', since the remaining features 1 and 4 are identical (agreement set).

3.3   Induction of Rules and Link with the Analogical Proportion-Based Inference

Obviously, one may consider that a particular disagreement pattern causes a particular change of class only if there is no contradiction when observing all the examples in TS, i.e. there should not exist 2 other examples z and z' having the same disagreement

---

[1] It is also worth mentioning that graded analogical proportions valued in $[0, 1]$ [5], agrees with the idea of interpolation between two points between $(a, f(a))$ and $(d, f(d))$ in $[0, 1]^2$. Indeed, then we have $a : b :: c : d = 1 - |(b - a) - (d - c)|$ for $a \leq b$ and $c \leq d$, and if $a : x :: x : d$ holds at degree 1, $f(a) : f(x) :: f(x) : f(d)$ holds at degree 1 for $f(x) = (f(a) + f(d))/2$.

pattern (i.e., $p_d(z, z') = p_d(x, x')$), but having the same class. Put in a formal way, this amounts to state the following: If it is the case that

- $\exists x, x' \in E$ such that $x \neq x'$, $cl(x) = c$ and $cl(x') = c'$ with $c \neq c'$
- $\nexists z, z' \in E$ such that $p_d(z, z') = p_d(x, x')$ and $cl(z) = cl(z') = c$

then the following *change rule* is inferred:

$$\forall y, y' \in E, \text{ if } p_d(y, y') = p_d(x, x') \text{ and } cl(y) = c \text{ then } cl(y') = c'.$$

Let us note that this rule is obviously valid on E and is just a generalization to the whole data set X. It is clear that $p_d(y, y') = p_d(x, x')$ implies $Dis(y, y') = Dis(x, x')$ but the reverse is false. Thus, when such a regularity is discovered, we generalize it by establishing a rule, assuming that the whole universe is not more sophisticated than the observable data. The application of this rule enables us to compute $cl(y')$ for $y' \in X \setminus E$, provided that there exists $y \in E$ such that $p_d(y, y') = p_d(x, x')$ and $cl(x) = cl(y)$. It is remarkable that the application of such a rule, which acknowledges the existence of regularities in the data set, is *nothing but* the application of the analogical proportion-based extrapolation rule. Indeed $x : x' :: y : y'$ holds feature by feature, since

- on $Dis(y, y') = Dis(x, x')$, due to $p_d(y, y') = p_d(x, x')$, for each feature, we have either the pattern $1 : 0 :: 1 : 0$ or the pattern $0 : 1 :: 0 : 1$;
- on $Ag(y, y') = Ag(x, x')$ for each feature, we have a pattern that corresponds to one of the 4 other lines of the truth-table of the analogical proportion;
- this leads to assume that $cl(x) : cl(x') :: cl(y) : cl(y')$ holds as well, i.e. $cl(y') = cl(x') = c'$ since $cl(x) = cl(y) = c$.

The exploitation of regularities in change that can be found in the data set is thus encapsulated in the analogical proportion-based extrapolation principle.

Clearly, the same reasoning can be applied when there is no label change, i.e. $cl(x) = cl(x')$: In that case, it means that the particular change in feature values does not induce a change of the class. Provided that the two conditions below hold:

- $\exists x, x' \in E$ such that $x \neq x'$, $cl(x) = cl(x') = c$
- $\nexists y, y' \in E$ such that $p_d(y, y') = p_d(x, x')$ and $cl(y) = cl(x) \neq cl(y')$

the following *no-change rule* is inferred:

$\forall y, y' \in E$, if $p_d(y, y') = p_d(x, x')$ and $cl(y) = c$ then $cl(y') = c$.

Due to the central permutation property, when $x : x' :: y : y'$ holds, which may allow us to generate a *change* rule, then $x : y :: x' : y'$ holds as well which may lead to a *no-change* rule. However, one cannot consider only one type of rule, since for a given training set, one may be applicable, while the other is not.

4   Completion Algorithm and Experiments

Now we are in position to complete a training set $\mathcal{TS}$ according to *change* or *no − change* rules in the above sense. Non classified data should behave in "conformity" with $\mathcal{TS}$, i.e. fit with these rules. We proceed as follows: first we extract all the rules from the available classified data $\mathcal{TS}$, then given a new element $y' \notin E$, we tentatively look for another element $y \in E$, to be paired with y' in such a way a rule applies to the pair (y, y'). We repeat this process with all suitable y and if *all* the applicable rules provide the same class c, we allocate this label c to y'. It is what we call unanimous classification. When it is not possible, we allocate the label c to y' as provided by the *majority* of the applicable rules. It remains undecided if there is no majority.

We have then implemented a simple classifier dealing with 6 dimensions Boolean vectors. Each vector $\vec{x} = (x_1, \ldots, x_6)$ is associated to a class. The class of $\vec{x}$ is a Boolean function of its attributes. In the experiment, we go from a dataset where we have a lot of useless information (or noise) (with $cl(x) = x_1 \vee x_2$) to a dataset where almost all the attributes are necessary to predict the class ($cl(x) = (x_1 \wedge x_4) \vee (x_2 \wedge x_3) \vee x_5$). We also consider $cl(x) = (x_1 \wedge x_4) \vee (x_2 \wedge x_3)$ and $cl(x) = x_1 \, XOR \, x_2$. We can then build the 5 corresponding data sets (each one has 64 lines). The testing protocol is:

- we extract 32 lines from each dataset: they constitute $\mathcal{TS}$, such that there is at least one case of the 2 classes 0 and 1; we avoid to have some feature equal for the 32 lines.

- using the generated rules, we complete $\mathcal{TS}$ on the 32 remaining instances.

The results of the experiments are given in Table 2 with the Boolean function to be estimated, the average number of correctly classified data (among 32), either unanimously or majoritarily, the average number of wrongly classified data (among 32), either unanimously or majoritarily, the average number of undecided cases. We can observe that all the instances are classified. Table 1 provides the average number of change rules and of no change rules generated from the training sets for each Boolean function. Up to 16% of instances may be wrongly classified. Correct classification is obtained mainly thanks to a majority-based choice (even if a number of instances are classified unanimously, except for the $x_1 \, XOR \, x_2$ example where this number is close to 0). The classification is almost never unanimously wrong. Undecided cases are quite rare. These results are really encouraging, but would need to be confirmed on a larger number of benchmarks.

It is clear that this simple algorithm may be the basis of a lazy learning technique: starting from our classified data E, a subset of $\mathbb{B}^n$, we first extract the set of rules as described in the algorithm. We then complete the set E as far as we can by just using the rules to classify the remaining data in $\mathbb{B}^n \setminus E$, then getting E_{ext}. Given a new piece of data to be classified, if $d \in E_{ext}$, then d is classified, else we cannot classify d this way.

Table 1. Average value (on 10 different training sets) of the number of rules generated

	$x_1 \vee x_2$	$x_1 \vee (x_2 \wedge x_3)$	$(x_1 \wedge x_4) \vee (x_2 \wedge x_3) \vee x_5$	$(x_1 \wedge x_4) \vee (x_2 \wedge x_3)$	$x_1 \, XOR \, x_2$
change rules gener.	32	28	30.2	33.9	32
no change rules gener.	30.9	44.1	27.5	21.3	30.9

Table 2. Results (cross-validation on 10 different training sets of 32 instances)

	$x_1 \vee x_2$	$x_1 \vee (x_2 \wedge x_3)$	$(x_1 \wedge x_4) \vee (x_2 \wedge x_3) \vee x_5$	$(x_1 \wedge x_4) \vee (x_2 \wedge x_3)$	$x_1 XOR x_2$
unanim. right classif.	7	11. 17	6.1	6.3	0.4
majorit. right classif.	24.2	16.8	20.3	22	26.6
unanim. wrong classif.	0	0.1	0	0	0
majorit. wrong classif.	0.9	3.1	5	3	4.1
undecided	0	0.3	0.4	0.6	0.9

5   Concluding Remarks

This paper constitutes a first attempt to understand the core logic underlying the analogy-based classifiers and why this common-sense reasoning may lead to successful procedures. Nevertheless, more investigations remain to be done to get a completely clear picture. Considering pairs of items (analogical proportions make parallels between such pairs), and observing regularities on pair's features (without counterexample) is the basis for reproducing both contextual permanence in particular changes or absences of change. This implicitly amounts to build analogical patterns. On the same basis, analogical proportions have been used [1] to solve IQ tests that amounts to find the contents of the 9th cell in a 3×3 table, positioning the ability to establish links between observed changes as an important feature of human mind.

References

1. Correa, W.F., Prade, H., Richard, G.: When intelligence is just a matter of copying. In: Proc. 20th Europ. Conf. Artific. Intellig (ECAI 2012), Montpellier, August 29-31 (2012)
2. Dubois, D., Prade, H.: An overview of ordinal and numerical approaches to causal diagnostic problem solving. In: Gabbay, D.M., Kruse, R. (eds.) Abductive Reasoning and Learning, pp. 231–280. Kluwer Academic Publ. (2000)
3. Miclet, L., Bayoudh, S., Delhay, A.: Analogical dissimilarity: definition, algorithms and two experiments in machine learning. JAIR 32, 793–824 (2008)
4. Miclet, L., Prade, H.: Handling Analogical Proportions in Classical Logic and Fuzzy Logics Settings. In: Sossai, C., Chemello, G. (eds.) ECSQARU 2009. LNCS, vol. 5590, pp. 638–650. Springer, Heidelberg (2009)
5. Prade, H., Richard, G.: Multiple-valued logic interpretations of analogical, reverse analogical, and paralogical proportions. In: Proc. 40th IEEE Int. Symp. on Multiple-Valued Logic (ISMVL 2010), Barcelona, pp. 258–263 (2010)
6. Prade, H., Richard, G., Yao, B.: Classification by means of fuzzy analogy-related proportions: A preliminary report. In: Proc. 2nd IEEE Int. Conf. on Soft Computing and Pattern Recognition (SocPar 2010), Evry, pp. 297–302 (2010)
7. Prade, H., Richard, G., Yao, B.: Enforcing regularity by means of analogy-related proportions - A new approach to classification. Int. J. of Comp. Information Syst. and Industrial Manag. Appl. 4, 648–658 (2012)
8. Stroppa, N., Yvon, F.: Du quatrième de proportion comme principe inductif: une proposition et son application à l'apprentissage de la morphologie. Trait. Autom. Lang. 47(2), 1–27 (2006)
9. Wojna, A.: Analogy-based reasoning in classifier construction. PhD thesis, Warsaw Univ., Faculty of Math., Info. and Mechanics, Poland (2004)
10. Zadeh, L.A.: The concept of a linguistic variable and its application to approximate reasoning - 1. Inf. Sci. 8(3), 199–249 (1975)

An Adaptive Algorithm for Finding Frequent Sets in Landmark Windows

Xuan Hong Dang[1], Kok-Leong Ong[2], and Vincent Lee[3]

[1] Dept. of Computer Science, Aarhus University, Denmark
dang@cs.au.dk
[2] School of IT, Deakin University, Australia
leong@deakin.edu.au
[3] Faculty of IT, Monash University, Australia
vincent.cs.lee@monash.edu

Abstract. We consider a CPU constrained environment for finding approximation of frequent sets in data streams using the landmark window. Our algorithm can detect overload situations, i.e., breaching the CPU capacity, and sheds data in the stream to "keep up". This is done within a controlled error threshold by exploiting the Chernoff-bound. Empirical evaluation of the algorithm confirms the feasibility.

1 Introduction

Many new applications now require compute of vast amount of data. The data is fast changing, infinitely sized and has evolving changing characteristics. Data mining them over limited compute resources is thus a problem of interest. As data rates change to exceed compute capacity, the machinery must adapt to produce results on time. The way to achieve this is to incorporate approximation and adaptability into the algorithm.

We introduce an algorithm that addresses approximation and adaptability for finding frequent sets. While this paper reports the method of finding frequent sets in landmark windows, we have in fact developed three methods of finding frequent sets for the landmark window, forgetful window, and sliding window [6]. All three methods are able to adjust their compute according to available CPU capacity and the data rate seen in the stream. All three methods achieved this outcome by shedding load to keep up with the data rate. A significant contribution of our approach is that we imposed an error guarantee on the load shed action using a probabilistic load shedding algorithm. In this paper, we report one of the three algorithms developed.

2 Related Works

The study of load shedding in data streams first begun in query networks as a formulation of query plans, where "drop operations" are inserted [3,7,10]. In data mining models for data streams, those that considers load shedding to

E. Hüllermeier et al. (Eds.): SUM 2012, LNAI 7520, pp. 590–597, 2012.

deal with overload situations include [5] and [2]. Nevertheless, these works do not deal with counting frequent sets. In works addressing frequent sets mining, we can classify them into three categories. Those that count frequent sets in a landmark window, e.g., [12]; those that count frequent sets in a forgetful (or decaying) window, e.g., [4,8]; and those that count frequent sets in a sliding window, e.g., [9]. These techniques all focus on summarising the data stream within the constraints of the main memory. In contrast, our work addressed CPU limits and also provided an error guarantee on the results.

The closest work to what we are proposing is found in [1] where sampling of the *entire* database is performed. Multiple samples are taken as [1] runs in a grid setting. The sampling is controlled via a probability parameter that is not a function of CPU capacity. Further, the approximate results are also not guaranteed by an error limit. Therefore, this contrasts with our proposed algorithm, where we 'sample' only in overload situations and the degree of sampling is controlled by the CPU capacity rather than a pre-set probability parameter. We operate our algorithm on a single compute – not over the grid, and we provide an error guarantee to our results.

3   Algorithm STREAML

We assume a data stream being transactions $\{t_1, t_2, \ldots, t_n, \ldots\}$, where $t_i \subseteq I = \{a_1, a_2, \ldots, a_m\}$ and a_j is a literal to a data, object, or item. We further denote t_n as the current transaction arriving in the stream. A data stream $\{t_1, \ldots\}$ is conceptually partitioned into time slots $TS_0, TS_1, \ldots, TS_{i-k+1}, \ldots, TS_i$, where TS_n holds transactions that arrived at interval (or period) n.

Definition 1. *A **landmark window** is all time slots between the given interval n and the current time slot i, i.e., $TS_n, TS_{n+1}, \ldots, TS_{i-1}, TS_i$. And in terms of transactions, it will be transactions in those time slots. A landmark window hence allows a analyst to perform analytic over a specific period. A data stream of call statistics for example, would benefit from a landmark window where analysis on the past 24 or 48 hours of calls can be obtained. In this case, the last 24 hours would be the landmark window's size.*

Definition 2. *An itemset X is a set of items $X \in (2^I - \{\varnothing\})$. An ℓ-itemset is an itemset that contains exactly ℓ items. Given $X \subseteq I$ and a transaction t_i, t_i supports (or contains) X if and only if $X \subseteq t_i$.*

Definition 3. *Given an itemset X of size ℓ (i.e., an ℓ-itemset), a set of all its subsets, denoted by the power set $\mathcal{P}(X)$, is composed of all possible itemsets that can be generated by one or more items of the itemset X, i.e., $\mathcal{P}(X) = \{Y | Y \in (2^X - \{X\}) \wedge (Y \neq \varnothing)\}$. An immediate subset Y of X is an itemset such that $Y \in \mathcal{P}(X)$ and $|Y| = (\ell - 1)$.*[1]

Definition 4. *The cover of X in $\mathcal{DS}$ is defined as the set of transactions in $\mathcal{DS}$ that support X, i.e., $cover(X, \mathcal{DS}) = \{t_i | t_i \in \mathcal{DS}, X \subseteq t_i\}$.*

---

[1] $|Y|$ denotes the number of items in Y.

Definition 5. *The frequency of X, denoted by $freq(X)$, is the number of trans-actions in $\mathcal{DS}$ that contain X, i.e., $freq(X) = |cover(X, \mathcal{DS})|$, and its support, $supp(X)$, is the ratio betweeb $freq(X)$ and $|\mathcal{DS}|$.*

3.1   Workload Estimation

As indicated, our algorithm has two unique characteristics. The first is adapt-ability to the time-varying characteristics of the stream. In this case, the issue is to detect overload situations. The first instance when an overload occur lie not in detecting the increased rate of transaction arrivals. Rather, it is a combina-tion of arrival rate and the number of literals present in each transaction. We denote the measure reflecting arrival rate and transaction size as L. Under an overloaded situation, figuring the exact L is not a pragmatic solution. Instead we favour an estimation technique for L that is computationally affordable in overloaded situations.

The intuition to estimate L lie in the exponentially small number of maximal frequent sets (MFS) compared to usual frequent sets [11]. Let m be the number of MFS in a transaction t and denote X_i as a MFS, $1 \leqslant i \leqslant m$. Then the time to determine L^2 for each transaction is given as

$$L = \sum_{i=1}^{m} 2^{|X_i|} - \sum_{i,j=1}^{m} 2^{|X_i \cap X_j|} \tag{1}$$

and for n transactions over a single time unit, we have the following inequality for load shedding decision

$$P \times r \times \frac{\sum_{i=1}^{n} L_i}{n} \leq C \tag{2}$$

where $P \in (0, 1]$ is repeatedly adjusted to hold the inequality. Hence, the LHS of the inequality also gives the maximum rate for transactions to be processed within a time unit, i.e., at $P = 1$. The process capacity of the system is denoted as C. Hence, the objective is to find all frequent sets while guaranteeing that the workload is always below C.

3.2   Chernoff Bound Load Shedding

Clearly when load is shed, the consequence is the loss of accuracy in the frequency of an itemset. Therefore, it makes sense to put an error guarantee on the results so that analysis can be made with proper context. We achieved this in our algorithm with the Chernoff bound.

Recall that $P = 1$ sets the arrival rate to its maximum. If this breaches C, then we must adjust P for $P < 1$ is when transactions are dropped to meet CPU

---

[2] Of course, computing L directly will be expensive and impractical. Instead, the implementation uses a prefix tree, where the transaction in question is matched to estimate the number of distinct frequent sets.

constraints. If N are all the transactions, then n represents the sampling of N transactions with $N - n$ the number of transactions dropped. For each itemset X, we compare the frequency for n sampled transactions against the frequency for the original N transactions.

The existence of X in a transaction is a Bernoulli trial that can be represented with a variable A_i such that $A_i = 1$ if X is in the i-th transaction and $A_i = 0$ otherwise. Obviously, $\Pr(A_i = 1) = p$ and $\Pr(A_i = 0) = 1 - p$ with $\Pr(.)$ being the probability of a condition being met. Thus, n randomly drawn transactions are n independent Bernoulli trials. Further, let r (a *binomial* random variable) be the occurrence of $A_i = 1$ in the n transactions with expectation np. Then, the Chernoff bound states that for any $\varepsilon > 0$ and $\epsilon = p\varepsilon$, we have

$$\Pr\{|r - np| \geq n\epsilon\} \leq 2e^{-n\epsilon^2/2p} \tag{3}$$

If $s_E(X) = r/n$ is the estimated support of X computed from n sampled transactions, then the Chernoff Bound tells us how likely the true support $s_T(X)$ deviates from its estimated support $s_E(X)$ by $\pm\epsilon$. If we want this probability to be $\leqslant \delta$, the sampling size should be at least (by setting $\delta = 2e^{-n\epsilon^2/2p}$)

$$n_0 = \frac{2p\ln(2/\delta)}{\epsilon^2} \tag{4}$$

3.3 Algorithmic Steps

In landmark windows, we compute answers from a specific time in the past to the latest transaction seen. For ease of exposition, we assume the first transaction within a landmark window to be t_1. Hence, we can conceptually compute as if the entire stream is seen, i.e., dropping the time boundary. Further, we *conceptually* divide the stream into buckets; each holding Δ transactions. To keep results within a given threshold, we set $\Delta = n_0$ (Equation. 4).

Due to resource limits, we do not count all itemsets. Rather, we identify the candidates that will become frequent and track those. With the conceptual buckets, we use a deterministic threshold $\gamma(< \sigma)$ to filter unlikely frequent sets. If an itemset's support falls below γ, then it is unlikely to become frequent in the near future. Since an approximate collection of frequent sets are produced, this must be within the promised error guarantee. As transactions are only read once, there is a higher error margin with longer frequent sets as a potential candidate is only generated after all its subsets are significantly frequent, i.e., the downward closure property.

To counter this, an array that tracks the minimum frequency thresholds is used to tighten the upper error bound. We set m to the longest frequent sets of interest to the user. The array thus must satisfy $a[i] < a[j]$ for $1 \leq i < j \leq m$, $a[1] = 0$ (for 1-itemsets) and $a[m] \leq \gamma \times \Delta$. A j-itemset is thus generated if its immediate subsets in the current bucket is $\geqslant a[j]$. During load shedding, we set each bucket to Δ transactions to bound the true support error probability to no more than ϵ. As will be discussed next, the support error of frequent sets discovered is guaranteed to be within γ and ϵ.

The STREAML algorithm uses a prefix tree $\mathcal{S}$ to maintain frequent sets discovered from the landmark window. Each node in $\mathcal{S}$ corresponds to an itemset X represented by items on a path from the root to the node, where $Item$ is last literal in X; $f_c(X)$ is the frequency of X in the current bucket; and $f_a(X)$ is the accumulated frequency of X in the stream.

Further, let b_c be the most current bucket in the stream. Hence, $b_c = 1$ upon initialisation and is incremented each time Δ transactions are processed. The algorithmic steps of STREAML is thus given below.

1. Periodically detect capacity C and if no overload, set $P = 1$. Otherwise, set P to maximal value identified in Equation 2.
2. Process transaction t_n with probability p as follows
   Increment if $X \subseteq t_n$ is in $\mathcal{S}$, then $f_c(X) = f_c(X) + 1$
   Insert, where $|X| = 1$ $\forall X \subseteq t_n \wedge X \notin \mathcal{S} \rightarrow$ insert X into $\mathcal{S}$ with $b(X) = b_c$
   and $f_c(X) = 1^3$;
   Insert, where $|X| > 1$ Insert X into $\mathcal{S}$ when *all* following conditions hold
   – All immediate subsets of X are in $\mathcal{S}$;
   – $\nexists Y$ (Y is any immediate subsets of X) s.t. $b(Y) < b_c$ and $f_c(Y) \leq$ $a[|X|]$, i.e., no Y is inserted into $\mathcal{S}$ from previous buckets but its frequency in b_c is insufficiently significant;
   – $\nexists Y$ s.t. $b(Y) = b_c$ and $f_c(Y) \leq (a[|X|] - a[|Y|])$, i.e., no Y is inserted into $\mathcal{S}$ in b_c and after which its frequency is $\leqslant (a[|X|] - a[|Y|])$.
   In cases where X is not inserted into $\mathcal{S}$, all its supersets in t_n need not be further checked.
   Prune $\mathcal{S}$ and update frequency After sampling Δ transactions, remove all itemsets (except 1-itemsets) in $\mathcal{S}$ fulfilling $f_a(X) + f_c(X) + b(X) \times a[|X|] \leq b_c \times a[|X|]$. If an itemset is removed, all its supersets are also removed. This does not affect itemsets insert into bucket b_c since their immediate subsets have became sufficiently frequent. For unpruned itemsets, their frequencies are updated as $f_a(X) = f_a(X) + f_c(X) \times P^{-1}$ and $f_a(X) = 0$. To compensate the effects of dropping transactions, X's frequency is scaled by P^{-1} to approximate its true frequency in the data stream. Finally, we update $b_c = n/\Delta + 1$.
3. At any point when results are requested, STREAML scans $\mathcal{S}$ to return 1-itemsets satisfying $f_a(X) + f_c(X) \times P^{-1} \geqslant \sigma \times n$ and ℓ-itemsets satisfying $f_a(X) + f_c(X) \times P^{-1} + b(X) \times a[|X|] \geqslant \sigma \times n$.

3.4   Error Analysis

We now present the accuracy analysis on the frequency estimates produced by our algorithm. For each itemset X^4, we denote its *true* frequency by $f_T(X)$ and

---

[3] The immediate subsets of a 1-itemset is $\varnothing$ which appears in every transaction. All 1-itemsets are therefore inserted into $\mathcal{S}$ without condition. For the same reason, they are also not pruned from $\mathcal{S}$.

[4] We note that we will not discuss the error for 1-itemsets as their support will be precisely counted in the absence of load shedding. Therefore the true and estimated support are the same for these itemsets.

estimated frequency by $f_E(X) = f_a(X) + f_c(X)$ using synopsis $\mathcal{S}$. Respectively, $s_T(X)$ and $s_E(X)$ denote its *true* and *estimated* supports.

Lemma 1. *Under no load shedding (i.e., $P = 1$), StreamL guarantees if X is deleted at b_c, its true frequency $f_T(X)$ seen so far is $\leqslant b_c \times a[\|X\|]$.*

Proof. This lemma can be proved by induction. Base case when $b_c = 1$: nothing is deleted at the end of this bucket since an itemset is inserted into $\mathcal{S}$ only if its immediate subsets are sufficiently significant. Therefore, any itemset X not inserted into $\mathcal{S}$ in the first bucket will not have their true frequency greater than $a[\|X\|]$; i.e., $freq_T(X) \leqslant a[\|X\|]$. Note that the maximal error in counting frequency of an inserted itemset X (at this first bucket) is also equal to $a[\|X\|]$. This is because X is inserted into $\mathcal{S}$ and started counting frequency *only after* all X's subsets gain enough $a[\|X\|]$ on their occurrences.

When $b_c = 2$: Likewise, itemsets inserted into this bucket are not deleted since they are in $\mathcal{S}$ only after their subsets are confirmed significant. Only itemsets inserted in the first bucket might be removed as they can become infrequent. Assume that X is such itemset, its error is at most $a[\|X\|]$. On the other hand, $f_E(X)$ is its frequency count since being inserted in the first bucket. Thus, $f_T(X) \leqslant f_E(X) + a[\|X\|]$. Combing with the delete condition: $f_E(X) + a[\|X\|] \leqslant b_c \times a[\|X\|]$, we derive $f_T(X) \leqslant b_c \times a[\|X\|]$.

Induction case: Suppose X is deleted at $b_c > 2$. Then, X must be inserted at some bucket $b_i + 1$ before b_c; i.e., $b(X) = b_i + 1$. In the worst case, X could possibly be deleted in the previous bucket b_i. By induction, $f_T(X) \leqslant b_i \times a[\|X\|]$ when it was deleted in b_i. Since $a[\|X\|]$ is the maximum error at b_i+1 and $f_E(X)$ is its frequency count since being inserted in b_i+1, it follows that $f_T(X)$ is at most $f_E(X) + b_i \times a[\|X\|] + a[\|X\|] = f_E(X) + (b_i+1) \times a[\|X\|]$; or $f_T(X) \leqslant f_E(X) + b(X) \times a[\|X\|]$. When combined with the deletion rule $f_E(X) + b(X) \times a[\|X\|] \leqslant b_c \times a[\|X\|]$, we get $f_T(X) \leqslant b_c \times a[\|X\|]$.

Theorem 1. *Without load shedding, StreamL guarantees that the true support of any $X \in \mathcal{S}$ is limited within the range: $s_E(X) \leqslant s_T(X) \leqslant s_E(X) + \gamma$.*

Proof. By definition $f_E(X)$ is the counting of X since it was inserted into $\mathcal{S}$. Therefore the true frequency of X is always at least equal to this value, i.e., $f_E(X) \leq f_T(X)$. We now prove that $f_T(X) \leq f_E(X) + b(X) \times a[\|X\|]$.

If X is inserted in the first bucket, its maximal frequency error is $a[\|X\|]$. Therefore, $f_T(X) \leqslant f_E(X) + a[\|X\|]$; In the other case, X is possibly deleted some time earlier in the first b_i buckets and then inserted into $\mathcal{S}$ at $b_i + 1$. By Lemma 1, $f_T(X)$ is at most $b_i \times a[\|X\|]$ when such a deletion took place. Therefore, $f_T(X) \leqslant f_E(X) + a[\|X\|] + b_i \times a[\|X\|] = f_E(X) + b(X) \times a[\|X\|]$. Together with the result above, we derive $f_E(X) \leqslant f_T(X) \leqslant f_E(X) + b(X) \times a[\|X\|]$. On the other hand, we define $a[i] < a[j]$ for $1 \leq i < j \leq m$ and $a[m] \leq \Delta\gamma$. And since $\Delta \times b(X)$ is always smaller than the number of transactions seen so far in the stream, i.e., $\Delta \times b(X) \leq n$, we have $a[\|X\|] \times b(X) \leq n \times \gamma$. Dividing the inequality above for n, the theorem is proven.

Theorem 2. *With load shedding, StreamL guarantees that* $s_E(X) - \epsilon \le s_T(X) \le s_E(X) + \gamma + \epsilon$ *with a probability of at least* $1 - \delta$.

Proof. This theorem can be directly derived from the Chernoff bound and Theorems 1. At periods where the load shedding happens, the true support of X is guaranteed within $\pm\epsilon$ of the counting support with probability $1 - \delta$ when the Chernoff bound is applied for sampling. Meanwhile, by Theorems 1, this counting support is limited by $s_E(X) \le s_T(X) \le s_E(X) + \gamma$. Therefore, getting the lower bound in the Chernoff bound for the left inequality and the upper bound for the right inequality, we derive the true support of X to within the range $s_E(X) - \epsilon \le s_T(X) \le s_E(X) + \gamma + \epsilon$ with a confidence of $1 - \delta$.

4   Empirical Evaluation

We summarise our results here but refer the reader to our technical reports for an in-depth discussion to meet space constraints.

We used the IBM QUEST synthetic data generator for our test. The first data set T10.I6.D3M has an averaged transaction length of T = 10, average frequent set length of $I = 6$ and contains 3 million transactions (D). The other two data sets are T8.I4.D3M and T5.I3.D3M. On all three datasets, we are interested in a maximum of $L = 2000$ potentially frequent sets with $N = 10,000$ unique items. As our algorithm is probabilistic, we measure the *recall* and *precision* on finding the the true frequent sets. The experiment is repeated 10 times for each parameter setting to get the average reading.

To test **accuracy**, we assume a fixed CPU capacity and varied the load shed percentage between 0% and 80%. A shed load of 50% for example corresponds to an input stream rate that is twice the CPU capacity. When $P = 1$, the algorithm behaves like its deterministic cousins with recall of 100% but its precision is not 100%. This lack of precision is due to false positives as a result of over estimating true occurrences by $\gamma \times n_w$. When load is shed, the false positives and false negatives increase as reflected in the precision and recall. This is as expected since less transactions are processed during a load shed and the action impacts accuracy. The important point though is that the accuracy of the results across our experiments show that it is highly maintained – in two of the three data sets, both measures were above 93%. We also observed that as load sheds increases beyond 60%, there is a general huge drop of accuracy so we note the practical extend of shedding possible.

To test the **adaptability** of our algorithm, we create a hybrid data set by concatenating the first 1 million transactions from T5.I3.D3M, T8.I4.D3M and T10.I6.D3M respectively. We note from the experiments that the time to compute the statistics for managing CPU capacity is negligible compared to finding frequent sets. The cost of maintaining the statistics is also linearly proportioned to the time across support thresholds experimented. From the results seen, we can conclude that we can use the statistics collected to identify the appropriate amount of data for shedding. On the results seen, the adaptability of our algorithm is practically reasonable as the cost to achieve a response to the load is relatively small.

5   Conclusions

We considered computing frequent sets in streams under unpredictable data rates and data characteristics that affect CPU capacity. Our unique contribution is a parameter to control the stream load imposed on CPU capacity through load shedding. Additionally, we bound the error to a Chernoff bound to ensure that the error is guaranteed to within a certain threshold.

References

1. Appice, A., Ceci, M., Turi, A., Malerba, D.: A Parallel Distributed Algorithm for Relational Frequent Pattern Discovery from Very Large Data Sets. Intelligent Data Analysis 15(1), 69–88 (2011)
2. Bai, Y., Wang, H., Zaniolo, C.: Load Shedding in Classifying Multi-Source Streaming Data: A Bayes Risk Approach. In: SDM (2007)
3. Babcock, B., Datar, M., Motwani, R.: Load Shedding for Aggregation Queries over Data Streams. In: ICDE, pp. 350–361 (2004)
4. Chang, J.H., Lee, W.S.: Finding Recent Frequent Itemsets Adaptively over Online Data streams. In: SIGKDD (2003)
5. Chi, Y., Yu, P.S., Wang, H., Muntz, R.R.: LoadStar: A Load Shedding Scheme for Classifying Data Streams. In: SDM, pp. 346–357 (2005)
6. Dang, X., Ong, K.-L., Lee, V.C.S.: Real-Time Mining of Approximate Frequent Sets over Data Streams Using Load Shedding. Deakin University, Technical Report (TR 11/06) (2011), http://www.deakin.edu.au/~leong/papers/dol.pdf
7. Gedik, B., Wu, K.-L., Yu, P.S., Liu, L.: Adaptive Load Shedding for Windowed Stream Joins. In: CIKM, pp. 171–178 (2005)
8. Giannella, C., Han, J., Pei, J., Yan, X., Yu, P.S.: Mining Frequent Patterns in Data Streams at Multiple Time Granularities. Next Generation Data Mining. AAAI/MIT (2003)
9. Lin, C., Chiu, D., Wu, Y., Chen, A.: Mining Frequent Itemsets from Data Streams with a Time-Sensitive Sliding Window. In: SDM, pp. 68–79 (2005)
10. Tatbul, N., Çetintemel, U., Zdonik, S.B.: Staying Fit: Efficient Load Shedding Techniques for Distributed Stream Processing. In: VLDB, pp. 159–170 (2007)
11. Yang, G.: The Complexity of Mining Maximal Frequent Itemsets and Maximal Frequent Patterns. In: SIGKDD, pp. 344–353 (2004)
12. Yu, J.X., Chong, Z., Lu, H., Zhou, A.: False Positive or False Negative: Mining Frequent Itemsets from High Speed Transactional Data Streams. In: VLDB, pp. 204–215 (2004)

Instantiation Restrictions for Relational Probabilistic Conditionals*

Marc Finthammer and Christoph Beierle

Dept. of Computer Science, FernUniversität in Hagen, Germany

Abstract. The role of instantiation restrictions for two recently proposed semantics for relational probabilistic conditionals employing the maximum entropy principle is discussed. Aggregating semantics is extended to conditionals with instantiation restrictions. The extended framework properly captures both grounding and aggregating semantics and helps to clarify the relationship between them.

1 Introduction

Relational probabilistic conditionals of the from *if A, then B with probability d,* formally denoted by $(B|A)[d]$, are a powerful means for knowledge representation and reasoning when uncertainty is involved. As an example, consider the common cold scenario from [3] that models the probability of catching a common cold, depending on a persons's general susceptibility and his contacts within a group of people. The knowledge base $\mathcal{R}_{CC}$ contains:

$$c_1 : (cold(U)|\top)[0.01]$$
$$c_2 : (cold(U)|susceptible(U))[0.1]$$
$$c_3 : (cold(U)|contact(U,V) \wedge cold(V))[0.6]$$

Conditional c_1 says that the general probability of having a common cold is just 0.01. However, if a person U is susceptible, then the probability of having a common cold is much higher (namely 0.1 as stated in conditional c_2), and c_3 expresses that if U has contact with V and V has a common cold, then the probability for U having a common cold is 0.6.

At first glance, $\mathcal{R}_{CC}$ makes perfect sense from a commonsense point of view. However, assigning a formal semantics to $\mathcal{R}_{CC}$ is not straightforward. In the propositional case, the situation is easier. For a ground probabilistic conditional $(B|A)[d]$, a probability distribution P over the possible worlds satisfies $(B|A)[d]$ iff $P(A) > 0$ and $P(B|A) = d$, i.e., iff the *conditional probability* of $(B|A)$ under P is d (see e.g. [7]). When extending this to the relational case with free variables as in $\mathcal{R}_{CC}$, the exact role of the variables has to be specified. While there are various approaches dealing with a combination of probabilities with a first-order language (e.g. [6,5], a recent comparison and evaluation of

---

* The research reported here was partially supported by the Deutsche Forschungsgemeinschaft (grant BE 1700/7-2).

E. Hüllermeier et al. (Eds.): SUM 2012, LNAI 7520, pp. 598–605, 2012.

some approaches is given [1]), in this paper we focus on two semantics that both employ the *principle of maximum entropy* [10,7] for probabilistic relational conditionals, the *aggregation semantics* [8] proposed by Kern-Isberner and the logic FO-PCL [4] elaborated by Fisseler. While both approaches are related in the sense that they refer to a set of constants when interpreting free variables, there is also a major difference. FO-PCL needs to restrict the possible instantiations for the variables occurring in a conditional by providing constraint formulas like $U \neq V$ or $U \neq a$ in order to avoid inconsistencies. On the other hand, aggregation semantics is only defined for conditionals without such instantiation restrictions, and it does not require them for avoiding inconsistencies. However, the price for avoiding inconsistencies in aggregation semantics is high since it can be argued that it may lead to counter-intuitive results. For instance, when applying aggregation semantics of [8] to $\mathcal{R}_{CC}$ and the set of constants $D_{CC} = \{a, b\}$, under the resulting maximum entropy distribution P^* the conditional $(cold(a)|contact(a, b) \wedge cold(b))$ has the surprisingly low probability of just 0.027.

In this paper, we elaborate the reasons for this behaviour, discuss possible work-arounds, and point out why instantiation restrictions would be helpful. We propose a logical system PCI that extends aggregation semantics to conditionals with instantiation restrictions and show that both the aggregation semantics of [8] and FO-PCL [4] come out as special cases of PCI, thereby also helping to clarify the relationship between the two approaches.

The rest of this paper is organized as follows: In Section 2, we motivate the use of instantiation restrictions for probabilistic conditionals. In Section 3, the logic framework PCI is developed and two alternative satisfaction relations for grounding and aggregating semantics are defined for PCI by extending the corresponding notions of [4] and [8]. In Section 4, the maximum entropy principle is emloyed with respect to these satisfactions relations; the resulting semantics coincide for knowledge bases that are parametrically uniform. Finally, in Section 5 we conclude and point out further work.

2 Motivation for Instantiation Restrictions

Simply grounding the relational knowledge base $\mathcal{R}_{CC}$ given in Section 1 easily leads to an inconsistency. Using $D_{CC} = \{a, b\}$ yields the four ground instances

$$
\begin{aligned}
c_3^{aa} &: \ (cold(a)|contact(a, a) \wedge cold(a)))[0.6] \\
c_3^{ab} &: \ (cold(a)|contact(a, b) \wedge cold(b)))[0.6] \\
c_3^{ba} &: \ (cold(b)|contact(b, a) \wedge cold(a)))[0.6] \\
c_3^{bb} &: \ (cold(b)|contact(b, b) \wedge cold(b)))[0.6]
\end{aligned}
$$

for c_3. E.g., the ground instance c_3^{aa} is unsatisfiable since for any probability distribution P, the conditional probability of $P(q|p \wedge q)$ must be 1 for any ground formulas p, q. As FO-PCL uses a grounding semantics requiring that all *admissible* ground instances of a conditional c must have the probability given by c, all FO-PCL conditionals have additionally a constraint formula restricting the admissible instantiations of free variables. For instance, attaching $U \neq V$ to c_3

and the trivial constraint $\top$ to both c_1 and c_2 in $\mathcal{R}_{CC}$ yields the knowledge base $\mathcal{R}_{CC}^{fo\text{-}pcl}$. Under FO-PCL semantics, $\mathcal{R}_{CC}^{fo\text{-}pcl}$ is consistent, where a probability distribution P satisfies an FO-PCL conditional r, denoted by $P \models_{\text{fopcl}} r$ iff all admissible ground instances of r have the probability specified by r.

In contrast, the aggregation semantics, as given in [8], does not consider instantiation restrictions, since its satisfaction relation (in this paper denoted by $\models_{\odot}^{no\text{-}ir}$ to indicate *no instantiation restriction*), is less strict with respect to probabilities of ground instances: $P \models_{\odot}^{no\text{-}ir} (B|A)[d]$ iff the quotient of the sum of all probabilities $P(B_i \wedge A_i)$ and the sum of $P(A_i)$ is d, where $(B_1|A_1), \ldots, (B_n|A_n)$ are the ground instances of $(B|A)$. In this way, the aggregation semantics is capable of balancing the probabilities of ground instances resulting in higher tolerance with respect to consistency issues. The knowledge base $\mathcal{R}_{CC}$ is consistent under aggregating semantics, but while the maximum entropy distribution P^{ME} for $\mathcal{R}_{CC}$ induced by aggregation semantics assigns probability 1.0 to both c_3^{aa} and c_3^{bb}, as is required for any probability distribution, we have $P^{\text{ME}}(c_3^{ab}) = P^{\text{ME}}(c_3^{ba}) = 0.027$. This low probability is enforced by that fact that P^{ME} aggregates the probabilities over *all* ground instances of c_3, including c_3^{aa} and c_3^{bb}. The satisfaction condition $P^{\text{ME}}(c_3) = 0.6$ requires

$$P^{\text{ME}}(c_3^{aa}) = \frac{P^{\text{ME}}(cold(a) \wedge contact(a,a) \wedge cold(a))}{P^{\text{ME}}(contact(a,a) \wedge cold(a))} = \frac{0.0057}{0.0057} = 1.0$$

$$P^{\text{ME}}(c_3^{ab}) = \frac{P^{\text{ME}}(cold(a) \wedge contact(a,b) \wedge cold(b))}{P^{\text{ME}}(contact(a,b) \wedge cold(b))} = \frac{0.0001}{0.0039} = 0.027$$

$$P^{\text{ME}}(c_3^{ba}) = \frac{P^{\text{ME}}(cold(b) \wedge contact(b,a) \wedge cold(a))}{P^{\text{ME}}(contact(b,a) \wedge cold(a))} = \frac{0.0001}{0.0039} = 0.027$$

$$P^{\text{ME}}(c_3^{bb}) = \frac{P^{\text{ME}}(cold(b) \wedge contact(b,b) \wedge cold(b))}{P^{\text{ME}}(contact(b,b) \wedge cold(b))} = \frac{0.0057}{0.0057} = 1.0$$

so that

$$P^{\text{ME}}(c_3) = \frac{0.0057+0.0001+0.0001+0.0057}{0.0057+0.0039+0.0039+0.0057} = 0.6$$

holds. In this example, we could add the deterministic conditional $(contact(U,U)|\top)[0.0]$, yielding a probability of $P^{\text{ME}}(c_3^{ab}) = 0.6$, and this is actually done in [11] for $\mathcal{R}_{CC}$. However, adding such sentences to a knowledge base may be seen to be not adequate, and moreover, such a work-around is not always possible.

Example 1 (Elephant Keeper). The elephant keeper example, adapted from [2] and also [3], models the relationships among elephants in a zoo and their keepers. $\mathcal{R}_{EK}$ consists of the following conditionals:

$$ek_1 : \ (likes(E,K) \mid \top)[0.9]$$
$$ek_2 : \ (likes(E, fred) \mid \top)[0.05]$$
$$ek_3 : \ (likes(clyde, fred) \mid \top)[0.85]$$

Conditional ek_1 models statistical knowledge about the general relationship between elephants and their keepers, whereas ek_2 represents knowledge about the exceptional keeper Fred. ek_3 models subjective belief about the relationship between the elephant Clyde and Fred. From a common-sense point of view, the knowledge base $\mathcal{R}_{EK}$ makes perfect sense: ek_2 is an exception from ek_1, and ek_3 is an exception from ek_2. However, $\mathcal{R}_{EK}$ together with a set of constants may be inconsistent under aggregation semantics. This is due to the fact that ek_3 must be considered when satisfying ek_2 and that both ek_2 and ek_3 influence the satisfaction of ek_1, but aggregating over all probabilities may not be possible if there are not enough constants. For $\mathcal{R}_{EK}$, there is no obvious way of adding conditionals in order to guarantee consistency under aggregating semantics.

Example 2 (Elephant Keeper with instantiation restrictions). In FO-PCL, adding $K \neq fred$ to ek_1 and $E \neq clyde$ to ek_2 in $\mathcal{R}_{EK}$ yields the knowledge base $\mathcal{R}'_{EK}$

$$ek'_1 : \quad \langle (likes(E, K) \mid \top)[0.9], K \neq fred \rangle$$
$$ek'_2 : \quad \langle (likes(E, fred) \mid \top)[0.05], E \neq clyde \rangle$$
$$ek'_3 : \quad \langle (likes(clyde, fred) \mid \top)[0.85], \top \rangle$$

that avoids any inconsistency.

3 PCI Logic

We develop a logical framework PCI (*probabilistic conditionals with instantiation restrictions*) that uses probabilistic conditionals with and without instantiation restrictions and that provides different options for a satisfaction relation. The current syntax of PCI uses the syntax of FO-PCL [3,4]; in future work, we will extend PCI to also include additional features like specifying predicates to be irreflexive on the syntactical level. In order to precisely state the formal relationship among $\models_{\odot}^{no\text{-}ir}$, $\models_{fopcl}$, and the satisfaction relations offered by PCI, we first define the needed PCI components.

As FO-PCL, PCI uses function-free, sorted signatures of the form $\Sigma = (\mathcal{S}, \mathcal{D}, Pred)$. In a PCI-signature $\Sigma = (\mathcal{S}, \mathcal{D}, Pred)$, $\mathcal{S}$ is a set of sorts, $\mathcal{D} = \bigcup_{s \in \mathcal{S}} \mathcal{D}^{(s)}$ is a finite set of (disjoint) sets of sorted constant symbols, and *Pred* is a set of predicate symbols, each coming with an arity of the form $s_1 \times \ldots \times s_n \in \mathcal{S}^n$ indicating the required sorts for the arguments. Variables $\mathcal{V}$ also have a unique sort, and all formulas and variable substitutions must obey the obvious sort restrictions. In the following, we will adopt the unique names assumption, i.e. different constants denote different elements. The set of all terms is defined as $Term_{\Sigma} := \mathcal{V} \cup \mathcal{D}$. Let $\mathcal{L}_{\Sigma}$ be the set of quantifier-free first-order formulas defined over Σ and $\mathcal{V}$ in the usual way.

Definition 1 (Instantiation Restriction). *An* instantiation restriction *is a conjunction of inequality atoms of the from $t_1 \neq t_2$ with $t_1, t_2 \in Term_{\Sigma}$. The set of all instantiation restriction is denoted by $\mathcal{C}_{\Sigma}$.*

Definition 2 (q-, p-, r-Conditional). *Let $A, B \in \mathcal{L}_{\Sigma}$ be quantifier-free first-order formulas over Σ and $\mathcal{V}$.*

1. *$(B|A)$ is called a* qualitative conditional *(or just* q-conditional*). A is the antecedence and B the consequence of the qualitative conditional. The set of all qualitative conditionals over $\mathcal{L}_{\Sigma}$ is denoted by $(\mathcal{L}_{\Sigma}|\mathcal{L}_{\Sigma})$.*

2. Let $(B|A) \in (\mathcal{L}_\Sigma|\mathcal{L}_\Sigma)$ be a qualitative conditional and let $d \in [0,1]$ be a real value. $(B|A)[d]$ is called a probabilistic conditional (or just p-conditional) with probability d. The set of all probabilistic conditionals over $\mathcal{L}_\Sigma$ is denoted by $(\mathcal{L}_\Sigma|\mathcal{L}_\Sigma)^{prob}$.

3. Let $(B|A)[d] \in (\mathcal{L}_\Sigma|\mathcal{L}_\Sigma)^{prob}$ be a probabilistic conditional and let $C \in \mathcal{C}_\Sigma$ be an instantiation restriction. $\langle(B|A)[d], C\rangle$ is called a instantiation restricted conditional (or just r-conditional). The set of all instantiation restricted conditionals over $\mathcal{L}_\Sigma$ is denoted by $(\mathcal{L}_\Sigma|\mathcal{L}_\Sigma)^{prob}_{\mathcal{C}_\Sigma}$.

Instantiation restricted qualitative conditionals are defined analogously. If it is clear from the context, we may omit *qualitative, probabilistic,* and *instantiation restricted* and just use the term *conditional*.

Definition 3 (PCI knowledge base). *A pair* $(\Sigma, \mathcal{R})$ *consisting of a PCI-signature* $\Sigma = (\mathcal{S}, \mathcal{D}, Pred)$ *and a set of instantiation restricted conditionals* $\mathcal{R} = \{r_1, \ldots, r_m\}$ *with* $r_i \in (\mathcal{L}_\Sigma|\mathcal{L}_\Sigma)^{prob}_{\mathcal{C}_\Sigma}$ *is called a PCI knowledge base.*

For an instantiation restricted conditional $r = \langle(B|A)[d], C\rangle$, $\Theta_\Sigma(r)$ denotes the set of all ground substitutions with respect to the variables in r. A ground substitution $\theta \in \Theta_\Sigma(r)$ is applied to the formulas A, B and C in the usual way, i. e. each variable is replaced by a certain constant according to the mapping $\theta = \{v_1/c_1, \ldots, v_l/c_l\}$ with $v_i \in \mathcal{V}$, $c_i \in \mathcal{D}$, $1 \leq i \leq l$. So $\theta(A), \theta(B)$, and $\theta(C)$ are ground formulas and we have $\theta((B|A)) := (\theta(B)|\theta(A))$.

Given a ground substitution θ over the variables occurring in an instantiation restriction $C \in \mathcal{C}_\Sigma$, the evaluation of C under θ, denoted by $[\![C]\!]_\theta$, yields *true* iff $\theta(t_1)$ and $\theta(t_2)$ are different constants for all $t_1 \neq t_2 \in C$.

Definition 4 (Admissible Ground Substitutions and Instances). *Let* $\Sigma = (\mathcal{S}, \mathcal{D}, Pred)$ *be a many-sorted signature. and let* $r = \langle(B|A)[d], C\rangle \in (\mathcal{L}_\Sigma|\mathcal{L}_\Sigma)^{prob}_{\mathcal{C}_\Sigma}$ *be an instantiation restricted conditional. The set of* admissible ground substitutions *of* r *is defined as*

$$\Theta_\Sigma^{adm}(r) := \{\theta \in \Theta_\Sigma(r) \mid [\![C]\!]_\theta = true\}$$

The set of admissible ground instances *of* r *is defined as*

$$gnd_\Sigma(r) := \{\theta(B|A)[d] \mid \theta \in \Theta_\Sigma^{adm}(r)\}$$

For a PCI-signature $\Sigma = (\mathcal{S}, \mathcal{D}, Pred)$, the *Herbrand base* $\mathcal{H}_\Sigma$ with respect to Σ is the set of all ground atoms constructible from $Pred$ and $\mathcal{D}$. Every subset $\omega \subseteq \mathcal{H}_\Sigma$ is a *Herbrand interpretation*, defining a logical semantics for $\mathcal{R}$. The set $\Omega_\Sigma := \{\omega \mid \omega \subseteq \mathcal{H}_\Sigma\}$ denotes the set of all Herbrand interpretations. Herbrand interpretations are also called *possible worlds*.

Definition 5 (PCI Interpretation). *The probabilistic semantics of* $(\Sigma, \mathcal{R})$ *is a possible world semantics [6] where the ground atoms in* $\mathcal{H}_\Sigma$ *are binary random variables. A PCI interpretation* P *of a knowledge base* $(\Sigma, \mathcal{R})$ *is thus a probability distribution* $P : \Omega_\Sigma \to [0,1]$. *The set of all probability distributions over* Ω_Σ *is denoted by* $\mathcal{P}_{\Omega_\Sigma}$.

Currently, the PCI framework offers two different satisfaction relations: $\models_{\mathcal{G}}^{\mathrm{pci}}$ is based on grounding as in FO-PCL, and $\models_{\odot}^{\mathrm{pci}}$ extends aggregation semantics to r-conditionals.

Definition 6 (PCI Satisfaction Relations). *Let* $P \in \mathcal{P}_{\Omega}$ *and let* $\langle (B|A)[d], C \rangle \in (\mathcal{L}_{\Sigma}|\mathcal{L}_{\Sigma})_{\mathcal{C}_{\Sigma}}^{prob}$ *be an r-conditional with* $\sum_{\theta \in \Theta_{\Sigma}^{adm}(\langle (B|A)[d],C \rangle)} P(\theta(A)) > 0$. *The two PCI satisfaction relations* $\models_{\mathcal{G}}^{\mathrm{pci}}$ *and* $\models_{\odot}^{\mathrm{pci}}$ *are defined by:*

$$P \models_{\mathcal{G}}^{\mathrm{pci}} \langle (B|A)[d], C \rangle \quad iff \quad \frac{P(\theta(A \wedge B))}{P(\theta(A))} = d \quad \begin{array}{l} for\ all \\ \theta \in \Theta_{\Sigma}^{adm}(\langle (B|A)[d], C \rangle) \end{array} \quad (1)$$

$$P \models_{\odot}^{\mathrm{pci}} \langle (B|A)[d], C \rangle \quad iff \quad \frac{\displaystyle\sum_{\theta \in \Theta_{\Sigma}^{adm}(\langle (B|A)[d],C \rangle)} P(\theta(A \wedge B))}{\displaystyle\sum_{\theta \in \Theta_{\Sigma}^{adm}(\langle (B|A)[d],C \rangle)} P(\theta(A))} = d \quad (2)$$

We say that P *(PCI-)satisfies* $\langle (B|A)[d], C \rangle$ *under* grounding *semantcis iff* $P \models_{\mathcal{G}}^{\mathrm{pci}} \langle (B|A)[d], C \rangle$. *Correspondingly,* P *(PCI-)satisfies* $\langle (B|A)[d], C \rangle$ *under* aggregating *semantcis iff* $P \models_{\odot}^{\mathrm{pci}} \langle (B|A)[d], C \rangle$.

As usual, the satisfaction relations $\models_{\triangle}^{\mathrm{pci}}$ with $\triangle \in \{\mathcal{G}, \odot\}$ are extended to a set of conditionals $\mathcal{R}$ by defining

$$P \models_{\triangle}^{\mathrm{pci}} \mathcal{R} \quad iff \quad P \models_{\triangle}^{\mathrm{pci}} r \text{ for all } r \in \mathcal{R}.$$

The following proposition states that PCI properly captures both the instantiation-based semantics $\models_{\mathrm{fopcl}}$ of FO-PCL [3] and the aggregation semantics $\models_{\odot}^{no\text{-}ir}$ of [8] (cf. Section 2).

Proposition 1 (PCI captures FO-PCL and aggregation semantics). *Let* $\langle (B|A)[d], C \rangle$ *be an r-conditional and let* $(B|A)[d]$ *be a p-conditional, respectively. Then the following holds:*

$$P \models_{\mathcal{G}}^{\mathrm{pci}} \langle (B|A)[d], C \rangle \quad iff \quad P \models_{\mathrm{fopcl}} \langle (B|A)[d], C \rangle \quad (3)$$

$$P \models_{\odot}^{\mathrm{pci}} \langle (B|A)[d], \top \rangle \quad iff \quad P \models_{\odot}^{no\text{-}ir} (B|A)[d] \quad (4)$$

4 PCI Logic and Maximum Entropy Semantics

If a knowledge base $\mathcal{R}$ is consistent, there are usually many different models satisfying $\mathcal{R}$. The *principle of maximum entropy* [10,7] chooses the unique distribution which has maximum entropy among all distributions satisfying a knowledge base $\mathcal{R}$. Applying this principle to the satisfaction relations $\models_{\mathcal{G}}^{\mathrm{pci}}$ and $\models_{\odot}^{\mathrm{pci}}$ yields

$$P_{\mathcal{R}}^{\mathrm{ME}_{\triangle}} = \arg \max_{P \in \mathcal{P}_{\Omega} : P \models_{\triangle}^{\mathrm{pci}} \mathcal{R}} H(P) \quad (5)$$

with $\triangle$ being $\mathcal{G}$ or $\odot$, and where

$$H(P) = - \sum_{\omega \in \Omega} P(\omega) \log P(\omega)$$

is the *entropy* of a probability distribution P.

Example 3 (Misanthrope). The knowledge base $\mathcal{R}_{MI}$ adapted from [3] models friendship relations within a group of people with one exceptional member, a misanthrope. In general, if a person V likes another person U, then it is very likely that U likes V, too. But there is one person, the misanthrope, who generally does not like other people:

$$mi_1 : \quad \langle (likes(U,V) | likes(V,U)))[0.9], U \neq V \rangle$$
$$mi_2 : \quad \langle (likes(a,V) | \top)[0.05], V \neq a \rangle$$

Within the PCI framework, consider $\mathcal{R}_{MI}$ together with constants $\mathcal{D} = \{a, b, c\}$ and the corresponding ME distributions $P_{\mathcal{R}_{MI}}^{\mathrm{ME}_{\mathcal{G}}}$ and $P_{\mathcal{R}_{MI}}^{\mathrm{ME}_{\odot}}$ under grounding and aggregation semantics, respectively.

Under $P_{\mathcal{R}_{MI}}^{\mathrm{ME}_{\odot}}$, all six ground conditionals emerging from mi_1 have probability 0.9, for instance, $P_{\mathcal{R}_{MI}}^{\mathrm{ME}_{\mathcal{G}}}(likes(a,b) \mid likes(b,a)) = 0.9$.

On the other hand, for the distribution $P_{\mathcal{R}_{MI}}^{\mathrm{ME}_{\odot}}$, we have $P_{\mathcal{R}_{MI}}^{\mathrm{ME}_{\odot}}(likes(a,b) \mid likes(b,a)) = 0.46$ and $P_{\mathcal{R}_{MI}}^{\mathrm{ME}_{\odot}}(likes(a,c) \mid likes(c,a)) = 0.46$, while the other four ground conditionals resulting from mi_1 have probability 0.97.

Example 3 shows that in general the ME model under grounding semantics of a PCI knowledge base $\mathcal{R}$ differs from its ME model under aggregating semantics. However, if $\mathcal{R}$ is *parametrically uniform* [3,4], the situation changes. Parametric uniformity of a knowledge base $\mathcal{R}$ is introduced in [3] and refers to the fact that the ME distribution satisfying a set of m ground conditionals can be represented by a set of m optimization parameters. A relational knowledge base $\mathcal{R}$ is parametrically uniform iff for every conditional $r \in \mathcal{R}$, *all ground instances* of r have the *same* optimization parameter (see [3,4] for details). For instance, the knowledge base $\mathcal{R}'_{EK}$ from Example 2 is parametrically uniform, while the knowledge base $\mathcal{R}_{MI}$ from Example 3 is not parametrically uniform. Thus, if $\mathcal{R}$ is parametrically uniform, just *one* optimization parameter for each conditional $r \in \mathcal{R}$ instead of one optimization parameter for *each ground instance of* r has to be computed; this can be exploited when computing the ME distribution. In [9], a set of transformation rules is developed that transforms any consistent knowledge base $\mathcal{R}$ into a knowledge base $\mathcal{R}'$ such that $\mathcal{R}$ and $\mathcal{R}'$ have the same ME model under grounding semantics and $\mathcal{R}'$ is parametrically uniform.

Using the PCI framework providung both grounding and aggregating semantics for conditionals with instantiation restrictions, we can show that the ME models for grounding and aggregation semantics coincide if $\mathcal{R}$ is parametrically uniform.

Proposition 2 ($\mathcal{R}$ parametrically uniform implies $P_{\mathcal{R}}^{\mathrm{ME}_{\mathcal{G}}} = P_{\mathcal{R}}^{\mathrm{ME}_{\odot}}$). *Let $\mathcal{R}$ be a PCI knowledge base. If $\mathcal{R}$ is parametrically uniform, then $P_{\mathcal{R}}^{\mathrm{ME}_{\mathcal{G}}} = P_{\mathcal{R}}^{\mathrm{ME}_{\odot}}$.*

5   Conclusions and Further Work

In this paper, we considered two recently proposed semantics for relational probabilistic conditionals. While FO-PCL uses instantiation restrictions for the free variables occurring in a conditional, aggregation semantics avoids inconsistencies by aggregating probabilites over the ground instances of a conditional. While the latter allows more flexibility, we pointed out some shortcomings of aggregating semantics that surface especially if the set of constants in the underlying universe is small relative to the number of exceptional individuals occurring in a knowledge base. Based on these observations, we developed the framework PCI that extends aggregation semantics so that also instantiation restrictions can be taken into account, but without given up the flexibility of aggregating over different probabilities. PCI captures both grounding semantics and aggregating semantics without instantiation restrictions as special cases. For the case that a knowledge base is parametrically uniform, grounding and aggregating semantics coincide when employing the maximum entropy principle. In future work, we will extend our investigations also to the *averaging* [11] and other ME semantics.

References

1. Beierle, C., Finthammer, M., Kern-Isberner, G., Thimm, M.: Evaluation and Comparison Criteria for Approaches to Probabilistic Relational Knowledge Representation. In: Bach, J., Edelkamp, S. (eds.) KI 2011. LNCS, vol. 7006, pp. 63–74. Springer, Heidelberg (2011)
2. Delgrande, J.: On first-order conditional logics. Artificial Intelligence 105, 105–137 (1998)
3. Fisseler, F.: Learning and Modeling with Probabilistic Conditional Logic. Dissertations in Artificial Intelligence, vol. 328. IOS Press, Amsterdam (2010)
4. Fisseler, J.: First-order probabilistic conditional logic and maximum entropy. Logic Journal of the IGPL (to appear, 2012)
5. Getoor, L., Taskar, B. (eds.): Introduction to Statistical Relational Learning. MIT Press (2007)
6. Halpern, J.: Reasoning About Uncertainty. MIT Press (2005)
7. Kern-Isberner, G.: Conditionals in Nonmonotonic Reasoning and Belief Revision. LNCS (LNAI), vol. 2087. Springer, Heidelberg (2001)
8. Kern-Isberner, G., Thimm, M.: Novel semantical approaches to relational probabilistic conditionals. In: Lin, F., Sattler, U., Truszczynski, M. (eds.) Proceedings Twelfth International Conference on the Principles of Knowledge Representation and Reasoning, KR 2010, pp. 382–391. AAAI Press (2010)
9. Krämer, A., Beierle, C.: On Lifted Inference for a Relational Probabilistic Conditional Logic with Maximum Entropy Semantics. In: Lukasiewicz, T., Sali, A. (eds.) FoIKS 2012. LNCS, vol. 7153, pp. 224–243. Springer, Heidelberg (2012)
10. Paris, J.: The uncertain reasoner's companion – A mathematical perspective. Cambridge University Press (1994)
11. Thimm, M., Kern-Isberner, G., Fisseler, J.: Relational Probabilistic Conditional Reasoning at Maximum Entropy. In: Liu, W. (ed.) ECSQARU 2011. LNCS, vol. 6717, pp. 447–458. Springer, Heidelberg (2011)

Artificial Intelligence for Identification of Material Behaviour Using Uncertain Load and Displacement Data

Steffen Freitag

Georgia Institute of Technology,
210 Technology Circle, Savannah GA 31407, USA
steffen.freitag@gtsav.gatech.edu

Abstract. A concept is presented for identification of time-dependent material behaviour. It is based on two approaches in the field of artificial intelligence. Artificial neural networks and swarm intelligence are combined to create constitutive material formulations using uncertain measurement data from experimental investigations. Recurrent neural networks for fuzzy data are utilized to describe uncertain stress-strain-time dependencies. The network parameters are identified by an indirect training with uncertain data of inhomogeneous stress and strain fields. The real experiment is numerically simulated within a finite element analysis. Particle swarm optimization is applied to minimize the distance between measured and computed uncertain displacement data. After parameter identification, recurrent neural networks for fuzzy data can be applied as material description within fuzzy or fuzzy stochastic finite element analyses.

Keywords: artificial neural network, particle swarm optimization, fuzzy data, finite element analysis, material behaviour.

1   Introduction

The application of new materials in engineering practice as well as the evaluation of existing structures require knowledge about their behaviour. Material tests can be performed. As a result of experimental investigations, data series (processes) for measured structural actions and responses are available. However, often only imprecise information can be obtained due to varying boundary conditions, inaccuracies in measurements and measurement devices, and incomplete sets of observations. Imprecise measurements can be described by the uncertainty model fuzziness, see e.g. [14]. Time-dependent structural actions and responses are quantified as fuzzy processes.

Commonly, constitutive models are used to describe material behaviour by means of stress-strain relationships. Optimization approaches can be applied to determine unknown parameters of predefined models using stress and strain patterns from experimental data. An alternative is the application of artificial intelligence for identification of constitutive material behaviour. Artificial neural

E. Hüllermeier et al. (Eds.): SUM 2012, LNAI 7520, pp. 606–611, 2012.

networks can be used to describe material behaviour, see e.g. [8]. Similar to constitutive models, they can be implemented as material formulation within the finite element method (FEM), see e.g. [7].

In [2], an approach for uncertain material formulations (stress-strain-time dependencies) is presented, which is based on recurrent neural networks for fuzzy data [3]. For parameter identification (network training), a modified backpropagation algorithm is used. In this paper, particle swarm optimization (PSO) [9] is applied for network training. The PSO approach presented in [4] is utilized, which can deal with fuzzy network parameters. This also enables to create special network structures with dependent network parameters considering physical boundary conditions of investigated materials.

Whereas in prior works stress and strain processes are required for network training, a new concept is introduced using measured uncertain load and displacement processes of specimens. This indirect training approach requires a numerical simulation of the experiment. A finite element analysis (FEA) is performed to compute displacements due to the applied forces. PSO is applied to minimize the distance between experimentally and numerically obtained uncertain displacements resulting in appropriate fuzzy or deterministic network parameters.

2   Artificial Neural Networks for Constitutive Material Behaviour

Artificial neural networks can be utilized to describe constitutive material behaviour, i.e. relationships between stresses and strains. Whereas feed forward neural networks can be used for nonlinear elastic behaviour, recurrent neural networks are suitable to capture time-dependent phenomena (viscoelasticity), see e.g. [18]. Recurrent neural networks consist of an input layer, a number of hidden layers and an output layer. The neurons of each layer are connected by synapses to the neurons of the previous layer. In contrast to feed forward neural networks, hidden and output neurons are also connected to context neurons to store the material history. Recurrent neural networks can be applied to map the components of the time-varying strain tensor onto the components of the time-varying stress tensor or vice versa. For 3D material formulations, six input and output neurons are required representing the six independent stress and strain components of the respective tensors. Accordingly, the number of input and output neurons is three for 2D formulations and one for 1D formulations. The number of hidden layers and neurons can be defined with respect to the complexity of the material formulation. In general, fully or partially connected network architectures with symmetric substructures can be created [4] in order to reflect special material characteristics, e.g. anisotropy, orthotropy or isotropy.

In order to deal with imprecise data, recurrent neural networks for fuzzy data have been developed, see e.g. [3]. The network outputs can be computed using α-cuts and interval arithmetic. If neural networks are applied as material formulation within a fuzzy or a fuzzy stochastic FEA, an α-level optimization (e.g. according to [16]) is required to calculate the fuzzy stress components (network outputs), see [2].

Determination of the tangential stiffness matrix of the neural network based material formulation requires to evaluate the partial derivatives of the network outputs with respect to the network inputs. This is obtained by multiple applications of the chain rule, see [2].

3   Identification of Network Parameters

The deterministic or fuzzy network parameters can be identified using imprecise load and displacement data obtained from experimental investigations. For direct network training, patterns of input and output data (i.e. components of the strain and stress tensors) are required. An optimization can be performed minimizing the averaged total training error

$$
E^{av} = \frac{1}{H} \sum_{h=1}^{H} \left[\frac{1}{N_h} \sum_{n=1}^{N_h} {}^{[n]}E \right] ,
\tag{1}
$$

where $h = 1, \ldots, H$ are the training patterns and N_h is the number of time steps per pattern. The error of each time step n can be computed by

$$
{}^{[n]}E = \frac{1}{K \cdot S} \sum_{k=1}^{K} \sum_{s=1}^{S} \left[\left({}^{[n]}_{sl}\sigma_k - {}^{[n]}_{sl}\sigma_k^* \right)^2 + \left({}^{[n]}_{su}\sigma_k - {}^{[n]}_{su}\sigma_k^* \right)^2 \right] ,
\tag{2}
$$

which contains a distance measure for neural network computed fuzzy stress components ${}^{[n]}\tilde{\sigma}_k$ and experimentally obtained fuzzy stress components ${}^{[n]}\tilde{\sigma}_k^*$ represented by their lower l and upper u interval bounds of each α-cut s. In Equation (2), K is the number of stress components and S is the number of α-cuts.

In general, it is difficult to determine all components of the stress and strain tensors from measurements of displacements and forces. Boundary conditions are often not fulfilled to assume homogeneous stress and strain fields in the specimens. As an alternative, an indirect training approach is presented.

3.1   Finite Element Simulation of Experiments

The load-displacement behaviour of specimens can be simulated numerically using the FEM. For constitutive material models, an inverse analysis can be performed to identify unknown material parameters, see e.g. [11] and [17]. Displacements at measurement points of real and virtual specimens can be used to solve an optimization task. The objective is the minimization of the difference between computed and measured displacements in order to find the material parameters of an a priory defined material model.

This concept can be adapted to identify deterministic or fuzzy network parameters. In [5], an indirect training approach (autoprogressive algorithm) is presented for feed forward neural networks. Deterministic stress and strain data are required within the used gradient based backpropagation network training.

They are computed adaptively within a FEA taking the difference between computed and measured displacements into account. This approach may be extended to recurrent neural networks for fuzzy data.

If optimization strategies without using gradient information are applied for network training, deterministic or fuzzy network parameters can be identified indirectly by minimizing the difference between computed and measured displacements. Swarm intelligence [10] can be used to solve this optimization task.

3.2   Particle Swarm Optimization

PSO [9] is a biosocial motivated search algorithm. It can be used for training of feed forward neural networks, see e.g. [12] and [13]. In [4], a PSO approach using fuzzy or interval numbers is presented for training of feed forward and recurrent neural networks. Using PSO for training of recurrent neural networks enables to modify all network parameters during training. This is an improvement compared to backpropagation training, where only parameters of the forward connections can be updated. Additionally, individual network structures with dependent parameters can be created in order to consider physical boundary conditions of investigated materials, e.g. anisotropy, orthotropy or isotropy.

The PSO approach presented in [4] can also be applied to determine deterministic or fuzzy network parameters indirectly using uncertain load and displacement data. A swarm consists of $p = 1, \ldots, P$ particles representing identical network architectures with different sets of network parameters. The following procedure is repeated until the functional value of the objective function (1) of a particle is less than a predefined value or a maximal number of runs is reached.

For each particle of the swarm, a numerical simulation of the experiment is performed to compute fuzzy displacement processes at selected measurement points $q = 1, \ldots, Q$. The fuzzy FEM is applied using a recurrent neural network for fuzzy data as material formulation, see Section 2. For each time step n and each component q, an α-level optimization according to [16] is carried out. The computed fuzzy displacement components $^{[n]}\tilde{v}_q$ are compared with experimentally obtained displacements $^{[n]}\tilde{v}_q^*$. In contrast to Equation (2), the error of time step n is defined by

$$^{[n]}E = \frac{1}{Q \cdot S} \sum_{q=1}^{Q} \sum_{s=1}^{S} \left[\left({}_{sl}^{[n]}v_q - {}_{sl}^{[n]}v_q^* \right)^2 + \left({}_{su}^{[n]}v_q - {}_{su}^{[n]}v_q^* \right)^2 \right] . \tag{3}$$

The objective function (1) is evaluated for each particle. The parameter set of the particle with the least value of the objective function is stored as global best of all runs. The individual best parameter sets of each particle are updated, if the value of the objective function is less than the current individual best.

The new parameter set of each particle is computed based on information of its own search history, information of other particles, and random influences. In general, information between particles in the swarm can be shared differently, see e.g. [1]. If a fully connected swarm topology is selected, each particle gets information from all other particles and also transmits its search experience to all other particles of the swarm. Details are presented in [4] for the update of deterministic, interval, and fuzzy parameters.

4   Application for Structural Analysis

Experimental investigations with different loading scenarios and boundary conditions are required to get realistic material formulations. Validation of neural network based material formulations is realized by additional load-displacement-time dependencies which are not used for parameter identification.

After identification of deterministic or fuzzy network parameters, the neural network based material formulation can be applied for several tasks in structural analysis, e.g. reliability assessment, structural design, and lifetime prediction.

Fuzzy response processes of arbitrary structures can be computed within fuzzy FEA. The same computational approach (α-level optimization) as used for indirect network training can be applied, see [16].

The proposed neural network based material formulation is based on the uncertainty model fuzziness. However, it can also be applied within the generalized uncertainty model fuzzy randomness [15]. This enables to perform fuzzy stochastic FEA, see e.g. [6] and [19] resulting in fuzzy stochastic structural responses such as fuzzy failure probabilities. Due to the high computational effort (e.g. Monte Carlo simulation with FEA), the fuzzy stochastic analysis can be realized with numerical efficient surrogate models.

5   Conclusion

A new concept has been presented for identification of uncertain material behaviour. Fuzzy load and displacement processes of real and virtual experiments can be used to create uncertain stress-strain-time dependencies based on recurrent neural networks. PSO is applied to determine fuzzy or deterministic network parameters. Recurrent neural networks for fuzzy data can be used as material formulation within fuzzy or fuzzy stochastic FEA.

Acknowledgments. The author gratefully acknowledges the support of the Deutsche Forschungsgemeinschaft (DFG – German Research Foundation) within the project (FR 3044/1-1) in the framework of a research fellowship.

References

1. Fontan, M., Ndiaye, A., Breysse, D., Bos, F., Fernandez, C.: Soil–structure interaction: Parameters identification using particle swarm optimization. Computers and Structures 89(17-18), 1602–1614 (2011)
2. Freitag, S., Graf, W., Kaliske, M.: Recurrent Neural Networks for Fuzzy Data as a Material Description within the Finite Element Method. In: Tsompanakis, Y., Topping, B.H.V. (eds.) Proceedings of the Second International Conference on Soft Computing Technology in Civil, Structural and Environmental Engineering, Chania, paper 28. Civil-Comp Press, Stirlingshire (2011)

3. Freitag, S., Graf, W., Kaliske, M.: Recurrent Neural Networks for Fuzzy Data. Integrated Computer-Aided Engineering 18(3), 265–280 (2011)
4. Freitag, S., Muhanna, R.L., Graf, W.: A Particle Swarm Optimization Approach for Training Artificial Neural Networks with Uncertain Data. In: Vořechovský, M., Sadílek, V., Seitl, S., Veselý, V., Muhanna, R.L., Mullen, R.L. (eds.) Proceedings of the 5th International Conference on Reliable Engineering Computing, Litera, Brno, pp. 151–170 (2012)
5. Ghaboussi, J., Pecknold, D.A., Zhang, M., Haj-Ali, R.M.: Autoprogressive training of neural network constitutive models. International Journal for Numerical Methods in Engineering 42(1), 105–126 (1998)
6. Graf, W., Sickert, J.-U., Freitag, S., Pannier, S., Kaliske, M.: Neural Network Approaches in Structural Analysis under Consideration of Imprecision and Variability. In: Tsompanakis, Y., Topping, B.H.V. (eds.) Soft Computing Methods for Civil and Structural Engineering, ch. 4, pp. 59–85. Saxe-Coburg Publications, Stirlingshire (2011)
7. Hashash, Y.M.A., Jung, S., Ghaboussi, J.: Numerical implementation of a neural network based material model in finite element analysis. International Journal for Numerical Methods in Engineering 59(7), 989–1005 (2004)
8. Jung, S., Ghaboussi, J.: Neural network constitutive model for rate-dependent materials. Computers and Structures 84(15-16), 955–963 (2006)
9. Kennedy, J., Eberhart, R.C.: Particle Swarm Optimization. In: Proceedings of the IEEE International Conference on Neural Networks IV, pp. 1942–1948. IEEE, Piscataway (1995)
10. Kennedy, J., Eberhart, R.C., Yuhui, S.: Swarm Intelligence. Morgan Kaufmann, San Francisco (2001)
11. Kučerová, A., Lepš, M., Zeman, J.: Back Analysis of Microplane Model Parameters Using Soft Computing Methods. Computer Assisted Mechanics and Engineering Sciences 14(2), 219–242 (2007)
12. Kuok, K.K., Harun, S., Shamsuddin, S.M.: Particle swarm optimization feedforward neural network for modeling runoff. International Journal of Environmental Science and Technology 7(1), 67–78 (2010)
13. Mendes, R., Cortez, P., Rocha, M., Neves, J.: Particle Swarms for Feedforward Neural Network Training. In: Proceedings of the 2002 International Joint Conference on Neural Networks, pp. 1895–1899. IEEE, Honolulu (2002)
14. Möller, B., Beer, M.: Engineering computation under uncertainty – Capabilities of non-traditional models. Computers and Structures 86(10), 1024–1041 (2008)
15. Möller, B., Beer, M.: Fuzzy Randomness – Uncertainty in Civil Engineering and Computational Mechanics. Springer, Berlin (2004)
16. Möller, B., Graf, W., Beer, M.: Fuzzy structural analysis using α-level optimization. Computational Mechanics 26(6), 547–565 (2000)
17. Novák, D., Lehký, D.: ANN inverse analysis based on stochastic small-sample training set simulation. Engineering Application of Artificial Intelligence 19(7), 731–740 (2006)
18. Oeser, M., Freitag, S.: Modeling of materials with fading memory using neural networks. International Journal for Numerical Methods in Engineering 78(7), 843–862 (2009)
19. Sickert, J.-U., Freitag, S., Graf, W.: Prediction of uncertain structural behaviour and robust design. International Journal for Reliability and Safety 5(3/4), 358–377 (2011)

On Cluster Validity for Fuzzy Clustering of Incomplete Data

Ludmila Himmelspach[1], João Paulo Carvalho[2], and Stefan Conrad[1]

[1] Institute of Computer Science,
Heinrich-Heine-Universität Düsseldorf, Germany
[2] INESC-ID, TULisbon - Instituto Superior Técnico,
Lisbon, Portugal

Abstract. In this study, we address the problem of finding the optimal number of clusters on incomplete data using cluster validity functions. Experiments were performed on different data sets in order to analyze to what extent cluster validity indices adapted to incomplete data can be used for validation of clustering results. Moreover we analyze which fuzzy clustering algorithm for incomplete data produces better partitioning results for cluster validity.

Keywords: fuzzy cluster analysis, cluster validity, incomplete data.

1 Introduction

Clustering is an unsupervised learning technique for automatic exploring the distribution of objects in a data set. A great number of well-performing algorithms for assigning data objects into a pre-defined number of hard or fuzzy partitions has been previously proposed in the literature. However, since the quality of the resulting partitioning of a data set produced by clustering algorithms strongly depends on the assumed number of clusters, the determination of that parameter is a crucial problem in cluster analysis. A widely used method for estimating the optimal number of clusters consists in comparing the clustering solutions obtained for different numbers of clusters using *Cluster Validity Indices* (CVIs).

In [1], we addressed the question of to what extent the optimal number of clusters can be found when validating partitionings obtained on data with missing values. We adapted different cluster validity functions to incomplete data and used them for validation of clustering results obtained by fuzzy clustering methods for incomplete data. The experimental results showed that the optimal number of clusters could only be found for small percentages of missing values in data. Since in that study both the clustering algorithms and the cluster validity indices were adapted for incomplete data, in this work we address the problem of finding what factors cause such poor performance: the adaption of the clustering algorithms, the adaption of the validity functions, or the loss of information in data itself.

E. Hüllermeier et al. (Eds.): SUM 2012, LNAI 7520, pp. 612–618, 2012.

2   Cluster Validity for Clustering of Incomplete Data

Since cluster validity indices are computed using partitioning results obtained by clustering algorithms, the determined optimal number of clusters depends on both clustering methods and CVIs. In this study, we focus on three approaches for adapting the fuzzy c-means algorithm (FCM) [2] that handle missing values in different ways and therefore produce different partitioning results. We validated the clustering results using three cluster validity functions that consider different aspects of an optimal partitioning.

Table 1. Cluster Validity Indices

$$\mathrm{NPC}(U) = 1 - \frac{c}{c-1}\left(1 - \frac{1}{n}\sum_{k=1}^{n}\sum_{i=1}^{c} u_{ik}^2\right)$$

$$\mathrm{FHV}(U,X,\mu) = \sum_{i=1}^{c}\sqrt{\det\left(\frac{\sum_{k=1}^{n}(u_{ik})^m i_{kp}i_{kl}(x_{kp} - \mu_{ip})(x_{kl} - \mu_{il})^T}{\sum_{k=1}^{n}(u_{ik})^m i_{kp}i_{kl}}\right)}, \; i_{kj} = \begin{cases} 1 & \text{if } x_{kj} \in X_{avl} \\ 0 & \text{else} \end{cases}$$

$$\mathrm{FS}(U,X,\mu) = \sum_{k=1}^{n}\sum_{i=1}^{c} u_{ik}^2 D_{part}(x_k,\mu_i) - \sum_{k=1}^{n}\sum_{i=1}^{c} u_{ik}^2 \|\mu_i - \bar{\mu}\|, \; \bar{\mu} = \frac{1}{c}\sum_{i=1}^{c} \mu_i$$

2.1   Approaches for Fuzzy Clustering of Incomplete Data

Partial Distance Strategy (PDS): Pursuing the available case approach, the membership degrees of incomplete data items to clusters are calculated using the partial distance function [3]. The cluster prototypes are calculated only on the basis of all available feature values of data items.

Optimal Completion Strategy (OCS): In this approach missing values are used as additional components to minimize the objective function [3]. They are estimated depending on all cluster prototypes in an additional iteration step of the algorithm.

Nearest Prototype Strategy (NPS): The idea of this approach is to completely substitute missing values of an incomplete data item by the corresponding values of the cluster prototype to which it has the smallest partial distance [3].

2.2   Cluster Validity Indices for Incomplete Data

Normalized Partition Coefficient (NPC) rates a partitioning of a data set as optimal if data items are clearly assigned into clusters [4]. Since it uses only membership matrix for calculation, NPC can be applied in case of incomplete data without any changes.

Fuzzy Hypervolume (FHV) considers the compactness within clusters. It rates a fuzzy partitioning as optimal if the clusters are of minimal volume [5]. We adapted FHV to incomplete data according to [6].

Fukuyama-Sugeno (FS) index combines compactness and separation between clusters [7]. We adapted this index using partial distance function.

Table 1 summarizes original and adapted versions of CVIs, respectively. Note that in case of complete data these CVIs reduce to their original versions.

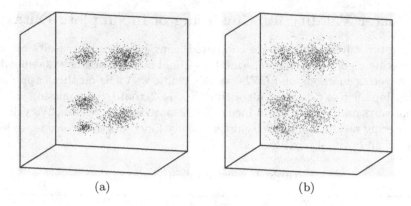

(a)   (b)

Fig. 1. Test data: (a) Data Set 1, (b) Data Set 2

3   Experimental Results

3.1   Test Data and Experimental Setup

We tested the described clustering algorithms and CVIs on several data sets
with different numbers of clusters. Due to the lack of space we only describe the
experimental results for two artificial data sets. Both data sets consist of 2000
data points generated by compositions of five 3-dimensional Gaussian distributions. Both data sets have the same mean values, and the second data set has
larger standard deviations. The five clusters build two groups of three and two
differently sized clusters. While all clusters in Data Set 1 are clearly separated
from each other, there is some overlap between clusters within two groups in
Data Set 2 (see Figure 1). We generated incomplete data sets by removing values in all dimensions with probabilities of 10%, 25% and 40% according to the
most common *missing completely at random (MCAR)* failure mechanism.

We first clustered the complete data sets with basic FCM with different numbers of clusters. Due to the hierarchical structure of clusters, many CVIs like
NPC determined two clusters in the data sets and only obtained their max-
ima/minima for $c = 5$. Only FHV and Fukuyama-Sugeno indices could correctly
determine the optimal number of clusters in both data sets (cf. Table 2 (left)).
In the first experiment we isolated and tested the performance of the adapted
CVIs by evaluating them on incomplete data using partitioning results produced
by FCM on complete data. In the second experiment, we validated partitioning
results obtained by three fuzzy clustering algorithms for incomplete data using
NPC and adapted FHV and Fukuyama-Sugeno indices.

3.2   Experimental Results

Table 2 (right) shows the performance results for the adapted FHV and
Fukuyama-Sugeno index on incomplete data using membership degrees and cluster prototypes obtained by FCM on complete data. Even for a large percentage

Table 2. Cluster validity results of clusterings produced by FCM and using complete (left) and incomplete data (right)

	Data set 1				Data set 1							
					10%			25%			40%		
CVIs	NPC	FHV	FS	CVIs	NPC	FHV	FS	NPC	FHV	FS	NPC	FHV	FS
c = 2	0.797	53.92	-201147	c = 2	0.797	53.69	-201057	0.797	53.04	-202566	0.797	53.28	-202774
c = 3	0.681	47.00	-226881	c = 3	0.681	46.96	-226893	0.681	46.35	-227429	0.681	46.25	-228392
c = 4	0.671	36.76	-223743	c = 4	0.671	36.75	-223765	0.671	36.34	-224061	0.671	37.44	-223602
c = 5	0.708	26.03	-273846	c = 5	0.708	26.02	-273874	0.708	26.17	-273982	0.708	26.25	-273696
c = 6	0.593	29.58	-224122	c = 6	0.593	29.49	-224104	0.593	29.49	-224225	0.593	29.95	-223999
c = 7	0.543	33.12	-216209	c = 7	0.543	33.04	-216195	0.543	33.24	-216272	0.543	33.47	-216121
c = 8	0.503	35.01	-210778	c = 8	0.503	34.87	-210756	0.503	35.25	-210845	0.503	35.56	-210695
	Data set 2				Data set 2								
c = 2	0.739	115.32	-170471	c = 2	0.739	115.26	-170322	0.739	114.49	-171419	0.739	113.38	-171207
c = 3	0.574	127.26	-190852	c = 3	0.574	126.87	-190913	0.574	125.50	-191817	0.574	125.71	-191694
c = 4	0.513	117.56	-172540	c = 4	0.513	117.67	-172435	0.513	117.02	-172921	0.513	116.67	-172399
c = 5	0.517	103.55	-202799	c = 5	0.517	103.52	-202719	0.517	104.46	-202751	0.517	102.89	-202688
c = 6	0.438	113.03	-168654	c = 6	0.438	112.90	-168610	0.438	114.17	-168645	0.438	112.17	-168549
c = 7	0.401	123.33	-163122	c = 7	0.401	123.64	-163081	0.401	123.99	-163137	0.401	121.72	-163050
c = 8	0.370	129.23	-148879	c = 8	0.370	129.90	-148835	0.370	129.43	-148922	0.370	126.94	-148835

Table 3. CV results of clusterings produced by PDSFCM on incomplete data

	Data set 1													
	10%					25%					40%			
CVIs / SM	NPC	FHV	FS	HR	DCP	NPC	FHV	FS	HR	DCP	NPC	FHV	FS	HR	DCP
c = 2	0.767	52.07	-202354	0.92	0.386	0.718	48.87	-204958	0.83	0.942	0.676	45.59	-208001	0.75	1.415
c = 3	0.663	44.83	-221928	0.92	0.264	0.638	40.53	-214374	0.83	0.834	0.608	35.14	-208274	0.74	1.500
c = 4	0.631	35.21	-217267	0.92	0.320	0.589	31.26	-212744	0.83	1.019	0.551	28.17	-205824	0.77	1.309
c = 5	0.655	25.28	-257279	0.93	0.131	0.582	24.30	-235732	0.85	0.490	0.544	21.98	-223380	0.79	0.777
c = 6	0.564	27.53	-217062	0.93	1.324	0.546	23.52	-215449	0.86	2.077	0.529	19.79	-211834	0.80	2.586
c = 7	0.535	28.72	-217460	0.93	0.370	0.510	24.80	-205263	0.86	2.984	0.522	18.99	-209022	0.81	3.484
c = 8	0.485	31.46	-192622	0.93	0.585	0.495	24.81	-196666	0.86	2.527	0.512	18.15	-201974	0.82	2.834
	Data set 2													
c = 2	0.704	112.26	-171335	0.93	0.328	0.664	106.23	-176452	0.83	0.803	0.613	95.68	-178915	0.76	1.648
c = 3	0.568	118.47	-190622	0.92	0.494	0.556	106.71	-191540	0.83	1.266	0.551	90.34	-194736	0.75	2.430
c = 4	0.494	111.05	-171356	0.92	0.688	0.483	99.94	-174912	0.84	1.481	0.522	77.33	-194253	0.77	2.901
c = 5	0.490	98.35	-195190	0.93	0.449	0.458	91.95	-189805	0.85	1.074	0.500	69.04	-203955	0.79	2.802
c = 6	0.432	103.47	-170006	0.92	1.495	0.440	87.91	-177594	0.85	2.696	0.499	63.86	-200461	0.80	5.145
c = 7	0.399	111.52	-162272	0.91	1.692	0.429	88.24	-175516	0.86	2.662	0.488	59.29	-196055	0.80	5.655
c = 8	0.371	116.83	-150911	0.93	1.051	0.408	88.00	-166817	0.86	2.539	0.478	57.29	-192358	0.81	5.133

Table 4. CV results of clusterings produced by OCSFCM on incomplete data

	Data set 1														
	10%					25%					40%				
CVIs / SM	NPC	FHV	FS	HR	DCP	NPC	FHV	FS	HR	DCP	NPC	FHV	FS	HR	DCP
c = 2	0.823	51.69	-210773	0.93	0.592	0.855	49.60	-219748	0.82	0.866	0.894	45.57	-238787	0.74	1.700
c = 3	0.723	44.48	-235427	0.92	0.428	0.776	40.31	-223027	0.81	0.817	0.822	36.63	-220472	0.70	1.439
c = 4	0.703	35.76	-229871	0.93	0.429	0.766	31.45	-252634	0.83	0.953	0.812	30.64	-257025	0.75	1.904
c = 5	0.727	25.77	-266899	0.92	0.331	0.764	25.41	-276824	0.84	0.875	0.796	26.26	-282493	0.79	1.510
c = 6	0.629	27.71	-229573	0.93	0.863	0.675	25.31	-251324	0.85	1.994	0.733	22.71	-262295	0.79	2.348
c = 7	0.565	31.34	-218060	0.92	1.268	0.627	26.47	-225223	0.85	2.532	0.692	23.73	-262888	0.79	3.438
c = 8	0.533	31.31	-202013	0.93	1.499	0.600	27.22	-236880	0.86	1.369	0.685	21.92	-238827	0.80	3.228
	Data set 2														
c = 2	0.769	111.23	-179969	0.92	0.578	0.811	106.30	-193870	0.82	1.106	0.854	96.01	-206988	0.72	1.836
c = 3	0.621	116.74	-197556	0.91	0.529	0.682	105.11	-192125	0.80	1.227	0.747	93.24	-217406	0.72	1.620
c = 4	0.555	110.89	-183545	0.91	0.603	0.624	102.55	-204446	0.82	1.686	0.696	82.51	-215742	0.75	3.593
c = 5	0.548	98.65	-205569	0.92	0.821	0.593	94.77	-200758	0.83	2.059	0.672	76.69	-220553	0.76	4.626
c = 6	0.485	102.08	-187271	0.92	1.708	0.553	91.57	-200363	0.84	2.421	0.639	72.47	-222156	0.77	4.636
c = 7	0.441	111.22	-167158	0.90	1.856	0.525	91.63	-202853	0.83	4.485	0.621	72.18	-221035	0.77	5.379
c = 8	0.415	116.64	-160721	0.92	1.707	0.494	94.11	-197417	0.84	3.840	0.593	71.99	-217658	0.79	4.223

Table 5. CV results of clusterings produced by NPSFCM on incomplete data

	Data set 1														
	10%					25%					40%				
CVIs / SM	NPC	FHV	FS	HR	DCP	NPC	FHV	FS	HR	DCP	NPC	FHV	FS	HR	DCP
c = 2	0.823	51.69	-210876	0.92	0.593	0.861	47.72	-227244	0.83	1.307	0.897	43.58	-243307	0.73	1.875
c = 3	0.721	44.60	-237804	0.93	0.560	0.789	39.54	-241932	0.79	1.463	0.850	32.30	-265685	0.71	2.342
c = 4	0.711	34.23	-236103	0.92	0.596	0.777	29.24	-255258	0.82	1.535	0.829	23.71	-277435	0.75	2.296
c = 5	0.739	24.74	-283561	0.93	0.538	0.784	22.59	-291786	0.85	1.314	0.834	18.82	-310182	0.78	2.391
c = 6	0.641	26.24	-239114	0.92	1.972	0.718	21.10	-266387	0.85	2.511	0.787	17.65	-298662	0.79	2.749
c = 7	0.586	28.70	-226936	0.91	2.542	0.673	22.55	-252513	0.85	2.695	0.773	16.72	-295092	0.79	4.018
c = 8	0.552	29.61	-212591	0.92	0.904	0.656	22.58	-263830	0.85	3.261	0.746	16.12	-281954	0.80	4.449
	Data set 2														
c = 2	0.769	111.09	-180745	0.92	0.624	0.816	102.95	-199409	0.81	1.334	0.861	90.60	-217180	0.74	2.313
c = 3	0.629	115.80	-199406	0.91	0.811	0.680	101.05	-226658	0.82	1.741	0.788	81.79	-239796	0.70	3.610
c = 4	0.565	107.06	-187282	0.90	1.268	0.653	89.67	-216146	0.81	2.513	0.740	68.93	-249989	0.74	3.899
c = 5	0.562	94.92	-213702	0.91	1.283	0.637	80.94	-236234	0.83	2.607	0.736	57.98	-279220	0.75	4.239
c = 6	0.496	98.50	-190611	0.91	1.905	0.605	76.24	-234962	0.83	3.917	0.714	53.74	-265106	0.76	5.960
c = 7	0.467	104.23	-184992	0.92	1.540	0.583	77.43	-231260	0.83	4.213	0.696	52.55	-269111	0.76	6.051
c = 8	0.436	108.10	-170956	0.92	1.847	0.556	76.29	-216932	0.84	3.755	0.685	48.31	-263907	0.77	6.604

of missing values in data sets, the values for both CVIs only slightly differ from the values obtained on complete data sets and both FHV and FS index could perfectly recognize the optimal number of clusters. That means that the adapted versions of FHV and FS maintain the properties of the original CVIs and the loss of values in data set does not have much effect on the performance of CVIs.

Tables 3, 4, 5 show validating partitioning results obtained by PDSFCM, OCS-FCM and NPSFCM on incomplete data. The additional columns give Hüller-meier/Rifqi index (HR) [8] between the membership matrices and Frobenius norm distance between the cluster prototyps (DCP) produced on complete and incomplete data. Although the values for CVIs differ from the original ones, all CVIs determined the same optimal number of clusters for small percentages of missing values as in the case of complete data. However, it is already possible to state some differences between the clustering algorithms: while PDSFCM slightly underestimates, OCSFCM and NPSFCM overestimate the values for NPC. This can be explained by the fact that OCSFCM and NPSFCM estimate missing values closer to the cluster prototypes achieving a compacter and clearer parti-tioning of a data set. Due to this fact, the values of NPC increase with increasing percentage of missing values in the data sets. Dealing with missing values in this way plays a central role in the case of overlapping clusters for a large percentage of missing values because the structure of the original data set can be changed for different numbers of clusters. The values for Hüllermeier/Rifqi index and DCP also show that the partitionings produced by PDSFCM on incomplete data are slightly more similar to the partitionings produced on complete data.

Comparing the performance of cluster validity indices the best results were obtained by Fukuyama-Sugeno index. It only failed for a large percentage of missing values on the data set with overlapping clusters. As we already stated in [1], FHV tends to overestimate the number of clusters with increasing number of missing values in a data set. That can be explained by the fact that clustering algorithms compute cluster prototypes close to available data items that are taken into account for calculation of FHV. Thus, with increasing number of clusters the distances between data items and cluster prototypes get smaller and the value for FHV as well. This property of FHV disqualifies this CVI for validating clustering results obtained on data with a large percentage of missing values unless the cluster prototypes can be well determined by clustering algorithms.

4 Conclusions and Future Work

In this study, we analyzed different kinds of cluster validity indices for finding the optimal number of clusters on incomplete data. In experiments, we showed that the partitioning results produced by clustering algorithms for incomplete data are the critical factor for cluster validity. We also showed that cluster validity indices that consider only one of the aspects of an optimal partitioning, e.g. only the compactness within clusters are more affected by the partitioning results produced by clustering algorithms on incomplete data. Because of that such CVIs fail to work for a high percentage of missing values in a data set.

As future work it is our intention to focus on the adapting CVIs to incomplete data that combine compactness and separation between clusters.

References

1. Himmelspach, L., Hommers, D., Conrad, S.: Cluster Tendency Assessment for Fuzzy Clustering of Incomplete Data. In: Procs. EUSFLAT 2011, pp. 290–297 (2011)
2. Bezdek, J.C.: Pattern Recognition with Fuzzy Objective Function Algorithms. Kluwer Academic Publishers (1981)
3. Hathaway, R.J., Bezdek, J.C.: Fuzzy c-means Clustering of Incomplete Data. IEEE Transactions on Systems, Man, and Cybernetics 31(5), 735–744 (2001)
4. Roubens, M.: Pattern classification problems and fuzzy sets. Fuzzy Sets and Systems 1, 239–253 (1978)
5. Gath, I., Geva, A.B.: Unsupervised Optimal Fuzzy Clustering. IEEE Transactions on Pattern Analysis and Machine Intelligence 11, 773–781 (1989)
6. Timm, H., Döring, C., Kruse, R.: Different Approaches to Fuzzy Clustering of Incomplete Datasets. Int. J. on Approximate Reasoning 35, 239–249 (2004)
7. Fukuyama, Y., Sugeno, M.: A new method for choosing the number of clusters for the fuzzy c-means method. In: 5th Fuzzy Systems Symposium, pp. 247–250 (1989)
8. Hüllermeier, E., Rifqi, M.: A Fuzzy Variant of the Rand Index for Comparing Clustering Structures. In: Procs. IFSA–EUSFLAT 2009, pp. 1294–1298 (2009)

Evaluation of the Naive Evidential Classifier (NEC): A Comparison between Its Two Variants Based on a Real Agronomic Application

Yosra Mazigh[1], Boutheina Ben Yaghlane[2], and Sébastien Destercke[3]

[1] LARODEC Laboratory, University of Tunis
[2] LARODEC Laboratory, University of Carthage
[3] CNRS UMR 7253 Heuristique et diagnostique des systèmes complexes

Abstract. We introduce the notion of naive evidential classifier. This classifier, which has a structure mirroring the naive Bayes classifier, is based on the Transferable Belief Model and uses mass assignments as its uncertainty model. This new method achieves more robust inferences, mainly by explicitly modeling imprecision when data are in little amount or are imprecise. After introducing the model and its inference process based on Smet's generalized Bayes theorem (GBT), we specify some possible methods to learn its parameters, based on the Imprecise Dirichlet Model (IDM) or on predictive belief functions. Some experimental results on an agronomic application are then given and evaluated.

1 Introduction

When modeling and processing uncertainty, computing on multivariate spaces is an important issue. If $X_1, \ldots, X_N$ is a set of variables assuming their values over some finite spaces $\mathcal{X}_1, \ldots, \mathcal{X}_N$, defining directly a joint uncertainty model over the Cartesian product $\mathcal{X}_1 \times \ldots \times \mathcal{X}_N$ and making some inferences with this model is often impossible in practice. The use of graphical models based on network architecture can solve this problem by decomposing the joint uncertainty model into several pieces of conditional models. This decomposition is possible thanks to conditional independence property. Note that outside their computational tractability, another attractive feature of such models is their readability for non-experts, thanks to their graphical aspects.

The aim of this paper is first to introduce the notion of the Naive Evidential Classifier (NEC) as the counterpart of the Naive Bayes (NB) in evidential theory. It uses the directed evidential network structure (DEVN) proposed by Ben Yaghlane in [1] and preforms inference by using the modified binary join tree (MBJT) algorithm, which uses the disjunctive rule of combination (DRC) and the generalized Bayesian theorem (GBT), both proposed by Smets in [2].

In Section 2, we briefly present the Basics of NEC. Section 3 then provides some details about the practical instanciation and use of the NEC structure. Finally, Sections 4 and 5 present some preliminary experiments on an agronomical problem and on an UCI data set. Due to lack of spaces, only the essential elements are provided, and the reader is referred to references for details.

E. Hüllermeier et al. (Eds.): SUM 2012, LNAI 7520, pp. 619–624, 2012.

2   Naive Evidential Classifier (NEC)

NEC is the TBM counterpart of the Naive Bayes (NB) classifier. Thereby, it is a graphical model having two parts: a qualitative and a quantitative one. Recall that the aim of a classifier is learn a mapping from input values $X_1, \ldots, X_N \in \mathcal{X}_1 \times \ldots \times \mathcal{X}_N$ to an output class $C \in \mathcal{C}$ from available (training) data, in order to predict the classes of new instances.

2.1   Graphical Structure

NEC maintains the same graphical presentation as NB. In the figure 1, the root C is class to predict and the leafs from X_1 to X_n present features.

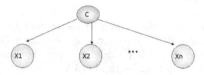

Fig. 1. A generalized presentation of the naive evidential classifier NEC

2.2   Quantitative Part of NEC

Each edge represents a conditional relation between the two nodes it connects. To each edge will be associated conditional mass distribution[1], while to each node will be associated a prior mass as follows:

- A prior mass distribution $m(C)$ in the root (class) node.
- Both a prior mass distribution $m(X)$ and a conditional mass distribution $m^X[C](x_i)$ in each leaf node associated to an edge.

Section 3 explains how these masses can be learnt. The propagation of beliefs for NEC, ensured by a Modified Binary Join Tree Structure, is illustrated in Figure 2 for nodes C and X_2. We refer to [1] for details.

3   Learning and Decision

3.1   Learning

Mass distributions in the model can be elicited from experts or constructed from observed data (the usual case in classification). We propose two methods to infer such distributions: the Imprecise Dirichlet model (IDM) [3] and Denoeux's multinomial model [4], obtaining respectively the NEC1 and NEC2 paremeterization.

The imprecision of mass distributions obtained by each methods is governed by hyperparameters, respectively the positive real number $v \in \mathbb{R}^+$ for the IDM and the confidence level $\alpha \in [0, 1]$ for Denoeux's model. The higher these parameters, the higher the imprecision of mass distributions.

---

[1] Recall that a mass distribution $m : 2^{|X|} \to [0, 1]$ on $\mathcal{X}$ is such that $m(\emptyset) = 0$, $\sum_{E \subseteq \mathcal{X}} m(E) = 1$ and induces a belief and a plausibility measure such that $bel(A) = \sum_{E \subseteq A} m(A)$ and $pl(A) = 1 - bel(A^c)$.

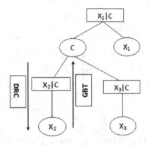

Fig. 2. The propagation process in the MBJT using the GBT and the DRC

3.2   Classification and Evaluation

As imprecision is an interesting feature of evidence theory, we propose an imprecise classification based on pairwise comparisons [5]. Using the mass on $\mathcal{C}$ obtained for an instance $x_1, \ldots, x_N$ by propagation on the NEC, a class c_1 is told to dominate c_2, denoted by $c_1 \succeq c_2$ if the belief of c_1 is larger than the belief of c_2 ($bel(c_1) \geq bel(c_2)$) and the plausibility of c_1 is larger than the plausibility of c_2 ($pl(c_1) \geq pl(c_2)$) for all the distributions. The retained set Ω_m of possible classes is then the one of non-dominated classes, i.e. $\Omega_m = \{c \in \mathcal{C} | \not\exists c', c' \succ c\}$. Evaluations are then made through the use of set accuracy and of discounted accuracy (see [6]). Roughly speaking, if $\hat{c}$ is the true class, set-accuracy counts 1 each time $\hat{c} \in \Omega_m$, while the discounted accuracy counts $1/|\Omega_m|$ with $|\Omega_m|$ the cardinality of Ω. Both counts 0 when $\hat{c} \notin \Omega_m$, and classical accuracy is recovered when $|\Omega_m| = 1$.

4   Application of the NEC in Durum Wheat Industry

In this section we present preliminary results on an agronomic application of NEC consisting in the prediction of the semolina milling value of the durum wheat. Being able to predict this value from easy to measure parameters would be very valuable both for farmers and industrials, the former because it would help in grain selection, the latter to quickly assess wheat quality.

The database contains 260 samples issued from IATE experimental mill, where the output class was the semolina value (discretized in four classes specified by experts) and the input parameters were the Hectolitre Weight (HLW), The Thousand Kernel Weight(TKW) and Vitreousness. NEC1 and NEC2 are both evaluated according to set and discounted accuracy measures, and a ten-fold cross validation method was used.

Figures 3 and 4 show the variations of set- and discounted- accuracy as a function of the parameters v (IDM) and α (Denoeux's model). A first remark is that NEC1 tends to have worse performances than NEC2, an observation that can be explained by the fact that the IDM is more "simple" than Denoeux's model. Hence NEC 1 is easier to compute, while NEC2 gives better predictions. Another noticeable fact is that set-accuracy does not necessarily decrease as the model gets more imprecise (Figure 4 left). This is due to the non-motonicity of

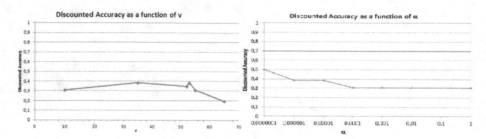

Fig. 3. Discounted Accuracy as a Function of v and α

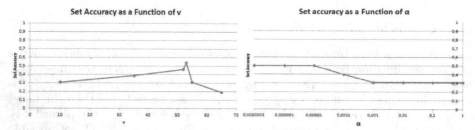

Fig. 4. Set Accuracy as a Function of v and α

the used decision rule (i.e., a more imprecise m does not mean a bigger Ω_m), and this could be solved by choosing another (more imprecise) rule such as interval-dominance [5].

Figure 5 displays the confusion matrix for both classifiers (where the precise classification was chosen as the most plausible, i.e., $c = \arg\max_{c \in C} pl(\{c\})$). Again, both classifiers appear to be at odds, and it is difficult to say which is better. However, as is supported by the relatively low level of well-classed items, this difficulty to differentiate probably also comes from the poor explanatory power of input variables, and more refined modelling (planned in the future) as well as tests on other data sets would be needed to make further conclusions.

Finally, to choose the best configuration, we propose to retain the best one according to discounted accuracy (as it reflects balance between imprecision and prediction quality), that is to say the NEC1 classifier with $v = 53$.

5   A Comparison between NEC1 and the Naive Bayes Classifier

The proposed comparison is based on the confusion matrix returned by NEC1 and the Naive Bayes Classifier. Figure 6 displays the confusion matrix for both classifiers applied to the IATE experimental mill database. Figure 7 displays the confusion matrix for both classifiers applied to LENSES database, from UCI repository, which contains 24 samples, where the output class is the type of lenses fitted to the patient (the patient should be fitted with hard contact lenses, soft contact lenses or should not be fitted with contact lenses) and the

Prédites / Vraies		[63;70] 1	[70;72] 2	[72;74] 3	[74;78,3] 4
[63;70]	1	13	43	13	0
[70;72]	2	11	61	14	0
[72;74]	3	12	31	27	0
[74;78,3]	0	7	19	9	0

(a)

Prédites / Vraies		[63;70] 1	[70;72] 2	[72;74] 3	[74;78,3] 4
[63;70]	1	36	24	9	0
[70;72]	2	21	65	13	0
[72;74]	3	9	36	25	0
[74;78,3]	0	6	10	19	0

(b)

Fig. 5. Confusion Matrix: (a)given by NEC1 with $v = 53$; and (b)given by NEC2 with $\alpha = 0.0000001$

Prédites / Vraies		[63;70] 1	[70;72] 2	[72;74] 3	[74;78,3] 4
[63;70]	1	13	43	13	0
[70;72]	2	11	61	14	0
[72;74]	3	12	31	27	0
[74;78,3]	4	7	19	9	0

(a)

Prédites / Vraies		[63;70] 1	[70;72] 2	[72;74] 3	[74;78,3] 4
[63;70]	1	36	32	7	0
[70;72]	2	13	45	26	2
[72;74]	3	5	41	17	7
[74;78,3]	4	1	22	9	3

(b)

Fig. 6. Confusion Matrix(IATE experimental mill database): (a)given by NEC1 with $v = 53$; and (b)given by Naive Bayes Classifier

Prédites / Vraies		hard contact lenses 1	soft contact lenses 2	No lenses 3
hard contact lenses	1	2	0	2
soft contact lenses	2	2	0	3
No lenses	3	0	1	14

(a)

Prédites / Vraies		hard contact lenses 1	soft contact lenses 2	No lenses 3
hard contact lenses	1	1	0	3
soft contact lenses	2	0	4	1
No lenses	3	2	1	12

(b)

Fig. 7. Confusion Matrix (LENSES database): (a)given by NEC1 with $v = 15.5$; and (b)given by Naive Bayes Classifier

input parameters are the age of the patient, the spectacle prescription, astigmatic and the tear production rate.

For the IATE experimental mill database, NEC1 is better than the Naive Bayes Classifier ($PCC = 38,85\% \geq 36,54\%$) while for the LENSES database the Naive Bayes Classifier is better than NEC1 ($PCC = 70,83\% \geq 66,66\%$).

6   Conclusion

In this paper, we have introduced the idea of Naive Evidential Classifier as well as tools to instantiate and evaluate it. A preliminary application has been achieved on an agronomical problem. Considering the quality of available data, first results are encouraging, however much work remains to be done:

- concerning the model itself, it would be desirable to realx the independence assumption and use an augmented tree model;
- concerning the application, a refined statistical analysis of the data may allow to extract more relevant information or identify subgroups of interest (e.g., differentiating big and small grains or varieties of wheat);
- concerning the general evaluation of the model, it remains to apply it to usual benchmarks and to confront it to other classical classifiers.

References

[1] Ben Yaghlane, B., Mellouli, K.: Inference in directed evidential networks based on the transferable belief model. International Journal of Approximate Reasoning 48, 399–418 (2008)
[2] Smets, P.: Belief Functions: the Disjunctive Rule of Combination and the Generalized Bayesian Theorem. International Journal of Approximate Reasoning 9, 1–35 (1993)
[3] Bernard, J.M.: An introduction to the imprecise dirichlet model for multinomial data. International Journal of Approximate Reasoning 39(2-3), 123–150 (2005)
[4] Denoeux, T.: Constructing belief functions from sample data using multinomial confidence regions. International Journal of Approximate Reasoning 42(3), 228–252 (2006)
[5] Destercke, S.: A Decision Rule for Imprecise Probabilities Based on Pair-Wise Comparison of Expectation Bounds. In: Borgelt, C., González-Rodríguez, G., Trutschnig, W., Lubiano, M.A., Gil, M.Á., Grzegorzewski, P., Hryniewicz, O. (eds.) Combining Soft Computing and Statistical Methods in Data Analysis. AISC, vol. 77, pp. 189–197. Springer, Heidelberg (2010)
[6] Corani, G., Zaffalon, M.: Jncc2: The java implementation of naive credal classifier 2. Journal of Machine Learning Research 9, 2695–2698 (2008)

An Approach to Learning Relational Probabilistic FO-PCL Knowledge Bases*

Nico Potyka and Christoph Beierle

Department of Computer Science, FernUniversität in Hagen, Germany

Abstract. The principle of maximum entropy inductively completes the knowledge given by a knowledge base $\mathcal{R}$, and it has been suggested to view learning as an operation being inverse to inductive knowledge completion. While a corresponding learning approach has been developed when $\mathcal{R}$ is based on propositional logic, in this paper we describe an extension to a relational setting. It allows to learn relational FO-PCL knowledge bases containing both generic conditionals as well as specific conditionals referring to exceptional individuals from a given probability distribution.

1   Introduction

In the area of learning, different motivations, scenarios, and approaches have been investigated, especially in classical Data Mining [6]. The field of Statistical Relational Learning [5] explicitly focuses on relational dependencies. Correspondingly, also different success and quality criteria are used.

In [7], Kern-Isberner advocates a view on learning from a knowledge representation point of view. There, knowledge representation with probabilistic conditionals of the form *if ψ then ϕ with probability ξ*, formally denoted by $(\phi|\psi)[\xi]$, are considered. A probability distribution $\mathcal{P}$ over the possible worlds satisfies $(\phi|\psi)[\xi]$ iff $\mathcal{P}(\psi) > 0$ and $\mathcal{P}(\phi|\psi) = \xi$. Among all distributions satisfying a knowledge base $\mathcal{R}$ consisting of a set of conditionals, the principle of *maximum entropy* [11,7,9] selects the uniquely determined distribution $\mathcal{P}^* = \text{ME}(\mathcal{R})$ having maximum entropy. Thus, $\text{ME}(\mathcal{R})$ inductively completes the knowledge given by $\mathcal{R}$ in an information-theoretically optimal way. Based on this scenario, in [7] it is argued that learning can be seen as an operation being inverse to inductive knowledge completion: Given a probability distribution $\mathcal{P}$, the objective of the process of *conditional knowledge discovery* [7,8] is to find a set $CKD(\mathcal{P})$ of probabilistic conditionals such that $\text{ME}(CKD(\mathcal{P})) = \mathcal{P}$, cf. Fig. 1. Conditional knowledge discovery has been implemented in the CondorCKD system [8,4]. Here, we propose an extension of the CKD approach to a relational setting. We use relational probabilistic conditionals as they are used in the logic FO-PCL developed in [2,3].

---

* The research reported here was partially supported by the Deutsche Forschungsgemeinschaft (grant BE 1700/7-2).

E. Hüllermeier et al. (Eds.): SUM 2012, LNAI 7520, pp. 625–632, 2012.

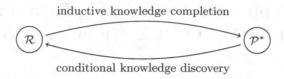

inductive knowledge completion

conditional knowledge discovery

Fig. 1. Conditional knowledge discovery as inverse to inductive knowledge completion

Example 1. Suppose we want to find out about the social behavior of a population of monkeys. A monkey X may be hungry, denoted by $h(X)$. At feeding time, hungry monkeys will eat their food, otherwise they might allow another monkey to eat it which will be expressed by $al(X,Y)$ with $X \neq Y$. Constraints like $X \neq Y$ on the variables restrict the variable bindings in terms of our intended meaning and avoid inconsistencies that would be obtained by simply considering all groundings of a conditional. The set $\mathcal{R}_{Mk}$ with

$$r_1: \quad \langle (al(X,Y)|\overline{h(X)})[0.9], X \neq Y, X \neq c \rangle$$
$$r_2: \quad \langle (al(c,Y)|\overline{h(c)})[0.2], Y \neq c \rangle$$

is an FO-PCL knowledge base where r_1 expresses generic knowledge, while r_2 involves knowledge about an exceptional individual monkey c who allows another monkey to eat his food only with probability 0.2, even if he is not hungry.

After briefly sketching the basics of FO-PCL and the idea of inverting inductive knowledge completion, we present the successive phases of our learning method and illustrate them using the scenario from Ex. 1. Basically, it follows a bottom-up approach. Starting with a set of most specific conditionals, redundancies between the conditionals are detected by algebraic means. These redundancies are used to generalize conditionals, thereby reducing the set of conditionals. Exceptional individuals are identified afterwards using statistical methods.

2   Background

2.1   FO-PCL

For the representation of relational knowledge we consider *FO-PCL* [2,3]. A restricted many-sorted first-order logic built over signatures of the form $\Sigma = (S, Const, Pred)$. S is a set of sorts, $Const$ a set of sorted constants, and $Pred$ a set of sorted predicate symbols. Formulas are built up over a signature Σ and a set of sorted variables $\mathcal{V}$ in the usual way using conjunction (where $\phi\psi$ abbreviates $\phi \wedge \psi$), disjunction and negation, but no quantifiers. An *FO-PCL conditional* $R = \langle (\phi \mid \psi)[\xi], C \rangle$ consists of two formulas over Σ and $\mathcal{V}$, the *consequence* ϕ and the *antecedence* ψ, a probability $\xi \in [0,1]$, and a *constraint formula* C using equality as the only predicate symbol. A set $\mathcal{R}$ of such conditionals is called an *FO-PCL knowledge base*. An instance of an FO-PCL conditional is called *admissible* if its constraint function evaluates to true. The *grounding operator* gnd maps each FO-PCL conditional R to the set of its admissible instances $\text{gnd}(R)$, in the following just called instances.

The Herbrand base $\mathcal{H}(\mathcal{R})$ of $\mathcal{R}$ includes all ground predicates appearing in grounded conditionals of $\mathcal{R}$. A Herbrand interpretation is a subset of $\mathcal{H}(\mathcal{R})$, and Ω denotes the set of all Herbrand interpretation, also called (possible) worlds. A probability distribution $\mathcal{P} : \Omega \to [0,1]$ satisfies a ground conditional $(\phi \mid \psi)[\xi]$ iff $\mathcal{P}(\phi\psi) = \xi \cdot \mathcal{P}(\psi)$, where the probability of a formula is the sum of the probabilities of Herbrand interpretations in Ω satisfying the formula, i.e., $\mathcal{P}(F) = \sum_{\omega \in \text{Mod}(F)} \mathcal{P}(\omega)$. $\mathcal{P}$ satisfies an FO-PCL conditional R iff it satisfies each ground instance in $\text{gnd}(R)$, and it satisfies an FO-PCL knowledge base $\mathcal{R}$ iff it satisfies each conditional in $\mathcal{R}$. The maximum entropy distribution $\text{ME}(\mathcal{R})$ is the uniquely determined distribution having maximum entropy and satisfying $\mathcal{R}$, where $\text{H}(\mathcal{P}) = -\sum_{\omega \in \Omega} \mathcal{P}(\omega) \log \mathcal{P}(\omega)$ is the entropy of $\mathcal{P}$.

Example 2. A possible signature Σ_{Mk} for $\mathcal{R}_{Mk}$ from Ex. 1 uses a single sort $S = \{Monkey\}$, three constants $Const = \{a, b, c\}$ of sort *Monkey* and two predicate symbols $Pred = \{h(Monkey), al(Monkey, Monkey)\}$. Then the ground instances of r_1 are $(al(a,b)\mid\overline{h(a)})[0.9]$, $(al(a,c)\mid\overline{h(a)})[0.9]$, $(al(b,a)\mid\overline{h(b)})[0.9]$, $(al(b,c)\mid\overline{h(b)})[0.9]$, and the ground instances of r_2 are $(al(c,a)\mid\overline{h(c)})[0.2]$, $(al(c,b)\mid\overline{h(c)})[0.2]$. Hence $\mathcal{H}(\mathcal{R}_{R_{Mk}}) = \{al(a,b), al(a,c), al(b,a), al(b,c), al(c,a), al(c,b), h(a), h(b), h(c)\}$.

Note that each of the ground predicates in $\mathcal{H}(\mathcal{R}_{R_{Mk}})$ can be viewed as a binary propositional variable from $\{h_a, h_b, h_c, al_{a,b}, al_{a,c}, al_{b,a}, al_{b,c}, al_{c,a}, al_{c,b}\}$. So the representation is essentially propositional.

2.2   Inverting Maximum Entropy Reasoning

Given a knowledge base $\mathcal{R}$ consisting of n propositional conditionals $(\phi_i \mid \psi_i)[\xi_i]$, one can apply the method of lagrange multipliers to show that there are $n + 1$ non-negative real values a_i such that the ME-optimal probability distribution $\mathcal{P}$ satisfying $\mathcal{R}$ is given by $\mathcal{P}(\omega) = a_0 \prod_{i=1}^{n}(\prod_{\omega \models \phi_i\psi_i} a_i^{1-\xi_i} \prod_{\omega \models \overline{\phi_i}\psi_i} a_i^{-\xi_i})$. Each $a_i \in \mathbb{R}_0^+$ is a factor corresponding to the i-th conditional with $0^0 := 1$, and a_0 is a normalizing constant. Depending on whether the world verifies or falsifies a conditional, the corresponding factor affects the probability with exponent $1 - \xi_i$ or $-\xi_i$. If the conditional is inapplicable in the world, i.e., $\omega \models \overline{\psi_i}$, it does not influence the probability at all. So basically there are three factors corresponding to each conditional, namely a positive, a negative and a neutral effect.

In [7] this idea is expanded to an algebraic framework. For each conditional two abstract symbols α^+, α^- representing the positive and negative effect are introduced. The *conditional structure* of worlds with respect to the knowledge base $\mathcal{R}$ is $\sigma_{\mathcal{R}}(\omega) = \prod_{i=1}^{n}(\prod_{\omega \models \phi_i\psi_i} \alpha_i^+ \prod_{\omega \models \overline{\phi_i}\psi_i} \alpha_i^-)$, which is an abstraction of the numerical representation above.

Example 3. Consider three binary variables A, B, C and a knowledge base $\mathcal{R} := \{(B \mid A)[0.8]\}$ consisting of a single conditional. For each world ω, the following table shows how the probability $\mathcal{P}(\omega)$ is determined by the factor a_1

corresponding to the conditional, and in the third column the conditional structure $\sigma_{\mathcal{R}}(\omega)$ of ω is given:

ω	$\mathcal{P}(\omega)$	$\sigma_{\mathcal{R}}(\omega)$	ω	$\mathcal{P}(\omega)$	$\sigma_{\mathcal{R}}(\omega)$
$A\,B\,C$	$0.18 = a_0 \cdot a_1^{0.2}$	α^+	$\overline{A}\,B\,C$	$0.14 = a_0$	1
$A\,B\,\overline{C}$	$0.18 = a_0 \cdot a_1^{0.2}$	α^+	$\overline{A}\,B\,\overline{C}$	$0.14 = a_0$	1
$A\,\overline{B}\,C$	$0.04 = a_0 \cdot a_1^{-0.8}$	α^-	$\overline{A}\,\overline{B}\,C$	$0.14 = a_0$	1
$A\,\overline{B}\,\overline{C}$	$0.04 = a_0 \cdot a_1^{-0.8}$	α^-	$\overline{A}\,\overline{B}\,\overline{C}$	$0.14 = a_0$	1

Worlds having the same conditional structure must have the same probability. If conversely the same probability implies the same structure, the distribution is called a *faithful representation* of $\mathcal{R}$ [7].

Example 4. We continue the previous example. The three different probabilities constituted are $0.18, 0.04$ and 0.14. In particular we have $\frac{\mathcal{P}(A\,B\,C)}{\mathcal{P}(A\,B\,\overline{C})} = 1$. Assuming faithfulness, we can conclude $\frac{\sigma_{\mathcal{R}}(A\,B\,C)}{\sigma_{\mathcal{R}}(A\,B\,\overline{C})} = 1$ for our searched knowledge base $\mathcal{R}$. Because of the independency of the abstract symbols, we can separate the known basic conditionals with respect to their consequence literal [7]. We consider the basic conditionals with consequence B: $b_1 := (B \mid A\,C)$, $b_2 := (B \mid A\,\overline{C})$, $b_3 := (B \mid \overline{A}\,C)$, $b_4 := (B \mid \overline{A}\,\overline{C})$. Evaluating the world product above with respect to their conditional structure σ_B yields $1 = \frac{\sigma_B(A\,B\,C)}{\sigma_B(A\,B\,\overline{C})} = \frac{b_1^+}{b_2^+}$, hence $b_1^+ = b_2^+$. Hence both conditionals b_1 and b_2 have the same effect on the probability distribution and therefore can be combined to a single conditional. We obtain $(B \mid A\,C \vee A\,\overline{C}) = (B \mid A)$ which is indeed the original conditional in $\mathcal{R}$. Other equations can be used to shorten and remove the remaining basic conditionals in a similar way. In general equations can be much more complex and new effects for the shortened conditionals have to be considered. We refer to [7] for a comprehensive analysis.

3   Learning FO-PCL Knowledge Bases

3.1   Learning Input

We describe our domain of interest by an FO-PCL signature Σ. As sketched in Sec. 1, our initial learning input is a probability distribution over the worlds induced by Σ.

Example 5. The relationships among the three monkeys from Ex. 2 are described by the nine grounded predicates

$$h(a),\ h(b),\ h(c),\ al(a,b),\ al(a,c),\ al(b,a),\ al(b,c),\ al(c,a),\ al(c,b).$$

Each observation is given by a Herbrand interpretation; e.g., $\{h(b), h(c), al(a,b), al(a,c)\}$ describes the situation where b and c are hungry, a is not hungry, and a allows b and c to eat his food. A set of such interpretations induces an empirical probability distribution like:

$h(a)$	$h(b)$	$h(c)$	$al(a,b)$	$al(a,c)$	$al(b,a)$	$al(b,c)$	$al(c,a)$	$al(c,b)$	$P(\omega)$
0	1	1	1	1	0	0	0	0	0.10
1	0	1	0	0	1	1	0	0	0.10
1	1	0	0	0	0	0	0	0	0.06
1	1	0	0	0	0	0	1	1	0.01

. . .

3.2 Language Bias and Background Knowledge

In principle, a structure over grounded predicates as in Sec. 3.1 can be viewed as propositional, and we could apply CondorCKD to find a set of propositional conditionals $\mathcal{R}$ representing $\mathcal{P}$. However, if we do not take information about the underlying relational structure into account, we might end up with conditionals like $(al(a,b)|al(c,d))$ which may be sufficient to represent the probability distribution, but that are not desirable from a learning perspective. Therefore we will restrict our language appropriately.

As in [7], we consider only *single-elementary conditionals*, i.e., conditionals whose consequence consists of a single literal and the antecedence is a conjunction of literals. Basically, this representation is as expressive as the Horn clauses usually considered in ILP [10]. Furthermore, in a first phase, we will learn *free conditionals* only, i.e. conditionals containing only variables; in a later phase, specific knowledge about individuals will be identified. As often done in ILP, we can consider *restricted* rules where each variable appearing in the antecedence has to appear in the consequence, too.

The relational learning systems CLAUDIEN and ICL use the template language $\mathcal{DLAB}$ instead [1]. If background knowledge about the structure of the conditionals to be learned is available, we can use a similar approach. A *template conditional* is a pair consisting of a non-probabilistic conditional and a constraint formula. Given a template $template((\phi|\psi), C)$ each FO-PCL conditional $\langle(\phi \mid \psi')[\xi], C\rangle$ with a sub-conjunction ψ' of ψ and arbitrary $\xi \in [0,1]$ is covered and so are its ground instances.

Example 6. The template $template((al(X,Y)|h(X)\,h(Y)), X \neq Y)$ covers the following four FO-PCL conditionals and all their ground instances:

$$\langle(al(X,Y)|h(X)\,h(Y))[\xi], X \neq Y\rangle \qquad \langle(al(X,Y)|h(X))[\xi], X \neq Y\rangle$$
$$\langle(al(X,Y)|h(Y))[\xi], X \neq Y\rangle \qquad \langle(al(X,Y)|\top)[\xi], X \neq Y\rangle$$

3.3 Relational CondorCKD

As explained in Section 2.2, CondorCKD starts with a set of most specific conditionals, the so-called *basic conditionals* [7]. Then $\mathcal{P}$ is examined to detect dependencies between conditionals by algebraic means, and equations on possible effects of conditionals are resolved. If a conditional turns out to be non-effective w.r.t. the observed distribution, it is removed. If conditionals similar to each other turn out to have the same effect w.r.t. the observed distribution, they

---

Algorithm 1. RCondorCKD

1: **procedure** RCKD($\mathcal{P}$, *Bias*)
2:     $\mathcal{R} \leftarrow createBasicConditionals(Bias)$
3:     $\mathcal{D} \leftarrow findDependencies(\mathcal{P})$
4:     **while** $d \in \mathcal{D}$ can be resolved **do**
5:         Resolve d and decrease $\mathcal{R}$
6:     **end while**
7:     $\mathcal{R} \leftarrow postprocess(\mathcal{R})$
8:     **return** $\mathcal{R}$
9: **end procedure**

---

are combined pairwise to more general conditionals. Algorithm 1 sketches our relational version of CondorCKD.

RCondorCKD takes an empirical distribution $\mathcal{P}$ and a language bias and computes a set of conditionals reflecting dependencies between the observed individuals. Starting with the basic conditional set generated with respect to our language bias, the initial set of conditionals is shortened by deleting or combining conditionals as sketched above. If no more shortenings are possible, we compute the probabilities and detect knowledge about specific individuals in a post-processing step. In the following we describe the steps in more detail.

Since the shortening operations of CondorCKD are based on algebraic equations and are defined on unquantified conditionals, we use a wild card symbol $*$ for probabilities. The basic conditional set is the set of most specific conditionals w.r.t. our language bias.

Example 7. Using the template $template((al(X,Y)|h(X)\,h(Y)), X \neq Y)$ from Example 6 and $template((h(X)|al(X,Y)\,h(Y)), X \neq Y)$ for our monkey example, yields the following basic conditional set:

$$\langle(al(X,Y) \mid h(X)\,h(Y))[*], X \neq Y\rangle \qquad \langle(h(X) \mid al(X,Y)\,h(Y))[*], X \neq Y\rangle$$
$$\langle(al(X,Y) \mid h(X)\,\overline{h(Y)})[*], X \neq Y\rangle \qquad \langle(h(X) \mid al(X,Y)\,\overline{h(Y)})[*], X \neq Y\rangle$$
$$\langle(al(X,Y) \mid \overline{h(X)}\,h(Y))[*], X \neq Y\rangle \qquad \langle(h(X) \mid \overline{al(X,Y)}\,h(Y))[*], X \neq Y\rangle$$
$$\langle(al(X,Y) \mid \overline{h(X)}\,\overline{h(Y)})[*], X \neq Y\rangle \qquad \langle(h(X) \mid \overline{al(X,Y)}\,\overline{h(Y)})[*], X \neq Y\rangle$$

Each ground instance g of a free conditional r corresponds to a propositional conditional and therefore can be handled similar to the propositional case. As a first approach, suppose *each* ground instance of c turns out to be redundant, i.e., can be removed; then removing c is perfectly justified. Similarly, if each ground instance of a free conditional r_1 can be combined with exactly one appropriate ground instance of another free conditional r_2, then combining the free conditionals r_1 and r_2 is justified.

Example 8. Suppose there are only two monkeys a, b and we discover that the ground instances $(al(a,b) \mid h(a)\,h(b))$ and $(al(b,a) \mid h(b)\,h(a))$ can be deleted. Then we can delete the free conditional $\langle(al(X,Y) \mid h(X)\,h(Y))[*], X \neq Y\rangle$ since all its ground instances are redundant.

Further, suppose we detect that $(al(a,b) \mid \overline{h(a)}\, h(b))$ can be combined with $(al(a,b) \mid \overline{h(a)}\, \overline{h(b)})$ to $(al(a,b) \mid \overline{h(a)})$. Additionally suppose $(al(b,a) \mid \overline{h(b)}\, h(a))$ can be combined with $(al(b,a) \mid \overline{h(b)}\, \overline{h(a)})$ to $(al(b,a) \mid \overline{h(b)})$. Then we can replace the corresponding free conditionals $\langle (al(X,Y) \mid \overline{h(X)}\, h(Y))[*], X \neq Y \rangle$ and $\langle (al(X,Y) \mid \overline{h(X)}\, \overline{h(Y)})[*], X \neq Y \rangle$ by $\langle (al(X,Y) \mid \overline{h(X)})[*], X \neq Y \rangle$, since this is justified for all their ground instances.

The condition for taking into account *each* ground instance in these reduction steps may be too restrictive. It can be replaced by a *bold* option being content with a justified shortening for a single instance, or by a *threshold-based* option. These options are similar to the heuristics already implemented in CondorCKD for shortening propositional conditionals [8,4].

3.4 Identifying Exceptional Individuals

We now have a knowledge base of unquantified, free conditionals. The probability of each of their ground instances is evaluated using the empirical distribution $\mathcal{P}$. While this might result in different probabilities for the ground instances, the probabilities for what can be considered prototypical elements will be within a small interval. It could also be possible that the probabilities separate the conditionals into different groups. In this case, it might be appropriate to introduce new types in the FO-PCL signature to realize a kind of clustering. However, for the time being we suppose there is one representative group and maybe some outliers.

In order to classify the probabilities, we can use standard measures as used in descriptive statistics. For each free conditional we compute the lower and upper quartile Q_1, Q_3 and the median Q_2 of the ground instances. The user may define a threshold $\tau > 0$, such that each instance with probability differing more than $(1+\tau) \cdot (Q_3 - Q_1)$ from the median will be considered exceptional. Specific individuals can be separated from the general conditional by appropriate constraint formulas. Then the probability of the general conditional can be computed as the mean or the median of the non-exceptional ground instances.

Example 9. Suppose there are five monkeys a, b, c, d, e and we learned the free conditional $\langle (al(X,Y)|\overline{h(X)})[*], X \neq Y \rangle$. Further suppose we computed the median $Q_2 = 0.9$ and the lower and upper quartiles $Q_1 = 0.8, Q_3 = 0.95$. For $\tau = 0.5$ each conditional with a probability lower than $0.9 - 1.5 \cdot (0.95 - 0.8) = 0.675$ is classified as exceptional. Suppose for monkey c we obtain $(al(c,t) \mid \overline{h(c)})[0.2]$ for $t \in \{a, b, d, e\}$. Since each ground conditional for $X = c$ is exceptional, we split c from the general conditional. Supposing that for the remaining ground instances we have a mean of 0.9, the learned FO-PCL knowledge base will contain the conditionals

$$\langle (al(X,Y)|\overline{h(X)})[0.9], X \neq Y, X \neq c \rangle$$
$$\langle (al(c,Y)|\overline{h(c)})[0.2], Y \neq c \rangle$$

which correspond to the conditionals in $\mathcal{R}_{Mk}$ from Example 1.

4 Conclusions and Further Work

Employing the view of learning as an operation inverse to inductive knowledge completion put forward by Kern-Isberner [7,8], we developed an approach how CondorCKD can be extended to the relational setting. We demonstrated how relational probabilistic FO-PCL knowledge bases containing both generic conditionals as well as conditionals referring to exceptional individuals can be learned from a given probability distribution.

Several of the intermediate steps which we could only briefly describe here can be optimized by exploiting the relational structure of the data. We will also investigate to which extent we can use this information to refine the neighborhood graph among worlds considered in [8]. While major parts of our approach have already been implemented, we will use a full implementation of RCondorCKD to experimentally evaluate our approach.

References

1. De Raedt, L., Blockeel, H., Dehaspe, L., Laer, W.V.: Three companions for data mining in first order logic. In: Relational Data Mining, pp. 105–139. Springer (2001)
2. Fisseler, F.: Learning and Modeling with Probabilistic Conditional Logic. Dissertations in Artificial Intelligence, vol. 328. IOS Press, Amsterdam (2010)
3. Fisseler, J.: First-order probabilistic conditional logic and maximum entropy. Logic Journal of the IGPL (to appear, 2012)
4. Fisseler, J., Kern-Isberner, G., Beierle, C., Koch, A., Müller, C.: Algebraic Knowledge Discovery Using Haskell. In: Hanus, M. (ed.) PADL 2007. LNCS, vol. 4354, pp. 80–93. Springer, Heidelberg (2007)
5. Getoor, L., Taskar, B. (eds.): Introduction to Statistical Relational Learning. MIT Press (2007)
6. Han, J., Kamber, M., Pei, J.: Data Mining: Concepts and Techniques. Morgan Kaufmann (2011)
7. Kern-Isberner, G.: Conditionals in Nonmonotonic Reasoning and Belief Revision. LNCS (LNAI), vol. 2087. Springer, Heidelberg (2001)
8. Kern-Isberner, G., Fisseler, J.: Knowledge discovery by reversing inductive knowledge representation. In: Proceedings of the Ninth International Conference on the Principles of Knowledge Representation and Reasoning, KR 2004, pp. 34–44. AAAI Press (2004)
9. Kern-Isberner, G., Lukasiewicz, T.: Combining probabilistic logic programming with the power of maximum entropy. Artificial Intelligence, Special Issue on Nonmonotonic Reasoning 157(1-2), 139–202 (2004)
10. Muggleton, S., De Raedt, L., Poole, D., Bratko, I., Flach, P.A., Inoue, K., Srinivasan, A.: ILP turns 20 - Biography and future challenges. Machine Learning 86(1), 3–23 (2012)
11. Paris, J., Vencovska, A.: In defence of the maximum entropy inference process. International Journal of Approximate Reasoning 17(1), 77–103 (1997)

An SMT-Based Solver
for Continuous t-norm Based Logics

Amanda Vidal[1], Félix Bou[2,1], and Lluís Godo[1]

[1] Artificial Intelligence Research Institute (IIIA - CSIC)
Campus de la Universitat Autònoma de Barcelona s/n,
08193 Bellaterra, Spain
[2] Department of Probability, Logic and Statistics
Faculty of Mathematics, University of Barcelona (UB)
Gran Via 585, 08007 Barcelona, Spain

Abstract. In the literature, little attention has been paid to the development of solvers for systems of mathematical fuzzy logic, even though there is an important number of studies on complexity and proof theory for them. In this paper, extending a recent approach by Ansótegui et al., we present ongoing work on an efficient and modular SMT-based solver for a wide family of continuous t-norm based fuzzy logics. The solver is able to deal with most famous fuzzy logics (including BL, Łukasiewicz, Gödel and Product); and for each of them, it is able to test, among others, satisfiability, tautologicity and logical consequence problems. Note that, unlike the classical case, these problems are not in general interdefinable in fuzzy logics. Some empirical results are reported at the end of the paper.

1 Introduction

In the literature, with a few exceptions mainly for Łukasiewicz logics [13,15,4,14], little attention has been paid to the development of efficient solvers for systems of mathematical fuzzy logic, even though there is an important number of studies on complexity and proof theory for them (see [12,9,10,1,6]). This is a problem that limits the use of fuzzy logics in real applications. In [2], a new approach for implementing a theorem prover for Łukasiewicz, Gödel and Product fuzzy logics using Satisfiability Modulo Theories (SMT) has been proposed. The main advantage of this approach based on SMT is the modularity of being able to cope with several fuzzy logics.

In this paper, we extend this approach in order to be able to cope with more logics (including Basic Fuzzy Logic BL): we study the implementation and testing of a general solver for continuous t-norm based fuzzy logics. We have generalized the solver so it can perform satisfiability, theoremhood and logical consequence checks for any of a wide family of these fuzzy logics. Also, we have changed the coding for product logic from the one of [2] to one based on Presburger Arithmetic (Linear Integer Arithmetic), and this has dramatically enhanced its performance.

E. Hüllermeier et al. (Eds.): SUM 2012, LNAI 7520, pp. 633–640, 2012.
© Springer-Verlag Berlin Heidelberg 2012

Structure of the paper. Section 2 introduces the propositional logics considered, and gives a brief introduction to SMT. Section 3 describes the SMT-based solver proposed in this paper. Section 4 starts with an explanation of the design of the experiments we ran on our solver, and then the results are analyzed in Section 4.1. Section 5 presents the conclusions and future work.

2 Preliminaries

2.1 Continuous t-norm Based Fuzzy Logics

Continuous t-norm based propositional logics correspond to a family of many-valued logical calculi with the real unit interval $[0, 1]$ as set of truth-values and defined by a conjunction $\&$, an implication $\rightarrow$ and the truth-constant $\overline{0}$, interpreted respectively by a continuous t-norm $*$, its residuum $\Rightarrow$ and the number 0. In this framework, each continuous t-norm $*$ uniquely determines a semantical propositional calculus L_* over formulas defined in the usual way (see [10]) from a countable set $\{p, q, r, \ldots\}$ of propositional variables, connectives $\&$ and $\rightarrow$ and truth-constant $\overline{0}$. Further connectives are defined as follows:

$$\neg\varphi \ \text{is} \ \varphi \rightarrow \overline{0},$$
$$\varphi \wedge \psi \ \text{is} \ \varphi \& (\varphi \rightarrow \psi),$$
$$\varphi \vee \psi \ \text{is} \ ((\varphi \rightarrow \psi) \rightarrow \psi) \wedge ((\psi \rightarrow \varphi) \rightarrow \varphi),$$
$$\varphi \equiv \psi \ \text{is} \ (\varphi \rightarrow \psi) \& (\psi \rightarrow \varphi).$$

L_*-evaluations of propositional variables are mappings e assigning to each propositional variable p a truth-value $e(p) \in [0, 1]$, which extend univocally to compound formulas as follows: $e(\overline{0}) = 0$, $e(\varphi\&\psi) = e(\varphi) * e(\psi)$ and $e(\varphi \rightarrow \psi) = e(\varphi) \Rightarrow e(\psi)$. Actually, each continuous t-norm defines an algebra $[0, 1]_* = ([0, 1], \min, \max, *, \Rightarrow, 0)$, called *standard L_*-algebra*.

From the above definitions it holds that $e(\neg\varphi) = e(\varphi) \Rightarrow 0$, $e(\varphi \wedge \psi) = \min(e(\varphi), e(\psi))$, $e(\varphi \vee \psi) = \max(e(\varphi), e(\psi))$ and $e(\varphi \equiv \psi) = e(\varphi \rightarrow \psi) * e(\psi \rightarrow \varphi)$. A formula φ is a said to be a *1-tautology* (or *theorem*) of L_* if $e(\varphi) = 1$ for each L_*-evaluation e. The set of all 1-tautologies of L_* will be denoted as $TAUT([0, 1]_*)$. A formula φ is *1-satisfiable* in L_* if $e(\varphi) = 1$ for some L_*-evaluation e. Moreover, the corresponding notion of *logical consequence* is defined as usual: $T \models_* \varphi$ iff for every evaluation e such that $e(\psi) = 1$ for all $\psi \in T$, $e(\varphi) = 1$.

Well-known axiomatic systems, like Łukasiewicz logic (Ł), Gödel logic (G), Product logic (Π) and Basic Fuzzy logic (BL) syntactically capture different sets $TAUT([0, 1]_*)$ for different choices of the t-norm $*$ (see e.g. [10,6]). Indeed, the following conditions hold true, where $*_L$, $*_G$ and $*_\Pi$ respectively denote the Łukasiewicz t-norm, the min t-norm and the product t-norm:

φ is provable in Ł   iff $\varphi \in TAUT([0, 1]_{*_L})$
φ is provable in G   iff $\varphi \in TAUT([0, 1]_{*_G})$
φ is provable in Π iff $\varphi \in TAUT([0, 1]_{*_\Pi})$
φ is provable in BL   iff $\varphi \in TAUT([0, 1]_*)$ for all continuous t-norms $*$.

Also, taking into account that every continuous t-norm $*$ can be represented as an ordinal sum of Łukasiewicz, Gödel and Product components, the calculus of any continuous t-norm has been axiomatized in [8]. All these completeness results also hold from deductions from a finite set of premises but, in general, they do not extend to deductions from infinite sets (see [6] for details).

2.2   Satisfiability Modulo Theories (SMT)

The satisfiability problem, i.e. determining whether a formula expressing a constraint has a solution, is one of the main problems in theoretical computer science. If this constraint refers to Boolean variables, then we are facing a well-known problem, the (propositional) Boolean satisfiability problem (SAT).

On the other hand, some problems require to be described in more expressive logical languages (like those of the first order theories of the real numbers, of the integers, etc.), and thus a formalism extending SAT, Satisfiability Modulo Theories (SMT), has also been developed to deal with these more general decision problems. An SMT instance is a first order formula where some function and predicate symbols have predefined interpretations from background theories.

The more common approach [16] for the existing SMT solvers is the integration of a T-solver, i.e. a decision procedure for a given theory T, and a SAT solver. In this model, the SAT solver is in charge of the Boolean formula, while the T-solver analyzes sets of atomic constraints in T. With this, the T-solver checks the possible models the SAT solver generates and rejects them if inconsistencies with the theory appear. In doing so, it gets the efficiency of the SAT solvers for Boolean reasoning, long time tested, and the capability of the more concrete T-oriented algorithms inside the respective theory T.

The current general-use library for SMT is SMT-LIB [3], and there are several implementations of SMT-solvers for it. For our experiments, we use Z3 [7], which implements the theories we need for our purposes:

- QF_LIA (Quantifier Free Linear Integer Arithmetic), which corresponds to quantifier free first order formulas valid in $(\mathbb{Z}, +, -, 0, 1)$,
- QF_LRA (Quantifier Free Linear Real Arithmetic), which corresponds to quantifier free first order formulas valid in $(\mathbb{R}, +, -, \{q : q \in \mathbb{Q}\})$,
- QF_NLRA (Quantifier Free non Linear Real Arithmetic), which corresponds to quantifier free first order formulas valid in $(\mathbb{R}, +, -, \cdot, /, \{q : q \in \mathbb{Q}\})$.

3   A SMT Solver for Continuous t-norm Based Fuzzy Logics

Inspired by the approach of Ansótegui et. al in [2], we aim at showing in this short paper that a more general solver for fuzzy logics can be implemented using an SMT solver. The main feature of the solver is its *versatility*, so it can be used for testing on a wide range of fuzzy logics (like BL, Łukasiewicz, Gödel, Product and logics obtained through ordinal sums) and also for different kinds of problems, like tautologicity and satisfiability but also logical consequence, or getting evaluations for a given formula (i.e., obtaining variable values that yield a formula a certain truth degree given some restrictions).

It is well-known that every continuous t-norm can be expressed as an ordinal sum of the three main continuous t-norms $*_L$, $*_G$ and $*_\Pi$. The fact that the three basic t-norms are defined using only addition and multiplications over the real unit interval was used in [2] to develop a solver for theoremhood in these three logics using QF_LRA and QF_NLRA. The case of BL was not considered in [2] because its usual semantics is based on the whole family of continuous t-norms, and not just on a single one. However, thanks to a result of Montagna [11] one can reduce proofs over BL, when working with concrete formulas, to proofs over the logic of an ordinal sum of as many Łukasiewicz components as different variables involved in the set of formulas plus one. This trick is the one we use in our solver for the implementation of BL.

We have implemented a solver that allows the specification, in term of its components, of any continuous t-norm; and the use of BL too. We also allow finitely-valued Łukasiewicz and Gödel logics, and all these logics can be also extended with rational truth-constants. Also, we have considered interesting to add to our software more options than just testing the theoremhood of a formula in a certain logic. In our solver, we can check whether a given formula (possibly with truth-constants) is a logical consequence of a finite set of formulas (possibly with truth-constants as well).

On the other hand, for the particular case of Product logic we have employed a new methodology. This is so because the previous approach, directly coding Product logic connectives with product and division of real numbers, has serious efficiency problems (inherited from QF_NLRA). Indeed, it is already noted in [2] that these problems appear with really simple formulas. To overcome these problems, we have used an alternative coding based on QF_LIA. We can do this because Cignoli and Torrens showed in [5] that the variety of Product algebras is also generated by a discrete linear product algebra: the one with domain the negative cone of the additive group of the integers together with a first element $-\infty$. Indeed, it holds that $TAUT([0,1]_{*_\Pi}) = TAUT((\widetilde{\mathbb{Z}^-})_\oplus)$, where $\widetilde{\mathbb{Z}^-} := \mathbb{Z}^- \cup \{-\infty\}$ endowed with the natural order plus setting $-\infty < x$ for all $x \in \mathbb{Z}^-$, and with its conjunction operation $\oplus$ defined as:

$$x \oplus y := \begin{cases} x + y, & \text{if } x, y \in \mathbb{Z}^- \\ -\infty, & \text{otherwise.} \end{cases}$$

Notice that its corresponding residuated implication is then defined as:

$$x \Rightarrow_\oplus y := \begin{cases} 0, & \text{if } x \leq y \\ y - x, & \text{if } x, y \in \mathbb{Z}^-, x > y \\ -\infty, & \text{otherwise.} \end{cases}$$

Therefore, for dealing with Product logic it is enough to consider this discrete algebra; and this particular algebra can be coded using just natural numbers with the addition (i.e. using Presburger Arithmetic). Our experiments have shown that, for concrete instances, this approach based on the discrete theory of integers with the addition (instead of the reals with product) works much better.

The interested reader can find in an extended version of this paper [17] an appendix with Z3-code examples generated by our software to solve several kinds of problems that clarify the methodology explained above.

4 Experimental Results

We consider the main advantage of our solver to be the versatility it allows, but we have also performed an empirical evaluation of our approach using only its theorem-prover option to compare it to [2].

We have conducted experiments over two different families of BL-theorems, see (1) and (2) below. First, for comparison reasons with [2], we have considered the following generalizations (based on powers of the & connective) of the first seven Hájek's axioms of BL [10]:

$$
\begin{aligned}
&\text{(A1)}\ \ (p^n \to q^n) \to ((q^n \to r^n) \to (p^n \to r^n))\\
&\text{(A2)}\ \ (p^n \& q^n) \to p^n\\
&\text{(A3)}\ \ (p^n \& q^n) \to (q^n \& p^n)\\
&\text{(A4)}\ \ (p^n \& (p^n \to q^n)) \to (q^n \& (q^n \to p^n))\\
&\text{(A5a)}\ (p^n \to (q^n \to r^n)) \to ((p^n \& q^n) \to r^n)\\
&\text{(A5b)}\ ((p^n \& q^n) \to r^n) \to (p^n \to (q^n \to r^n))\\
&\text{(A6)}\ \ ((p^n \to q^n) \to r^n) \to (((q^n \to p^n) \to r^n) \to r^n)
\end{aligned}
\tag{1}
$$

where p, q and r are propositional variables, and $n \in \mathbb{N} \setminus \{0\}$. It is worth noticing that the length of these formulas grows linearly with the parameter n.

In [2] the authors refer to [13] to justify why these formulas can be considered a good test bench for (at least) Łukasiewicz logic. In our opinion, these formulas have the drawback of using only three variables. This is a serious drawback at least in Łukasiewicz logic because in this case tautologicity for formulas with three variables can be proved to be solved in polynomial time[1].

To overcome the drawback of the bounded number of variables, we propose a new family of BL-theorems to be used as a bench test. For every $n \in \mathbb{N} \setminus \{0\}$,

$$
\bigwedge_{i=1}^{n} (\&_{j=1}^{n}\, p_{ij}) \ \to\ \bigvee_{j=1}^{n} (\&_{i=1}^{n}\, p_{ij})
\tag{2}
$$

is a BL-theorem which uses n^2 variables; the length of these formulas grows quadratically with n. As an example, we note that for $n = 2$ we get the BL-theorem $((p_{11} \& p_{12}) \wedge (p_{21} \& p_{22})) \to ((p_{11} \& p_{21}) \vee (p_{12} \& p_{22}))$. We believe these

---

[1] This polynomial time result is outside the scope of the present paper, but it can be obtained from the rational triangulation associated with the McNaughton function of the formula with three variables. It is worth noticing that the known proofs of NP-completeness for Łukasiewicz logic [12,10,1] need an arbitrary number of variables.

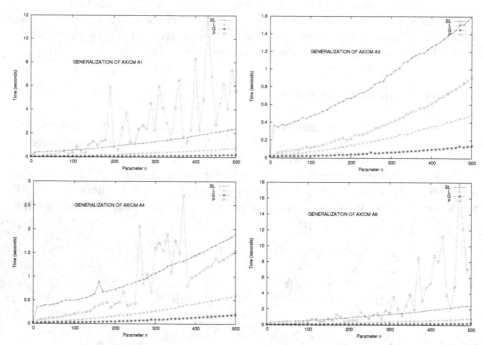

Fig. 1. Generalizations of BL-axioms given in (1)

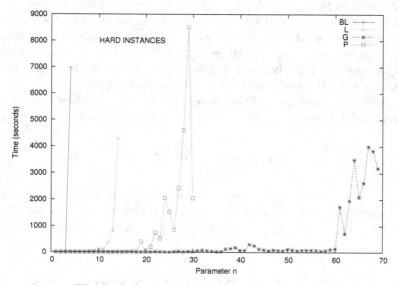

Fig. 2. Our proposed BL-theorems given in (2)

formulas are significantly harder than the ones previously proposed in [13]; and indeed, our experimental results support this claim[2].

4.1 Data Results

We ran experiments on a machine with a i5-650 3.20GHz processor and 8GB of RAM. Evaluating the validity in Łukasiewicz, Product and Gödel logics of the generalizations of the BL axioms (1), ranging n from 0 to 500 with increments of 10, throws better results than the ones obtained in [2], but since our solver is an extension of their work for these logics, we suppose this is due to the use of different machines. For Product Logic, we obtained really good timings. Actually, they are worse than the ones for Łukasiewicz and Gödel logics in most of the cases, since the Presburger arithmetic has high complexity too, but the difference with the previous approach is clear: complex formulas are solved in a comparatively short time, whereas in [2] they could not even be processed. In Figure 1 one can see and compare solving times (given in seconds) for some of the axioms of the test bench for the cases of BL, Łukasiewicz, Gödel and Product logics. It is also interesting to observe how irregularly the computation time for Product Logic varies depending on the axiom and the parameter.

The experiments done with the other family of BL-theorems (2) (see Figure 2 for the results) suggests that here the evaluation time is growing non-polynomially on the parameter n. We have only included in the graph those answers (for parameters $n \leq 70$) obtained in at most 3 hours of execution (e.g. for the BL case we have only got answers for the problems with $n \leq 4$). The high differences in time when evaluating the theorems were expectable: Łukasiewicz and Gödel are simpler than BL when proving the theoremhood because of the method used for BL (considering $n^2 + 1$ copies of Łukasiewicz, where n is the parameter of the formula). On the other hand, the computation times for Product logic modeled with the Presburger arithmetic over $\mathbb{Z}^- \cup \{-\infty\}$ are also smaller than for BL.

5 Conclusions

We have extended the use of SMT technology to define general-use logical solvers for continuous t-norm logics, considered two test suites for these logics, and performed empirical evaluation and testing of our solver. Also, we have provided a new approach for solving more efficiently problems on Product logic.

There are a number of tasks and open questions that we propose as future work. Firstly, solving real applications with SMT-based theorem provers: the non-existence of fast and modern theorem provers has limited so far the potential of fuzzy logics to real applications. Secondly, using Presburger arithmetic has been very useful for our solver to deal with product t-norm, but it is still missing an

---

[2] We point out that the natural way to compare our formula with parameter n is to consider the formulas in [13] with the integer part of $\sqrt{n}$ as parameter.

implementation where this trick is used for ordinal sums where one of the components is the product t-norm. Finally, we would like to point out a more challenging problem: to design an SMT solver for MTL logic (i.e., the logic of left-continuous t-norms), since no completeness is currently known using just one particular t-norm.

Acknowledgments. The authors acknowledge support of the Spanish projects TASSAT (TIN2010-20967-C04-01), ARINF (TIN2009-14704-C03-03) and AT (CONSOLIDER CSD2007-0022), and the grants 2009SGR-1433 and 2009SGR-1434 from the Catalan Government. Amanda Vidal is supported by a JAE Predoc fellowship from CSIC.

References

1. Aguzzoli, S., Gerla, B., Haniková, Z.: Complexity issues in basic logic. Soft Computing 9(12), 919–934 (2005)
2. Ansótegui, C., Bofill, M., Manyà, F., Villaret, M.: Building automated theorem provers for infinitely valued logics with satisfiability modulo theory solvers. In: ISMVL 2012, pp. 25–30. IEEE Computer Society (2012)
3. Barrett, C., Stump, A., Tinelli, C.: The SMT-LIB standard: Version 2.0. Technical report, Department of Computer Science, The University of Iowa (2010), http://www.SMT-LIB.org
4. Bobillo, F., Straccia, U.: Fuzzy description logics with general t-norms and datatypes. Fuzzy Sets and Systems 160(23), 3382–3402 (2009)
5. Cignoli, R., Torrens, A.: An algebraic analysis of product logic. Multiple-Valued Logic 5, 45–65 (2000)
6. Cintula, P., Hájek, P., Noguera, C.: Handbook of Mathematical Fuzzy Logic. Studies in Logic, 2 volumes 37, 38. Mathematical Logic and Foundation, College Publications (2011)
7. de Moura, L., Bjørner, N.: Z3: An Efficient SMT Solver. In: Ramakrishnan, C.R., Rehof, J. (eds.) TACAS 2008. LNCS, vol. 4963, pp. 337–340. Springer, Heidelberg (2008)
8. Esteva, F., Godo, L., Montagna, M.: Equational Characterization of the Subvarieties of BL Generated by t-norm Algebras. Studia Logica 76(2), 161–200 (2004)
9. Hähnle, R.: Automated deduction in multiple valued logics. Oxford Univ. Pr. (1993)
10. Hájek, P.: Metamathematics of fuzzy logic. Kluwer Academic Publishers (1998)
11. Montagna, F.: Generating the variety of BL-algebras. Soft Computing 9(12), 869–874 (2005)
12. Mundici, D.: Satisfiability in many-valued sentential logic is NP-complete. Theoretical Computer Science 52(1-2), 145–153 (1987)
13. Rothenberg, R.: A class of theorems in Łukasiewicz logic for benchmarking automated theorem provers. In: Olivetti, N., Schwind, C. (eds.) TABLEAUX 2007, pp. 101–111 (2007)
14. Schockaert, S., Janssen, J., Vermeir, D.: Satisfiability checking in Łukasiewicz logic as finite constraint satisfaction. Journal of Automated Reasoning (to appear)
15. Schockaert, S., Janssen, J., Vermeir, D., De Cock, M.: Finite Satisfiability in Infinite-Valued Łukasiewicz Logic. In: Godo, L., Pugliese, A. (eds.) SUM 2009. LNCS, vol. 5785, pp. 240–254. Springer, Heidelberg (2009)
16. Sebastiani, R.: Lazy satisfiability modulo theories. Journal on Satisfiability, Boolean Modeling and Computation 3(3-4), 141–224 (2007)
17. Vidal, A., Bou, F., Godo, L.: An SMT-based solver for continuous t-norm based logics (extended version). Technical report, IIIA - CSIC (2012), http://www.iiia.csic.es/files/pdfs/ctnormSolver-long.pdf

Seismic Hazard Assessment on NPP Sites in Taiwan through an Observation-Oriented Monte Carlo Simulation

Jui-Pin Wang[*] and Duruo Huang

The Hong Kong University of Science and Technology, Hong Kong
{jpwang,drhuang}@ust.hk

Abstract. This paper uses Monte Carlo Simulation to evaluate the seismic hazard in two nuclear power plants sites in Taiwan. This approach is different from the two commonly-used methods, featuring the direct use of observed earthquakes to develop magnitude and distance probability functions. Those earthquakes with larger sizes and closer to the site are considered capable of causing damage on engineered structures. The result shows that even though two NPP sites in Taiwan are only 30 km away from each other, NPP site 4 is situated with the seismic hazard four-time as large as NPP site 1. Given our limited understanding and the complicated, random earthquake process, none of a seismic hazard analysis is perfect without challenge. The accountability of seismic hazard analysis relates to analytical transparency, traceability, etc., not to analytical complexity and popularity.

Keywords: Seismic hazard analysis, Monte Carlo simulation, NPP, Taiwan.

1   Introduction

Earthquake prediction is by all means one of the most controversial subjects in earth science that has been debated. A perfect prediction for engineering design and hazard mitigation needs to pinpoint "when" "where" and "how large" of next earthquakes, which is just difficult given the complicated earthquake process and inevitable natural randomness. Alternatively, the science community has developed several practical solutions to earthquake mitigation, such as earthquake early warning and seismic hazard analysis.

Seismic hazard analysis has become the state-of-the-art approach for estimating the best ground motion that would occur within a certain period of time. Probabilistic seismic hazard analysis (PSHA) and deterministic seismic hazard analysis (DSHA) are the two methods that have been generally accepted and that have been codified in technical references for critical structures [1-2]. However, the fight and defense about their pros and cons are nothing new in the history of their development. We agree with Mualchin's footnote on seismic hazard analysis [3]: "None of them is going to

---

[*] Corresponding author.

E. Hüllermeier et al. (Eds.): SUM 2012, LNAI 7520, pp. 641–646, 2012.

be perfect without challenge." This presents much analogy to other engineering analyses, such as finite element analysis versus discrete element analysis, Terzaghi's bearing capacity equation versus Meyerhof's expression [4], Monte Carlo simulation versus Taylor probabilistic analysis [5], Bishop's method and Janbu's method of slice in slope stability analysis [4].

As a result, this study develops another approach in estimating seismic hazard for the nuclear power plants (NPP) in Taiwan, where is "infamous" with high seismicity. The approach aims to use less subjective judgments involved during input characterization. Those details and estimated seismic hazard at NPP sites are provided in this paper.

2   Probabilistic Seismic Hazard Analysis

The essential of PSHA is to account for the uncertainty in earthquake magnitude, earthquake location, and ground motion attenuation [6]. It must be noted that there must be other probability-oriented approaches in seismic hazard assessment. But when it comes to PSHA,.it is recognized as the Probabilistic Seismic Hazard Analysis or as the Cornell-McGuire method.

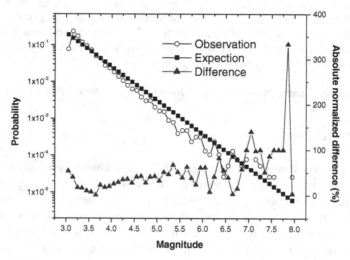

Fig. 1. Observed and expected magnitude probability distribution around NPP site 4 in Taiwan

For characterizing earthquake magnitude and source-to-site distance probabilities, recurrence parameters through the empirical Gutenberg-Richter relationship [7] and the seismic zoning model are needed in advance. According to more than 20,000 earthquakes, Fig. 1 shows the observed and expected magnitude probability distributions within 200 km from NPP site 4 in the northeastern of Taiwan. The observed magnitude distribution is based on a published earthquake catalog around Taiwan since year 1900 [8].

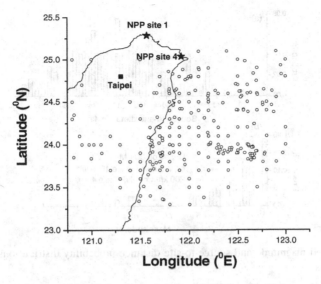

Fig. 2. Spatial locations of $M_w \geq 5.5$ earthquakes from NPP site 4 in Taiwan since year 1900.

3   Observation-Oriented, MCS-Based Seismic Hazard Analysis

3.1   Observation Directly Used as Inputs

Fig. 2 shows the spatial distribution of 301 $M_w \geq 5.5$ earthquakes since year 1900 within 200 km from NPP site 4. Such moderate to large earthquakes are considered complete during the period of recording in the earthquake catalog used [8]. Fig. 3 shows their statistics of magnitude and source-to-site distance. With the statistical information from abundant observations in the past hundred and plus years, the magnitude and distance probability of such earthquakes for the site are considered statistically convincing when used in seismic hazard analysis. Without other judgments needed, this analysis is on the condition that the magnitude and distance thresholds are set at 5.5 M_w and 200 km. (PSHA needs this too.) Those quakes with a lower magnitude and higher distance are considered not to cause damage to engineered structures.

3.2   Monte Carlo Simulation for Estimating Seismic Hazard

The essential of Monte Carlo simulation (MCS) is to generate random numbers with prescribed probability distributions. For magnitude, distance, and ground motion attenuation, their distributions are given or developed. We used the observed distribution for the first two variables as described earlier, and adopted the lognormal distribution for ground motion attenuation that was commonly accepted [8].

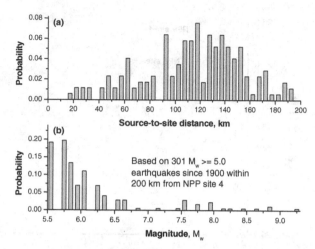

Fig. 3. Observed magnitude and source-to-site distance probability distributions based on the 301 earthquakes

In addition to magnitude, distance, and ground motion uncertainties, this proposed approach accounts for the variability of earthquake occurrence in time. Apparently, given a specific earthquake with its mean annual rate equal to v, it is very unlikely that such an event will recur v times every year, but should follow a Poisson process with general acceptance [5-6]. Note that PSHA does not account for such uncertainty as the mean seismic hazard is governed by $v \times H$, where H denotes the seismic hazard induced by a single earthquake.

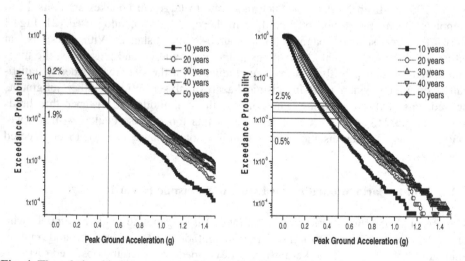

Fig. 4. The relationship between exceedance probability and PGA under five different periods for NPP site 4 (left), for NPP site 1 (right)

As a result, we can generate a series of random maximum ground motion through MCS given the best-estimate statistical attributes of earthquake magnitude, etc. Therefore, the probability that the seismic hazard (i.e., maximum motion) exceeds a given motion is the ratio of trials N $(Y_{max} > y^*)$ to total simulation or known as MCS sample size n, as follows:

$$Pr(Y_{max} > y^*) = \frac{N(Y_{max} > y^*)}{n} \tag{1}$$

It is worth noting that the use of maximum motion as seismic hazard estimation is logical for earthquake resistant design. As long as engineered structures can withstand the maximum motion, they are considered reliable with a deterministic failure criterion being adopted.

4   Seismic Hazard Assessments on NPP Sites 1 and 4

Fig. 4 (left) shows the relationship between exceedance probability and maximum PGA for NPP site 4 through the proposed MCS method with sample size of 50,000. Such a large size ensures the reliability of this MCS result. Given the design motion at 0.5 g in PGA, the exceedance probabilities in 10 and 50 years are 1.9% and 9.2%. Following the same procedure, Fig. 4 (right) shows the seismic hazard estimation for NPP site 1 located at the north tip of Taiwan (Fig. 2), around 30 km away from NPP site 4.   Lower seismic hazard is situated at the site in contrast to NPP site 4 closer to the "hot zone" of seismicity relating to plate subduction zones. Given the same design motion design, the exceedance probabilities in 10 and 50 years reduce to 0.5% and 2.5%, around one fourth as large as those in NPP site 4. As a result, this approach is also site-specific, as the-state-of-art earthquake resistant design for safety-related structures at nuclear power plants demanding site-specific design inputs.

5   Discussions

This paper by no means attends to degrade the PSHA methodology. In fact, the fundamental ideas of PSHA to take earthquake uncertainty into account are acknowledgeable and used in this study. But we here like to refer to Klugel's points [9] on robust seismic hazard analyses, which must feature "transparency" and "traceability" with model assumptions and analytical results being supported and validated. "The best way of improving analytical transparency is to keep the analysis as simple as possible."

Combining Mualchin's amd Krinitzsky's points [3], it is worth noting that the accountability of seismic hazard estimation relates to those factors, but not including the complexity, or even popularity, of the method adopted. "Among a variety of imperfect approaches, it is more critical for decision makers and reviewers to fully understand the fundamentals of the selected method in seismic hazard analysis, to ask hard questions about every assumption and input data, and to be open-minded during a decision-making process."

6   Conclusions

This paper presents a new site-specific, observation-oriented, and probability-based seismic hazard analysis, in which the uncertainties of earthquake magnitude, location, rate, and motion attenuation are accounted for. This approach features less judgments needed during analysis, and directly develops magnitude probability distribution from abundant earthquakes rather than using empirical relationships to indirectly develop the distribution. This approach was demonstrated during the seismic hazard evaluation on two nuclear power plants in Taiwan. The site-specific analysis shows that the seismic hazard is situated at NPP site 4 is four-time as large as that at NPP site 1 in terms of exceedance probability. As pointed out, given our still limited understanding on the high uncertain earthquake process, not a seismic hazard analysis is perfect without challenge. The accountability of a seismic hazard analysis relates to analytical transparency, traceability, etc., not relates to complexity.

References

1. U.S. Nuclear Regulatory Commission, A performance-based approach to define the site-specific earthquake ground motion, Regulatory Guide 1.208, Washington D.C. (2007)
2. Caltrans, Seismic Design Criteria 1.4, California Department of Transportation, State of California (2012)
3. Mualchin, L.: Seismic hazard analysis for critical infrastructures in California. Eng. Geol. 79, 177–184 (2005)
4. Das, B.M.: Principles of Foundation Engineering. PWS Publishing Co., Boston (1995)
5. Ang, A.H.S., Tang, W.H.: Probability Concepts in Engineering: Emphasis on Applications to Civil and Environmental Engineering. John Wiley & Sons, New York (2007)
6. Kramer, S.L.: Geotechnical Earthquake Engineering. Prentice Hall Inc., New Jersey (1996)
7. Gutenberg, B., Richter, C.F.: Frequency of earthquakes in California. Bull. Seism. Soc. Am. 34, 1985–1988 (1944)
8. Wang, J.P., Chan, C.H., Wu, Y.M.: The distribution of annual maximum earthquake magnitude around Taiwan and its application in the estimation of catastrophic earthquake recurrence probability. Nat. Hazards. 59, 553–570 (2011)
9. Klugel, J.U.: Seismic hazard analysis – Qua vadis? Earth. Sci. Rev. 88, 1–32 (2008)

Author Index